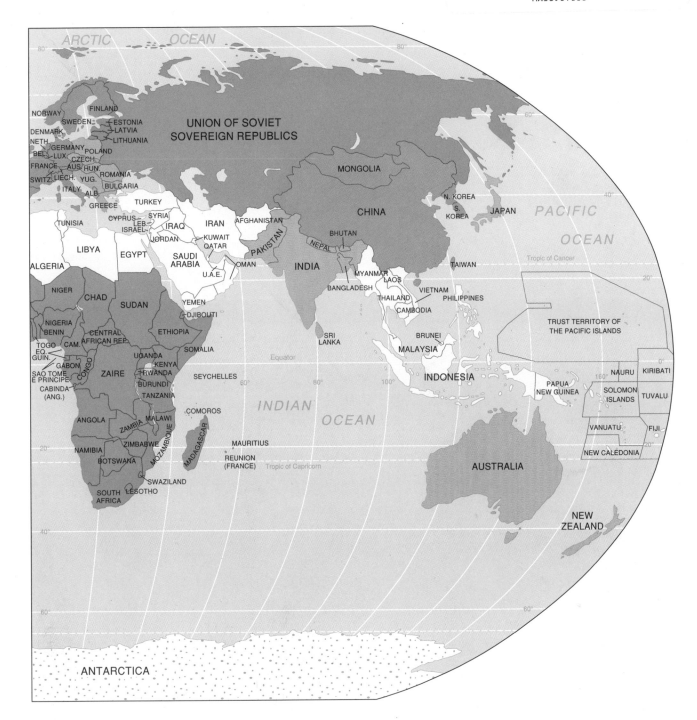

WORLD REGIONAL GEOGRAPHY

A Global Approach

WORLD REGIONAL GEOGRAPHY

A Global Approach

GEORGE F. HEPNER
University of Utah

JESSE O. McKEE
University of Southern Mississippi

West Publishing Company
St. Paul ■ New York ■ Los Angeles ■ San Francisco

Production credits

Copyediting: *Patricia Lewis*
Cartography: *Maryland Cartographics, Inc.*
Illustrations: *Visual Graphic Systems, Ltd., Rolin Graphics, Inc.*
Composition: *Parkwood Composition Service, Inc.*
Cover design: *Diane Beasley Design*
Cover photograph: *Joe Rossi, St. Paul, Minnesota*

COPYRIGHT ©1992   By WEST PUBLISHING COMPANY
610 Opperman Drive
P.O. Box 64526
St. Paul, MN 55164-0526

Printed in the United States of America

99 98 97 96 95 94 93   8 7 6 5 4 3 2

LIBRARY OF CONGRESS CATALOGING-IN-PUBLICATION DATA

Hepner, George F.
  World regional geography: a global approach / George F.
Hepner, Jesse O. McKee
    p. cm.
  Includes index.
  ISBN 0-314-84672-7
    1. Geography.   I. McKee, Jesse O.   II. Title.
G128.H45 1991   CIP
910—dc20   91-4254

Klaus J. Meyer-Arendt is an assistant professor of geography at Mississippi State University. He received his undergraduate degree from Portland State University and his Ph.D. from Louisiana State University. His areas of specialty include cultural/historical geography, coastal environments, and Latin America.

Andrew Burghardt is a professor emeritus of geography at McMaster University in Hamilton, Ontario, Canada. Professor Burghardt received his undergraduate degree in geography from Harvard University and his Ph.D. from the University of Wisconsin in 1958. He has lived and done research in various parts of Europe and has held visiting teaching positions in Vienna, Berlin, and Stuttgart. In addition to numerous articles, Professor Burghardt has written two books on European geographical issues.

Gary Hausladen is an associate professor of geography at the University of Nevada–Reno. His Ph.D. is from Syracuse University. Professor Hausladen did undergraduate work at Stanford University and was a former Soviet Union pilot in the Vietnam War. His areas of specialty include urban, economic, and political geography and East Asia.

George F. Hepner is an associate professor and chair of the Department of Geography at the University of Utah in Salt Lake City. He received his Ph.D. from Arizona State University in 1979. His research has focused on trends in urbanization geographic information systems and land-use change, particularly in the western United States.

Arthell Kelley is a professor emeritus of geography at the University of Southern Mississippi. He received his undergraduate degree from the University of Southern Mississippi and his Ph.D. from the University of Nebraska. He has traveled extensively throughout Europe and Australia and New Zealand. His areas of specialty include climatology, political geography, Europe, and Australia/New Zealand.

Martin S. Kenzer is an associate professor of geography at Florida Atlantic University. He received his undergraduate degree from California State University at Chico and his Ph.D. from McMaster University. His areas of specialty include historical and cultural geography and Latin America, and he has also published several articles and books on geographic thought and the history of geography.

Ian M. Matley is a professor of geography at Michigan State University. He received his masters degree from the University of Edinburgh and his Ph.D. from the University of Michigan. He has traveled extensively throughout Eastern Europe and the Soviet Union and has published several articles about these geographic areas. His areas of specialty include cultural geography, Eastern Europe, and the Soviet Union.

Jesse O. McKee is professor and chair of the Department of Geography at the University of Southern Mississippi. He received his undergraduate degree from Clarion University of Pennsylvania and his Ph.D. from Michigan State University. His areas of specialty include cultural/historical geography, population and human resources, and Africa. Professor McKee has been the recipient of a Fulbright-Hayes travel grant to Africa and has published several articles and books on cultural/historical and ethnic geography.

Basheer K. Nijim, late professor of geography at the University of Northern Iowa, received his undergraduate degree from Augustana College and his Ph.D. from Indiana University. His areas of specialty included political geography, the Middle East, and population. He traveled extensively in the Middle East and published several articles on this area.

Gwenda Rice, an assistant professor in geographic education at Western Oregon State College in Monmouth, Oregon, was born and raised in India. She received her undergraduate and masters degrees from the University of Northern Colorado and her Ph.D. from the University of North Carolina at Chapel Hill in 1983. Her specialty areas include Africa, South Asia, and geographic education.

Richard Ulack is a professor of geography and departmental chair at the University of Kentucky in Lexington, Kentucky. He received his Ph.D. from Pennsylvania State University in 1972. His primary research interests involve urban structure in Southeast Asia. In 1991, he was a Fulbright Scholar at the University of the South Pacific in Fiji.

BRIEF CONTENTS

CONTENTS

LIST OF MAPS

Recent world events, such as the changes in Eastern Europe and the crisis in the Gulf, are intricately related to the geography of these areas—past, present, and future. One of Iraq's goals in invading Kuwait was the acquisition of a coastal outlet to the Persian Gulf. Eastern Europe had acted as a geographic buffer zone for the Soviet Union. Thus, in both of these situations, geographic issues lay beneath the ideological and political rhetoric.

As a result of these and similar events that indicate the value of geography, the discipline is undergoing a renewal in the United States and other parts of the world. In the scientific/high-technology arena, geographic studies using computers for cartography, satellite remote sensing, spatial analysis, and information systems are becoming indispensable to regional economic development, resource exploitation, military planning, and environmental management. In addition, the impact that geographic illiteracy can have on international business opportunities and the peace and welfare of the world is finally gaining attention in the United States. Consequently, geography is increasingly being viewed as a fundamental discipline necessary for comprehension of the history, politics, and economies of the nations and regions of the world.

Geographic educators play an important role in introducing students to fundamental, nontrivial knowledge about the variety of places in the world. Implicit in the study of world places and regions is an understanding of one of the essential concepts of geography: that relationships and interactions between humans and their environments influence the evolution of specific geographic places, regions, and landscapes. At the same time, geographic education must help the student understand the global interrelationships and dependencies among countries and regions.

This text addresses the variations in the regions of the world while focusing on the commonalty of linkages in the increasingly interconnected world. The regional chapters in this text have been prepared by leading regional specialists. All contributors have lived and worked in their respective regions and are thus able to provide the text with unmatched geographical knowledge and insight. Continuity between the regional chapters is maintained by the use of a consistent topical structure. At the same time, flexibility is achieved by leaving the style, presentation, and emphasis given to individual topics to the contributor's discretion. We feel this is an optimal way of ensuring a measure of predictability in the discussions of the regions, which will aid student comprehension, while avoiding the monotonous, sterile presentations that may result when diverse regions are forced into a rigid structure.

It is the intention of the authors that the student recognize that each region has a distinctive character, but that global processes bind all regions together into an interdependent web. The increasing number of socioeconomic, political, and natural environmental connections among the regions of the world is an undeniable reality of the modern age. The desire to produce a text that focuses on these evolving connections was the major impetus for this book. The theme of global interconnectivity and interdependency is the frame of reference through which the regional chapters have been prepared. Each of the regional chapters includes discussion of interconnectivity topics common to all chapters, such as trade and alliances. The chapter contributors further enhance the theme of interconnectivity by examining issues unique and significant to the particular region. Each regional chapter strives to address both interregional and intraregional connectivity by highlighting those connections most significant to the particular region.

World Regional Geography: A Global Approach is divided into three parts. Part One, composed of Chapters 1, 2, and 3, provides the framework used in the regional discussions in Parts Two and Three. These chapters are an explication of foundational ideas and global patterns that provide the students with a basis for understanding and comparing specific regional patterns. Chapter 1 provides an overview of the basic concepts and ideas in regional, physical, and human geography. Chapter 2 discusses the elements and processes of the global natural environment as a foundation for human activities. Global patterns of culture, population, technology, and economic development are introduced in Chapter 3. These spatial variations in technological achievements and applications contribute significantly to the division between the economically developed and the developing worlds.

The introductory student needs an organizational structure for comprehending the role of both individual countries in a region and of regions in the world. The level of economic development is the criterion most commonly used to provide this structure. Categorizing countries and regions by the level of economic development is at best an inexact undertaking, however. It glosses over inequities within countries and regions, minimizes the recognition of economic development as a continuum and ignores noneconomic criteria in the definition of development. These important considerations and subtleties are all addressed in the initial chapters.

The regional chapters presented in Part Two and Three are designed with a consistent format: (1) Physical Environment, (2) Human Environment, (3) Spatial Connectivity, and (4) Problems and Prospects. Part Two discusses the regions that are part of the economically developed world. This category includes regions having both predominantly capitalistic, market economies and predominantly socialistic, planned economies. Although the radical changes taking place in the Soviet Union and Eastern Europe make firm ideological and economic classification impossible at this time, it appears that a form of democratic socialism will be the dominant political-economic theme in these regions. The inertia of socialistic control and centralized economic planning is likely to be felt for many years to come in these countries in spite of the recent changes. Part Three examines regions having economic levels significantly below those of the developed world. These developing regions encompass a mixture of planned and free market economies but are similar in their modest to low standard of living.

All of the chapters are illustrated with maps, graphs, tables, and photographs. Each chapter begins with a list of terms and a statement of important concepts and issues. Each regional chapter includes a table of statistical information for each country in the region and a base map of the region. Special feature boxes are included in the chapters to delve further into important geographic concepts, connectivity issues, and special interest topics. Chapters conclude with a list of suggested readings. The text material is supported by a comprehensive glossary at the end of the text.

Supplements

A full supplement package accompanies this text.

■ *The Instructor's Manual with Test Bank* by Gwenda Rice, contributor, contains lecture outlines and enhancements and student map exercises. Two versions of chapter tests that include objective and essay questions are provided for each chapter. The test bank is also available on Westest, a computerized testing program that includes editing features for instructors.
■ *The Student Study Guide* contains learning objectives, conceptual and map review exercises, and sample test questions that correspond to the test questions in the test bank.
■ Full-color transparency overhead acetates of maps and key figures in the text are provided.
■ Slides are available to qualified adopters. See your West representative for additional media.

Acknowledgments

We wish to acknowledge the invaluable comments of the following reviewers of the several versions of the manuscript. Their input not only contributed to the accuracy of this book, but also helped the discussions of the world regions to achieve a richness that the individual authors could not have attained.

W. A. Bladen
University of Kentucky

Jorge Brea
Central Michigan University

Alvar Carlson
Bowling Green State University

Fred Day
Southwest Texas State University

James Delehanty
University of Wisconsin at Madison

Dennis Dingemans
University of California at Davis

George Erickson
St. Cloud State University

John Florin
University of North Carolina at Chapel Hill

Orville E. Gab
South Dakota State University

Charles Gritzner
South Dakota State University

Harold Gulley
University of Wisconsin at Oshkosh

Ray Henkel
Arizona State University

Dave Icenogle
Auburn University

Jim Kenyon
University of Georgia

Paul Lehrer
University of Northern Colorado

Elliot McIntire
California State University at Northridge

E. Joan Wilson Miller
Illinois State University

Mary Lee Nolan
Oregon State University

Lee A. Opheim
South Dakota State University

Richard Pillsbury
Georgia State University

Steve Pontius
Radford University

Milton Rafferty
Southwest Missouri State University

Gregory Rose
Ohio State University at Marion

There is no substitute for the knowledge and insight of the contributing authors who have lived and worked in their respective regions. They are to be greatly commended for their dedication to and hard work on this book. Sadly, Basheer Nijim passed away near the end of the writing of this text. Prior to his death, however, he had successfully completed and revised his contributions to the book.

Many people also contributed to this book at the production stage. Pat Lewis is to be highly praised for her outstanding copyediting skills and care. Gregory Rose provided a keen cartographic sense in his reviews of the maps and graphics. Peter Marshall, Mark Jacobsen, Becky Tollerson, Kara ZumBahlen, and the other talented people at West Publishing did a superb job of taking a rough manuscript and turning it into a polished text.

Our inspirational debt for this book goes to those attendees at a Wingspread Conference several years ago who fostered a desire to prepare a world regional book that had both regional expertise and an explicit global focus. Also to be thanked is John Poth for his many discussions of the geography of the multitude of places in the world. Our colleagues at Florida State University, the University of Utah, and the University of Southern Mississippi and staff members Carolyn Robbins and Betty Blackledge were supportive and helpful in our efforts.

Our wives, Prudy Hepner and Janet McKee, and our children, Evan Hepner and Pamela and Russell McKee, deserve thanks for their support and understanding through the several years of this project.

George Hepner
Jesse McKee

WORLD REGIONAL GEOGRAPHY

A Global Approach

PART ONE

Geographic Concepts in Understanding World Regions

CHAPTER 1

Geography: Structure and Concepts

IMPORTANT TERMS

Homogeneous or formal region
Functional or nodal region
Spatial connectivity
Spatial distribution
Relative location
Site
Situation
Global interdependence

CONCEPTS AND ISSUES

- Interrelationships between humans and their natural environment.
- Spatial connectivity and interrelationships within and between places.
- Quality-of-life and economic differences between the developed and the developing world.

CHAPTER OUTLINE

The Geographic Viewpoint
Geographic Regions
World Regions
Geographic Analysis of Regions
Concepts in Geography

T he word *geography* may mean different things to different people, depending on whether they are citizens, students, or academicians. All of these groups would probably define geography differently. But if the various definitions were compiled and analyzed, certain key words would very likely emerge. This list would probably include places, earth, people, space, distribution, connectivity, and the natural and human environments. Indeed, geography is a study of the earth, places, people, natural environment, human-environment interaction, and the movement of people, goods, and ideas. It is also a study of connections and linkages, focusing on how people, goods, and ideas are interconnected. Thus, geographers seek to describe places on the earth's surface and to explain relationships among people, places, and events.

Knowledge and understanding of ourselves, our culture, and the places we live in and interact with are important if we are to function in today's interconnected world where time and space are shrinking. The flow of information is quickening, and events occurring in other parts of the world may have global as well as local consequences. A nuclear explosion resulting in atmospheric contamination can cause severe environmental damage in specific places, and oil spills and other natural and human-initiated hazards often damage fragile environments. Ethnic strife and religious conflicts may instigate wars; the control and monopolization of a natural resource by cartels may result in economic gains or losses for organizations and/or nations and cause monetary, social, and political disparities between peoples and places.

At a more local level, communities may be struggling with how to cope with urban growth and sprawling suburbs, stagnation in the inner city, dwindling employment opportunities, mass transit and transportation gridlock, or environmental degradation. An understanding of the key ideas, concepts, and themes of geography helps us describe, explain, and predict many of the problems and issues at the local, regional, and national scale in the world today. Not only does geography seek to further knowledge about our planet earth, but it attempts to apply this knowledge to solving problems affecting humans and their environment.

THE GEOGRAPHIC VIEWPOINT

Geography comes from a Greek word meaning "to describe the earth" or "description of the earth." Today, geographers are still interested in describing and ana-

lyzing planet earth. They investigate both the various elements of the **physical environment,** such as climate, vegetation, soils, landforms, water resources, energy sources and minerals, and wildlife resources, and the **human environment,** which includes people and their cultures, their settlement patterns, and economic activities. As geographers study planet earth, they are concerned with certain fundamental questions; *What* is it? *Where* is it? *Why* is it there? More specifically, they are interested in all of the following:

1. **Location.** Where is the phenomenon or phenomena being studied located?
2. **Spatial distribution.** How is the phenomenon distributed on the earth's surface, and is there a pattern to the distribution?
3. **Human-environment interactions.** How do people respond to and modify their environment?
4. **Movement.** What are the kinds of interaction and types of movement that occur among people, goods, ideas, and places on the earth's surface?
5. **Place.** What are the physical and cultural features of a particular place that give it its character and identity?
6. **Regions.** How can the earth be divided into smaller parts in order to gain perspectives on the character and relationships between geographic areas and places?

Defining geography is a difficult task, and many specific definitions currently exist. Nevertheless, we can say that geography is concerned with describing and analyzing the physical and human environmental landscapes and the interaction between humans and their physical environment in order to understand the differences, similarities, and relationships between places on the surface of the earth. There are many kinds of places such as neighborhoods, cities, counties, states, nations, mountains, valleys, plains, river systems, and deserts. Most places have distinctive features that give them identity and character and distinguish them from other places. Certainly, places can be viewed as being different in many respects, perhaps even unique. One might say, for example, that there is only one New York City, Singapore, or Nairobi. There is only one Texas, Cameroon, China, or Soviet Union. Because places are unique and different, describing and analyzing these differences is a major part of geography.

Despite their unique qualities, however, places also have similarities. Cities, states, and countries often share certain characteristics. One reason for this is that human

A view from Glendhu Bay, Lake Wanaha, New Zealand, reflecting the natural beauty of the physical environment.

beings in society have common needs, such as food, clothing, and shelter, that must be met. The attempt to meet these needs may lead to similarities in rural or urban human settlement patterns, architectural styles, economic activities such as agricultural and industrial practices, and transportation and communication networks, as well as in political and religious ideologies. Thus, although New York City, Singapore, and Nairobi are unique in some ways, they also share many similarities. Each has a central business district (CBD), residential areas, business offices, and a transportation network. Cameroon, China, and the Soviet Union are unique in certain respects, yet all three are organized under a political system and have cities, agricultural areas, and industrial regions. Since a science cannot be based solely upon the uniqueness of the phenomena studied, geographers examine similarities in order to bring rigor to the discipline of the study of places. Thus, rather than just emphasizing differences between places, geographers look for similarities and patterns. To do so, they must classify data and theorize about the physical and human characteristics of a place and the interaction within and between places.

GEOGRAPHIC REGIONS

One way to classify data and theorize about places is through the conceptual framework of geographic regions. A region is a concept that is used to classify certain areas of the earth to help us comprehend the similarities and differences on the earth's surface. Hemispheres (e.g., northern or southern, eastern or western), continents (e.g., South America, North America, Eurasia), or countries are frequently used as an organizational framework for the study of regions or places. Such regions are often referred to as *general regions*.

But the earth's surface can also be divided into regions of similar climate, landforms, vegetation and soil types, political or economic systems, racial groups, or cultural characteristics such as languages spoken or religious beliefs, to name only a few of the numerous variables geographers use to regionalize the complex whole we call earth. Generally, a region exhibits some form of uniformity or commonality in one or more of those variables that distinguishes it from other regions. Such regions are known as **homogeneous** or **formal**

FIGURE 1.1

Regions of the World

regions. These regions may be delimited by *single factors* (e.g., climate, types of crops, or per capita income) or by *multiple factors* (e.g., religion, language, and political ideology). Frequently, however, multiple factors are used in delimiting homogeneous or formal regions. For example, most Americans would consider the American South to be a distinctive geographic region—but based upon what criteria? If we use multiple factors to define the South, the concept of a homogeneous region becomes clearer. For example, the South has a warm mild climate that differentiates it from the cold North and arid West, religiously it is overwhelmingly Protestant Baptist, its dialect is characterized by a distinctive Southern drawl, and politically it has tended to be conservative. So by examining multiple factors or variables, and seeing uniformity in them expressed upon the landscape, a distinctive Southern cultural region can be defined. Similar variables could be used to delimit other regions in the United States such as the Midwest, Southwest, or the Northeast.

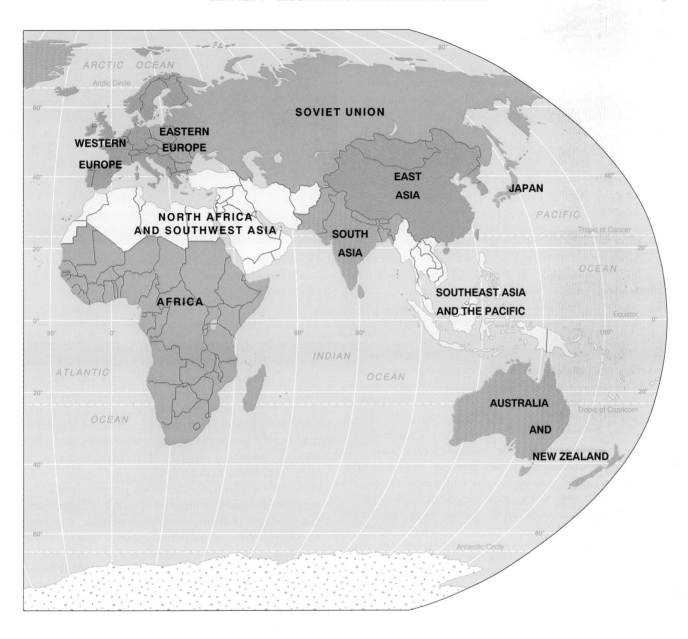

The other basic type of region is a **functional** or **nodal region.** This is an operational unit that is interdependent and operates over space. It may have a node or core area that directs and coordinates the function. Examples of functional regions include voting precincts, retail trade areas, school districts, and newspaper distribution areas. Each of these functional regions centers on a node such as the polling place, shopping center or CBD, school, or newspaper printing office. Usually, functional regions are continually changing and dy-

namic with considerable spatial movement of ideas and products, whereas formal regions are more static with less spatial movement.

WORLD REGIONS

The focus of this text is on the geography of world regions. Thirteen specific geographic regions have been identified (Figure 1.1). We discuss individual countries and specific places and their characteristics as integral components of one of the thirteen regions. To understand how these countries and specific places interact within a particular region is important, but it is also necessary to see how these larger regions and specific places within them interact with other places and regions to form a larger global network.

Taking a global perspective, a convenient and popularly accepted way of regionalizing the earth is to divide it into two broad regions: (1) a modern or ***developed world,*** which relies heavily upon technology, is highly mechanized, consumes a large amount of inanimate energy, maintains intensely interconnected transportation and communication networks, and has a tremendous amount of wealth as measured by gross national product (GNP) and per capita income; and (2) a traditional or ***developing world,*** which is linked in various ways to the developed world but in which the majority of people follow a more traditional way of life independent from the many conveniences of the developed world and its level of living (Figure 1.2).

The developed world, in turn, can be further divided into Western-style ***commercial economies,*** such as those found in the United States, Canada, Western Europe, Japan, Australia, and New Zealand, and ***centrally planned economies*** which include primarily the communist-dominated Soviet Union and, to a lesser degree, those countries formerly dominated by communism in Eastern Europe. In 1989 most of the Eastern European countries moved toward multiparty socialist democracies and began experimenting with ways to reduce centralized planning. In the future, as the region

Rice planting in the Philippines typifies parts of the developing world where the majority of the people follow a traditional way of life.

assumes more characteristics of free market commercial economies, these countries may no longer be classed as centrally planned economies. Similar changes are also occurring in the Soviet Union as it increases its experimentation with a more free market–oriented economy. The developing world includes both countries that have commercial or free market economies (such as South Korea), and countries that have planned systems (such as China and North Korea).

Although we refer to the developing world as a single category, individual countries within that category are in various stages of socioeconomic development as they progress toward modernization. Certain countries of the developing world have industrialized, and some, through income from oil exports, have accumulated surplus capital. Figure 1.3 reflects the economic transition that is taking place in some of these developing countries. Several Middle and South American countries have recently industrialized as have Algeria and South Africa in Africa; Turkey, Iraq, and Iran in Southwest Asia; Taiwan, Hong Kong, and South Korea in East Asia; and Singapore in Southeast Asia. Although income from oil has helped Saudi Arabia, Libya, Bahrain, and the United Arab Emirates to acquire a capital surplus, the general quality of life, as measured by literacy, life expectancy, birth and death rates, health care, housing, and sanitation conditions, is still appreciably lower in these countries than in the developed countries of the world. At the same time, many developing countries are poor, and some are extremely poor. Some of these nations lack a mineral resource base to provide a source of income,

The skyline of Chicago, Illinois, showing Lincoln Park and Lake Michigan. As part of the developed world, the United States and its cities maintain intensely interconnected transportation and communication networks that rely heavily upon modern technology.

FIGURE 1.2

Developed and Developing World

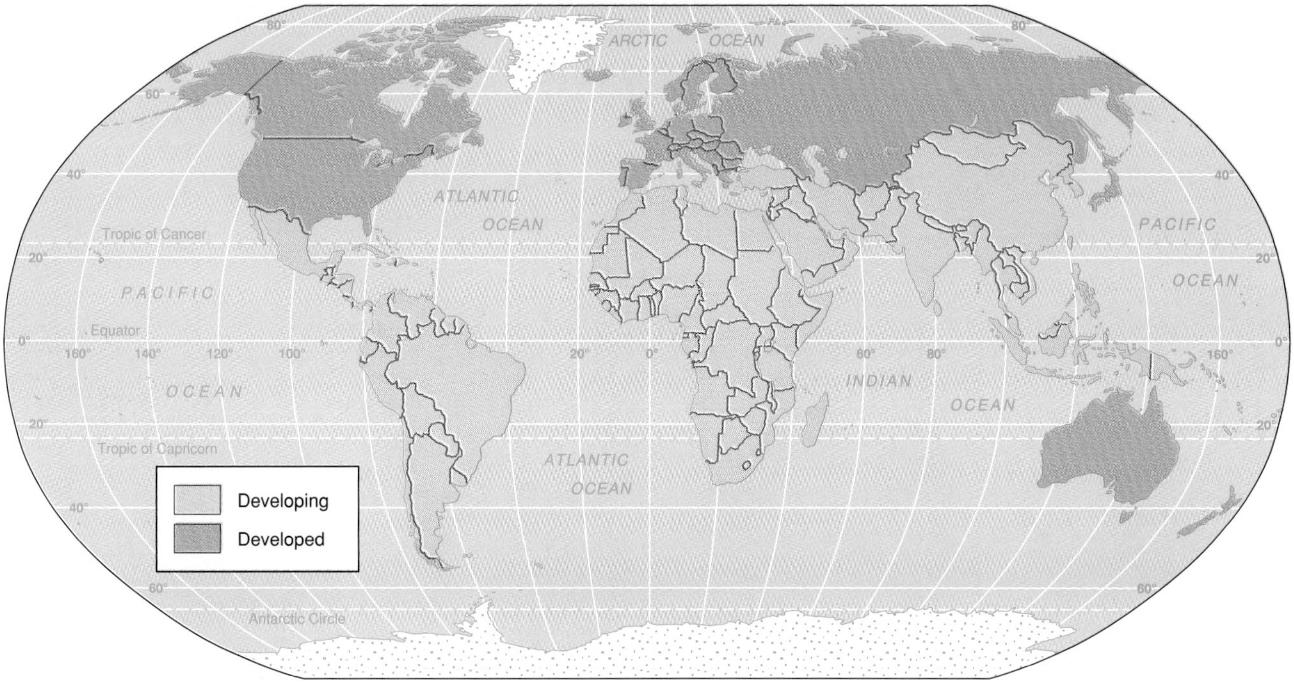

are unable to produce enough food to feed themselves, or lack the technology and foreign investment that may be necessary to improve standards of living. Many of these poor and least developed countries are located in Africa, while others are found in Latin America and Asia.

As this brief introduction suggests, the developed and developing worlds are not static, but rather are changing economically, socially, and politically. Change is occurring extremely rapidly in the United States and the developed world, as well as in the Soviet Union and parts of Eastern Europe. In the developing world, the rate of change, as measured by certain types of socioeconomic development, varies markedly between countries and regions. Therefore, one of the focal points of the specific regional chapters of this text will be to evaluate the change and progress that is occurring in specific geographic regions of the world.

GEOGRAPHIC ANALYSIS OF REGIONS

Geographers use a variety of approaches in describing and analyzing regions and countries, including the topical approach and the concept of spatial connectivity.

Here we will examine both these approaches as well as the techniques used in applying them.

Topical Approach

The topical or systematic approach assists greatly in understanding regions and countries. The topical approach considers two elements of the environment: physical and human (Table 1.1). The key elements of the physical environment include climate, vegetation, soils, landforms, water resources, energy resources and minerals, and wildlife resources. Frequently asked questions about a place often include; What is the climate of that area? What kind of vegetation exists? How fertile is the soil? What is the topography like? These and many other questions are critical to understanding a region or place.

When focusing on the human environment, attention is given to the people and their characteristics. First, we look at the biological characteristics of a population. Important biological components of a population include race, age structure, and gender composition. Geographers also examine population characteristics such as the total population of a country or region and the dis-

FIGURE 1.3

World Regions Based on the Level of Industrial Development

SOURCE: Modified from Richard H. Jackson and Floyd E. Hudman, *World Regional Geography: Issues for Today*, (New York: John Wiley & Sons, 1990), p. 10.

Legend:
- ◼ Industrialized countries
- ◼ Newly industrialized countries
- ◼ Developing countries
- ◻ Surplus capital (oil) countries

tribution, density, and spatial settlement pattern of the people. Migration trends and population growth rates are also significant.

Another aspect of the human environment is culture, the learned traits that people possess. The key elements of culture include language, religion, political systems and ideology, and technology. From the time of birth, individuals are constantly exposed to the cultural traits around them as they learn to speak a language, to believe or not believe in some form of religion, to live under a political system, and to function in society at a

particular technological level and quality of life.

Therefore, in learning about places, we need to ask questions about the inhabitants. What language do they speak? What form of religion do they practice? What type of government do they live under? At what level of technological development is their country? Technology, the key element in cultural change and quality of life, plays a major role in housing construction and quality, occupations of the population, types of transportation and communication facilities, agricultural productivity, industrial output, health care, degree of ur-

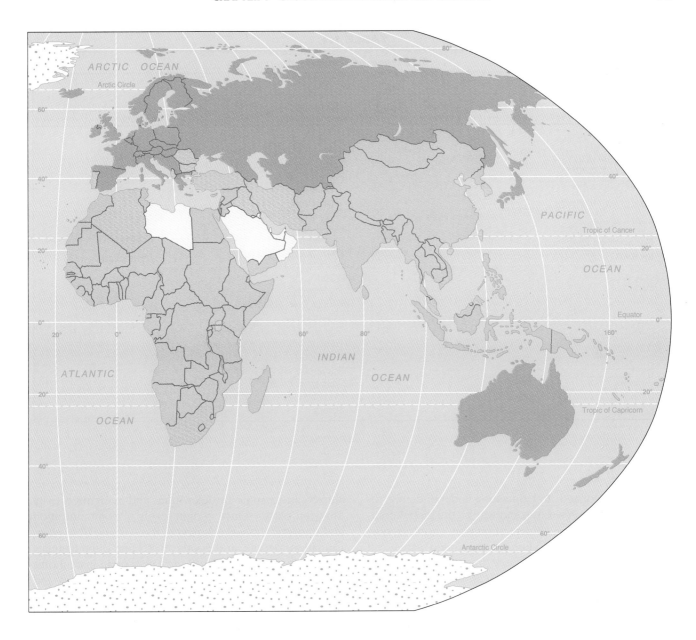

banization, food preferences and caloric intake, and a host of other socioeconomic variables. All these characteristics distinguish places and regions and give them their own identity. The regional chapters in this book use the topical approach and discuss various physical and human topics for each major region.

Spatial Connectivity

Having specific information about particular topics (e.g., climate, landforms, culture) on individual countries or regions is indeed critical to the geographic analysis of an area. But geographic inquiry about places and regions requires further investigation into how these elements of the physical and the human environment are spatially interconnected within and between countries and regions. The organization of space requires a strong emphasis upon *direction* (e.g., north, south, west, east), *distance* (e.g., absolute—kilometers or degrees; or relative—the nonmathematical measurement of distance using reference points), and **spatial connectivity** (e.g., spatial linkages between phenomena). No city, country,

TABLE 1.1

Structure of Geography

TECHNIQUES	TOPICAL/SYSTEMATIC		REGIONAL
	Physical Elements of the Environment	Human Elements of the Environment	
1. Cartography	1. Climate	1. Population	1. North America
2. Remote sensing Aerial photographs Imagery from spacecraft	2. Vegetation	Biological/demographic components Race, age structure, sex ratio Population Size	2. Latin America
	3. Soils	Number	3. Europe
3. Quantitative methods	4. Landforms	Distribution Density	4. Africa
4. Field methods	5. Water resources	Growth	5. Asia
5. Geographic Information Systems	6. Energy and mineral resources	2. Culture Language	6. Australia/New Zealand
	7. Wildlife resources	Religion Political system	7. Antarctica
		Technological capabilities Urban/rural Agriculture/industry/services Transportation/communication facilities Occupational characteristics Education/literacy Medical/health care Others	

or region is isolated and operating independently to-day—all are connected in some way to the global community. The spatial movement of oil from countries bordering the Persian Gulf to Europe, the United States, and Japan is one indication of how interconnected these parts of the world have become with regard to oil production and consumption. Economies are interconnected in other ways as well, whether by trade alliances or financial investments. Transactions on the New York Stock Exchange may cause reactions in Tokyo or London and vice versa. Cultural, political, and military alliances also promote interconnectivity.

Although people and places are indeed intricately connected today, the precise level of the connectivity is frequently dependent upon technological achievements in communication and modes of transportation in a particular society. Advances in the production, distribution, and utilization of communicative devices (e.g., radios, televisions, newspapers, satellites, and computer hardware and software) and transportation modes (e.g., bicycles, motorcycles, automobiles, trains, planes, and

ships) can promote or deter intra- and interconnectivity between places. Once certain levels of communication and transportation are achieved, the ensuing flow of information, people, and goods affects and often changes the society's social, political, and economic institutions. It should be remembered, however, that technology and culture are intertwined. Thus, a society's social, economic, and political structure can advance or impede technological progress, thereby encouraging or discouraging the production and consumption of goods, the flow of information, or the movement of people.

As this brief overview indicates, spacial connectivity is having an impact on virtually every society on earth. Because of spatial differences in the distribution of people, political ideologies, economic systems, and environmental resources, decision makers within various countries and regions are forced to communicate and interact with each other. Therefore, each of our regional chapters includes a discussion of spatial connectivity giving a geographic perspective and analysis of the spatial interaction between people, places, ideas, and goods.

Techniques and Skills

In studying the physical and human environments, their interrelationship and spatial connectivity, and how they can produce distinguishable regions and landscapes, geographers rely on certain tools and techniques. These include maps, aerial photographs, statistical analysis, and computer modeling (refer back to Table 1.1).

Maps are a fundamental means of communication and a vital tool for the geographer engaged in teaching and research. Just as musicians express themselves through music and artists through paintings, geographers rely on maps. They are the discipline's shorthand. All kinds of information can be mapped, including distributions of phenomena. The creation of maps is the concern of the art or science of cartography.

Aerial photographs and images generated from spacecraft are a second tool used in geography. Geographers use them to analyze land use and human interactions with, and modifications of, the physical environment. Using aerial photographs and stereoscopic devices, geographers can more accurately map various features on the earth's surface such as coastlines, rivers, mountains, and cities. Images taken from orbiting satellites, such as those in the *U.S. Landsat* series, are able to portray such information as seasonal variation in vegetation, areas of drought, and environmentally polluted areas. Satellite images are important, but sometimes geographers must go directly to the field site for more direct observation.

Much of the data geographers work with are statistical and can be quantified. Consequently, basic statistical procedures used in descriptive statistics, such as the mean, median, mode, standard deviation, locational quotient, and variance, are commonly used. More advanced procedures such as regression analysis and coefficients of correlation are also routinely used in geographic teaching and research.

The use of the computer has helped to revolutionize cartography and the handling and managing of geographic data. PREVU, CALFORM, GEOSYS, SPANS, ATLAS GRAPHICS, and ARC/INFO are but a few of the many computer mapping and geographical information systems programs used today to help manage and display data. Thus, the computer has become an essential tool that helps the geographer develop modeling techniques of spatial analysis.

In Parts Two and Three of this text, which discuss the regions of the world using a topical approach with an emphasis on spatial connectivity, numerous maps, graphs, and charts will assist the reader in understanding and interpreting data concerning countries and regions.

CONCEPTS IN GEOGRAPHY

Geography is a generalizing discipline with a distinctive approach and methodology, and it has developed many important theories and concepts. Some of the more prominent theories in geography have focused on location, the importance of central places, and spatial structure. Numerous concepts have also been developed to help the researcher, teacher, and student better understand the world we live in. Six of the more important concepts are discussed below.

Location

Where are the people, places, or things located or positioned in earth space? Location can be measured in three ways—nominal, relative, and absolute—depending upon the degree of accuracy required.

Nominal location is the most generalized and imprecise; basically, it uses names of places to gain an initial understanding of where a place is located. For example, when one hears or reads about a particular place, be it Brazil, California, or Los Angeles, one is able to give it a nominal location in his or her mental map. We mentally picture Brazil's location in South America, California in the United States, and Los Angeles in California. Most of us experience almost daily the need to locate some place mentally on the earth's surface. However, we may not always have an extremely accurate mental image of the location of a particular place, and frequently (depending upon how well known a place is) we may need more nominal response to locate it accurately in our mind. Most of us can locate a well-known place like Chicago or San Francisco in Illinois and California, respectively. But, when we are confronted with a place like Ouagadougou, we may need more nominal responses until we can mentally locate it. To locate Ouagadougou, we may need to know that it is in the country of Burkina Faso and that Burkina Faso is in West Africa. Now that we have been given some more nominal locations, we can more accurately imagine where Ouagadougou is located.

Relative location is a nonmathematical way to locate something. For example, let us say we want to locate the town of Clarion, Pennsylvania, but we do not know

New Orleans, located near the mouth of the Mississippi River, is on a difficult *site* that helped determine the physical layout of the city. However, its *situation* is excellent since its hinterland is comprised of much of the Mississippi River Valley drainage basin and it is centrally located in the Gulf Crescent Region from Brownsville, Texas, to Tampa, Florida.

where in Pennsylvania this community is located. Therefore, to answer this question, we choose another place or nominal location to serve as a reference point. If Pittsburgh can be used as a reference point, and if we know where it is located, then Clarion can be located "relative" to Pittsburgh. In this case, Clarion is about 80 miles north of Pittsburgh, and it is about midway between Chicago and New York City. Now we have a clearer idea of the relative location of Clarion. Note that relative location refers to a position with respect to other locations. It implies a relationship and can be a key concept in understanding the interaction between places.

When we wish to be more precise, we can use ***absolute location***, which utilizes a mathematical grid system. The latitude and longitude system is one of the most widely used reference grid schemes. To give an absolute location of our sample city of Clarion, one would say that it is located 41°10′ north latitude and 79°25′ west longitude. With this information, Clarion can be located precisely on the surface of the earth.

Site and Situation

Site refers to the characteristics of a place or location, whereas **situation** refers to how that particular place is related to other places. Site looks at "internal" factors, whereas situation is "external." Scale is also important in the application of these terms. At a smaller or local scale, a house and its lot may represent the site. Many questions can be asked about the characteristics of the lot: Is it level or sloping? How many trees are on the site? What is its size? Next the individual may want to examine how this site, consisting of the house and lot, is situated in relation to neighborhood features such as the nearest school, shopping center, and place of employment. At a larger scale, one might choose the United States as the site; then examination of its situation would include its spatial relationship and interconnectivity to neighboring countries, surrounding water bodies, and more distant countries.

President Thomas Jefferson once stated that "New Orleans is destined to be the greatest city the world has ever seen." After the purchase of the Louisiana Territory in 1803, it seemed rather obvious that New Orleans, located on a swampy site near the mouth of the Mississippi River, with a situation such that its hinterland was comprised of the Missouri, Mississippi, and Ohio river drainage basins, would definitely serve as the port for future development and settlement in the Middle West. Even though the site characteristics of New Orleans may have been poor, its situation was excellent. However, the building of the Erie Canal and the eventual predominance of an east-west alignment in railroad construction meant that the natural link to New Orleans had been modified by a human-made, artificial link to New York City. Since much of the U.S. trade was with Europe, New York became the "front door" of the United States, and New Orleans became a "side door," despite its excellent situation. Geographers studying places usually examine both the site and situation characteristics of a particular location.

Spatial Distribution

Spatial distribution refers to the distribution and arrangement of physical and human geographic phenomena on the earth's surface. People traditionally think of historians as being interested in time and geographers in space. Thus, the historian might ask the questions:

(continued on page 20)

aps are used to represent part or the whole of the earth's surface on a flat surface. Maps can portray many natural and human-made features such as continents, oceans, rivers, terrain features, cities, roads, and political boundaries. Almost anything can be mapped to reveal the location of places or events and the spatial distribution and spatial interaction, or flow, of people, goods, and ideas. A map communicates facts, concepts, and ideas and can help us understand our local environment as well as the world around us.

UNDERSTANDING MAPS

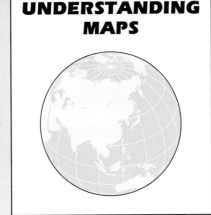

maps show a larger area (e.g., the world or a continent), and the information they contain is usually less detailed.

Map Properties and Projections

One cannot project a curved surface onto a flat surface and accurately maintain all of the major properties of a map such as (1) area, (2) shape, (3) distance, (4) direction, and (5) proximity. Therefore, cartographers have designed certain types of map *projections* that display at least one but not all of these properties on a specific map (Figure 1). Some map projections called *equal-area* or *equivalent* projections display equal areas of a region on a map but distort the shape of the areas within the region. Maps made from a *conformal* projection portray accurate shape but distort area. *Equidistant* projections may show true distance from one point to another but distort other properties.

The three major types of map projections are *cylindrical, conic,* and *planar* (or azimuthal), but there are numerous other projections. Some show the world in an oval or heart shape, while others, like the equal-area Goode's Homolosine projection, interrupt the oceans or continents. The Goode's projection can be likened to peeling an orange so the orange peel can be placed on a flat surface. Imagine trying to place the continents on the peeled surface leaving the oceans to be interrupted. Projections like Goode's Homosoline try to preserve

Scale

Maps have several characteristics and properties. Since the earth's surface must be reduced to portray it on a *map, most maps have a map scale* that shows the ratio of the distance on the map to the distance on the earth. Several methods are used to represent scale: (1) representative fraction, (2) graphic scale, or (3) verbal scale. A representative fraction is usually expressed by a figure such as 1:5000 or 1/5000 or 5000, which indicates that the linear unit on the map (e.g., 1 inch) represents 5000 such units on the earth's surface. A graphic scale is similar, except that a graph is used to indicate the linear unit. A verbal scale is given in words, such as 1 inch represents 1 mile, or 1 centimeter represents 1 kilometer. Thus, the following are all ways of stating 1 inch represents 1 mile:

Word statement. "1 inch represents 1 mile"

Graphic scale. 0 ⌞_____⌟ 1 mi.

Representative fraction.

$$\frac{1}{63,360} \quad \text{or} \quad 1:63,360$$

The number utilized in the representative fraction usually indicates if the map is a large- or small-scale map. Larger fractional values, usually less than 1:100,000, indicate a large-scale map. For example, 1:1200 indicates that 1 inch on the map represents 1200 inches or 100 feet on the ground. Large-scale maps commonly contain more detail and cover a smaller geographic area. Small-scale maps have a smaller representative fraction, usually greater than 1:1,000,000. For example, 1:63,360,000 indicates that 1 inch on the map equals 1000 miles. Small-scale

(continued on page 20)

FIGURE 1

Map Projections

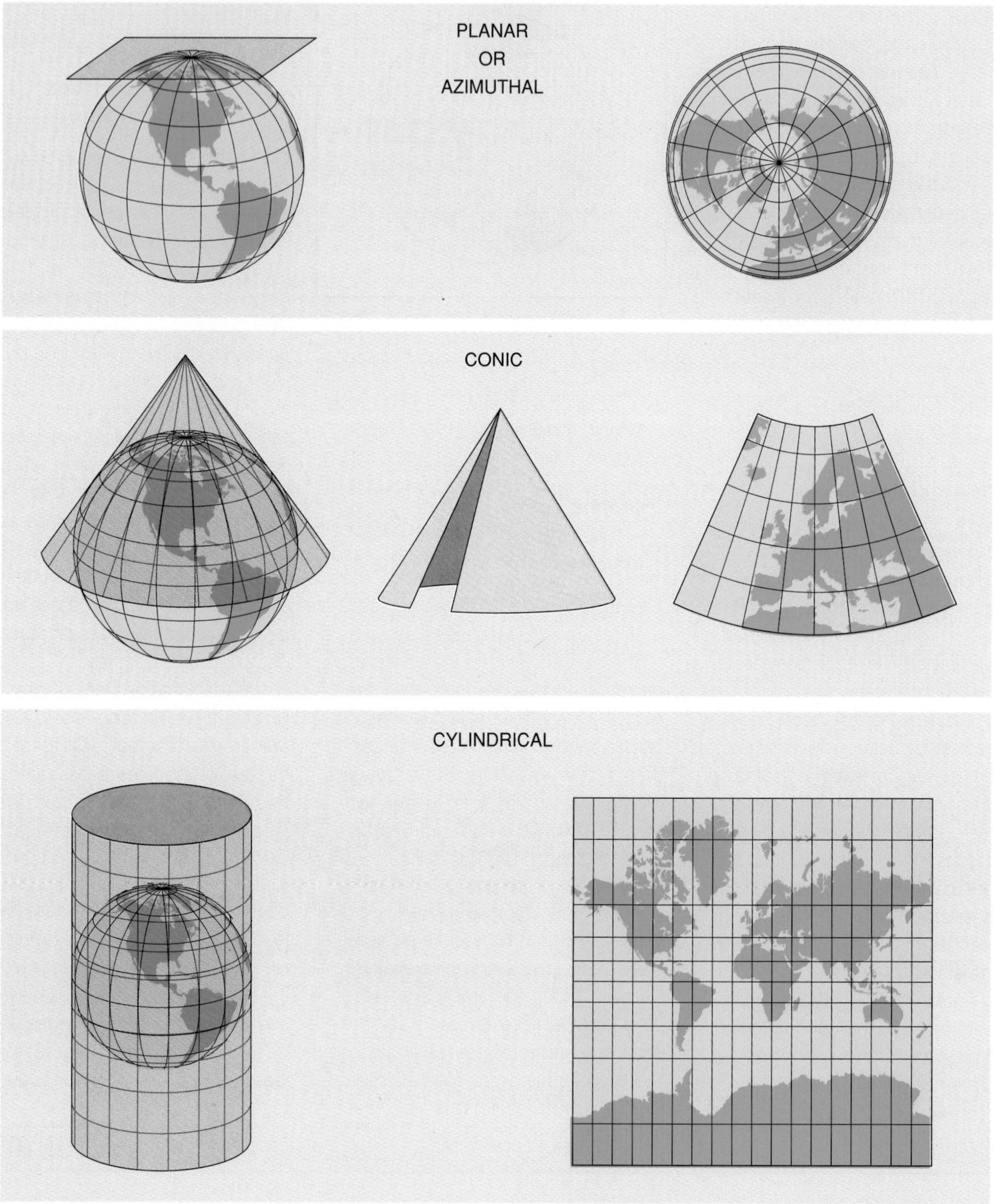

PLANAR
OR
AZIMUTHAL

CONIC

CYLINDRICAL

FIGURE 1 (continued)

Map Projections

GOODE'S HOMOLOSINE EQUAL AREA PROJECTION

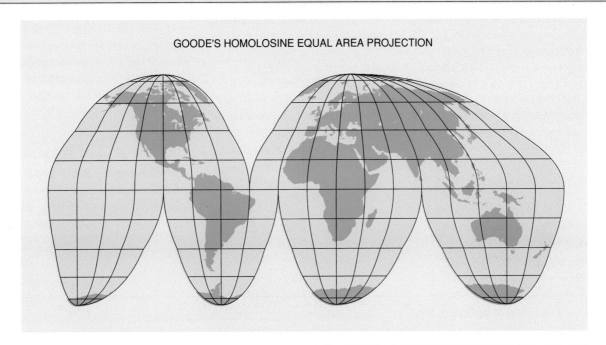

ROBINSON'S PROJECTION

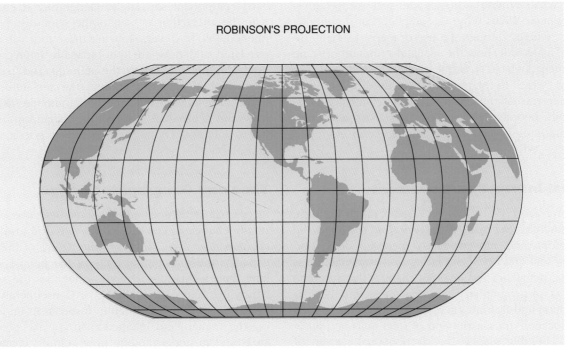

one of the map properties—in this instance, equal area. Other projections in which no particular map property is preserved, but rather all of the properties are distorted, are referred to as compromise projections. The Robinson's projection, originally developed for Rand McNally and recently adopted by the National Geographic Society, is a compromise projection with an oval shape but with some exaggeration near the poles.

The Mercator projection is one of the most commonly used cylindrical projections. In cylindrical projections, the meridians are equally spaced vertical lines that do not converge at the poles. This projection tends to distort the land areas located in the higher latitudes.

Conic projections are good for showing half of the globe, and the map properties of area and shape are not seriously distorted. Planar projections are useful for showing equidistance, facilitating the measurement of distance from the point of tangency to another point. Thus, mapmakers are careful to choose the type of projection that best displays the data being mapped and to use the proper scale. Just as statistics can be misused to prove a particular point, map projections can misrepresent the shape or size of part of the earth's surface. Selection of the proper projection to display specific kinds of data or portray a specific geographic region or the earth as a whole is critical.

Classification and Symbolization

Most maps have a key or *legend* and frequently use symbols such as circles, squares, or dots to represent specific kinds of data. After determining the type of data to be presented on a map, the cartographer usually classifies the data (e.g., population size of cities or countries) into ranked interval categories (i.e., 0–999, 1000–2499, 2500–4000, and so on). Usually, the interval classes utilized are placed in the legend. Depending on the type of map, various symbols such as lines, dots, and stars are also used to represent map features such as rivers, cities, or population density.

When? What? Why? The geographer will ask the questions: Where? What? Why? Geographers can frequently answer "where" through the use of maps. In many instances, maps will show the spatial distribution of some phenomena, which is frequently the first step in answering "why." By mapping the spatial distribution of such elements of the environment as climate, vegetation, soils, people, cities, mineral resources, and manufacturing sites, geographers are more readily equipped to answer "why."

each component separately. Geography is concerned with the interaction as well as the static distribution of phenomena, however. How are these four variables interrelated and how are they spatially interconnected? For example, measuring the amount and pattern of movement of coal and iron ore to the steel mills, of steel to the automotive plants, and of automobiles to the various auto dealerships displays the dynamism of spatial interaction and connectivity between goods, people, and places.

Spatial Interaction/Movement

The distribution of phenomena is static, rather uniform, and immobile. But *spatial interaction* concerns the connections and movements of people, ideas, and goods within and between places. Spatial interaction is dynamic, not static, and involves changes through time and space. A geographer might be interested in the spatial location and distribution of coal and iron ore as well as the location and distribution of steel mills and automobile assembly plants. The distributional pattern of these four specific variables results from considering

Time and Geographic Change

Places and landscapes are not static but rather are constantly changing. The rate of the physical and cultural changes may vary, however. The nature of human occupancy and the cultural traits of the people, particularly their technology, greatly influences the speed at which the landscape is changed. Consequently, a historical dimension is frequently essential in understanding the human-land interaction in specific geographic areas and in understanding how cultures change over space and through time.

People are very mobile, and their spatial movements and migration patterns are of interest to geographers. Through the years, the United States has attracted a large number of immigrants from all over the world, including these young women from India.

Global Interdependence

It is difficult for an individual, city, state, or nation to be wholly independent; in fact, it is almost impossible. Individuals and nations are interconnected politically, economically, and socially—a phenomenon known as **global interdependence**. As improved methods of transportation and communication help to shrink the world in time and space, it is becoming increasingly apparent that the global community is interconnected and interdependent.

An understanding of the concepts mentioned above will enhance your ability to understand the geographic viewpoint and to think geographically. The next two chapters provide a topical introduction to the world that examines the ways in which the physical environment, the human environment, and the role of technology differentiate the developed from the developing world. The remaining chapters discuss the major regions of the world by examining the physical and human environments of these regions, the interaction between the human and physical environments, the resulting cultural landscape that distinguishes each region from other regions, and the spatial connectivity and interaction between regions.

Suggested Readings

Cole, J. P. *Development and Underdevelopment: A Profile of the Third World*. London: Methuen, 1987.

Gaile, G. L., and C. J. Willmott. *Geography in America*. Columbus, Ohio: Merrill Publishing Company, 1989.

Geography and International Knowledge (Report). Washington, D.C.: Association of American Geographers, 1982.

Guidelines for Geographic Education: Elementary and Secondary Schools. Macomb, IL: The National Council for Geographic Education and the Association of American Geographers, 1984.

Hart. J. F. "The Highest Form of the Geographer's Art." *Annals of the Association of American Geographers* 72 (1982): 1–29.

Kenzer, M. *On Becoming a Professional Geographer*. Columbus, Ohio: Merrill Publishing Company, 1989.

Pattison, W. D. "The Four Traditions of Geography." *Journal of Geography* 63 (1964): 211–216.

Murphey, R. *The Scope of Geography*. New York: Methuen, 1982.

Robinson, J. "A New Look at the Four Traditions." *Journal of Geography* 75 (1976): 520–530.

CHAPTER 2

The Natural Environment

IMPORTANT TERMS

Lithosphere

Tectonic processes

Gradational processes

Climate

Desertification

Biomes

Mineral reserves

CONCEPTS AND ISSUES

■ Interaction of all components of the earth, including humans, in a dynamic system.

■ Influence of the earth's natural system on the location and character of human activities.

■ Definition of resources by their usefulness to humans within the existing cultural, economic, and technological situation.

CHAPTER OUTLINE

Landforms
Soils
Climate
Water

Vegetation
Energy and Mineral Resources

G eographers believe that a knowledge of the natural environment is critical to the analysis and understanding of the spatial distribution of human activities and cultures. As in all systems, an alteration in one component of the earth's natural environment affects other components and the overall functioning of the system. Humans are a critical component of the earth's environmental system and are dependent on the earth for food, shelter, and the luxuries of life. Unlike other components of the environmental system, humans have the explicit ability to dramatically alter the natural environment. Events continually demonstrate the power of humans and our technology to alter the natural environment. They also indicate the profound interrelationships between natural systems and human well-being.

The nuclear power plant accident at Chernobyl in 1986 in the Ukrainian Republic of the Soviet Union released harmful radiation into the atmosphere. Spread by the wind, this radiation contaminated land, agricultural products, and water across a large area of the Soviet Union, Eastern Europe, and Scandinavia (Figure 2.1). In less harmful concentrations, the radiation was transported by the air and water to the soils, wildlife, and human food systems of other regions of the world. Thus, local human activities can have regional and even global impacts.

Recent scientific evidence also suggests that human actions may be altering the global climate. Over time, the water and carbon dioxide components of the atmosphere have allowed the earth's surface to warm to habitable temperatures. Fossil fuel combustion, however, is introducing additional carbon dioxide into the atmosphere. In addition, deforestation and ocean pollution have reduced the vegetative mass that performs the necessary task of transforming the carbon dioxide gas to carbon in plant tissue and oxygen gas for the atmosphere. Many atmospheric scientists believe that these practices will cause the earth's surface to heat up from 1°C to 5°C during the next century. The possible global

FIGURE 2.1

Spread of Nuclear Radiation from the Chernobyl Accident

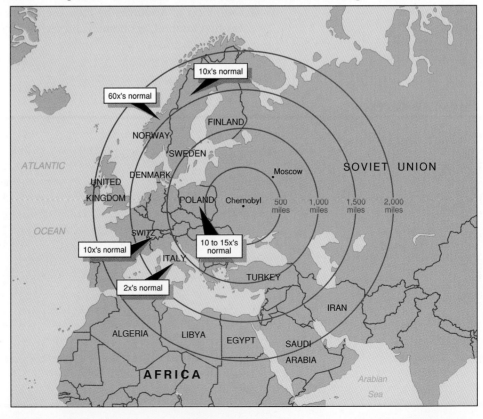

FIGURE 2.2

Drift of Pacific Plate over Hot Spot to Form Hawaiian Island Chain

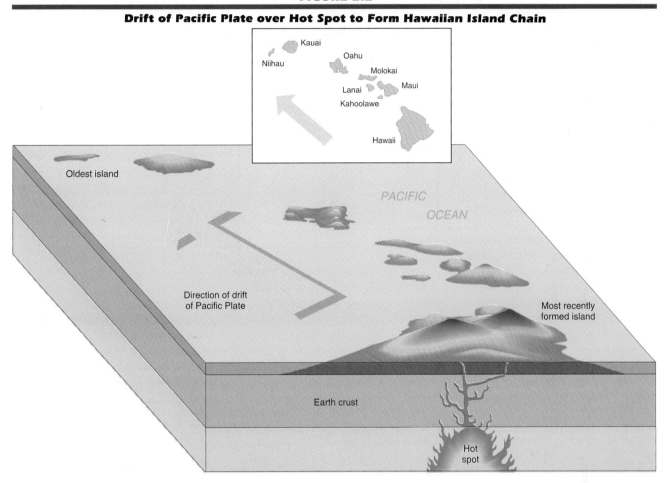

effects include melting of the polar ice caps, coastal flooding, and the alteration of entire regional agricultural economies due to climate change. This chapter introduces the primary components of the earth's natural environment such as landforms, climate, water, soils, vegetation, and energy and mineral resources with an emphasis on their systematic interactions with people.

LANDFORMS

The earth is composed of a central core of molten rock surrounded by a thick layer called a mantle. Upon this mantle is the earth's crust or **lithosphere** composed of rigid plates 6–75 miles (10–120 kilometers) thick. The continental landmasses and oceanic basins of the earth's surface are laid upon this interlocking arrangement of tectonic plates.

A Hawaiian legend illustrates the fundamental theory of crustal plate tectonics. According to this legend, the fiery goddess of volcanoes, Pele, lived originally on the island of Kauai at the western end of the Hawaiian island chain. Pele was forced to flee by the god of the sea to Oahu, then Maui, and finally to the Crater at Kilauea on the island of Hawaii. In reality, the Pacific crustal plate is moving over a relatively stationary "hot spot" that has opened in the earth's mantle. The heat and molten rock are penetrating through the crustal plate from deep in the earth's core to form volcanoes on the surface. Over millions of years as the Pacific plate moved, the intense volcanic activity (Pele) has moved eastward creating each of the Hawaiian Islands in succession. This conclusion is reinforced by estimations of the ages of the various islands. The greater amount of erosion on the islands of Kauai and Oahu (Figure 2.2) indicates that

RESOURCE CREATION

Resources do not simply appear on earth, but are "created" by a perceived human need or desire, science and technology, and economic feasibility. For example, during the period of westward expansion in the United States, ranchers in Texas and Oklahoma cursed the black, foul-smelling ooze that soiled their grazing lands. In the early 1800s, scientific knowledge and technology had not advanced to the point that humans could use this black ooze we now call petroleum. The development of distillation technology that could convert raw petroleum into kerosene and gasoline transformed petroleum from a nuisance into a valued resource. Resource creation to satisfy a human function is an important concept for examining human use of the natural environment. Many natural materials that appear unimportant at the present may become of great value if a function is found for them in the future. The relatively recent recognition that uranium can be used to produce electricity and warheads and the development of technology to do so have greatly altered human existence.

Resource creation is affected by culture as well as technology. Many people, particularly in Asia, believe eating the powdered horn of the rhinoceros will enhance their sexual potency. This cultural manifestation of rhinoceros horn as an aphrodisiac resource threatens the existence of the species. The few remaining wild rhinoceroses are being poached solely for their horns. The concept of resource creation also provides insights into the solution to saving the rhinoceros. Using technology to create a substitute that really works as an aphrodisiac would diminish the killing of the rhino for its horn. Rhino horn would be "decreated" as a resource much as whale oil was supplanted by petroleum products for lighting and heating functions. As these examples illustrate, most human and natural resource issues can be examined using ideas associated with cultural values, economic cost, technology, and resource creation.

they were formed prior to the formation of Maui and Hawaii.

The movement of tectonic plates accounts for much of the volcanic activity, mountains, earthquakes, and rift valleys on the earth's surface. These actions have taken place over hundreds of millions of years, compared to the short three to five million years that humans have inhabited the earth. Approximately 500 million years ago (late Cambrian), the earth's surface landmasses on the various plates were separate and far different from their present arrangement. By 250 million years ago (Permian), the landmasses had merged into one massive supercontinent called *Pangaea* (Figure 2.3). Evidence of Pangaea is found in the apparent interlocking shape of the continents and the fossils of similar land-living and shallow-water organisms that have been found on continents widely separated by deep oceans. A reasonable explanation is that these organisms lived, died, and became fossilized when the continents were joined together, but later were separated by the drift of the continents.

Recent evidence indicates that tectonic plates are in continuous motion. The African plate is moving northward and colliding with the Eurasian plate. The earthquake and volcanic activity around the Mediterranean Sea is the result. The sliding action of the Pacific plate against the North American plate accounts for the earthquake and volcanic activity along the Pacific coast of the United States.

The landforms on the earth's surface include mountains, hills, valleys, plateaus, and plains. **Tectonic processes,** such as diastrophism—the folding and faulting of the earth's crustal plates—and volcanism, work to build up the earth's surface. **Gradational processes,** such as mechanical and chemical weathering, glaciation, and erosional and depositional forces, work to weather and wear down the earth's surface. Running water, wind, glaciers, and ocean waves and currents

FIGURE 2.3 (right)

Crustal Plate Movement

are gradational agents that affect the formation of landforms.

SOILS

Tiny rock particles mixed with organic mater, water, air, and living organisms form the thin veneer of soil that covers the earth's surface. The regional differences in soils are based upon variations in bedrock or parent material, topography, climate, and vegetation. The parent material is the primary source of the chemical elements in a soil and is a governing factor in the fertility of a given soil. The vegetation, temperature, and moisture influence the rate at which soil is developed from the parent material. The topography affects soil composition and retention. Mountainsides are areas of shallow soil with low productivity because gravity and water tend to scour the soil from the mountainside. The valleys of the Nile River in Egypt, the Huang He (Yellow) River in the People's Republic of China, and the Mississippi River in the United States are well known for their deep, rich soil and agricultural productivity. Topography has aided in the deposit of nutrient-rich soils in these valley areas.

Table 2.1 lists the major soil orders or classes and the approximate percentage of the earth's surface that they cover. The classes in the table are based on the United

An eroded butte in the desert of northern Arizona. The butte was formed primarily by water erosion of less-resistant rock.

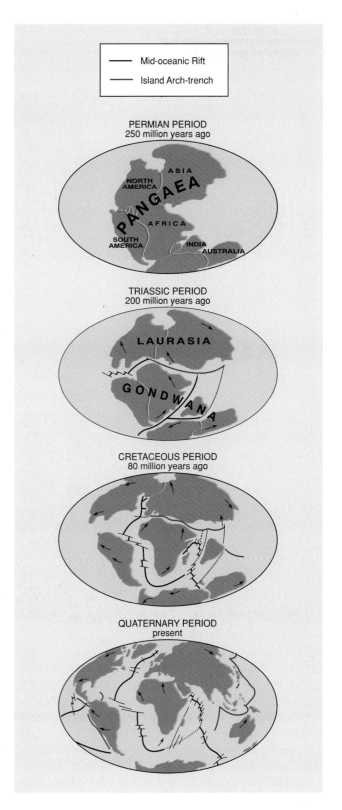

——	Mid-oceanic Rift
——	Island Arch-trench

PERMIAN PERIOD
250 million years ago

ASIA
NORTH AMERICA
PANGAEA
AFRICA
SOUTH AMERICA
INDIA
AUSTRALIA

TRIASSIC PERIOD
200 million years ago

LAURASIA
GONDWANA

CRETACEOUS PERIOD
80 million years ago

QUATERNARY PERIOD
present

FIGURE 2.4

Soils of the World

SOURCE: Adapted from U.S. Soil
Conservation Service, 1972.

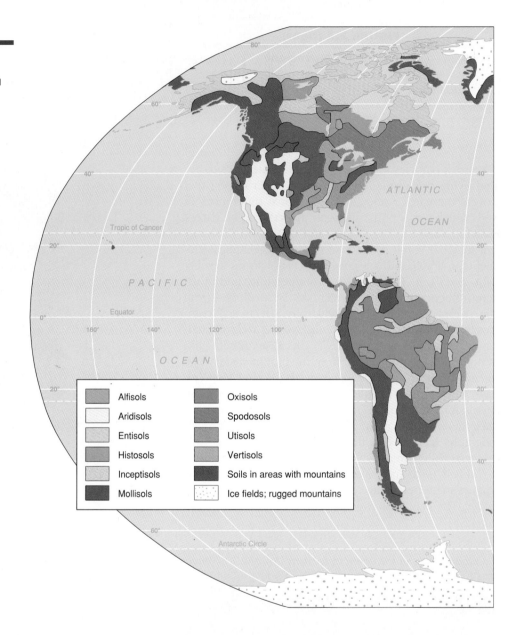

Alfisols		Oxisols	
Aridisols		Spodosols	
Entisols		Utisols	
Histosols		Vertisols	
Inceptisols		Soils in areas with mountains	
Mollisols		Ice fields; rugged mountains	

States Comprehensive Soil Classifications System (often called the 7th Approximation System), which focuses on the specific properties of a given soil rather than on its environment or how it was formed. This classification is more logically consistent, but makes regionalization more difficult. The third column of Figure 2.4 and Table 2.1 provide an overview of the geographic distribution of the various soils across the earth's surface.

Human use of these various soils is dependent on the climate and the level of available technology as much as on the properties of a particular soil. For example, the vertisol soil order is found in subtropical and tropical areas, such as east central Africa, central India, and eastern Australia (Figure 2.4). These soils retain nutrients and water necessary for agriculture, but they are difficult to till when wet. Expensive, fossil-fuel–consuming tractors and other modern farm equipment are necessary to plow these heavy, packed soils to achieve maximum production. Cultivation by human or animal power, as is the case in Africa and India, results in diminished productivity. The mollisol order of soils is located in the Great Plains region of the United States, the

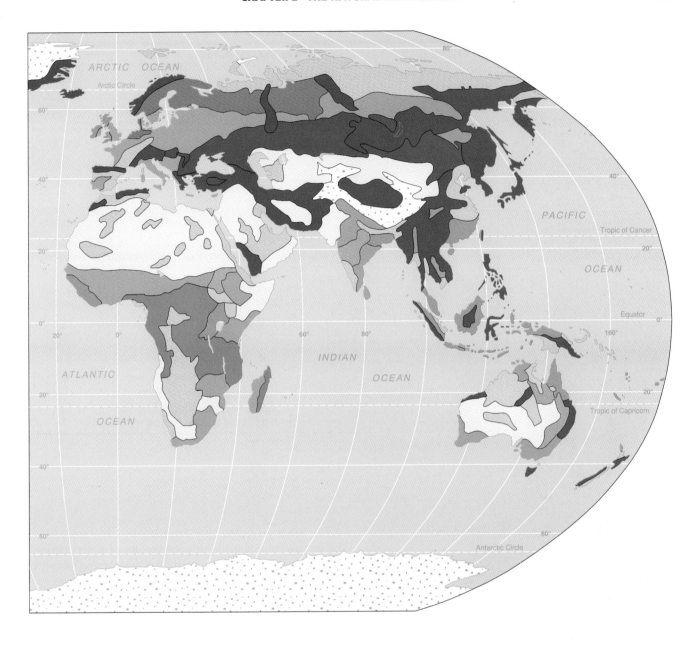

Pampas of Argentina and Uruguay, and the steppes of the Soviet Union, Mongolia and northwestern China. They are some of the most naturally fertile soils in the world and produce most of the world's commercially sold grain. Many areas where mollisols occur have variable precipitation, so maximum sustained production requires irrigation. Intensive farming in these areas invites wind and water erosion should the tilled soil become too dry or be left unvegetated for a time. As Table 2.1 indicates, the aridisols and inceptisols cover large portions of the earth's surface. Figure 2.4 shows that these generally fertile soils occur in areas where intensive agriculture is limited by climatic and technological factors.

CLIMATE

Several earth processes and human activities are directly linked to the variabilities of temperature, moisture, and wind. The long-term temporal and geographic pattern of these atmospheric and weather characteristics is termed **climate.**

TABLE 2.1

Global Soil Orders

SOIL ORDER	PERCENTAGE OF TOTAL WORLD SOILS*	CHARACTERISTICS
Inceptisol	16%	Young soils found on glacial, alluvial, or volcanic deposits; productive for agriculture.
Aridisol	19	Soils of dry areas having little organic matter and shallow depth.
Mollisol	9	Mature soils often found under midlatitude grasslands; most agriculturally productive.
Spodosol	5	Infertile, acidic, sand soils that occur mostly in cold-winter forested areas.
Alfisol	15	Soils of wide geographic distribution, containing concentrations of aluminum (Al) and iron (Fe) compounds.
Ultisol	9	Red soils due to concentrations of iron, acidic with poor fertility.
Entisol	13	Soils with limited profile development due to erosion, youth, or human disturbance.
Vertisol	2	Clay-rich soils of tropical and subtropical areas.
Oxisol	9	Soil of the wet tropics; cultivation results in severe nutrient depletion.
Histosol	1	Very organic soils; found extensively in higher latitude tundra and bog environments.

* Ice fields and unclassified areas make up 2–3 percent.

Lake Stelli and the Matterhorn in the Swiss Alps. High elevation results in a tundra environment.

The foremost factor affecting climatic variation across the earth's surface is the uneven receipt of solar energy. The spherical earth (actually an oblate spheroid) rotates once every 24 hours on an axis tilted at 23½ degrees. The earth revolves around the sun once every 365¼ days (Figure 2.5). The axis of the earth tilts the northern half of the earth (northern hemisphere) toward the sun during the period from late March to late September creating summer in the northern hemisphere. This relationship is reversed in December (winter in the northern hemisphere). Without the axis tilt, the earth would not have seasonal change. The surface of the earth near the equator (0° latitude) receives the greatest amounts of sun energy (insolation) because the most direct rays (perpendicular to the surface) are received between the Tropics of Capricorn (23½° south latitude) and Cancer (23½° north latitude). These areas tend to have the warmest climates on an annual basis with more poleward locations being colder and more variable.

Latitudinal position is not the only reason for climatic variations. The abundance of sun energy in the equatorial area is the generating force behind the circulation patterns of the atmosphere and ocean currents. The relative abundance of heat energy creates convectional currents in the oceans and atmosphere that transfer heat toward the midlatitudes and poles. Areas of the northern United States and southern Canada lying at 40–45° north latitude are much different climatically than the relatively warm areas of southern France and Italy also at 40–45° north latitude. The difference is attributable to the transfer of heat from the Mediterranean Sea and the Atlantic Ocean and the prevailing westerly winds at this latitude. Water bodies retain heat longer than the adjacent land areas. The winter winds blowing across the water transfer heat from the water to the land areas keeping the land temperatures much more moderate.

Another example is found in the southern hemisphere. The cold ocean current (Humboldt Current) moving northward from the Antarctic is the return flow from the convectional flow of heat energy from equatorial waters toward the poles. This cold ocean current accounts for the extreme dryness of the coast of northern Chile in South America. The winds blowing across the cold water carry very little moisture. When moving onshore, the winds are warmed, decreasing the chance of rainfall still further. The result is the Atacama Desert where in some locations human inhabitants have never recorded rainfall.

As Figure 2.6 indicates, climatic regions can be identified by considering variations in temperature, precipitation, and the seasonality of precipitation. Variation in vegetation types is closely associated with climatic variation. The tropical wet climates encompass the rainforest vegetation zone, the tropical **monsoon,** and the **savanna** vegetation areas near the equator. The rainforest areas of central Africa and South America are warm and wet with no dry season. Ironically, the seeming lush productivity of the tropical rainforest proves to be an illusion when humans attempt to convert these areas to

FIGURE 2.5

Earth-Sun Relationship

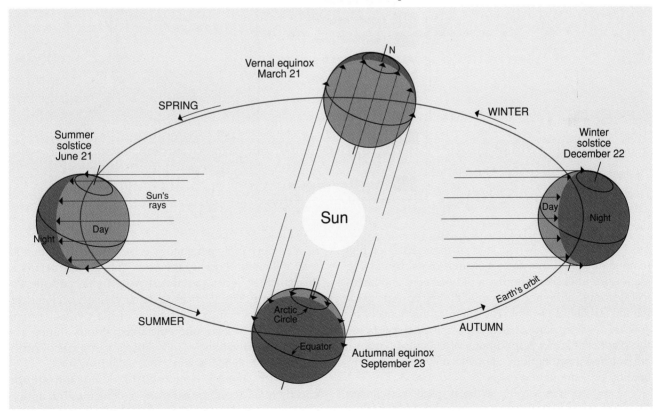

FIGURE 2.6

Climates of the World

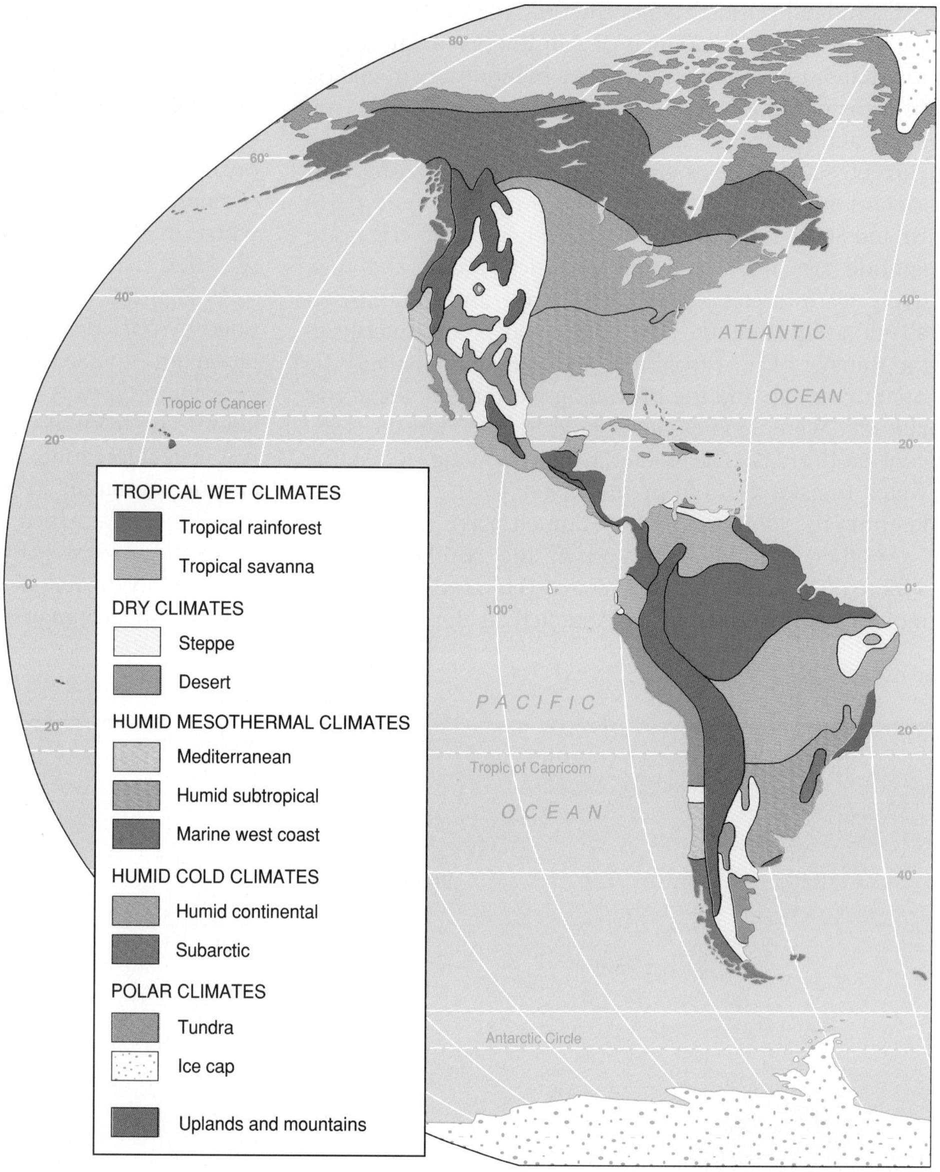

TROPICAL WET CLIMATES
- Tropical rainforest
- Tropical savanna

DRY CLIMATES
- Steppe
- Desert

HUMID MESOTHERMAL CLIMATES
- Mediterranean
- Humid subtropical
- Marine west coast

HUMID COLD CLIMATES
- Humid continental
- Subarctic

POLAR CLIMATES
- Tundra
- Ice cap

- Uplands and mountains

modern agriculture. When land in these areas of high temperatures and moisture is cleared of its natural vegetation for intensive farming, rapid bacterial and chemical decomposition and high rainfall rapidly deplete the natural fertility of the soil. The chemical nutrients of the soil are released and carried deep into the soil below the plant root zone. Applications of chemical fertilizers are washed away rapidly, allowing only a minimal uptake by the agricultural crops.

The monsoon is a form of the tropical wet climate. Monsoon areas result where the general wind patterns in a region shift, generating large amounts of rainfall over a limited period of the year. This climate occurs in coastal West Africa, Southeast Asia, India, and the Pacific islands. The tropical savanna climate is characterized by a dry winter with much of the rainfall coming during one or two unpredictable periods during the year. The characteristic landscape of vast areas of Africa and South America is grassland, called savanna.

The dry climates composed of desert and **steppe** constitute a second broad class of climates. The annual rainfall in the desert climate areas, which include por-

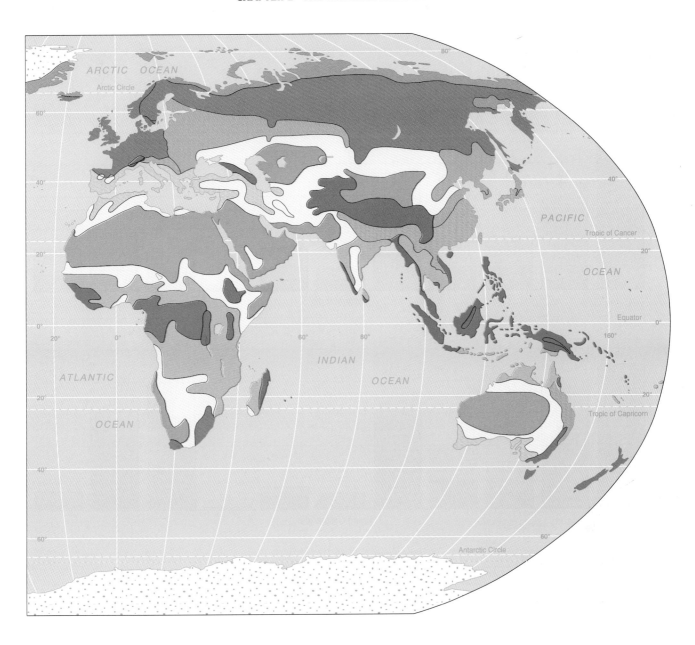

tions of North Africa, the southwestern United States, central Asia, and Australia, ranges from zero to 10 inches (25 centimeters). Receiving slightly more precipitation than the desert is the semiarid steppe (10–20 inches; 25–50 centimeters). These steppe areas border on the desert areas. It is in steppe areas that accelerated **desertification** is taking place due to human activities. The United Nations Environmental Program estimates that desertification threatens 35 percent of the earth's land surface and 19 percent of its population. Land is being reduced to desertlike conditions at a rate of 23,160 square miles (60,000 square kilometers) annually. The causes include natural drought conditions aggravated by overpopulation, overgrazing by livestock, farming of marginal lands, and the lack of effective countermeasures. As with all problems of the physical environment, the spread of desertification has a human cost. Figure 2.7 indicates the number of rural people affected by desertification. The effects of desertification are markedly different between regions and between types of dryland uses. The most severely impacted regions are rangelands in Sahelian Africa, rainfed croplands in Africa and

FIGURE 2.7

Populations Affected by Desertification (in Millions)

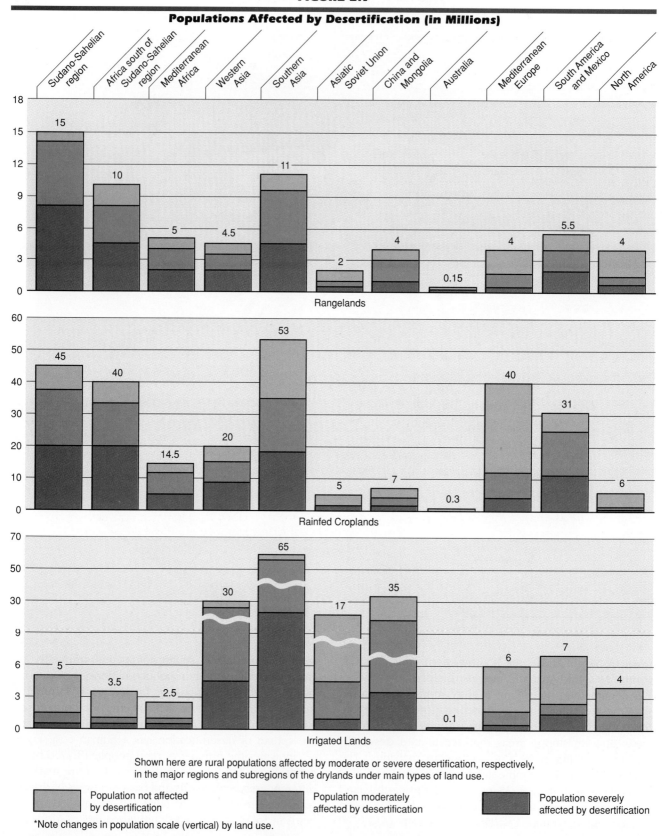

Shown here are rural populations affected by moderate or severe desertification, respectively, in the major regions and subregions of the drylands under main types of land use.

Population not affected by desertification

Population moderately affected by desertification

Population severely affected by desertification

*Note changes in population scale (vertical) by land use.

SOURCE: *Desertification Bulletin,* United Nations Environmental Program, 11 (December 1984): 3.

southern Asia, and irrigated lands in southern and western Asia.

The humid mesothermal (moderate temperature) climates occur in the mid-latitudes. This broad climate type encompasses the dry, sunny ***Mediterranean climate*** of southern California and the coastal areas of the Mediterranean Sea, the humid subtropical areas of the southeastern United States, and the mild, wet marine west coast climates of the northwestern coast of the United States, southern Alaska, and a portion of Europe.

The humid microthermal climates include the humid continental climate located in the north central and northeastern portions of North America and the western portion of the Soviet Union, and the subarctic climate located in the northern portion of North America and the Soviet Union. These climates occur only on the large landmasses of the northern hemisphere. The effects of continental size (great distance from a moderating ocean to the west) and high latitude make the subarctic areas extremely cold and dry.

The last of the major climate classes is the cold polar climate. The average temperature of the warmest month of the year is below 50°F (10°C). The polar ice caps, the ***tundra*** climate-vegetation zone, and permanently frozen soil (***permafrost***) exist in the polar region. Buildings, petroleum pipelines, and other heated structures must be built above ground to avoid thawing the permafrost. Because permafrost does not thaw uniformly when heated by a house or a pipeline, the thawing soil provides uneven ground support for these structures, causing them to fracture and disintegrate.

WATER

The water environment of the earth, termed the ***hydrosphere,*** is intricately related to climate. The total amount of water in the hydrosphere is fixed. The primary variables in the human use of the hydrosphere are as follows:

1. In what form is the water? Only freshwater is directly useful for human and animal consumption and farming.
2. What is the geographic distribution of the available water?
3. What is the quality of the water in terms of contaminants?

Approximately 97 percent of the earth's water is found in the oceans. While this is a tremendous amount of water, natural salinity makes ocean water unusable

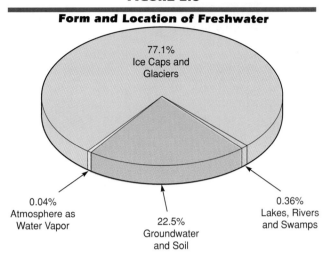

FIGURE 2.8

Form and Location of Freshwater

77.1%
Ice Caps and
Glaciers

0.04%
Atmosphere as
Water Vapor

22.5%
Groundwater
and Soil

0.36%
Lakes, Rivers
and Swamps

for direct human consumption or agriculture. That leaves only 3 percent of the earth's water as freshwater. As Figure 2.8 shows, even this 3 percent is not totally accessible to people. The majority of the freshwater (77.1 percent) is contained in the polar ice caps and glaciers of the world far from the concentrations of population. The second major source is groundwater; this is more widely available, but in many areas only limited amounts are easily accessible. Obtaining groundwater from below the earth's surface requires money and technology to construct wells and pumps. But money and technology are not available in many of the poorer areas of the world where the need for water is greatest. Thus,

Clear-cut forest harvesting in the Pacific Northwest. This practice leads to erosion and stream sedimentation if it is undertaken on sloped terrain.

FIGURE 2.9

Per Capita Water Availability in the Year 2000

SOURCE: Council on Environmental Quality, *Global 2000 Report* (Washington, D.C.: U.S. Government Printing Office, 1981).

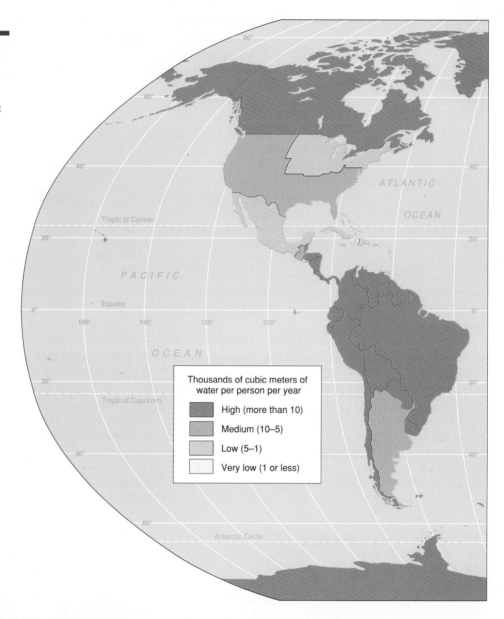

Thousands of cubic meters of water per person per year

High (more than 10)
Medium (10–5)
Low (5–1)
Very low (1 or less)

only a small amount of the earth's water is in an accessible location and an easily usable form.

Precipitation and water available for human use are not the same thing. A comparison of Africa and North America points this out very vividly. Africa receives approximately 5040 cubic miles (21,000 cubic kilometers) of precipitation per year. North America receives approximately 3840 cubic miles (16,000 cubic kilometers) per year. In Africa, however, an estimated 82 percent of this precipitation is lost to human use by evaporation, while only 60 percent is lost to evaporation in North

America. As a result of its higher latitudinal position, North America has lower temperatures, more atmospheric humidity, and consequently a higher net availability of liquid freshwater for human use—1536 cubic miles (6400 cubic kilometers). Areas in hotter and drier Africa generally will have less water available for human use—907 cubic miles (3780 cubic kilometers)—due to evaporation and evapotranspiration through plants. An examination of projected water availability for the year 2000 indicates a vast global problem affecting most regions of the world (Figure 2.9). The most severe prob-

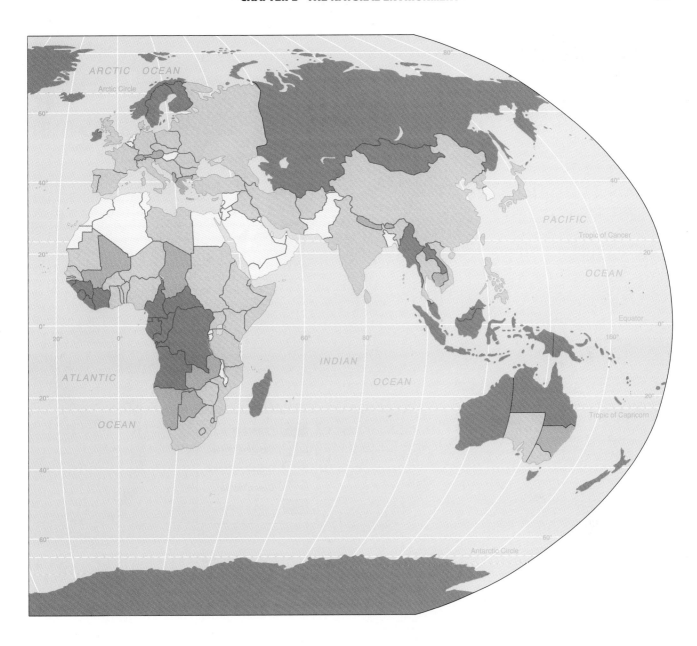

lems will be in the semiarid and arid areas of West and East Africa, the central United States, Mexico, and central Asia. In the United States, the problem will result from the great increase in commercial irrigated agriculture and urban growth in arid areas. In Mexico, Africa, and Asia, increased agricultural usage and the growth in rural and urban populations will tax the available supplies.

In many areas where water is sufficient, contamination by urban sewage, chemicals, heavy metals, and agricultural waste products has rendered the water unfit for many uses. Portions of the Great Lakes in North America, Lake Baikal in the Soviet Union, and the Rhine River basin in Western Europe are major examples of valuable freshwater resources that have become so contaminated that their use for human consumption is limited.

VEGETATION

Natural vegetation can be classified into five general vegetative **biomes:** forest, savanna, grassland, desert,

FIGURE 2.10

FIGURE 2.10

Vegetative Biomes of the World

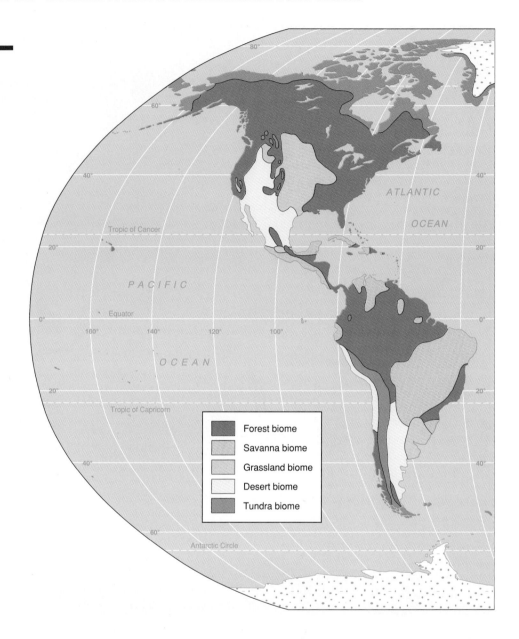

and tundra (Figure 2.10). The characteristics of the predominant vegetation in an area are a reflection of the climate and, to a lesser degree, the soils of that region. Figure 2.11 shows the approximate zonation of vegetation biomes by latitude. The natural grassland and bush savanna biomes are in areas too dry or too variable in precipitation for tree growth. Forest vegetation is located in both the moist equatorial latitudes and the mid-latitudes, but the species vary.

The conditions that create these biomes also control the type of crops and agricultural methods available for human use in these areas. In general, shifting agriculture is practiced in the tropical rainforest areas where the depletion of soil nutrients forces the farmers to relocate periodically. The predominately pastoral areas and irrigated areas correspond to the semiarid savanna and grasslands and the arid desert areas. Intensive, row crop agriculture cannot be utilized in these areas without a massive investment in irrigation and farming technology. People in these areas in Africa and Asia are facing food shortages, but are unable to afford the investments necessary for more intensive agriculture. Much of the

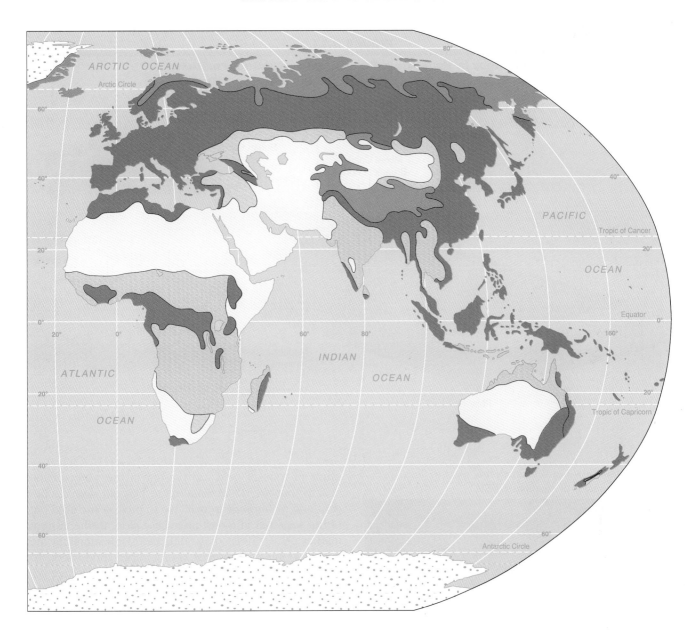

earth's surface that has adequate, available water, moderate climate, and suitable soils for more intensive agriculture lies in regions of high economic development, low population growth, and abundant food, such as the United States and Europe.

ENERGY AND MINERAL RESOURCES

Human success on earth results in large part from our abilities to use inanimate sources of energy. The evolution of the use of wood, water, coal, petroleum, and

uranium as energy sources was critical to the development of modern civilization. Associated with energy sources and use are the other mineral resources, such as iron, copper, and diamonds, that are required for modern commerce and industry. Estimation of how much of a resource, such as petroleum, coal, or gold, actually exists below the earth's surface in an area is difficult. This difficulty is heightened because the **mineral reserves** of a resource are estimated as the amount of a resource obtainable at the current market price using the available technology. For example, thou-

FIGURE 2.11

Changes in Vegetation Relative to Latitudinal Position

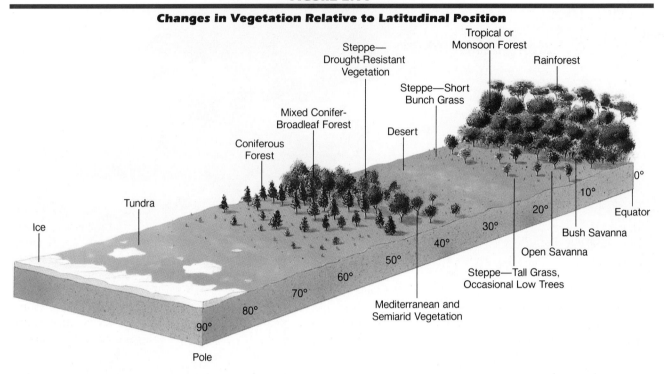

A small output (low-head) hydroelectric plant in the Pyrenees Mountains of southern France.

sands of tons of gold are present in the water of the oceans, but are not considered to be gold reserves because no technology is available to process the gold out of the seawater at a cost anywhere close to the current market price. Given these difficulties in estimating the total amounts of resources, useful information is obtainable by examining the general patterns of resource acquisition and use. These patterns provide insights into the geopolitical and economic development future of the world.

The ideal situation is to have energy and mineral resources. Many of the wealthy, developed nations of the world have large domestic supplies of these resources (i.e., the United States and Canada). In the United States, the dwindling of the domestic resource base has resulted in the necessity to import a high percentage of resources to meet consumption demands (Table 2.2). The Soviet Union retains a large and relatively untapped domestic resource base. If a nation does not have these resources, then it is important to maintain relations with those nations that do have them. For a time, many developed countries controlled colonial territories that they exploited for resources to build their own econ-

TABLE 2.2

Mineral Imports as a Percentage of Mineral Consumption, 1988

	UNITED STATES	EUROPEAN COMMUNITY	JAPAN	SOVIET UNION
Bauxite	93%	100%	100%	35%
Copper	32	100	92	0
Iron ore	37	82	98	0
Manganese	100	100	100	0
Chromium	100	100	100	0
Platinum group	98	98	98	0

SOURCE: U.S. Central Intelligence Agency, *Research Aid: Handbook of Economic Statistics*, (September 1990).

omies. Others such as Japan lacked significant colonial holdings, but have been able to acquire the necessary resources to supply a rapidly expanding economy by maintaining good relations and trade with resource-rich countries.

Table 2.3 indicates the countries in which the major minerals reserves are located. The United States and Canada control significant reserves of many of these minerals, but many other countries, particularly the So-

viet Union and the Republic of South Africa, are major suppliers. Geopolitical strategies are very evident in resources supply and demand relationships. The United States and the countries of Western Europe have found it difficult to sever relations with the Republic of South Africa in spite of South Africa's apartheid situation. This is because the Republic of South Africa is a major supplier of uranium, manganese, chromium, diamonds, phosphates, gold, and platinum. The only other major

The quest for energy often conflicts with other human activities. Here a surface (strip) coal mine is encroaching on a farm residence in the midwestern United States.

TABLE 2.3

Major Mineral Reserves

MINERAL	LOCATION OF MAJOR RESERVES
Antimony	Bolivia, South Africa, Mexico
Bauxite	Guinea, Australia, Brazil, Jamaica
Beryllium	Brazil
Cadmium	Canada, USA, Australia
Chromium	South Africa, Zimbabwe, Finland
Cobalt	Zaire, Zambia, Morocco
Copper	Chile, USA, Zambia, Canada, USSR
Diamond	Zaire, Botswana, Australia
Gold	South Africa, USSR, USA
Iron ore	USSR, Brazil, Australia, India
Lead	USA, Australia, Canada
Lithium	Chile, USA, Zaire, Canada
Manganese	USSR, South Africa, Australia
Mercury	Spain, USSR, Algeria
Molybdenum	USA, Chile, Canada
Nickel	New Caledonia, Canada, Cuba
Platinum group	South Africa, USSR, Zimbabwe
Selenium	USA, Canada, Chile, Peru
Silver	USA, Canada, Mexico
Tin	Malaysia, Indonesia, Thailand, China
Tungsten	China, Canada, USA, Korea
Vanadium	South Africa, USSR, China
Zinc	Canada, USA, Australia

supplier for several of these resources so critical to the Western industrialized countries is the Soviet Union. For example, the Republic of South Africa and the Soviet Union control approximately 70 percent of the world's gold reserves and over 90 percent of the platinum reserves.

The key to high-production agriculture, industrialization, and corresponding economic development is energy. Until the 1950s the major industrial energy sources were wood and coal. At present, petroleum and natural gas have surpassed coal as the major sources of energy while nuclear, biomass, and solar energy are of increasing but lesser importance. Coal, petroleum, and natural gas (*fossil fuels*) are required to power industry, generate electricity, heat buildings, and fuel the

transportation system. Regions that have access to fossil-fuel energy sources will have the economic development advantage for the foreseeable future (Figure 2.12). Several of the alternative energy sources are too costly, require extensive use of high technology, or produce the wrong form of energy. This is the situation with nuclear fission energy. Nuclear technology is very expensive, requires a high level of technological skill to operate, and produces only electricity.

As the activities of the Organization of Petroleum Exporting Countries (OPEC) have demonstrated in recent years, the world's economy is tied to the price and availability of petroleum. Figure 2.13 shows the convoluted nature of the petroleum transport system from the countries of supply to the countries of demand. The United

FIGURE 2.12

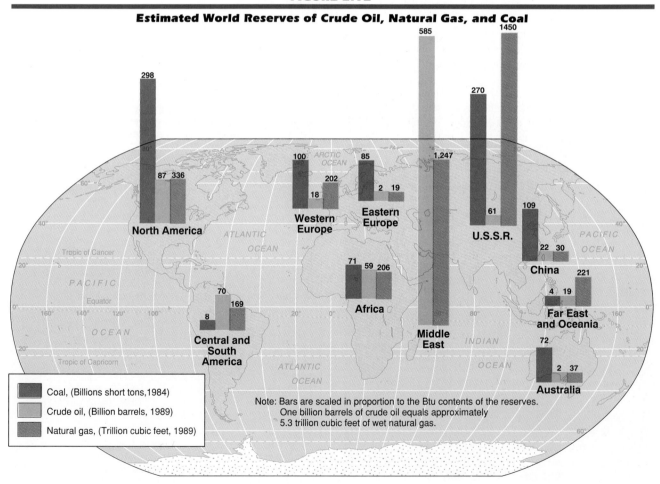

Estimated World Reserves of Crude Oil, Natural Gas, and Coal

Note: Bars are scaled in proportion to the Btu contents of the reserves. One billion barrels of crude oil equals approximately 5.3 trillion cubic feet of wet natural gas.

Legend:
- Coal, (Billions short tons, 1984)
- Crude oil, (Billion barrels, 1989)
- Natural gas, (Trillion cubic feet, 1989)

SOURCE: U.S. Energy Information Administration, *Annual Energy Review* (1989).

FIGURE 2.13

International Crude Oil Flow (Thousand Barrels per Day), 1986

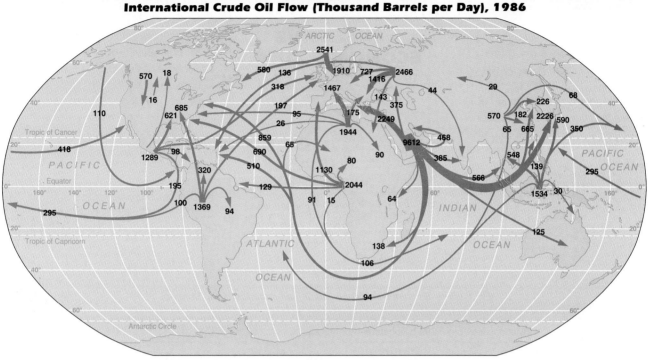

Note: Arrows indicate origins and destinations but not necessarily specific routes. Several minor routes and quantities are not displayed.

SOURCE: U.S. Energy Information Administration, *Annual Energy Review* (1986).

States, Japan, and Western Europe are the major consumers of all forms of energy, especially petroleum fuels (liquid and natural gas), which are used primarily for transportation, heating of buildings, and generation of electricity. Several countries with large populations do not have significant petroleum resources, nor do they have large reserves of coal, as do China, the United States, and Germany. These countries include India, Brazil, Japan, Bangladesh, Pakistan, Italy, and France. Developing countries, such as Bangladesh and India, are restrained from more rapid development by this lack of fossil fuel. Developed countries, such as Italy and Japan, can afford to import fuels, but are in a precarious dependency situation to maintain their economic growth.

The natural environment provides the materials necessary for human existence. Although people can alter the natural environment dramatically, the environment greatly influences the substance and form of human endeavors. A very close association exists between use of

the natural environment and cultural evolution and economic development.

Suggested Readings

Brown, L. R. *State of the World*. New York: W. W. Norton, 1991.

Cutler, S. L. *Exploitation, Conservation, Preservation,* 2nd ed. New York: John Wiley & Sons, 1991.

Marsh, W. M. *Earthscape: A Physical Geography*. New York: John Wiley & Sons, 1987.

Miller, G. T. *Living in the Environment,* 5th ed. Belmont, Calif.: Wadsworth Publishing, 1988.

National Research Council. *The Effects on the Atmosphere of a Major Nuclear Exchange*. Washington, D.C.: National Academy Press, 1985.

Strahler, A. N., and A. H. Strahler. *Modern Physical Geography,* 3d ed., New York: John Wiley & Sons, 1987.

EXCLUSIVE ECONOMIC ZONES

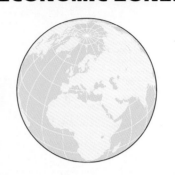

F or centuries the open oceans have been recognized as a common property resource available for use by people from all nations. In recent years increasingly large areas of ocean adjacent to national coastlines have been placed under sovereign control. The most widely recognized geographic zones of sovereign coastal control include (1) the internal coastal area, such as coastal bays and river mouths; (2) the territorial zone, defined as a continuous zone bounded by a 3-nautical-mile (5.56 kilometers) limit outward from the coastal baseline; and (3) the contiguous zone, which is recognized as having a 12-mile (22.2 kilometers) outward limit by most nations. These zones are encompassed by (4) the exclusive economic zone (EEZ), which has a 200-mile (370.4 kilometers) limit. (5) the continental margin is the farthest outward zone varying in width between 700 miles (1296.4 kilometers) in the Barents Sea and 20 miles (37 kil-

FIGURE 1

Exclusive Economic Zone (EEZ) of the United States

Acreage deemed within the U.S. EEZ includes the United States proper, 2.787 billion acres; the Commonwealth of the Northern Mariana Islands, 0.299 billion acres; territories and possessions, 0.839 billion acres.

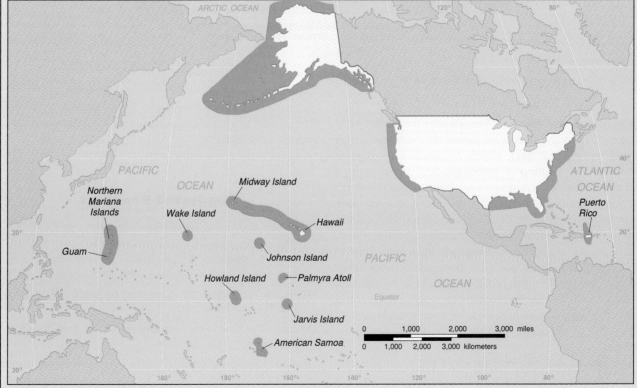

SOURCE: R. W. Rowland, M. R. Goud, and B. A. McGregor, *The U.S. Exclusive Economic Zone—A Summary of its Geology, Exploration and Resource Potential* (U.S. Department of the Interior, Geological Survey Circular 912, 1983)

ometers) west of the Niger River delta depending on the continental geology of the ocean floor.

Control of these zones is complicated by the lack of agreement of all nations on 3-, 12-, and 200-mile limits for control of fishing, minerals, navigation, and environmental regulation. Compounding the situation further are definitional problems involving the location of the coastal baseline and the outer edge of the continental margin. Overlapping claims by nations in close proximity to each other have led to both legal and military action.

Currently, the geographic definition and use of the EEZs are the most controversial issues. The nation claiming an EEZ has exclusive control over the use of the ocean areas and the resources contained on the sea-floor. This control includes access to the fish, petroleum, and seabottom mineral deposits located in the zone. For example, the United States established an EEZ in 1983 for an area that encompasses 3.9 billion acres. This EEZ is much larger than the 2.3 billion land acres of the United States and its territories (Figure 1). Deposits of manganese nodules, cobalt-ferromanganese crusts on seamounts, and polymetallic sulfides formed by seafloor volcanism are massive and, in some cases, far richer than land deposits of these minerals.

For all nations control over the coastal zones has numerous implications. A nation may impose limits on foreign fishing in the richer near-coastal areas. National control of the rich energy and mineral resources in these zones gives coastal nations a major advantage in the future quest for these resources. Exploitation of these undersea resources requires high technology, however. Developing nations may control their EEZs, but without the needed technology, they may be unable to realize their resource and economic development potential. As a result, the developing nations will be dependent on the developed countries for exploitation of their resource frontier. Landlocked nations are precluded from any access to the wealth of the oceans and the seafloor. Thus, control of the oceans—the last great resource frontier on earth—will affect all nations in the future.

CHAPTER 3

Cultures and Economies in a Global Perspective

CONCEPTS AND ISSUES

■ Awareness and appreciation of cultural differences in a global setting.

■ Interrelationship of population dynamics, the demographic transition, and economic development.

■ The tempering of political ideologies with practical concerns to yield different paths toward development.

■ The role of heightened expectations in motivating socioeconomic development or, if unrealized, revolution.

This generation is witnessing the collision of traditional values and cultures with a high-technology, economically well-developed world. Although recent history indicates that massive population and cultural change is inseparable from modern economic development, not all authorities believe that this must be the rule. Many are optimistic that a people can undergo sustainable, ecologically sound economic development and, at the same time, maintain their cultural integrity. This chapter will introduce the features of culture, cultural identity, and population and will examine the ways in which they interact with economic development.

CULTURAL CHARACTERISTICS

When examining specific places and regions, an understanding of the people, their ethnic character, and their culture is crucial to comprehending the geography of the region. When we visit another country or a strange or new place, our senses are bombarded by a rush of images showing how the inhabitants have modified or imprinted the landscape. In the cities and urban areas, we may be impressed by the general atmosphere, the architectural style of the buildings, the city plan, the clothing styles of the people, their language, religion, varieties of food, modes of transportation, and types of businesses and industries. The rural landscape may present an image of tidiness and order or of poverty and overcrowding caused by population pressure and overutilization of the land.

We must learn to examine, evaluate, and critique different landscapes, but in doing so we must be aware that our own background and culture affect the manner in which we view another culture. Seeing another culture objectively, or from that culture's viewpoint, can be difficult. Yet being cognizant of the specific traits, achievements, and accomplishments of other cultures is vital to obtaining a world view and to understanding and comprehending the complexities of our earth.

Culture is learned behavior; it comprises the total way of life of a people including their religious and political beliefs, their language, and their technological knowledge and achievement. It includes material traits, such as clothing and buildings, and nonmaterial traits, such as ideas and values. These material and nonmaterial traits together with behavior patterns are learned, shared, and transmitted by members of a society and passed on to future generations.

Language

Language is defined as a systematic means of communication between people. Most societies transmit their

Indian temple dancers in Rajasthan, India. Although geographers normally consider language, religion, political organization, and technological achievements to be the key cultural traits, the arts including music, painting, the theater, and dance can also help us understand or interpret a geographic region.

TABLE 3.1

Principal Languages

LANGUAGE FAMILY	LANGUAGE	SPEAKERS (MILLIONS)
Indo-European	English	355
	Spanish	250
	Hindi	235
	Russian	215
	Bengali	160
	Portuguese	140
	German	100
	Punjabi	80
	French	75
	Italian	60
Sino-Tibetan	Chinese	1000
Japanese-Korean	Japanese	120
	Korean	60
Afro-Asiatic	Arabic	150
Dravidian	Telugu	65
	Tamil	60
Malay-Polynesian	Indonesian	110

SOURCE: Modified from H. J. deBlij and P. O. Muller, *Human Geography: Culture, Society, and Space* (New York: John Wiley & Sons, 1986), p. 187.

language in vocal and written form, but people do communicate in other ways. Sign language has been used by American Indians and is a principal means of communication among the hearing impaired. Facial and body expressions, such as a smile, frown, or hand gesture, can also transmit a desired message.

Language is the "data bank" of a culture, recording, storing, and dispensing information. Moreover, language is not static; as a society changes, new words must be added to its language. As a society becomes more complex, the ability of its citizenry to communicate becomes more sophisticated.

An estimated 2500 to 3500 languages exist in the world; they vary in grammar, vocabulary, and phonology from place to place. A dialect, which is a variation in the way words are pronounced and used within a language, should not be confused with a language. Dialects and accents are generally associated with specific geographic regions.

There are several existing linguistic classifications, which group languages according to their characteristics into families, subfamilies, and finally language groups. At a broad level of generalization, the world can be divided into about 14 major linguistic families. Chinese, which is part of the Sino-Tibetan language family has the largest number of speakers, followed by English, Spanish, Hindi, and Russian (Table 3.1; Figure 3.1). Twelve other languages have more than 50 million speakers.

Language is certainly a key element of culture, and a culture group will attempt to maintain its language to help further group identity. With trade and finance as well as politics increasingly being conducted on a global basis, and with more spatial interaction and interconnectivity occurring, it has become obvious, however, that the world's inhabitants cannot communicate with each other in 2500 to 3500 different languages. As a result, certain international languages, or *lingua francas,* have emerged as media of communication among persons of different languages. English, Spanish, Russian, Portuguese, French, and Arabic are among the most important languages that serve as lingua francas.

Religion and Belief Systems

Geographers are interested not so much in the theological aspects of religion, but in how religious beliefs affect human spatial behavior. Specifically, geographers

FIGURE 3.1

Language Families of the World

SOURCE: Modified from H. J. deBlij and P. O. Muller, *Human Geography: Culture, Society, and Space* (New York: John Wiley & Sons, 1986), pp. 184–85.

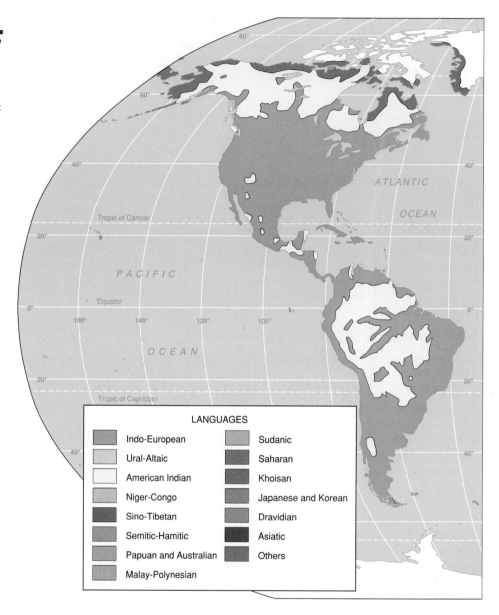

LANGUAGES

- Indo-European
- Ural-Altaic
- American Indian
- Niger-Congo
- Sino-Tibetan
- Semitic-Hamitic
- Papuan and Australian
- Malay-Polynesian
- Sudanic
- Saharan
- Khoisan
- Japanese and Korean
- Dravidian
- Asiatic
- Others

ask, what is the relationship between religious beliefs and activities and the physical and human environment, and how does this relationship help us to interpret the landscape and human behavior?

Religion, or at least some type of belief system, pervades almost all aspects of daily living from eating and working habits to marital relationships, child rearing, views regarding fertility and mortality, and attitudes concerning utilization and management of natural re-

sources. Thus, religions provide most of the standards of basic human conduct. Religion can serve as a catalyst to encourage change in a society, or conversely it can be conservative and promote traditional customs and morals. Religion has affected the environment and vice versa. Not only do churches, mosques, synagogues, and temples dot the world landscape, but the activities of these religious institutions may affect land-use patterns and influence the location of other economic, social,

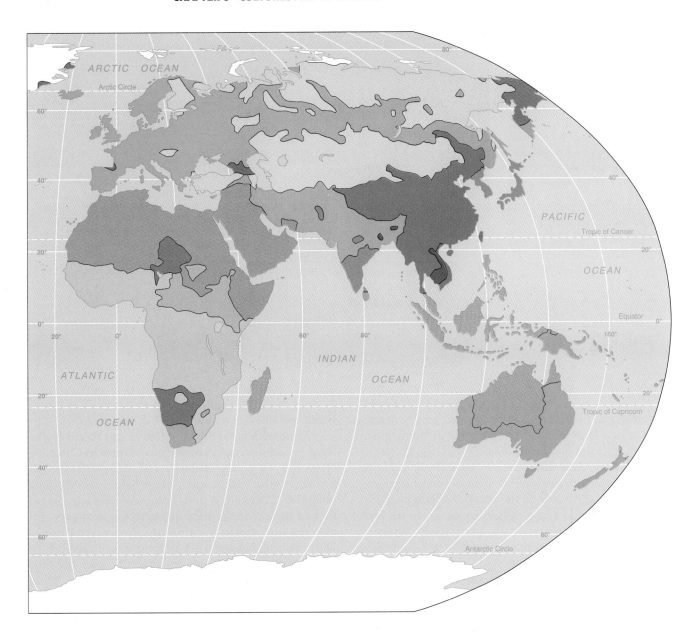

and political institutions. Thus, the nature of the world's major religions, the number of their adherents, and their distributional patterns are important to understanding how and why cultures differ.

Religions tend to be either universalizing or traditional/ethnic. Most universalizing religions attempt to proselytize and appeal to a large number of persons, and normally they are rather widespread geographically. In contrast, traditional/ethnic religions are more regionally concentrated and do not have widespread appeal. Christianity, Islam, and Buddhism are the three main universalizing religions, and they each have a large number of adherents. Hinduism, although not a universalizing religion, is also one of the four largest religions of the world.

Judaism is also considered a major religion even though it does not have a large following compared to the universalizing religions. Unlike other traditional/eth-

A multilingual sign at the Church of the Nativity in Bethlehem. Many cities that receive a large number of foreign visitors each year for religious, recreational, or business purposes frequently print such things as brochures and signs in various languages as an aid to the visitors.

nic religions, however, its followers are widely distributed with a large number of adherents in North America, Europe, and Southwest Asia.

Statistics on religious affiliations are difficult to obtain and are approximations rather than accurate accounts. They are usually based on the number of adherents not membership. To confuse matters more, some denominations, such as Protestantism, count only adults as members, whereas statistics for other religions include all constituents, even infants.

Of the world's 4.4 billion people in 1980, approximately 33 percent adhered to Christianity, 16.5 percent to Islam, 13.3 percent to Hinduism, 6.3 percent to Buddhism, and 0.4 percent to Judaism. More than 70 percent of the world's population adheres to one of these major religions (Table 3.2). Christianity, primarily composed of Roman Catholic, Protestant, and Eastern Orthodox branches, is found mainly in the western hemisphere, Europe, Australia, and New Zealand. Islam is centered in Southwest Asia, North Africa, and Indonesia, whereas Hinduism is found in India and Buddhism is in Southeast and East Asia (Figure 3.2).

Of the remaining 30 percent, most people are nonreligious and atheist (21 percent) with the balance de-

voted to believers in traditional/ethnic religions. Many important traditional/ethnic religions are found in East Asia, including Confucianism and Taoism in China and Shintoism in Japan. Sikhism and Zoroastrianism are found in India. Africa, south of the Sahara, is religiously complex. Christianity and Islam are making inroads there, but traditional religions are also prominent. It is also not uncommon for Africans to practice a Christian faith but also maintain ties to their ethnic religion. For example, two funeral services—one Christian, the other ethnic—might be conducted for an individual. Ethnic religions are also found among peoples in the tropical climates of South America and the tundra climate areas of North America and Eurasia.

Political Systems

The population of the world is divided into more than 200 political units, most of which are referred to as states or nations. The word *state* refers to a political structure, whereas the term *nation* refers to a people. A **state** is a geographical area that is organized into a political unit. In international reference, state and country are often used synonymously; thus one should not confuse the term state at the international level with, or limit it to, the meaning of state as it is used in the United States. A ***sovereign state*** is a country that controls its internal affairs and is not governed by a foreign power. A **nation** encompasses the citizens of a state, but it refers to more than the country's political structure. It is concerned with the people and their strong sense of cultural and national unity. This sense of shared cultural unity is commonly expressed by similarities in language, religion, ethnic composition, and other cultural attributes. The term ***nation-state*** is used to denote a state where there is internal homogeneity within a specific political unit. Such homogeneity exists within very few states, however. Thus, a state is not necessarily a nation or a nation-state, and many states today may have difficulty achieving nation-state status because they include so many diverse linguistic, religious, political, and ethnic groups.

With the demise of colonialism (see the discussion later in this chapter), individual states have sought independence and are promoting nationalism. In several newly created sovereign states, the political boundaries appear to correspond to the region inhabited by a people, a measure that encourages national unity. Many boundaries that have been established on the landscape, however, actually divide or separate people of a common heritage and cultural background. North and South Korea is an excellent example. Conversely, the placement of some boundaries has resulted in diverse ethnic, linguistic, and religious groups being contained within the same political unit, making it difficult to create a "national spirit" or national identity. Many African states fall into this category, and since achieving independence, most of them, including Nigeria, Zaire, and Cameroon, have struggled to obtain national unity. Admittedly, some diversity within a state is probably in-

TABLE 3.2

Major World Religions

RELIGION	1900 Adherents (Millions)	1900 Percentage of World Population	1980 Adherents (Millions)	1980 Percentage of World Population	2000 Adherents (Millions)	2000 Percentage of World Population
Christianity	558	34.4%	1433	32.8%	2020	32.3%
Roman Catholic	272	16.8	809	18.5	1169	18.7
Protestant and Anglican	153	9.4	345	7.9	440	7.0
Eastern Orthodox	121	7.5	124	2.8	153	2.4
Other	12	0.7	155	3.6	258	4.1
Nonreligious and atheist	3	0.2	911	20.8	1334	21.3
Islam	200	12.4	723	16.5	1201	19.2
Hinduism	203	12.5	583	13.3	859	13.7
Buddhism	127	7.8	274	6.3	359	5.7
Chinese folk religion	380	23.5	198	4.5	158	2.5
Traditional and shamanist	118	7.3	103	2.4	110	1.8
"New religions"	6	0.4	96	2.2	138	2.2
Judaism	12	0.8	17	0.4	20	0.3
Other*	13	0.8	36	0.8	61	1.0
World Population	1620		4374		6260	

* including Sikh, Confucian, Shinto, Baháí, Jain, Spiritist, and Parsi.
Due to rounding off, percents may not equal 100.

SOURCE: *World Christian Encyclopedia*, (Oxford: University Press, 1982).

FIGURE 3.2

Religions of the World

SOURCE: Modified from H. J. deBlij
and P. O. Muller, *Human Geography:
Culture, Society, and Space* (New York:
John Wiley & Sons, 1986) pp. 200–201.

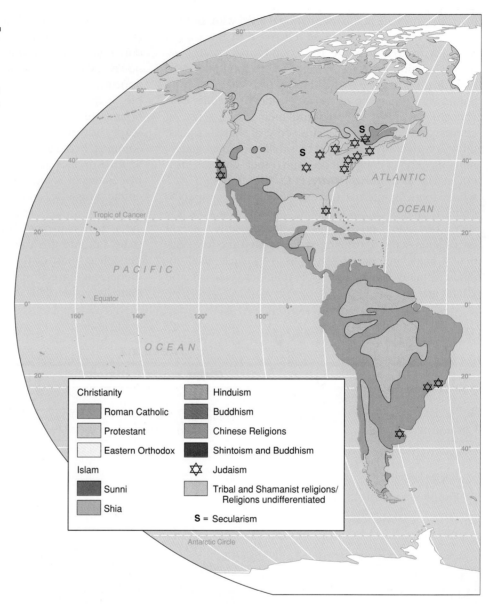

evitable, but too much diversity may lead to divisive-
ness. At the same time, too much homogeneity may
foster stagnation and the perpetuation of the status quo.

Currently, there are about 170 sovereign states in the
world. They vary immensely in population size and land
area. In addition, there are about 35 states that are not
independent and have colony, territory, or protectorate
status.

Many types of governmental structures and legal systems can be found in modern states. Most states in the

developed world hold competitive elections and are
presidential republics such as France. There are several
exceptions, however, such as Belgium, Canada, Denmark, the Netherlands, Norway, Spain, and the United
Kingdom, which have parliamentary constitutional
monarchies. Other developed nations such as Japan,
Australia, and New Zealand also have parliamentary
constitutional monarchies.

Most Eastern European countries and the Soviet
Union are socialist republics that hold elections, but un-

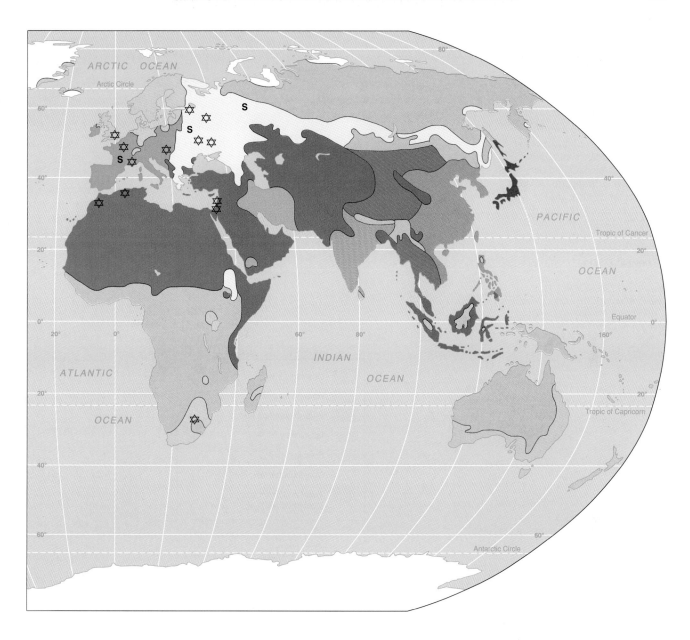

til the recent political changes, which began to occur in 1989, there was no political competition and these countries were controlled by a single political party. All that is changing now as legitimate, competing political parties are beginning to emerge and assume power. The developing world is quite complex and reflects a mixture of presidential republics, such as Bolivia and Colombia in South America and Cameroon and Gabon in Africa; parliamentary presidential republics in South America, Africa, and in India; and parliamentary consti-

tutional monarchies in such countries as Jordan in Southwest Asia. Probably one of the distinguishing features of the developing world is the large number of countries with military regimes. This is especially true in much of Africa.

The legal systems of the world today have been affected by the impact of the European powers. Through the establishment of their respective empires, they frequently modified the traditional political systems they encountered. Spanish settlers were responsible for es-

FIGURE 3.3

Major Legal Systems of the World

In the major legal systems of the world there has been considerable blending and overlapping of legal tradition. This blending is partly because European powers established overseas empires. (In part, after John H. Wigmore, "A Map of the World's Law," *Geographical Review*, 19 (1929), 114–20, and Ernst S. Easterly III, "Global Patterns of Legal Systems," *Geographical Review*, 67 (1977), 209–20.)

SOURCE: Modified from T. G. Jordan and L. Rowntree, *The Human Mosaic: A Thematic Introduction to Cultural Geography* (New York: Harper & Row, 1986), p. 112.

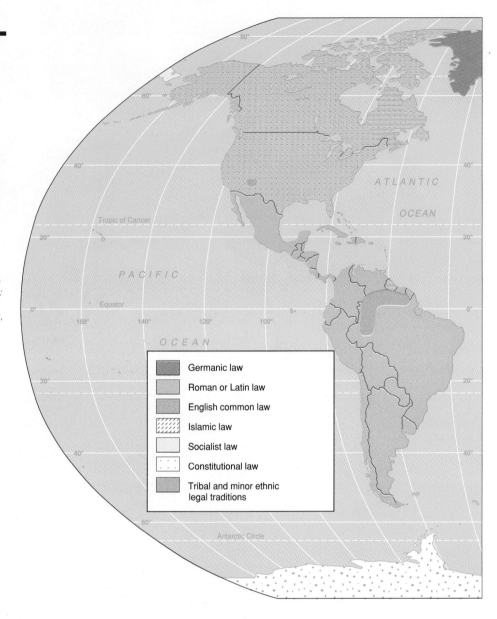

Germanic law

Roman or Latin law

English common law

Islamic law

Socialist law

Constitutional law

Tribal and minor ethnic legal traditions

tablishing their largely Roman legal system in Latin America, while the British had a similar effect on the United States and Canada where they implanted English common law. There are basically six major types of legal systems in the world: Germanic law, Roman or Latin law, English common law, Islamic law, socialist law, and varied ethnic traditions (Figure 3.3). In many areas, particularly Africa, the systems overlap considerably. In several instances, Islamic, English common, or Roman law coexists with ethnic legal traditions. In a court trial, for example, depending on which system prevails in the country in question, the individual might take an oath swearing on the Bible (English common law), Koran (Islamic law), or a gun (traditional or ethnic law). In parts of Cameroon, for example, a person who took an oath on a gun and did not tell the truth on the witness stand would die sometime in the not too distant future by some divine force.

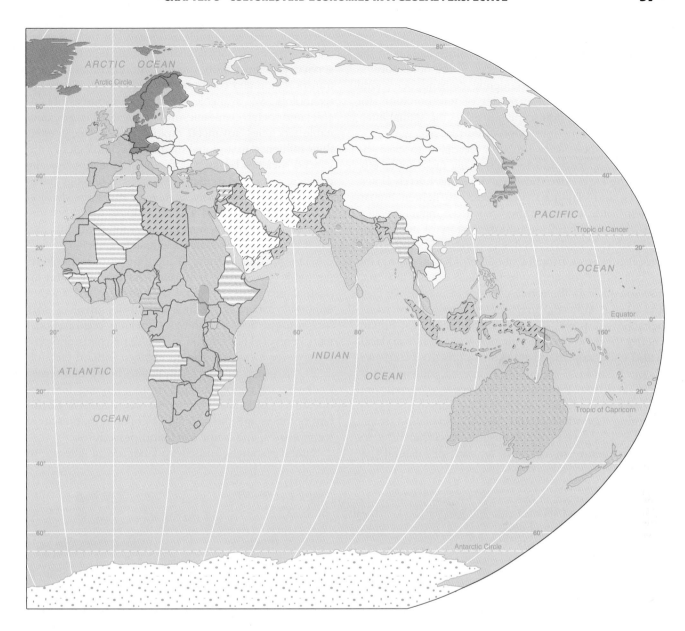

Technology

Of the four major traits of culture (e.g. language, religion, political systems, and technology), technology is the most important source of change in the interconnectivity of societies. It is the key element that differentiates the developed from the developing world. Throughout history those societies possessing the highest technological capabilities and achievements have usually been the most advanced economically and politically.

Technology refers to the ideas, procedures, tools, and systems that use natural and other resources to support the needs of a population. Although all people have certain basic needs, such as food, clothing, and shelter, the type, complexity, and quality of the provisions for meeting these needs vary widely around the globe. Society is composed of interrelationships between people, their social organization and physical environment, and their technological capabilities. Usually, changes in the technological base cause reactions in the other subsystems of a society that frequently demand behavioral changes. Changes in the modes of transportation, for instance, alter the landscape and affect the spatial interaction between people, goods, and places.

The first tools were probably designed to ease the human work load and to make daily tasks more efficient. The early adze (hoe), carved stone knives, bows, baskets, and clay-fired pots helped to lighten human chores. Later, energy sources such as animals, water, and wind were harnessed. In the history of humankind there are two easily identifiable periods where technological advances brought about rapid change: (1) the agricultural revolution, which began about 7000 B.C. and spread from the fertile crescent in the Middle East, and (2) the industrial and scientific revolution, which began about A.D. 1770 and diffused from Great Britain. During these two periods, the physical and human landscape underwent rapid transformations. As economies changed, land-use patterns were altered, settlement patterns were transformed, new life-styles emerged, and the level of living increased for most persons.

During the industrial and scientific revolution, scientific knowledge was applied to the development of inanimate energy sources such as coal, oil, natural gas,

electricity, and later nuclear energy. As these energy sources were developed, changes in manufacturing and service activities, food production, urbanization, and transportation and communication occurred. Where the technology was available, the natural resources obtainable, and social, economic, and political structures amenable, societies advanced rapidly and began to exert their influence in a global manner. Technologically advanced societies began to influence the technologically deficient ones, and in many instances technology transfer did occur. Conversely, in several cases, the division between the developed and developing world became more pronounced when recipient societies were unable to adopt new technologies adequately.

When most of us think of the transfer of technology, we think of a society acquiring new machines and industrial processes, or **hard technology;** but another important aspect of technology transfer involves **soft technology.** Soft technologies include organizational management skills, education, health practices, and procedures for increasing the acceptance of hard technologies. Examples of soft technologies being transferred to developing nations include contour farming, farm cooperatives, programs employing breast feeding for infant nutrition and the spacing of children, reforestation technologies, and home sanitation procedures.

Natural resources and cheap labor are of little use to

A Sudanese classroom in Africa. Education plays an important role in the development of developing countries. Although many Africans once traveled to Europe, the United States, or the Soviet Union for advanced education, as the various countries improve their educational systems, Africans are being educated at home by Africans trained in their own country's schools.

TECHNOLOGY DIFFUSION

Access to technological innovation has always been critical to human success. The use of human intellect to control fire and construct primitive weapons allowed early humans to survive in the harsh natural environment. Technological improvements are spread from area to area by a process called **technology diffusion**. The diffusion of technology is impeded by barriers, such as physical obstacles, cultural differences (languages, ideologies), and psychological predispositions toward technology acceptance. For example, the Chinese invented techniques for producing paper prior to the eighth century, but it took until the thirteenth century for the first paper factories to be built at Xativa and Fabriano in Europe. The barriers to the geographic diffusion of paper-making technology included distance, cultural differences, language, and the isolationist view of the Chinese toward the rest of the world. The grand duke of Florence in the 1500s is another example of bar-

riers to diffusion. The duke held a monopoly on the technology for producing brocade, fine cloth embroidered with raised ornamentation. Not wanting to lose his lucrative monopoly, the duke created a very tangible barrier to the spread of brocade-making knowledge. He offered a reward for the return of any brocade craftsman who left Florence. The craftsman could be returned dead or alive.

In recent times, the diffusion of innovations to developing nations has been

aided by international assistance programs. The direct transfer of technology from developed nations has not always been successful, however. In some cases the needed technology is unavailable, too costly, or not appropriate for a developing country. Large farm tractors, well suited for American farms, are used inefficiently in the small irregular farm plots of many developing nations. These machines are left idle due to a shortage of fuel, repair skills, and spare parts. Fortunately, many important innovations, such as improved strains of crops and vegetables, have been introduced successfully in developing nations. Nevertheless, major obstacles to widespread adoption of such innovations within a developing region must be faced. The primary obstacles involve resistance to alteration of the traditional farming, educational, and cultural patterns. The geographic diffusion of technology will become more critical as technology's role in development increases.

---

many developing nations without the hard and soft technologies and financial assistance to create and sustain domestic economic development. Unfortunately, many of the advances in food production, medicine, energy production, industrial processing, and electronics are not available or are too expensive for the poorer countries to obtain. In some cases the technology that is provided to developing countries is too costly or complex to be used effectively and maintained.

Even when technology is affordable, it is not the total panacea for society's problems. It can solve problems, but it can intentionally or unintentionally create new problems. Thus, while a culture must cope with economic, social, and political changes brought about by new technology, it must also concern itself with the

problems technology creates. One such problem is environmental degradation, including air, water, and noise pollution, destruction of forests and extinction of animals, soil erosion, and inadequate treatment and disposal of toxic wastes. Societies must use technology adequately and appropriately for the betterment of humankind not its destruction and extinction. Technology must be used responsibly and its attributes and shortcomings must be recognized.

POPULATION CHARACTERISTICS

In addition to examining the culture of a specific geographic area, geographers wish to learn about the composition of the population and its characteristics. One way of describing a population is to look at its biological traits: race, age, and gender.

Race

A key biological trait of humans is race. Race refers to the physical and/or biological characteristics of a group of persons that are transmitted genetically. Although race is a term full of misunderstandings and myths, people do differ physically. Different racial groups can be recognized based upon their distinctive physical traits. The problem for the academic and scientific communities is to decide what criteria to use to differentiate humans (e.g., skin color or pigmentation, blood chemistry, physical appearance, or geographical proximity). None of the classification systems suggested thus far has been adequate to explain all of the differences that exist between people. There are no clear-cut racial types into which all humans can be placed. Because race is so difficult to define, the use of the term "ethnic" has become popular since ethnic includes both physical and cultural traits.

Although it is generally believed that humans evolved from a common stock, migration, genetic drift, and spatial and social isolation have worked through the centuries to produce particular racial groups in specific geographical areas. But with increased migration more racial mixing occurs. Traditionally, racial groups have been defined as Caucasoid, Mongoloid, and Congoid, with other groups emerging between these cores or physical extremes (e.g., American Indians).

Age-Gender Structure

Age-gender structure is the composition of a population according to the number or proportion of males and females in each age category at a specific point in time. The key factors affecting the age-gender structure of a country are past trends in fertility and mortality rates and migration patterns. Much of a country's economic, social, and even political behavior is influenced by the proportion of the population found in the various age groups. For example, the need for schools, jobs, housing, and health care is directly affected by a country's age structure. Many of the developed countries have lower birth rates, higher median ages for their populations, and a larger percentage of elderly persons than do the developing countries, whose populations are frequently younger or more youthful.

Age-gender structure is best described by use of the population pyramid, which shows the percentage of the male and female population according to five-year age intervals. The overall shape of the pyramid indicates the age-gender distribution of a country's population and its potential for future population growth. Usually, countries make the transition from youthful/rapid growth populations to older/negative growth populations (Figure 3.4). In the *rapid growth* model, the pyramid has a wide base and a narrow top indicative of a young population with a high birth rate. Mexico is a good example of a rapid growth country (Figure 3.5).

With regard to the *declining growth* or *constrictive* model, the pyramid has an indentation at the base that

FIGURE 3.4

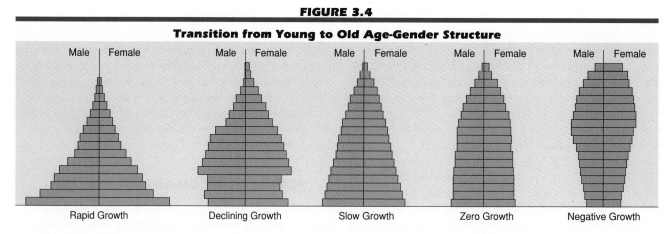

Transition from Young to Old Age-Gender Structure

Rapid Growth   Declining Growth   Slow Growth   Zero Growth   Negative Growth

SOURCE: Population Reference Bureau, *World Population Data Sheet* (Washington, D.C.: Population Reference Bureau, 1984).

FIGURE 3.5

Population Composition: Age and Gender

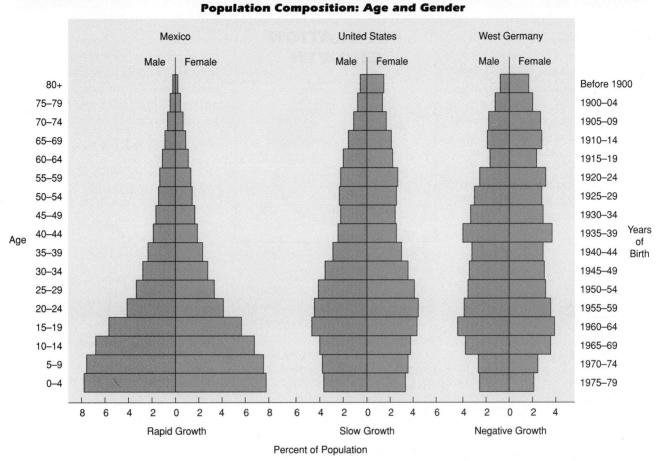

SOURCE: Population Reference Bureau, *World Population Data Sheet* (Washington, D.C.: Population Reference Bureau, 1984).

is narrower than the middle of the pyramid indicative of the fact that the country has just recently started to experience rather rapid decreases in its birth rate. It still does not have a large proportion of older persons, and the upper portions of the pyramid are still relatively pyramidal in shape. In time, if the trend continues, the population will begin to assume slow growth characteristics. Japan is a country that has experienced this kind of transition.

The *slow growth* model, of which the United States is an example, reflects a decline in fertility and mortality rates, infant mortality decreases, and life expectancy increases; the median age of the population rises, and the population begins to age. With the ***zero population***

growth (ZPG) model, there are further decreases in the fertility and mortality rates, and the growth of the population begins to stabilize. Sweden is frequently cited as an example of this model.

In the *negative growth* model, such as in former West Germany, the country experiences a very low birth rate, and the proportion of the population in the childbearing years (15–44 years) may start to become smaller than the proportion above 45 years of age. The upper portion of the pyramid begins to bulge, and death rates begin to rise as the population ages, resulting in the death rate exceeding the birth rate; the net effect is a natural decrease in the population (Figure 3.4). Quite often countries experiencing negative growth also have minimal

POPULATION GROWTH

Populations are not static, but instead are continually changing in size. The population of the United States today is different from what it was yesterday and what it will be tomorrow. The population of your hometown also changes daily. These changes are caused by birth, death, and migration. Births and in-migrants add to the total and deaths and out-migrants subtract. This is easily expressed by the basic demographic equation:

$$TP = SP + B - D + I - O,$$

where

TP = total population, some time interval beyond SP
SP = starting population
B = births during the interval
D = deaths during the interval
I = in-migration during the interval
O = out-migration during the interval

A population's rate of *crude natural* increase is determined by using the crude birth rate (i.e., number of annual births per 1,000 persons) minus the crude death rate (i.e., number of annual deaths per 1,000 persons). For example, in 1990, the United States had a crude birth rate of 16 and a crude death rate of 9, meaning that its crude natural increase was 7 per 1,000, or an annual rate of 0.7 percent. However, the **population growth rate** differs from the rate of crude natural increase in that net mi-

gration must be calculated. Thus, if the rate of natural increase is 7 per 1,000, and the rate of net migration adds 8 persons per 1000, the population growth is 15 per 1000, or a rate of 1.5 percent annually.

Rate of Natural Increase

$$\text{Birth rate} = \frac{\text{births per year}}{\text{population}} \times 1000$$

$$\text{Death rate} = \frac{\text{deaths per year}}{\text{population}} \times 1000$$

$$\text{Rate of natural increase} = \frac{\text{birth rate} - \text{death rate}}{10}$$

Rate of Population Growth

$$\text{Annual population growth rate} = \frac{\text{natural increase} \pm \text{net migration}}{10}$$

increases or even losses in net migration. Since most migrants tend to be young in age and in their fertile childbearing years, a country with low in-migration cannot expect to increase its population through migration.

Although these models are useful in explaining general population growth trends through utilization of the age-gender structure of a population, these broad trends can be interrupted by particular events in a specific country. Wars, economic depressions, migration flow changes as a result of political upheavals or natural disasters, epidemics, or changes in immigration laws, and social and or religious changes in attitudes regarding birth control or abortions can all contribute to periodic "bulges" and "contractions" in the age-gender pyramids.

Thus, one should keep in mind that changes in the age-gender pyramids are ongoing and that the population structures of countries are always in transition. When specific countries are examined through time, sometimes these changes become more apparent. Colombia, for example, in 1964 reflected the rapid growth model, by 1973 it had begun to lower its birth rate, and by 1985, it had started to assume certain characteristics of a slow growth model (Figure 3.6).

Thus, the age-gender structure of a country is important for understanding the population composition and its potential for population growth. It also provides an insight into the possible economic, social, political, and health problems of a particular geographic area.

FIGURE 3.6

Colombia: Changing Demographic Structure (Percent of Total Population)

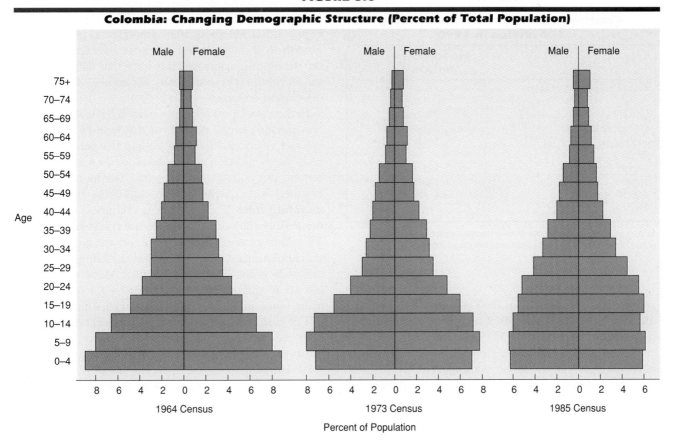

SOURCE: DANE (Departamento Nacional de Estadistica).

POPULATION DYNAMICS

As we have seen, populations are constantly changing. The study of population dynamics looks at several aspects of that change, including changes in the number of persons in the population and the population growth rate; changes in the spatial distribution of the population; and changes in population density.

Population Size and Growth

More than 5.3 billion persons live on earth. Asia (excluding the Soviet Union) with 3.1. billion persons has 58 percent of the world's population, Africa has 661 million (12 percent), Europe (excluding the Soviet Union) has 501 million (10 percent), Latin America 447 million (8 percent), the Soviet Union 291 million (6 percent), North America 278 million (5 percent), and Oceania 27 million. More than 3.3 billion live in just 10 countries, of which 6 are located in Asia (Table 3.3).

The average annual natural increase in the world's population is 1.8 percent. At this rate, the world's population will double in 39 years, so that sometime near the year 2030, the world could be supporting 10 billion people. This rather rapid increase in the world's population is a rather recent phenomenon. In A.D. 1, the world's population was only about 300 million. It was not until after the beginning of the industrial revolution in the middle 1700s that population growth began to accelerate, particularly in some of the more advanced countries that began to urbanize and improve living standards and health care. The world's population

TABLE 3.3

Countries with Populations Exceeding 100 Million in 1990

	1990 POPULATION (MILLIONS)
People's Republic of China	1119.9
India	853.4
Soviet Union	291.0
United States	251.4
Indonesia	189.4
Brazil	150.4
Japan	123.6
Nigeria	118.8
Bangladesh	114.8
Pakistan	114.6

SOURCE: Population Reference Bureau, *World Population Data Sheet* (Washington, D.C.: Population Reference Bureau, 1990).

reached its first billion in 1800, the second billion in 1930, and the fourth in 1975 taking only 45 years to double between 1930 and 1975 (Figure 3.7).

The growth of the world's population is not evenly distributed around the globe. Although the world's rate of natural increase is currently 1.8 percent, Africa and Latin America exceed this rate, Asia is at the world av-

erage, and North America, Europe, the Soviet Union, and Oceania are below the world average (Table 3.4; Figure 3.8). Until about 1925, annual rates of population growth in the developed countries exceeded those of the developing countries. But since that date, growth in the developing countries has far exceeded that in the developed countries.

In the developed countries, which have about 1.2 billion persons or 24 percent of the world's population, the rate of natural increase is only 0.5 percent. At this rate, the population of the developed countries would require 140 years to double in size. The number of years required for a population to double in size is called ***doubling time***. The developing countries, exclusive of the People's Republic of China, are comprised of about 2.9 billion persons (54 percent) and are growing at 2.4 percent annually. The population doubling time at this rate is 29 years. China has about 1.1 billion persons and has an average annual growth rate of 1.4 percent, with a doubling time of 54 years (Table 3.5; Figure 3.9).

The easiest way to estimate population doubling time is to use the *law of 70,* which is calculated by dividing the percentage of growth into 70. Therefore, at an annual natural increase of 1 percent, the population takes 70 years to double in size, at an increase of 2 percent only 35 years are required, and so forth.

FIGURE 3.7

World Population Growth through History

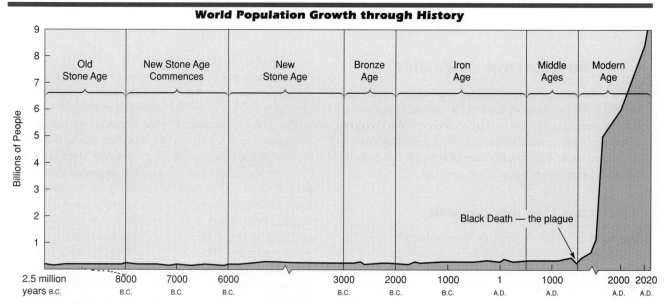

SOURCE: United Nations Population Division.

FIGURE 3.8

Population Growth through Natural Increase: Developing and Developed Countries, 1775–1985

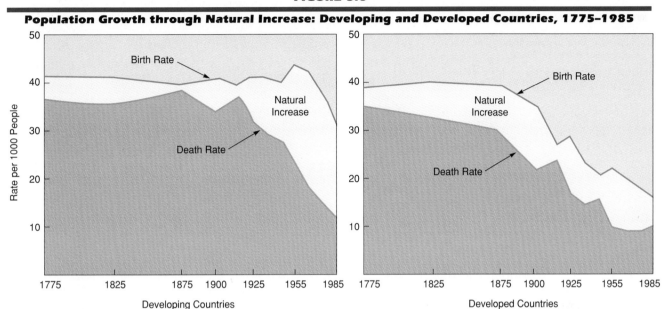

SOURCE: Population Reference Bureau.

If we project the earth's population to the year 2020, it could reach 8.2 billion. If so, 14 countries are expected to exceed 100 million in size (Table 3.6).

Although no one has a crystal ball, population growth models can be built around selected trends and projections can be made. If the two-child family (replacement fertility) assumption model is used, the world's future population growth would tend to level off (Figure 3.10). But if the three-child assumption or the current average of 3.8 children per family is projected to the year 2100, the world's population could be more than 20 billion and 50 billion, respectively. Future population growth trends are always based upon a number of "ifs," however: if the rate of population growth continues at this figure, then the population will be this amount; if the developed world continues to have low birth rates, then its population total will be this size; if families in the developing world continue to have a rather large number of children, then this outcome will happen; or if the developing world begins to lower its birth rate, then this result will occur. Many assumptions and many ifs must be made, but the bottom-line question is still how many people can the earth support at an adequate level of living.

Population Distribution

More than 90 percent of the earth's population lives north of the equator, where 80 percent of the land surface of the world is located. The disparity is much greater than this suggests, however, because about 90 percent of the people actually live on only 10 percent

TABLE 3.4

Growth Rates for World Regions: 1980–1985

REGIONS	AVERAGE ANNUAL GROWTH RATE
World	1.7%
Africa	3.0
Asia	1.7
Latin America	2.3
North America	0.9
Europe	0.3
Soviet Union	0.9
Oceania	1.5

SOURCE: United Nations, World Population Prospects as Assessed in 1982, U.N. Population Division, December 1983.

FIGURE 3.9

World Population Growth

SOURCE: Data compiled from
Population Reference Bureau.

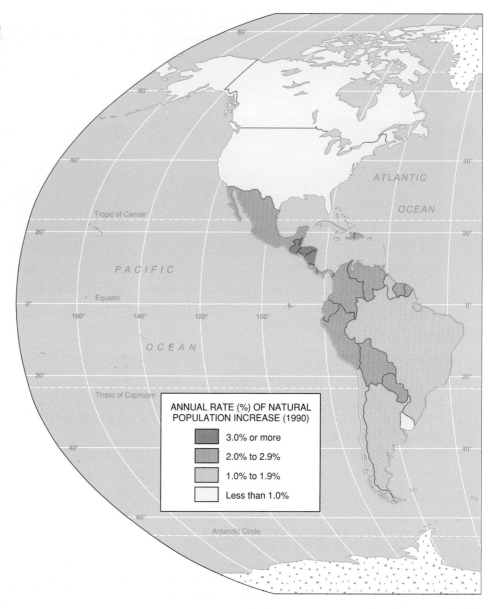

of the world's land. People tend to inhabit low plains areas, many of which are located along margins of the continents. Thus, more than 60 percent of the population lives within 300 miles of the coastline and below 700 feet in elevation. Eurasia has more than 70 percent of the world's population in thee major concentrations: (1) East Asia—China and Japan; (2) South Asia—India; and (3) Europe—Germany, the United Kingdom, and France (Figure 3.11).

Why are people spatially distributed in this manner on the earth's surface? An examination of two basic factors may help answer this question: (1) the natural environment and its associated elements such as climate (e.g., precipitation and temperature), vegetation, soils, landforms, water resources, and other natural resources (e.g., animals, energy sources, and minerals), and (2) the human environment and its cultural elements such as religion, political systems, socioeconomic insti-

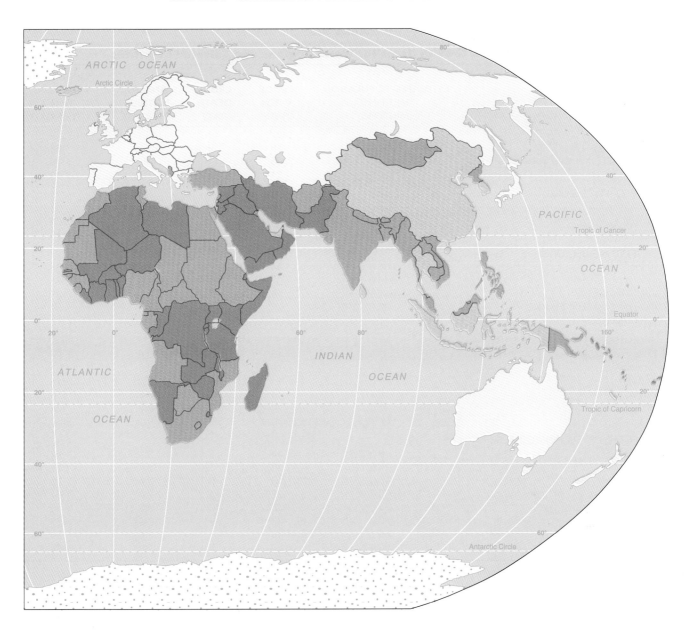

tutions, and technical development. Initially, when humans had limited technological capabilities, they and the natural environment were rather closely intertwined. As technology improves, natural environmental factors have less influence upon human behavior. Nevertheless, despite today's technology, vast areas of dry and cold land and high-altitude lands are still incapable of supporting large human settlements. Sections of the earth that are basically uninhabited are referred to as

the *nonecumene,* whereas the **ecumene** is the permanently inhabited portion of the earth.

Population Density

Population density (the number of persons per square mile or kilometer), a standard measure of population concentration, can be determined in several ways. The *arithmetic density* is the total population of a place

TABLE 3.5

Average Annual Growth Rates for World Regions and Selected Countries

	ESTIMATED POPULATION, MID-1990 (MILLIONS)	CRUDE BIRTH RATE	CRUDE DEATH RATE	ANNUAL NATURAL INCREASE	POPULATION DOUBLING TIME (YEARS)
WORLD	5321.0	27	10	1.8%	39
Developed	1214.0	15	9	0.5	128
Developing (excluding China)	2987.0	35	11	2.4	29
China	1119.9	21	7	1.4	49
DEVELOPED REGIONS					
North America	278.0	16	9	0.7	93
Europe	501.0	13	10	0.3	266
Soviet Union	291.0	19	10	0.9	80
DEVELOPING REGIONS					
Africa	661.0	44	15	2.9	24
Asia	3116.0	27	9	1.9	37
Latin America	447.0	28	7	2.1	33
Oceania (including Australia, New Zealand)	27.0	20	8	1.2	57
SELECTED DEVELOPED COUNTRIES					
United States	251.4	16	9	0.8	92
Denmark	5.1	12	12	0.0	(−)
Sweden	8.5	14	11	0.2	311
Bulgaria	8.9	13	12	0.1	630
Poland	37.8	16	10	0.6	122
Italy	57.7	10	9	0.1	1155
SELECTED DEVELOPING COUNTRIES					
Ghana	15.0	44	13	3.1	22
Kenya	24.6	46	7	3.8	18
Saudi Arabia	15.0	42	8	3.4	20
India	853.4	32	11	2.1	33
Iran	55.6	45	10	3.6	20
Vietnam	70.2	33	8	2.5	28
Mexico	88.6	30	6	2.4	29
Brazil	150.4	27	8	1.9	36

SOURCE: Population Reference Bureau, *World Population Data Sheet* (Washington, D.C.: Population Reference Bureau, 1990).

compared to the size of the land area. The world's arithmetic density in 1988 was 88 persons per square mile or 34 persons per square kilometer. The figure for the United States in 1990 was about 70 per square mile or 28 per square kilometer, whereas the comparable figures for Japan are 860 and 334, respectively (Table 3.7).

The ***physiological*** or ***nutritional density*** is the ratio of population to arable land—land that is suitable for tillage. In using this measure, however, some potential arable land is excluded if it is in pasture, forest, or some other related use. Some scholars believe this method is a little more indicative of population pressure

on the land. The physiological density for the United States is 350 persons per square mile whereas for Japan it is 6,595 (Table 3.7).

The question then arises as to whether there are optimum arithmetic and physiological densities that can be used as indicators to tell when a country is overpopulated or when there is too much population pressure on the land. Can optimum numbers be derived to serve as guidelines to ensure adequate living standards? It is generally agreed that these two types of density measurements do not tell us much about population pressure or levels of living, and that optimum numbers are difficult to derive. For example, densely settled areas like the Netherlands and some other Western European countries and Japan have high standards of living, whereas similarly densely settled areas in East Asia and the Caribbean do not possess a comparably high standard of living. Thus comparisons of population densities to levels of living are difficult to make.

One of the early theorists to examine the relationship between population growth and food production was Thomas R. Malthus, an English clergyman and economist. He believed that population increases geometrically and food production arithmetically; thus, the human species would increase in units such as 1, 2, 4, 8, 16, 32, 64, . . ., and food in units such as 1, 2, 3, 4, 5, 6, 7, ... He argued that population growth is limited by food production, and that if diseases, epidemics, and wars did not limit population growth, then hunger and famine would. Many neo-Malthusians feel that population growth will still outstrip food production, and that increasing population pressure hastens environmental deterioration. Although some of today's scientists concur with these conclusions, others believe that advances in technology will avoid some of the calamities proposed by Malthus. On the other hand, some scientists see technology as the major contributor to environmental degradation and do not believe it is the ultimate solution to the problem. What is clear from the debate is that the relationship between population and food is still an important question. The problem is that no adequate answer has been found.

The ***carrying capacity*** of the land is a measure of the number of people that can be supported by a given

TABLE 3.6

Countries with Populations Exceeding 100 Million in the Year 2020

	POPULATION (MILLIONS)
People's Republic of China	1496.3
India	1374.5
Soviet Union	355.0
Nigeria	273.2
United States	294.4
Indonesia	287.3
Brazil	233.8
Bangladesh	201.4
Pakistan	251.3
Mexico	142.1
Japan	124.2
Ethiopia	126.0
Vietnam	119.5
Philippines	117.5

SOURCE: Population Reference Bureau, *World Population Data Sheet* (Washington, D.C.: Population Reference Bureau, 1990).

FIGURE 3.10

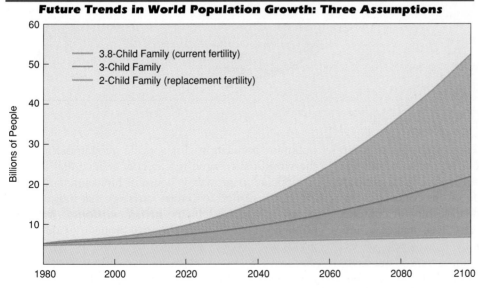

Future Trends in World Population Growth: Three Assumptions

- 3.8-Child Family (current fertility)
- 3-Child Family
- 2-Child Family (replacement fertility)

SOURCE: Population Reference Bureau.

FIGURE 3.11

Distribution of World Population

SOURCE: Modified from *Goode's World Atlas* (Chicago: Rand McNally, 1990).

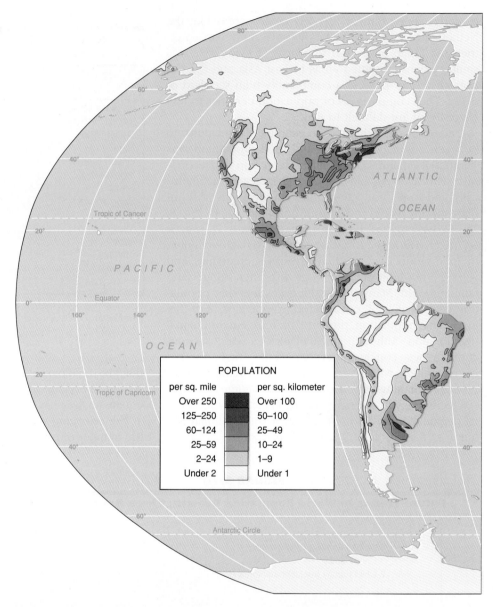

POPULATION

per sq. mile		per sq. kilometer
Over 250		Over 100
125–250		50–100
60–124		25–49
25–59		10–24
2–24		1–9
Under 2		Under 1

area of land. This measure focuses on the relationship or potential relationship between population and the environmental content of an area, with emphasis upon self-sufficiency. The shortcoming of this concept is that it fails to consider the trading relationships or complex exchange systems between areas. This makes it less meaningful when countries with high levels of exchange such as Singapore, Taiwan, Japan, or the Netherlands are considered.

Although no agreed-upon measurements for defining optimum populations for specific geographic areas of the world exist, some researchers believe that the rate of population growth is a useful indicator particularly when it is related to certain variables that indicate economic growth such as *gross national product (GNP),* which refers to the total monetary value of all goods and services produced by a country, typically in one year. Other variables include industrial or agricultural production and consumption of food and other resources. Different countries have different goals and

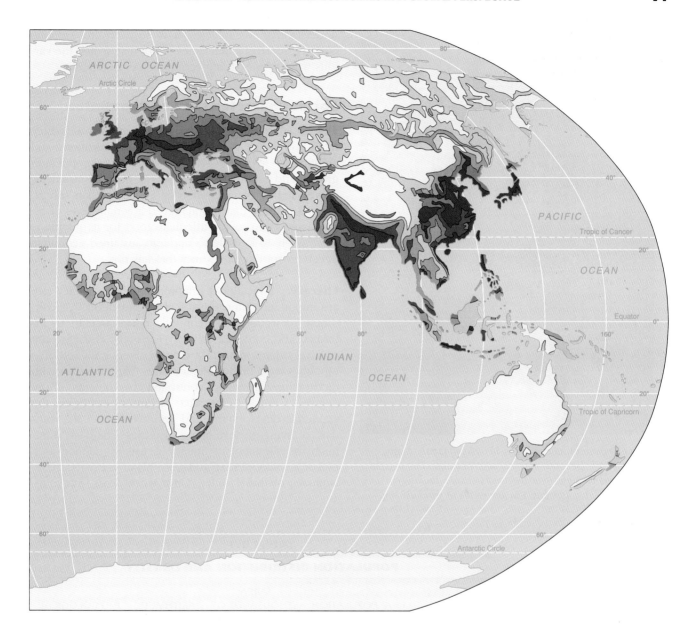

different perceptions as to the standard of living they wish to achieve. Once these goals and perceptions are devised, governmental officials can then use various measurements to gauge their country's productivity against population growth. If the population growth rate is exceeding rates of production, or vast shortages exist in food, housing, education, and social services, officials can gain a clearer view of the problem areas and are in a better position to determine the optimum population or decide if overpopulation does indeed exist. Thus

there are no definite measurements to determine optimum population for a geographic area, and just how many people specific areas of the globe can or should contain is still a matter for conjecture.

POPULATION AND SOCIOECONOMIC DEVELOPMENT

A country's population dynamics and its level of socioeconomic development are interconnected. In order to

A crowded scene on Ginza Street in Tokyo, Japan. Many Asian cities have become very densely populated as people have migrated to them over the past decades seeking economic opportunities and the amenities of urban living.

understand these linkages, it is necessary to examine both the social characteristics of population and the various ways of measuring economic development.

Social Characteristics of Population

The social characteristics of a population—its literacy rate, nutritional status, degree of urbanization, and amount of migration—all are related to population growth and economic development. In some instances, however, this connection is more direct than others. Thus, a country that has achieved a certain level of literacy and can provide adequate food for its people is more likely to be able to undergo sustained economic development than a country that has not.

Literacy.   Educational attainment and the literacy rate are important characteristics of a population that contribute to its level of economic development. An informed and educated citizenry is usually crucial to the political organization of a country. Frequently, elective republics have higher literacy rates than countries with dictatorships or military-controlled governments. A low degree of literacy can place serious obstacles on the economic development and improvement of the standard of living of developing countries. The basic minimum measurement of literacy is usually defined as a person's ability to read and write his or her name in the language of that country. On a world basis, this mini-

TABLE 3.7

POPULATION DISTRIBUTION AND DENSITY

COUNTRY	1990 POPULATION (MILLIONS)	AREA (THOUSANDS OF SQUARE MILES)	ARITHMETIC DENSITY	PHYSIOLOGICAL DENSITY
Japan	123.6	143.7	860	6595
Egypt	54.7	386.7	141	6348
Netherlands	14.9	14.4	1033	4362
Bangladesh	114.8	55.6	2064	3277
Colombia	31.8	439.7	72	1345
India	853.4	1269.3	672	1294
Iran	55.6	636.3	87	1006
Ethiopia	51.7	471.8	110	828
Nigeria	118.8	356.7	333	812
Argentina	32.3	1068.3	30	592
United States	251.4	3615.1	70	350

SOURCES: Population Reference Bureau, *World Population Data Sheet* (Washington, D.C.: Population Reference Bureau, 1990); United Nations, Food and Agriculture Organization, *Production Yearbook* (1990); United Nations population data; World Bank reports.

mum qualification is probably an indication of how literate a population is; but it is not as meaningful in the developed countries since close to 100 percent of the populations of the developed nations would be literate by this low standard. There, functional literacy involves more than this minimum qualification and includes the ability to read and write adequately in the language of one's country.

Increases in the rate of minimal world literacy have been accelerating during the past decades. As early as 1930, only about 40 percent of the earth's population was literate, by 1955 the figure had risen to 50 percent, and today it is about 67 percent.

Nevertheless, regional variations in literacy rates are still quite striking, and the dichotomy between the developed world and the developing world still exists. Although several countries in the developing world, mainly in Latin America, have increased their literacy rates, rates for many countries in Africa, Southwest Asia, and South Asia are still below 50 percent (Figure 3.12).

Nutrition. Nutrition or food intake usually involves two aspects: an energy-yielding aspect, as measured by calories, and a health-protecting aspect, as measured by vitamins and minerals. Calorie deficiency (quantity) creates a condition referred to as *undernourishment,* whereas deficiencies in health-protecting aspects (quality) create *malnourishment.* Quite often the question is not a matter of eating enough food, but rather one of a balanced diet.

An average calorie intake of 2360 is considered to be the minimum daily requirement. It is estimated that about half of the world's population lives on less than 2500 calories a day (Table 3.8). Frequently, the countries with low caloric intake also cannot provide a balanced diet. Consequently, much of the population in many countries is both undernourished and malnourished. Thus, world hunger is a major problem, particularly in the developing countries (Figure 3.13). Of course, many argue that the world is producing enough food to feed its population, but that the real problem is one of food distribution and regional disparity. Obviously, many countries cannot feed themselves, but when attempts from the outside are made to ease the situation, most of these same countries lack adequate transportation, communication, and distribution facilities to properly disperse the food. Often political and economic constraints regarding trade and other issues prohibit the flow of food supplies between and within countries. Thus, many of the developing countries that

TABLE 3.8

Daily Calorie Supply per Capita in Selected Countries

HIGH CALORIES	
New Zealand	3480
United States	3647
United Kingdom	3322
U.S.S.R.	3328
Argentina	3405
Romania	3337
Italy	3716
ADEQUATE CALORIES	
Brazil	2529
South Africa	2825
Korea (South)	2931
Mexico	2805
Cuba	2766
Venezuela	2642
Costa Rica	2686
Food and Agriculture Organization (United Nations) minimum for adequate nutrition: (2360) calories	
LOW CALORIES	
Peru	2183
Nigeria	2360
Thailand	2303
Colombia	2521
Tanzania	1985
Zaire	2135
Laos	1986
VERY LOW CALORIES	
Bolivia	2179
Algeria	2433
India	1906
Iran	2795
Somalia	2119
Haiti	1879
Indonesia	2342

SOURCE: World Bank, *World Development Report 1984* (New York: World Bank, 1984).

are experiencing rapid population increases are unable to increase their food supply proportionally to meet the food needs of their burgeoning populations.

Coupled with calorie inadequacies and protein and vitamin deficiencies is the incidence of diseases. Kwashiorkor, which is a disease caused by a diet deficient in protein, makes a child's belly grow disproportionately and is found in famine-ridden portions of Africa and other malnourished countries. Marasmus, caused by a

SOURCE: Modified from *Goode's World Atlas* (Chicago: Rand McNally, 1990).

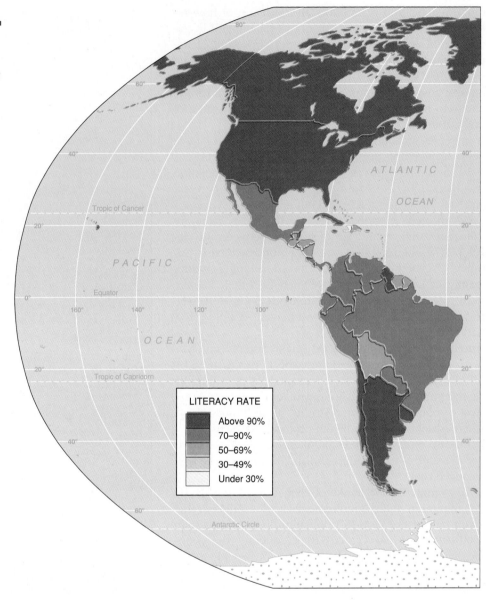

FIGURE 3.12

Literacy of World Population

LITERACY RATE

Above 90%
70–90%
50–69%
30–49%
Under 30%

lack of protein and calories, causes the body to be thin and bony, and the eyes of the individual often appear huge in comparison to other facial features. Not only can malnourishment and undernourishment cause skin diseases, gum problems, and eye disorders, but they can lower the body's ability to fight off other disorders such as cholera, hookworm, yellow fever, malaria, beriberi, scurvy, and rickets.

The rate of infant mortality and years of life expectancy are also related to food intake. Life expectancy at birth in much of Africa averages only about 51 years, and in southern Asia the average is 54 years. In contrast, much of the developed world, such as North America and Europe, averages about 75 years.

Urbanization.   During this century, the world's population has been urbanizing at a rather rapid rate considering the fact that in 1800 only 3 percent of the world's population lived in urban areas and by 1900 the urban population had only reached 14 percent. But

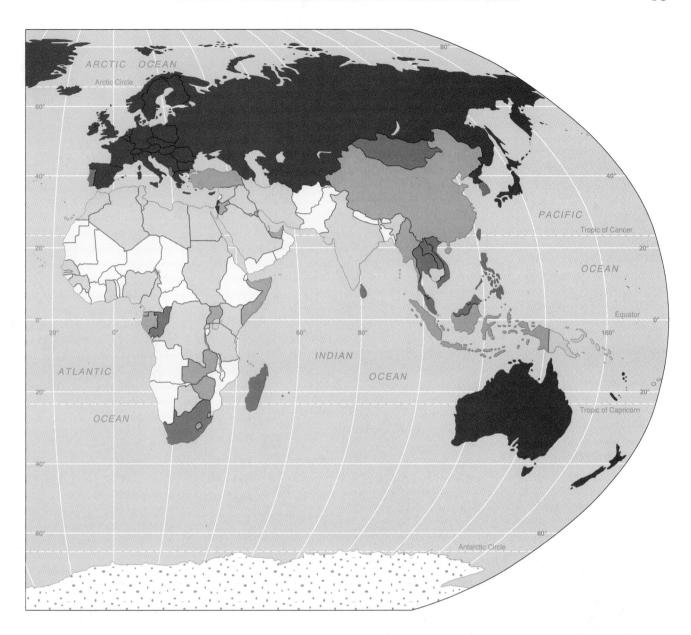

since 1900 it has increased dramatically so that the world today is approximately 41 percent urban (Table 3.9). Thus the degree of **urbanization** (which is the proportion of the total population living in urban areas) in the world has been increasing, and the urbanization process of rural to urban migration continues at a rather rapid pace.

The concept of urbanization should not be confused with *urban growth,* which is the increase in the size of urban populations. For example, certain cities may grow

more through natural increase than through migration or vice versa, but this urban growth may be faster or slower in specific cities than the rate of urbanization for the country as a whole.

Not only are urban growth and the degree of urbanization important indicators in examining the urbanization process, but another factor to consider is the percentage of the population that is living in selected large cities. Some countries tend to have much of their urban population concentrated in a single "super" city while

FIGURE 3.13

World Average Daily per Capita Calorie Consumption

SOURCE: Data from United Nations, Food and Agriculture Organization Statistics. Figure modified from H. J. deBlij and P. O. Muller, *Human Geography: Culture, Society, and Space*, 3rd ed. (New York: John Wiley & Sons, 1986), pp. 60–61.

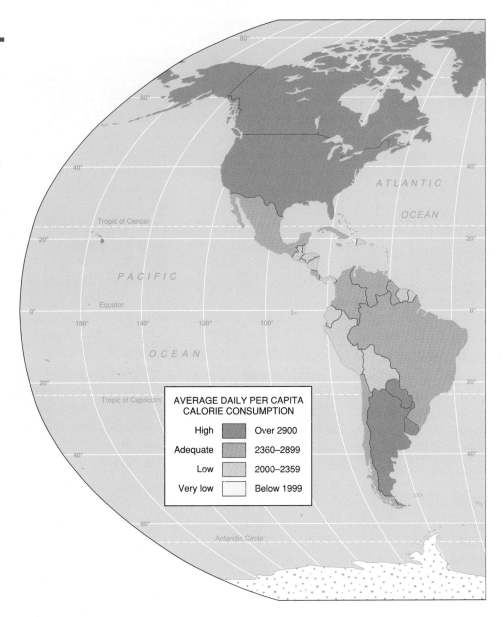

AVERAGE DAILY PER CAPITA CALORIE CONSUMPTION

High		Over 2900
Adequate		2360–2899
Low		2000–2359
Very low		Below 1999

others have their urban populations distributed among many cities. This super city, termed a **primate city,** is often the national capital and the city most identified with a particular country. For example, 7 million of England's 42 million urban dwellers live in London; in contrast only about 7 million of the United States' 200 million urban persons live in New York City.

The definition of an urban place is not standard around the world, varying from 1000 to 30,000 in population size; therefore, making international comparisons is difficult. Although the criteria may vary, it is generally agreed that an urban place is an area where the dwellers are usually not dependent upon agriculture for a living and where the population is more densely concentrated. As cities spread spatially, the boundary between the incorporated city, the *urbanized area* or built-up area around the city, and the adjacent rural area becomes increasingly blurred. Therefore, for most urban

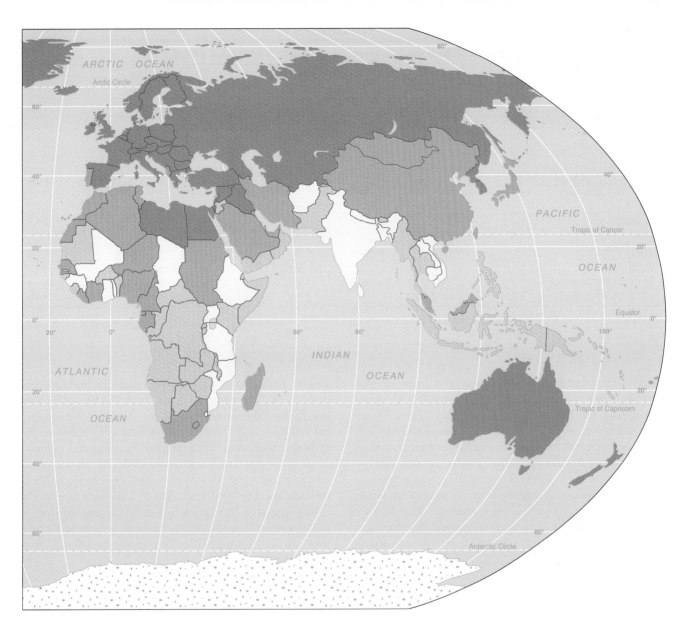

places, separate statistics exist for the city population, the urbanized area, and/or the **metropolitan area,** which includes territory beyond the urbanized area that is still integrated with the central city. As cities spread and begin to merge along transportation corridors, the term **conurbation** or **megalopolis** is frequently used to refer to urban coalescence. In some countries, such as the United States, cities begin to spread, merge, and interconnect with others to form several vast intercon-

nected megalopoli. The urban corridor in the United States extending from Boston to Washington, D.C. (Bo-wash) is just one example of a megalopolis in the United. States.

In the past and to some degree still today, urbanization in the developed nations appeared to be accompanied by industrial and economic development, rises in per capita incomes, and a slowing of population growth as birth rates declined. Even though cities in the

TABLE 3.9

World Urban Population, by Region: 1950 and 1990

WORLD REGION	1950 URBAN POPULATION		1990 ESTIMATED URBAN POPULATION	
	Millions	Percent	Millions	Percent
North America	105	64%	205	74%
Western Europe	177	60	294	77
Oceania	8	64	19	70
Latin America	67	41	308	69
Eastern Europe/Soviet Union	108	39	263	65
North Africa/Middle East	26	26	136	48
East Asia	112	17	388	29
Southeast Asia	23	13	123	27
South Asia	69	15	310	26
Subsaharan Africa	17	10	140	27
World	715	29%	2288	41%
Developed countries	457	53	886	74
Developing countries	257	16	1314	32

SOURCE: Compiled from Population Reference Bureau, *World Population Data Sheets*, (Washington, D.C.: Population Reference Bureau).

developed world have had and still have incidences of poverty, poor housing, high unemployment, and poor health and sanitary conditions, a higher degree of urbanization has generally meant a higher quality of life. But does industrial and economic development always accompany urbanization? The answer is not always. This question is fundamental in many developing countries. The process of urbanization is increasing rather rapidly in some countries, yet higher incomes and increased industrialization and economic development have not necessarily kept pace with the rate of urbanization. Thus, many high expectations go unfulfilled as persons flee the rural areas and move to the cities hoping to find their dreamed-of prosperity. All too often many cities in the developing world are unable to absorb the multitude who migrate, causing the cities to strain their resources and have difficulty in delivering adequate services. As a result, cities are often unable to provide the life-style anticipated. Although the rate of urban growth is frequently lower in developed than in developing countries, some cities in developed countries are also having difficulty meeting the demands of

TABLE 3.10

Ten Largest Urban Areas Projected for the Year 2000

RANK	URBAN AREA	PROJECTED POPULATION (MILLIONS)
1	Mexico City	26.3
2	São Paulo	24
3	Tokyo	17.1
4	Calcutta	16.6
5	Bombay	16
6	New York City	15.5
7	Seoul	13.5
8	Shanghai	13.5
9	Rio de Janeiro	13.3
10	Delhi	13.3

FIGURE 3.14

Patterns of Urbanization, 1900–2020

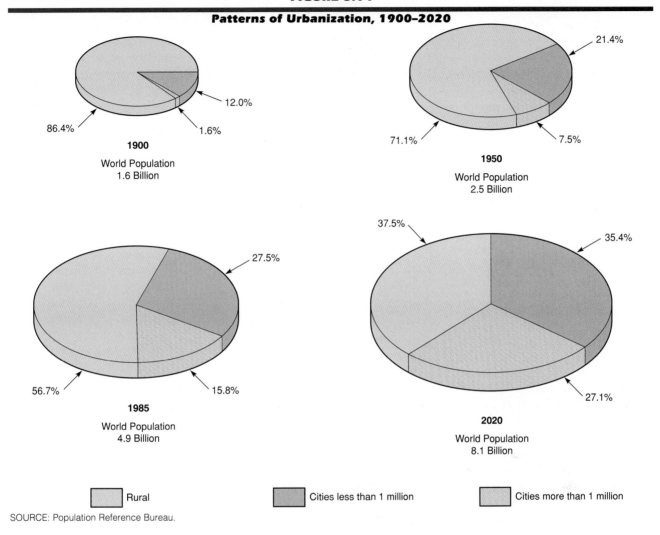

SOURCE: Population Reference Bureau.

their citizenry as resources become strained and certain environmental problems begin to mount. Current trends indicate that the world will continue to urbanize; by the year 2000 eight of the ten largest urban areas in the world will be in nations that we categorize today as developing (Table 3.10). By the year 2020 approximately 63 percent of the world's population will be urban with more than 27 percent living in cities with populations larger than 1 million (Figure 3.14).

Migration.   Migration for both permanent residence and employment is a major ramification of population

growth, development, and global interdependency. The movement of individuals results in a transfer of ideas, social customs, and financial resources between countries and regions. Today, migration primarily involves the movement of peoples from the developing countries to the developed nations. Between 1976 and 1980, the United States received over 2,000,000 legal permanent immigrants with approximately 44 percent coming from South America, Central America, and Canada; 40 percent from Asia; and 2.2 percent from Africa.

Labor imigration has been a favored method of combating labor shortages in certain sectors of developed

DEMOGRAPHIC TRANSITION MODEL

Theories are helpful in explaining and/or analyzing phenomena. One of the most notable theories that helps to explain population growth is the **demographic transition model.** It is based on historical observation and postulates that there is a causal link between modernization and fertility decline and slow population growth. The demographic transition is from high birth and death rates to low birth and death rates (Figure 1). Although the model is based upon the demographic experience of northwestern Europe, it does have implications for other regions of the world.

The model is divided into four time stages and assumes that every country is in one of the four stages:

Stage 1. The first stage consists of high crude birth and death rates that result in a low natural increase in the population. The crude birth and death rates frequently average above 35 per 1000, with the crude natural increase averaging less than 10 per 1000; this results in an annual increase in the population of less than 1 percent. Many countries might average less than 0.5 percent.

Stage 2. The crude death rate declines while the crude birth rate remains basically constant, resulting in a rather dramatic expansion in the crude natural increase of the population. Annual rates of population growth may average greater than 2.5 percent. Usually, technological assistance in the areas of food supply and health care helps to produce a decline in the death rate, particularly in infant mortality.

Stage 3. The crude death rate continues to decline, but the key characteristic of stage 3 is a decline in the crude birth rate. The population continues to grow but at a slower rate; often the growth rate is below 2.5 percent. In this stage, many social, economic, and political changes occur as technology increases within the society. Social mores may change regarding the ideal family size, and structural changes may occur in the composition of the labor force as well as in views on the role of women in society.

Stage 4. Countries in stage 4 have low crude birth and death rates. The crude natural increase is frequently below 1 percent and approaches zero population growth. Thus, the cycle is almost complete, returning to a stage of low population growth as experienced in stage 1. Whereas birth and death rates averaged above 35 per 1000 in stage 1, they are frequently below 20 in stage 4.

The demographic transition model is useful for explaining demographic changes in Europe, but how applicable is it to other regions of the world? If the developing regions of the world experience the effects of industrialization and urbanization and assume other attributes of modernized society, then optimists believe that these countries will indeed proceed through the

nations' economies. Between World War II and the early 1970s, Western European countries had a sustained foreign worker flow in the millions, mainly from southern Europe, Turkey, and North Africa. Since the 1970s, restrictions on foreign workers have reduced the flow into Western Europe. It is noteworthy that in spite of these restrictions, foreign populations have increased in Western Europe. This is due to the liberalization of laws regulating family reunion, higher natural population increase among immigrants, and illegal residency in these countries.

The oil-rich countries of the Middle East and North Africa are another major region of labor importation. Between 1975 and 1980, Kuwait had an 82.1 percent increase in temporary foreign workers, Bahrain a 131.1 percent increase, while Saudi Arabia, the lowest, had a 32.2 percent increase. These global migrations have several implications. Domestic labor forces feel threatened

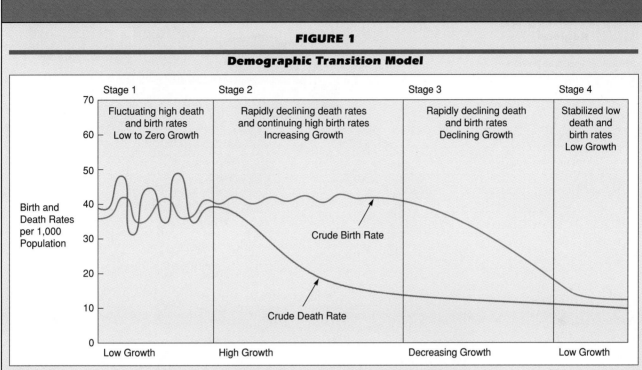

FIGURE 1

Demographic Transition Model

SOURCE: James M. Rubenstein and Robert S. Bacon, *The Cultural Landscape: An Introduction to Human Geography* (Saint Paul, Minn.: West Publishing Company, 1983), p. 54. Reproduced with permission. All rights reserved.

demographic transition as outlined here and that their populations will level off sometime in the future. If the demographic transition happens, population growth trends by the mid-twenty-first century could start to resemble the low growth rates experienced prior to 1750. But this will occur only *if* the developing countries follow the European model.

Although the four stages of the demographic cycle are generally accepted, the question arises, does the cycle end with stage 4? Recent research suggests it will not. If Europe is to be used as the model, evidence is already available that suggests that societies may move from low birth and death rates to low birth and increasing death rates. Already Denmark, Austria,

former West Germany, and Hungary are experiencing decreases in their natural population growth. Thus, stage 5 may prove to be a transition period when death rates exceed birth rates as a result of higher fertility rates of an earlier generation. After going through this stage, a stage 6 may emerge having characteristics similar to those of stage 4.

by newly arriving workers. This may result in civil unrest like that which led to restrictions on immigration into Western Europe. The closed and cohesive societies of countries, such as Kuwait and Saudi Arabia, risk being undermined by the ideas and aspirations of the foreigners. The increase in Islamic fundamentalism and political terrorism in these countries can be traced, in part, to these foreign workers. Another implication is a massive transfer of people out of developing countries

and a massive transfer of money back to the migrants' families in the source countries.

Measures of Economic Development

When the characteristics of culture and population dynamics are viewed by region, a linkage is apparent between the population situation and the level of socioeconomic development of that region. Population

FIGURE 3.15

GNP per Capita by Major Regions

SOURCE: Population Reference Bureau, *World Population Data Sheet* (Washington, D.C.: Population Reference Bureau, 1991).

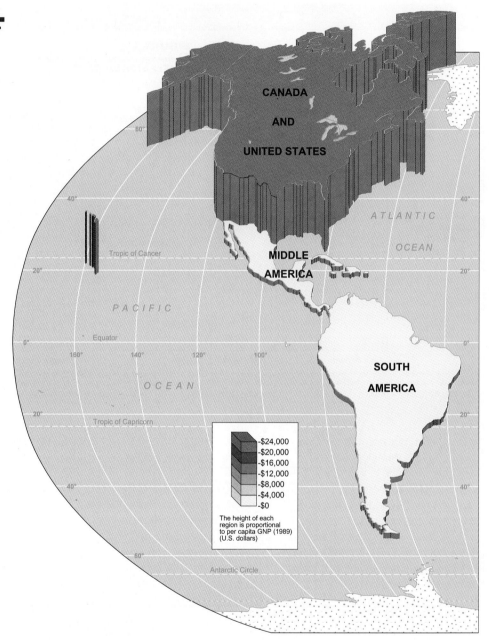

growth, the level of human welfare, urbanization, and a host of other cultural features are related to economic development. The quest for economic development, the path to obtain it, the roles of government and private corporations, and its cultural and ecological costs are major social as well as economic issues. In recent years modern global communications, trade, and tourism have permitted increasing numbers of people to make comparisons between other cultures and their own. The knowledge of a better life has heightened expectations, leading many people to reevaluate their situation and to demand the higher quality of life that only economic development can provide. Ideas and information on well-being, liberation, resource consumption, and global events are being diffused to all parts of the world via radio, television, fax, and a host of other avenues.

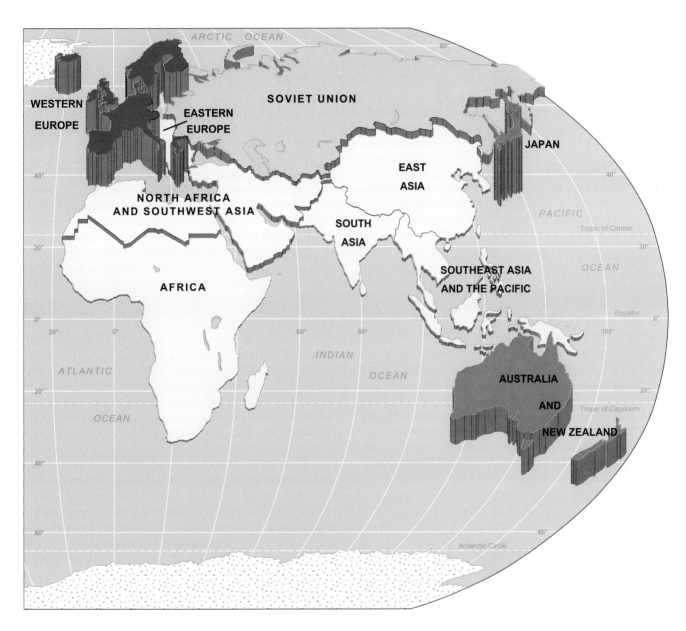

This growing awareness of other cultures is the basis for much of the cultural change and turmoil that exists in the present world.

In classifying countries by level of economic development, geographers usually rely on statistical measures. At the broadest level, countries can be defined as developed and developing using the gross national product (GNP) or ***gross domestic product (GDP),*** which is the total value of goods and services produced domestically by citizens and businesses of a country, per capita per year. Figure 3.15 shows quite dramatically the differences in GNP per capita, with North America, Europe, the Middle East, and Japan having much higher per capita GNP than South America, Asia, and Africa.

The GNP statistics are useful in classifying nations and regions, but do not offer a complete picture. The oil-

TABLE 3.11

Selected Measures of Development

WORLD AVERAGE	DEVELOPED NATIONS	DEVELOPING NATIONS
GNP per capita/per year: $3470 (1988 U.S.$)	High GNP per capita/per year: $15,830 (1988 U.S.$)	Low GNP per capita/per year: less than $3000; average $710 (1988 U.S.$)
Fertility rate: 3.5 children per woman	Low fertility rate: average 2.0 children per woman	High fertility rate: average 4.0 children per woman
Urbanized population: 41 percent	Highly urbanized: 73 percent	Low level of urbanization: 32 percent
Infant mortality: 73 deaths per 1000 live births per year	Low infant mortality rate: 16 deaths per 1000 live births per year	High infant mortality rate: 81 deaths per 1000 live births per year
Life expectancy: 64 years	High life expectancy: 74 years	Low life expectancy: 61 years

SOURCE: Population Reference Bureau, *World Population Data Sheet* (Washington, D.C.: Population Reference Bureau, 1990).

producing countries of the Middle East are wealthy because of a great many petrodollars and limited populations. Measured solely on GNP per capita, the United Arab Emirates (GNP per capita, $15,720) is more developed than Austria ($15,560). However, the high GNP per capita for the Middle East does not encompass all of the aspects of economic development. Although wealthy, the oil-producing countries of the Middle East do not have a high level of economic development when their industrialization, technological development, and educational and social systems are compared to those of Europe and Japan. Table 3.11 shows a selected list of more comprehensive measures of economic development, including population growth, urbanization, infant mortality, and life expectancy. These are indicative of the social and medical systems, as well as the economic standard of a country or region.

The classification of nations into developed and developing countries shows the vast differences between them. These differences seem all the more dramatic when one considers that over 75 percent (4.1 billion) of the world's people live in the developing countries, and 800 million people are in desperate poverty. Figure 3.16 suggests that the gap between the developed and developing nations is widening. In 1975 the average GNP per capita for the developed nations was over eleven times greater than in the developing countries; by the year 2000 the GNP per capita for the developed countries will be almost 15 times greater than the developing countries. Future GNP per capita projections by region predict significant increases in the United States, Western Europe, the Soviet Union, and Eastern Europe. Asia and Africa will undergo limited increases, although percentage increases in selected countries of Asia may be sizable.

At the most complex and comprehensive level of classification, measures of economic development include the previously mentioned factors of urbanization, health, and GNP per capita. An additional element is the ideological orientation of the government and people. The range is from free enterprise with limited government intervention in the economic marketplace to centralized control of most farms and factories by the government. In practice, ideology is tempered by the practical concerns of equity, human incentives, and global interdependency.

Petrodollars have enhanced the purchasing power of Saudi Arabians. These shoppers are in the Diplomatic Quarter in Riyadh, Saudi Arabia.

Other considerations in economic development include a nation's future development potential based on its ability to use monetary and technological assistance, its domestic resource base, population control, and internal commitment to development. Arranging countries in this fashion produces five general classes or worlds with subclasses in the second and third worlds (Table 3.12). The map of the five worlds (Figure 3.17) shows the spatial distribution of economic development as measured in this manner.

The first world includes the most affluent, industrialized and urbanized nations in the world. Gross national product per person is above $10,000 with modest economic growth rates. The second world is characterized by countries having a major portion of their economies under central government control. The second world can be broken into two subclasses to reflect their relative economic level and location in Asia and Eastern Europe. Communist Asia is comprised primarily of the People's Republic of China, North Korea, and Southeast Asia; these nations have GNP per capita of $1000 or less. The middle-income nations are the Soviet Union and the transitional socialist nations of Eastern Europe. These nations have a GNP per capita in the $2000–$6000 range.

The third world includes the greatest number of nations. It is politically and economically distinct from the first and second worlds and is only sporadically and impermanently aligned with either group. Three subclassifications of this group are apparent. The "significantly developed" countries (approximate GNP per capita of $4000 include the emerging industrial countries

and/or countries with a sizable natural resource base to generate income. The "transitional" group of third world countries have GNPs between $1000 and $2000 per capita, but have a potential for economic growth with the infusion of capital and technological investment. The third group of "net oil exporters" have a

FIGURE 3.16

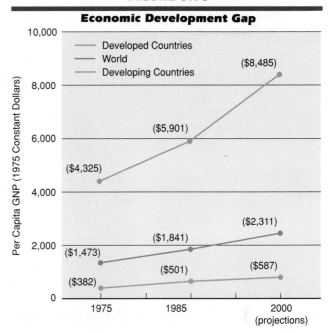

SOURCE: Council on Environmental Quality, *Global 2000 Report* (Washington, D.C.: U.S. Government Printing Office, 1981).

TABLE 3.12

Five Worlds of Development

	CLASS		REPRESENTATIVE COUNTRIES
	First World:	Affluent, market oriented	United States, France, Australia
Developed	Second World:	Socialist	
		Communist Asia	China, North Korea
		Transitional socialist	Soviet Union, Romania
	Third World:	Nonaligned middle income	
		Significantly developed	Portugal, Mexico
		Transitional	Tunisia, Peru,
		Net oil exporters	Bahrain, Saudi Arabia
Developing	Fourth World:	Managing poor	Tanzania, India
	Fifth World:	Extremely poor	Burkina Faso, Haiti, Bangladesh

FIGURE 3.17

Five Worlds of Development

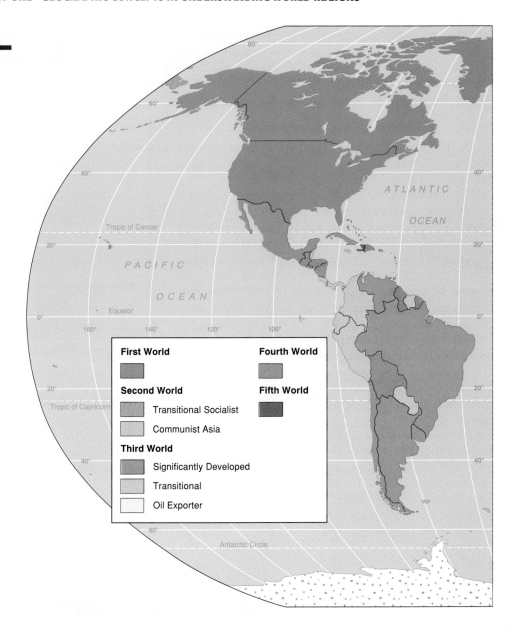

First World

Second World
Transitional Socialist
Communist Asia

Third World
Significantly Developed
Transitional
Oil Exporter

Fourth World

Fifth World

much higher GNP per capita ($10,000–$25,000) than other nations of the third world, yet in terms of the development of their social, medical, and economic systems and their nonaligned stance, these nations are part of the third world.

The nations of the fourth world can be characterized by low GNP per capita (approximately $750). Most have population growth rates exceeding their economic growth rates. However, they have a source of raw materials, an industrial base, or a growing middle class that offers potential for increased development. With a strong and self-reliant program for development, wise use of foreign aid, and access to markets, technology, and capital, these countries have the long-term potential for development.

The fifth world is comprised of countries with little hope for economic development. These countries have an inadequate resource base for their expanding populations and a seeming inability to cope with their multitude of problems. The GNP per capita is in the $100–

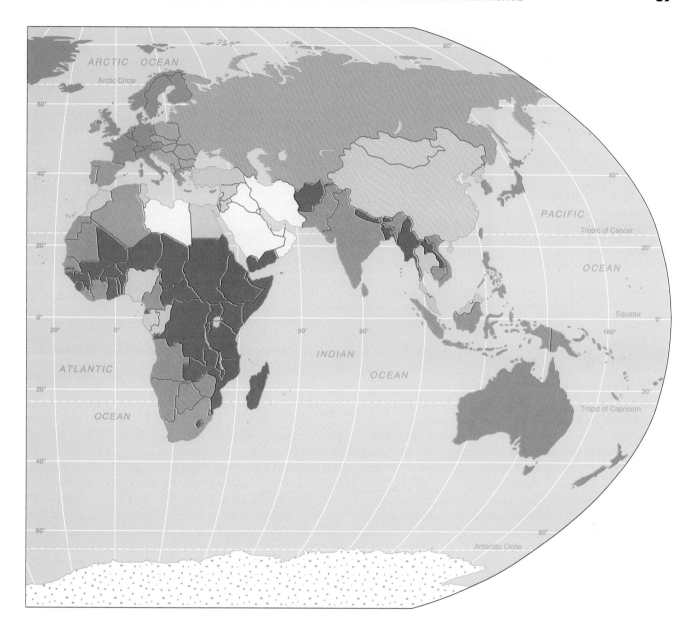

$500 range, which is barely enough to survive in the modern world.

COMPARATIVE VIEWS OF DEVELOPMENT

Many of the differences in the levels of economic development around the modern world can be traced to the period of European colonization. European exploration led to the "discovery" and colonization of the Americas, Asia, and Africa. Spain and Portugal led the way in colonizing Middle and South America in the late 1400s. Other European nations soon followed with extensive colonial acquisitions. Several European countries took control of large portions of North America and Asia. In the mid-1800s the Europeans assumed control of the interior of Africa. At this time the British, Russian, and French Empires had expanded to their greatest proportions covering much of North America, Africa, and Asia.

An automated control room overlooking the production areas of an aluminum mill in the United States. As many manufacturing plants in the United States began to age and become outdated, some firms modernized their facilities while others closed them. Many others, however, relocated to other parts of the country such as the South and the West Coast or established plants in foreign countries.

The colonizers justified their colonial acquisitions and their efforts to Europeanize the indigenous people by claiming that it was their duty to Christianize the people, that they were stopping the slave trade, and that they were fulfilling their rightful destiny as superior beings. One should also note, however, that the colonies provided raw materials for European industry, cheap labor, and a captive market for European manufactured goods.

The nineteenth century was the beginning of the decline of colonialism as well as its high point. Spain's colonial holdings contracted as almost all of its colonies in Middle and South America became independent nations. By the early twentieth century, decolonization of the British Empire had accelerated with the independence of Australia, Canada, New Zealand, and South Africa. After World War I and World War II, the demands for independence from European control resulted in the demise of traditional colonial rule for most of the world. The number of independent nations in the world increased approximately threefold between 1945 and 1980. Most of these new nations were former European colonies in Africa and Asia.

The heritage of colonial rule lingers in these nations. In many cases the Europeanization of these colonies fostered the decline of the indigenous structures of government, economic enterprise, and social customs. In spite of their political independence, many of the former colonies have maintained a dependency relationship with their former masters. For example, Italy remains the main trading partner of Libya; English governmental organization is a model for many former colonies, like Canada, Australia, and India; several African countries look to France for military assistance; and the Philippines have maintained strong ties to the United States.

Allied with colonialization in fostering the present dichotomy between developed and developing countries was the spread of technology. The industrial revolution began in Western Europe in the 1700s and diffused at a variable pace throughout the world. Those nations having the advantages of access to a large resource base, capital, and industrial technology, plus internal cohesion, were able to join the European nations in the present-day group of highly developed nations. The majority of the countries and peoples of the world do not possess these advantages and have been left behind in the move toward development.

From the viewpoint of the developing countries, the following are the critical economic development issues in the modern-day world:

1. The legacy of colonialism, which exploited the best natural resources, altered traditional social relationships, and still pervades the outlook and operations of many countries.
2. The existing world economic system, which is structured to ensure the dominance of the already developed nations and the continued dependency of the poorer nations.
3. Misdirected assistance from the developed countries, which has encouraged high-consumption societies based on goods exported from the developed countries to the developing countries. Technology has been transferred to poorer nations, but not the information or skills necessary for domestic technological creation to take place in these nations. Foreign assistance programs have hindered the development of internal mechanisms in these countries for dealing with their problems and becoming self-sufficient.

These issues are valid in many cases, but the reasons for them are not simple. They do not have a single cause nor is a single group, such as the developed nations, *multinational corporations* (private corporations with facilities and interests in many nations), or the developing nations themselves, responsible.

Ideological Perspectives

Economic development is a complicated process that may mean different things to different groups of people. Just as two map projections may show the world differently, ideological projections present dissimilar realities of the same world. The *Marxist* view of economic development requires class struggle to overthrow capitalism and institute an organization of the working classes.

Supposedly, this will ensure that resources will be allocated on the basis of need, not profit. Achievement of these ends is defined as economic development.

At the other end of the ideological spectrum is free market *capitalism.* This view stems from Adam Smith's ideas for a free market economy unfettered by government intervention. Supposedly, this approach will ensure sufficient economic growth for all people to prosper while freeing them from governmental control.

In the ideological middle are ideas descended from David Ricardo, John Maynard Keynes, and John Kenneth Galbraith. This view, often called the cost-of-production school, questions whether the developing nations can ever achieve development in the present world market economy. Proponents of this ideological view call for the maintenance of capitalism with the creation of a new international economic order to redistribute wealth and technology to developing countries.

Economic Interdependency

Regardless of the level of economic development, the economies of the world are interlinked and interdependent through trade, multinational corporations, migration, and a host of other factors. Table 3.13 shows the distribution of world trade for 1976 and 1983.

As the table shows, the developed countries have a majority of the total, although their share declined from 70.3 percent in 1976 to 66.0 percent in 1983. This decline was compensated for by an increase in the shares of the developing and planned economies. All national economies have become more open economies because of the establishment of flexible exchange rates, the worldwide impact of the oil price rise, and the general expansion of trade and international financing since

TABLE 3.13

Distribution of World Trade

	SHARE (PERCENTAGE OF TOTAL)	
DEVELOPMENT CLASS	1976	1983
Developed market economies	70.3%	66.0%
Developing economies	20.3	23.5
Centrally planned economies (Soviet Union and Eastern Europe)	8.0	9.8

SOURCE: International Monetary Fund, *International Financial Statistics* and Direction of Trade Yearbook.

TABLE 3.14

Number of Countries with Growth Rates of Real GDP at or below the Rate of Growth of Population, 1979–1984

	TOTAL SAMPLE SIZE	1979	1980	1981	1982	1983	1984
World	112	26	37	50	66	63	46
Developing economies	83	24	29	37	54	55	44
South and East Asia	14	2	3	0	4	3	2
West Asia	10	3	4	4	6	7	5
Western Hemisphere	23	6	6	13	21	20	13
Africa	32	12	15	20	22	23	23
Mediterranean	4	1	1	0	1	2	1
Developed market-oriented economies	22	1	6	11	10	8	2
Centrally planned economies	7	1	2	2	2	0	0

SOURCE: Department of International Economic and Social Affairs of the United Nations Secretariat.

World War II. No country or bloc of countries, be they communist, capitalist, or in between, north or south, developed or developing, operates in isolation. The necessities of economic survival and growth override most other concerns.

The economic events of recent years point this out quite vividly. The economic stagnation, inflation, and recession in the developed countries in the late 1970s and early 1980s resulted in a lack of demand for natural resources and manufactured goods from the developing countries. High interest rates stemmed the flow of capital and technology to these developing countries, thus limiting their expansion plans. The period resulted in high debt loads for several developing nations. They had borrowed from financial institutions in the developed countries only to have the world economic decline make debt servicing and repayment impossible. Lastly, this period brought forth a wave of sentiment for increased protectionism in the developed nations.

One aspect of the current world situation is the failure of the economies of many nations to meet the expectations of their citizens. Modern mass media and travel have exposed the people of poorer nations to the lifestyles of the wealthier nations to a greater degree than ever before in history. As a result of this exposure, heightened expectations have developed that often cannot be met given the existing economic order. The frustration of the people may lead to revolution as in Eastern Europe or a more bloody upheaval, such as in South America.

Without access to markets in the developed nations, developing nations can have little hope of sustained economic development. Job creation, income increases, and overall development in the poorer countries are tied directly to the consumption patterns in the developed countries. For example, Bangladesh is one of the poorest countries on earth with a 1985 GNP per capita of $150, a population of 107 million, a fertility rate over 6.0, and wages for unskilled industrial workers averaging 8 to 12 cents per hour. Due to the low labor costs, Bangladesh has become a haven for clothing manufacturers with over 500 new textile companies being formed there between 1981 and 1985. This phenomenon resulted in 150,000 new jobs and an increase in the value of clothing production from $3.2 million in 1981 to $116.2 million by 1985. Recent quotas on imports of clothing imposed by Western Europe and the United States have reduced Bangladesh's access to these markets, however, and dealt a severe blow to a country whose only substantial resource is its cheap labor.

Table 3.14 shows the results of United Nations sample of 112 countries that compared growth rates of GDP with rates of population growth. In 1979, 26 countries had GDP growth rates at or below their rate of population growth. In the recession years of 1982 and 1983, 66 and 63 countries, respectively, had population growth rates higher than their economic growth rates. These data reveal several aspects of the world economy. Many developing nations, particularly in Africa and Latin America (western hemisphere in the table), are losing the race to maintain economic growth ahead of population growth. The situation in the developing countries parallels the developed countries where 10 out of 22 developed countries in 1982 had economic growth rates below their relatively low population growth rates. In contrast, in 1979 only one developed country had an economic growth below its population growth rate. Essentially, this indicates the dependency of the developing nations on the economic health of the market-oriented developed nations.

Suggested Readings

Austin, C. M., R. Honey, and T. C. Eagle. *Human Geography.* Saint Paul, Minn.: West Publishing Co., 1987.

Brown, L. "World Food Resources and Population: The Narrowing Margin." *Population Bulletin* 36 (1982): 1–43.

Brunn, S. D., and J. F. Williams. *Cities of the World.* New York: Harper & Row, 1983.

Clark, D. *Post-Industrial America.* New York: Methuen, 1985.

Cole, J. P. *The Development Gap: A Spatial Analysis of World Poverty and Inequality.* New York: John Wiley & Sons, 1981.

deBlij, H. J., and P. O. Miller. *Human Geography: Culture, Society, and Space,* 3rd ed. New York: John Wiley & Sons, 1986.

Glassner, M. I., and H. J. deBlij. *Systematic Political Geography.* New York: John Wiley & Sons, 1980.

Gritzner, C. F. "The Scope of Cultural Geography." *Journal of Geography* 65 (1980): 4–11.

Johnston, R., and P. J. Taylor, eds. *A World in Crisis? Geographical Perspectives.* New York: Basil Blackwell, 1986.

Jordan, T. G., and L. Rowntree. *The Human Mosaic: A Thematic Introduction to Cultural Geography.* New York: Harper & Row, 1990.

Kahn, H. *World Economic Development.* Boulder, Colo.: Westview Press, 1979.

Kirk, J., et al, eds. *Studies in Linguistic Geography.* Dover, N.H.: Longwood, 1985.

Knight, D. B. "Identity and Territory: Geographical Perspectives on Nationalism and Regionalism." *Annals of the Association of American Geographers* 72 (1982): 514–31.

McKee, J. O., ed. *Ethnicity in Contemporary America: A Geographical Appraisal.* Dubuque, Iowa: Kendall/Hunt, 1985.

Peters, G. L., and R. P. Larkin. *Population Geography: Problems, Concepts, and Prospects.* Dubuque, Iowa: Kendall/Hunt, 1989.

Population Reference Bureau. *World Population Data Sheet,* Washington, D.C.: Population Reference Bureau, 1990.

Rubenstein, J. M., and R. S. Bacon. *The Cultural Landscape: An Introduction to Human Geography.* Saint Paul, Minn.: West Publishing Co., 1983.

CHAPTER 4

The United States and Canada

IMPORTANT TERMS

Postindustrial era

Push-pull migration factors

Economies of scale

Comparative advantage

Industrial core

Agglomeration

Hinterland

Exurban

CONCEPTS AND ISSUES

■ Evolution of the economies of the United States and Canada toward service economies with less reliance on natural resource extraction and industry.

■ Redistribution of people, economic resources, and political power from the traditional core region urban centers to the "sunbelt" states and suburban and exurban communities.

■ Modification of the leadership role of the United States in the global economic and political arena.

CHAPTER OUTLINE

The United States and Canada as a Region

PHYSICAL ENVIRONMENT
Landforms
Climates
Soils and Vegetation
Water Resources
Mineral and Energy Resources

HUMAN ENVIRONMENT
Population
Cultural Aspects
Agricultural Production
Manufacturing
Urban Settlement

SPATIAL CONNECTIVITY
Connections within the Region
Global Connections
U.S. Transnational Corporations as Interconnections

PROBLEMS AND PROSPECTS

THE UNITED STATES AND CANADA AS A REGION

T hose nations with developed economies play a pivotal role in the economic development, social welfare, and peace of the world. They provide the technology, financial resources, and political leadership that influence all nations. At the same time, these developed nations rely on other countries and regions for the natural resources, labor, and political support necessary to maintain their position in the world community. Examination of the United States and Canada, which are the most familiar to most of us, will provide the backdrop necessary to compare the many settings, actions, and world views of the other peoples and nations of the world. It is to that purpose that we examine the United States and Canada in the first regional chapter of this text.

Canada and the United States have many commonalities that make it easy to view them as a homogenous region. The two nations share many natural environmental characteristics, such as climate and vegetation zones, landforms, and environmental problems. They have similar highly developed, capitalistic economic systems limited only modestly by government regulation of key activities. Both countries have progressed far into the **postindustrial era**, with the majority of the population urbanized and employed in white-collar jobs.

The open, pluralistic nature of both societies has encouraged the continuous infusion of new ideas and motivated people, both of which have stimulated social and economic progress. Although they have significant minority populations, the United States and Canada are predominately English speaking, with the majority of the citizens descended from European ancestors. For this reason, the region formed by the United States and Canada combined is termed Anglo-America. This European heritage has fostered similar democratic ideals, legal systems, and religious and cultural traditions in both countries. Due to this heritage, both countries have been assured early and continued access to the scientific and technological advances that have occurred in Europe since the advent of the industrial revolution. Combined with this technology, a large domestic resource base for both countries has created a high standard of living. Canada is the world's second largest country in areal size, and the United States is the fourth largest. This large land area, much of it suitable for agriculture, has provided the food, energy, and other natural resources necessary to support the urbanization and industrialization of Anglo-America.

Regional and Global Issues

The key issues facing the region involve the transition from an industrial to a postindustrial economy and the role of the region in the world political and economic situation. The systematic transition to a postindustrial economy is creating very significant regional alterations within both Canada and the United States. Jobs, people, and economic and political power are leaving the once prosperous Great Lakes region and the Northeast. They are relocating in the southern and western states of the United States and the central and western provinces of Canada.

Canada and the United States are relatively recent national entities compared to some nations in Europe and Asia, whose national identities can be traced back a thousand years or more. In spite of their relative youth, Canada and the United States have become the focal points of economic, political, and social activities for the entire world. The economies of many nations depend on trade and economic assistance from North America. The New York Stock Exchange sets the trend for stock markets and economic activities all over the world. Millions of people around the world eat grain, dairy products, and meat produced in the region. The American version of the English language is diffusing throughout the world via North American business, arts and music, and technology. It is becoming the most widely adopted second language in the world. The United States, in particular, has military, political, and economic influence over many other nations of the world.

The size and affluence of the American population make the United States the world's largest consumer market. An increasing share of this market is being dominated by foreign automobiles, electronics, clothing, and a multitude of other products. Citizens from other regions of the world are purchasing homes, investments, businesses, and land at a rapid rate. Thousands of acres of farmland, as well as whole blocks of Los Angeles and Manhattan, are owned by foreign investors. By some estimates more than 15 percent of the U.S. federal budget deficit is owed to foreigners who have purchased U.S. government securities. People from other regions of the world, often the most educated and motivated, are relocating in Canada and the United States

STATISTICAL PROFILE: UNITED STATES AND CANADA									
REGION OR COUNTRY	**Population Estimate mid-1990 (Millions)**	**Natural Increase (Annual)**	**Population Projected to 2000 (Millions)**	**Infant Mortality Rate**	**Life Expectancy at Birth (Years)**	**Urban Population (%)**	**Per Capita GNP, 1988 (US$)**	**Area, Thousands of Square Miles (Km²)**	**Population Density, No./mi² (No./km²)**
Canada	26.6	0.7	29.3	7.3	77	77	16,760	3852(9977)	7(3)
United States	251.4	0.8	268.3	9.7	75	74	19,780	3615(9363)	70(27)

to take advantage of the opportunities offered. This has created a "brain drain" problem in the developing world. Like the production of many products, leadership in economic and political matters has begun to shift from the United States and Canada to other parts of the world. These changes are modifying the manner in which the United States and Canada and all other regions of the world relate to each other.

Boundaries of the Region

Canada is comprised of 10 provinces and two territories. The federal government is based in Ottawa, Ontario. The United States is comprised of 50 states with the federal government located in Washington, District of Columbia (Figure 4.1). Although Hawaii is one of the 50 states of the United States, its location in the Pacific Ocean makes it an exception to many of the generalizations relevant to the rest of the country. Mexico shares a common boundary with the United States and is part of the landmass of North America in common with the United States and Canada. Nevertheless, despite this adjacency, Mexico and Anglo-America exhibit little similarity in economic development, language, and cultural factors. Mexico is considered part of Hispanic Middle America and is discussed in chapter 10.

PHYSICAL ENVIRONMENT

The United States and Canada comprise a region of varied environmental conditions. Most of the region is in the midlatitudes. The southernmost portion of Florida is positioned at approximately 25° north latitude. The northernmost portion of Canada extends past the Arctic Circle to 75–80° north latitude. Being midlatitude, no portion of the region receives direct rays of the sun during the year. Seasonal variation is thus very pronounced. Landforms vary from the high mountains of the Cana-

dian Rocky Mountains and the Alaska Range to the coastal lowland of Louisiana. The climate varies from tropical in Florida to polar tundra climate in northern Alaska and Canada. Renewable resources, such as forests, and nonrenewable resources, such as copper and coal, exist in great abundance in this region. All of these varied environmental conditions provide a complex of ecosystems supporting a great variety of animals and vegetation. Human activities in the region are varied and complex as well, reflecting this environmental diversity.

LANDFORMS

The region is composed of four major landform regions: the Western Mountains, the Interior Plains, the Appalachian Mountains, and the Atlantic Ocean–Gulf of Mexico Coastal Lowland. Within these major regions are significant subregions, including the intermontane Basin and Range area, the Hudson Bay Lowland, the Canadian Shield, and the Ozark-Ouachita Highlands (Figure 4.2).

The Western Mountain Region

The Western Mountain Region is composed of the Pacific Coastal, Sierra Nevada, and Rocky Mountain systems of mountains and highlands. The Pacific Coastal system is made up of the Cascade Mountains extending from California to Washington, the Coast Mountains in British Columbia, and the Alaska Range. The Pacific Coastal system is geologically young and is still undergoing changes (as was evidenced by the Mt. St. Helens's eruptions). These mountains are near the contact zone of the Pacific and the North American tectonic plates. Crustal movement is the reason for the volcanic and earthquake activity prevalent from California northward to Alaska. The mountains of the Pacific Coast are known for their steep sides, which rise up to create some of the highest mountain peaks in North America,

FIGURE 4.1

Provinces of Canada and the States of the United States

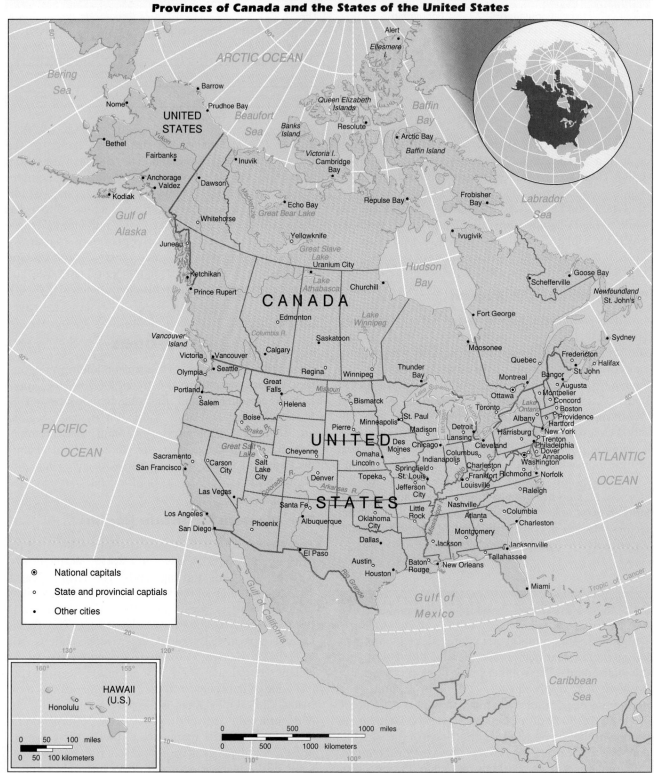

such as Mt. Denali (McKinley) in Alaska, Mt. Rainier in Washington, and Mt. Shasta in California.

The Sierra Nevada range parallels the Pacific Coastal system through much of eastern California and extreme western Nevada. These mountains were formed by the upward thrust of blocks of the earth's crust along cracks (faults) in the crust.

The third mountain system in the Western Mountain Region is the Rocky Mountain system. The Rocky Mountain system is a connected group of three mountain ranges: the Rocky Mountains in the United States and Canada, the Mackenzie Mountains in the Yukon Territory of Canada, and the Brooks Range of northern Alaska. The Rocky Mountain range is the largest of the three stretching from the Mexican border to northern British Columbia with peaks reaching 13,000 feet (3962 meters) in elevation.

Lying between the Sierra Nevada and the Rocky Mountain systems is the intermontane Basin and Range subregion. This area is noted for isolated ranges of fault block mountains and plateaus surrounded by sediment-filled valleys containing old lake beds with no outlet to either the Pacific or the Gulf-Atlantic drainage basins. The most famous of these internal drainage areas are the Great Salt Lake in Utah and Death Valley in California. Death Valley is the lowest area in the United States with an elevation of 282 feet (86 meters) below sea level. The U.S. portion of the intermontane area has primarily a desert climate created by the blockage of water-laden clouds by the mountain ranges on both sides of the area. The plateaus in this intermontane area are best known for the dramatic effects of rivers cutting downward through the uplifted sedimentary rock layers. The most famous example of this phenomenon is the Grand Canyon in northern Arizona formed by the incision of the relatively small Colorado River through the uplifted Colorado Plateau.

The Interior Plains

The Interior Plains encompass a vast area of flat to rolling topography from Texas to northern Canada paralleling the Western Mountain Region. The plains stretch from the Rocky Mountains eastward to the Appalachian Mountains in Ohio and Tennessee and to the Gulf-Atlantic Coastal Plain in Texas and Mississippi. The elevation of the region ranges from 6000 feet (1829 meters) along the western margin to sea level in the Hudson Bay area of Canada. This region contains large areas of wind-deposited soil, called *loess*. Soils formed by glacial deposition cover large portions of the region. Given adequate water from nature or irrigation, the fertile soils and level topography yield great harvests of wheat, corn, and soybeans.

The Valley of Ten Peaks, Banff National Park, in the Canadian Rocky Mountains.

FIGURE 4.2

Landform Regions and Subregions

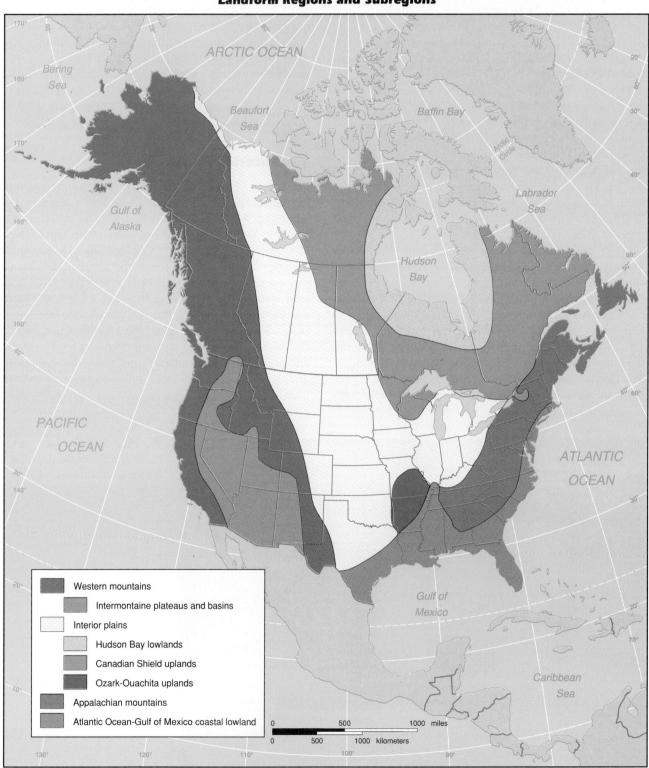

Western mountains

Intermontaine plateaus and basins

Interior plains

Hudson Bay lowlands

Canadian Shield uplands

Ozark-Ouachita uplands

Appalachian mountains

Atlantic Ocean-Gulf of Mexico coastal lowland

The Appalachian Mountains

The Appalachian Mountains comprise the third major landform region in Anglo-America. Like the Western Mountains and the Interior Plains, the Appalachian Mountains are a north-south trending region. They extend from Alabama northward to southern Quebec. These mountains are much older and more eroded than the Rocky Mountains. In the past they were as high as the Rockies, but now average only 3000–6000 feet (914–1829 meters) with the highest mountain being Mt. Mitchell in North Carolina at 6684 feet (2037 meters). The eastern portion of the Appalachian Mountains from Alabama to New Jersey is known as the Piedmont Plateau. Elevation declines very sharply where this plateau meets the Gulf-Atlantic Coastal Plain. Rivers flowing off the Piedmont Plateau toward the ocean have waterfalls and rapids at this physiographic line. This so-called Fall Line had an important impact on river navigation, water power supplies, and therefore industrial and urban location in the early settlement of the eastern United States.

The Atlantic Ocean–Gulf of Mexico Coastal Lowland

The Atlantic Ocean–Gulf of Mexico Coastal Lowland extends continuously from Texas to New York and reappears further north at Cape Code in Massachusetts. Much of the region is low-lying coastal beach and swamp areas with the more interior portions being gently rolling hills ranging up to 300 feet (91 meters) in elevation. The inland delta of the Mississippi River is part of the region, but it extends hundreds of miles northward from the Gulf of Mexico to southern Illinois. The presence of vast areas at or near sea level in this region have created tidal swamps and marshes in the southeastern United States unequaled in other parts of the country. These areas are the nursery for many species of wildlife and ocean inhabitants. The Florida *peninsula,* like the Mississippi Delta, is a distinctive portion of the coastal lowland. Geological evidence, including magnetic fields, deep bedrock profiles, and fossils, indicates that this peninsula is a remnant of the African continental landmass. The Florida peninsula remained attached to the North American continental landmass when the two landmasses separated some 200 million years ago.

Within the plains and coastal lowland regions of eastern Anglo-America are three separate and distinctive subregions: the Ozark-Ouachita Highlands, the Laurentian Uplands or **Shield**, and the Hudson Bay Lowland. The Ozark-Ouachita Highlands are located in the south-central United States in Missouri, Arkansas, and Oklahoma. This is a heavily eroded highland area whose age and form are similar to the Appalachian Mountains. It is separated from the Appalachian region by the Mississippi River Delta area of the Gulf Coastal Lowland. The Laurentian Upland subregion, termed the Canadian Shield, stretches in a semicircle from the Northwest Territories south and eastward around Hudson Bay to Newfoundland. This is an area of crystalline bedrock heavily scoured by continental glaciation. The result is a landscape with large areas of exposed bedrock with very shallow soils. The area is noted for its abundance of large freshwater lakes and rich, near-surface deposits of metallic ores. The Hudson Bay Lowland is the swampy, poorly drained area surrounding the vast Hudson Bay. Being so far north, the area is cold and supports only tundra grasses, scrub, and sparse forests. The rivers flowing off the Shield into the Hudson Bay Lowland provide great potential for hydroelectric power.

CLIMATES

The United States and Canada have great extremes in climate varying from desert to very wet, and from tropical to polar. Much of the region is exposed to Arctic blasts of cold air in the winter and tropical storm systems in the summer. The midlatitude position creates hot, moist summers and cold, drier winters for most of the region. In spite of these extreme variations within the region, a substantial portion of Anglo-America enjoys moderate temperatures with adequate precipitation for human habitation and activities.

Latitudinal position is the primary control of the climate. The climate and the geographic variation of climate within the region are also dependent on the air pressure systems, prevailing winds, effects of water bodies, and the physiographic features of the land surface. The prevailing winds across most of North America are "westerlies" coming off the Pacific Ocean (from the west) and moving toward the east. The high and low pressure systems tend to follow the same west to east movement. More permanent pressure cells, such as the Bermuda high, reside over portions of the region, following the sun northward in the summer and southward in the winter.

Across such an extensive landmass as North America, the distance to large bodies of water has a significant effect on the climate. Those areas downwind (east) of water bodies, such as the Pacific Ocean and the Great Lakes, have more precipitation and different temperature regimes than areas located upwind (west) of the water body. The effects of these winds blowing west to east over water explains the mild climate of the Pacific coast of Alaska and the heavy snowfalls in Buffalo, New York. Many areas of the interior of the region are far away from the moderating effects of water. These interior areas tend to be hot in the summer and very cold in the winter.

Landform features are also important to climatic variation. Because the mountains in North America follow a north-south trend, they play a minimal role in blocking the frigid polar air masses that move across the region in winter. This is in marked contrast to Europe where the east-west trend of the mountains helps to block the polar air and prevents it from reaching southern Europe. The mountains of the United States and Canada do impede the predominant east-west movement of moisture-laden air across the region. As this air is pushed over the mountains, it is forced further aloft. The rising air is cooled to its condensation temperature, and precipitation is formed. This physiographic effect explains the higher precipitation amounts on the western sides of the Coastal Mountains and Rocky Mountains and the relative dryness in the areas east of these mountains. The areas east of the mountains are in the **rain shadow.** The State of Washington is a good example of this phenomenon; the portion of the state west of the Cascade Mountains receives over 100 inches (254 centimeters) per year, while the area east of this mountain range averages only 10–20 inches (25–51 centimeters) per year.

Climatic Regions of Anglo-America

Defining climatic regions on the basis of similarities in climatic characteristics is a difficult task in a region as varied as Anglo-America. Nevertheless, it is useful to form a general impression of climate patterns for the region (Figure 4.3). Most of the Pacific coastal area has a marine west coast climate. This classification reflects the important impact the ocean and the prevailing westerly winds have on the temperature and precipitation of the area. This climate type extends as a narrow band from northern California northward to Alaska. Precipi-

tation occurs almost daily for much of the year, and the temperature rarely goes below freezing.

South of the marine region is the Mediterranean climate of southern California. This climate region is the result of a zone of high atmospheric pressure positioned off the California coast for much of the year. The Mediterranean climate is characterized by sunny, warm summers with very little rainfall. The winter is the season of maximum precipitation. This accounts, in part, for the massive brush fires that occur in southern California almost every autumn. The vegetation grows during the moist winter months, but dries out during the hot summer months. Over a period of years, dried vegetation accumulates on the hillsides. The natural fire cycle in which the dead vegetation is periodically burned off has been broken by fire prevention programs. Eventually, a fire does start from lightning or arson. The hot, dry weather and the large accumulation of vegetative fuel create fearsome fires that engulf large areas.

To the east of the Mediterranean climate is the desert climate of the southwestern United States. The summers are long and hot with many days over 100°F (52°C). The precipitation is minimal and erratic, amounting to 10–20 inches (25–51 centimeters) per year. A significant amount of the yearly rainfall comes during the so-called monsoon season. Although most people associate the term *monsoon* with torrential rains in parts of Asia, technically, the term means a shift in the prevailing winds in a region. This is what happens in the southwestern United States during July and August. The prevailing winds, usually from the west, shift to the southeast bringing moisture from the Gulf of Mexico across Mexico to this desert region. The winters are short and mild. In this area elevation is a major factor controlling temperature and precipitation. The many smaller mountain ranges in this area provide a welcome refuge from the hot, dry summers in the lower elevations.

East and north of the desert is the semiarid steppe climate, which extends from Texas to central Canada. This climate in Anglo-America is the result of the great distance from the steppe areas to a large body of water and the blockage of moisture to these areas by the Pacific Coastal and Rocky Mountain ranges. The areas farther east of the Rocky Mountains become more moist as the rain shadow effect decreases and more moisture is introduced from the Gulf of Mexico.

The central portion of the region is a transition zone between the desert and steppe climates and the humid subtropical and humid continental climates of eastern

FIGURE 4.3

Climatic Regions

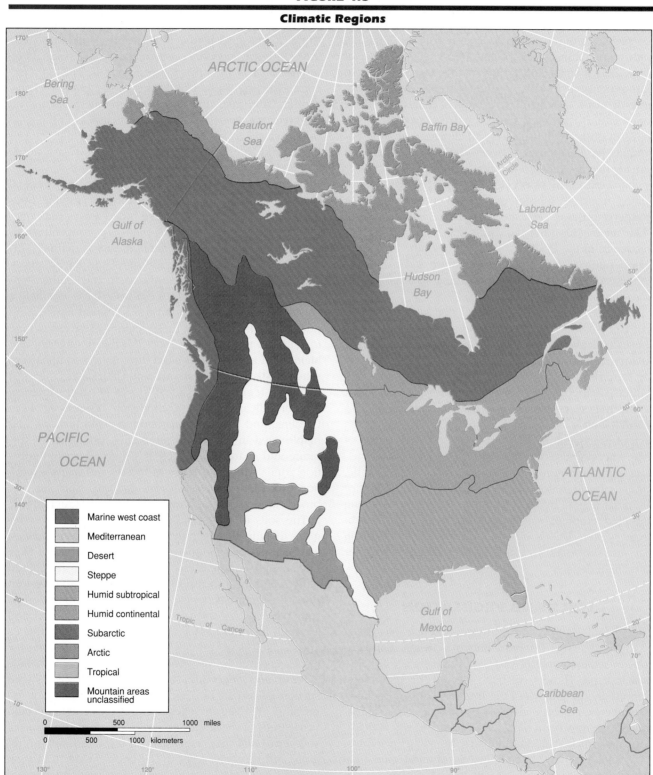

Legend:
- Marine west coast
- Mediterranean
- Desert
- Steppe
- Humid subtropical
- Humid continental
- Subarctic
- Arctic
- Tropical
- Mountain areas unclassified

0   500   1000 miles

0   500   1000 kilometers

Anglo-America. The southeastern United States has a humid subtropical climate with hot, humid summers and mild winters. The warm waters of the Gulf of Mexico provide the moisture and energy for the humidity, high temperatures, and violent storms characteristic of the region. A small climatic area in the extreme southern part of penisular Florida and the Florida Keys is the only area of tropical climate. This area is characterized by temperatures that never fall below freezing (32°F; 0°C).

North of the humid subtropical climate is an area encompassing the border between the United States and Canada that is humid continental in climate. This climate is characterized by warm summers and cold winters resulting from the interior continental location and the northerly position. The area is the interaction zone of polar air masses from the north and warm, humid air masses from the tropical areas to the south. This interaction often results in violent summer storms, heavy winter snowfall, and rapidly changing weather conditions throughout the year.

North of the humid continental region is the broad band of subarctic climate extending from the Pacific coast of Alaska to the Atlantic coast of Canada. In the subarctic, the summer period is short with the days having long daylight periods due to the high latitude. The winter is extremely cold with little sunshine. For most of the region, precipitation is minimal, averaging less than 20 inches (51 centimeters) per year. In spite of the lack of precipitation, the soil is moist due to the lack of evaporation and transpiration in this cold climate.

Extreme northern Alaska and Canada have an arctic, tundra climate, reflecting the vegetative environment of the polar region. In this region the winters are very long and cold although, surprisingly, not as cold as the subarctic winter. This is due to the proximity of Hudson Bay and the Arctic Ocean, which moderate the winter to a small degree.

SOILS AND VEGETATION

The natural vegetation of Anglo-America has been changed to a large extent by human activities. Early European settlers wrote of vast, dense forests across Ohio and Indiana. These have been cleared for agriculture. Reportedly the grasses of the Interior Plains grew taller than the height of a man on horseback. These grasslands have now been replaced by corn and wheat fields. In spite of major modifications by humans, an impression of the general geographic distribution of natural vegetative and soil types across the region can be ob-

tained by examining the vegetation in relation to soil characteristics and human activities.

The three general classes of vegetation in Anglo-America are grasslands, forest, and shrublands. The grasslands are those areas where the available soil moisture is too low to support forest growth. This area encompasses the semiarid portions of the Interior Plains, central California, and the northern intermontane region. For the most part, the soils are alfisols in the prairie provinces of Canada and mollisols in the Interior Plains and intermontane regions. The soils are fertile and well developed.

The forest areas can be classified into two general classes based on their leaf structure. Classifying trees according to whether they have broad leaves (oak, maple) or needlelike leaves (fir, pine) is a convenient, although not completely scientific, means of defining forest regions. Generally, the forests of the Pacific coast, the Rocky Mountains, the northeastern and southeastern United States, and most of Canada have needlelike leaves. These forests support the softwood lumber and pulp/paper industries in these regions. Between a pine forest band through Minnesota, Michigan, Ontario, and Maine and a pine forest band through the southeastern states of Mississippi, Alabama, and Georgia lies the most definitive region of broadleaf forests in Anglo-America. This broadleaf region encompassing Tennessee, North Carolina, and West Virginia supports oak, hickory, and maple used in hardwood lumbering, primarily for furniture and charcoal production. There are many transition zones with both types of forests and many exceptions to these general forest patterns, such as the broadleafed aspen growing in the Rocky Mountains.

The soils of the broadleaf and needleleaf forest regions include alfisols in the Great Lakes region, the wet, leached spodosols in the poor agricultural areas of New England and southeastern Canada, and the iron-rich, red ultisols of the southeastern United States. In the western portion of Anglo-America, the forest areas are underlain by mollisols that have a rich organic composition and are high in fertility. Much of the western United States and Canada is covered with soil patterns too complex to be classified in this generalized discussion.

Shrubland vegetation is comprised of bushes and small trees usually biologically limited by the lack of soil moisture. The shrub vegetation includes cacti and palo verde trees of the southwestern U.S. desert, chaparral of southern California, sagebrush in the intermontane region, and mesquite trees in the arid-semiarid transition zone of New Mexico and Texas. The predominant soil class in the shrubland region is the aridisol. As its name implies, it is an arid land soil having a low organic content. This class of soils has a high mineral content, making it a fertile agricultural soil in areas where irrigation is practiced.

The United States and Canada have large areas of good, fertile soils capable of supporting abundant natural and agricultural vegetation. This is one of the few global regions that is a surplus food producer. Great demands are made on these areas to provide the food, fiber, wood products, grazing land, and other necessities to maintain a highly economically developed society. Inattention to management of the soil and vegetative resources can have severe consequences, such as the "dust bowl" of the 1930s.

WATER RESOURCES

About two million years ago, massive continental ice sheets formed in northern Canada around Hudson Bay and expanded across the northern United States. These continental glaciers excavated the basins that are now the Great Lakes. The melting ice helped to fill the Great Lakes basins and the thousands of other lakes created by glacial action. The glaciers modified the drainage patterns of the Mississippi River and many of the other streams and rivers of the region. The glaciers left North America with one of the earth's greatest freshwater resources. The Great Lakes *watershed* which encompasses several states and provinces, drains out the St. Lawrence River to the Atlantic Ocean. The Great Lakes are a joint-use water resource supplying water for cities, industry, and agriculture in both the United States and Canada.

The other major water system in the region is the Missouri-Mississippi-Ohio River system, which drains much of the interior of Anglo-America from southern Canada to its mouth in the Gulf of Mexico in Louisiana. The western limit of this drainage area is the continental divide running along the spine of the Rocky Mountains through Montana, Colorado, and New Mexico. The eastern limit is the western side of the Appalachian Mountains through the states of West Virginia, Tennessee, and Alabama. East of the Mississippi River basin are a great number of large-volume rivers draining this humid subtropical, area. These rivers include the Savannah River, the Roanoke River, and the James River.

In the western United States, several river systems drain into the Pacific Ocean. The Columbia and Snake rivers are the principal rivers in the northwestern United

States and a portion of southern British Columbia. The Sacramento River drains much of California, and the Colorado River transports water from the west slopes of the Rockies through the arid desertlands of Utah and Arizona. These western rivers do not have the volume of flow of many of the eastern rivers. Since they flow through the most arid areas in the United States, their relative importance to the natural environment and to human activities is greater, however.

Canada has many large lakes and rivers that make it one of the most water-abundant nations on earth. Great Bear Lake and Great Slave Lake are part of extensive drainage systems that include the Athabasca River, the Mackenzie River, and the Peace River. Much of the Hudson Bay Lowland is made up of rivers and marshes transporting water into Hudson Bay. Northwestern Canada and much of Alaska are drained by the Yukon River.

Anglo-America's abundant supply of high-quality freshwater has played an important role in ensuring the prosperity of the region. Major issues for the future include the need to reduce pollution in order to maintain the quality of the water and the regional transfers of water from the northern part of the region to the arid, but growing southwestern United States. In past years, massive federal water projects have transferred water to allow growth in Texas, Arizona, and California. More projects, such as the diversion of water from western Canada and the Great Lakes to water-deficit areas, are envisioned by some. Significant environmental and political problems and the unwillingness of many taxpayers in states of limited growth to subsidize development in the rapid growth areas may force these grandiose plans to be abandoned.

MINERAL AND ENERGY RESOURCES

The United States and Canada have some of the largest deposits of mineral resources of any region on earth. These deposits are the basis for many local and regional economies within the two nations. Portions of the Canadian Shield, the prairie provinces, and the western mountains are the major mineral- and fossil-fuel–producing areas of Canada. The prairie provinces of Manitoba, Saskatchewan, and Alberta accounted for 71 percent of the $36 billion mineral and energy production in Canada in 1983. Alberta alone accounted for 62 percent of the total. Most of this was the result of escalating petroleum and natural gas prices that benefited these petroleum-rich provinces. Other resources of the prairie provinces include potash, gold, copper, and uranium. Many of the mines for these minerals are located in the Shield and the Rocky Mountain portions of these provinces. The petroleum deposits are located on the prairies with the major fields in the zone extending from Calgary to Edmonton, Alberta.

The province of Ontario is also a producer of minerals with about 10 percent of the total value of Canadian mineral production. Ontario's relative importance in the resource economy of Canada has declined in recent years due to the drop in prices of its major mineral outputs: nickel, uranium, and copper. Increases in the production and price of gold have helped offset these declines. Among the other provinces, Quebec is a leading producer of gold and iron. British Columbia is the leading zinc-producing province. In addition, large quantities of gold, silver, copper, and coal are mined.

In the United States, the western mountains are a major mineral-producing area. Large deposits of gold, copper, silver, molybdenum, and other metallic ores exist in the mountain states of Arizona, Utah, New Mexico, and Montana. Copper and molybdenum are mined by open-pit methods. Other metallic ores are mined using tunnel mines or water separation of placer deposits. As in Canada, the increase in the price of gold in recent years has reinvigorated the gold mining industry in Nevada, Colorado, California, and Idaho.

The iron ore deposits of the upper Midwest are the other significant metallic ore deposits in the United States. Minnesota and Michigan are the leading iron-ore–producing states. Copper is also found in this area, often in the relatively uncommon metallic form. These deposits were a major impetus to the development of the iron and steel industry in the Great Lakes region.

Significant deposits of nonmetallic minerals are found in various areas of the United States. Florida leads the nation in the mining of phosphate rock, followed by Idaho, North Carolina, and Tennessee. Phosphate rock is an important component of agricultural fertilizers and a source of industrial phosphorous compounds. Asbestos is mined in Arkansas, Ohio, and Wisconsin.

Coal, natural gas, and petroleum are the most important fossil-fuel energy resources in the United States. Major coal deposits are located in the Appalachian Mountains, the Midwest, and the eastern parts of the Rocky Mountain region. Kentucky and Wyoming are the leading coal-producing states, followed by West Virginia, Pennsylvania, and Illinois. An important distinction between the coal produced in the different states

is its sulfur content. The coal of Wyoming and Arizona is low in sulfur compared to Appalachian coal. The governmental restrictions on industrial air pollution have increased the demand for low-sulfur coal. The remoteness of these low-sulfur coal mines has resulted in a relocation of coal-fired electrical generation plants to the western coal mine sites. It is cheaper to transport electricity via transmission lines than to haul coal via railroad to the generation plants in urban areas.

The largest petroleum-producing areas are in the south central states of Texas, Louisiana, and Oklahoma. Texas had the highest production in 1983. Much of the production is on the coast and offshore areas of the Gulf of Mexico. The other producing area within the south central United States is northern Texas and Oklahoma. The current leader in petroleum production is Alaska. The north slope area of extreme northern Alaska is the most recently developed deposit. In spite of the long shipment necessary, Alaskan petroleum deposits will remain an important part of the petroleum future of the United States. Recent discoveries in the Beaufort Sea are likely to provide additional Alaskan petroleum for the United States, especially as the price rises in the 1990s following the decline of the mid-1980s.

A primary issue associated with these resources is the depletion of the resource base within the United States. The United States is becoming more dependent on other countries for the raw materials and energy resources necessary to maintain its high level of economic development. Figure 4.4 demonstrates the decline in the productivity of the U.S. natural resource base relative to the total world production of selected resources between 1970 and 1986. The increasing U.S. dependency on foreign petroleum is portrayed in Figure 4.5. After a peak in consumption and importation in oil in the late 1970s, both dropped, with the average daily consumption of imported oil reaching a low of 5,051,000 barrels in 1983. It appears that the trend in the 1990s will be for consumption and importation to continue to rise. Since the domestic economy depends on reliable and inexpensive supplies of petroleum and other natural resources, this situation has implications for the opening of environmentally sensitive areas in the United States for resource exploration and development. Foreign relations between the United States and other countries will also be affected to a large degree in the future by natural resource considerations.

The natural environment has fostered the high level of growth and development in the United States and

FIGURE 4.4

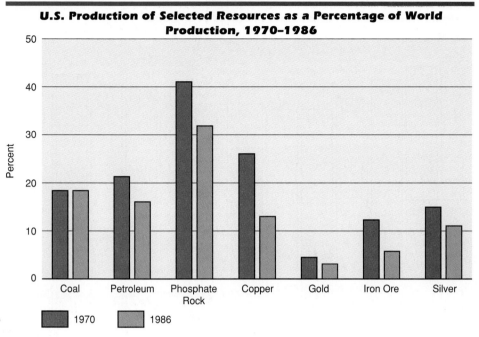

U.S. Production of Selected Resources as a Percentage of World Production, 1970–1986

1970   1986

SOURCE: U.S. Bureau of the Census, *Statistical Abstract of the United States*, 1989 (Washington, D.C.: U.S. Government Printing Office, 1989).

Canada. It has provided the setting and raw materials for the development of the human landscape of people and culture in Anglo-America.

HUMAN ENVIRONMENT

POPULATION

The population of the United States is about 10 times larger than that of Canada. The U.S. population as of 1990 was over 250 million people while Canada's population was in excess of 26 million people. The major population concentrations are in the northeastern and north central portions of the United States and the adjacent areas of Ontario and Quebec in Canada. The major areas of population growth are along the margins of both countries. The largest absolute population growth is occurring in Texas, California, and Florida in the United States and the near-border areas of Ontario, Alberta, and British Columbia in Canada. The northern in-

terior of Canada is virtually uninhabited. Many parts of the central region of the United States are experiencing population declines.

Many of the population problems that plague other regions of the world, such as a large population relative to the resource base or standard of living, are not major problems in the United States and Canada. In addition, rapid population growth is not a problem in Anglo-America as it is in less economically developed regions of the world. Since 1972, the average number of children born to each American woman has been two or less, resulting in an average increase in natural population growth of about 1 percent per year between 1965 and 1985. In 1990 the natural increase was 0.8 percent for the United States. Canada's natural population growth has dropped steadily from 2.5 percent in 1961 to 1.1 percent in 1981 to 0.7 percent in 1990. These natural increases are well below the developing countries of Asia, Africa, and Latin America where natural increases average 2–3 percent per year.

This is not to say that the United States and Canada do not have issues and problems associated with their populations. The most important population issues in-

FIGURE 4.5

U.S. Petroleum Consumption and Importation, 1973–2000

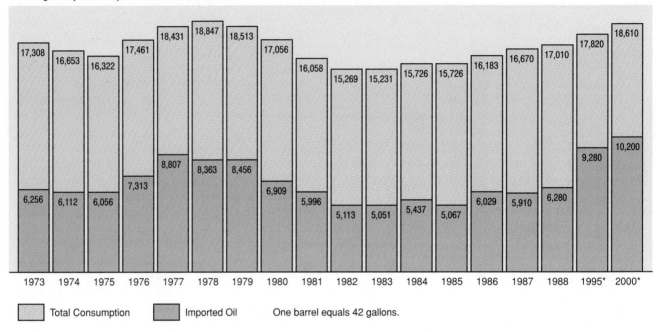

Average Daily Consumption in Millions of Barrels

☐ Total Consumption   ☐ Imported Oil   One barrel equals 42 gallons.

*Base estimates Energy Information Administration, *Annual Energy Outlook 1989*.

SOURCE: U.S. Energy Information Administration, *Annual Energy Outlook*, 1989.

---

volve the regional redistribution of population within Anglo-America, legal and illegal immigration, urbanization, and the role of minorities within the demographic and socioeconomic mix of the region.

The Effects of Regional Redistribution of Population

The populations of the United States and Canada are experiencing a geographic shift in areas of concentration. The populations of both countries were overwhelmingly nonurban and agricultural in the nineteenth century. The United States had become more than 50 percent urban by 1917. As of 1990, it was 74 percent urban. Canada has followed a similar trend with over 77 percent of its population being urban in 1990.

Canada. In Canada, the Atlantic Maritime Provinces of Newfoundland, Nova Scotia, and New Brunswick, the provinces of Quebec and Manitoba, and the Yukon and Northwest Territories all had net out-migrations in recent years. The provinces of British Columbia and Alberta had the largest influxes of migrants relocating from other provinces, followed by Ontario, Saskatchewan, and Prince Edward Island. As a percentage of the total population of Canada, the provinces of Alberta and British Columbia are growing most rapidly. This growth is a result of migration, immigration from other countries, and the natural increase of the resident population. During the period from 1951 to 1981, Alberta's share of the total population of Canada increased from 6.7 percent to 9.2 percent. British Columbia's share increased from 8.3 percent to 11.3 percent during the same period. The percentages of the total population for the Atlantic Provinces and Quebec declined during this time period. Ontario peaked at 35.9 percent of the Canadian population in 1976. It has declined slightly since that year.

Population shifts to other locations result from **push-pull migration factors.** Provincial shifts are the result of a decline in opportunities in the poorer, less industrial areas of eastern Canada—a push factor. The shifts have been fostered by the emergence of natural resource–oriented industrialization, urbanization, and opportunities in the western areas of Canada—pull factors. The economic impacts of this geographic redistribution have led to significant political and social impacts. A large number of residents of eastern Canada depend on federal governmental monies to subsidize their housing, food, and medical care. The eastern provinces spend more federal governmental dollars on such programs than they contribute to the federal treasury. This difference is made up by the western provinces, which contribute more to the treasury than they get back from Ottawa. This regional inequity in governmental payments and support has aggravated the conflicts associated with the dual language policy and the desire of some of the French-speaking people in Quebec to secede from Canada.

The United States. In the United States, regional redistribution of population and, therefore, political power is occurring as well. The geographic center of population for the United States has been moving steadily westward; it was located in eastern Maryland in 1790, in southern Ohio in 1870, and in eastern Missouri in 1980 (Figure 4.6). Between 1970 and 1980, the populations of the South and West regions, as defined by the U.S. Census Bureau, increased 20 and 23.9 percent, respectively.

This trend of the last several years runs counter to the migration patterns in the United States since the late 1800s when millions of people were leaving the rural areas of the South and Midwest to move to the cities of the upper Midwest and Northeast in search of industrial jobs and an urban life-style. Now the jobs that were available in the northern cities until the 1950s have declined as the American economy no longer depends on heavy industrial output as its mainstay. The social opportunities offered in the North have declined relatively along with the jobs. The competition for limited jobs, housing, schools, and political power has diminished the advantages of relocation to northern urban areas, especially for blacks, Hispanics, and rural southern whites.

In the 1960s and 1970s, a "new" economic thrust evolved in the United States. The economy became focused on light manufacturing, production of services, and high technology. Almost 70 percent of the jobs available in the United States are in services production

with only 25 percent of the jobs in industrial production. Agricultural employment accounts for less than 5 percent of total U.S. jobs. One of the results of this change in the American economy has been the creation of employment opportunities in the South and the West, mainly in the cities of these regions. The restrictions on political and social freedoms, based on race and religion, have lessened in the South and the West due to the civil rights movement and federal legislation. Along with these substantive changes, the perception of these areas has improved; they are now viewed as comfortable and rewarding living places. This has further stimulated the migration to the South and the West.

The development of transportation, energy, water transfer, and air conditioning technology has allowed large urban areas to be created in the West and the South. Ironically, the taxpayers of the Midwest and North subsidized the construction of the energy, transportation, and water projects that were necessary to foster the growth in the South and the West. Without large federal government projects, such as the Colorado River Projects and the Tennessee Valley Authority, the infrastructure required for urban, commercial, and industrial growth in these regions would be limited.

The age structure of the population has an important role is this regional redistribution of population. The median age of the American population is increasing. In 1971 the median age was 27.9 years. In July 1986, it had increased to 31.8 years. Twelve percent of the U.S. population was 65 years of age or older in 1990. In Canada the figure is about 11 percent. These percentages have increased markedly in recent years and will continue to do so in the future. More elderly persons are living longer due to advances in nutrition and medical technology. The life expectancy in the United States is now 75 years. Many people retire between the ages of 55 and 70. These numerous retirees are moving to the "sunbelt," particularly Florida, Arizona, and California. In the retirement areas, the economic system has been modified to meet the needs of these migrants, many of whom are on fixed incomes. The political complexion of many areas is also being altered to a great extent, reflecting the views and life-styles of these new residents. The retirees are less likely to vote for school levies and more likely to support transportation, social service, and medical programs that assist the elderly. These changes can embroil the longtime residents of these retirement areas in conflicts with the more recent, elderly citizens.

The future trends of regional growth appear to be a

FIGURE 4.6

The Westward Shift in the U.S. Center of Population, 1790–1980

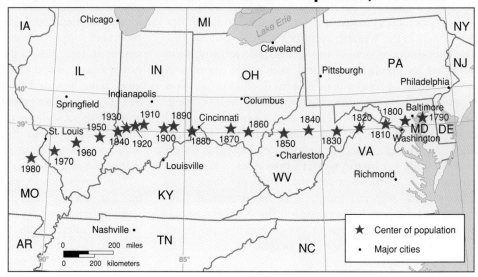

SOURCE: U.S. Bureau of the Census, *1980 Census of the Population,* vol. 1 (Washington, D.C.: U.S. Government Printing Office, 1980).

continuation of the existing patterns with some significant exceptions. The migration to the South and the West will continue. Most of the growth, however, will be directed toward certain coastal areas of California, Florida, and Texas and the larger metropolitan areas of these states, such as Los Angeles, Tampa, and Dallas. The more remote small towns of the regions have not and will not be the primary recipients of the growth and economic development. This is not to suggest that population growth will not occur in the Northeast and upper Midwest. In recent years, several northern cities have been experiencing renewed growth and vitality due to a new and relatively unique set of circumstances. Several cities in New England have undergone a dramatic rebound from the economic stagnation of the 1960s and 1970s. Most of this growth can be attributed to the presence of institutions of higher education and technology. These institutions wedded to billions of dollars in federal spending, primarily in the defense sector, provided the impetus for a renewal of the regional economy. Many other areas of the North are attempting to duplicate this model for postindustrial economic renewal with mixed results.

CULTURAL ASPECTS

The culture of Anglo-America is a rich blend of cultures brought by settlers from a multitude of nations. These cultural infusions, which have overlain the native American Indian culture, originated in the nations of Western Europe. This cultural mix was supplemented by the religions, languages, and beliefs of the thousands of Africans brought to America as slaves in the eighteenth and nineteenth centuries. More recently, the North American cultural milieu has been influenced by southern and eastern Europeans, Latin Americans, and Asians.

The Ethnic Melting Pot?

The United States and Canada have a diverse mixture of cultural and ethnic groups. This diversity and the integration of these groups into the American and Canadian cultural mainstreams are important issues. As of 1980, the ethnic composition of the United States was 80 percent non-Hispanic whites, mainly of European extraction. The largest minority are African Americans who comprise 11.7 percent of the population. The next largest minority group is composed of Hispanic Americans. This group accounts for 6.4 percent of the population, of which the Mexican-American group is the largest. Two percent of the population is composed of Asians, Native Americans, and others. Figure 4.7 shows the geographic distribution of several major ethnic groups within the United States.

The ethnic diversity of Canada reflects the predominately European heritage of the country. As of 1981, the largest European group was of British ancestry (40.2 percent of the population) followed by French (26.7 percent), German (4.7 percent), and Italian (3.1 percent). The native Indians and Inuits represent only a declining 2 percent of the population. They are located mostly in the Northwest Territories.

Minorities in Canada. Ethnic diversity underlies the fundamental division in Canadian society, which is based on language. Over 68 percent of Canadians have English as their first language. A large minority (24.6 percent), reflecting their ethnic proportion of the population, use French as their first language. For this reason, Canada has two official languages that are used in all official governmental transactions. A significant percentage (7.2 percent) of the population uses nonofficial languages corresponding to the significant ethnic minorities in Canada. People with Indo-Pakistani languages as their mother tongue doubled between 1976 and 1981 to 117,000. The population with Chinese as the mother tongue reached 224,000 in 1981.

Cree is the most common of the Canadian native languages. It is the mother tongue of over 67,000 Canadians of aboriginal descent. The 1981 census indicates a total of 491,460 native Indians and 25,390 Inuits in Canada. There are 576 bands of Indians who occupy or have access to 2,251 reserves distributed among all of the provinces and territories with the exception of Newfoundland. Most of the Inuits (15,910) reside in remote communities in the Northwest Territories with smaller groups in Quebec (4875), Labrador (1850), and Ontario (1095).

African Americans. The African-American or black population in the United States amounts to over 26 million people or about 12 percent of the total population. Most of these people share a common African racial heritage with an ethnic identity developed in America. Most of the black population was confined to Maryland, Vir-

ginia, and the Carolinas from the first introduction of blacks as indentured servants in the early 1600s until slaves began to be used on the cotton plantations in the late 1700s. This change in cropping and farm management resulted in the redistribution of blacks via the slave trade to the other states of the South from Georgia to Texas. By 1900, approximately 90 percent of the black population resided in the U.S. South.

After the Civil War and emancipation, the geographic distribution of African Americans began to reflect their adaptation to the American economy. As the economy became less agricultural, more industrialized, and more urbanized, so too did the distribution of blacks in America. The changing geographic pattern of blacks corre-

The City Hall in Pasadena, California, is a good example of Hispanic influence on the culture and architecture of the southwestern United States.

FIGURE 4.7

Distribution of Selected Ethnic Groups in the United States, 1980

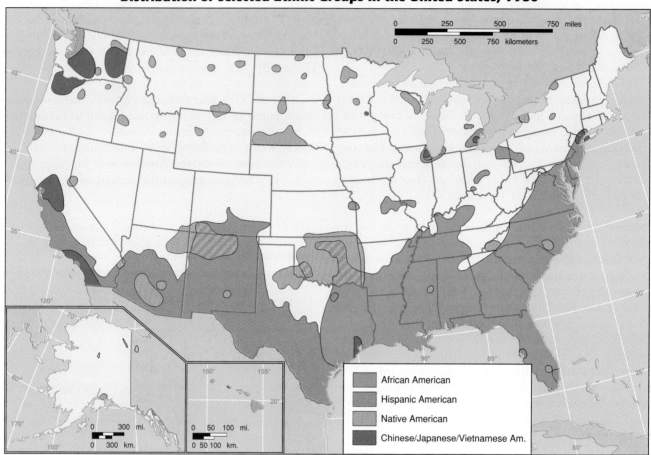

African American

Hispanic American

Native American

Chinese/Japanese/Vietnamese Am.

SOURCE: U.S. Bureau of Census Maps, GE Series; modified from Jesse McKee, *Ethnicity in Contemporary America* (Dubuque, Iowa: Kendall-Hunt, 1985).

sponds to the regional redistribution of people in general in the United States. The largest migration of blacks was to northern urban areas beginning about 1910. By 1920, New York, Philadelphia, and Chicago had the largest concentrations of blacks of any urban areas in the United States. By 1940, 1.5 to 2.0 million blacks had migrated from the South to the Northeast and the North central portions of the United States. This trend continued after World War II as black people increasingly moved to the urban areas of the West Coast.

A return migration of blacks to the South began in the 1960s and was firmly established by 1970. The changing economy created employment opportunities in the South. The improved racial, social, and political situa-

tion in the South also assisted this reverse migration. Most of the blacks returning to the South have opted for the major urban areas, such as Atlanta, Memphis, and Houston. Western migration has continued in the 1970s and 1980s. San Diego and Oakland, California, have been leading destination cities for migrating blacks during this period.

Hispanic Americans. The Hispanic population of the United States is distributed throughout the Southwest, Florida, and most large northern cities. The largest groups comprising the Hispanic population are of Mexican, Cuban, and Puerto Rican descent. Their geographic distribution is largely dependent on their ancestry. The people of Mexican ancestry are concentrated in states bordering Mexico that were formerly part of New Spain and then Mexico. Of the almost 9 million Mexican Americans in the United States, over 3.5 million live in California followed by Texas, Arizona, New Mexico, and Colorado. Some of these people are descendants of Mexicans who settled in these areas before they became U.S. territory in the mid-nineteenth century. Many more trace their descent to refugees from the Mexican Revolution (1910–1920) when thousands of Mexicans sought safety in the United States. In addition to the people in the border states, significant numbers of Mexican Americans live in the Midwest, particularly Illinois where over 400,000 reside, mostly in Chicago. As with the black population, many were drawn to this area by the industrial job opportunities available after World War I. Others came to the Midwest as seasonal migrant farm labor and took up permanent residence.

The second largest Hispanic group living in the 50 states of the United States is of Puerto Rican heritage. Their numbers are estimated to be close to 2 million or approximately 14 percent of the Hispanic population. The over 3 million residents of the island of Puerto Rico are American citizens as Puerto Rico has commonwealth status with the United States. The combined total number of Puerto Rican–American citizens is approximately 5 million. Within the 50 states, the largest concentrations of Puerto Ricans are in the large cities of the Northeast and Midwest. New York City is the largest with close to 1 million Puerto Rican residents. Other large concentrations are found in Chicago, Newark, Philadelphia, Los Angeles, and Miami.

The flow of Cubans to the United States has been modulated largely by the political and economic conditions in Cuba. Many of the approximately 1 million Cuban Americans can trace their heritage back to immigrants coming from Cuba in the 1800s. The largest immigration of Cubans has occurred in the last 30 years, however. The defeat of the Batista dictatorship by the revolution of Fidel Castro in 1959 led to a massive exodus to the United States. Over 200,000 Cubans left Cuba for the United States between 1959 and 1962. Nearly 800,000 have migrated during the postrevolution period. Cuban Americans are concentrated in a few states: Florida has almost half of the total with New York, New Jersey, and California being the other states of greatest numbers. The Cuban Americans are located in a few large urban areas within these states. Miami is the home of the largest number of Cuban Americans with New York, Los Angeles, and Chicago being far below Miami in numbers of residents.

Native Americans.   According to the 1980 census, the Indian, Eskimo, and Aleut population in the United States is an estimated 1,418,195. This minority group has distinctive languages, religions, and traditions far different from other American minority groups. This non-European and, initially, non-Christian culture has been hard-pressed by an antagonistic, dominant culture since the days of the first white settlements. The Eskimos and Aleuts comprise 66 percent of the Native American population in Alaska. These groups along with the separate American Indian group comprise approximately 16 percent of the total population of the state.

The western states contain the majority of the American Indian population in the conterminous United States. California, Arizona, New Mexico, and Oklahoma have significant percentages of the Indian population. These states have the largest reservations in the nation, such as the Navajo and Hopi Reservations in Arizona and New Mexico. These states also contain the urban areas to which a large segment of the Indian population has migrated in recent years. Although California has no large reservations, the cities of Los Angeles and San Francisco are leading migration destinations for Indians leaving the reservations in other states. Surprisingly, as of 1980 over 50 percent of the American Indian population is living in cities with populations greater than 2500. Lesser numbers are present in the states of North Carolina, Michigan, South Dakota, New York, Montana, and Washington and in smaller numbers throughout the country. These populations reflect the presence of community enclaves or reservations within these states.

Isolated Ethnic Groups.   Isolated ethnic groups occur in all regions of Anglo-America. These groups have a basis in ethnic origin, language, or religion that has allowed them to maintain their separation from the dominant culture. Generally, these groups can function within the dominant culture of America, but have not been totally assimilated by that culture. The French-speaking Acadians fled Canada in the mid-1700s. These refugees from British domination became the Cajuns of southern Louisiana. Until recently, this small group has maintained its French language and culture in the relative isolation of the bayou country.

The Amish are a Swiss Anabaptist sect concentrated in central Ohio and Pennsylvania with small groups in

A traditional Navajo Indian house or hogan on the Navajo Indian Reservation in Arizona.

several other states. Their numbers total about 90,000 spread throughout rural areas of the United States and Canada. They are known for their strict adherence to their religion and their refusal to own or use modern technological trappings, including electrical devices and motorized vehicles. They maintain their cultural isolation in the midst of many non-Amish through their religious teachings and their use of a German dialect as their primary language. A high birth rate among the Amish has led to an increase in their numbers despite a 25 percent dropout rate from the group.

Several other ethnic groups have also maintained their cultural identity and language within the North American melting pot. Significant numbers of Finns, Swedes, Germans, and Norwegian descendants reside in the rural areas of Minnesota and Wisconsin. Groups of Middle Eastern, Asian, and eastern and southern European ancestry reside in the large cities of Canada and the United States. Many of these people still use their ancestral language as a second language to English in social settings. Ethnic neighborhoods within these cities are easily defined by their distinctive spatial arrangements, types of businesses, churches, and other cultural symbols. In recent years, the trend has been for ethnic groups to enhance their common bond and the identifiable characteristics of their ethnicity, such as language, religious practices, architecture, and social ceremonies.

Ethnic groups in both cities and rural areas highlight their culture with festivals, traditional ceremonies, and instruction in language and culture for the young.

This examination of Anglo-American ethnic groups suggests that the "melting pot," is not the best culinary analogy for the dissolution of ethnic and cultural identities into the homogeneous mass of American culture. The uniform society suggested by the melting pot is not present, nor has it ever been present in the United States and Canada. A more accurate analogy is probably a "lumpy stew" in which various groups regard themselves as Americans while retaining their ethnic and cultural traditions.

Religious Composition

The religious diversity of the United States and Canada is an important part of the cultural diversity of these nations. In Canada as of 1981, the 11.4 million Catholics formed 47.3 percent of the population; the Protestants number 9.9 million or 41.2 percent of the population; and Eastern Orthodox and Jewish faiths accounted for 1.5 percent and 1.2 percent, respectively. Of the remaining 8–9 percent, 7.4 percent expressed no religious preference, and 1.3 percent were affiliated with other minor religions. The geographic pattern of religions is closely associated with the patterns of ethnicity and language. The provinces that are predominately Catholic are French-speaking Quebec with 88.2 percent and New Brunswick with 53.9 percent. Ontario and the provinces farther west are Protestant. Most of the Jews in Canada live in either Ontario or Quebec, primarily in the urban areas. In Ontario, Jews number nearly 150,000 or about 1.7 percent of the population. Although small in numbers, the Buddhists are the fastest growing religious group in Canada. This group recorded a 223 percent increase in numbers since 1971 to a 1981 total of 51,955. Pentecostals increased by 54 percent, with Mormons increasing by 36 percent, Catholics by 13 percent, and Jews by 8 percent over the same period.

In the United States, the patterns of religion are not regionally uniform. Overall, 70 percent of the population identify themselves as Protestant. The southeastern United States is predominately Baptist followed by Methodist. The Upper Midwest is an area with a strong Lutheran following due to the relatively large numbers of people of German, Swedish, and Norwegian ancestry. The northern areas of the West are Protestant as well. The fundamentalist and evangelical groups within the major Protestant faiths, particularly the Baptists and Assemblies of God, are the fastest growing Protestant groups. Roman Catholics comprise approximately 25 percent of the U.S. population. The major concentrations are in the large cities of the Northeast and Midwest reflecting the large Hispanic, Italian, and other southern European ethnic groups. The southwestern states also have a large Catholic population reflecting the large Hispanic population. In the Southeast, an island of Catholic beliefs exists among the people of French heritage who live in southern Louisiana.

The people of the Jewish faith number nearly 6 million or about 3 percent of the U.S. population. Almost 50 percent of this total (2.4 million) live in the New York City area. Other significant concentrations are in the Miami and south Florida area and the Los Angeles area with approximately 500,000 Jews living in each area. Philadelphia, Boston, and Chicago have sizable Jewish populations as well.

Other distinctive religious groups having a presence in the United States include the Mormons, Muslims, and Buddhists. The Mormon church began in New York in the 1800s. Today the Mormons are centered in Utah with churches spreading to all portions of the United States and Canada. Their social service network and practice of actively recruiting new members has made this religion one of the fastest growing in the region.

Muslims represent a significant religious group in Anglo-America for two reasons. Several thousand African Americans have converted to Islam as a response to the acceptance of slavery by the Christian church before the Civil War and as a means of ethnic identity. These American converts follow the teachings of the Koran as do Muslims around the world. In addition, immigrants from Muslim countries of the Middle East, Africa, and Asia have established groups of Muslims in several urban areas. New York, California, and Michigan have sizable Muslim populations. Many of the immigrants from Asia have maintained Hinduism and Buddhism as their religion in the United States and Canada. These religions are centered in large urban areas with large Asian populations, particularly in California, Washington, Texas, and New York.

Immigration to Anglo-America

Immigration to Anglo-America has ebbed and flowed according to the political and economic situations in other regions of the world. Immigration is the single most important factor in the limited amount of population growth that is occurring in Anglo-America at the present time. Net immigration is calculated by subtracting the number of people leaving the region (emigration) from the total number of legal and illegal immigrants entering the region. Net immigration accounted for almost 30 percent of the population growth in the United States between 1980 and 1985. The region of origin of most of the immigrants has shifted dramatically over the last several years. In the late 1800s, Europeans made up 90 percent of the immigrants to the United States. In 1985 only 11 percent of the 570,000 legal immigrants to the United States were from Europe. Asians accounted for 46 percent of the total of legal immigrants in 1985 while Latin Americans comprised 37 percent. Illegal immigration, resulting from the civil strife and depressed economies of Latin America, has risen greatly in the last several years. The U.S. Immigration and Naturalization Service (INS) located 231,100 deportable aliens in 1970 and 1,138,600 deportable aliens in 1984. Given that the INS admits that it locates only a small percentage of the total number of illegal aliens, the total entering and living in the United States must be in the millions.

Immigration into the United States is the result of push factors in the immigrant's home country, such as unemployment and political and social unrest. The decision to leave one's country, family, and culture to emigrate to Anglo-America also results from pull factors, the perceived and real economic opportunities and freedoms offered by the United States and Canada.

Immigration represents a complex proposition for the United States and Canada. New immigrants often perform difficult, menial, and dangerous labor for less pay than do resident workers. Compared to the jobs and pay in their home countries, even the most menial jobs in the United States may be quite acceptable to the immigrants. Competition between immigrants and resident workers for jobs and the fear that the immigrants' willingness to work for less will lower wages for all workers are major sources of conflict, however. Conflict also arises due to language, cultural, and social differences between residents and the immigrants. This conflict intensifies when newcomers and residents are competing for housing, education, social services, and cultural identity. It is important to note that this competition and conflict is not new or unique to Anglo-America. It happens in all countries and regions faced with a large influx of immigrants. The conflicts have not arisen just in the last few years, but rather have been present in Anglo-America for the last 150 years.

Immigration has also had many positive aspects for the United States and Canada. Much of the scientific, technological, and cultural vitality of Anglo-America is the result of the infusion of ideas, innovation, and human energy from the waves of immigrants over the last 300 years. Since the immigrants are often well educated, talented, and motivated, they have made important contributions to the economic and cultural development of the region. Furthermore, the cheap immigrant labor in agriculture, clothing production, construction projects, and other industrial activities allows some American companies to compete with foreign firms. A Rand Corporation study in 1985 reported that Mexican immigration to California has enabled many industries in the southern California area to flourish in spite of vigorous foreign competition. The study contended that the preservation of these industries has resulted in the maintenance of higher skilled, higher paying jobs for resident Americans throughout the region.

On a personal level, illegal aliens know that a month's pay at even the most menial job in the United States may exceed their earning potential for an entire year in an Asian or Latin American nation. Added to economic incentives are the turmoil, instability, and uncertainty that dominate life in many developing countries. As a result, no barrier, such as a fence or the Rio Grande, will keep individuals from crossing the border to benefit

Amish farmers in Pennsylvania harvest corn with horse-drawn machines and wagons. Religion and cultural practice prohibit the Amish from owning motorized vehicles, but not from using labor-saving machines.

themselves and their families. Any measures to stem the illegal flow of aliens must therefore deal with the push-pull factors that underlie the immigration. This means limiting the employment and social service opportunities available to illegal aliens in the United States and improving the economies and political situations in the source countries so that people will not want to leave.

The Future Impact of Immigration in the United States.   The role immigration will play in the U.S. population in the future is not clear. Current legal immigration levels will not significantly increase the total population in the future. Although immigrants are generally younger than the existing U.S. population, the trends toward an aging population will not change. These conclusions are firm unless a marked increase in fertility occurs in the general population. However, the ethnic composition of the U.S. population may change a great deal. Given the current levels of fertility and legal immigration, the percentage of non-Hispanic whites will decrease from the 80 percent of 1980 to 60 percent by the year 2080. Hispanics will increase from about 6 percent in 1980 to 16 percent by 2080, Asians from 2 percent to 10 percent, and blacks will have only a small increase from 12 to 15 percent.

The amount of illegal immigration is a key factor in the ethnic composition question. A continuation of the current high rate of illegal immigration will make non-Hispanic whites the largest minority in a nation with no clear majority by the twenty-second century. The percentage of Hispanics in the U.S. population would be close to 25 percent with Asians comprising about 12 percent and blacks accounting for 15 percent of the U.S. population. Most legal and illegal immigration is directed toward the states of California, Texas, Florida, New York, and Illinois. These states will undergo the most significant impact with other states experiencing minimal changes. The most dramatic effects would be in states having large Hispanic populations, such as California and Texas.

AGRICULTURAL PRODUCTION

Canada and the United States are very productive agricultural nations. The combination of large tracts of flat terrain, suitable climate, fertile soils, and technologically advanced farming methods makes this region one of the few net exporters of food in the world. The climate tends to be the primary limiting factor on the types of agricultural products produced in various areas of Anglo-America.

Canada

Only the very southern portion of Canada has significant agricultural activity. Canada, with its colder climate and shorter growing season, produces a smaller variety of products than does the United States. In the eastern portion of Canada, dairy farming is prevalent because the rocky, shallow soil inhibits intensive crop cultivation. Along the shores of the Great Lakes and the St. Lawrence River, some specialized agriculture takes place using the deeper, richer soils and the lake-induced microclimates offered in these areas. Grapes, tobacco, and vegetable crops are grown in these areas. In terms of farm receipts, Ontario is the leading agricultural province in Canada. The production of corn, hay, apples, grapes, beef, and dairy products accounts for this dominance. Farther west, wheat and small grain framing predominates. Wheat is the single most important crop in terms of total cash receipts in all of Canada. The interior plains of Manitoba, Saskatchewan, and Alberta are the major producing areas. Smaller areas of general farming, dairying, and irrigated farming occur in the river valleys of southern Alberta and British Columbia. The Fraser Valley in western Canada is known for its poultry, eggs, dairy products, and fruits. Coastal British Columbia and Vancouver Island have specialized fruit, vegetable, and flower industries.

The United States

In the United States, the much wider range of climate presents opportunities for a greater diversity of agricultural production. The actual geographic pattern is determined by physical factors, such as climate, soils, and terrain, the historical agricultural patterns of the settlers, and transportation cost factors, such as the distance to markets and the bulk of the products (Figure 4.8). The shallow, infertile soils and harsh winters of New England make dairying the primary agricultural activity in this area. Exceptions occur in the fertile river valleys, such as the valley of the Connecticut River, where tobacco and truck crops of fruits and vegetables for the nearby urban centers are produced. In the mid-Atlantic area along the coastal plain, general farming of many crops and truck farming of garden fruits and vegetables are practiced. This production of summer vegetables for the large urban centers of the East Coast is the reason New Jersey is called the "Garden State."

In the southeastern United States, the historical cotton belt has receded, but is still apparent in isolated production areas stretching from South Carolina to Texas. The dominance of cotton production has now largely been replaced by a more varied mix of soybeans, corn, and other grains. Within this area, other favored crops are peanuts in Georgia and Alabama and tobacco in North Carolina, Kentucky, and Virginia. On the southern periphery of the southeastern region, the production of specialty crops is very important. The winter harvesting of citrus, exotic fruits, and garden crops is the leading agricultural activity in the southern areas of Texas and Florida.

The northcentral United States is characterized by dairy production. The states of Wisconsin and Michigan are known for their milk and cheese. The physical limitations of climate and soil make livestock grazing the best use of much of the land in these states. The proximity of this region to the large urban areas of the Midwest makes it very suitable for producing relatively high-value, but quite perishable dairy products.

South of this area is the more fertile, milder climatic agricultural region called the corn belt or, perhaps, more appropriately, the cash grain belt. This area, comprised of Ohio, Indiana, Illinois, and Iowa is noted for its vast fields of soybeans and corn used to fatten cattle grazed on the grasslands farther west. The corn is used to supplement the diets of the large numbers of pigs raised in the corn belt, as well as the dairy herds to the north. The agricultural activities in this region are dependent

FIGURE 4.8

Agricultural Regions of the United States and Canada

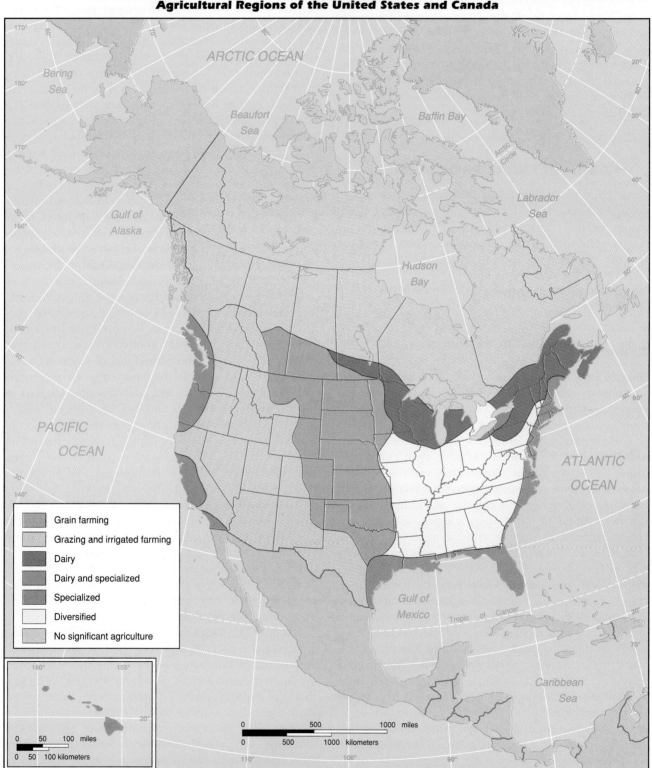

on its location at the nexus of the western grazing region, the dairy region, and the major urban concentration in the United States.

West of the corn belt lie the drier grassland areas suitable for winter and spring wheat and small grain production. This area is too arid to support crops, such as corn or soybeans. Along the eastern edge of the Rocky Mountains, the climate becomes so arid that no crops will grow without irrigation. Supplemental irrigation has extended the variety and range of crops grown in the area from Montana to Texas. Despite increased irrigation, much of this area and the mountain areas to the west are used solely for grazing cattle and sheep.

In the Mediterranean and desert climatic areas of the southwestern United States, irrigation is a necessity for intensive agriculture. This area offers mild temperatures and a long growing season that allows the production of a variety of crops. The production of more highly valued crops, such as citrus, cotton, avacados, and garden vegetables, overcomes the extra expenses of irrigation and transportation to the markets in the eastern United States and overseas. Early settlers from the wine-growing regions of Europe introduced grape cultivation in this area, particularly in California. Ironically, California grape rootstocks were used to replant many European vineyards wiped out by disease. The wine production of California now rivals that of Europe in amount and quality.

The Pacific Northwest has dairy and fruit production oriented toward the urban centers of the region. The region is noted for its production of apples and potatoes that are consumed throughout the United States. Grain farming in the central and eastern portions of Washington and Oregon is directed toward the Pacific Basin export market.

The American agricultural system is changing dramatically. The number of people employed in farming has decreased steadily since the 1800s. The number of individual farms is decreasing, while the remaining farms are becoming larger (Figure 4.9). The role of the farm unit in the farm economy is also changing as many farms are now owned by large diversified corporations or are contractually and financially bound to a corporate entity.

Smaller and less efficient farm operations have been forced out of business in recent years. The costs of fossil-fuel–based fertilizers and farm chemicals, equipment, and interest rates have increased while farmland values and crop prices have declined or remained stable over the last 15 years. This has brought about a decline in the number of farms and farmers. Much of the land has been purchased by nonfamily, publicly held corporations.

Publicly held corporations, as opposed to private family corporations, account for a very large share of total U.S. agricultural sales, especially in certain com-

Natural grassland, such as this area on the interior plains of Canada, is ideal for growing wheat.

modities. The share of sales controlled by corporations is greater than 25 percent in the vegetable, fruit and nuts, nursery products, poultry, and cattle industries. The corporate control of these specialized agricultural activities is concentrated in Texas, Florida, and California. These trends are continuing in the 1990s with more of the food supply being managed by fewer farming units, many of which are corporate. The corporate control of U.S. agriculture ofen improves efficiency, capital availability, and **economies of scale**—the increased efficiency of operation and profitability realized by a larger farm as opposed to a smaller farm. Economies of scale involve the increased use of mechanization, volume discounts for seed and fertilizer purchases, and less dependency on minor market changes. The possible negative side to corporate control is that many of these corporate units do not depend on agriculture for their primary earnings. As a result, these corporate producers may be less responsive to the food system and the consumer.

Foreign control of U.S. agriculture is becoming more widespread. The value of foreign currencies relative to the dollar has made investments in the United States very attractive. In the last few years, the decline in farm-land prices has provided excellent opportunities for foreign investors with the funds to buy one of the world's safest investments. As Table 4.1 indicates, 12,860,000 acres of land are owned by foreigners. Over 50 percent of the land owned by foreigners is forest land, about 20 percent is pasture, and about 21 percent is cropland and other agricultural land. The leading countries of foreign ownership either directly or through a U.S. corporation are Canada, the Netherlands, the United Kingdom, and Germany. The vast majority of these purchases are through a corporate entity of one form or another. Whether this trend in ownership will have a significant effect on the forest and agricultural systems remains to be seen.

The decline of farming as a way of life for individuals and families in the United States and Canada is having the effect of depopulating and decreasing the economic viability of many rural areas. The congressional report, "The Bicoastal Economy," concludes the economic growth in the noncoastal states was only about one-third of the growth in the coastal states from 1981 to 1985. This figure would be even lower for the central farm states if several of the resource-rich interior western states were excluded from the comparison. The

FIGURE 4.9

Changes in Farm Population, Number, and Size, 1950–1987

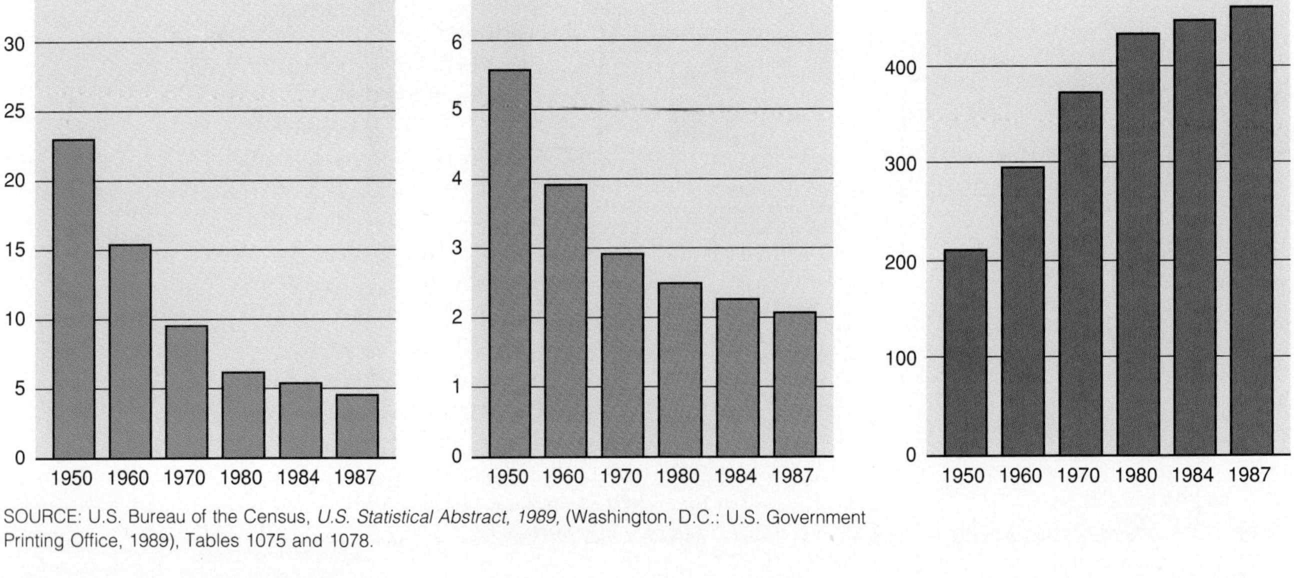

SOURCE: U.S. Bureau of the Census, *U.S. Statistical Abstract, 1989,* (Washington, D.C.: U.S. Government Printing Office, 1989), Tables 1075 and 1078.

TABLE 4.1

Foreign Ownership of U.S. Land

	TOTAL ACREAGE AS OF DECEMBER, 1987
Total	12,860,000
TYPE OF OWNER	
Corporation	10,329,000
Partnership	1,326,000
Individual	1,006,000
Other	199,000
LAND USE	
Forest	6,397,000
Pasture	3,086,000
Cropland	2,182,000
Other agriculture	566,000
Other nonagriculture	628,000

SOURCE: Department of Agriculture, Economic Research Service, *Foreign Ownership of U.S. Agricultural Land through December 1987* (1988), AGES 880314.

coastal states experienced nearly 70 percent of the growth in personal income along with the majority of the new jobs created during this period.

MANUFACTURING

Manufacturing in Anglo-America has undergone a number of changes in recent years, in particular, a regional shift away from the early centers of industrialization and a structural shift away from heavy industry toward more service-oriented activities.

The Industrial Core

Industrial production in Anglo-America has been centered in the Great Lakes region eastward to the Atlantic Coast since the 1800s. In the early days of industrialization, this region had the **comparative advantages** to become the **industrial core** for both Canada and the United States. The Great Lakes and the connecting river systems provided the cheap transportation necessary for hauling the bulk raw materials and the finished products of industrial activity, such as coal and steel. In addition to transportation, the natural resources essential for industrial development were present in the area. The iron ore of the Mesabi Range in Minnesota and the coal from the mines of Pennsylvania, Ohio, and Illinois were the

essential ingredients in steel production. The steel industry provided the material upon which many of the other industries in the region depended. These factors led to the emergence of this area as the leading industrial **agglomeration** in Anglo-America.

The growth of cities in this region provided the labor, capital, and infrastructure, such as rail systems and roads, for industrial growth to be sustained. The surrounding rich agricultural area of the upper Midwest provided the food to support this growing urban-industrial concentration. This industrial core region experienced impressive growth during the 1920s and again in the 1940s when it produced the majority of the industrial goods for all of Anglo-America. As a result, several corporations in this region, such as General Motors, U.S. Steel, and General Electric, became among the world's largest corporations during this period.

At present this region centered on the Great Lakes retains its dominance in industrial manufacturing. The traditional core region states of Illinois, Ohio, Michigan, New York, and New Jersey are still the major industrial states in the United States. Ontario has the largest manufacturing base in Canada, producing almost 50 percent of the manufacturing output for the nation. Quebec is second with approximately 25 percent of the national output. However, the industrial manufacturing trends in the United States and Canada are changing in type of output and location of production. The postindustrial era is very apparent across the two nations, particularly in the traditional industrial core region.

Postindustrial Change

Several changes have occurred to create stagnation and decline in this industrial core region and and move toward a postindustrial economy in the region as a whole. The increased cost of energy has made the old heavy industrial factories more costly to operate, particularly in the energy-intensive metals industries. Other costs of operation, such as labor and taxes, in the Midwest and Northeast have increased to make operations in these areas less profitable. Faced with antiquated, inefficient plants, foreign competition, and high operating costs, many industries have found alternative locations for new facilities. These alternative locations have often been outside the traditional industrial core area. The construction of rail and truck facilities, particularly the Interstate Highway System, has diminished the transportation advantages of the Great Lakes and connecting river systems. States in the South and the West offer

AGGLOMERATION

The spatial clustering of people, businesses, or industries for their mutual benefit is called agglomeration. Agglomeration minimizes the effort of overcoming distance. This minimizes costs and maximizes interaction and productivity. Cities are agglomerations of people, businesses, and industries. Shopping centers are agglomerations of retail stores. Often industrial firms are agglomerated so they can share a common raw material supply or use the outputs of one industry as the inputs for another without the costs of extensive transportation. The companies in the steel industry agglomerated near the Great Lakes to be close to the coal, iron ore, water transport, and markets for steel in the area. The automobile industry developed in this same area to be close to the steel industry upon which it is so dependent for raw materials.

The "Silicon Valley" of California (Figure 1) was spawned in large measure by the spin-offs of Fairchild Semiconductor Company. The primary scientists and managers of this first computer chip producer split off to form the nuclei for the other major U.S. producers of computer chips, such as Intel, National Semiconductor, Raytheon, and

Advanced Micro Devices. From this agglomeration of firms, dozens of other high-technology firms were formed in the Palo Alto–San Jose area. These firms located close to each other to share scientific innovations, managerial expertise, and computer chips and devices. Their outputs are quite different from the steel industry, but the concept of agglomeration is the same.

FIGURE 1

The Silicon Valley Area of California

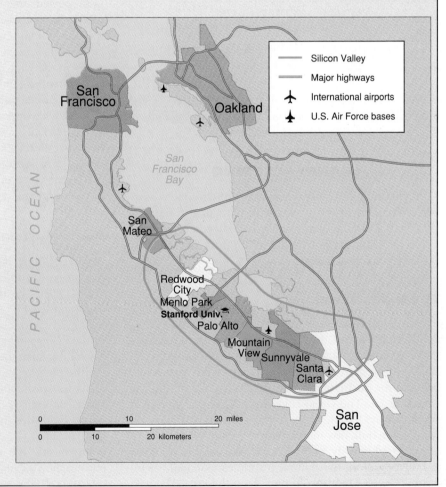

lower taxes, lower labor costs, fewer regulations, and large incentives for plants relocating to their states. The federal government has fostered the attractiveness of the South and the West through subsidized water and energy projects and by making federally controlled coal and petroleum lands available for development. These factors along with major structural changes in the American economy have redesigned the geography of industrial activity and urban settlement.

The major structural changes in the economy involve the shifting of the economic activity of Anglo-America from the primary and secondary sectors to the tertiary and quaternary sectors. This means that the greatest number of jobs and greatest income production have shifted from mining, agriculture, and industrial activities to wholesaling, retailing, and services-related activities, such as insurance, banking, and information processing. At the turn of the twentieth century, about 25 percent of the work force of the United States was engaged in tertiary and quaternary activities. In the 1980s, about 80 percent of the work force was involved in these sectors of the economy. Thus, the economy has evolved beyond the advantages of the Great Lakes industrial core region.

Within the secondary sector, a major change took place as well. The nature of industrial processing and products has changed in the last 20 years. In general, industrial processing is becoming less labor intensive as automation has taken over many of the jobs once performed by workers. The growth industries have shifted from heavy industry, such as steel and automobiles, to lighter industry, such as electronics. This has resulted in fewer industrial jobs, especially in those states reliant on heavy industry. The states of the Great Lakes region and the Northeast no longer have a comparative advantage over other regions because the industrial focus of the U.S. economy has changed. Lighter industry is more "footloose" in general. In these lighter industries, the operational costs become much more important than proximity to raw materials or market transportation costs, which are so crucial to heavy industry. Figure 4.10 shows the regional shift of manufacturers to the South and West during the period from 1963 to 1982. Although the industrial base remains very large in the traditional industrial core, these percentages indicate the trends in manufacturing relocation.

In high-technology industries, such as computer products and services, a primary location factor is the

FIGURE 4.10

Regional Shifts in Manufacturing, 1963–1982

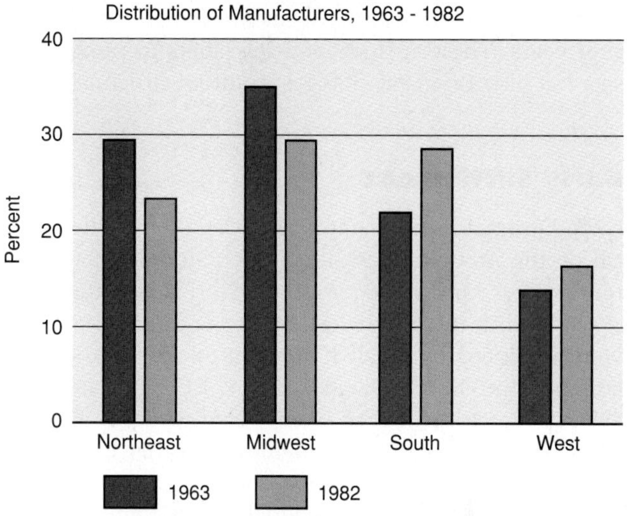

Distribution of Manufacturers, 1963 - 1982

SOURCE: U.S. Bureau of the Census.

proximity to educational institutions and centers of scientific innovation. Few of the traditional industrial location factors, such as transport, natural resources, or market proximity, are relevant to high-tech industry. This explains the emergence of the Research Triangle in North Carolina, Silicon Valley in California, and the Boston area as high-technology centers.

The regional shift in industry in the United States is likely to continue for the near term. It corresponds to the similar shifts in population and urban settlement. The factors that are encouraging these shifts are likely to be maintained into the early twenty-first century. Beyond that time, prediction is very difficult. Some evidence suggests that southern states that strived to induce northern industry to relocate have sacrificed their tax base without attracting enough industry to compensate for the tax losses. In some cases, they attracted undesirable and declining industries as well. Some observers believe that the operational costs for industry, such as taxes and labor, will equalize and other factors, such as the higher educational levels and greater water availability in the Northeast and Midwest, will reverse the current trends.

URBAN SETTLEMENT

Like the manufacturing patterns, urban settlement patterns in Anglo-America have undergone changes in recent years. These include not only shifts in population size but also dramatic changes in urban structure.

Early Settlement

Early European settlement was undertaken by the British on the Atlantic Coast, the French along the St. Lawrence River, and the Spanish in Florida, on the Gulf Coast, and in the Southwest. Alaska and the Northwest were explored by smaller numbers of Russian settlers crossing the Bering Straits. Figure 4.11 indicates the years that various areas were obtained from France, Mexico, and Spain as well as the years that U.S. statehood was extended to portions of these areas. In Canada early settlement was undertaken predominately by the French and later by the British. The settlement of Canada was an east to west movement similar to that of the United States.

Although they may have not expressed it as such, the early settlers attempted to satisfy both site and situational requirements in their selection of settlement locations. The first white settlers were explorers and agents of colonizing European countries. They selected sites for settlements based on their situational needs of access to the natural resources of the interior and to ocean waters to ship the resources back to Europe. Often other needs for defense and internal transportation in the colonies corresponded to this export focus. This led to the establishment of the settlements destined to become the major urban centers of Philadelphia, New York, New Orleans, and Montreal.

In many cases the settlers were not successful as is evidenced by the Jamestown settlement of 1607. This settlement had such a poor site in a swamp area that it was abandoned. Some cities, such as New Orleans, prospered in spite of a poor site. New Orleans is a terrible site for a city. Because much of the city lies below the level of the Mississippi River, high levees must be built to keep the river out of the city. The surface runoff after a rainfall must be pumped over the levees because of the higher elevation of the river. Much of the city is constructed on moist, river deposit soils that tend to shift and subside under the weight of urban development. However, the situation of New Orleans overcomes even these severe site limitations. New Orleans is located near the mouth of the extensive Mississippi River system where it enters the Gulf of Mexico. It is the major ocean access for international trade for much of the interior United States. This ensured the early growth of New Orleans and continues to make it one of the fastest growing centers of commerce in the United States.

The situation often changes for settlements causing them to either prosper or decline. As early settlers moved across the Appalachian Mountains through the interior, many settlements were established. Villages and towns that were very important trade centers declined as the settlement wave moved farther to the west. In the mid 1800s, towns that were bypassed by the railroad died while other cities on rail lines grew into major interior cities. The cities of Chicago, St. Louis, Kansas City, and Denver thrived as transfer and commerce points for the agricultural products and natural resources from the surrounding **hinterland**. Their importance was linked to their location on a water route or rail line and their situation relative to a productive hinterland.

Several cities on the West Coast were established for the same reasons as New York and New Orleans. San Francisco, Seattle, and Vancouver had natural harbor sites and acted as ocean shipping outlets for the interior west. Transportation-related site and situational factors explain the growth of the major urban centers in Anglo-America during the early settlement period until the early 1900s.

Modern Urban Settlement

Changes stimulated by postindustrialization, advances in communications and transportation technology, and changing regional perceptions have added to and revised the list of site and situational characteristics important to urban settlement. As was discussed in the previous sections on population and industry, the emphasis of the economy has undergone a major shift toward tertiary activities. The regional economies in the United States and Canada have responded to this postindustrial emphasis with growth moving out of the traditional industrial core area and into the South and West. The climate and the addition of recreational and cultural amenities in many cities have enhanced their growth potential.

The effect on the population size of many urban areas has been dramatic. Table 4.2 shows the 10 largest cities in total population in the United States in 1986. Several of the cities of the upper Midwest and Northeast remain among the largest in the nation. However, cities of the West and the South comprised the majority of the largest cities in 1986. The growth rates indicate the trends with Detroit having the largest percentage decline in population (−9.7 percent), and San Antonio, Texas, showing a vigorous 16.3 percent increase between 1980 and 1986. Selected smaller cities in the South and West have experienced unprecedented growth. The population of Arlington, Texas, located outside Dallas increased 56 percent between 1980 and 1986. At the same time, cities such as Youngstown, Ohio, and Peoria, Illinois, have

experienced declines in population in the 9–11 percent range. These numbers reflect the changes taking place across the entire United States. Important changes are also occurring at the micro-scale level within the urban areas across the United States.

Changes in Urban Structure

Within the individual urban areas of the United States and Canada, significant changes in the structure of transportation systems, land use, economic activities have occurred over the last 150 years. These changes paral-

FIGURE 4.11

Territorial Expansion of the United States

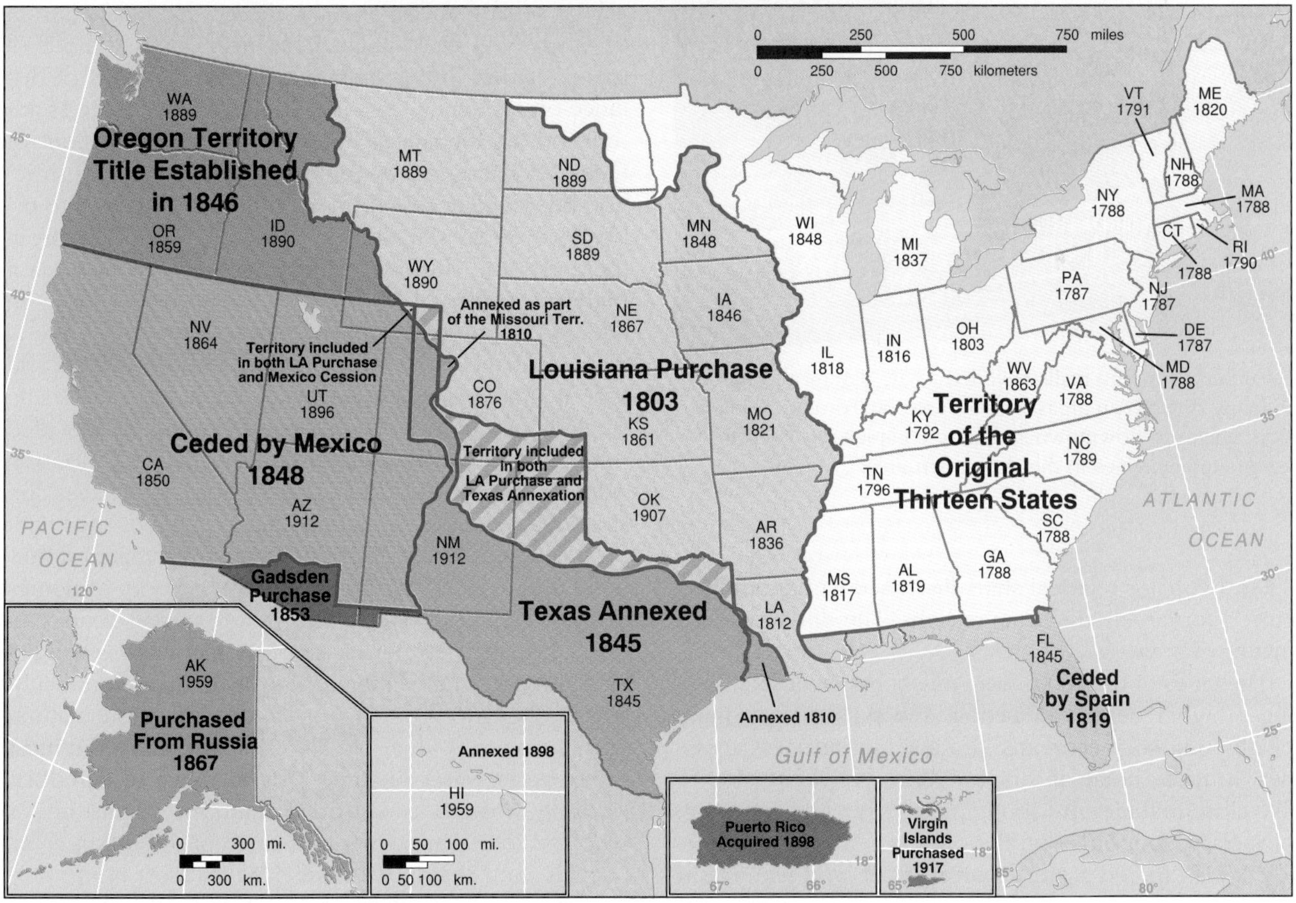

SOURCE: U.S. Bureau of the Census, *U.S. Statistical Abstract, 1989* (Washington, D.C.: U.S. Government Printing Office, 1989).

Suburban shopping malls have led to the demise of retailing in many downtown areas. Eaton Centre in Toronto, Canada, is an enclosed downtown shopping complex designed to keep retail activities in the downtown area.

leled the major regional shifts discussed earlier, and occurred for largely the same reasons. The difference is a matter of scale.

The cities of the 1800s were much less extensive, having a higher density of people and a closer proximity of differing socioeconomic groups and land-use activities. In these horse and buggy days, people needed to live close to their jobs, shopping, and services. This lack of mobility put limits on the outward expansion of cities. It also forced a greater diversity of land use within the central urban area. Businesses were adjacent to industry, and both were close to the residential neighborhoods of both workers and managers. While the railroad

tracks or a street may have acted as a barrier between poor residential areas and wealthy areas, the actual distance in most cities was minimal.

The advent of the streetcar altered the structure of many cities. This mode of transportation allowed people to live farther out from the central business district (CBD) of the city and still retain relatively rapid access to it. The shape of city expansion reflected this increase in accessibility from the outlying areas. Figure 4.12 shows the generalized changes that have occurred in urban shape as transportation technology has advanced. The land uses and housing stock in the central portions of many cities reflect these periods. Many of the largest, finest homes of wealthy urbanites, built in the mid and late 1800s, are very close to either the CBDs or the old streetcar lines.

The availability of inexpensive automobiles and an expanded road network led to the most dramatic change in urban structure in Anglo-America. Suburbanization was the urban and social form that resulted from this improved accessibility. This individualized mode of transportation further minimized the limitations of distance on residential location. In the 1950s, Congress authorized the Interstate Highway System. The billions of federal dollars spent on multiple-lane highway construction improved access from even greater distances to urban core areas. Workers could live many miles from their jobs and still have a degree of accessibility, via a highway, comparable to others living much closer. In addition to improved transportation technology, several factors enhanced the growth of the suburbs. After World War II, millions of people were intent on purchasing single-family homes. The federal government provided Veterans Administration (VA) and Federal Housing Administration (FHA) financing for home purchases. Restrictions on these programs, such as a 30-year age limit on the homes financed, virtually assured that a suburban home would be purchased. The majority of homes in most central city areas are more than 30 years old. The residential preferences of homebuyers changed from a focus on the extended family in the urban neighborhood to the desire for a larger yard, more natural setting, and individual privacy. The land market in most urban areas was structured for the move to suburbia. Not only was the lowest priced land forest, pasture, or cropland located on the rural fringe of the urban areas but this rural setting was preferred by most home purchasers. These factors combined to make the suburbanization of America inevitable (Figure 4.13). The commercial and industrial facilities in many cities relocated from the central cities to the suburban areas as well. Communications technology allowed businesses and industries to establish spatially decentralized units. This placed the jobs, services, and shopping in the suburbs adjacent to the residential areas. The decline of the central cities is both a cause and an effect of the move of the middle and upper economic classes to the suburbs. The more the central city declines, the greater the incentive for people and businesses to flee to the suburbs. This lowers the tax base and political power of the central city, which in turn makes further decline almost inevitable. The result is an economic vacuum in the central city.

The resulting spatial structure in most urban areas has become one of multiple activity centers located in the suburbs. These residential developments, shopping centers, and industrial parks provide the housing, shopping, and employment that only the central city could offer in past years. Rather than a single CBD and employment center in an urban region, most regions have several outlying centers interconnected by the communication and transportation systems. Many of these surpass the central city district in importance.

A modest trend toward **exurban** relocation began in the 1970s. Significant numbers of people and some industry are moving outward beyond the suburbs to the rural areas called the "exurbs." Many of the same mo-

Scene of filled Busch Stadium in St. Louis indicates the increasing importance of sports and the recreation industry to the American economy and society. The geographic location of professional sports teams is determined by criteria including geographic equity across the nation, urban population size and relative location of television markets.

TABLE 4.2

Ten Largest U.S. Cities, 1986

CITY	1986 POPULATION	PERCENTAGE CHANGE, 1980–1986
1. New York, NY	7,263,000	+ 2.7
2. Los Angeles, CA	3,259,000	+ 9.8
3. Chicago, IL	3,010,000	+ 0.1
4. Houston, TX	1,729,000	+ 8.4
5. Philadelphia, PA	1,643,000	− 2.7
6. Detroit, MI	1,086,000	− 9.7
7. San Diego, CA	1,015,000	+ 16.0
8. Dallas, TX	1,004,000	+ 10.9
9. San Antonio, TX	914,000	+ 16.3
10. Phoenix, AZ	894,000	+ 13.1

SOURCE: U.S. Bureau of the Census, *Statistical Abstract of the United States, 1989* (Washington, D.C.: U.S. Government Printing Office, 1989).

FIGURE 4.12

Changes in Urban Area Shape Due to Innovations in Transportation and Communication

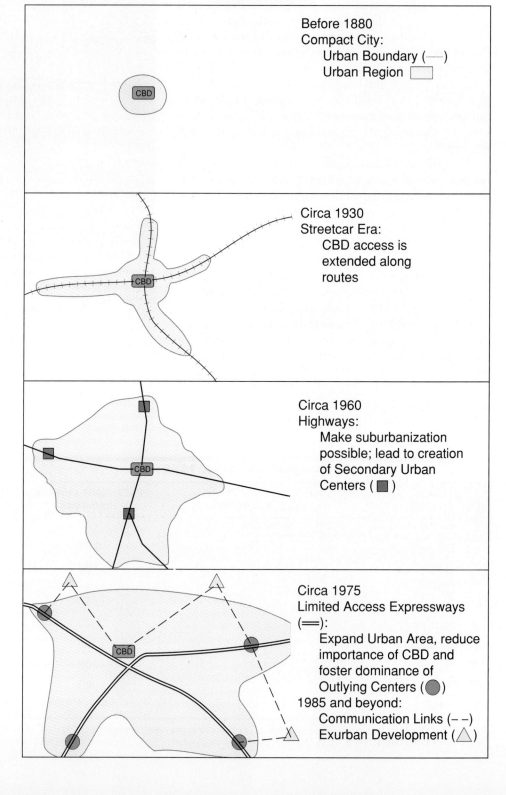

In recent years, some urban slum areas have been targeted for redevelopment projects such as this shopping center in central Philadelphia.

urban-oriented residential and commercial development have led to the coalescence of many urban and suburban areas. This phenomenon has created the megalopolis. This super city is composed of several different political jurisdictions, tied together by a dense web of transportation and communication networks crossing city, county, and state boundaries. The megalopolis contains large tracts of open land, but even these are oriented functionally to the urban activity centers. The earliest megalopolis, identified by geographer Jean Gottman in the early 1960s, is in the northeastern United States. This megalopolis extends from Boston southward through the New York City and Washington, D.C., metropolitan areas to Richmond, Virginia. Other megalopoli are forming in coastal California, southern Florida, the southern shore of the Great Lakes, and a corridor from Toronto to Montreal in Canada (Figure 4.14).

The changes in agriculture, manufacturing, and urban settlement in the United States and Canada are inevitable in the changing world. The domestic frontiers have been surpassed and the people, cities, economic activities, and ideas of the United States and Canada are being extended to areas outside the region of Anglo-America. An understanding and vision of the region in the future is dependent on knowledge of this extension to the rest of the world.

tivations for suburban relocation are involved in the exurban location decision. This move has been fostered by technological advances in telecommunications and isolated residential sewage and water supply equipment. The trend in Anglo-America is toward more workers being employed in services, such as sales, insurance, computer processing, and financial analysis. These occupations do not require a daily visit to a central business office. An exurban existence is possible under these circumstances. As a result many smaller towns in the outlying fringes of large metropolitan areas are becoming the homes of persons connected functionally to the large urban center many miles away.

These recent trends toward the areal expansion of

FIGURE 4.13

Relationship of Land Values and Residential Preference

The correspondence of low land values and high residential preference encourages outward expansion from the central city.

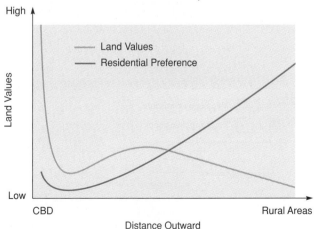

FIGURE 4.14

Megalopoli Formations

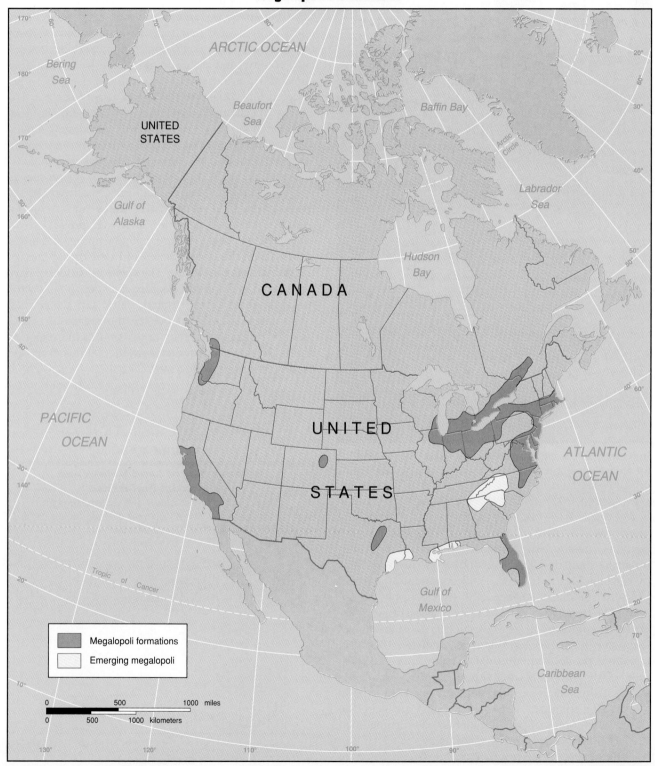

SPATIAL CONNECTIVITY

CONNECTIONS WITHIN THE REGION

The United States and Canada share interconnected social, political, and economic systems. The magnitude of the systematic and spatial connectedness of these two countries is beyond that found in most other regions of the world. Much of the future of both countries is dependent on their relationship with each other.

Adjacency and Trade

The United States and Canada share a cordial relationship unmatched in the world. The common border between the two countries is approximately 5,500 miles (8900 kilometers) long. If the much larger and more powerful United States were ever inclined to expand territorially, domination of Canada would be quite possible. This border is, however, the longest nonmilitarized border between two nations in the world. Military aggression between the two countries is not an issue on either side of the border. There are disagreements on trade, foreign policy, and environmental issues to be sure, but never the hostilities that plague many other adjacent nations.

This geographic adjacency and relationship of trust have resulted in the United States becoming Canada's single most important trading partner. Canada sells 70 percent of all of its exports to the United States and buys approximately 20 percent of all U.S. exports. The value of Canadian exports to the United States over imports from the United States has increased in recent years due to the U.S. reliance on Canadian natural resources as domestic supplies have become less available and more expensive. These resources include wood, fish, energy sources, and minerals. Canada is one of the top three foreign suppliers to the United States of nickel, tungsten, zinc, gold, copper, titanium, and cadmium among many other mineral resources. Canada is a major supplier of fossil fuels, particularly natural gas, to the United States. Figure 4.15 indicates how this relationship has grown through the 1970s and 1980s. This trade with Canada is very advantageous for the United States because it provides a reliable and close source of the resources necessary for U.S. industry. The free trade agreement signed by both governments in 1987 allows Canadians more access to the U.S. market. However, the agreement heightens the concern of many

Canadians that this more open relationship will lead to further economic domination of the Canadian economy by American firms.

The economies are linked by more than trade. Many U.S. corporations have installations in Canada. These corporate operations are intricately related in terms of management, employment, and production. In 1985 a strike by U.S. members of the United Automobile Workers union forced the shutdown of Canadian automobile plants. The Canadian workers were not on strike, but operations in the Canadian plants depended on the production of parts in the U.S. plants. Along with extensive U.S. interest in the Canadian economy, Canadian investors have found the United States a good place to purchase land and businesses. Many of the new residential and commercial developments in the sunbelt states, particularly Florida and Arizona, have been financed and controlled by Canadians.

The management of the common property natural resources of the two nations has been a major arena of interaction. The resolution of the serious **acid precipitation** problem requires a concerted action by both countries. The Canadians contend that most of the pollutants (over 3 million tons) responsible for the problem come from the U.S. manufacturing area south of the Great Lakes. The United States has taken limited action on the problem due to concern that these already depressed industrial areas will be further hurt by more stringent pollution enforcement. Scientific specification of the source areas of pollutants and the magnitude of the impacts on eastern Canada and the United States are making cooperation possible (Figure 4.16). Resolution of such problems is the responsibility of the International Joint Commission. This is an independent agency established in 1909 to handle the transboundary air and water supply and pollution problems between Canada and the United States.

The two nations also have a cooperative approach to regional defense. They participate in the North American Air Defense Command (NORAD), which shares facilities, duties, and management of the defense of the United States and Canada. Both nations are members of the North Atlantic Treaty Organization (NATO), which also includes many of the nations of Western Europe.

GLOBAL CONNECTIONS

The United States and Canada have interconnections and interdependencies with all regions and nations of

FIGURE 4.15

U.S. Dependency on Canadian Fossil Fuels

Crude Petroleum (millions of dollars)

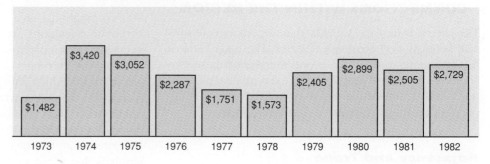

Natural Gas (millions of dollars)

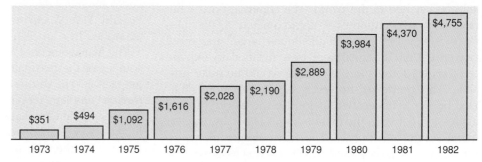

SOURCE: *Canada Yearbook, 1986* (Ottawa: Statistics Canada, 1986).

the world. The relationships involve mutual and reciprocal trade, travel, economic assistance, military arms and supplies, and investments. They are maintained by official governmental organizations, private citizens, and **transnational corporations** based in the United States and Canada. These relations are indicative of the increasing level of interdependence that exists among all nations of the world for defense, resources, technology, and economic opportunities.

Travel and Trade

Travel by individuals is a major interconnection between the United States and other nations. Money, goods, and services are moved across national boundaries. Ideas, culture, and attitudes are transmitted between nations through the travel of their citizens. Figure 4.17 shows the number of travelers between the United States and selected foreign regions in 1987. In terms of travel, Western Europe is the region most connected to the United States. In the "Other" category, most of the

foreigners visiting the United States are Japanese tourists. The combined total number of travelers was over 23 million in 1987.

Trade between the United States and Canada and other regions of the world is one of the more apparent interconnections. In 1983, Asia surpassed Western Europe as the second leading trading partner of Canada after the United States. This is surprising given the close historic and political relationship of Canada with Western Europe. It is only in the last few years that the trade role of the Asian countries, particularly Japan, has taken on such significance to Anglo-America.

The situation is similar in the United States. Taiwan, South Korea, and Japan are the major U.S. trading partners in Asia, exporting automobiles and electronic equipment to the United States. This has created a major trade imbalance since the United States does not export products to these countries in quantities equal to the U.S. imports of their products. One effect of the trade imbalance, combined with the declining value of the U.S. dollar, has been to spur the construction of auto-

FIGURE 4.16

Acid Precipitation Impact Areas

The acidity (pH) of natural rain is 5.0–5.6. Acid rain can cause harm to ecosystems when the pH becomes less than 4.6.

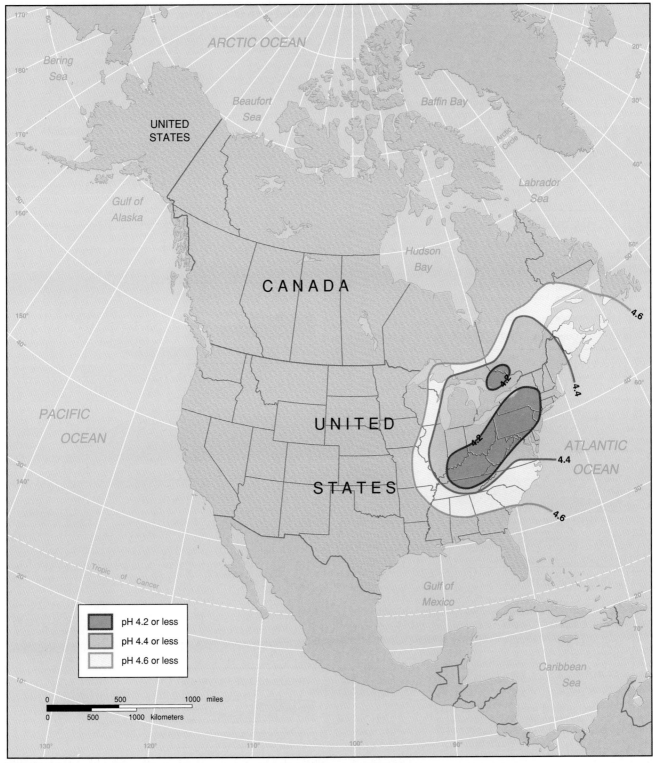

SOURCE: U.S. Congress, Office of Technology Assessment, *Acid Rain and Transported Air Pollutants: Implications for Public Policy* (New York: UNIPUB, 1985), p. 66.

FIGURE 4.17

Travel between the United States and Other Regions, 1987

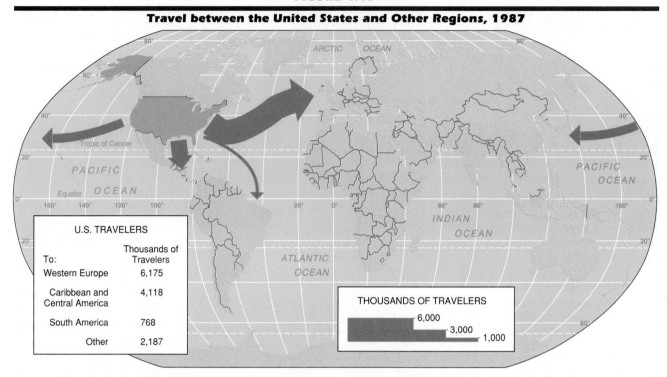

U.S. TRAVELERS	
To:	Thousands of Travelers
Western Europe	6,175
Caribbean and Central America	4,118
South America	768
Other	2,187

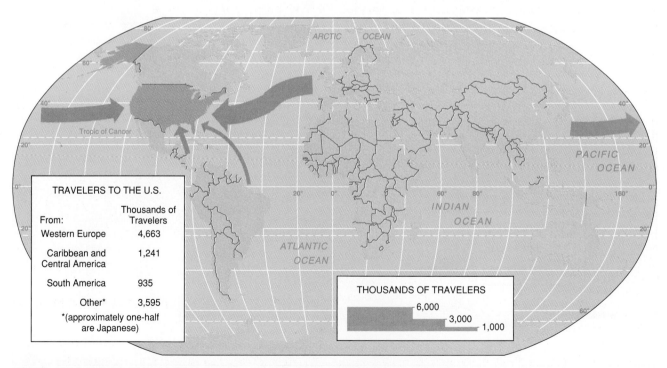

TRAVELERS TO THE U.S.	
From:	Thousands of Travelers
Western Europe	4,663
Caribbean and Central America	1,241
South America	935
Other*	3,595
*(approximately one-half are Japanese)	

SOURCE: U.S. Bureau of the Census, *Statistical Abstract of the United States, 1989* (Washington, D.C.: U.S. Government Printing Office, 1989) Tables 403, 404.

mobile plants in the United States by the Asian automakers.

The trade in certain commodities has an unusually significant effect on global relations. In particular, the United States takes a special interest in petroleum and military supplies for the global community. The United States controls about 5 percent of the global petroleum reserves according to 1986 estimates by the U.S. Department of Energy. Anglo-America as a whole controls 12 percent (Figure 4.18). The greatest reserves of petroleum are in the Middle Eastern nations, particularly Saudi Arabia and Kuwait. Although consumption and importation decreased in the early 1980s, it is expected that the United States will be importing 50 percent of its petroleum by the early 1990s. The Middle Eastern embargoes and the formation of the monopolistic Organization of Petroleum Exporting Countries (OPEC) forced the United States to diversify its sources in the 1970s. Increased reliance on Mexico and Canada have eased supply problems for the short term. The long-term conclusion is inescapable as these friendly, nearby sources deplete their deposits and U.S. consumption continues to increase. The United States may become increasingly dependent on the nations of the Middle East for the petroleum necessary to fuel its transportation and industrial systems. The competition of the United States and the other highly industrialized nations to obtain petroleum from a limited group of suppliers will cause a political and economic realignment in the global community.

Military and Economic Assistance

A major connection of the United States with the rest of the world is through military and economic assistance, which includes sales, grants, and loans of funds and equipment. The United States is one of the world's leading arms suppliers. The sale of military arms and supplies is a major global business with several important implications. The sale of expensive military hardware is a means of reducing the balance of trade problems between countries. This has been important in maintaining the trade balance between the developed nations and the OPEC petroleum countries in recent years. The manufacture of arms for the export market has helped to maintain industrial production and jobs in developed nations whose manufacturing sectors have begun to decline. Arms sales are a tangible means of assisting the political and ideological struggles around the world. For all of these reasons, the United States is the second lead-

ing arms exporter in the world as of 1986; in first place is the Soviet Union. Other major arms exporters are France, Germany, Great Britain, the People's Republic of China, Czechoslovakia, and Poland. With the exception of the People's Republic of China, all of these nations are members of either NATO or the dismantled Warsaw Pact.

As Figure 4.19 illustrates, the strategic interests of the United States and the other Western democracies influence the disbursal of military aid. The Middle East and South Asia include several strategically important nations. Israel is the single largest recipient of aid with $1.8 billion of assistance in 1987 and over $8.0 billion between 1983 and 1987. Reflecting the attempt by the United States to enlist the cooperation of moderate Arab nations, Egypt is a major recipient ($1.3 billion in 1987). The geopolitical importance of Turkey and Pakistan is also reflected on this map. Both countries are major recipients of aid due to their proximity to the Soviet Union, Iran, and Afghanistan. In Africa, Tunisia and Morocco are the leading 1987 recipients receiving 34 million and $46 million, respectively, in 1987. El Salvador was provided $112 million in military aid in 1987 to lead the Latin American region. The Philippines was the leading recipient in East Asia in 1987, and Spain and Portugal received $108 million and $83 million, respectively, for the European region.

The United States is the world's leading provider of non-military assistance to other nations. Humanitarian and strategic motivations have led to the creation of aid programs, such as the Agency for International Development, Food for Peace, the Peace Corps, and the Foreign Assistance Act allocations. The United States is also a primary benefactor of several international organizations, such as the World Bank. The 1987 total for economic aid was $9.386 billion; this total does not include military aid. Much of the 1987 U.S. foreign aid went to the Middle East and South Asia ($2.8 billion). Other amounts for 1987 were $1.3 billion and $620 million to Latin America and Africa, respectively. This assistance has been the salvation for many nations experiencing a natural disaster, a long-term food supply problem, or a capital shortage to undertake a development project.

U.S. TRANSNATIONAL CORPORATIONS AS INTERCONNECTIONS

Multinational or transnational corporations provide an exchange system of natural resources, products, and financial resources matching the interconnections of the

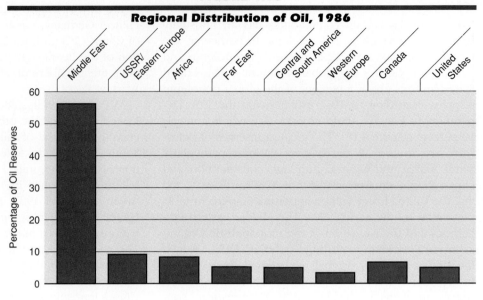

FIGURE 4.18

Regional Distribution of Oil, 1986

governments of the world. Many Americans think of transnational corporations in terms of the auto and electronics firms of Europe and Asia penetrating the American market and creating domestic unemployment and a U.S. trade deficit. In fact, the United States is the home country for many of the world's largest transnational corporations, which have been involved in other countries for many years. Table 4.3 indicates the foreign content of selected American-based transnational corporations. It is apparent that many of these corporations have substantial assets and employees in foreign countries. The growth of U.S. corporate investments in manufacturing in foreign countries has averaged 10–15 percent per year over the last 30 years. This large degree of participation by many U.S. firms in other nations is surprising to many Americans.

In many developing nations, these corporations are the most recognized image of the United States. The most obvious are Kentucky Fried Chicken or the Coca Cola and Pepsi soft drinks, but Ford is a leading auto producer in Europe and South America. Union Carbide has chemical plants around the world, including the infamous one in Bhopal, India. IBM is the leading computer manufacturer in the world, not just the United States.

PROBLEMS AND PROSPECTS

The future of the United States and Canada will involve an evolution of the structure that has existed for the last century. The changes will be dramatic and wrenching for many citizens. The most apparent changes will likely involve the role of the United States in the world political and economic situation. To a greater degree, the United States will share control of the world economic, political, and military balance of power. Indications are that the U.S. economy cannot support, nor are other nations willing to accept, such a strong U.S. influence in world affairs. The post–World War II strategy of containment of communist influence in the world will be revised further as the U.S. military presence in Europe, the Philippines, and elsewhere is reduced. The moves of the Soviet Union and the Eastern European countries toward less aggression and more democracy and capitalism are altering the posture that the United States has maintained over the last 45 years. Simplistic dichotomous views of geopolitics in which a country is either a friend or an enemy are not appropriate in the new context. The rise of other major economic powers in the world, such as Japan, Brazil and the People's Re-

FIGURE 4.19

U.S. Economic and Military Assistance to Other Regions, 1987

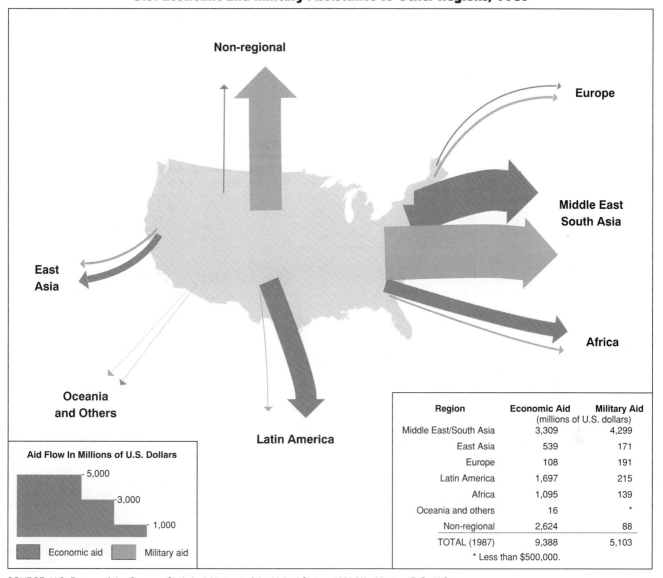

Region	Economic Aid	Military Aid
	(millions of U.S. dollars)	
Middle East/South Asia	3,309	4,299
East Asia	539	171
Europe	108	191
Latin America	1,697	215
Africa	1,095	139
Oceania and others	16	*
Non-regional	2,624	88
TOTAL (1987)	9,388	5,103
* Less than $500,000.		

Aid Flow In Millions of U.S. Dollars

5,000
3,000
1,000

Economic aid   Military aid

SOURCE: U.S. Bureau of the Census, *Statistical Abstract of the United States, 1989* (Washington, D.C.: U.S. Government Printing Office, 1989), Table 1366.

public of China, will alter the approach taken to trade, market access, and foreign economic assistance.

Internal changes in the region will be pronounced as the economy shifts more toward a service orientation. The geographic shifts in population, economic resources, and political power will redefine the economic and cultural landscapes of the region. Over the last sev-

eral years, the Northeast and Midwest regions of the United States have experienced an erosion of political power. At present, taxpayers in these regions pay approximately 47 percent of federal taxes, but the regions receive from the federal government only about 40 percent of federal expenditures to the states. After the 1990 census, these regions will lose approximately 12–15 seats in the U.S. House of Representatives due to population shifts and redistricting. Federal programs that are relatively more important to the older industrial states of the Midwest and Northeast, such as highway construction, urban renewal, economic development, and social programs, allocate their funds on the basis of census data. This likely means reductions in federal funds to these regions. At the same time many non-census-related programs including defense, the savings and loan bailout, nuclear waste clean-up at weapons plants, and the giant atom smasher to be built in Texas will direct additional federal dollars to the South and West.

TABLE 4.3

U.S. Transnational Corporations: Foreign Assets and Employment, 1980

CORPORATION	FOREIGN CONTENT AS A PERCENTAGE OF	
	Assets	Employment
General Motors	12%	31%
Ford Motor Company	40*	58
IBM	46	43
General Electric	28	29
ITT	33	53
Tenneco	25	—
United Technologies	21	32
Procter & Gamble	19*	33*
Dow Chemical Company	49	39
Union Carbide	31	46
Eastman Kodak	24	35
Caterpillar Tractor	40*	24
Goodyear	58	48
General Foods	34	45
Monsanto	28	27
International Harvester	48	35
3M	33	40
Pepsico	—	27
Consolidated Food	88	26
Coca Cola	36	52

NOTE: This table lists corporations with foreign assets or employment greater than 20 percent, but does not include petroleum corporations.
* 1976.

SOURCE: Based on United Nation Conference on Transnational Corporations, *Transnational Corporations in World Development: Third Survey* (New York: United Nations, 1983) Table II-31; *Transnational Corporations in World Development: A Re-examination* (New York: United Nations, 1978), Table IV-1.

THE INTERTWINING OF TRANSNATIONAL AUTOMOBILE FIRMS

he American automobile industry is viewed by many Americans as the major victim of foreign competition. To be sure, many autoworkers have been displaced by foreign competition and cheaper labor. However, the U.S. automaking firms as transnational corporations have lived by the old maxim, "if you can't beat them, join them." The largest U.S. automakers have gone into business with the Japanese to offset the effects of the Japanese control of a large share of the U.S. market. Figure 2 shows the tangle of connections between these companies as of 1986. These connections take the forms of technology transfer and ac-

tual equity ownership. General Motors (GM) has several ventures with different Asian auto producers. GM shares ownership of Nummi with Toyota and also owns sizable shares of Suzuki and Isuzu in Japan and Daewoo in South Korea. Chrysler owns 20 percent of Mitsubishi of Japan, which owns 15 percent of Hyundai, the largest auto producer in South Korea. South Korea is the next Asian source country for large numbers of automobiles after Japan. Both the U.S. and the Japanese firms are developing production facilities in South Korea because of the lower labor costs. In the mid-1980s, American autoworkers cost $20–25 an hour. The strong Japanese yen relative

FIGURE 2

U.S. Automakers' Ownership in Asian Automobile Firms, 1986

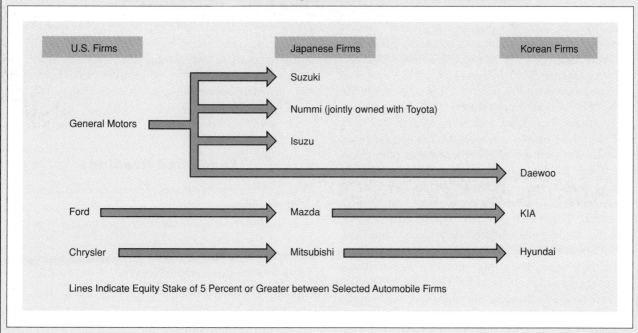

to the U.S. dollar makes a Japanese autoworker cost about $18–20 an hour. The Korean autoworker earns the equivalent of about $2–3 per hour. The fact that GM, Ford, and Chrysler actually own portions of these Japanese and Korean competitors changes the picture of the U.S auto market.

Many foreign auto firms have found it in their best interests to locate manufacturing facilities in the United States or Canada (Table 1). This enables them to cope more effectively with import restrictions, currency exchange problems, and rising transportations costs. The operation of these firms is having a profound effect on the management of auto plants and industrial plants in general in the United States and Canada. The Japanese are introducing new technology, management approaches, and employer/employee relationships. Given their success, the Japanese innovations are bound to alter the operations of many American businesses.

TABLE 1

Major Foreign Automobile Plants in the United States and Canada, 1988

FOREIGN AUTOMOBILE COMPANY	LOCATION OF PLANT
Toyota	Kentucky
Toyota (with GM)	California
Nissan	Tennessee
Honda	Ohio
Suzuki (with GM)	Ontario, Canada
Mazda	Michigan
Mitsubishi (with Chrysler)	Illinois
Volkswagen	Pennsylvania
Subaru-Isuzu	Indiana

SOURCE: *Foreign Investment—Growing Japanese Presence in the U.S. Auto Industry* (GAO/NSIA D-88-111, 1988).

Many foreign firms are finding the United States a good place to invest in manufacturing. As of 1988, Great Britain was the largest investor in U.S. assets with direct investment of over $102 billion. Japan was second in direct investment with $53 billion, followed by the Netherlands ($49 billion) and Canada ($27 billion). Figure 4.20 shows the spatial pattern of foreign direct investment in the United States over the last several years. Most states show an increase in the number of employees working for foreign-owned manufacturing firms. The most dramatic rate of increase in these employees has occurred in Massachusetts, New York, New Mexico, and Nevada. These changes may require great adaptation of the Anglo-American attitude and economic system, but are not necessarily detrimental to the two nations of the region. These investments counterbalance the investments of U.S. firms in foreign nations. The foreign investments provide an infusion of capital to many U.S. firms that allows them to reconfigure their operations to the altered reality of business in the 1990s.

Clearly, the United States and Canada have many important interactions upon which the global economy, global welfare, and world peace are dependent. Other developed, commercial economies in regions such as Europe and Japan play an equally important role in the world. The next chapters will define and explore the role of these regions in the interdependent world.

Suggested Readings

Axtell, J. *The Cultural Origins of North America.* New York: Oxford University Press, 1985.

Berry, B. *The Changing Shape of Metropolitan America.* Cambridge, Mass.: Ballinger, 1977.

Brunn, S., and J. Wheeler, eds. *The American Metropolitan System: Present and Future.* New York: V. H. Winston, 1980.

Canada Yearbook 1986. Ottawa: Statistics Canada, 1986.

Chang, K. "Japan's Direct Manufacturing Investment in the U.S." *The Professional Geographer* 41 (1989): 314–28.

Guinness, P. *North America: A Human Geography.* Totowa, N.J.: Barnes and Noble, 1985.

FIGURE 4.20

Foreign Direct Investment by State: Percentage Growth Rate, 1981–1987

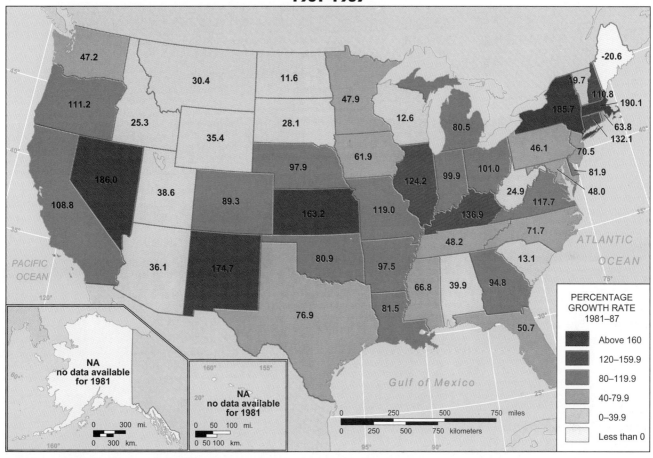

PERCENTAGE
GROWTH RATE
1981–87

- Above 160
- 120–159.9
- 80–119.9
- 40-79.9
- 0–39.9
- Less than 0

SOURCE: U.S. Bureau of Economic Analysis, *Direct Investment in the United States*, May 1989.

Jordan, T. "Preadaptation and European Colonization in Rural North America." *Annuals of the Association of American Geographers* 79 (1989): 489–500.

McHugh, K. "Hispanic Migration and Population Redistribution in the U.S." *The Professional Geographer* 41 (1989): 429–39.

McKee, J. *Ethnicity in Contemporary America*. Dubuque, Iowa: Kendall-Hunt, 1985.

The U.S. Population Data Sheet. Washington, D.C.: Population Reference Bureau, 1990.

Rand McNally Commerical Atlas and Marketing Guide. Chicago: Rand McNally, 1989.

CHAPTER 5

Western Europe

IMPORTANT TERMS

Industrial revolution

Conurbation (megalopolis)

Transhumance

Centralization

Core-periphery

Distance-decay

Growth poles

CONCEPTS AND ISSUES

■ Advantages of location in the core area relative to location on the periphery.

■ Long and vigorous industrialization of the region resulting in pollution problems that span the boundaries of Europe.

■ Major population issues of population decline and the role of "guest workers."

■ European anxiety regarding regional security, the presence of U.S. troops and weapons in Europe, and the future intentions of the Soviet Union.

■ The advancement of European economic cooperation conflicts with the protection of the agricultural and industrial sectors in several countries, but underlies necessary changes to meet international competition.

CHAPTER OUTLINE

Why Europe?
Kinship with Anglo-America

PHYSICAL ENVIRONMENT
The Seas
The Land
The Rivers
The Climate
Vegetation
The Soils
Mineral Resources

HUMAN ENVIRONMENT
Population Density
Population Change
Historical Background
Languages
Religions
Political Institutions
Agriculture
Manufacturing
The Great Cities

Regionalization: Core and Periphery
Internal Regionalization

SPATIAL CONNECTIVITY
Trade
The Movement of People
Political Ties

PROBLEMS AND PROSPECTS
Energy and Pollution
Toward a Unified Europe

No region of the world is as committed to or as dependent upon the connectivity of places as Western Europe (Figure 5.1). Western Europeans have been among the greatest seafarers, explorers, colonizers, and traders of the past thousand years. This small area, housing less than 8 percent of the world's population, accounts for some 46 percent of the total trade by value of the world. Of the world's leading trading countries, six are in Western Europe: Germany, France, the United Kingdom, Italy, the Netherlands, and Belgium. Granted, some three-quarters of this trade is within Europe itself, but that only proves the high value these small countries place upon open borders and connectivity between nations and regions. The economic unification of European nations into the European Community is a strategy for regional integration to foster European competitiveness in the world marketplace. The European community is an experiment in placing European regional advantages above nationalism and benefits for an individual country.

One must not think only of trade. As will be described later, the movement of people as workers, tourists and business people is intense. If one wishes to fly from North America to Africa or the Middle East, one will almost certainly pass through London, Paris, Frankfurt, Amsterdam, or Zürich. Much of the money of the world passes through those same cities, and four of the six major gold markets are there. Many worldwide organizations have their headquarters in such cities as Geneva and Vienna.

Europe has been a great teacher of the world. Almost every vital political principle active in the world today had its origin in Europe or its offspring, European North America. These principles include democracy, socialism, communism, capitalism, republicanism, and all the other "isms"; liberty, freedom, equality, fraternity, and feminism; parliaments and representative government; and all the varied economic models that are now applied around the world.

For better or worse, the same is true of the arts. Even though other parts of the world have produced rich folk arts, the culture of the West has become dominant. Europe has also been an effective teacher of technology. Modern science is a European invention, as is the modern university. The industrial revolution began in Western Europe and has now, within two centuries, spread to the whole world.

WHY EUROPE?

The question that arises is, why Europe? What was so special about this area that it could produce this incredible wealth, this variety of ideas, this amazingly rich culture? Even in terms of raw military power, it is amazing that in 1800 a mere 10 million Britons, while trying to defeat Napoleon, should have been able at the same time to control India and many other parts of the world, or that a few thousand Spaniards should have been able to conquer the entire Pacific coast of the Americas from Cape Horn to Oregon. The European soil and climate are no better than in many other parts of the world, and the mineral resources are modest. Europe had no population advantage, and, in fact, politics and languages split that population into scores of separate units. It is clear that if one is to understand the reasons for European cultural and economic predominance, one must look at the ideas and attitudes that motivated these people.

Several strains of thought have combined to create the European (and American) world view. First, from the ancient world, came the Greek love of abstract thinking. Greek philosophy formed the basis for European education. The result has been an emphasis on being "rational," making distinctions, and engaging in testing and analysis. The scientific method originated with the Greeks. The Greeks were also important architects, city planners, political scientists, poets, sculptors, athletes, historians, and playwrights. Although they had slaves, they gave us our democratic ideas.

The second major input came from Rome. The Roman Empire controlled all of Western Europe except its northern fringes for at least three centuries. The Romans had a genius for organization and management. From them have come our systems of law, our constitutions, and our administrative structures. Rome gave the world the two contrasting ideals, the Republic and the Empire. We could say that the United States is a union of idealized views of Roman republicanism and Greek democracy. Although the practice of dividing the land into squares for settlement was utilized in the Indus Valley prior to the Romans, the Romans advanced the practice and took over and expanded the "gridiron" street pattern to become the greatest city planners in history. Thus, the Greco-Roman traditions supplied Europeans with a rich base of ideas.

The force that moved European ideas and ambitions

FIGURE 5.1

Western Europe

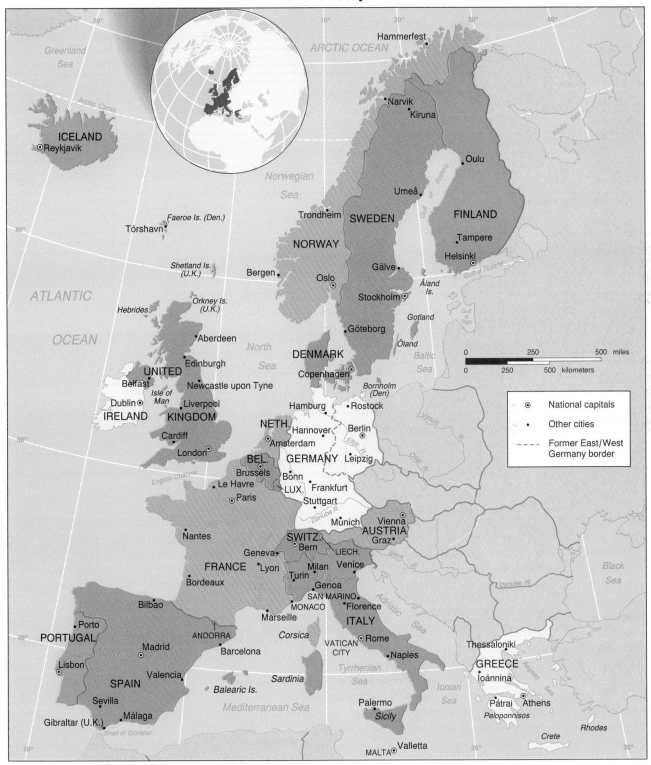

STATISTICAL PROFILE: WESTERN EUROPE

REGION OR COUNTRY	Population Estimate mid-1990 (Millions)	Natural Increase (Annual)	Population Projected to 2000 (Millions)	Infant Mortality Rate	Life Expectancy at Birth (Years)	Urban Population (%)	Per Capita GNP, 1988 (US$)	Area, Thousands of Square Miles (Km²)	Population Density, No./mi² (No./km²)
Austria	7.6	0.1	7.7	8.1	75	55	15,560	32.3(84)	235(91)
Belgium	9.9	0.2	9.9	9.2	74	95	14,550	11.8(31)	842(325)
Denmark	5.1	0.0	5.2	7.8	75	84	18,470	16.6(43)	309(119)
Finland	5.0	0.3	5.0	5.9	75	62	18,610	130.1(337)	38(15)
France	56.4	0.4	57.9	7.5	77	73	16,080	211.2(547)	267(103)
Germany, West	63.2	−0.0	65.7	7.5	76	94	18,530	96(249)	659(254)
Germany (unified)	79.5	0.0	81.2	—	—	—	—	137.8(357)	577(222)
Greece	10.1	0.2	10.2	11.0	77	58	4,790	50.9(132)	197(76)
Iceland	0.3	1.1	0.3	6.2	78	89	20,160	39.8(103)	7(3)
Ireland	3.5	0.6	3.5	9.7	74	56	7,480	27.1(70)	130(50)
Italy	57.7	0.1	58.6	9.5	75	72	13,320	116.3(301)	496(192)
Luxembourg	0.4	0.2	0.4	8.7	75	78	22,600	1.0(3)	374(144)
Malta	0.4	0.8	0.4	8.0	75	85	5,050	.12(.3)	2849(1100)
Netherlands	14.9	0.4	15.3	7.6	77	89	14,530	14.4(37)	1033(399)
Norway	4.2	0.3	4.3	8.4	76	71	20,020	125.2(324)	34(13)
Portugal	10.4	0.2	10.7	14.9	74	30	3,670	35.6(92)	291(112)
Spain	39.4	0.3	40.7	9.0	77	91	7,740	194.9(505)	202(78)
Sweden	8.5	0.2	8.8	5.8	77	83	19,150	173.7(450)	49(19)
Switzerland	6.7	0.3	6.8	6.8	77	61	27,260	15.9(41)	420(162)
United Kingdom	57.4	0.2	59.1	9.5	75	90	12,800	94.5(245)	607(234)

forward came out of the third main element, the Judeo-Christian religions. Judaism taught that God was a spirit above and infinitely greater than the world. The Israelites had no sacred trees, or springs, or rocks. Therefore, the earth could be freely used and studied, and humans were in a special position above nature. These beliefs helped to produce notions of progress, development, and evolution. Unfortunately, the idea that humans are dominant over nature led to overexploitation of the natural environment to further the economic development of Europe. Many areas of the world were despoiled to provide the natural resources necessary for European development during the colonial period.

Perhaps the most important element in the rise of Europe was the view of work as good, as a means of participating in God's creation. The monastic ideal of *ora et labore* ("pray and work") led to the famous "Protestant work ethic." Although the elite within Christianity accepted the work ethic with great reluctance, the ultimate result, when joined to the idea of the equality of all people before God, was to diminish the acceptance of slavery and feudalism.

KINSHIP WITH ANGLO-AMERICA

Western Europe is the part of the world with which Americans and Canadians feel the closest kinship. Even if our ancestors did not come from there, we have grown up within a culture that was largely created in Europe. Anglo-America is the cultural, spiritual, and technological heir of Europe. In studying Europe, we come to understand ourselves better.

The two sides of the Atlantic are growing closer together. The London-New York (or Toronto) flight has become a four- to eight-hour commuter run. It is not unusual for artists or athletes to be of northern European background, earn most of their money in the United States, and live in the south of France. The great transnational companies straddle the Atlantic. Ford and IBM may have their headquarters in the United States, Philips in the Netherlands, Nestlé in Switzerland, and Michelin in France, but in truth they are everywhere in the Western world. In addition, the North Atlantic Treaty Organization (NATO) has linked most of Western Europe militarily with North America for over 40 years. We have even coined phrases to express this unity: the West and the Atlantic Community.

PHYSICAL ENVIRONMENT

In its broad outline, Europe is a large peninsula extending southwestward from the main mass of the Eurasian continent. This peninsula is cut up by mountain chains and surrounded by arms of the ocean (Figure 5.2).

THE SEAS

Two great arms of the Atlantic Ocean extend into the Eurasian landmass, and each reaches to Russia. The northern arm, the Baltic Sea, extends around the Scandinavian peninsula. The furthest part is called the Gulf of Bothnia and lies between Sweden and Finland. All the ports of Finland, Poland, and former East Germany, and many of the ports of Sweden and the Soviet Union as well, are located on the Baltic. All the shipping to and from these ports must pass through the narrow straits of Denmark to reach the open ocean; Copenhagen, on the most used passage, has great commercial and naval importance.

Although not nearly as contained as the Baltic, the North Sea is still not the open ocean; it is separated from the Atlantic by Great Britain. Except for fishing ships, however, almost no vessels pass through the wide gap to the north. Instead almost all ships crowd through the narrow Straits of Dover and the English Channel. This bottleneck is the world's busiest seaway, and several of the greatest ports of the world are located in its vicinity: Rotterdam, Antwerp, London, Le Havre, Hamburg, and Bremen. As long as the British navy controlled the seas,

Britain was able to control all sea traffic to and from northern Europe in war time.

The Mediterranean Sea

The southern extension of the Atlantic is remarkable both for its length and its number of subparts. The Mediterranean Sea (the name means "in the middle of the lands") is a self-contained unit because its only natural contact with the oceans of the world is through the narrow Strait of Gibraltar. Very few large rivers empty into its basin, and evaporation is high. The flow of water at Gibraltar is therefore primarily in one direction, inward. As a result, the water at the Mediterranean is unusually blue because of the high salt content, and pollutants are not carried out. The heavy industrialization of northern Italy and southern France since 1950 and the massive development of tourism along the Spanish and Italian beaches have created major pollution problems, which are very difficult to solve.

The Mediterranean is about 3000 miles (4827 kilometers) long. It can claim to be the birthplace of Western culture, since Greece, Rome, Israel, and Egypt all developed along its shores. It divides roughly into two major portions; at its "waist", where Sicily reaches almost to North Africa, the sea is nearly cut in two. This location was the site of ancient Carthage (now Tunis) and the former British naval and air base at Malta, an indication of the area's strategic importance.

At its northeastern end, the Mediterranean receives the waters of the Black Sea, which pour in through the Dardanelles. Turks, Greeks, Russians, and British have often fought over these straits and over Istanbul (Constantinople), the historic city on them. The northeasternmost portion of the Mediterranean itself is called the Aegean Sea and is almost totally enclosed by Turkey, Greece, and the Greek islands.

West of Greece is the Ionian Sea, enclosed on three sides by mainland Greece, the southern end of the Italian peninsula, and Sicily. At its northern end, the Straits of Otranto lead from the Ionian Sea into the long Adriatic. For centuries Venice, connecting central Europe with the Adriatic Sea, was the richest port city of southern Europe. The Adriatic is very shallow, which has encouraged the rapid buildup of pollution, and its great length has made it difficult for the pollution to escape.

The principal subpart of the western Mediterranean is the Tyrrhenian Sea, which lies between the Italian peninsula and the large islands of Sicily, Sardinia, and

FIGURE 5.2

Principle Land and Sea Features of Western Europe

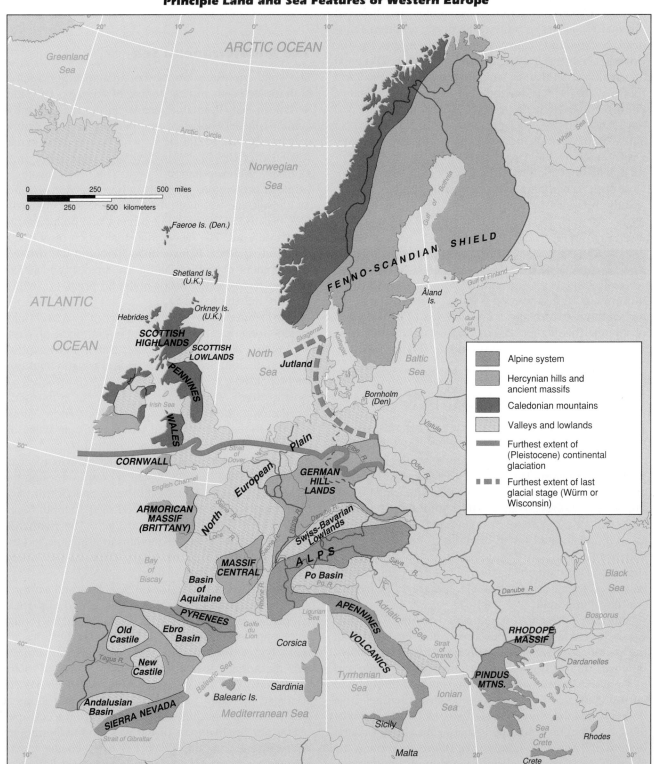

Corsica. One of the more unusual features of this sea is the presence of several active volcanoes in its southeastern corner. Along the north coast of the western Mediterranean Sea are the major ports, Marseille and Genoa. The mountainous coast between them is perhaps the most famous tourist area in the world, the Riviera. A chain of islands runs outward from the Spanish coast; these are the Balearic Islands (Majorca, Menorca, and Ibiza). Finally, at its western end, the long sea narrows down to a mere 6 miles (9.7 kilometers) at the Strait of Gibraltar, watched over by the British base on "the Rock."

These seas have had an importance in European life far beyond the transport routes they supplied. They were the source of one of Europe's most precious food resources, fish. Cities like Amsterdam, Lübeck, Germany, and Bergen, Norway, became wealthy on the herring trade. Many Norwegians, Danes, British, Greeks, Spaniards, and Portuguese still obtain their income and their food from these seas. All Europeans are conscious of the proximity of the ocean; no place in Western Europe is more than 300 miles (483 kilometers) from the sea.

THE LAND

Four roughly parallel bands of terrain make up the land parts of Europe. From south to north, these bands are composed of high mountains, clumps of old hill lands, then a long plain, and finally another band of mountains. We call these four bands the Alpine System, the Hercynian uplands or hills, the North European Plain, and the Caledonian Mountains. Each of these is interrupted or broken into segments by arms of the sea or great rivers, so that Western Europe is made up of a great number of separate small regions.

The Alpine System

The Alpine System includes those chains of mountains that were formed in recent geologic time. The Alps themselves are shaped like a hook lying on its side. They begin at the Riviera, then form an arc separating France from Italy. Turning eastward, they lie between Italy and Switzerland; then, widening out, they occupy most of eastern Switzerland and Austria, to end at Vienna.

In northwestern Italy, the Alps double back on themselves and then turn southeastward to form the backbone of Italy, the Apennines. Although not as high as

the Alps, the Apennines are rugged, and do reach 9554 feet (2912 meters). In southern Italy, the system turns back toward the west to form the northern coast of Sicily and, after a gap, the Atlas Mountains of North Africa. Further west, Spain is bordered both north and south by outliers of the Alpine system. The highest peaks in Spain are in the south in the Sierra Nevada ("snowy mountains"), but the biggest barrier to movement is formed by the Pyrenees between Spain and France. The Pindus Mountains, which form the rugged mainland of Greece, are also alpine. The drowned portions of the Pindus, with just the tops above water, have created the hundreds of Greek islands.

The Alps are mainly folded mountains. They resemble long high wrinkles that have been carved by glaciers and rivers. In the middle chain are the older, crystalline rocks that have been more resistant to erosion. They form the highest mountains: Mont Blanc (France) (15771 feet; 4807 meters) and the Monte Rosa and the Matterhorn (Italy-Switzerland). North of the crystalline area is a belt of limestone that weathers to form cliffs and peaks. These cliffs and peaks form the spectacular mountain walls overlooking Bavaria and northern Austria and Switzerland. They also contain the lakes; therefore the limestone belt is a popular tourist area (for example, Luzern, Interlaken, and Salzburg). South of the crystallines is another belt of limestones, the Dolomites of northern Italy.

Between the chains are long, narrow valleys, which are remarkably level and heavily used for transportation. Innsbruck, the capital of the Tyrol region (Austria), lies in such a valley. The rivers in these valleys all turn sharply north and cut through the limestone ranges in deep gorges. This is true of the Rhone, the Rhine, and several Austrian rivers like the Inn that flow into the Danube. The beautiful lakes occur in clusters at the edge of the piedmont plains; they were carved out by the glaciers surging out of the mountains.

Unlike our Rockies, the Alps have been densely settled for thousands of years. Much of the beauty of the landscape is produced by the settlements and the managed approach to nature. The word *alp* does not mean "mountain" but rather "upland pasture," which suggests the importance that dairying has long had in these mountains. Despite their elevation, the Alps are crossed by many passes, and most of the highways and rail lines now tunnel through well below the crests. Enclosed within the loops of the Alpine system are a number of important basins: The Po Valley in northern Italy and the Ebro and Andalusian Basins alongside the moun-

A MARITIME ORIENTATION

More than any other major world region, Europe has looked to the seas and oceans for its economic well-being. Luxembourg, Switzerland, and Austria are the only nations in Western Europe that do not have a coast on an ocean or sea. The ancient Greeks, the Vikings, the medieval Italians, the Spanish, Portuguese, and Dutch during the Age of Exploration, and the British and French of the colonial era all lived in close contact with the sea.

Several of the European nations have limited natural resources. The oceans allow the importation of the high-bulk relatively low value metallic ores, fuels, animal fibers, and lumber, which form the basis for a high standard of living. Maritime activities are the foundation of an export-based economy so important in the modern era.

The oceans have also helped give Europeans a worldwide perspective in foreign affairs and cultural matters. The knowledge of the world also allowed for emigration to many parts of the world. The interchange of goods over long distances spurred local manufacturing and encouraged the creation of sophisticated international financial institutions to link this local activity with international commerce.

These trends continue as European nations and businesses retain a focus on maritime activities. Several European countries are leaders in commercial fishing and ocean aquaculture. The port of Rotterdam is the world's leading market for noncontract petroleum (spot market) trading. Lloyds's of London is the leading insurer of ships and cargoes. European industry depends on the ocean shipment of petroleum from the Persian Gulf and North Africa. Many aspects of the European example have been replicated in the successful economies of Taiwan, Japan, and Hong Kong.

---

tains of Spain. Because they are well sheltered and can use the mountain waters for irrigation, these basins are of great agricultural importance.

The Hercynians

Around the western and northern edges of the Alpine ranges are the hill lands we call the **Hercynians.** "Hercynian" is the Latin name for one of the most prominent groups of these hills, the Harz (or Hartz) Mountains of Germany. Although often close to the Alps, these hills are usually separated from them by an important lowland belt. The Hercynians include a great variety of geologic forms, including raised blocks, volcanics, and eroded massifs. Compared to the Alps, they are very old, worn down, and rounded off. The individual ranges tend to be short (the Harz is only 60 miles (96 kilometers) long) and heavily forested. In fact, they are usually called forests rather than hills or mountains, as in the Black or Bohemian Forests. The major rivers, especially the Rhine, have cut gorges through them.

One of the largest of the raised blocks is the Meseta, which forms most of Spain and Portugal. This plateau is flanked by Alpine ranges, with a few old ranges of its own. Sardinia and Corsica are, for the most part, also parts of an ancient massif. Most of south central France is included in the Massif Central, a tilted block with a sharp edge facing to the south and east and numerous old volcanic plugs. Further north, the principal concentration of the Hercynians stretches from eastern France across Germany into Czechoslovakia. This is the home of some of Europe's best wines, and many of Germany's most famous fairy tales, set in the forests with their legendary hunters, witches, and wolves. Another grouping of Hercynians occurs off to the northwest, where they form the western points of France (Brittany), England (Cornwall), and Ireland (Kerry).

The North European Plain

Beyond the principal arc of the Hercynians lies the North European Plain. This broad lowland extends from the Pyrenees Mountains around the Atlantic and North Sea shorelines into Eastern Europe. It slopes gradually and gently to the sea and is crossed at right angles by the major rivers of Western Europe. A very high proportion of the population of Western Europe lives on this lowland.

The plain has been divided into two main parts by the actions of the ice sheets. The furthest extent of glaciation, the **terminal moraine,** runs roughly east-west across southern England, the Netherlands, and along the hill-plain contact zone in Germany. Because this was an early stage of glaciation, the terminal moraine has been largely worn away. It is most noticeable in the Netherlands where it appears as a line of low, sandy hills just north of the Rhine River.

South of that line, which is shown on Figure 5.2, the plain was not glaciated. Due to the absence of glaciation, the tops of hills were not scraped off and the valley bottoms filled in; the soil is therefore based on the underlying rocks. Just south of the terminal moraine, one can find broad areas of very fertile **loess,** made up of the fine silt blown by the winds from the moraine area and deposited miles away. From northern France, a belt of loess soils extends across Belgium and Germany at the base of the hills.

North of the moraine, all the land has been glaciated, which resulted in the landscape, except for the mountain chains, being leveled out. The dumping of rock debris by the glaciers interrupted drainage across the area. Therefore, much of northern Germany is made up of swampy areas that have had to be drained to become agriculturally productive.

One further distinction must be made within the glaciated area. The last ice sheet, called the Wisconsin in North America and the Würm in Europe, moved over Scandinavia, but did not reach the rest of Western Europe.

The terminal moraine for this ice sheet forms the Jutland peninsula of Denmark. Fortunately for Denmark and Sweden, there is an area of rich clayey soils behind this moraine; these form the base for the rich Danish agriculture.

The Caledonian Mountains

The northernmost of the four principal terrain bands is called the Caledonian Mountains, or the Northwestern Highlands. "Caledonia" was the Roman name for Scotland, and it was in Scotland that these mountains were

first studied. In elevation they fall between the Alps and the Hercynians, but they are far wilder and more barren than either. They also are old mountains and are now recognized to be of the same time period as the Appalachians of North America. They are between 2000 and 4000 feet (610–1220 meters) high in Ireland and Scotland, but considerably higher in Norway.

Perhaps their most distinguishing characteristic has been the severity of their glaciation. Whereas in the Alps the glaciers have moved down the valleys, here the ice sheets covered virtually the entire land surface. In Scotland the valleys ("glens") have often filled with water to form the long narrow lakes called "lochs." In Norway and Scotland, the valleys were gouged out into the sea;

The Cuillin Hills on the Island of Skye in Scotland are part of the northwestern highlands physiographic region that includes much of extreme western and northern Europe.

Narrow, but deep fjords allow large ships to reach inland villages in Norway. Note the necessity for zigzagged roads (switchbacks) to traverse steep mountain slopes.

when the ice melted away, these valleys filled with seawater to form steep-sided saltwater inlets called *fjords.* In both countries the mountains with their fjorded coasts face directly onto the wild North Atlantic. The frequent storms and the constant wave action have produced steep headlands and thousands of rock islands.

East of the Norwegian mountain wall is an extensive region of ancient hard rock. As in Canada, this is called a *shield,* here the Fenno-Scandian Shield. In northern Sweden, it slants downward toward the Gulf of Bothnia and hence is well drained. In Finland there are thousands of lakes upon the shield, much as there are under

the same conditions in northern Ontario and northeastern Minnesota.

THE RIVERS

Because of their significance to European life, the rivers deserve special mention. The most important are in northern Europe. They rise in the Alps or the Hercynians, cut through the hill lands, cross the plain at right angles, and empty into the North Sea or the open Atlantic. These rivers tend to carry enough water to permit transportation all year round. They form well-developed systems, which come quite close to each other, so that joining them with canals has proven to be relatively easy.

The main river of northern France is the Seine, which flows through Paris. The valley of one tributary, the Marne, can be followed eastward to the Meuse, the Moselle, and the Rhine. Just south of Paris, the Loire River flows west to the ocean. The Rhine, a bigger river, is even better for transportation and trade connections. The river and canal systems are so well linked that one can go by boat from Paris to Warsaw, Poland, without touching the seas.

The rivers of Great Britain and Ireland are much shorter than those on the European continent. Despite their small size, several are quite famous: the Shannon in Ireland, the Clyde in Scotland, and the Thames in England.

In the south of Europe, the rivers tend to be short and steep. The mountains lie close to the coasts, and the streams rush swiftly. The Po River is important to the industry of northern Italy. Due to the long, hot and dry summer, many of these streams become very low by late summer. The rivers of Spain and Portugal tend to be longer than the other southern rivers. The rivers crossing the Spanish Meseta have cut deep valleys, which were serious barriers to land transportation. The English word *canyon* derives from the Spanish *cañon*.

There is one great exception to the simple division of the rivers into northern and southern; that is the largest of them all, the Danube. This mighty river begins at the southern end of one of the Hercynian blocks, the Black Forest. It flows eastward along the northern edge of the alluvium washed down from the Alps. As the Danube flows to the southeast, it pulls in all the rivers flowing northward out of the Austrian Alps. It is already the greatest river of Europe by the time it leaves Western Europe at Vienna, and it still has more than half its course to go.

The most important single river of Europe is undoubtedly the Rhine. It forms the north-south axis of the European economy. It is heavily used for freight, and its valley through the Hercynians is followed by railroads and highways. As will be explained later, its delta is regarded as the economic core of the continent.

The rivers are also important as sources of water and of electricity. They are the obvious source of drinking water for the millions of people in the great urban areas, and the water in these rivers may be used several times before it reaches the sea. Pollution is thus obviously of great concern; the condition of the Rhine in particular has been the focus of international study. Some fortunate cities, notably Vienna, have reservoirs of mountain water to satisfy their needs. The availability of water is of greatest concern in southern Europe. Athens is in a difficult situation; it is almost surrounded by saltwater and has no freshwater river nearby. In Spain the major rivers have been dammed to form long, narrow reservoirs, hemmed in by the canyon walls. These have become very important as sources of irrigation water.

In the Alps, water is used more for the production of hydroelectric power than for irrigation or drinking water. Many of the upper valleys have been dammed for this purpose. Several of the longer rivers, notably the Danube, Rhine and Rhône, have also been tapped for power purposes.

THE CLIMATE

The climate of Europe is created by the interplay of four major physical influences: the westerly winds, the North Atlantic Ocean, the Azores (Bermuda) high pressure cell that creates the Sahara Desert, and the Eurasian landmass. Their interaction produces two major types of climate, the marine west coast and the Mediterranean (Figure 5.3). A third type, the continental, exerts an influence, but is felt mostly in Eastern and Central Europe. The passage of the weather patterns is usually from west to east, and the west winds are felt the most keenly, especially along the exposed shorelines.

The North Atlantic coast has a relatively mild climate for its latitude. The warm North Atlantic Drift, or Gulf Stream, flows along the western shores of Europe. Except for those of the Baltic Sea, none of the northern ports of Europe freezes in the winter. However, the waters never become very warm either, and the winds off the ocean are always cool and raw. Fog is also common. Swimming in the ocean is not a popular sport in Scotland or Norway.

The Eurasian landmass has an influence opposite to that of the ocean. Land heats and cools far more rapidly than water; therefore, continental locations tend to be both hotter in summer and colder in winter than seashore locations (this is known as ***continentality***). Also while maritime locations tend to have constant rainfall in winter, when storms are strongest, continental locations tend to have heavy summer rainfall because the heating of the land produces many showers. The high pressure cell that accounts for the Sahara Desert is

FIGURE 5.3

The Climates of Western Europe

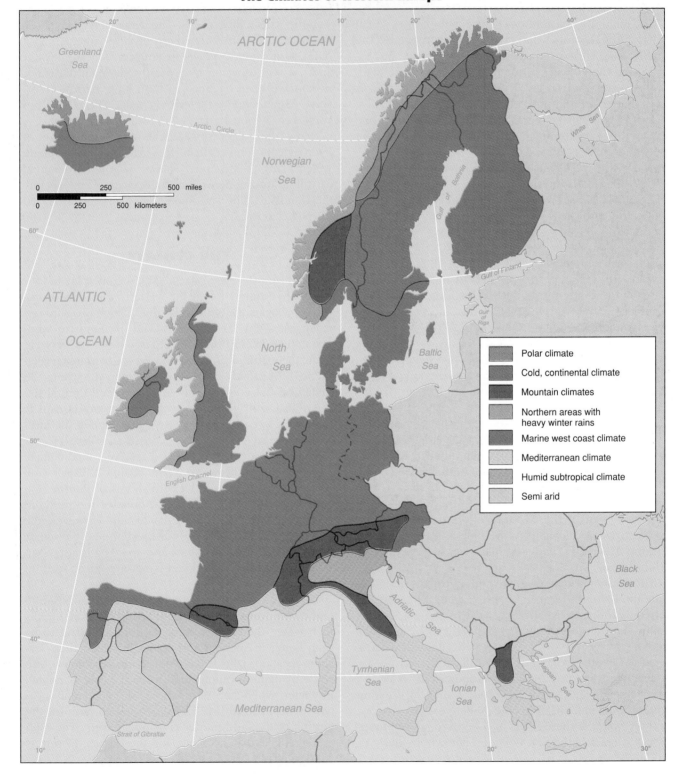

Polar climate

Cold, continental climate

Mountain climates

Northern areas with heavy winter rains

Marine west coast climate

Mediterranean climate

Humid subtropical climate

Semi arid

called the subtropical high. It moves north and south with the sun. During the northern summer, it covers most of the Mediterranean Sea; in the winter it moves well to the south of the sea into North Africa.

The two major climates of Western Europe are sharply separated from each other by the east-west line of high mountains, the Alps and the Pyrenees. For a north European, "south of the Alps" has a definite meaning in terms of climate, culture, and life-style. The lands of the Italians, Greeks, and Spaniards are felt to be very different from the lands of the Germans, English, and Swedes.

The Marine West Coast Climate

The marine west coast type is the prevailing climate for two-thirds of the West Europeans; it is probably the type of climate we associate most with Europe. The eastward extension of both the seas and the plain allows for the oceanic influence to be felt southward to the Alpine mountains and eastward to the boundaries of Eastern Europe. At times in winter, cold blasts can come from Russia or even across from Canada, but the duration of such cold spells is short. In summer the weather is mild in the sense that temperatures are well below those in eastern or central North America. In fact, in the more maritime portions of western Ireland there is very little difference between summer and winter.

Cloudy days are frequent, especially toward the northwest. Although total precipitation in cities like London or Paris is actually less than that in New York or Toronto, it falls on many more days. Drizzle is the most common form of rainfall; London business people are famous for carrying their umbrellas with them everywhere. Winter storms are both more frequent and more violent than those in summer. From October to February the days are very short (because of the high latitude) and are usually overcast and drizzly, all of which creates an atmosphere of gloom. It is no wonder that northern Europeans welcome spring ecstatically when it begins to arrive in March.

As one moves inland from the coast, the amount of cloud and drizzle decreases. The amount of sunshine, plus the reliability of rainfall, have allowed France to become one of the great wheat-growing countries of the world. Overall, the climate is good for agriculture in that the length of the growing season makes up for the lack of high temperatures. Grapes can be harvested in November; and animals can be kept outdoors all year.

Sweden and Finland lie adjacent to Russia and are cut off from the oceanic influence by the Norwegian mountains. This lack of an oceanic influence results in humid continental and subarctic climates, characterized by colder winters and less dampness and precipitation than in other areas of northern Europe.

The Mediterranean Climate

The Mediterranean climate, the other great climatic type characterizes most of southern Europe. In brief, it consists of winter rains and summer drought. It is a result of the seasonal movement of the sun, and the accompanying pressure belts. In the summer, relatively high pressure, arid air covers the area. In the winter the high pressure area moves southward, fostering the movement of Atlantic storms through the region.

The summer months are rainless and almost cloudless. The sky is bright blue; the sun is intense. Tourists from the cloudy north love this, but for the local people the months without rain can be a disaster. The people around the Mediterranean Sea have adjusted their lifestyles to the powerful sun. They have learned to divide the active hours of their day into morning and evening portions, with a long gap, often a siesta, in between. The Madrileños (persons from Madrid) of dry, central Spain commonly have their evening dinner between 10:00 P.M. and midnight; begin their evening theater performances at 10:30; and yet are up for work the next morning.

Winter can be more uncomfortable than summer because few of the homes are adequately heated. Mist and rain can seem extra cold in stone buildings or on marble floors. Of course, it does not rain every day, and between storms the winter days can be delightful in the sun, even if bone chilling in the shade. Winter is also a time when strong winds whip across the open sea or pour down the Alpine valleys. When storms pass through the Mediterranean basin, cold air is drawn in from northern Europe. The most famous of these winds are the *mistral* rushing down the Rhone Valley in France and the *bora* along the Adriatic coasts.

Although this description provides a general profile of the Mediterranean climate, there actually is a great deal of variation within the region because of the many mountain ranges. In Greece, the Athens area, east of the mountains, is far hotter and drier than the western, Ionian coast. In Spain, the northern coast is much like northern Europe with adequate rainfall all year round, whereas central and southern Spain are very hot and dry. The greatest local contrasts are in Italy where the mountains divide the country into many distinct areas. Only in the extreme south does one find the true Mediterranean climate. In contrast, the Po Basin of the north, which is almost encircled by high mountains, has a semicontinental climate. Milan is colder than Edinburgh, Scotland, in winter and receives rain every month of the year.

VEGETATION

It is difficult to say what the natural vegetation of Europe was because so little of it is left. Through cutting, planting, and overgrazing, humans have almost totally modified the vegetation of Europe. The only truly natural vegetation we can find probably would be in the north of Scandinavia, on some mountains, and in a few of the drier regions of the south. Even areas that now look wild and natural to us do not have the original vegetation. The peaty glens of northwestern Scotland once had many more trees than they have now. Many of the hills of central Germany, which were once covered with deciduous trees, are now being planted in evergreens because they are more profitable for construction and pulp uses and are more popular with hikers.

The Fenno-Scandian Shield appears to be a natural area for coniferous trees, much like the Canadian Shield. In northernmost Scandinavia, the trees are replaced by the northern shrub growth we call tundra. Most of the continental Europe and the British Isles were covered by deciduous forests, (analogous to the broadleafs of Anglo-America) although the Hercynians carried conifers (needlelike leaves) as well. We think of Ireland as a natural pastureland, but that is after thousands of years of deforestation and grazing.

In southern Europe, the "natural" vegetation can endure the summer drought. *Maquis,* a French word, is the name commonly used, but we are more accustomed to the Spanish word *Chaparral* for the same type of vegetation. This is a scrubby, thorny family of plants. They are **sclerophytic**, which means that the leathery leaves can hold in the precious moisture. Trees are relatively rare in the south largely because of the climate, but also because generations of sheep and goats have eaten away the young saplings. The most common southern trees are the oak and a cultivated tree, the olive. In summary, one may think of the far north and the

mountains as the home of pines and firs, of the North European Plain as a land of beeches and oaks, and of the south as an area of oaks and olives. But, almost everywhere humans have totally modified the environment and its vegetation.

THE SOILS

Western Europe has the full range of soils, as one would expect from an area that is open to oceanic winds and extends from the margins of the world's largest hot desert to the edges of the polar ice cap.

Finland and most of the Scandinavian peninsula are covered with spodosols, thin, ash-colored soils that normally develop under coniferous trees in northern locations. Other spodosols have developed on the glacial material that covers northern Germany, the Netherlands, and the western strip of Denmark.

The alfisols are the most important agricultural soils of northern Europe. These soils occupy the southeastern halves of both Great Britain and Ireland, the northwestern half of France, most of Belgium, southern Sweden, eastern Denmark, the fertile belt along the edges of the hills, in northern Germany, southwestern Spain, southern Portugal, and most of Italy.

The broad hill areas of southern and central France and southern Germany are covered by inceptisols, relatively thin soils that have formed recently on the slopes and in the valleys. In fact, this name includes a great variety of local soil types. Entisols, soils without clear horizons, are common in eastern and central Spain. At the other end of Western Europe, the histosols, or bog soils, filled with deep layers of peat, cover most of Scotland, Wales, and the northern and western coasts of Ireland.

In addition, the mountainous regions include a number of undifferentiated soils. The elevations and slopes of these regions are so varied that it is hard to characterize the soils precisely. These dominate in the Alpine chain in Austria and Switzerland and the adjacent strips of France and Italy, in Greece, in northern Spain, and, most extensive of all, along the full length of Norway.

Through great diligence and wisdom, European farmers have made most of these soils productive. The alfisols are the premier soils of Europe, but the Dutch and Germans have achieved great success on the spodosols. Perhaps the most intensive productivity of all is on the alluvial entisols of the Netherlands, Italy, and Spain. Through centuries of care, the Europeans have almost remade their native soils, making them among the most productive in the world.

MINERAL RESOURCES

Western Europe is not well endowed with mineral resources and is therefore a major importer of them. However, it does have sufficient amounts of the two minerals necessary for the development of heavy industry, coal and iron ore. Coal fueled **the industrial revolution**. The availability of coal helped to determine which countries industrialized early and which did not. In the power politics of the last century, the possession of good coal deposits was felt to be essential.

The principal outcrops of coal have been along the margins between the hill lands and the North European Plain. The most important coal deposit begins in northernmost France and then follows the northern edge of the Ardennes Forest (Hercynian), along the line of the Sambre and Meuse rivers, through Belgium. It touches the southernmost tip of the Netherlands and then enters Germany. East of the Rhine it becomes the famed Ruhr Valley. Further east it continues on and off through Germany and Poland. The other great coal area lies around the Pennine hills in England. It extends from Newcastle in the northeast, past Leeds and Sheffield, around Birmingham, and back northward to Manchester. In addition, Great Britain has had important coal mines in southern Wales and in the Scottish Lowland. Outside of these two major coal belts, only a few small scatterings of coal occur: in the Saar Basin along the German-French boundary, in northern Spain, and in the Massif Central of France.

The most highly prized hydrocarbon now is petroleum; Europe has little of it. The biggest producing field is in the northern half of the North Sea, shared by Great Britain and Norway. As of now, it looks as if the North Sea oil will begin to run out by the year 2000. Elsewhere there are only very minor oil fields in Germany and Austria. Natural gas is more plentiful. It occurs in southern France and the Po Basin, but the principal deposits have been in and around the North Sea: in the Netherlands, Great Britain, and northern Germany. The supply of natural gas is not nearly sufficient for the demands of the European economy, however. As in oil, Western Europe is a major importer of natural gas, much of it arriving by pipeline from Siberia.

Peat is in plentiful supply in Ireland and Scotland. However, it is a bulky, heavy product that has a low caloric (energy) value. Its high water content makes it extremely heavy to dig out, and then it has to be allowed to drain and be compacted before it can be used. Only in central Ireland is it used on a large scale for heating and the generation of electricity.

Iron ore was the second mineral essential for the industrial revolution. The largest body of high grade ore in Europe is around Kiruna in northern Sweden. Besides supplying the Swedish steel industry, this ore is exported in large quantities, most through the nearby ice-free Norwegian port of Narvik. A second large deposit is in eastern France. This ore is relatively low grade with a high level of impurities. Despite its shortcomings, this ore is used because of its location in the middle of the European manufacturing belt, close to the coal of the Saar Basin and alongside the navigable Moselle River. Lesser outcrops of iron ore occur widely in the East Midlands of England, at Eisenerz in Austria, in northern Spain, and in central Sweden.

There are only small deposits of other minerals. Spain and Greece have the greatest variety: copper, lead, zinc, manganese, gold, and even mercury, but all in small deposits. Portugal has tungsten; Sicily is famous for sulfur. In most of these, Europe remains a major importer.

The use of canals and levees in the Netherlands is the only way stable, dry land can be maintained in this densely populated country where much of the land area is at sea level.

HUMAN ENVIRONMENT

POPULATION DENSITY

Western Europe is one of the most densely populated regions of the world. Over 350 million people live in an area equal to about one-third of the United States. The result is a population density of 255 per square mile (100 per square kilometer).

This population is not evenly distributed. Europe is characterized by both great concentrations of population and sparsely populated stretches. As Figure 5.4 shows, the major concentration is along the two sides of the southern end of the North Sea. The core of Europe's population is located near the mouth of the Rhine River. The Netherlands has the highest density of Europe. The zone of over 750 per square mile (300 per square kilometer) includes most of the Netherlands and Belgium, the lower Rhine region of Germany, and the northern strip of France. Across the North Sea, the English portion is centered on London but extends northward to include Birmingham, Manchester, Liverpool, and Leeds.

Away from this primary core of the population, we can pick out a few other clusters. Most of these fall within a broad north-south corridor ending in northern Italy. The western part of the Po Basin, around Milan, can be thought of as the southern anchor of the main population belt. In addition, there are major concentrations along the Rhine, around Paris, and in a narrow belt at the edge of the hills in northern Germany.

A major boulevard in Madrid, Spain.

A density of 125 per square mile (50 per square kilometer) would include almost all of Germany, Italy, and Great Britain, as well as the coastal strips of France, Spain, and Portugal and the southern parts of Scandinavia. Finally, we can use a figure of 50 per square mile (20 per square kilometer) the density of Iowa, as marking the limits of the settled areas of Europe. Outside are the sparsely populated areas of the Scottish Highlands, the western Alps, and large parts of Northern Scandinavia. Perhaps because of the high densities, Europeans are very conscious of the need for open space. Urban sprawl is limited, so that even in the Netherlands, one can easily bicycle into the countryside.

Despite the recent actions toward the economic unification of Europe, Europeans still think in terms of the individual countries. The Statistical Profile shows the population and density of each country. Unified Germany has the largest population size followed by three large countries of roughly equal size: Italy, the United Kingdom, and France. In an intermediate position is Spain. After the Netherlands at 14 million, all the other major countries fall between 5 and 10 million. It is intriguing to note that in Europe there appears to be no correlation between the physical size and population of a country and its prosperity.

Most West Europeans live in cities; the urban population exceeds 80 percent of the total in almost every country. Table 5.1 lists in order the largest cities or complexes of cities (**conurbations**). The population given are estimates for the entire metropolitan area since official city limits rarely coincide with the extent of the urban area. In addition, there are many cities around or above 1 million people each: the Cardiff area in south Wales; Newcastle, England; Hamburg, and Munich, Germany; Naples and Turin, Italy; Brussels, Belgium; Vienna, Austria; Copenhagen, Denmark; Stockholm, Sweden; Helsinki, Finland; Lyon, Marseille, and Lille, France; and Lisbon, Portugal.

POPULATION CHANGE

As can also be seen from the Statistical Profile, Europe has had a very low rate of population increase in recent years. In several countries, the population is at zero population growth or lower. Before the recent inflow of East Germans and the unification of Germany, the West German population had been declining since 1981. Were it not for the high birth rates among the *migrant or guest workers*, imported into the former West Ger-

FIGURE 5.4

Population Density of Western Europe

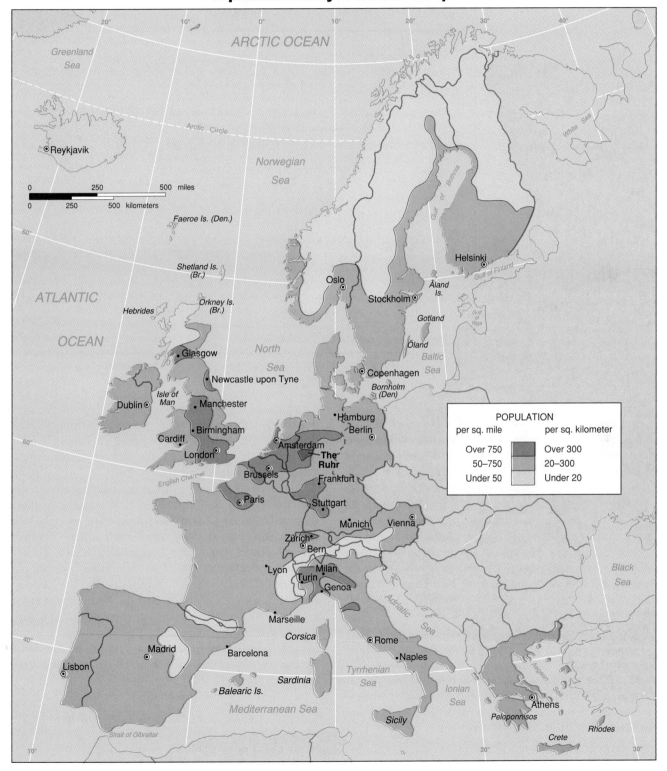

TABLE 5.1

Population of the Principal Urban Complexes of Western Europe

	POPULATION
1. Ruhr Region, Germany (Cologne, Essen, Dortmund, Düsseldorf, Duisburg, etc.)	10,000,000
2. Paris, France	8,700,000
3. London, United Kingdom	8,000,000
4. Randstad, Netherlands (Rotterdam, Amsterdam, The Hague, Utrecht, Haarlem, etc.)	5,000,000
5. South Lancashire, United Kingdom (Greater Manchester and Merseyside [Liverpool])	4,100,000
6. Madrid, Spain	3,188,000
7. Athens, Greece	3,027,000
8. Milan plus satellite cities, Italy	3,000,000
9. Barcelona plus satellite cities, Spain	3,000,000
10. Berlin, Germany	3,000,000
11. Rome, Italy	2,831,000
12. West Midlands (Birmingham), United Kingdom	2,658,000
13. Strathclyde (Glasgow), United Kingdom	2,383,000
14. West Yorkshire (Leeds, etc.), United Kingdom	2,059,000

SOURCE: United Nations, *1987 Demographic Yearbook* (New York: United Nations, 1989)

many as less expensive laborers, Germany's decline would be even more rapid. Even countries that were formerly considered to have high birth rates, like Italy, now have low birth rates by world standards.

Several possible explanations have been advanced for the fall of the birth rate in recent years. Europeans, especially the Germans, are far more prosperous than they were even 30 years ago. Programs for medical and old age care are in place in most countries. In addition, people who, in a nuclear age, find themselves between the great powers may have a fear of the future. Whatever the reasons may be, families with more than two children are rare in Europe.

Some European leaders are becoming quite concerned. Soon many countries could experience a shortage of workers to pay taxes and of service personnel to care for the aging population. Berlin and Vienna are already noted for their concentrations of older people—mostly older women because so many men were killed in World War II. In both of these cities, the total population has been steadily declining because of the high death rate among the aged. Some governments have been offering various inducements to young couples (money, living space, maternity leave) to have more children, but with little effect so far.

The labor shortages of the recent past have been met principally with the importation of migrant workers. Originally, these persons were regarded as only temporary or seasonal, and many were under contract. These workers come from Spain, Italy, Greece, Turkey, Algeria, and a host of other less economically developed countries. They do the kinds of work that the more affluent Europeans do not care to do. The migrant laborers live in the poorer sections of the cities or in camps. Rather than being temporary, they have become permanent parts of the urban scene and are highly visible in the urban slums and blue-collar working class districts.

As a result of this immigration, Europeans have had to face the prospect of large numbers of foreigners living among them. North Americans are used to the periodic arrival of immigrants, Europeans are not. The English, Germans, French, and others have been proud of their heritage to the point of snobbery and beyond. Consequently, a profound culture clash has occurred, reinforced by the prejudice often felt toward migrant people who do not act in the expected way. In some European countries, birth within the borders does not automatically confer citizenship; one must be the child of a citizen. The presence of the migrant workers raises many questions of social justice: To what level of social benefits are these people entitled? Unemployment insurance? Retirement? Old age pensions? Citizenship? If such benefits are granted, will the system be abused by people using it as an easy way to move from the developing world? The Europeans want and need the less expensive labor, but are troubled by the other social and economic consequences of these migrants.

Table 5.2 shows the country of origin and the number of such workers in the major West European countries. Until the 1960s the Irish were the only large group in Great Britain. Since then, however, a great number of people have arrived from the countries that once were parts of the British Empire. Indian restaurants and Pakistani corner stores are now common in British towns, and several cities have large mosques.

For many years France was the goal of southern Europeans, but in recent years the great mass of workers has come from France's former colonies in Africa, especially Algeria and Morocco. In Switzerland the adjacent Italians have been the largest group for almost a century, and repeated attempts have been made to force most of them to leave. Germany has the largest number of migrant workers, with the Turks being by

TABLE 5.2

Migrant Workers in Western Europe, May 1980 (Thousands)

COUNTRY OF ORIGIN	Total	HOST COUNTRIES							
		Austria	Belgium	France	West Germany	Luxembourg	Netherlands	Sweden	Switzerland
Algeria	363.7	—	2.4	361.0	—	—	—	0.3	—
Austria	101.2	—	3.7	—	76.0	—	—	2.0	19.5
Finland	107.4	—	—	—	2.9	—	—	104.5	—
Greece	177.8	—	9.6	—	153.3	—	2.0	8.1	4.8
Italy	853.3	2.0	106.4	175.8	305.1	10.8	10.4	2.7	240.1
Morocco	235.2	—	22.2	181.4	—	—	31.0	0.6	—
Portugal	472.9	—	3.9	385.0	59.9	12.9	5.4	0.9	4.9
Spain	390.9	0.2	27.3	184.5	95.8	2.2	17.6	1.7	61.6
Tunisia	77.2	—	1.9	73.7	—	—	1.2	0.4	—
Turkey	686.3	26.2	17.0	36.3	540.5	—	45.1	4.5	16.7
Yugoslavia	606.9	121.1	3.0	43.1	380.2	0.6	8.1	24.6	26.2
Other EEC	487.2	12.7	87.7	66.2	131.6	20.7	52.6	32.0	83.7
Other, Non-EEC	484.1	14.5	18.2	135.8	216.6	1.9	23.0	42.2	31.9
Total	5047.1	176.7	306.3	1642.8	1961.9	49.1	196.4	224.5	489.4

SOURCE: *Organization for Economic Co-operation and Development (OECD) Observor*, 104 (May 1980).

In the census of 1981, the United Kingdom listed 3,359,825 persons as born outside the country. The leading home countries were:

1. Irish Republic	606,851	4. Jamaica	164,119	7. Italy	97,848	9. West Germany	92,789
2. India	391,474	5. USA	118,079	8. Poland	93,369	10. Cyprus	84,327
3. Pakistan	188,198	6. Kenya	102,144				

TABLE 5.3

Foreigners Living in West Germany (Thousands)

	TOTAL	TURKS	YUGOSLAVS	ITALIANS	GREEKS	AUSTRIANS	SPANIARDS	DUTCH	PORTUGUESE
1982	4666.9	1580.7	631.7	601.6	300.8	175.0	173.5	109.0	106.0
1984	4363.6	1425.8	600.3	545.1	287.1	172.1	158.8	108.6	83.0
Percentage change	−6.5	−9.8	−5.0	−9.4	−4.6	−1.7	−8.5	−0.7	−21.7
Number of residents per worker, 1980	2.27	2.65	1.66	2.03	1.94	2.27	1.88	Not available	1.87

SOURCE: *Statistisches Jahrbuch 1985* (and 1981) *für die Bundesrepublik Deutschland.*

far the most noticeable. The Italians and the Yugoslavs (mostly Croats) as European Christians, fit into the German population fairly easily, but the Turks are so different in culture that they stand out. As Table 5.3 indicates, the number of workers is gradually decreasing, because of a recent recession, increased mechanization, and government efforts to encourage the workers to return home.

HISTORICAL BACKGROUND

The geography of Europe is tied closely to portions of its history. At the time of Christ, which is the beginning of our system of numbering years, European civilization was centered on the Mediterranean Sea. The Roman Empire included all the lands around the sea, all the lands south of the Danube and west of the Rhine, and England. By A.D. 400 the Roman Empire was assaulted by waves of Germanic warrior tribes, some of whose names are still on the map of Europe: the Franks (France), the Bajuvari (Bavaria), the Angles (England), and the Saxons (Saxony), to name only a few. These attacks shattered the western half of the Empire, and Rome itself fell in 476. The eastern half, which was Greek speaking, became known as the Byzantine Empire; it remained a great power until the Christian Crusades (1100–1250) and was finally wiped out by the Turkish conquest of Constantinople in 1453.

Around the year 1000, Europe underwent major changes. Austria was established in 976, the French monarchy in Paris in 987, England was united by 1066, and central Spain reconquered by Christian forces in 1085. Further east the kingdoms of Poland and Hungary were established in 963 and 1000, respectively. In 1091 Norman descendants of the Vikings recaptured Sicily

from the Arabs. By 1100 the Italian city-states were winning back the Mediterranean, and Christian soldiers had conquered the Holy Land in the Crusades.

The centuries after 1000 were a time of great expansion and growth. Most of our patterns of international trade, banking methods, and much of the basic industry of Europe began at this time. Cities grew rapidly, notably London, Paris, Cologne, Lübeck, Bruges, Frankfurt, Augsburg, Milan, Florence, Venice, Genoa, and Barcelona. Flanders (Belgium) became a center of industry. Famous universities, such as Bologna, Paris, Oxford, Cambridge, Heidelberg, and Salamanca, were established. These centuries from 1000 to 1400 are misleadingly called the Middle Ages, or medieval. Many scholars believe they were the beginning, not the middle of a civilization.

The vast expansion of trade and travel (a Venetian called Marco Polo was in China in 1280) led to a renewed fascination with the ancient Greek and Roman writings and art forms. Italy led the way in what we call the **Renaissance** ("rebirth"), which characterized the century after 1400. The continued development of navigational skills led to the epic voyages of Columbus to the Americas in 1492 and of the Portuguese around Africa to India in 1498.

"Modern" times began with the French Revolution of 1789. These years coincided with the beginning of the industrial revolution in Great Britain. These two forces, technology from Britain and ideology from France, swept across Europe and then the world. The democratic ideals led directly to **nationalism,** which became murderous when joined to technology. Between 1789 and 1945, Europe endured an almost continual series of struggles. A set of wars between 1860 and 1871 led to the unification of Italy and Germany. Finally came the gigantic struggles of 1914–1918 (World War I) and 1939–1945 (World War II) among the Great Powers. Out of these came the extremist political philosophies called communism and fascism, which have made their way into many countries throughout the rest of the world. Within Western Europe, the near suicide of the two World Wars has led to a strong revulsion against war. Nationalism is still strong, but there has been a partial return to the older ideal of a united Europe.

LANGUAGES

The abundance of languages has been both a blessing and a curse to Europe. Each group has developed a rich literature, but each language has become the focus for nationalism. The larger groups have often felt that they were doing the others a favor by imposing their language on them, including the English on the Welsh, the Germans on the Slavs, and the Parisians on any persons who did not speak their version of French. At the same time, intense feelings have been aroused whenever a language has been threatened.

Dialects are common in Western Europe. Those of the south and north of Italy are at least as far apart as are Swedish, Danish, and Norwegian, but the first set are all called Italian because they are in the same country, whereas the others are separated because they are in different countries. All countries have had to overcome the problem of dialects by choosing one to be the official language. Anyone who learns Hochdeutsch, the official German, will probably find Wienerisch (Viennese) impossible to understand. Despite its long unification, England is rich in local dialects: "Oxford English," Cockney, and Yorkshire among them. And, of course, "Scots" English is unique in its own way.

Figure 5.5 shows the languages of Western Europe. The Romance languages are those that have developed from Latin (Roman). Besides the major tongues—French, Italian, Spanish, and Portuguese—there are a number of others. Perhaps the most important is Catalan, the language of Barcelona, in northeastern Spain. The official Spanish is really the language of Castile, the area around the capital; therefore it is often called Castilian. In the south of France, the original Provençal is still used, but the official Parisian French has become dominant.

The Germanic, or Teutonic, languages form the other main grouping. These include all forms of German, the Scandinavian languages (except Finnish), and the various forms of Dutch. Even so small a country as the Netherlands has its regional dialects, and the Frisians of the north speak what could be called a separate language. The Dutch category also includes the Flemish of Belgium.

English is perhaps the most interesting language in Europe, because it is a hybrid, a combination of Romance and Germanic. It is usually classified as Germanic, but its grammar and vocabulary are closer to French than to German. The small everyday words in English are close to German; therefore American tourists often find it easier to get around in Germany than in France. Almost all the longer words are close to French, however, which makes it easier for Americans to read French than German.

The Celtic languages constitute a third family. These languages were once dominant over most of Europe, but have been pushed to the "Celtic fringe" in the northwest. The healthiest appears to be Welsh, which is still spoken by about one-third of the people of Wales. On the western fringes of Scotland and Ireland, one finds Gaelic, which is required in the Irish schools. The northwestern cape of France is the home of Breton, the Celtic language of Brittany. The desire to maintain the Breton language and culture has led to the rise of a strong nationalistic movement in this area of France.

The other languages tend to be unique and hence incomprehensible to other Europeans. The Basques of northern Spain have resorted to violence in their attempts to gain cultural and linguistic independence for their language, which appears to be a remnant from pre-Roman times. Greek was once the dominant language all through the eastern Mediterranean and western Asia, but is now limited to Greece and part of Cyprus. Maltese, peculiar to its small island republic, is based on Arabic. Finnish and Lapp are loosely related to the languages spoken by the northern tribes of the Soviet Union; Finnish is also close to Estonian.

The medieval merchants who moved between Italy and Flanders needed a language, which was actually spoken as opposed to Latin, the standard written language of the time, to conduct their business. Since many of the fairs and routes were in France, a form of French was accepted. The Italians called this the lingua franca ("language of the French"); we still use that term to describe an international language. As France continued to rise in wealth and power, French became the lan-

FIGURE 5.5

Languages of Western Europe

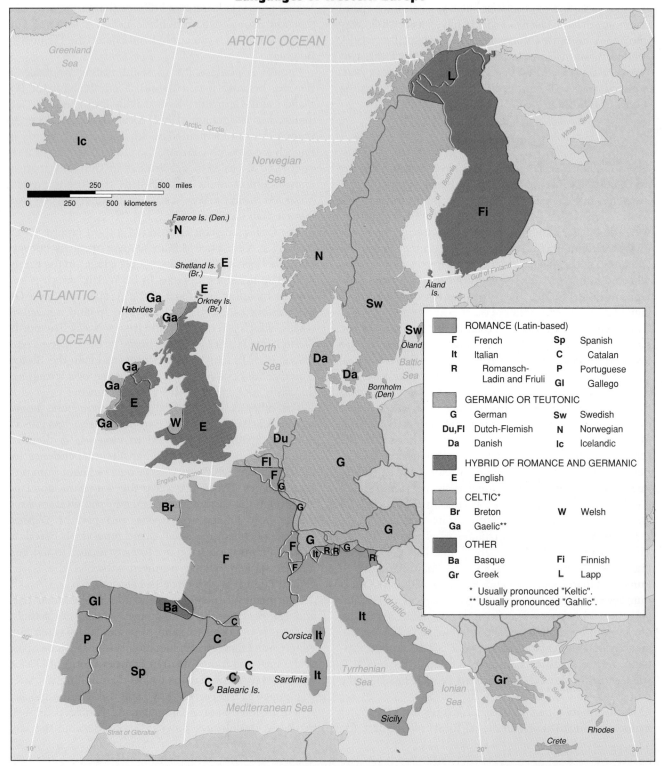

guage of culture and diplomacy and the language of the nobility everywhere. When the English duke of Wellington was fighting Napoleon, he spoke to his German allies in French; during the French invasion of Russia in 1812, the Russian nobility continued to speak French—they hardly knew Russian. In music and the arts, Italian was the accepted language; most of Mozart's operas were written in Italian, even though they were for a German audience.

English did not become important until the nineteenth century when works of Shakespeare, Lord Byron, and Sir Walter Scott became very popular in Europe. At the same time, the industrial revolution made the British the technicians of the world. To learn the new techniques, one had to know English. Then, as the British Empire circled the globe, English became worldwide. In the decades since 1945, North American science and medicine have made English dominant in the universities and laboratories. Academics and researchers everywhere must be able to read English if they are to keep up with developments. At another level, popular music came to be dominated by the Americans and the British. At present, if one deals with students, technicians, and expensive hotels, English can be found everywhere in Western Europe. Two countries are impressively multilingual. Switzerland has four official languages, and every educated Swiss will know at least German and French, perhaps Italian, and also English. The Netherlands has only one language, Dutch, but most educated Dutch will also know the big three: English, French, and German. Nothing seems more ridiculous to a European than the American belief that learning two or more languages is confusing.

RELIGIONS

Figure 5.6 shows the distribution of religions in Western Europe. It is, of course, only an approximation because much mixing has occurred in the cities. Protestantism shows a general correspondence with the Germanic language groups, especially if that includes English; the Romance language countries are Catholic. The Greeks are Orthodox. There are also some interesting splits. The Rhineland, Bavarian, and Austrian Germans are Catholics; the other Germans are mostly Lutheran. The northern Dutch are Calvinist Protestant, but the southern Dutch and Flemish are Catholic. In Switzerland the cities tend to be Protestant and the rural areas Catholic. Germany, the Netherlands, and Switzerland are each half

and half, although the Protestants have been the more influential group in the latter two.

The pattern on the map is a result of historical political factors. After the Reformation, each local ruler saw to it, through persecution or otherwise, that his people followed his choice in religion. Not to obey was considered a form of treason. It is no accident that Sweden was 100 percent Lutheran and Spain 100 percent Catholic. As we know, many of the smaller groups fled. The only large groups that stayed and held out despite all attempts by the crown to force them to change seem to have been the Irish Catholics and, to a lesser degree, the Scottish Presbyterians.

POLITICAL INSTITUTIONS

All countries of this region are democracies, although several are also monarchies. The king or queen has only persuasive power in most cases, but is a source of loyalty and stability. Those countries that do not have a monarch, have a president who is a dignified ceremonial figure, much like a monarch. Every country has a prime minister (or the equivalent) who is the leader of the major party; the legislative and executive branches are, therefore, more united than they are in the United States. France is somewhat of an exception in having a powerful president and a relatively weak prime minister.

Political parties tend to be more ideological—that is, committed to big ideas—than they are in North America. This dates back to the last century when Marxist parties represented labor and were opposed by conservative or Christian parties. The usual two-party system in Western Europe thus has a Socialist (or Labour) party versus a conservative or Christian-Democratic party. Some countries, however, especially France and Italy, have many smaller parties, as well. Having lived through the horrors of the dictatorships of the 1930s, Europeans are strongly in favor of democracy and often seem to be afraid to give any one party too large a majority.

As Table 5.4 shows, European governments have accepted much of the responsibility for human welfare in their countries. The lowest figures are for Greece (a poor country) and Switzerland (a very wealthy country!). *Cradle-to-the-grave* support programs are usually associated with Sweden, but the figures show that most of the countries have similar policies. In most countries, health care is a part of the social system and is available to all. Of course, all of this requires massive funding, and tax rates tend to be high in Europe.

FIGURE 5.6

Religions of Western Europe

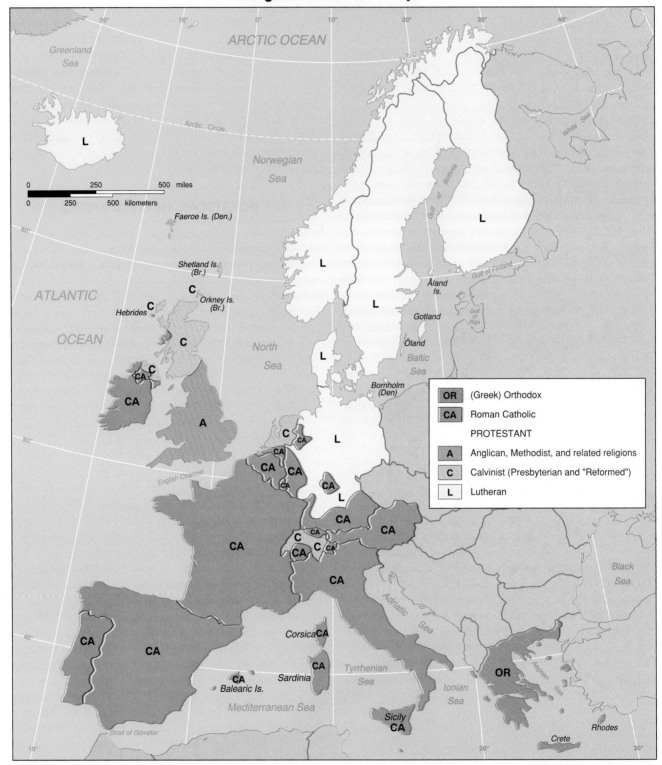

TABLE 5.4

Social Expenditures of Western European Countries

	SOCIAL EXPENDITURES AS A PERCENTAGE OF GDP, 1981	TAX RECEIPTS AS A PERCENTAGE OF GDP, 1980
United States	21.0%	30.7%
Canada	21.7	32.8
Austria	27.9	41.5
Belgium	38.0	42.5
Denmark	29.0	45.1
France	23.8	Not available
West Germany	31.5	37.2
Greece	12.8	Not available
Ireland	27.1	37.5
Italy	29.1	30.1
Netherlands	36.1	46.2
Norway	27.1	47.4
Sweden	33.5	49.9
Switzerland	14.9	30.7
United Kingdom	24.9	35.9

SOURCE: *Organization for Economic Co-operation and Development (OECD) Observor*, 114 (January 1982), 126 (January 1984).

AGRICULTURE

The modern world economy, with its emphasis on industrialization, research, and trade, began in Western Europe. Although it suffered a severe setback at the end of World War II, Europe recovered quickly with American help and is again one of the most active, productive, and prosperous parts of the world.

Agriculture remains relatively more important in Europe than in North America. Many Europeans vividly remember the starvation they had to endure during the final years of both World Wars. Parts of the continent have been repeatedly crossed by armies, and many people view the land as the only base of security. Further, there has been a strong tradition for the family holding to be passed on from one generation to another. Governments have supported these ties to the land not only because of the need for food in war time, but also because the peasantry has been seen as a conservative force, true and loyal to state and church, and as the bearers of the national folk culture.

During the past three decades, the agricultural policy of the European Community (EC) or Common Market has been to subsidize farmers and to set guaranteed prices for most crops. One result has been that Europeans pay relatively more for food, especially meat, than do North Americans. When this policy led to over-production, the EC bought up the surplus and stored it. Terms such as "the butter mountain" and "the wine lake" are sarcastically used now to describe the huge piles of stored foods. Occasionally, these are dumped at low prices. The EC seems to be slowly changing its policy in order to eliminate some of the worst aspects of the support system.

The core of the problem is that, overall, the Europeans are the world's most productive farmers. In an area about one-third the size of the United States or Canada, with a population much higher than the two combined, European agriculture is suffering from the problem of overproduction of certain commodities. France and Germany produce more than four times as much wheat per acre as that of the United States or Canada. This productivity is due to the intensity of land use, skill, and experience. Although machinery has replaced horse and oxen almost everywhere, much work is still done by hand, especially where steep slopes are used for fruit trees and vineyards.

Statistics given for the number of people engaged in farming are very misleading because many European farmers combine farming and factory work. According to the official statistics, fewer than 5 percent of the Dutch and only 3 percent of the Belgians are farmers; in truth, the numbers engaged in farming are much higher. Great Britain is the exception because there the

Dairy farming for the production of cheeses and other dairy specialty items that are higher in value, but lower in bulk than milk is an important activity in the Turbach Valley of the Swiss Alps.

division between city and farm is sharp, and fewer than 2 percent of the people are engaged in agriculture. That separation may help explain the fact that no nation romanticizes its landscape more than the English; in no other country do city people try to prevent farmers from cutting down hedges on the grounds that it would spoil the landscape.

European food is generally of very high quality. Europeans endeavor to have fresh, minimally processed food. Viennese whipped cream must be whipped cream, not some concoction of chemicals; butter must be pure butter; eggs must be truly fresh; cheeses are rarely processed and are loved in their hundreds of varieties. The governments enforce standards and geographical designations for food production. The farmers' cooperatives often not only handle marketing, but also grade the produce and guarantee the standards.

The location of the differing types of agriculture is influenced by the climate and soil, the taste preferences of Europeans, and the presence of large urban markets. The basic divide in Western Europe is between north and south, corresponding to the two main climatic regions, the marine west coast and the Mediterranean. The north is characterized principally by dairying, hardy grains, and apples; the south by wheat, olives, grapes, and other fruit.

In the British Isles, grazing of animals takes up most of the land. Dairying is important because of the home market and the dampness which precludes extensive crop agriculture. Cattle are plentiful, but the peculiarity of Great Britain is the huge number of sheep on the hill lands of England, Wales, and Scotland. Alone among northern Europeans, the English have loved to eat lamb, and Britain has long been involved in the production of woolens. The southeast of England is relatively sheltered from clouds and is used for the growing of wheat and barley. (Barley is the principal ingredient of beer, the most popular drink of northern Europe.) The southwest has vast areas of apple orchards, grown as much for cider as for fruit. A few favored areas are devoted to berries for the production of the much loved jams and preserves. Contrary to common opinion, Irish potato production is quite low compared to other European countries.

The northwestern coast of the continent is devoted mainly to dairying. Since the French drink very little milk, most of the milk goes into some of the world's most famous cheeses. Apples are grown widely, and an alcoholic cider is the basic local drink in Normandy. Inland, the climate is drier, allowing for the productive wheatlands of France, the sugar beets of Belgium, and the bacon, ham, and egg farms of Denmark. The pig is favored in Germany and Denmark as an efficient source of meat, much of it made into sausages. The Netherlands is famous for its miles of flower farms and greenhouses.

Further south in the Hercynian hills, one finds the northern edge of grape production. The world's most highly prized and priced wines come from here, from Burgundy and Champagne in France, and the Rhine and Moselle valleys in Germany. In a somewhat similar climate are the Bordeaux wines of France and those produced on the Alpine foothills in Austria and Switzerland. Many kinds of fruit are grown as well—for the table, for jams and perserves, and for liqueurs.

The polar peripheries of Europe have both poor soils and poor climate. The northern two-thirds of Norway, Sweden, and Finland are almost without agriculture.

Despite their ruggedness, the Alps are intensively used. In North America similar areas have been left as wilderness, but in Europe the mountain valleys have been used for centuries. The use of mountain slopes and valleys has led to **transhumance**, the practice of moving the dairy cattle up and down the slopes with the seasons. Since the fluid milk cannot easily be brought down the mountains, cheese making has been the rule. The hard facts of modern economics have made Alpine farming difficult, and many of the more remote Swiss mountain farms have been abandoned in recent years.

For thousands of years, the farmers in southern Europe have relied upon those crops that could survive the dry, hot summer. The three traditional crops have been wheat, the olive, and the grape. Wheat and barley are planted in the fall, grow with the winter rains, and then mature in the early summer. Both the olive tree and the grape have long roots, and the olive has sclerophytic leaves. The olives are pressed for cooking oil, and the grapes supply the basic drink of the south, red wine. Because of the hot sun, the wines tend to be heavier or sweeter than those in the north, and two higher alcohol or fortified wines, sherry and port, are major exports of Spain and Portugal, respectively. Grazing favors sheep and goats, which can survive on scrubby vegetation and need little water. Southern Europe also has a number of rich oases, which supply rice, early vegetables, citrus and other fruits, and melons to the northern markets.

MANUFACTURING

The earliest manufacturing centers of Europe had developed by 1100 in Flanders and in the Italian cities. The industrial revolution (1770–1900), however, began in Great Britain. Industry concentrated on the source of the power needed to turn the machines and smelt the metals, relying first on water power and later on coal. Both of these were located around the edges of the hills in England and Scotland. Textile manufacturing led the way, using first native wool and then cotton imported from the southern United States. The invention of the steam engine and the steam hammer led to the growth of heavy manufacturing and of railroads, which transformed transportation. England became the world's greatest industrial nation, Liverpool one of its busiest ports, and Manchester rivaled even London in economic importance.

In the twentieth century when manufacturing has come to stress consumer products, industries have increasingly located in or near the largest cities. For many products, proximity to market is more important than closeness to resources, and many of the older regions, located near the coal mines, are depressed. Further, with steadily improving transportation, especially by

Due to the climate and the shallow, rocky soils, agricultural productivity is limited in much of Ireland. The grazing of livestock in stone-fenced areas is often the best use of the land and the stones.

trucks on the superb superhighways, the cost of shipping has become increasingly less important. Many industries therefore tend to locate in areas where the owners and the highly trained employees prefer to live.

Within the British Isles, London is now the principal manufacturing center because of its command of port industries, auto assembly, publishing, clothing, and the host of industries catering to a large rich market. To the northwest, a major industrial belt extends through Birmingham, specializing in hardware and other metal products, and Stoke-on-Trent, famous for its chinaware, on to the older textile areas straddling the Pennine hills, this area produces the woolens of Leeds to the east and cottons around Manchester to the west. Despite a shift to electrical machinery and communications, "the North" is a severely depressed area. When Britain's principal ties were overseas, these manufacturing centers were close to their markets via their port city, Liverpool; now that ties to the European continent are stressed, the cities of the North find themselves at the back of England, and both they and their port, Liverpool, have suffered badly. Further north or northwest yet are Newcastle, England, Glasgow, Scotland, and Belfast, Northern Ireland. These three cities had all stressed shipbuilding; Glasgow built the huge *"Queens"*; Belfast the *Titantic*. The decline of the passenger ship trade and severe competition from more modern shipyards in

other countries have closed these yards, except for a few government contracts.

The greatest concentration of manufacturing in Europe is along and close to the lower Rhine River. One of the largest industrial concentrations in the world, the Ruhr Valley, developed a century ago on Germany's principal coal field. Iron and steel still form the base, but the manufacturing has diversified into all sorts of metallurgical products, transport equipment, chemicals, and textiles. Cologne, Düsseldorf, Essen, and Dortmund are the principal centers, and the western capital, Bonn, is only a few miles to the south. Downstream are the many industrial cities of the Netherlands, and just to the west, in Belgium and northern France, is the descendant of medieval Flanders, now engaged in steel, chemicals, and textiles.

Branching out from this core concentration are three areas of high manufacturing and population. An area to the southwest extends through northern France to Paris, which is by far the largest manufacturing center of France. Southward from the core, an extension follows the Rhine and its tributaries to Frankfurt, Mannheim, Stuttgart, and the Saar-Lorraine-Luxembourg steel region. Eastward an industrial area follows the edge of the hills into former East Germany and to Berlin, once the greatest single manufacturing city of Germany and still important for electrical machinery. Now that Ger-

The Rhine River is the major industrial corridor and commercial route from western Germany to the North Sea.

GERMAN REUNIFICATION

The demise of the Iron Curtain between Eastern and Western Europe is resulting in events and ideas that were only fantastic notions a short time ago. One idea that fits this characterization is the reunification of West and East Germany. Germany was partitioned in 1949 as part of the aftermath of its defeat in World War II. East Germany became the satellite and buffer zone of the Soviet Union. This ensured that any war between East and West would occur on German rather than Russian land. The line between the two Germanys has been the focus of superpower alliances (NATO and the Warsaw Pact) and animosities. In spite of the rhetoric to the opposite, many in the West were comfortable with this arrangement because it diminished the long-standing problem of German militarism that had existed since the Teutonic Knights made war on the peoples of Eastern Europe in the thirteenth century. This sentiment is captured in the statement by the French novelist, Francois Mauriac, "I love Germany so much, that I am glad there are two of them."

The relaxation of Soviet hegemony over the entire region in 1989 made reunification possible. The unified Germany shares a common currency and national election. In less than a year East Germany became an integral part of the German Federal Republic. Berlin is the national capital with Bonn remaining as the home of the parliament.

The unified Germany has the largest population in Europe and the highest GNP. It ranks first in steel production and second in both grain and energy production.

In spite of the favorable response to reunification by many Germans, many obstacles exist. The staggering cost of unification to the German economy could counter the prospect of increased economic dominance in Europe for many years to come. East Germany had only one-quarter the population and one-half the land area of West Germany. Even with political unification the former East Germany is tied functionally in many ways to Eastern Europe. The people of the two Germanys had developed their own cultural perspectives that may not always be compatible. Europeans worry that the united Germany could dominate much of the European scene from the economic arena to the Olympic Games. Europe is concerned not only about economic and military domination by a unified Germany, but also that the new Germany will focus on Central Europe rather than fostering the unity and economic strength of Western Europe and the EC.

many is unified, manufacturing in former East Germany will undoubtedly change substantially. Although many of the manufacturing establishments may be considered outdated, engineering and chemical industries are two of the largest industries in terms of employment of the labor force and exports by value. Engineering plants making such items as heavy industrial equipment and machine tools are found in major cities throughout former East Germany, but concentrations can be found in Magdeburg, Leipzig, and Berlin. Hopefully, in a six to eight year period former East Germany's economy and industrial capacity can be integrated into West Germany and can become as competitive and modern.

One of the more interesting aspects of Germany's manufacturing is its lack of concentration. The automobile industry is perhaps the best example. Volkswagen's home is at Wolfsburg, a small city in the middle of northern Germany, but the assembly plants are in several cities, and parts manufacturing seems to be everywhere. The Opel is manufactured outside Frankfurt, the Mercedes-Benz and Porsche at Stuttgart, and the BMW at Munich. Although most manufacturing is still in the north, the greatest expansion has been around the attractive cities of Stuttgart and Munich in the south.

The Alpine region has a surprising amount of industry. Steel and chemicals are manufactured in the Austrian valleys and along the Danube at Linz. Zürich and Basel, Switzerland, are both prospering, the first in electrical machinery, the second in pharmaceuticals and chemicals. The famous Swiss watch industry is in the valleys next to the French border; it has suffered badly from Japanese competition, but is now recovering.

Southern Europe has one major industrial region, the Po Valley of western Italy. Milan is the dominant center; it is surrounded by satellite cities manufacturing almost everything exported from Italy. Nearby are Turin, the home of the Fiat car, and Genoa, the port and steel center. Similar to Milan in Italy is Barcelona in Spain. Like Milan it began in textiles, but has branched out into a variety of products. Spain also has some steel and metallurgy around two northern cities, Bilbao and Oviedo. Neither Portugal nor Greece has much manufacturing except for that associated with shipping.

St. Paul's Cathedral and the older section of the city are in sharp contrast to the modern skyscrapers in the background in this view of the London skyline.

There are two important manufacturing centers in Scandinavia, Copenhagen and Göteborg. The former, as the only large city in Denmark, includes most of that country's manufacturing; the latter is Sweden's main port and a shipbuilding, lumber, and automobile center. A line of smaller industrial cities lies between Göteborg and Stockholm, the capital, which is the publishing and consumer-products center. In Finland, manufacturing is concentrated in the capital, Helsinki, and in the lumber and paper town, Tampere.

THE GREAT CITIES

The larger European cities are far more than local economic or political centers. They are *world cities,* serving the entire globe. Because of their attractiveness, culture, and wealth, they draw in millions of people from around the world. Two cities, London and Paris, are preeminent in this regard.

London was founded by the Romans at the lowest possible bridging point on the Thames River. London Bridge was the point to which most of the major English roads led. It was also the furthest point on the river that the North Sea boats could reach. London became the center of government, the home of the monarchy, the busiest port, and one of the world's most important financial centers. Despite the end of the British Empire, London remains a great and dynamic city, a major attraction to those hundreds of millions of people around the world who treasure British traditions and literature and the pageantry of monarchy. The city structure shows a strong differentiation east to west. In the old "City," near London Bridge, is the financial center. To its west are the main shopping area, the hotels and theaters, the government buildings, parks, and palaces; to the east along the Thames are the old docking areas with their port industries and poor housing.

Paris was also founded by the Romans at a bridging point of the Seine River, and the old cathedral and the city hall arc still at the old bridge site. The **centralization** of France under both the monarchy and the Republic brought almost all the major highways and railroads of France to the capital. These factors made Paris a primary industrial center in France as well. The wealth of the monarchy established a level of culture that was copied all over Europe. Paris became the center for fashions in clothing, cuisine, art, and manners. It remains a magnet for all those peoples who speak French and is one of the most attractive great cities of the world.

A panoramic view of Paris showing the geometric patterns of tree-lined boulevards designed by Napoleon III.

Like London it shows strong internal differentiation. The famous shops and hotels and the new office towers are generally to the west of the old city core, whereas an old industrial belt forms a broad semicircle from the northwest, by north to east, along the Seine River and the railroad yards. South of the Seine River is the university district, also known as "the Latin Quarter" and "the Left Bank."

Several other cities perform special functions. Rome is one of the most historic cities in the world because of its wealth of ancient, medieval, and Renaissance structures. As Italy's national capital, it ties together the north and south. Vatican City, on the western outskirts, is the headquarters of the Roman Catholic church and makes Rome one of the principal pilgrimage centers of the world. To a lesser extent, Athens plays a similar historical role. Vienna is accepted in much of the world as the ultimate center for classical and operetta music and as a meeting place for Western and Eastern Europe. Despite crowding, pollution, and dangerous traffic, European cities and towns remain very attractive to natives and tourists alike. Europeans love their cities and are proud to be known as Parisians or Berliners.

Among the more interesting facts of urban life are the rivalries between two cities for dominance in a country. Usually, one city is the capital and the other the economic center. The latter's citizens resent that they work and pay the taxes while the capital collects the taxes and spends the money. The best example is in Italy where Milan, the industrial and banking center, challenges Rome. Similarly in Spain, Barcelona challenges Madrid, and in Portugal to a lesser extent, Porto challenges Lisbon. In Sweden, Göteborg is the worker, Stockholm the spender; in Scotland, Glasgow works while Edinburgh puts on the grand show. A century ago even London had a challenger in Manchester.

REGIONALIZATION: CORE AND PERIPHERY

From the preceding discussions of population, agriculture, and manufacturing, it is clear that there is a marked regional pattern to the economic organization within Western Europe. This pattern resembles a **core-periphery** model and also shows a **distance-decay** pattern. In this model, the "core" concentrates the decision making, the wealth, and much of the productive capacities; the periphery is treated as a kind of colonial territory. For Europe, the distance-decay concept implies that wealth and power steadily decline as distance increases from a center. Essentially, it means that the further a place is from the economic core, the poorer it will be.

Western Europe clearly has its core area around the mouths of the Rhine and Meuse rivers (Figure 5.7). Included are the Ruhr Valley; Rotterdam, the world's busiest port; Brussels, the headquarters of both the European Community and NATO; three major international airports; three capitals; other major ports such as Antwerp; and the greatest concentration of canals, railways, and superhighways in the world. Over 35 million people live here in an area of less than 25,000 square miles (64,750 square kilometers), half the size of Pennsylvania. To delimit the full core, however, we have to expand outward slightly to include London, Paris, and Frankfurt, the three principal financial centers of Europe. This core controls most of the European economy. In this small area, most of the great international companies have their head offices or European headquarters.

There is a secondary core for southern Europe. This is in northwestern Italy, with extensions eastward in northern Italy and along the Mediterranean coast (the Riviera) to Marseille, the principal port of France. A dense network of roadways links these northern and southern cores. Along the broad corridor between them one finds several important cities: Lyon, Stuttgart, Munich, and Zürich.

Beyond these cores and extensions of high productivity lie the less favored parts of Western Europe. As an overall generalization, the further outward one travels from the core, the poorer the area becomes. The poorest countries of Western Europe are Portugal and Greece, followed by Ireland and Spain. Although the Scandinavian countries are on the margin of Europe, they have high standards of living. These countries have low population densities and industrial economies, and their mineral and forest products are in high demand.

Great Britain is probably the best example of the distance-decay function in operation. London dominates the British government, culture, economy, and even tourist trade to the extent that the rest of the country seems to be out of the mainstream. There has been a gradual flow of people from the north toward London not only to seek employment, but also for retirement. The cities within 150 miles of London are rather prosperous and manufacture most of Britain's machinery and auto and aircraft parts. Further out to the north are the old factory towns. Ultimately, one reaches the northwestern highlands of Scotland and troubled Northern Ireland. The British government has tried to encourage industries, especially branches of foreign companies, to settle in these peripheral locations, but has had only slight success. A small computer components "Silicon Valley" has developed east of Glasgow, but attempts to make Belfast an auto manufacturing center (the DeLorean car) have been failures.

Such problems are far less severe in Germany and France than they are in Great Britain. However, the belt along the former Iron Curtain has been a zone of economic decline. This could be changed by German unification, however. In France the Massif Central in the south and Brittany in the northwest are relatively underdeveloped, and the steel belt of Lorraine has been suffering the closure of mills and mines.

The most serious problem areas are in southern Europe, where one also finds the greatest contrasts. The coastlines of Spain swarm with vacationers, but only a few miles inland are some of the poorest farming areas of Europe. Among the poverty areas of Europe are the interior of Portugal, the southern and western interior of Spain, southern Italy, and, perhaps the most remote of all, the interior of Greece. In all these areas, a poor version of subsistence agriculture is still the way of life, and landholding is a problem. Many of the villages are almost devoid of young men as they have gone off to Milan, Barcelona, Athens, or northern Europe in search of work. True to the distance-decay principle, the northern areas are the most prosperous and the southern the poorest in Portugal, Spain, and Italy.

The various governments and the European Community have made attempts to solve their regional problems. The strongest efforts have been made by Italy. **Growth poles** have been selected, and industries established, among them the steel and chemical complexes at Naples and Taranto. This has been accompanied by land reform, the improvement of roads, the construction of service buildings, and an emphasis on education and sanitation. The national banking and business are still controlled from the north, however, and young people continue to move toward Milan or other countries. This movement is eased by the European Community policy of allowing all its people to work in any member country.

INTERNAL REGIONALIZATION

In common perception, Western Europe divides into three superregions: the West (or Northwest), the North, and the South. The West has most of the plains and the

FIGURE 5.7

The Economic Structure of Western Europe

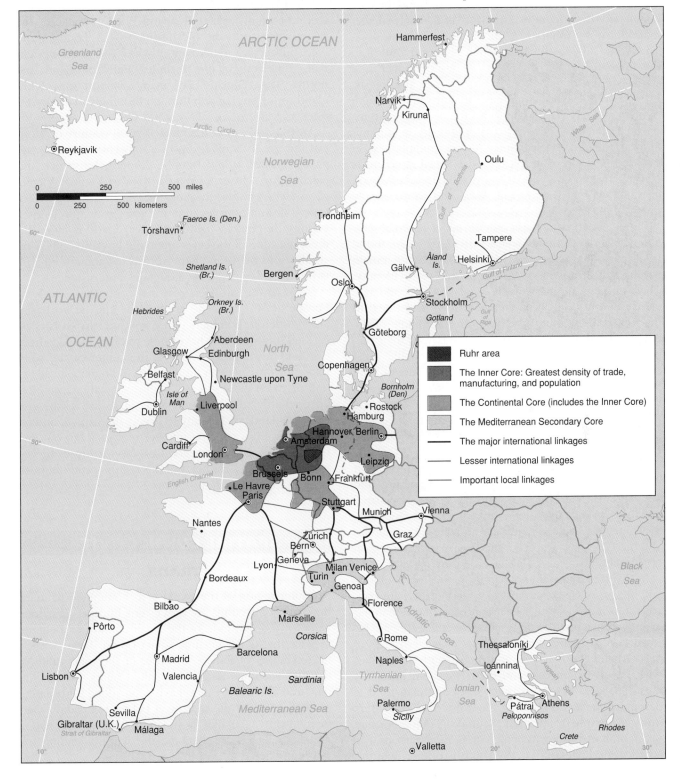

Legend:
- Ruhr area
- The Inner Core: Greatest density of trade, manufacturing, and population
- The Continental Core (includes the Inner Core)
- The Mediterranean Secondary Core
- The major international linkages
- Lesser international linkages
- Important local linkages

marine west coast climate and faces the North Atlantic. It has two-thirds of the population and most of the industry and wealth of Europe and is the most heavily committed to world trade. The North is distinctly smaller in population, although it encompasses a large area. The North supplies lumber, paper, fish, ore, and oil to the huge market of the West. Its northern half is a sparsely populated area of hard rock, lakes, and forests. It has a cold climate with great variations in the length of night and day. The South forms the northern margins of the Mediterranean Sea. It is hot and dry in summer and cool and damp in winter. It is incredibly rich in history but poor economically. The social structure is conservative and traditional and is still based heavily on agriculture. Tourists from the West and North supply much of the income.

Individual countries also exhibit major regional variations. Even in Germany where thriving areas are found in all parts of the country, regional differences remain strong in dialect, way of life, and people's local identity. The north has the plain with its glacial soils and raw climate. Giant population clusters have formed next to the lower Rhine and at the ports. In the middle of the country are the forested Hercynian hills with their picturesque towns, castles, and vineyards and the great corridor of the Rhine gorge. The south is interior with a more continental climate. The broad Bavarian plateau, noted for its agriculture, leads to the tourist towns in the Alps. Although industry and agriculture are found everywhere, the common perception remains of an industrial, Lutheran north and a peasant, agricultural, Catholic south. Frankfurt is the principal center of the transport network, but the largest cities are Berlin, Hamburg, and Munich at the furthest ends, and the core of industrial life is in the Ruhr (Figure 5.8). In former East Germany, the impact of the communist period on ideas, culture, and the economy will make this area a separate region of Germany for years to come.

In France, almost all the head offices of corporations and banks, as well as almost half the industry, are concentrated in and around Paris. Most of the remaining manufacturing and the most intensive agriculture are located north of Paris. France is thus characterized by a major difference between north and south and somewhat less between east and west. Northern France is mostly flat and has the marine west coast climate. It is densely populated, although the western coast suffers from its peripheral location. Southern France is quite different. It is warmer and much more mountainous. Except for the Riviera coast, the south is poorer than the north. The principal transport axis runs from Paris through Lyon to Marseille.

Great Britain divides readily in two. The southeastern half, commonly called "lowland Britain," includes the richest farmlands, the old historic towns, London, and the south coast retirement towns. The northwestern half, "highland Britain," includes Wales, Scotland, and northwestern England. The main transport axis of England runs from London through Birmingham to Manchester and Liverpool.

Austria is the most fragmented small country of Western Europe. It is made up of a collection of mountain valleys, joined to the Danube corridor. Innsbruck and Salzburg are closer to Munich than to Vienna. Vienna is, in fact, in a fine location for links with Eastern Europe, but in a poor location for linkages within its own country. In contrast to Austria, Switzerland is more unified, because all the cities are in the lowland with good connections between them. Norway, on the other hand, is heavily fragmented by mountains and distance.

Some of the strongest regional contrasts in Western Europe are in Italy. The north is the alluvial Po Basin, an area rich in both agriculture and industry. It is crisscrossed by routes that have come through the Alps from Germany and France. Milan is the center of the web, but is far off-center in the country. Bologna acts as the link between the north and peninsular Italy. From Florence southward, one major route leads across the volcanic hill lands of Tuscany to Rome, the capital, and on to Naples, from which a number of routes fan out into the south. Rome is in a good position to try to tie the north and south together, but it is not the center of business and transportation.

SPATIAL CONNECTIVITY

TRADE

No other region of the world is as committed to international connections as is Western Europe. The total value of foreign trade of the small country of Belgium is twice that of the gigantic Soviet Union. Table 5.5 lists the countries of Europe in order of the total value of their international trade. The United States, Canada, Japan, and the Soviet Union have been added for the sake of comparison. Note that even before unification Germany, with a smaller population than Japan, had a higher total and has become the world's leading exporting nation.

FIGURE 5.8

Internal Connections: France, the United Kingdom, Spain, West Germany, Italy, and Austria

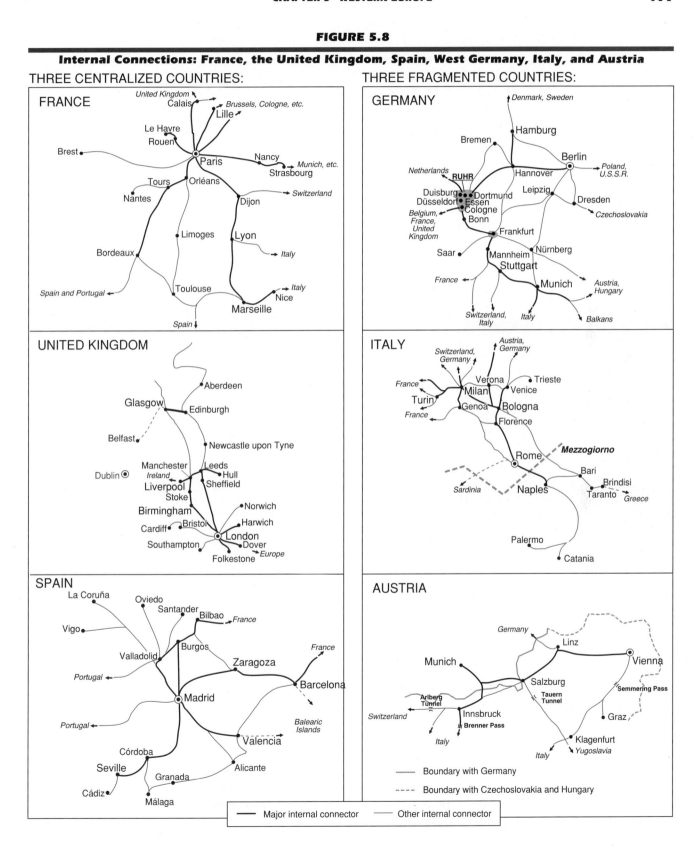

THREE CENTRALIZED COUNTRIES:

THREE FRAGMENTED COUNTRIES:

TABLE 5.5

Trade of Western European Countries, 1988

	TOTAL (MILLIONS OF U.S.$)	EXPORTS (MILLIONS OF U.S.$)	IMPORTS (MILLIONS OF U.S.$)	TOTAL TRADE AS A PERCENTAGE OF EUROPEAN EXPORTS PLUS IMPORTS	PER CAPITA (U.S.$)
EC					
1. West Germany	$573,939	$323,374	$250,565	22.84%	$ 9,409
2. France	346,634	167,783	178,851	13.80	6,234
3. United Kingdom	334,505	145,166	189,339	13.31	5,889
4. Italy	267,120	128,529	138,591	10.63	4,654
5. Netherlands	202,638	103,194	99,444	8.06	13,879
6. Belgium-Luxembourg	184,353	92,103	92,250	7.34	17,898
7. Spain	100,872	40,341	60,531	4.01	2,587
8. Denmark	54,425	27,877	26,548	2.17	10,672
9. Ireland	34,302	18,736	15,566	1.37	9,801
10. Portugal	26,664	10,556	16,108	1.06	2,587
11. Greece	19,701 *	6,533*	13,168*	0.78	1,970
NON-EC					
1. Switzerland	107,114	50,624	56,490	4.26	16,229
2. Sweden	95,374	49,747	45,627	3.80	11,354
3. Austria	67,024	31,022	36,002	2.67	8,819
4. Norway	45,241	22,086	23,155	1.80	10,772
5. Finland	43,066	21,930	21,136	1.71	8,789
6. Iceland	3,021	1,424	1,597	0.12	12,231
7. Cyprus	2,643	763	1,880	0.11	3,887
8. Malta	2,065	711	1,354	0.08	5,975
United States	781,170	321,600	459,570	—	3,273
Canada	225,583	116,841	108,742	—	8,812
Japan	452,234	264,856	187,378	—	3,699
Soviet Union	92,913	41,328	51,585	—	327

* 1987.

SOURCE: International Monetary Fund, *Direction of Trade Statistics, Yearbook* (1989).

Perhaps more meaningful is the final number, which shows the value of trade per capita for each country. Here the small countries of the north stand out. Of course, it is easy for an active small country to attain a high figure since any goods that move over one hundred miles from Rotterdam, Brussels, or Zürich are almost sure to cross an international boundary, whereas many goods that travel the same distance in Germany or France stay inside the country and hence are not counted. Nevertheless, the extraordinarily high figures for Belgium and the Netherlands show quite clearly that the Low Countries are the focal area of the European trading system. Among the larger countries, Germany is the leader. Denmark, commanding the exit from the

Baltic Sea and the crossing over to Sweden and Norway, is a major center of interchange.

The European Free Trade Association (EFTA) nations also achieve very high values. Switzerland, on the major north-south routes of Europe, has long been famous for being a "free trader." Norway's figure has been raised recently by the export of oil from the North Sea. Sweden is noted as an exporter of quality products. In contrast, note the low values for the poorer countries of the south. Greece imported twice as much as it exported. Unfortunately, the value of farm products can rarely equal the cost of imported automobiles and machinery.

With a few exceptions, notably Germany, most of the countries import more than they export. They are not afraid to do this because it is the level of the total trade as much as any surplus gained that accounts for the prosperity of the trading countries. In any case, the difference is made up in shipping earnings, bank charges, dividends on investments, and, in some cases, tourism earnings.

A breakdown of the value of the trade among the European countries is given in Table 5.6. The importance of the proximity of these countries to each other is clear from these numbers. Europe may be small, but countries do tend to trade heavily with their closest neighbors. All the Scandinavian countries trade heavily with each other. Similarly, Austria's trade is dominated by Germany. The removal of the Iron Curtain and the changes in economic policy of the Eastern European countries have opened new trading possibilities for Western Europe.

The developing world group includes both the major oil-exporting nations and the truly poor countries. Much of the trade of the European countries with this part of the world involves the importation of petroleum. This helps explain the high numbers for Italy and Spain and the fact that imports exceed exports for the Netherlands and Germany. In contrast, the United Kingdom, which has its own oil and obtains both oil and gas from Norway, has a low percentage of imports from the developing world. Former colonial ties are significant too because in most cases the colonial power still has trading advantages in the former colonies and may also favor imports from those ex-colonies. This is clearly the case of Italy with Libya, France with North Africa, and of the United Kingdom with the Commonwealth countries. The high percentages for Spain may be due to the fact that all the other Spanish-speaking countries are part of the developing world.

Probably the most obvious fact to be seen in this table is the importance of Germany to the economy of Western Europe. Of the 12 European countries listed, 10 of them import more from Germany than from any other country in the world. Germany has often been called the "locomotive" of the European Community. Considering its enormous importance and potential, it could become the "locomotive" of Western and Eastern Europe.

German industry is extremely varied and highly skilled. Automobiles are probably the most important single export, but the range of heavy or intricate machinery is impressive. The pipe for the huge pipeline carrying Soviet natural gas from Siberia to Central Eu-

rope was made in Germany, and Germans built both the new Moscow airport and the hotel for the 1980 Moscow Olympics. The Germans are experts in monorail systems and roller coasters. They produce some of the world's heaviest mining equipment and are noted for their electrical machinery and medical technology. Chemicals, shipbuilding, heavy textiles, automobiles, and communications and computer technology all rate highly. Germany is a major partner in the European rocket and satellite program. At one time, the Germans and French were the bitterest of enemies, but they are now very close economic partners; much of the success of the European Community is due to their fruitful cooperation.

THE MOVEMENT OF PEOPLE

The European Community is as devoted to the free movement of its citizens as it is to the movement of goods. Any citizen of an EC member country may work in any of the member countries. In a sense this has merely made easy what was already a pattern. Italians, Spaniards, and Greeks had long been working in France and in Germany. It has become common for young Britons to work in the Bordeaux wine harvest. Perhaps the best sign of this freedom of movement has been the appearance of Italian restaurants all over Western Europe.

Another form of human movement of great importance is tourism. Tourism allows some of the smaller or poorer European nations to balance their budgets and has supplied the funds for economic development and social improvement.

It is important to realize two facts, which are not fully understood in North America. First, most tourists in Europe are Europeans. Americans form only a small minority of the visitors and are no longer the most conspicuous; the Germans have become the archetypical tourists. Second, Europeans want the sun when they go on their summer vacations. August is the main vacation time in Europe. In that month, literally tens of millions of north Europeans head south for the Mediterranean. For several decades, Spain has been the favored holiday country of Europeans. In particular, they pour onto the island of Majorca and the Costa del Sol in the south. The south coast of France west of the Riviera has become the site of thousands of condominiums. In Italy the most crowded areas are along the continuous beach that lies along the Adriatic south of Venice.

TABLE 5.6

Trade among West European Countries, 1988

	West Germany		France		United Kingdom		Italy		Netherlands		Belgium-Luxembourg		Spain		Denmark	
	EXP to (%)	IMP from (%)	EXP to (%)	IMP from (%)	EXP to (%)	IMP from (%)	EXP to (%)	IMP from (%)	EXP to (%)	IMP from (%)	EXP to (%)	IMP from (%)	EXP to (%)	IMP from (%)	EXP to (%)	IMP from (%)
EC																
West Germany			15.5	19.3	11.5	16.7	16.8	21.7	25.0	25.8	19.0	24.9	12.3	16.4	17.7	22.9
France	12.3	12.9			10.4	8.7	16.4	15.0	10.8	7.4	19.9	15.6	18.5	13.7	5.6	4.9
United Kingdom	9.2	6.8	9.3	7.5			8.4	5.3	10.9	7.8	9.8	7.2	10.2			
Italy	9.1	8.9	12.0	11.1	5.2	5.9			6.4	4.0	6.6	4.2	9.2	9.7	4.6	3.9
Netherlands	8.2	10.2	5.1	5.1	6.5	7.6	3.0	5.6			14.3	18.0	4.7	3.1	3.7	6.0
Belgium-Luxembourg	7.0	6.9	8.6	9.0	5.3	4.7	3.1	5.0	14.4	14.1			3.3	3.4	1.9	3.2
Spain	3.2	2.1	5.1	4.3	3.7	2.5	4.4	2.4	2.2	1.4	2.5	1.6			1.6	1.1
Denmark	1.9	2.0	—	—	1.3	1.9	—	1.0	1.9	1.1	—	—	—	—		
NON-EC																
Switzerland	5.7	4.5	3.9	2.5	2.1	3.8	5.0	4.4	2.0	1.4	2.2	1.7	1.6	1.5	2.3	2.0
Sweden	3.1	2.5	1.4	1.6	2.9	3.2	1.3	1.5	2.0	2.3	1.7	2.3	1.1	2.1	11.9	12.5
Austria	5.6	4.3	1.0	—	—	—	2.3	2.3	1.1	—	1.0	—	—	—	—	1.2
Norway	—	1.3	—	1.0	1.3	2.5	—	—	1.1	1.1	—	—	—	—	6.6	4.8
United States	8.7	6.5	6.2	7.0	13.0	10.2	8.9	5.3	4.6	8.0	4.9	4.4	8.1	9.9	6.0	6.4
Canada	—	—	—	—	2.5	1.9	1.1	—	—	1.0	—	—	1.2	—	—	—
Japan	2.6	6.4	1.6	4.1	1.9	5.8	1.9	2.5	—	3.2	1.1	2.2	1.3	5.5	3.6	4.2
Soviet Union	1.6	—	1.5	—	—	—	1.5	—	—	—	—	—	—	—	—	—
Developing world	11.6	13.4	16.3	12.9	17.6	12.1	15.5	17.0	9.3	13.5	9.7	10.4	14.1	16.7	10.5	9.5

Values less than 1.09 percent of Total Trade are not included. Only those countries having greater than $45 billion in trade are included.

SOURCE: International Monetary Fund, *Direction of Trade Statistics* (June 1989).

In contrast, relatively few North Americans travel to Europe only for the beaches. Rather, they head for the famous cities, with the result that they think Europe is swarming with nothing but North Americans and Japanese. Standing, crowded together in the Louvre or Notre Dame de Paris, they may wonder where all the French are—they are all on the Mediterranean shore or in the Alps.

Table 5.7 lists the tourist arrivals in Europe in 1986. Unfortunately, not all countries report this figure every year, and definitions of tourist vary as well. Nevertheless, it is clear that France, Italy, and Spain receive the greatest number, although which is highest varies from year to year. Note that the number for Austria is higher than the total for Switzerland. The United States and Canada are added for comparison.

The second column shows the gross receipts from tourism for each country. The citizens of these countries also spend money in other countries, however. The more significant figure, therefore, is the net receipts in the third column. Spain and Italy have the highest "net" totals by far; these billions of dollars give their governments huge amounts of money to use for development. Note the disparity between northern and southern Europe. Every country north of the Alps, except Ireland, runs a deficit in tourist expenditures. The Germans, in particular, spend billions more per year on tourism in other countries than is taken in by the German tourist industry. In a way then, tourism is an important way of redistributing money from the wealthy to the poorer countries of Europe.

In the final column, the net receipts have been divided by the population of each country to obtain a per capita receipts figure. This gives us the best idea of the value of tourism to the economy. Given these statistics, Austria becomes the leader in tourism. Austria is blessed with both summer and winter tourism. In summer Austria has the Alps and their lakes, the great city, Vienna,

TABLE 5.6 (continued)

Trade among West European Countries, 1988

| | NON-EC | | | | | | | |
| | Switzerland | | Sweden | | Austria | | Norway | |
	EXP to (%)	IMP from (%)	EXP to (%)	IMP from (%)	EXP to (%)	IMP from (%)	EXP to (%)	IMP from (%)
EC								
West Germany	20.5	34.6	11.8	21.7	33.8	46.5	11.4	12.3
France	9.4	10.7	5.2	4.9	4.6	4.3	7.0	3.0
United Kingdom	7.4	5.4	11.4	9.1	4.5	2.6	23.2	7.3
Italy	8.6	10.3	4.0	3.9	10.1	9.2	3.4	2.8
Netherlands	2.7	4.4	4.6	4.1	2.6	2.8	6.6	4.0
Belgium-Luxembourg	2.0	3.3	3.9	3.4	2.3	2.7	2.4	2.5
Spain	1.9	1.3	2.3	1.2	1.9	—	—	—
Denmark	1.1	1.1	6.5	6.5	—	—	5.8	7.3
NON-EC								
Switzerland			2.5	2.1	6.9	4.4	1.0	1.5
Sweden	1.9	2.2			2.1	2.1	12.8	16.2
Austria	3.6	3.9	1.3	1.4			—	1.0
Norway	—	—	8.6	6.1	—	—		
United States	8.9	5.4	10.1	7.5	3.4	3.6	5.9	5.5
Canada	1.2	—	1.7	—	—	—	1.4	1.9
Japan	3.8	4.9	1.9	6.0	1.3	4.8	2.1	4.0
Soviet Union	1.1	—	—	—	3.4	—	—	—
Developing world	17.1	8.7	10.2	6.4	7.7	7.0	8.7	16.9

and the exquisite small cities, Salzburg and Innsbruck; in winter it has become the premier skiing country of Europe. Spain ranks second, followed by Greece, Switzerland, Italy, and Portugal. The Scandinavian countries, Germany, and the Netherlands run the highest deficits.

There is another interesting type of human movement, the training of foreign students in European schools. Table 5.8 shows the number of foreign students who were studying in each country. France has by far the highest number of foreign students, mostly because it is the undisputed center for learning in the French language. The citizens of most of the former French colonies in Africa come to France for their advanced education. The United Kingdom is second highest, because it is the obvious place for students from its former colonies to go. However, Britain faces competition from the United States, Canada, and Australia and hence falls far behind France. Considering that it has had no colonies since 1914, Germany's high position is surprising,

and may be due to its good reputation and to its willingness to give grants to foreign students. Italy has a smaller colonial heritage, but is an attractive place for persons studying the arts; Florence has several "branch plants" of American universities. The other countries are well behind. Spain has, of course, its ties with Latin America. Thus, France, Great Britain, and Germany are performing a major service to the developing world by helping to train their brighter young people.

POLITICAL TIES

Since the end of World War II in 1945, Europe has steadily tended toward unity. At the present time, every major country of western Europe is a member of one of two economic groups, the European Community (Common Market) or the European Free Trade Association (EFTA). The beginnings can be traced back to the end of World War I when Belgium and Luxembourg decided

TABLE 5.7

Tourism in Western Europe, 1986

	ARRIVALS AT FRONTIER OF TOURISTS FROM ABROAD	RECEIPTS (MILLIONS OF U.S. DOLLARS)		NET PER CAPITA (U.S. DOLLARS)
		Gross	Net	
Austria	15,000,000*	$ 6,076	$2,819	$373
Belgium	7,000,000*	1,970*	−870	−88
Denmark	4,000,000*	1,759	−360	−70
Finland	500,000*	596	−464	−95
France	36,080,000	9,704	3,200	58
West Germany	12,000,000*	7,826	−12,838	−210
Greece	7,024,779	1,834	1,340	135
Ireland	2,378,000	634	133	37
Italy	53,314,936	9,855	7,097	125
Netherlands	3,500,000*	2,228	−2,742	−189
Norway	1,250,000*	991	−1,435	−346
Portugal	5,409,201	1,574	1,242	123
Spain	46,664,622	9,680*	8,480	217
Sweden	3,500,000*	1,540	−1,271	−152
Switzerland	10,000,000*	3,730*	870	135
United Kingdom	13,844,000	7,921	−765	−14
United States	25,358,501	12,913	−4,714	−20
Canada	15,660,500	3,860	−434	−17

* Estimate, based on 1982 data

SOURCE: World Tourism Organization, *Yearbook of Tourism Statistics* (1987).

to form a customs union, the Belgium-Luxembourg Economic Union, or BLEU. During World War II, the Belgian and Dutch governments-in-exile in London agreed to work toward a union. This was achieved after the end of the war and became known as Benelux. At the end of the war, a few statesmen of Germany and France came together to try to prevent any future French-German conflict. They hoped for ultimate unity, and the economic route seemed the easiest to take. To begin, they agreed to drop all tariffs and restrictions on the movement of coal, iron, and steel. Six countries agreed: the three Benelux countries plus France, Germany, and Italy. The European Coal and Steel Community, or Schumann Plan, was a great success, and the six partners were soon looking for ways to expand the plan. At this point, Great Britain, as the leading country of Europe, was invited to join. Britain refused, for it felt that its ties were principally overseas and that it was a Great Power in its own right. The six therefore established the European Economic Community (EEC or Common Market) on January 1, 1958. Great Britain gathered the other countries of Europe into a loose organization called the

European Free Trade Association. Three neutrals—Sweden, Switzerland, and Austria—joined as did Denmark, Norway, and Portugal because of their close ties to Britain.

Economically, the EEC (now EC) prospered enormously, since Germany, France, and Benelux formed an obvious economic core. Within a decade, Great Britain recognized that it had made a mistake; it decided to try to leave EFTA and join the EEC. In 1973 Britain entered the EEC along with Denmark and Ireland. Norway was also supposed to enter, but in a heated referendum, the Norwegians voted against entry; they feared that their identity would be swamped in the larger community. Subsequently, Greece entered in 1980, and Spain and Portugal in 1986. The EC is advancing toward complete economic unification in 1992. Despite Britain's leaving, EFTA continued and now includes Sweden, Norway, Finland, and Iceland (all the Scandinavian countries except Denmark) and the two Alpine neutrals Switzerland and Austria. Relations between the two groups have been friendly, and both Switzerland and Austria have special relationships that make them almost semimembers of the EC. Other EFTA members are moving toward membership in the EC.

The unity of Western Europe has also had a military dimension. Most of the countries are joined together with the United States and Canada in the North Atlantic Treaty Organization or (NATO), an alliance whose purpose is to protect Western society from a possible Soviet threat. Except for the neutrals, Sweden, Finland, Switzerland, and Austria, all of the Western European countries have been in NATO, although the degree of their participation has varied from time to time.

On the broader international scene, Europe has played a large role in the United Nations (UN). Despite the fact that Switzerland has refused to join the UN, Geneva has been the principal European headquarters of the organization. A host of related international organizations also have their headquarters in Geneva. During the past 30 years, a second UN city has grown in Europe, Vienna. The Austrian capital has the advantage of being located right on the boundary between "the West" and the former Communist Bloc. Both cities are used frequently for international meetings.

Being one of the wealthiest regions of the world, Europe has felt the responsibility to assist in the development of the poorer countries of the globe. Table 5.9 lists the total aid given by the Western Europeans to the developing world. The pattern matches that of the foreign student numbers, reflecting the link to former colonies. Again France is first, Britain second, and Germany third. France and Great Britain clearly feel an obligation to help their former colonies advance. Belgium's total is high for so small a country, but it too had a large colony; Switzerland is the wealthiest country, hence its high per capita figure.

PROBLEMS AND PROSPECTS

In world terms one would have to say that the prospects for Europe look good. The subcontinent has weathered the oil crises of the 1970s and the recession of the 1980s and has bounced back well. None of the countries is seriously in debt. Rather the problem is in the opposite direction. Many of the larger banks have lent huge amounts of money to countries that will evidently not be able to repay them. Overall, the prosperity of Europe

TABLE 5.8

Foreign Students in Western European Universities and Technical Colleges, 1978

HOST COUNTRY	NUMBER	PERCENTAGE OF THE WORLD TOTAL
World	842,705	100.0%
Europe	360,456	42.8
Austria	11,416	1.3
Belgium	17,822	2.1
Denmark	2,893	0.3
France	108,288	13.3
West Germany	55,337	6.6
Greece	8,647	1.0
Holy See (The Vatican)	7,867	0.9
Ireland	2,079	0.2
Italy	26,663	2.9
Netherlands	3,851	0.5
Spain	9,853	1.2
Sweden	about 4,000	0.5
Switzerland	13,176	1.6
United Kingdom	59,625	6.9
United States	263,940	31.3
Canada	26,285	3.1
Soviet Union	62,942	7.5

SOURCE: UNESCO, *Statistics of Students Abroad, 1974–1978.*

TABLE 5.9

Assistance Given by Western European Countries

	AMOUNT (MILLIONS OF U.S. DOLLARS)	PERCENTAGE OF WORLD TOTAL	GIVEN PER CAPITA (U.S. DOLLARS)
World total	$82,182		
Western Europe total	40,737	49.3%	
Austria	401	0.5	$ 53.09
Belgium	2,794	3.4	283.57
Denmark	871	1.1	170.38
Finland	142	0.2	29.16
France	10,451	12.6	191.62
West Germany	7,342	8.9	119.54
Italy	3,323	4.0	58.37
Netherlands	1,969	2.4	136.78
Norway	620	0.8	149.98
Sweden	1,349	1.6	161.93
Switzerland	1,822	2.2	283.67
United Kingdom	9,653	11.7	172.81
United States	24,996	30.2	110.34
Canada	3,449	4.2	141.68
Japan	12,128	14.7	107.00

SOURCE: United Nations, *U.N. Yearbook* (New York: United Nations, 1982).

depends heavily on the free movement of trade, especially the importation of raw materials and fuels.

Europe's problems and prospects center on its energy dependency and associated problems, such as pollution. It appears that its ability to cope with these problems and enhance its prospects lies in the continuation of the unification of the European countries.

ENERGY AND POLLUTION

Of the countries of Western Europe, only Norway and Great Britain are major exporters of energy, both of oil from the North Sea. The Netherlands, which possesses natural gas, has a slight excess of exports over imports. All the other nations are importers. The three leading importers are Germany, France, and Italy, followed by Spain and Belgium.

Most of the potential hydroelectric locations in Europe have been used; damming still more valleys would upset the ecology of the Alps and interfere with both farming and tourism. The burning of coal is becoming an increasing concern because of the resulting air and water pollution. The importation of oil from the Arab countries or natural gas from the Soviet Union makes

the Europeans uncomfortable. They all suffered badly when the OPEC cartel suddenly cut off the flow of oil and raised prices in 1973. European countries would love to reduce their dependency on other nations for energy. But how? The only other known way of generating electricity cheaply on a large scale is through nuclear energy. As a result, Western Europe has made a large investment in nuclear generating stations. Despite the accidents in Pennsylvania and the Soviet Union, and the terrible question of what to do with the waste materials, some European countries continue to expand their nuclear capacities. Table 5.10 lists the countries according to their use of nuclear power; France is clearly first. Even Sweden, famous for its social conscience, adopted nuclear power. In a bitterly fought referendum, however, the Swedes voted to phase out all their nuclear plants by 2010. How they will replace the lost power is still unclear.

Unfortunately, the burning of coal, peat, and oil leads to water and air pollution. This pollution along with the other industrial sources of pollution has become a major worry for Europeans during the past 20 years. Probably the most serious water pollution problem now is the Mediterranean Sea. Industrialization and tourism are

TABLE 5.10

Use of Nuclear Power in Generating Electricity, 1986

	TOTAL ELECTRICAL POWER (MILLION KW HOURS)	NUCLEAR (MILLION KW HOURS)	NUCLEAR PROPORTION OF THE TOTAL, 1986	NUCLEAR PROPORTION, 1983
Europe	2,529,594	671,491	26.5%	17.8%
1. France	343,045	241,400	70.4	48.7
2. West Germany	406,386	119,580	29.4	17.7
3. Sweden	138,023	70,243	50.9	37.7
4. United Kingdom	298,156	59,079	19.8	18.1
5. Belgium	57,621	39,394	68.4	46.4
6. Spain	127,713	37,446	29.3	7.8
7. Switzerland	54,857	21,303	38.8	29.1
8. Finland	46,856	17,998	38.4	41.4

SOURCE: United Nations, *Energy Statistics Yearbook, 1986* New York: United Nations, (1988).

being expanded in an effort to develop the poorer areas of the south, but sewage treatment facilities are inadequate. In addition, huge tankers cross the sea from Africa and the Suez Canal. As noted earlier, the pollutants cannot escape from the sea, and the high evaporation tends to increase their concentration. Fourteen countries border the Mediterranean and three more border the Black Sea, which feeds into it. International cooperation will be hard to achieve, considering the massive investment that will be needed.

In the northern countries, air pollution is the greatest concern. Unfortunately, the acid precipitation often begins in another country. The fumes of the British thermal generating stations blow across the North Sea, much to the distress of the Swedes, who deplore the deaths of their lakes and trees. Other serious offenders have been the East Germans and Czechoslovaks who burned huge amounts of low-quality coal. Shutting down the factories and generators of Europe would mean a slowdown of industry and a decline in the economies of many nations. International cooperation is needed between all of the countries of Western and Eastern Europe.

TOWARD A UNIFIED EUROPE

Less than a half century after World War II, Western Europe has achieved a remarkable degree of unity. Yet, like all regions, Western Europe suffers from regional disparities.

First of all, there are the problems within the individual countries. Many have areas that suffer from high unemployment and have little hope for the future. Most of them are on the peripheries of the European economy. Among the best known are the north and west of the British Isles, the west of France, the southern Interiors of Spain and Portugal, and the Interior of Greece. In addition, the coal, iron, and steel areas present problems. Many people choose to stay in such areas because they are reluctant to leave home and because of their love of local folkways. The energetic young do leave, but the local dialects and outlooks they have learned often make it difficult for them to fit into the life-styles of the sophisticated cities.

In addition, inequalities in the various countries' standard of living present difficulties. Unquestionably, there are major differences in income per capita and in social services. As Table 5.11 shows, however, the tendency in Europe has been for the poorer countries to catch up. In particular, Spain, Norway, and Austria have shown spectacular gains; in 1955 Spain was still shunned because of the Franco dictatorship, while Austria was just freeing itself from military occupation. In both cases, earnings from tourism have been a major factor in the rise in the economic level. Norway has benefited from North Sea oil. Greece and Italy have also improved. Switzerland, which has been the richest country of Europe for several decades, has remained close to the European average. Sweden, which also avoided both World Wars, has fallen slightly behind. Britain, which

TABLE 5.11

Increase in GDP Per Capita, 1955–1985

1. Spain	25.5 times
2. Norway	19.8
3. Austria	19.6
4. Italy	17.5
5. Finland	15.5
6. West Germany	14.8
7. Greece	14.5
8. Netherlands	14.4
European Average	14.4
9. Denmark	14.3
10. Ireland	14.0
11. Switzerland	13.1
12. United Kingdom	12.6
13. Portugal	12.5
14. Sweden	12.4
15. France	11.3
16. Belgium	9.4

had been the leading country of Europe in 1945, dropped steadily for several decades, but has rebounded. The ports and industries of Belgium were relatively untouched by World War II, but in recent years the country has suffered from a depression in the coal and steel industries. Overall then one can say that Europe is becoming equalized. Anyone who has traveled there will know the sad news that few places are bargains any more.

Nevertheless, the countries of the south do have weaker economies than those of the north. To help them weather the shock of facing the competition, they have been given special safeguards for a number of years to smooth their entry into the EC. The Spanish, Portuguese, and Greek industries are to gain strength gradually with the hope that in the end, those countries will be able to compete equally with their efficient northern partners.

All the European countries will soon face the prospect of declining and aging populations. The result will be fewer local workers and fewer taxpayers, and yet more people requiring expensive social services. The needed number of workers will increasingly have to be drawn from countries with non-European cultures. The direction and management of the businesses will remain in European hands and will have to be supplied from the shrinking population base. If this situation were to continue for a long time, a two-class system composed of

wealthy local people and poorer imported workers could result and produce serious tensions. Most European governments are frightened at the prospect and are offering various incentives for couples to have children.

Europeans also know too that they must keep up to date on technological changes. They had been left far behind by the computer revolution in the United States and Japan. The Japanese have led in the use of industrial robots. In both computers and robots, the Western Europeans have been gaining on the leaders. Similarly, rocketry and space research seemed for a time to belong only to the Americans and Soviets, but a European consortium, based on German-French cooperation, has entered the field and has been sending satellites into space. The *SPOT* earth-imaging satellite and the Ariane and Exocet rocket programs are examples of progress in this area. Competition is fierce on the frontiers of technology, but Europeans know that is where the future lies.

A number of other problems must be faced as well. All countries recognize that the EC agricultural policy must be altered to allow both for a decrease in surpluses and for an increasing standard of living for (fewer) farmers. Defense cooperation also proves difficult. Few Europeans like the idea of having foreign troops and nuclear weapons, particularly U.S. weapons, on their territory over which they have limited control. The proximity of Soviet nuclear weapons places the Western Europeans in a quandary. Every time the issue arises, protest marches and peace demonstrations abound, and most governments try to find compromises between public opinion and defense needs. The dramatic changes in Eastern Europe and the 1990 Conventional Forces in Europe treaty may lessen these apparent problems, but may create new ones such as the hostility of ethnic factions igniting warfare as happened in World War I. The unification of Germany has not been welcomed by some Europeans who fear it will be an economic and perhaps military monolith.

A potential solution lies in the emergence of the Conference on Security and Cooperation in Europe (CSCE). This organization is composed of most of the Eastern and Western European nations, the Soviet Union, and also includes the United States. CSCE is a response to the need for a pan-European security structure and may render NATO obsolete relative to the new realities of Europe.

Suggested Readings

Abromeit, H. *British Steel: an Industry between the State and the Private Sector*. New York: St. Martin's Press, 1986.

Bowler, I. R. "Intensification, Concentration, and Specialisation in Agriculture: The Case of the European Community." *Geography* 71 (January 1986): 14–24.

Burtenshaw, D., M. Bateman, and G. J. Ashworth. *The City in West Europe*. New York: John Wiley & Sons, 1981.

Clout, H. *Regional Variations in the European Community*. New York: Cambridge University Press 1986.

Clout H. M. Blacksell, and D. Pinder. *Western Europe: Geographical Perspectives*. New York: Longmans, 1985.

Diem, A. "Are the Alps Dying?" *The Geographical Magazine* 60 (June 1988): 2–11.

El-Agraa, A. M., ed. *The Economics of the European Community*, 2d ed. New York: St. Martin's Press, 1985.

Hughes, M. "The Channel Tunnel." *The Geographical Magazine* 60 (April 1988): 36–43.

Ilberry, B. W. *Western Europe: A Systematic Human Geography*, 2d ed. New York: Oxford University Press, 1988.

Jordan, T. G. *The European Culture Area: A Systematic Geography*, 2d ed. New York: Harper & Row, 1988.

Mead, W. R. "Sweden in Perspective." *Geography* 70 (January 1985): 36–44.

Morris, A. and G. Dickinson. "Tourist Development in Spain: Growth versus Conservation on the Costa Brava." *Geography* 72 (January 1987) 16–26.

Mounfield, P. R. "Nuclear Power in Western Europe: Geographic Patterns and Policy Problems." *Geography* 70 (October 1985): 315–27.

Spooner, D. J. "The Southern Problem, the Neapolitan Problem and Italian Regional Policy." *The Geographical Journal* 150 (March 1984): 11–26.

Straubhaar, T. "International Migration within the Common Market: Some Aspects of EC Experience." *Journal of Common Market Studies* 27 (September 1988) 45–62.

CHAPTER 6

The Soviet Union

IMPORTANT TERMS

COMECON

Glasnost

Perestroika

Collectivization

Autonomous republic

Oblast

Rayon

Kolkhoz

Sovkhoz

CONCEPTS AND ISSUES

■ Reconciliation in a multiethnic society.

■ Reform of the party and governmental administrative structure.

■ Decentralization of the planning and control of the economy.

■ Effects of large areal size on movement and governance.

■ External relations with Europe.

CHAPTER OUTLINE

PHYSICAL ENVIRONMENT
Climate
Natural Zones, Vegetation, and
 Soils
Topography
Rivers and Water Bodies

HUMAN ENVIRONMENT
Historical Background
Population
Administrative Structure
Agriculture
Fishing
Manufacturing and Industry
Urbanization
Transportation

SPATIAL CONNECTIVITY
Foreign Trade
Foreign Aid
Military Aid

PROBLEMS AND PROSPECTS

A mong the countries of the world with planned economies, the Soviet Union was the most important (Figure 6.1). Although the People's Republic of China has a large population, it did not develop a socialist economy until the early 1950s, whereas the Soviet Union was already experimenting with planning in the 1920s. The Soviet economy has served as a model not only for China during its early stages of socialist development, but also for the countries of Eastern Europe, Cuba, Mongolia, and Vietnam. The Soviet Union has not only influenced the economic systems of these countries over the past four decades, but through the mechanism of **COMECON** (Council of Mutual Economic Assistance or CMEA) was able to control trade and other economic links not only within the Soviet Bloc itself, but also with the outside world. The Soviet Union effectively tied the COMECON member countries together into an organization that tended to maximize trade and economic cooperation between members and minimize contacts with capitalist countries. However, as a result of the failed coup in August of 1991, momentous reforms have been initiated in the Soviet Union which will restructure the political union of the Soviet Republics and alter the economic relationships within and outside the Soviet Union.

Soviet leader Mikhail Gorbachev's policies of **glasnost** (openness, frankness), **perestroika** (restructuring), and ***demokratizatsiya*** (democratization) are aimed at increasing economic productivity, reducing incompetency and inefficiency in the Soviet economy, and increasing cooperation and trade with the Western industrial nations while forsaking the more traditional policy of autarky (economic self-sufficiency). Although these policies were initiated in the mid-1980s, it took an attempted coup in 1991 to accelerate these reforms. At present, COMECON has ceased to exist, and it is likely that the whole trade pattern of the Soviet Union and its allies as well as the internal economic and political structure of the Soviet Union will undergo considerable changes in the aftermath of the failed coup.

PHYSICAL ENVIRONMENT

One of the most important facts to realize about the Soviet Union is its great size. With an area of 8.6 million square miles (22.4 million square kilometers), it is the largest country in the world. It has an area greater than

that of the entire North American continent. From its western border with Poland to the Bering Strait, the country extends for 6000 miles (9700 kilometers) and crosses 11 time zones. At its maximum—from the Arctic coast to the heart of Central Asia—it is 3000 miles (4800 kilometers) wide. The Soviet Union has the longest coastline of any country, most of it lying north of the Arctic Circle.

CLIMATE

Not only is the Soviet Union much larger than the United States, but its physical environment is considerably different. Much of the Soviet Union lies in the latitude of Canada. Even the most southerly part of the country is on the same latitude as Wisconsin, while the Black Sea is located as far north as the southern Great Lakes. Moscow lies farther north than Edmonton, Alberta, and Vladivostok on the Pacific coast has the same latitude as Halifax, Nova Scotia. This northerly location has an influence on the climate of the country, which is marked in general by cold winters, cool summers, and a short growing season.

The climate of the Soviet Union is also affected by the location and extent of its major mountain ranges (Figure 6.2). The southern borders of the country are fringed by several ranges of high mountains, from the Caucasus in the west, through the Central Asian mountains to the complex ranges of East Siberia and the Far East. These ranges act as a major barrier to the movement of warm tropical air masses from the Indian Ocean and prevent their penetration of the landmass of the Soviet Union. No similar barrier, however, prevents the movement of Arctic air masses over virtually the whole country in winter, resulting in below-freezing temperatures as far south as Central Asia. The result is cool summer temperatures throughout most of the northern and central parts of the country and very cold temperatures over most of the country in winter. For such a large area, the climate of the Soviet Union is remarkably uniform. Its distance from major oceans, except for the frozen Arctic Ocean, results in a very weak maritime influence with a relatively low level of precipitation in all areas except in the west. These western areas receive some precipitation from the Atlantic Ocean as well as from the Baltic and Black seas.

The vast size of the landmass of the Soviet Union, much of it far from the sea, results not only in relatively dry conditions in the center of the country, but also in

STATISTICAL PROFILE: SOVIET UNION

	Population Estimate mid-1990 (Millions)	Natural Increase (Annual)	Population Projected to 2000 (Millions)	Infant Mortality Rate	Life Expectancy at Birth (Years)	Urban Population (%)	Per Capita GNP, 1988 (US$)	Area, Thousands of Square Miles (Km²)	Population Density, No./mi² (No./km²)
Soviet Union	291	0.9	312	29	69	66	7400	8649(22,402)	34(13)

a wide range between summer and winter temperatures. This continental influence is more extreme than on the North American continent.

The pattern of air masses in North America is quite different, due to the north-south alignment of the major mountain masses, which block most Pacific Ocean air from penetrating inland. However, the lack of any east-west barrier means that tropical air masses from the Gulf of Mexico can penetrate as far north as southern Canada in summer, while Arctic air masses can reach southern Florida in winter. This interplay of cold and warm air masses brings the changeable weather so typical of North America, particularly in the spring. The phenomena that accompany this weather, such as tornadoes, severe thunderstorms, and hurricanes, are little known in the Soviet Union, where weather is much more stable. The alignment of the mountain barriers, the degree of penetration of air masses, and the differing latitudes explain why the United States has large areas of subtropical climate and why even the Great Lakes area has warm summers and favorable conditions for agriculture. The climate of the Soviet Union, on the other hand, is not a major asset in terms of agriculture.

The effects of these climatic controls can be seen in the temperature ranges and annual precipitation for different areas in the Soviet Union. Moscow, in the western part of the country, has an average temperature in January of 14°F (−10°C) and in July of 66°F (19°C) with total annual precipitation of 21 inches (53 centimeters). This winter temperature is similar to that of Minneapolis, Minnesota, but the latter has an average July temperature of 72°F (22°C) and total precipitation of 27 inches (68 centimeters). To find summer temperatures similar to those in Moscow, one has to go as far north as Winnipeg, Manitoba. In other words, summer temperatures in much of the Soviet Union are more similar to those of the prairie provinces of Canada than to those in the continental United States. Only in the most southerly parts of the European area of the Soviet Union and in the Caucasus and Central Asia do summer temperatures exceed 70°F (21°C) on the average (Figure 6.2).

The Soviet Union lacks any climatic zone similar to that of the subtropical American South or southern California. Only along the southern coast of the Crimean Peninsula and on the eastern coast of the Black Sea and the adjacent Colchis Lowlands are winter temperatures above freezing on the average. The mild winters are largely due to the shelter afforded by the Crimean Mountains and the western ranges of the Caucasus against the cold northern air masses. The Colchis Lowlands have the highest average precipitation total of any part of the Soviet Union with over 100 inches (250 centimeters). The northern coast of the Kola Peninsula also has relatively mild summer temperatures for its latitude. The North Atlantic current skirts the west coast of Norway and rounds the North Cape before vanishing in the cold waters of the Barents Sea off the Kola Peninsula. The port of Murmansk on the Kola coast has an average January temperature of 14°F (−10°C), compared with 8°F (−13°C) at Arkhangelsk on the White Sea, 300 miles (480 kilometers) to the south. Murmansk can be kept open in winter, while Arkhangelsk is generally closed by ice from November to May and Leningrad from December to May. This fact helps to explain the vital role played by Murmansk as a Soviet naval base.

NATURAL ZONES, VEGETATION, AND SOILS

Due to the cool, short summers and long, cold winters, much of the eastern and northern areas of the Soviet Union has a short growing season. Most of the country that lies to the north and east of a line drawn from Leningrad on the Baltic Sea to the southern tip of Lake Baykal in eastern Siberia has a growing season averaging under 90 days annually. Most agricultural crops require a longer growing season to reach maturity. This area corresponds with two major natural zones: the *taiga,* or coniferous forest, which extends from the western borders with Finland to the Pacific Coast and covers most of Siberia, and the tundra, which stretches along the Arctic coast from Scandinavia to the Bering Strait.

FIGURE 6.1

The Soviet Union

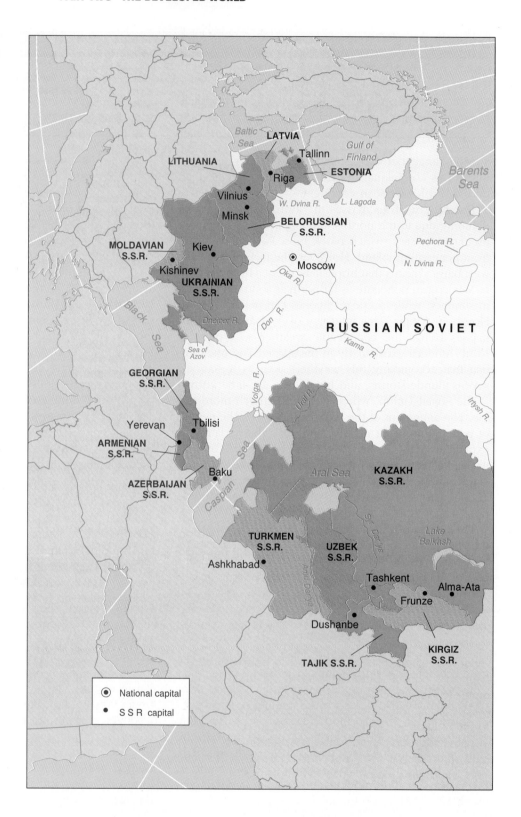

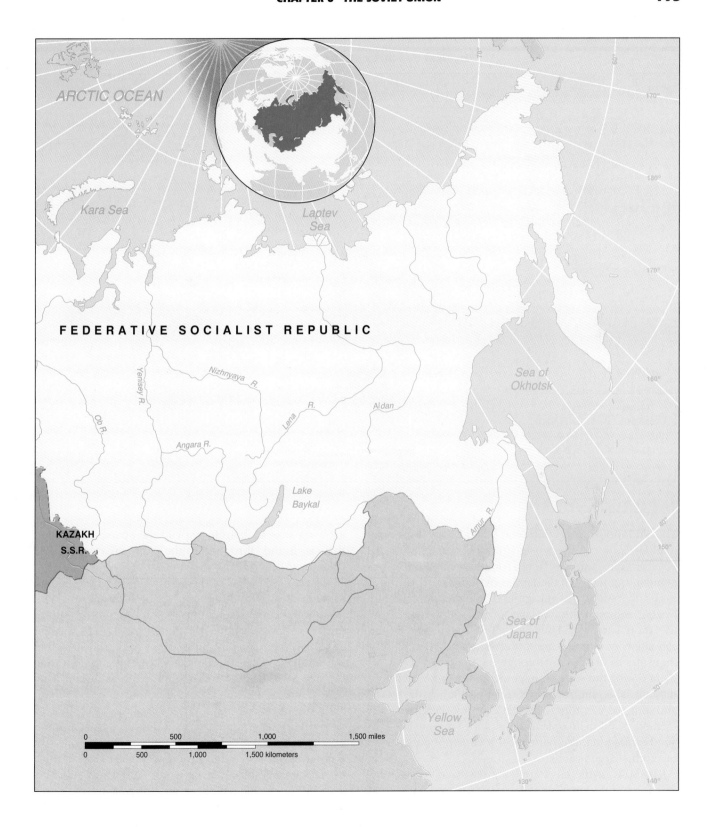

ARCTIC OCEAN

Kara Sea

Laptev
Sea

FEDERATIVE SOCIALIST REPUBLIC

Sea of
Okhotsk

Yenisey R.

Nizhnyaya R.

Ob R.

Lena R.

Aldan

Angara R.

Lake
Baykal

Amur R.

KAZAKH
S.S.R.

Sea of
Japan

Yellow
Sea

| 0 | 500 | 1,000 | 1,500 miles |

| 0 | 500 | 1,000 | 1,500 kilometers |

FIGURE 6.2

Physical Features of the Soviet Union

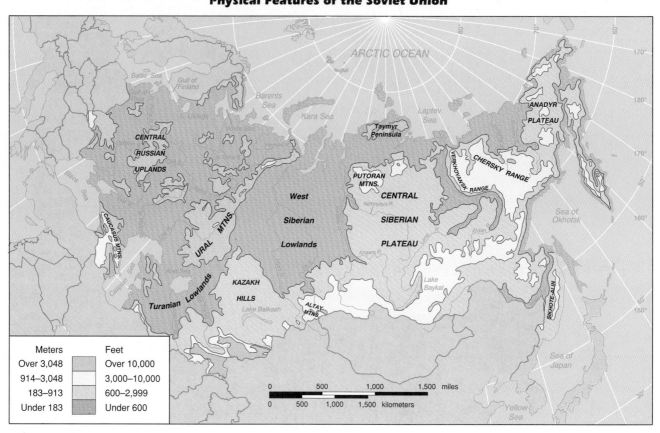

Meters		Feet
Over 3,048		Over 10,000
914–3,048		3,000–10,000
183–913		600–2,999
Under 183		Under 600

This enormous area, which comprises about half of the total land area of the Soviet Union, is not only unsuitable for agriculture and hence for large-scale human settlement, but much of the land is in permafrost. Moisture under the surface of the ground remains permanently frozen, although the top layer of the soil thaws during the short summers. However, this thaw results in the formation of swamps and ponds due to the inability of the water to drain away through the frozen layer. Movement during the summer in permafrost areas is therefore difficult. The construction of buildings, railroads, roads, and airfields is a formidable task as any disturbance or melting of permafrost leads to sinking or landslides. In southern Siberia, the permafrost area is not continuous but forms large islands.

The soils of the taiga zone consist mainly of spodosols formed under a coniferous forest cover. These acid soils are poor for farming and, combined with the short growing season, help to explain the sparse population that is found throughout the vast area of Siberia (Figure 6.3).

Another large area of the Soviet Union where human settlement is difficult lies to the south of a line drawn from the port of Odessa on the Black Sea to the southern tip of Lake Baykal. It includes Kazakhstan, Central Asia, and the plains to the north of the Caucasus. Here the annual average precipitation is generally less than 12 inches (30 centimeters), while parts of Central Asia have less than 8 inches (20 centimeters). In this southern area, temperatures in July average around 88°F (31°C) with maximum temperatures around 115°F (46°C). However, freezing temperatures occur in winter. Much of this area consists of desert and semidesert with a zone of steppe (grassland) in the north. The warm summers and the sunshine make this region suitable for the growing of a number of crops. The main limiting factor is the lack of water, but a number of large rivers, including the Syr Dar'ya and the Amu Dar'ya, fed by the glaciers and snows of the Central Asian mountains, provide sufficient water for a number of large irrigation schemes. The alluvial soils along the river banks and the large areas of loess soils are used for the growing of cotton on a large scale, while the mollisols of the northern steppes are good for the cultivation of wheat. Because of the large amounts of water used for irrigation along the Amu Dar'ya and Syr Dar'ya, however, little remains to empty into the Aral Sea to counteract evaporation. Therefore, it is predicted that by the year 2000 the Aral Sea may only be two-thirds of its present size. The steppes of Kazakhstan and southern Siberia experience

considerable variations in precipitation from year to year, and grain production is irregular. This region was known as the "Virgin Lands" in the early 1950s, when large areas of grass were plowed up and sown with wheat in an attempt to create a new area where grain could be produced without the use of scarce fertilizer. The government also hoped to release land in the more humid regions of the west for growing more valuable crops than wheat. The Virgin Lands Scheme did not meet the ambitious targets set for it. Underestimation of the drought problem created "dust bowl" conditions in some areas. Nevertheless, the scheme resulted in a useful addition to the total grain production of the country. The steppes of the European part of the Soviet Union

FIGURE 6.3

Natural Regions of the Soviet Union

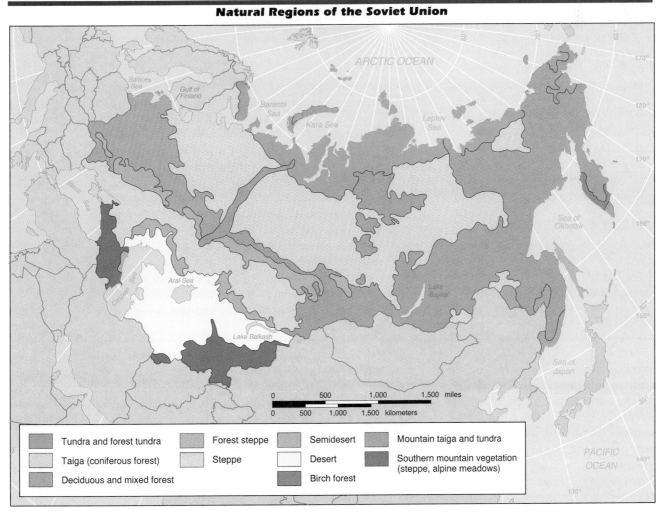

Tundra and forest tundra	Forest steppe	Semidesert	Mountain taiga and tundra
Taiga (coniferous forest)	Steppe	Desert	Southern mountain vegetation (steppe, alpine meadows)
Deciduous and mixed forest		Birch forest	

LOCATION AND SIZE

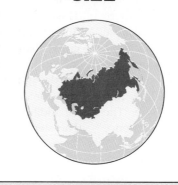

wo important concepts in geography are the location and size of a particular geographic place. These are particularly important in the case of the Soviet Union.

The northerly location of this country and the configuration of its terrain have a considerable effect on its climate. Much of the country is too cold, too dry, or too mountainous to present ideal conditions for agriculture, although a sizable portion of the European part of the country has a milder climate and is more suitable for agriculture. Access to the world oceans is greatly limited by the freezing of the waters along much of the coastline during the winter months, and the landlocked nature of the Baltic and Black seas creates serious problems for the Soviet Union as a maritime power.

The size of the country has made it difficult to develop an efficient and extensive transportation network. This problem is especially significant when the irregular distribution of natural resources and industrial raw materials is taken into account. Long hauls of coal and other materials by rail are characteristic of the Soviet transport system. The lack of a well-developed highway network to move goods and people is also a factor limiting economic devel-

opment. Recent statistics show that the Soviet Union has a total highway network of 1,000,344 miles (1,609,900 kilometers), of which 74 percent consists of paved roads; for comparison, the United States has a highway network of 3.9 million miles (6.3 million kilometers). A similar disparity exists between the two countries with regard to automobile ownership. Whereas the United States averages 570 automobile registrations per 1000 persons, the Soviet Union averages only 42 per 1000. Although many Soviet people use various forms of mass transit and do not rely on the automobile as much as Americans do, the country still needs a more modern highway system to increase accessibility and interconnectiv-

ity between places, people, and goods.

The vast size of the Soviet Union also creates considerable problems for a government that traditionally has aimed at centralized control of the political and economic system from the capital city. Frequently, regional and local problems receive little attention from authorities in Moscow. Similar difficulties are affecting the implementation of Gorbachev's policies of glasnost and perestroika. Not only may implementing national economic reform or restructuring prove to be difficult, but since most of the various nationalities are located in the country's borderlands far away from Moscow, enforcing a nationalities policy will prove more difficult now that the Baltic states have bolted from the Soviet Union and others have declared their independence.

At the same time, many positive attributes, such as the Soviet Union's extensive natural resources, are associated with being the largest nation in the world. Whether in developing those resources or making other decisions related to political, economic, and social affairs, however, the Soviet government and its people must always consider the factors of location and size.

have a more humid climate than those of Asia and are more reliable as producers of grain.

The zone with the best conditions for human settlement lies between the coniferous forests to the north and the dry steppes and deserts to the south. The northern part of this zone consists of areas of deciduous and mixed forest, with grey and brown forest soils of moderate fertility. To the south the more humid areas of the steppes, mainly in the Ukraine, have fertile mollisols and constitute some of the best areas for farming in the

country. This zone forms a triangle or wedge with its base along the western boundary of the country and its apex at the southern tip of Lake Baykal in Siberia. Because of its relatively good conditions for farming, this zone is often known as the "Agricultural Triangle". However, it must be realized that conditions for farming in this zone are no better than those in much of Wisconsin or Minnesota, and in no way can it be compared to the Corn Belt of Iowa, Illinois, and Indiana in the United States.

TOPOGRAPHY

Although the vast plain extending from the western borders to the heart of Siberia is one of the major geographical features of the country, there are also considerable areas of mountains and hills (Figure 6.2). The Urals, the long range of mountains that cut across the great plain from north to south, do not pose a major barrier either to climate or to the movement of people. The Urals are crossed by a number of major transportation routes, and a number of mining and industrial settlements are located in the southern part. The highest range in the Soviet Union, the Pamirs, is found in the Central Asian mountains along the borders with Afghanistan and China. They contain the highest peak in the Soviet Union, the Peak of Communism, at 24,590 feet (7495 meters). Other ranges radiate from the Pamirs, of which the most important are the Tien Shan Mountains, which reach over 10,000 feet (3000 meters) in height. Further east, the Altay Mountains mark the beginning of a series of complex ranges that cover much of eastern Siberia and the Far East. In the latter region, the large peninsula of Kamchatka contains 22 active volcanoes.

In the European part of the Soviet Union, the highest mountains are the Caucasus with the peak of Mount Elbrus reaching over 18,400 feet (5630 meters). The Caucasus extend some 700 miles (1100 kilometers) from the eastern coast of the Black Sea to the Caspian Sea. Further west are the Crimean Mountains, a continuation of the Caucasus, while part of the large loop of the Carpathian Mountains, extending from Czechoslovakia to Romania, is located within the western borders of the Ukraine. They rise to over 6500 feet (2000 meters).

RIVERS AND WATER BODIES

The Soviet Union contains some of the largest rivers in Eurasia. The Volga is the longest river in Europe and flows 2300 miles (3690 kilometers) from the Valday Hills northwest of Moscow to the Caspian Sea. Its main tributaries are the Oka and Kama rivers. These rivers also have a number of tributaries. In all, the Volga river system has a navigable length of over 10,000 miles (17,000 kilometers). However, ice restricts navigation between eight and nine months of the year. The construction of large hydroelectric dams has created several large reservoirs on the river.

Other large rivers in the European region are the Dnieper and the Don, both of which flow to the Black Sea. In the north, the Northern Dvina ends in the White Sea and the Pechora in the Arctic Ocean. The main river flowing to the Baltic Sea is the Western Dvina (Daugava).

Central Asia is too dry to support many large rivers, but the Syr Dar'ya and the Amu Dar'ya carry enough water to reach the Aral Sea. Further east, the vast area of Siberia is crossed by three large rivers, the Ob, the Yenisey, and the Lena. These rivers flow across the whole width of Siberia to empty into the Arctic Ocean. From the source of its major tributary, the Irtysh, it is nearly 3400 miles (5400 kilometers) to the mouth of the Ob. The mouths of these rivers are frozen for over seven months each year. During the spring, the water flowing from the south brings ice floes, which pile up on the ice of the still frozen mouth of the Ob, causing ice jams and vast floods. The rivers are navigable, but the season is short. The Yenisey and its tributary, the Angara River, which rises in Lake Baykal, are particularly suitable for hydroelectric power development. The value of the Siberian rivers for navigation and electric power production is somewhat limited by their distance from the major areas of population and the fact that they end in the Arctic Ocean.

The major river of the Far East is the Amur, which flows to the Pacific Ocean. It is as long as the Siberian rivers, but carries less water. Navigation is restricted by the shallow mouth of the river, which remains frozen for several months.

The largest lake in the Soviet Union is the Caspian Sea. It receives most of its water from the Volga, and since it has no outlet, it is salty enough to be called a "sea." Until recently, the level of the Caspian was dropping by about two and a half inches (6.3 centimeters) a year. This lowering of the water level created problems for navigation and fisheries. This problem has resulted from a decline in precipitation in the Volga Basin, increased use of water from the Volga for irrigation, and a high rate of evaporation from the surface of reservoirs on the river. Plans for building a dam across the northern part of the Caspian or diverting water from northern rivers into the Volga systems have never materialized. Recently, however, there have been signs of a gradual rise in the level of the sea.

In Central Asia, the Aral Sea and Lake Balkhash are shallow lakes with a low salt content. Lake Baykal in eastern Siberia is the deepest lake in the world and contains a volume of fresh water almost equal to that of all the American Great Lakes combined. European Russia contains the two largest lakes in Europe, Lakes Ladoga (Ladozhskoye Ozero) and Onega (Onezhskoye

Ozero), which are connected by the River Svir. Lake Ladoga drains into the Baltic Sea through the short Neva River, on whose delta the city of Leningrad is located.

The only world oceans to which the Soviet Union has direct access are the Arctic and the Pacific. The Soviet Union controls parts of the coasts of the Black Sea and the Baltic Sea, but both of these seas are connected to other seas by narrow straits controlled by foreign powers. Although the Black Sea is a southern sea by Soviet standards, it has a colder climate than the Mediterranean Sea with which it is connected. To the northeast, the Sea of Azov freezes over in winter. The Baltic Sea freezes in the north, especially in the Gulfs of Riga and Finland. The southern part is ice-free. Neither the Black nor the Baltic Sea has good fishing grounds.

The Arctic Ocean is divided into a number of seas, of which the most westerly, the White Sea, penetrates far inland. It is frozen for six months of the year, but in summer ships can sail from the port of Arkhangelsk to the Atlantic Ocean around northern Norway without encountering any ice. The Barents Sea is the meeting place of the warm currents from the north Atlantic and the cold currents of the Arctic and is an ideal environment for supporting a large fish population. Further east, the Arctic Ocean has severe icing conditions for much of the year, and navigation is possible only with icebreakers.

The Pacific coast is plagued by ice and storms and even the port of Vladivostok is obstructed by ice for over three months of the year. In the northern Pacific, the meeting of cold and warm climates results in large supplies of fish.

HUMAN ENVIRONMENT

HISTORICAL BACKGROUND

For centuries under the rule of the czars, Russia remained primarily an agrarian society, with the vast majority of the population living under autocratic rule in conditions of poverty and illiteracy. A small portion of the population controlled the wealth, and what might be called the middle class was comprised of small groups of merchants, farmers, and professional people. Everyday life varied little under the czars, but change did occur for better or worse, depending on which czar was in power. In the sixteenth century, Ivan IV, or Ivan

the Terrible as he was later called, was successful in expanding the Russian Empire from its early origins in the state of Muscovy centered on Moscow (Figure 6.4).

From this core in Muscovy, the Russian Empire expanded into the Volga River basin and into the regions north of Moscow. Expansion continued eastward across the Ural Mountains and into Siberia reaching the shores of the Pacific Ocean by 1640. Under Peter the Great (1672–1725), the empire expanded toward the Baltic Sea, and by the end of the eighteenth century, the Russians controlled the lands north of the Black Sea. During the nineteenth century, Russian expansionism spread into the southern Caucasus Mountain area and into Central Asia east of the Caspian Sea.

The Russians came in contact with numerous ethnic groups as a result of expanding their borders. Some groups were Slavic peoples closely related to the Russians, like the Belorussians and Ukrainians, while others were ethnically and culturally different from the Russians like the Turkic-speaking Muslims in Central Asia or the Mongols in Siberia.

By the early twentieth century, Russia had begun to change. It had experienced its own industrial revolution in the 1890s, and its population had doubled from 60 million in 1850 to 130 million in 1900, reaching 160 million in 1913. Increased population pressure on the farmland caused many persons to flee to the urban centers. But chronic food shortages, failure to achieve meaningful land redistribution, and poor working conditions in the factories and mines triggered several riots in 1905. Even though Nicholas II, the last of the Romanov czars, was aware that he needed to initiate more industrial growth and transform the country's economy, the peasants were still repressed. The loss of former territories in Eastern Europe including Finland, Poland, and part of the Ukraine during World War I came as a further setback to the Russians. Continued pressure from the urban workers and rural peasantry forced Nicholas II to abdicate the throne in February 1917. Although Alexander Kerensky was able to establish a provisional government for a few months, eventually Vladimir Ilich Lenin and the Bolsheviks wrested control from Kerensky and his provisional government in November 1917.

The Bolsheviks (later called communists) were faced with many problems: a nationwide civil war, peasant unrest, famine, and the loss of agricultural land, factories, and coal and iron ore mines during the war. The young communist state also had to gain control of the cities and rural areas within its boundaries. Following

FIGURE 6.4

The Growth of the Russian Empire

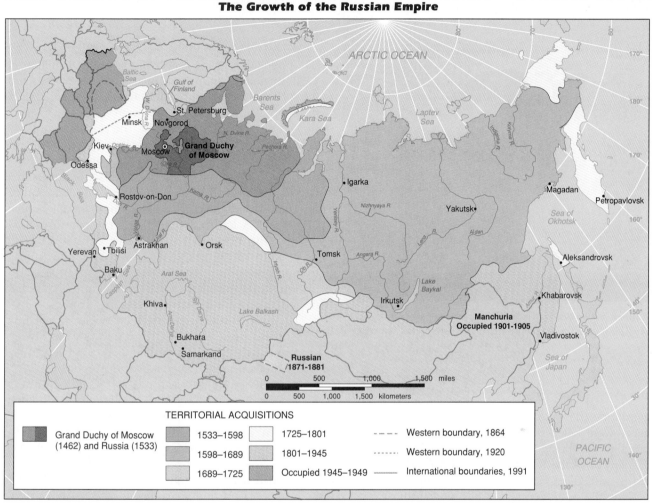

SOURCE: Modified from H. J. DeBlij and Peter O. Muller, *Geography: Regions and Concepts*, 6th ed. (New York: John Wiley & Sons, Inc., 1991), p. 140.

Marxist theory, Lenin nationalized important segments of the economy and seized private and church property, making it state property. It was hoped that through such measures class differences based on land ownership would disappear, exploitation of the workers by the capitalists would cease, and a classless society of social equals would emerge.

But economic and social problems persisted despite Lenin's push for a new order. Sensing that more reforms were needed, he instituted a New Economic Policy in 1921, encouraging private and foreign investment to assist in the development of agriculture and industry. Although this infusion of capitalism went against the grain of orthodox Marxists, Lenin's decision helped to restore production levels to prewar levels. In 1922, the communists established the Union of Soviet Socialist Republics.

Following the death of Lenin in 1924, Joseph Stalin gradually gained control of the Communist party. Continuing to reform and improve the Soviet economy, Stalin felt that the agricultural land should be collectivized; that is, many small holdings would be grouped together into a larger unit that would be worked by the peasants as a group under government supervision. **Collectivization** of agriculture, he believed, would increase grain production, and the payments received for agricultural products could be used for investment in industry. Beginning in 1929, Stalin inaugurated his policy of collec-

Soldiers at Red Square during the coup attempt of August 1991 show the force but not the will to crush the resistance.

tivization. The peasants resisted the new policy, and many were killed during the ensuing years while the program was being implemented. But within 10 years, 93 percent of the Russian peasants lived on collective farms, and Soviet industry was expanding.

Upon the death of Stalin, Nikita Khrushchev came to power. He concentrated on improving agriculture and industry in the Soviet Union in an attempt to move the country toward world power status. Still the Soviet Union was isolated from much of the West as it focused its attention on domestic development and on extending trade with communist bloc countries—mainly those in Eastern Europe. During this time, it also expanded its influence in developing countries. The ouster of Krushchev and the subsequent rise to power of Leonid Brezhnev did little to change basic Soviet ideology or attitudes toward the West, as the Soviet Union and the United States continued to compete with each other to gain influence in emerging developing nations around the world.

The rise of Mikhail Gorbachev, however, caused drastic shifts in Soviet ideology and thinking. Gorbachev's attempts to open up Soviet society and increase inter-

action with the West resulted in the policies of glasnost and perestroika.

POPULATION

The population of the Soviet Union was 291 million in 1990. Beginning with World War I, the Revolution, and the subsequent civil war with its famines and epidemics, the population has suffered from losses that have slowed its growth considerably. The population of 147 million in 1926 was some 25 to 30 million less than it would have been under normal circumstances. However, these losses were somewhat offset by an annual rate of increase of 2.4 percent. The collectivization of agriculture, deportations to labor camps, and the famine of 1932–1933 led to a further loss of some 5 million people. By 1939, the annual rate of increase had dropped to 1.9 percent, due to lower birth rates arising from increased urbanization. In that year the total population was 171 million, some 20 million less than expected, taking natural growth rates into account.

The most pronounced effect on Soviet population growth was caused by World War II. It is generally accepted that 15–20 million people died in between 1941 and 1945 as a result of the war and purges by Joseph Stalin. Little information is available for the period immediately following the war, but a figure of 198 million for 1956 indicates that the population of the country did not reach its prewar level again until 1955 (taking boundary changes into account). The census of 1959 gave the annual rate of population increase as 1.7 percent and revealed an imbalance of males and females in the population due to wartime losses. By about 1967, the annual rate of natural increase had dropped to under 1 percent per annum due to falling birth rates and a rise in the death rates of the male population. In 1979, the census showed an annual rate of increase of 0.8 percent and a total population of 262 million. In 1990, the population was 291 million with an annual rate of natural increase of 0.9 percent.

Recent trends in Soviet population dynamics place the Soviet Union in the same category of growth rate as the United States. However, the figures for growth trends of the total population do not reveal two important characteristics of the Soviet population: (1) ***internal migration*** of the population and (2) the great differential in birth rates between the European and Asian regions of the country.

The greatest concentrations of population in the Soviet Union are found in the western base of the Agricultural Triangle (Figure 6.5) especially in the Ukraine, with other concentrations in the Moscow, Volga, and Ural regions. Outside the Triangle, there are minor concentrations in parts of the Caucasus region and Central Asia and the Kuznets industrial region of Western Siberia. Further east, most of the population is scattered along the route of the Trans-Siberian Railroad and in the Far East.

Many of the people living in the eastern parts of the country moved there during the early period of industrial expansion, when new cities in the Urals, Kazakhstan, and the Far East attracted workers from the west. During the 1926–1939 period, almost 3 million people moved to the Urals and beyond. This movement accelerated after 1939. During World War II, many workers were moved to the factories of the Urals, Central Asia, and Western Siberia, and more followed after the war. In recent years, however, the eastward movement of the population has slowed.

Internal migration of the population has not only involved a west-east movement, but also a major rural-urban migration in all parts of the country. Although no period has matched that of 1926–1939 in the rapidity of increase of urban population, the percentage of the total population living in cities increased from 56 percent in 1970 to 66 percent in 1990. The number of cities with a population of over one million increased from 10 in 1970 to 23 in 1987. The largest is Moscow with a 1987 population of 8.8 million. As the capital and major industrial city in the country, Moscow has shown continuous growth since 1939 when its population totaled 4.5 million. St. Petersburg on the Baltic has experienced a less dramatic growth rate. In spite of its importance as a port and as the second manufacturing center of the country, the population of St. Petersburg grew from a 1939 total of 3.4 million to only 4.9 million in 1987. This slow growth is mainly due to the population losses incurred during the siege of the city during World War II. Two other

FIGURE 6.5

The Agricultural Triangle and Major Industrial Regions

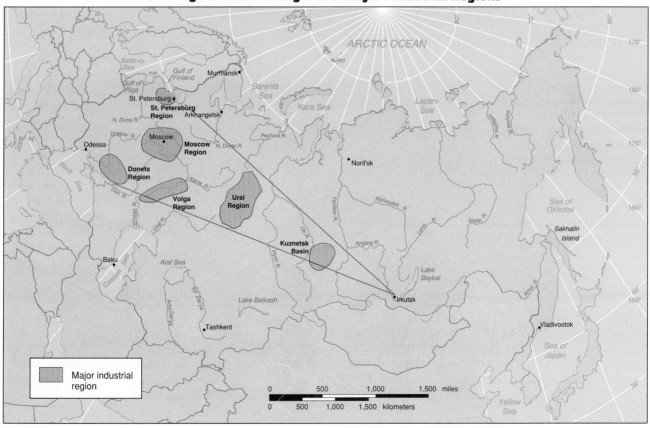

The new Moscow district of Troparyovo. Moscow, like many other cities in Europe, has been expanding its environs to accommodate its growing population.

cities had over two million inhabitants in 1987: Kiev, the capital of the Ukrainian republic (2.5 million), and Tashkent, the largest city of Central Asia and capital of the Uzbek republic (2.1 million). Other cities with over one million inhabitants include Baku, Kharkov, Minsk,

Gorky, Kuybyshev, Sverdlovsk, and Novosibirsk. The last is the largest city in Siberia and owes its importance to its location at the crossing point of the Trans-Siberian Railroad and the Ob River and its proximity to the Kuznets Basin. Omsk, the second city of Siberia, also has over one million inhabitants, but most of the other Siberian cities are relatively small in size. Vladivostok on the Pacific coast had 615,000 inhabitants in 1987. The major concentration of urban population still remains at the western base of the Agricultural Triangle.

Birth rates vary considerably between different parts of the country. This difference is mainly accounted for by the existence of more than one hundred ethnic groups with differing cultures, levels of education, and standards of living (Figure 6.6). The largest group are the Russians who account for about 52 percent of the total population, followed by other Slavic speakers, such as the Ukrainians (16 percent) and the Belorussians (4 percent). In all, about 72 percent of the population are of Slavic origin. Other peoples of European origin are the Estonians, Latvians, and Lithuanians in the Baltic states, the Moldavians along the Romanian border, as well as scattered groups of Finnic peoples such as the Udmurts and Mari in the Volga and Ural regions. Most of these people have relatively low birth rates, and their annual rate of increase is on the average below the 0.9 percent average for the country as a whole.

A teahouse in Dushanbe, Tajikistan. Teahouses are a major part of the social environment in Soviet Central Asia where many persons of Turkic and Iranian descent reside.

FIGURE 6.6

Peoples of the Soviet Union

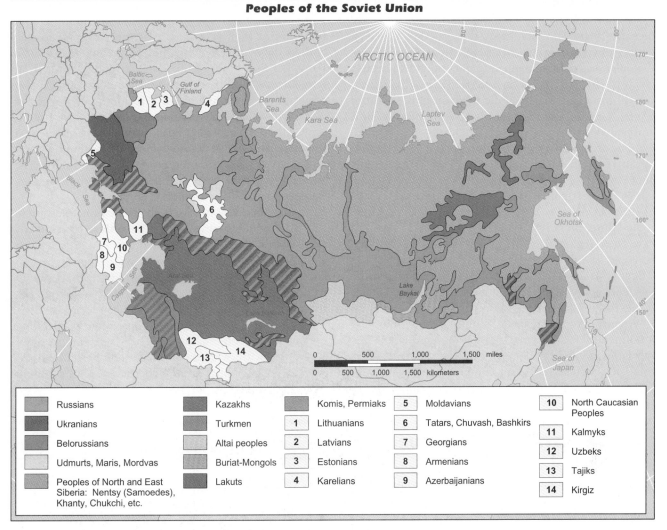

In the southern regions of the Soviet Union are a number of ethnic groups with a variety of cultures and languages. The most numerous are the Central Asian peoples, such as the Uzbeks, Kazakhs, Turkmens, Kirgiz, and Tajiks. They are Muslims and are mostly Turkic speaking. The 1979 census showed an average annual rate of growth of over 3 percent for these peoples. Even today these groups exhibit a higher growth rate than the non-Asian groups in the Soviet Union. The peoples of the Caucasus region, such as the Azerbaijanians, Georgians, and Armenians, had growth rates of between 1 and 2 percent per annum in 1979. Although these growth rates show some signs of decreasing, a major shift in the ethnic composition of the Soviet population is taking place. Between now and the year 2000, about half of the country's population growth will take place in Central Asia. The 43 million Muslims in the Soviet Union at present comprise about 16 percent of the total population. The Muslim population increased some 25 percent between 1970 and 1979, compared with 5 percent for non-Muslims. In view of the rise of religious feeling in such neighboring countries as Iran and Pakistan, the Soviet government must have some misgivings about the rate of increase of its own Muslim population.

During the past two decades, the Soviet Union has suffered from a labor shortage in its urbanized and industrial areas. This has been largely due to a drop in the birth rates in these areas, leading to a marked decrease in young people entering the labor force since the early 1960s. The problem is aggravated by the large number of young men in the armed forces. Employment of women has partially helped to alleviate this shortage, which is especially acute in Siberia. There is as yet no sign of any migration from the labor-surplus region of Central Asia to these regions. This lack of movement is attributable to several factors. The people of Central Asia are predominantly rural and so far have been reluctant to move to the cities of their own region, let alone to cities in other regions. In addition, Central Asians differ culturally and racially from most of the other Soviet peoples, and their educational level is low. In the future, however, it may be necessary for them to leave their region to work unless more local employment can be found for the growing population.

Central Asians thus do not play the same role in the Soviet economy as the Mexicans do in the United States or the Turks, Yugoslavs, North Africans, and others do in Western Europe. Menial jobs in the Soviet Union are performed by local people, not by imported labor. The only labor to enter the country from abroad are workers from such Eastern European countries as Bulgaria, Poland, or former East Germany, who are employed on special projects, as well as a few from Vietnam.

ADMINISTRATIVE STRUCTURE

In 1922 the Union of Socialist Soviet Republics was formed, with the idea of giving each major ethnic group its own republic within a federal system. Smaller ethnic groups have their own administrative units, known as **autonomous republics (ASSRs).** There were 15 full republics, known as *union republics (SSRs),* of which the largest is the Russian Soviet Federated Socialist Republic. It contains a number of non-Russian peoples, each with its own administrative unit. It is, in fact, a federation within a federation. The smallest ethnic groups in the Russian republic have their own administrative units, known as autonomous oblasts or *national okrugs.* All these units, from union republic to national okrug, sent representatives to the Soviet of Nationalities, which in theory shared legislative power with the Soviet of the Union, whose deputies were chosen on the basis of population rather than ethnic group. In practice, however, these Soviets had no legislative power, but met only occasionally to ratify decisions

made by the Presidium of the Supreme Soviet, which in turn was controlled by the Politburo of the Communist party. The federal system gave little autonomy to the republics or their subordinate divisions. Each republic had its own government, ministries, and Communist party, but they exercised little initiative, and virtually all decisions of importance were made in Moscow. The administrative divisions formed a hierarchy of command and at the same time gave an illusion of ethnic autonomy. All of this central control is now being challenged by various republics whose leaders are beginning to initiate moves toward more autonomy for the republics. Several republics have seceded and the union now shows signs of disintegration (Figure 6.7).

Apart from divisions based on ethnic principles, there are other administrative units. The **oblast** corresponds roughly to a county in the United States, while a **rayon** is somewhat similar to a township. In sparsely populated areas of Siberia a large division known as a *kray* is found. These administrative units have no ethnic significance and serve purely administrative purposes.

To assist the planning process, a number of planning regions have been established. Their boundaries do not always coincide with those of the administrative divisions. Some republics are divided into several planning

FIGURE 6.7

The Baltic Republics: From Democratization to Independence

1986	Mikhail Gorbachev announces *demokratizatsiya* (democratization).
1989	Gorbachev is freely elected president of the new Soviet Parliament.
	In November the Berlin Wall falls.
1990	Lithuania declares independence. Latvia and Estonia take steps toward independence.
1991	In January Soviet troops move into Lithuania to stop the independence movement.
	On August 19 Soviet hardliners oust Gorbachev in a coup d'état.
	On August 22 the coup d'état fails. Gorbachev returns and dismantles the Communist party.
	By early September the U.S. and several foreign countries recognize the independence of Lithuania, Estonia, and Latvia.
	On September 4 the Soviet Parliament recognizes the independence of Lithuania, Estonia, and Latvia.
	On September 17 Lithuania, Estonia, and Latvia are admitted to the United Nations.

FIGURE 6.8

Administrative Divisions of the Soviet Union Based Ethnic Groups

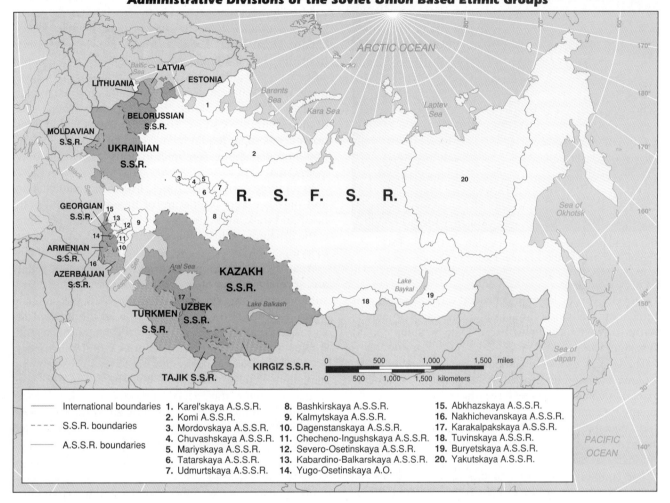

	International boundaries	**1.** Karel'skaya A.S.S.R.	**8.** Bashkirskaya A.S.S.R.	**15.** Abkhazskaya A.S.S.R.
- - - -	S.S.R. boundaries	**2.** Komi A.S.S.R.	**9.** Kalmytskaya A.S.S.R.	**16.** Nakhichevanskaya A.S.S.R.
		3. Mordovskaya A.S.S.R.	**10.** Dagenstanskaya A.S.S.R.	**17.** Karakalpakskaya A.S.S.R.
———	A.S.S.R. boundaries	**4.** Chuvashskaya A.S.S.R.	**11.** Checheno-Ingushskaya A.S.S.R.	**18.** Tuvinskaya A.S.S.R.
		5. Mariyskaya A.S.S.R.	**12.** Severo-Osetinskaya A.S.S.R.	**19.** Buryetskaya A.S.S.R.
		6. Tatarskaya A.S.S.R.	**13.** Kabardino-Balkarskaya A.S.S.R.	**20.** Yakutskaya A.S.S.R.
		7. Udmurtskaya A.S.S.R.	**14.** Yugo-Osetinskaya A.O.	

regions. Their boundaries are periodically adjusted as the pattern of economic activity changes. These regions have no political or administrative function, but are used mainly for planning and statistical purposes (Figure 6.8).

The government of the Soviet Union is still somewhat centralized, but reforms being initiated are helping to decentralize authority. However, attempts to decentralize economic control by giving more power to regional management were not very successful before the 1991 coup attempt, but now this economic power is being diffused. With certain republics having declared independence, and others desiring it, a whole new economic and political arrangement between the former republics and the Soviet government is being negotiated.

AGRICULTURE

Soviet agriculture is characterized by several features that are to a great extent explainable in terms of the environment, but also owe something to Soviet organization and planning. In particular, the selection and distribution of the main agricultural crops are partly in response to the limits of the environment and partly to the perceived requirements of the economy.

In terms of crop distribution, there are three major zones in the country (Figure 6.9). The northern zone occupies the northern part of the Agricultural Triangle. This zone is characterized by cool summers and a relatively humid climate. The main crops are flax, used for

The opening session of the Nineteenth All-Union CPSU Conference held in the Kremlin Palace of Congresses on June 28, 1988.

the production of linen and linseed oil, and some oats and barley. Dairying is an important activity, particularly in the Baltic region, which forms the western end of this zone.

The central zone occupies most of the deciduous forest area and extends eastward to the Urals. The predominant agricultural activities are the growing of grain, mainly rye, and potatoes, along with suburban dairying and vegetable growing for the markets of the cities of the region. In the southern parts of this zone, wheat is gradually replacing rye.

The southern zone corresponds with the area of the steppes. In this dry grassland environment with its chernozem/mollisol (black earth) soils, wheat grows well, especially in the southern Ukraine where precipitation is greater and more regular than in the steppes of Siberia and Kazakhstan. Wheat is the most important grain crop in the Soviet Union and accounts for about half of the total area sown with grain. Other grains, such as barley and oats, are also grown in this region. Corn is, however, an unsuitable crop for the dry steppes, and attempts to introduce corn as a crop throughout the Soviet Union in the 1950s were unsuccessful, mainly due to the lack of a suitable environment, but also partly due to the inexperience of Soviet farmers in growing and processing corn. In addition to grain, the most important crop is sunflowers. This plant is the major source, along with flax, of vegetable oil. The lack of access in the past to tropical sources of vegetable oils, along with the

more recent unwillingness to import raw materials, has encouraged the growing of sunflowers. They are also used to supplement protein in livestock fodder.

The southern zone produces most of the Soviet Union's sugar beets. They are grown in a belt extending from the central Ukraine to the Volga. This belt coincides with the area known as the forest steppe, where the grasslands merge with the deciduous forests. Again, the lack of tropical colonies promoted the growing of a domestic supply of sugar. Today, the Soviet Union does not meet its sugar requirements from its own production and imports about 5–6 million tons a year, mainly from Cuba.

In all three agricultural zones, livestock, mainly cattle and pigs, are kept. Cattle are of breeds that give both meat and milk, and there are no distinct regions where cattle are kept exclusively for either meat production or dairying.

Apart from the three major agricultural zones, there are several regions of specialized farming. The warm summers and long growing season in Central Asia have encouraged the growing of cotton with the help of irrigation. Cotton is grown in this region to the virtual exclusion of all other crops. Most of Central Asia's grain comes from outside the region in order to reserve the maximum acreage for cotton. In other countries with irrigated farming, the area of land devoted to food crops is generally much higher. In Pakistan grain crops occupy an acreage about five times that of cotton, while in the United States much of the best irrigated land is used to grow high-value crops, such as off-season vegetables and fruits for the urban markets of the north. Fruit, especially melons and peaches, are grown in Central Asia, but mainly for local consumption. The virtual monoculture of cotton in Soviet Central Asia is a good example of how the Soviet planners attempt to utilize the resources of a region, in this case climatic, to develop a particular specialization that will benefit the Soviet Union as a whole. It also demonstrates the desire for self-sufficiency in raw materials by the maximum domestic development of a major ingredient of the textile industry.

The Caucasus region has warm summers and, in general, more precipitation than Central Asia. Some cotton is grown there, but in the areas of higher precipitation along the Black Sea coast and in the Colchis Lowlands, citrus fruits (oranges, tangerines, and lemons) and tea are grown. The latter crop is important as the Soviet peoples are predominantly tea drinkers, coffee being relatively unpopular. Grapes for the manufacture of wine are also grown there, although grapes are also found in Moldavia, the southern Ukraine, and Central Asia.

The lack of sufficient land with climate and soils of the quality of the U.S. Midwest or parts of the South has restricted the choice of crops in Soviet agriculture and limited the use of such a valuable fodder crop as corn. The desire of the Soviet government to be as self-sufficient as possible has also dictated the priorities for certain crops such as cotton and sugar beets. Another factor has been the lack of experience on the part of Soviet farmers with certain possible crops and their traditional adherence to others. It is probable that soybeans, which are versatile plants, could be grown successfully over wider areas of the country than at present, but as a crop they are unknown to most Soviet farmers. In the United States, the expansion in soybean produc-

tion in recent decades has enabled livestock production to meet not only the requirements of the domestic market, but also those of other countries. Through its large exports of soybeans and corn, the United States has become a major supplier of livestock fodder to the world, including the Soviet Union.

Environmental problems, inefficient farming methods, and poor organization have forced the Soviet Union to import large quantities of grain since 1963. In 1985, the Soviet Union bought a record 55 million tons of grain on the world markets, of which about one-third came from the United States and much of the rest from Canada, Argentina, and Australia. The imports of wheat, corn, and soybeans by the Soviet Union are more of an attempt to improve supplies of fodder for meat production than to obtain grain for the production of bread or other foods. In 1984, the per capita meat consumption in the Soviet Union was 134 pounds (61 kilograms),

FIGURE 6.9

Agricultural Regions of the Soviet Union

compared with 247 pounds (112 kilograms) in the United States. Attempts to achieve a goal of 165 pounds (75 kilograms) per capita by 1980 were not successful. Since that time, some increase in production has taken place, but inefficient use of fodder and poor breeds of livestock have prevented any dramatic achievements.

Organization of Agriculture

It is impossible to judge the performance of agriculture in the Soviet Union without some understanding of the way in which farming in that country has been organized. The collective farm, or **kolkhoz,** is the basic unit of agricultural production. Organized on the basis of a village or a number of smaller villages, the kolkhoz owns the land, which is given to it by the state, the ultimate owner of all land in the Soviet Union. The kolkhoz also owns most of the livestock and farm machinery. Each family owns a small ***private plot*** on which it can grow vegetables, fruit, and other crops for its own consumption or for sale at a nearby town market. Each family may keep a cow and a small number of other livestock and poultry. The kolkhoz is run by a committee with an elected chairman nominated by the Party and a clerical staff. Thus, the kolkhoz is under strong political control. The members are paid for the work that they put in on the kolkhoz lands. The income of the kolkhoz is decided by the prices paid by the state for its products. As the state is the only purchaser, it can control the level of prices and the standard of living of the farmers as well as influence the kinds of crops grown.

In practice, the private sector of collective farming is very important. Over 60 percent of the national production of potatoes and about one-third of the vegetables, milk, eggs, and meat come from the private plots, which occupy less than 3 percent of the total sown area of the country. Almost one-third of the nation's cows are pri-

An example of housing on a collective farm. Shown here is Molodyozhnaya (youth) Street in the village of Novosyolki, at the Suvorov Collective Farm, in the Vitebsk Region.

vately owned. However, less than one-quarter of these products are marketed since the majority are consumed largely by the kolkhoz families themselves. This private farming is intensive and produces much higher yields than do the lands of the kolkhoz.

Apart from the kolkhoz and the private plot, there is another form of agricultural enterprise, the state farm or **sovkhoz.** The sovkhoz, which is run by the state, not by a collective, is regarded as the ideal form of socialist farming organization. Its employees are paid wages like factory workers, irrespective of the yearly variations in farm income. Sovkhozes often have specialized roles as research or experimental farms. In general, they have a higher level of mechanization and employ more technicians than kolkhozes. Attempts to amalgamate kolkhozes into larger units and convert them into sovkhozes were abandoned some years ago, and the kolkhoz is still an important unit in Soviet agricultural production.

The problems of Soviet agriculture arise from a combination of environmental and organizational factors. Environmental limitations have already been noted above. Organizational problems are tied to ideology and the planning approach to farming. It is difficult, if not impossible, to organize farming along the lines of industry, with yearly plans and targets. Attempts to convert peasants into factory workers did not succeed. The kolkhoz remains a halfway house between private and socialist farming. It has, however, turned farmers into shift workers who drop their work at five o'clock irrespective of the urgency of completing the harvesting, sowing, or plowing. The continuing existence of the private plot remains an ideological anomaly in a communist country.

The kolkhoz system provides little incentive for its workers, and productivity is low. Wastage on a large scale, inefficient use of machinery, and poor transportation and storage facilities are common. Large quantities of foodstuffs are lost every year for these reasons. Many young people have left farming for city jobs, and it is often difficult to find an expert manager for a kolkhoz. In spite of the movement of people from agriculture to industry in recent decades, about 20 percent of the labor force is still active in farming, a much higher proportion than in most Western countries.

If the recent reforms are put into action, the structure and organization of the farming system will be modified. The reforms established a single large ministry to handle agriculture and strongly support the idea of private plots as an essential part of the system. It is possible that families on the kolkhoz may be permitted to farm land given to them by the kolkhoz under contract. It is too early, however, to predict how far these suggested changes may go. It is likely that the kolkhoz will undergo considerable organizational changes, but whether private farming will return on a wide scale is difficult to predict. There is considerable resistance on the part of kolkhoz families to the idea of private farming as they have little experience in choosing or marketing crops.

A lack of investment in agriculture has also hampered production. Until recently, the production of chemical fertilizers was insufficient for the requirements of agriculture. Although this situation has been rectified by large investments in the chemicals industry, the use of fertilizer is inefficient. The Soviet Union produces some 6 tons of grain per ton of fertilizer compared with 16 in the United States. Supplies of the right types of pesticides and herbicides are lacking. In spite of a large investment program for agriculture currently in operation, it is unlikely that major advances in production of farm products will take place without some major reforms in the organization of the system of farming.

FISHING

The lack of protein in the Soviet diet due to poor supplies of meat has been countered to a certain extent by the development of the fishing industry. In 1984, per capita consumption of fish was 39 pounds (18 kilograms) compared with 14 pounds (6.4 kilograms) in the United States. The Soviet Union and Japan lead the world in terms of their fish catch, and both countries send their fishing fleets far afield. The Atlantic Ocean accounts for about one-third of the Soviet catch. The major port of the Atlantic fleet is Murmansk. Many fish are caught in the nearby Barents Sea, but the Atlantic coast of North America is also a favorite fishing ground. The Baltic Sea also produces some fish. About half of the Soviet catch comes from the waters of the Pacific coast and from the Amur River. Salmon and crab from the Sea of Okhotsk are exported, mainly to Western Europe. Vladivostok is the main fishing port for the region.

MANUFACTURING AND INDUSTRY

If the development of Soviet agriculture has been less than spectacular, the same cannot be said of industry. Since the Revolution, the Soviet Union has changed from a predominantly agrarian country into an industrial giant comparable with the United States and Japan.

Before the Revolution, Russia had developed an iron and steel industry near Moscow and in the Ukraine based on coal supplies in the latter region. The Ural region also had a metallurgical industry based on local supplies of iron ore and other metals. The engineering industry was largely concentrated in St. Petersburg (now Leningrad) and Moscow. Textile manufacturing, using imported cotton, was located in the Moscow region. Much of this industry was financed by foreign capital. It was on the basis of this existing pattern of industry that Stalin began in his First Five-Year Plan (1928–1932) to transform the Soviet Union into a major industrial power. New iron and steel and engineering plants were built, and priority was given to the rapid development of heavy industry at the expense of consumer goods and housing. In subsequent years, other five-year plans followed these trends and resulted in the construction of new plants, railroad lines, canals, and power stations. New mineral and energy deposits were exploited and industrial regions established. World War II accelerated the development of the industry of the Urals and Western Siberia. By 1984, the Soviet Union was first in the world in the production of oil, natural gas, iron and steel, iron ore, chemical fertilizers, and cement. In spite of these successes, it must be remembered that this industrialization took place at a forced pace with great suffering to the population and that it has not provided the Soviet population with the range of consumer goods, necessities as well as luxuries, that is found in most of the countries of the industrialized West. A high price was paid for industrial success and economic power, both by the industrial workers and by the peasants, who were forced into the collective farms where the state could organize supplies of food for the cities.

The planning of the Soviet economy is handled by a number of organizations at national, republic, and local levels. Most industrial enterprises are run by a ministry that develops plans for the enterprises under its control. Coordinating the plans at various levels is a complex task, and few plans, in fact, ever attain their goals. It must also be remembered that the state determines prices and wage levels and controls the production and allocation of all goods, from raw materials to finished products. It is thus not surprising that Soviet planners have been anxious to obtain modern computers to aid them in the development of more sophisticated planning techniques.

The planning system leads to the establishment of priorities. For many years, the Soviet government gave the highest priority to heavy industry, including armaments and equipment for the space program, and low priorities to consumer goods, housing, and agriculture. In recent years, greater attention has been given to agriculture and, to a lesser extent, consumer goods. The scale of priorities also affects the allocation of raw materials, investment, and the employment and training of the labor force. Increasing demands from the public for more consumer goods has led government agencies to revise these priorities, and it is even suggested that some armament factories may be turned over to the production of consumer goods. Various schemes for the conversion of the planned economy to a market-oriented one have been considered, but so far no definite decisions have been made.

Industrial Resources

The Soviet Union possesses a large variety of raw materials for industrial development, including energy and metallic and nonmetallic minerals. The variety and, in many cases, the quantity of these resources are impressive and compare very favorably with those of other industrial countries. For the Soviet Union, the problem is not one of variety and volume, but of location. Soviet industrial development has been made more difficult by the fact that many of the deposits of the most essential raw materials are located in inaccessible areas or far from the major regions of population and industry. It is important to realize that a comparison of the number and size of mineral deposits of the Soviet Union and of countries like the United States or the People's Republic of China is not very meaningful if the factor of availability is not taken into account. These remarks apply especially to Soviet energy supplies. Much of the country's coal, oil, natural gas, and hydroelectric power resources are located in Siberia, while those in the European region are somewhat unevenly distributed (Figure 6.10).

Coal. Coal is the most important energy resource in terms of quantity, with reserves of some 9000 billion tons. About two-thirds of this coal is bituminous, while the rest is lignite or brown coal. These reserves are enormous by world standards, and the Soviet Union faces no problem of exhaustion of supplies. However, about half of these reserves are located in the Tungus and Lena coal basins east of the Yenisey in Siberia, far from transportation routes.

As far as production is concerned, the most important

FIGURE 6.10

Energy Resources of the Soviet Union

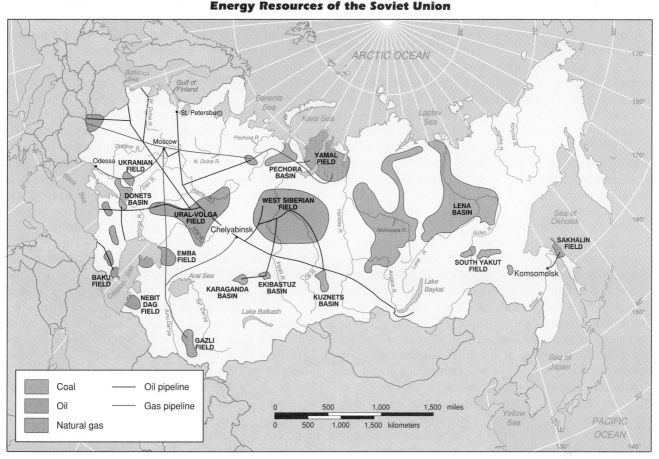

field is the Donets Basin (Donbass) in the eastern Ukraine. In 1985, it produced about 27 percent of total Soviet coal and about 45 percent of coking coal for the iron and steel industry. The quality of the coal is generally good, but the sulfur content is high. The complex geology and the great depth of the best seams make mining expensive. In addition, many seams are thin and difficult to work. In spite of these problems, the proximity of the Donbass to the industrial centers of the Ukraine and European Russia, and to the other raw materials required by the iron and steel industry, has helped to maintain its importance.

The Kuznets Basin or Kuzbass of Western Siberia follows the Donbass in volume of production. It accounted for some 20 percent of total coal production in 1985 and about 30 percent of coking coal. The coal is of good quality and can be cheaply extracted from thick seams at no great depth; some is mined by open-cut methods.

Kuzbass coal is transported considerable distances to supply Ural industry and also for export to Japan. Kuzbass coke is much cheaper than Donbass coke and may be shipped as far west as the Moscow region.

Other important coalfields are located in Kazakhstan. Karaganda produced about 7 percent of the total coal mined in 1985. This field was developed in the 1930s to supplement and, if possible, replace deliveries of Kuzbass coal to the Urals. Karaganda ia situated about 650 miles (1050 kilometers from the Urals, compared with 1100 miles (1800 kilometers) in the case of the Kuzbass. The coal has a high ash content and has to be mixed with Kuzbass coal for metallurgical purposes. As well as being sent to the Urals, it is also used in a local iron and steel plant and to produce electricity. Another Kazakhstan coalfield at Ekibastuz is easily mined by open-cut methods and now outproduces Karaganda. It accounted for 11 percent of total production in 1985.

This coal is not of good enough quality for making coke for industry, but is used for electric power production in Kazakhstan and the Urals.

The Pechora coalfield produced less than 4 percent of total national coal production in 1985, but is important as a supplier for Cherepovets, the iron and steel center for the northern European region of the country. The Pechora field lies north of the Arctic Circle and only became of real importance when the Donbass was occupied by the Germans in World War II. A railroad was built to bring the coal to the industrial centers in the south. The coal is of good quality, but the cost of extraction is high, and the cost of transportation also limits the further growth of the field. There has been talk of building a railroad from the mining town of Vorkuta directly sought to the Urals or developing a water route by diversion of rivers, but so far nothing has come of these ideas.

Other coalfields are relatively minor or produce low-grade coal. Potentially, the most important of these minor fields are the South Yakut field and the Kansk-Achinsk field, both located in Siberia. The former lies to the south of the Lena Basin and was reached in 1978 by a branch line from the new Baykal-Amur railroad, which is currently under construction. Coking coal from this field should shortly replace Kuzbass coal for export to Japan. The Kansk-Achinsk field is a vast lignite field located east of Krasnoyarsk on the railroad. The lignite is mined by open-pit methods. Due to its softness and low heat-producing capacity, it cannot be transported far. Therefore, large thermal power stations are being built on the field to provide power for the region. This will release Kuzbass coal for use in the Urals. The cost of extraction of Kansk-Achinsk lignite is 50 percent less than that of coal mined in the Kuzbass by open-cut methods and is the lowest in the Soviet Union. However, the poor quality of the coal seems to be the reason for recent delays in the development of the Kansk-Achinsk project.

Another major lignite field is located near Moscow. In the past, it was used to provide fuel for heating and for thermal power stations in the Moscow region, but has increasingly been replaced by natural gas. A little coal is produced in Georgia and in southern Central Asia, as well as in the western Ukraine. This production is of some importance for local industry and power production.

The Soviet Union, thus, relies mainly on the Dunbass, Kuzbass, Karaganda, and Pechora fields to supply coal of coking quality that can be transported for use in other regions of the country. In 1985, about 60 percent of coal mined came from these four fields.

One of the main problems of the Soviet coal industry is the depletion of the older mines in the European part of the country and the necessity of sinking ever-deeper shafts to reach new seams. The Siberian fields are not being developed fast enough to replace them due to locational, technical, and economic factors and labor shortages. The share of coal in the Soviet fuel balance is, at present, about 23 percent compared with 66 percent in 1950. The Soviet authorities would like to raise this proportion to about 30 percent. Development of new mines, most of them in Siberia, will take time, however, and it is unlikely that coal production will rise dramatically in the near future.

Oil.   In 1950, coal accounted for 66 percent of total fuel production and oil 17 percent. In 1987, the proportion of coal fell to 21 percent and oil rose to 40 percent. The production of petroleum has grown dramatically to a total of 624 million tons in 1988, compared with 38 million tons in 1950. Since 1976, the Soviet Union has been the world's largest producer, accounting for about 20 percent of world output. The Soviet Union has proven petroleum reserves about twice those of the United States. It has become a major oil exporter; about 25 percent of its production is sold abroad. About half goes to hard-currency countries and the rest to Eastern Europe and the developing world. Petroleum is thus an important earner of hard currency for the Soviet Union. One of the limiting factors in increasing production has been the lack of modern equipment, which the Soviet Union has been attempting to obtain in larger quantities from the West.

Apart from exports to the West to earn hard currency, the Soviet Union has sold petroleum to developing countries at reduced prices in order to reduce their reliance on oil from Western companies. A tendency prior to the Iraqi invasion of Kuwait in 1990 had been to increase exports to such countries as India and Brazil, while selling reduced amounts of oil to the West for higher prices. The Soviet Union is also determined to be self-sufficient in petroleum, a fact that will affect the level of exports in the future if production falls or consumption within the Soviet Union increases.

The earliest location of the oil industry was at Baku on the Caspian Sea and in the foothills of the North Caucasus. By World War II, 90 percent of the country's petroleum was still coming from this region, but by 1965, 72 percent of oil production came from a new field located between the Volga and the Ural Mountains. The cost of production from the new field was low, and a system of pipelines was built from it to various parts of the country and to Eastern Europe. In the 1960s, exploitation of newly discovered fields in Western Siberia began.

In spite of difficult problems posed by the remoteness of the area, the climate, the swamps, and the permafrost, the development of West Siberian petroleum has taken place at a rapid rate. In 1965, West Siberia produced 1 million tons of oil. Ten years later production reached 148 million tons and in 1985, 372 million tons. In 1987, the region produced 67 percent of total national production. The proportion of petroleum from the Ural-Volga field has dropped to 20 percent and from the Baku field to 2 percent. West Siberian oil is purer than that from the Ural-Volga field. Conditions in the region are favorable for the discovery of further oil deposits. Other potential fields may lie along the Arctic coast, providing further environmental and locational challenges.

There are several minor fields in scattered locations. The largest is in northwestern Kazakhstan, while there are some small fields in the western Ukraine. Fields on Sakhalin Island have a small production, amounting to 3 million tons annually. This petroleum is of regional importance as it supplies about 40 percent of the Far East region's requirements. In Estonia oil is produced from oil shale.

At present, it is impossible to predict any further major shifts in the location of Soviet oil production. It is probable that the major fields in relatively favorable areas have all been located and that further discoveries will probably be in increasingly unfavorable regions such as the Arctic. Already the exploitation of the West Siberian fields has placed the major part of the Soviet Union's oil production in a relatively undesirable region. This trend is not likely to change in the future.

Natural Gas. Natural gas has emerged as the most recent addition to Soviet energy resources. In 1950, production was only 204 billion cubic feet (5.8 billion cubic meters), or 2.3 percent of total fuel production. By 1970, it had reached 7000 billion cubic feet (198 billion cubic meters), and in 1988, 25,356 billion cubic feet (718 billion cubic meters), or 38 percent of total fuel production. The largest fields are in the Arctic, especially in the area around the estuary of the Ob, on the Yamal Pen-

insula to the west of the estuary, and at Urengoy to the east. These fields are said to have reserves greater than any in the world. In 1985, 59 percent of Soviet natural gas production came from Siberia, with 19 percent from Central Asia, 7 percent from the Ukraine, and the rest from the Ural-Volga field and northern European Russia.

In spite of having about 35 percent of world reserves, the Soviet Union has lagged behind the United States in the production and use of natural gas and has not yet developed the dense network of gas pipelines found in the latter country. Nevertheless, the Soviet Union has begun to use natural gas as a substitute for other fuels in any number of industries, especially in the fuel-deficit Moscow region. Gas is used in the chemical industry and the iron and steel industry, as well as for domestic heating. This rapid growth in gas production has resulted in a surplus that can be exported abroad. The main problem facing the utilization of the natural gas fields has been the remoteness of the largest reserves and the difficulty of manufacturing or procuring large quantities of pipes and other equipment for long-distance transmission. In fact, the fuel-deficit Transcaucasus region imported Iranian gas from 1967 to 1980, when the agreement was not renewed because of disputes over prices. In 1986, deliveries were resumed. This importation is necessary due to the difficulty of bringing Siberian or Central Asian gas to the region. It also helps to release gas from the northern fields for export. The Soviets hope to use the vast reserves of the West Siberian fields as a source of hard-currency earnings by transporting the gas through a pipeline from the Urengoy field to Western Europe, completed in 1984. The plan was controversial, in that it would result in former West Germany obtaining about 30 percent of its gas supplies from the Soviet Union. However, the changing political atmosphere in Europe makes this reliance on Soviet supplies less of a problem. The construction of the 3300-mile (5500 kilometer) pipeline probably cost between $10 and $15 billion. The Soviet Union hopes to earn some $9 billion a year in hard currency from the sale of this gas.

Five other major pipelines are under construction to bring more Siberian gas to the industrial regions of the country. This drive to exploit natural gas reserves has already made the Soviet Union the second largest producer in the world after the United States, and in the near future it may surpass the latter. There is no doubt that the Soviet government has given the highest priority to the production of natural gas as the fuel of the future to replace the sagging production of coal and oil. Natural gas is one of the few success stories in recent Soviet industrial development, even though this has had to be accomplished with considerable Western technical aid.

Electricity.   From an early period, the Soviet government put a high priority on the development of electric power and its transmission to both urban and rural areas. In particular, emphasis was placed on the building of hydroelectric power plants and dams. In the eyes of the Soviet planners, power produced from water was virtually free. In a system where interest on capital was not taken into account, the large amount of capital tied up for lengthy periods during the construction of a dam and power plant did not appear in cost calculations. Labor was cheap, and in some cases forced labor was used. The first major power project was the Dneproges plant on the Dnieper River. Completed in 1932, it had an eventual capacity of 550,000 kilowatts, making it the biggest in the world, until the construction of the Grand Coulee dam in the United States in 1942. The major period of hydroelectric power plant construction started in 1950, when work on two big dams on the Volga began. The first completed was the Kuybyshev plant in 1957 with a capacity of 2.3 million kilowatts. The Volgograd plant began operating in 1958 with a capacity of over 2.6 million kilowatts, making it the world's largest at that time. Much of the electric power produced by these dams is transmitted by high-voltage lines to Moscow and the Urals. The large reservoirs formed on the Volga by the dams are also used as sources of irrigation water. The main problem encountered by the Volga plants is the uneven flow of the river, which leads to fluctuations in power production. After the opening of these plants, the Soviet planners began to have some doubts about the huge amounts of capital invested in them. It was claimed that the same amount of capital invested in the construction of thermal power plants over the seven-year period required to build the Volgograd plant would have produced a total capacity of over 12 million kilowatts. However, the temptation to harness the hydropower resources of the country was too great, and work on hydroelectric plants continued.

Before the opening of the Volga power stations, work had already begun on a major dam on the Angara River in Eastern Siberia. The Angara is a tributary of the Yenisey and has its source in Lake Baykal. The Angara and the Yenisey together have the greatest hydropower potential of any major river system in the country, totaling

about 30 million kilowatts. In all, the East Siberian and Far East regions have about 40 percent of the total hydropower potential of the country. It was this vast unused potential that was so attractive to Soviet planners, in spite of the distance of the region from any major industrial centers. The Bratsk dam and power plant were completed in 1961, with a capacity of 4.1 million kilowatts. In 1956, construction had also started on an even larger dam at Krasonyarsk on the Yenisey. It went into operation in 1967 and has a capacity of 6 million kilowatts. Another major plant on the Angara at Ust-Ilimsk was completed in 1975 with a capacity of 3.5 million kilowatts, while the giant of them all, the Sayan dam, was completed in 1978 with a capacity of 6.4 million kilowatts. Ultimately, seven hydroelectric plants are planned for the Yenisey and four for the Angara.

This huge capacity for power production has posed another problem for the planners: finding uses for the power. Some of this power has been used for the electrification of the Trans-Siberian Railroad, while power-hungry industries, such as aluminum and pulp and paper industries, have been developed in the region. The aluminum plants at Krasnoyarsk and Bratsk are the largest in the Soviet Union. The development of the long-distance transmission of electric power has received a high priority, and work on a national power grid is under way, with the aim of transmitting Siberian power to the industrialized regions of European Russia, where 70 percent of the power is consumed.

Other hydropower plants have been built on the Dnieper and other rivers in the European part of the Soviet Union and in the Caucasus and Central Asia. In spite of this present and past large-scale investment in hydropower, only 13 percent of total Soviet electric power production in 1987 came from water power, compared with 19.7 percent in 1958. In 1988 hydropower accounted for 10 percent of total electric power production in the United States. The spectacular nature of huge hydroelectric projects, plus the great publicity given to them by the Soviet government, have exaggerated their importance and their role in Soviet energy production.

After 1958, although construction of hydroelectric projects was not curtailed, a greater emphasis was placed on the building of thermal power plants. In the 1960s, a major increase in the building of large thermal plants took place, mainly on the Siberian coalfields. The idea is to use low-grade coal in mineside power plants to produce electricity that can be used locally or transmitted to other regions. A number of large thermal stations have been built or are under construction on the Kansk-Achinsk lignite field and at the Ekibastuz coalfield in Kazakhstan. The Berezovskaya plant at the Kansk-Achinsk field will have a capacity of 6.4 million kilowatts, as large as the Sayan hydropower plant. Several similar plants are to be built at the field. The construction of such mammoth plants in a region where large hydroelectric plants already are in operation will help even out the fluctuations in electric power production due to seasonal changes in the flow of the rivers or during periods of maximum icing. Not all the giant thermal plants are located in Siberia or Kazakhstan. Near Kostroma, northeast of Moscow, a large power complex is being expanded to have a capacity of 4.8 million kilowatts, while one at Konakovo, north of Moscow, has a capacity of 2.4 million kilowatts. These plants, and others in the European part of the country, use natural gas and oil as fuel. Several Central Asian power stations also operate on natural gas.

The Soviet Union has been a pioneer in the field of nuclear power development. One of the first nuclear power plants in the world went into operation at Obninsk, near Moscow, in 1954 with a capacity of 5000 kilowatts. Since that time, a number of nuclear power stations have been built in most parts of the country, although, at present, priority is given to the energy-

deficit regions west of the Urals. In 1988, 56 nuclear power stations were producing 12.6 percent of total electric power, compared with 1.9 percent in 1975. In 1988, the United States had 108 nuclear plants, producing 20 percent of total electric power. Soviet planners have placed great hopes on the rapid development of nuclear plants, although the disaster at the Chernobyl power plant in the Ukraine near Kiev must give them much food for thought. Soviet engineers have shown no concern about the potential dangers of nuclear plants, even though other serious accidents are known to have taken place. In the past, the lack of adequate shielding and backup cooling systems presented potential problems. An added problem is that most of the plants are in populated areas near cities, as was the case with Chernobyl, and any accident may affect a relatively large number of people. The advantage of an urban location is that the waste heat can be used for heating or in local industry. As in other countries, storage of nuclear waste is a problem. Old coal and salt mines have been used. In the winter of 1957–1958, strontium-90 escaped from an explosion in a nuclear waste storage area in the Urals and contaminated a considerable area, causing hundreds of casualties.

The first major nuclear plant went into operation in the 1960s near Sverdlovsk in the Urals and near Voronezh. In the 1970s and 1980s, plants were built at St.

Construction of the protective wall of unit 4 at the Chernobyl nuclear plant in August 1986. Chernobyl became the site of one of the world's major nuclear accidents.

Petersburg, in Armenia, on the Kola Peninsula, and in the Ukraine and Belorussia. A small nuclear plant operates in northeastern Siberia, and a plant produces plutonium at a secret location somewhere in Siberia. A number of new plants are under construction or on the drawing board, but it doubtful that all will be completed. Soviet planners clearly hope that the greater use of nuclear power will reduce their reliance on conventional fuels. The Soviet Union has developed a reactor export program based on the large new $4 billion Atommash factory at Volgodonsk near the Black Sea. These mass-produced reactors will be sold abroad. Soviet nuclear plants have been built in Eastern Europe and Finland, and others are planned for Cuba and Libya.

Soviet energy policy has undergone some changes in the postwar period as old sources of energy have become depleted and new sources have been found. As in the case of other industrial raw materials, the location of major new reserves in regions east of the Urals creates problems for the future development of coal, natural gas, and hydroelectric resources. In spite of great efforts, the Soviet Union, in 1987 produced only 5881 kilowatt-hours of electricity per capita compared with 11,313 kilowatt-hours in the United States. Since the Chernobyl accident, a policy of energy conservation has been announced, with the possibility of reduced electric power supplies, not only to Soviet cities but also to some sectors of industry. The Soviet Union exports electricity to Finland and Turkey, as well as Eastern Europe, but it is not clear how these reduced supplies will affect export.

Metallic Minerals.   The Soviet Union has a rich variety of metallic minerals and, with the exception of a few metals such as tin and tungsten, is self-sufficient in most of them. Since the Revolution, the Soviet government has launched major geological explorations in remote areas of the country, resulting in the discovery of many, and in some cases, major deposits of a wide variety of minerals.

With a major emphasis on the development of heavy industry, the discovery and exploitation of ferrous metals have received much attention. The first supplies of iron ore for early Russian industry came from the Tula area to the south of Moscow and from the Ural region. The discovery of the large deposit at Krivoy Rog in the Ukraine close to the Donbass coalfield, and its exploitation on an increasing scale after 1881, enabled a major new industrial region to develop. In 1987, 47 percent of total Soviet iron ore production came from the Ukraine,

of which about 90 percent was from Krivoy Rog. The Urals accounted for 9 percent of total production, compared with almost the total national production in 1860. In the 1930s, the exploitation of a major iron ore deposit near Kursk, known as the Kursk Magnetic Anomaly or KMA, began. In 1987, the field supplied 20 percent of total production. Much of the ore is mined by open-pit methods and is of good quality. Most of the ore is used in the iron and steel industry in the Moscow region, but some goes to the Urals. In 1987, 10 percent of iron ore came from Kazakhstan, where it is mined southwest of Karaganda and sent to the iron and steel plants at Karaganda and also to the Urals. Other sources of iron ore are the Kola Peninsula and Karelia, near the Kuzbass in West Siberia, and the Transcaucasus. In 1987, Soviet production was seven times that of the United States and is sufficient to meet all the country's requirements. Iron ore is also exported, much of it to Eastern Europe.

The Soviet Union has good supplies of metals for use in the production of steels of various types. Manganese ore is mined in quantities that make the Soviet Union the largest producer in the world, and about one million tons a year are exported. Production figures for other metallic ores are generally not published, but the output of nickel is known to be large, most coming from Norilsk in West Siberia, the Urals, and the Kola Peninsula. Molybdenum is mined mainly in Armenia, East Siberia, and Kazakhstan, while cobalt comes not only from Norilsk, but also from the Kola Peninsula and the Urals. Chromium comes mainly from Kazakhstan and a little from the Urals. About half is exported, mainly to the West.

Supplies of nonferrous metals are also ample. The Soviet Union is the third largest producer in the world of copper, which comes mainly from northern Kazakhstan, the Urals, and East Siberia. It is probably the second largest producer of lead and zinc in the world, with mining in Kazakhstan, the Urals, and other locations. Both metals are exported. The Soviet Union has supplies of low-grade bauxite and nepheline for the manufacture of aluminum. Over 70 percent of aluminum production comes from bauxite and a quarter from nepheline. Some bauxite is imported from Hungary and Greece. The Soviet Union also has good supplies of metals such as titanium and vanadium and rarer metals, including selenium and tellurium, which are used in the manufacture of rockets, jet engines, aircraft, and nuclear plants. The only major metal in short supply is tin, which has to be imported. Most Soviet tin comes from the Far East region and East Siberia. It is hoped that the new Baykal-

Amur railroad will facilitate the exploitation of deposits of tin as well as those of other metals, such as copper, lead, zinc, nickel, and titanium, which have been inaccessible until now.

A metal that has been of great importance to the Soviet Union, although not as a raw material for industry, is gold. Although no official information on Soviet gold production or reserves is available, the Soviet Union is known to be the world's second largest producer after South Africa. Gold is used for foreign exchange requirements, and, along with sales of oil and natural gas, it forms a very important means for obtaining needed food and technology from abroad. About 80 percent of production is probably sold abroad, and in some years gold from reserves is also exported. Two-thirds of supplies come from northeastern Siberia and the Far East, with the rest from the Urals, Kazakhstan, Central Asia, and the Caucasus.

Nonmetallic Minerals. The Soviet Union has adequate supplies of most nonmetallic minerals, although mica, fluorspar, and barite have to be imported. The chemical industry uses large quantities of sulfur, which is available in large reserves of relatively poor quality. Apatite, a mineral associated with nepheline, is used as a material for the manufacture of super-phsophate fertilizer. Most supplies come from the Kola Peninsula. Phosphorite is also mined in Kazakhstan. Potassium for the manufacture of potash fertilizer comes mainly from the Urals and Belorussia.

As in the case of gold, the Soviet Union is fortunate in possessing large deposits of another rare and valuable mineral, diamonds. Production is about 10 million carats a year, of which 80 percent are industrial and the rest of gem quality. Part of this production will be sold by DeBeers, a South African company. The main area of diamond mining is in the Yakut ASSR, with large open-pit mines at Mirny on the Vilyuy River and placer mines to the north. The Soviet Union is one of the largest producers in the world, along with Zaire, South Africa, and Botswana. Along with gold and fossil fuels, diamonds form a major source of foreign currency earnings.

The Iron and Steel Industry

In line with the high degree of priority given to the development of heavy industry under the series of five-year plans beginning in 1928, it is not surprising that the iron and steel industry has seen rapid development (Figure 6.11). Soviet planners have regarded the iron and steel industry as a major catalyst for other industries. In their plans for regional development, the establishment of an iron and steel plant was seen as an important first step in creating a supply of metal for a variety of other manufacturing industries.

The first center of iron production in Russia was at Tula, south of Moscow. The discovery of iron ore deposits in the Urals resulted in the development of the Ural region as the major producer until the development of Donbass coal mining and the rise of the Ukrainian iron and steel industry. By 1913, the Ukraine produced almost three-quarters of the country's iron.

After the Revolution, some major changes in the rate of development and the location of the iron and steel industry took place. The First Five-Year Plan (1928–1932) saw the creation of the Ural-Kuznets Combine, a scheme for the erection of large iron and steel plants in the Urals and in the Kuzbass. A shuttle service by rail brought Kuzbass coal to the Ural plants and Ural iron ore to the Kuzbass, a distance of some 1200 miles (1940 kilometers). The basic idea, apart from obtaining a coal supply for the Ural industry, was to decentralize the Soviet iron and steel industry by establishing new metallurgical bases far from the border in the Urals and Siberia. The Donbass was perceived as an insecure region, as it had almost been detached from Russia in 1918, when the Ukraine became briefly independent, and was in the direct path of an invader from the west or via the Crimean Peninsula. In 1933, Magnitogorsk produced its first steel and, with expansion of its facilities, became the largest iron and steel plant in the country with a capacity of 17 million tons of pig iron in 1980.

The Ural-Kuznets Combine was a great aid to the Soviet Union in World II as it enabled steel to be produced after the occupation of the Donbass by the Germans. After the war, however, a search for alternative sources of raw materials was intensified. The high costs of the operation of the combine, including the tying up of large numbers of railroad cars for lengthy periods and the intensive use of part of the Trans-Siberian Railroad, began to worry the government. Iron ore from the southern fringe of the Kuzbass, as well as from East Siberia, was used increasingly in the Kuzbass plants, while coal from Karaganda was brought to the Urals with the aim of reducing both the use of Kuzbass coal and the distance of hauling coal. At present, Kuzbass coal is still coming to the Urals, but the combine no longer operates at its past scale. A large new iron and steel plant was built near Novokuznetsk in the 1960s,

FIGURE 6.11

The Soviet Iron and Steel Industry

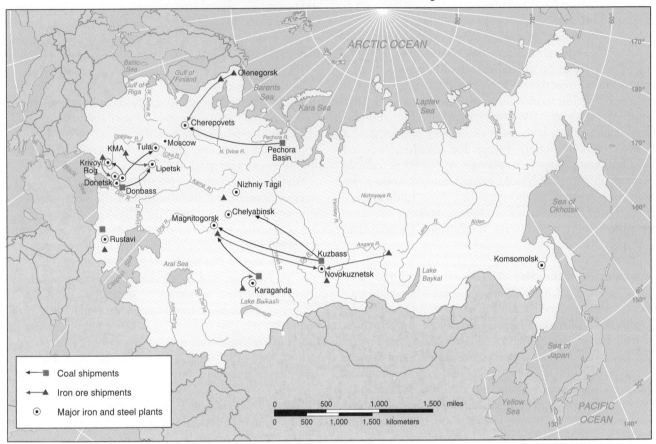

← ■ Coal shipments

← ▲ Iron ore shipments

⊙ Major iron and steel plants

and further expansion is planned with the aim of supplying most of Siberia's steel requirements from the Kuzbass.

Since World War II, investment in iron and steel plants in the Ukraine and European Russia has again increased. The most important development has taken place at Lipetsk, near the iron ore of the Kursk Magnetic Anomaly (KMA). Coal comes mainly from the Donbass. At present, the Lipetsk plant is one of the largest in the country. Another major plant in European Russia began operation in 1955 at Cherepovets, located equidistant between St. Petersburg and Moscow. It was built primarily to supply the heavy industries of St. Petersburg and the northwest. It uses coal from the Pechora Basin, which is brought over 900 miles (1500 kilometers) to the plant, and iron ore from the Kola Peninsula, some 750 miles (1200 kilometers) to the north. Some iron ore

also comes from the KMA, 620 miles (1000 kilometers) to the south. Because of these long and expensive hauls of raw materials, iron ore deposits in Karelia have been developed; these are located at half the distance of those on the Kola Peninsula. The plant also uses large amounts of scrap metal for steel production. Like the Ural-Kuznets Combine, the Cherepovets plant is a high-cost operation where strategic requirements outweigh economic ones and the state is willing to carry the losses.

Although the iron and steel industry of the Ukraine has mainly expanded around the original centers in the Donbass, the plant at Krivoy Rog, located at the iron ore field, has emerged as the largest. Founded in the 1930s, it has been expanded, especially in the 1970s, to become the second largest plant in the country after Magnitogorsk.

In 1987, the European part of the country, including the Urals, produced about 85 percent of Soviet steel, the Ukraine accounting for 34.8 percent and the Urals for 28.4 percent. Siberia, mainly the Kuzbass, produced 10.8 percent. Production in other regions was small. About 4 percent of total steel production came from the plants at Karaganda in Kazakhstan, which use local coal and iron ore. In 1960, a large new plant was added to an older one established in the 1930s with the aim of supplying the needs of Kazakhstan and Central Asia. In the Caucasus, a small iron and steel plant was established in the 1950s at Rustavi in Georgia using local coal and iron ore. Coal from the Donbass has to supplement local supplies, and the plant is expensive to maintain. However, it supplies most of the steel requirements of the region. Small plants where steel is produced from scrap metal or pig iron from other plants are found in a number of locations. Between the Kuzbass and the Pacific coast, there are no iron and steel plants of any significance. At Komsomolsk in the Far East, a steel plant was built in the 1930s to provide this isolated region of the country with steel in case of a war with Japan. The town had been founded in 1932 by members of the Young Communist League who came east to settle this uninhabited site. In 1987, Komsomolsk had 316,000 inhabitants. The plant uses scrap and pig iron from the Kuzbass and the Urals to supply about 40 percent of the region's steel requirements. Some local coal is used, but most of the steel is produced in electric furnaces. The plant is in fact an uneconomic operation and clearly owes its existence to strategic necessity.

Nonferrous Metallurgical Industries

Much of the smelting and other processing of nonferrous metal ores takes place at or near the locations where the ore is mined. However, because of the special requirement of the industry, the manufacture of aluminum takes place in a number of locations distant from raw material supplies. Large amounts of electric current are needed to produce the metal from alumina, which in turn is derived from bauxite, nepheline, or alunite. Thus, a location near a source of cheap and plentiful electric power is preferable for the final stage of aluminum production. The largest aluminum plants in the Soviet Union went into operation at Krasnoyarsk in 1964 and Bratsk in 1966, using the large amount of surplus electric power in the region. Other Siberian plants are at Novokuznetsk and near Irkutsk. Much of the alumina

comes from the Urals and from a plant at Achinsk, west of Krasnoyarsk. Other aluminum plants are located at Volgograd, in the Ukraine, and on the Kola Peninsula. Some aluminum is also produced in the Transcaucasus region.

The Engineering Industry

Mechanical engineering is a key industry in the Soviet Union and accounts for over 30 percent of the total volume of industrial production. Machine tool manufacture is particularly emphasized. Most of the production of machine tools comes from the European part of the country, especially the Ukraine and Belorussia. The manufacture of railroad equipment, including locomotives and rolling stock, is located in a variety of places in the European part of the country. The automobile industry, however, is located in only a few places. The first production of automobiles after the Revolution took place in Moscow, but in 1932 a large factory was built at Gorky with the aid of the American Ford Company. It produced mainly trucks, but at present, it also produces automobiles, as do plants in Moscow and at Izhevsk in the Urals. The two latter plants were equipped and designed with the help of the French Renault Company. Small cars are made at Zaporozhye in the Ukraine. Since 1966, the Soviet Union has invested about $5 billion in an attempt to develop a modern automobile industry. In that year the Soviet government signed a contract with the Fiat Company of Italy to build a large passenger car plant at Togliatti on the Volga. By 1973 the plant had reached its capacity production of 600,000 a year. It also manufactures cars for export. In spite of these major efforts to increase production, annual output of passenger cars is low. In 1987, it amounted to 1,332,000 compared with 7,085,000 in the United States. Nevertheless, this represented a large increase over the 344,200 produced in 1970.

The manufacture of trucks has also lagged in the Soviet Union in spite of their importance to the economy and the military. In an attempt to increase production of heavy-duty trucks, in 1971 the Soviet government began the construction of the world's largest heavy truck plant at Nabereznyye Chelny on the Kama River, some 600 miles (970 kilometers) east of Moscow. This Kamaz plant was built with Western help, including engineering and designing work from the United States, France, West Germany, Italy, and Japan. It began operation in 1976. Trucks are also manufactured at a variety of other locations, including Gorky, Moscow, Minsk, and Kutaisi in Georgia. At present, several foreign automotive companies are negotiating with the Soviet authorities with a view to entering the Soviet market. It is likely that the Soviet automobile industry will see considerable restructuring in the future.

The location of the agricultural machinery industry exhibits some interesting features. At an early date, the necessity to produce large numbers of tractors and other equipment for the modernization of agriculture and the efficient working of collective and state farms became apparent. In particular, equipment was needed to plow and cultivate the vast areas of the steppes. Thus, many of the major tractor and combine plants were located on the fringes of the steppe regions, reducing the necessity to transport bulky machinery any distance. The major concentration is in the Ukraine, with the largest farm machinery plant in the country at Rostov, located between the Ukraine and the steppe lands of the North Caucasus region. Tractors are manufactured at Volgograd, Kharkov, and Chelyabinsk in the Urals. The largest producer is the factory at Minsk, which makes light tractors, some of which are exported to the West, including the United States. Other farm machinery plants serve the farmers of the Siberian and Kazakh steppes and are located at Omsk, Tselinograd, and Pavlodar. A plant at Tashkent produces cotton picking machines for Central Asia.

Other branches of engineering, such as the construction of ships and aircraft, have received a high degree of priority. In general, the mechanical engineering industry has seen a rapid rate of growth since the Revolution, and, in spite of the specific location of certain branches of industry, most cities have one or more factories producing engineering products.

A tractor works in Minsk, Belorussia. Many of the manufacturing plants producing heavy construction and farm equipment are located in the Ukraine and Belorussia.

The Chemicals Industry

The chemicals industry has not shown the rapid rate of growth of metallurgy or engineering. In fact, the Soviet Union lags behind much of the industrialized West in the development of this important industry. This is partly due to a rather outdated view of industrialization, whereby the iron and steel industry was viewed as the essential ingredient in regional development at a time when other countries were investing in their chemical and, particularly, their petrochemical industries. Only in the 1950s did the Soviet Union give a higher priority to the chemicals industry, especially in view of the lack of fertilizers for agriculture. In general, the industry is located with reference to raw material supplies, but in cases where the final product is difficult or dangerous to transport, manufacturing plants are located near the market. The production of synthetic rubber, petrochemicals, and synthetic fibers has received considerable expansion in recent years, and these branches of the industry are located in a wide variety of places.

The Textile Industry

The textile industry still retains the basic locational pattern that it had before the Revolution, when it was one of the largest branches of Russian industry. Until the conquest of Central Asia in the 1860s, most of Russia's cotton supply came from the United States. The American Civil War and the drop in cotton exports from the South were partially instrumental in hastening Russia's desire to control cotton-producing areas of its own. The cotton textile industry was established in a number of small towns to the northeast of Moscow, particularly in Ivanovo, and the industry is still mainly located here. Recently, new factories have been opened in the cotton-growing areas of Central Asia and the Transcaucasus. Linen is also manufactured in the same centers that produce cotton textiles, as well as in the flax-growing regions of Belorussia and the Baltic republics. Some wool and silk are also produced.

The Timber, Pulp, and Paper Industry

With about 30 percent of the country covered by forests, it is not surprising that the Soviet Union has industries based on the processing of timber. During the early days of the Soviet regime, the export of timber helped to earn foreign currency, and, at present, Soviet timber is exported to Western European countries and Japan. Although 75 percent of the forest resources are located east of the Urals, most of the timber is cut and processed in European Russia. Apart from the problem of transporting timber long distances from Siberia to the industrial centers of the west, the quality of Siberian timber is not always good. In particular, the larch, which is the prevalent tree in Siberia, is not very suitable for lumber. The best-grade lumber comes from northern European Russia and is processed in a number of centers located near the market. The major port for export of timber is Arkhangelsk, where the largest sawmill in the Soviet Union is located. In Siberia, the large amounts of available hydroelectric power have been used for the production of wood pulp and cellulose. The largest wood pulp plants in the country are at Bratsk and at Ust-Iimsk, while other plants are located near Lake Baykal. The major port for the export of Siberian timber is Igarka on the Yenisey. Lumber and pulp from the forests of the Far East region are exported to Japan.

The problem of distant resources and high transportation costs is particularly obvious in the case of Soviet timber. Whenever possible, water transportation is used to move bulky unprocessed timber, while rail is used to move lumber and pulp from the processing centers to the market. Fortunately, the forest resources of European Russia are still extensive, but an increase in the rate of cutting will make further exploitation of the Siberian forests necessary.

Light Industry

Until recently, the production of consumer goods has received a very low priority, and although output of clothing, footwear, household equipment, and other goods of this type has increased considerably over the years, the standard of living of average Soviet citizens lags noticeably behind that of their counterparts in the West. Gorbachev's program of perestroika is intended to correct this problem, but attempts to vitalize the consumer goods and housing construction industries have so far met little success, and the standard of living of the average worker has, in fact, deteriorated.

Industrial Location

Soviet economic geographers and planners have been critical of what they consider the irrational location

Boats loaded with lumber at moorage at Cape Astafyev. The Soviet Union exports lumber to Japan, Cuba, and other trading partners.

of industry under the capitalist system. According to their ideas, the profit motive does not lead to the optimum location of industry from the point of view of the national interest, nor does it lead to the development of the industries most needed for the country or the region. They claimed that under the communist system the planning of the economy in all its aspects leads to a rational and efficient distribution of industry and an even, balanced growth in industrial production. In particular, the development of the regions of the country were planned so that economic activities would be evenly distributed over the whole country. Any region endowed with a particular natural resource or other advantage would receive special attention for development.

In practice, these principles have led to the establishment of regional iron and steel industries in many of the major regions, along with associated industries, such as engineering. In other regions the production of a specialized crop, such as cotton, or a source of energy, such as oil or natural gas, has been emphasized. In Siberia and the Far East, a number of "territorial-production complexes" have been developed. The size and specialization of such a complex are determined by the resources at its disposal, the size of the population and labor force of the area, and the available transport facilities. The largest of these complexes consists of a combination of the Kuzbass and the Novosibirsk regions. The city of Novosibirsk has become a major center of various branches of the engineering industry using Kuzbass steel. A territorial-production complex has arisen around the West Siberian oilfields, including engineering plants, building materials plants, and thermal power stations. The Krasnoyarsk complex includes engineering, hydroelectric power, aluminum production, timber processing, and the Kansk-Achinsk brown coal basin with its large thermal power stations. Other complexes have or are being developed in the Irkutsk, Bratsk, and Sayan areas. The concept of the territorial-production complex has to a great extent replaced older ideas for the planned development of the regions of Siberia, the Far East, Kazakhstan, and Central Asia.

It is difficult to predict what factors will influence the location of industry in the future. Abandonment of centralized planning is probable, but so far no new system of industrial location has been unveiled.

Industrial Regions

In spite of earlier attempts to spread industrialization more evenly over the territory of the Soviet Union, the major concentration of industry still remains within the boundaries of the Agricultural Triangle. Even within the Triangle, however, industry is not evenly distributed, and there are some clearly identifiable concentrations forming industrial regions, some old, some modern.

The most important of these regions consists of the city of Moscow and its surrounding cities and towns, which form the Central Economic region. Poor in natural resources, the region has a history of industrialization beginning in the eighteenth century. Starting with the ironworks at Tula, the region became a major producer of metal and, later, of engineering products. Textiles and chemicals are also of importance. Variety is the keynote of the region, which contains about 20 percent of Soviet manufacturing. Close to the Moscow region, but much smaller, is the St. Petersburg region, which consists mainly of the city itself. St. Petersburg also has a variety of industries, the most important being engineering industries; most of those produce specialized machinery with the aid of a skilled labor force.

The most important center of heavy industry still remains the eastern Ukraine or Donets-Dnieper economic region, with the Donbass and the industries along the Dnieper River. The major products are iron and steel, machinery, and chemicals. The region contains about 15 percent of Soviet manufacturing. A close competitor in terms of heavy industrial production is the Ural region, which received a major boost during World War II with the evacuation of plants from the west and the operation of the Ural-Kuznets Combine. The Urals account for about 12 percent of Soviet industry.

One of the newer industrial regions is the Volga region, which grew rapidly after the opening of the Volga power plants and the exploitation of the Ural-Volga oilfields in the 1950s. Major industries are engineering, oil refining, and chemicals, especially petrochemicals.

The only other industrial region within the Agricultural Triangle is the Kuzbass-Novosibirsk region, now designated as a territorial-production complex. It is the major supplier of iron and steel and engineering products to Siberia, but is small compared with the other heavy industrial regions. The only other industrial regions in Siberia are the small concentrations of manufacturing around Irkutsk and the southern end of Lake Baykal and at Komosomolsk-na-Amure and in the Ussuri Valley in the Far East. Kazakhstan has a concentration of metallurgical industries, mainly based in Karaganda. The Transcaucasus and the North Caucasus regions have some industry, including a small iron and steel capacity, while there is a small Central Asian industrial region, based on Tashkent and the steel plant at Bekabad. These peripheral industrial regions play a relatively unimportant role, however, since over 75 percent of Soviet manufacturing is concentrated in the European part of the country, including the Urals.

URBANIZATION

Rapid industrialization is linked with a steady growth in the number and size of cities. Some of these cities are old centers of population and industry, such as Moscow and St. Petersburg. Other cities, such as Gorky, Kharkov, Riga, or Tashkent, were urban settlements before the Soviet period, but have seen rapid growth during the industrialization drives of the five-year plans of the 1930s and later. Other cities are virtually new. During the Soviet period, no less than 1284 cities have been founded. Many of these were developed for specific purposes. Magnitogorsk, Lipetsk, Cherepovets, and Novokuznetsk were established as iron and steel centers, while a number of mining settlements, such as Norilsk, north of the Arctic Circle, were built from scratch.

The authorities have not always been able to provide adequate urban housing and amenities to meet this demand. Cheaply constructed apartment blocks, assembled from prefabricated sections, provide poor-quality housing in the new districts that arose around the older cities or formed the centers of new ones. Apartments are small and often house too many people. Plumbing and other fixtures sometimes do not work properly. Older houses are divided into flats, and the inhabitants must often share kitchens and bathrooms. In recent years, housing has been given a higher priority by the government, but finding a suitable home is still a major problem for many Soviet families. Restrictions on settling in cities such as Moscow also create problems for people who wish to change their jobs or residences.

TRANSPORTATION

A study of the transportation network of the Soviet Union and its growth is essential for a complete understanding of the history of the country and its economic development. The vast size of the Soviet Union and the large areas of difficult terrain and climate have made the development of an extensive and efficient transportation system difficult. The location of many of the major reserves of raw materials in relatively isolated areas of the country has added to the problems of providing accessibility to them by rail or road.

The first major routes used for movement over any distance were the rivers. Without the river routes, the spread of the Russian people over such a large area at a relatively early date would have been impossible. In the early eighteenth century, Peter the Great constructed several canals to link the rivers into an inland waterway system.

Nevertheless, transportation in Russia remained somewhat primitive until the arrival of the railroad (Figure 6.12). The St. Petersburg–Moscow line was opened in 1851, and the basic railroad network in European Russia was laid down in the 1860s and 1870s. Lines to the Ukraine for the transportation of Donbass coal and Ukraine grain and to the Urals and Caucasus were built during this period. Many of these lines radiated from Moscow, which became the center of the rail network. Railroad construction reached its peak before World War I with the building of the Trans-Siberian Railroad between 1891 and 1916. The line to Vladivostok crossed Russian-controlled Manchuria, but after the loss of that province to Japan in 1905, a further line was laid between 1908 and 1916 along the Amur and Ussuri valleys on Russian territory. During World War I, a major effort to increase the railroad network took place, and 6300 miles (10,900 kilometers) of new track was laid. One major achievement of this period was the railroad from St. Petersburg to the north coast of the Kola Peninsula where the port of Murmansk was founded.

After the Revolution, railroad construction continued. The Turkestan-Siberian railroad, or Turksib, was completed in 1930. It joined the Central Asian network with the Trans-Siberian at Novosibirsk and helped to bring cotton to the factories of the north and to develop agriculture in areas along the line. A line was laid from Karaganda to bring coal to Magnitogorsk.

The 1930s also saw the building of the White Sea–Baltic Canal, which linked the rivers and lakes of Karelia to provide a water route from Leningrad to the Arctic Ocean. The canal was finished in just over one and a half years using forced labor and local construction materials. It was substantially rebuilt in the 1970s. Its construction was linked to the development of the Great Northern Sea Route along the Arctic coast. This route can only be kept open with difficulty during the summer months by the use of icebreakers, aerial ice patrols, and meteorological stations. Its aim is more strategic than economic. Along with the White Sea–Baltic canal, it enables Soviet naval vessels to sail from the Baltic to the Pacific through Soviet territorial waters. Since the 1930s, the period of operation of the Northern Sea Route has been extended by the use of large ice breakers and improved ice and weather observations.

By World War II, the basic transportation network of the country had been developed, but the war caused great damage to railroads, roads, and canals in the areas of the country occupied by the enemy. Some new rail lines were built, notably the line to the Pechora coal basin. The postwar period was marked mainly by reconstruction of damaged lines and by electrification of some lines. Two new lines were laid in Kazakhstan, the so-called South-Siberian and Central-Siberian lines, with the aim of serving the newly settled Virgin Lands and facilitating the transport of grain.

Railroads

As the above historical review of Soviet transportation suggests, the railroads have played an increasingly important role in the total transport system. The vast distances and the relatively flat terrain have encouraged railroad construction. In 1940, 87 percent of freight went by rail.* In recent years, however, the role of railroads has decreased with the rising importance of other forms of transportation. In 1988, the proportion of total freight moved by rail was 48 percent. In the case of freight movement, the main competition came from the rapid increase in the number of pipelines. In 1988, 35 percent of freight (oil, oil products, and natural gas) went by

*Figures given for the movement of freight by different forms of transportation are based on Soviet totals given in ton-kilometers.

FIGURE 6.12

The Soviet Railroad System

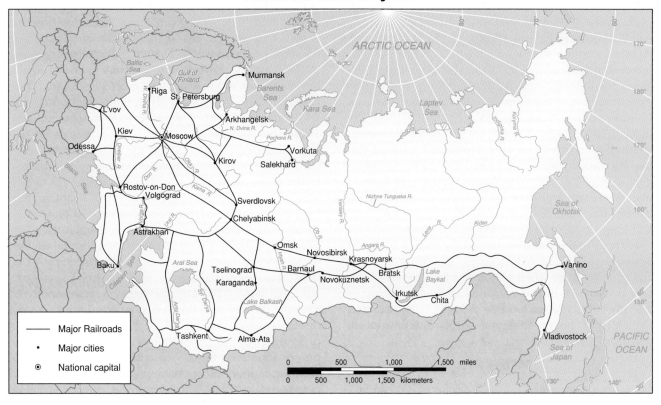

pipeline, compared with 0.8 percent in 1950. If we omit pipelines as being a specialized form of transportation for a limited number of products, then the railroads in 1988 moved some 74 percent of other freight. The railroad is, thus, still by far the most important form of freight transportation. Passenger transportation by rail has been affected by the increase in bus transportation; in 1988, 42 percent of passengers went by road, compared with 5 percent in 1950.

The Soviet rail system is plagued by long hauls of bulky goods, such as coal, iron ore, timber, and grain. In particular, long hauls of coal are becoming an increasing burden. The use of pipelines for the transportation of coal has been discussed, but so far none exists. In 1984, coal accounted for about 19 percent of the freight hauled by the railroads. With an increasing necessity for more hauls of coal from the Kuzbass and Ekibastuz fields as production in the Donbass declines, these long hauls will remain a major problem for Soviet railroads form some time to come.

Compared with other countries, the Soviet Union moves a huge volume of freight by rail. The Soviet Union has a freight density (ton-miles divided by miles of route) six times that of the United States. The railroads of the Soviet Union carry about 70 percent more tonnage than U.S. railroads on a rail network less than half the size of that of the United States. In order to handle this intensive traffic, the operation of the Soviet railroad system is coordinated and organized to a much greater extent than that of the United States. The turnover time for railroad cars is shortened as much as possible to get the maximum use out of them. Frequent services, automatic signaling, and electrification have helped the efficient operation of rail services. In contrast with the United States, the Soviet Union has moved from steam to electric power rather than to diesel. In 1987, about 35 percent of the Soviet network was electrified, and further work on the electrification of the eastern section is proceeding, using surplus regional hydroelectric power.

The most important development of the Soviet railroad system in recent years was the building of the Baykal-Amur Mainline or BAM. Although there were plans in the 1932–1937 period for building a new line parallel with the eastern section of the Trans-Siberian Railroad, nothing was done until 1974. The reasons for the construction of this 2000-mile (3200-kilometer) line are partly strategic and partly economic. The present eastern section of the Trans-Siberian line runs along the Chinese border, while the BAM is located about 200 miles (320 kilometers) north of the border throughout much of its length. In view of recent poor relations with the People's Republic of China, this new line offers greater security. The new line runs through areas of mineral wealth with deposits of iron ore, copper, manganese, tin, lead, zinc, asbestos, oil, and natural gas. It will assist in the transportation of coal from the South Yakut coal field and also timber for export to Japan. Several tunnels and numerous bridges were required, and the total cost, when completed in 1983, was probably over the $8 billion estimated by Soviet authorities. The line will be electrified with power from two large hydroelectric plants on the Zeya and Bureya Rivers, both tributaries of the Amur. A large new port at Vanino, near Sovetskaya Gavan, is under construction to serve as the terminus of the BAM. Large numbers of workers came from all parts of the country to take part in the much-publicized "project of the century." There is no doubt that this was a massive and difficult feat of railroad engineering, involving a struggle with rugged terrain, permafrost, and severe winter conditions. However, it is the only practical solution to the further utilization of the mineral and other resources of this remote area of eastern Siberia.

Rivers and Canals

In 1913, 23 percent of freight went by rivers and canals, compared with only 3 percent in 1988. This comparison is somewhat misleading, however, since the actual freight traffic carried by water routes increased over eight times during the same period. The development of the railroad as the main mover for a variety of relatively high bulk goods and materials during the early 1930s had already considerably reduced the role of internal waterways in the total transportation system.

Although the Soviet Union possesses a navigable river network about 310,000 miles (500,000 kilometers) in length, many of these rivers flow into the Arctic Ocean or are too remote for effective use. At present, about 89,000 miles (143,000 kilometers) of inland waterways are in use, of which 13,000 miles (21,000 kilometers) are artificial. The main problem in using these waterways is their seasonality. All the major waterways freeze for some period in the winter. For example, the Dnieper at Kiev and the lower Volga are frozen for about 100 days a year on the average. The great Siberian rivers are frozen for about half the year in their middle stretches and for considerably longer periods at their mouths. These short navigation seasons reduce the value of this large waterway system, as do the slow speed of river traffic and the circuitous routes followed by many rivers.

The largest waterway system is the Volga and its tributaries, which carry 60 percent of all inland waterway freight. Its value has been increased by a number of canals. The Volga-Don Canal, completed in 1952, gives Volga shipping access to the Black Sea instead of only to the landlocked Caspian Sea. In 1964, the Volga-Baltic Canal was opened along the route of Peter the Great's old canal system. The Moscow Canal, built in 1937, joins Moscow directly with the Volga. Moscow is thus linked by water with the Baltic, Black, White, Azov, and Caspian seas. Navigation on the Volga has not been improved, however, by the construction of the dams and their large reservoirs, which become very difficult to navigate during storms. Other important waterways are the Dnieper system and the Northern Dvina. In Siberia the Ob, Irtysh, Yenisey, and Lena play an important role as links between the Trans-Siberian Railroad and the territories to the north. In the Far East region, the Amur and its tributaries form the main water routes.

By far the most important freight moved by rivers and canals consists of bulky products with low value, such as timber and construction materials. In many cases, timber is floated on rafts. Some grain, coal, and oil are also moved by water.

Highways

In comparison with the United States and other Western industrialized countries, the Soviet Union moves very little freight on its highways. In 1988, only 2 percent of freight went by road compared with about 24 percent in the United States. Although the volume of freight moved had increased more than fourfold over the 1960 total, it still remained small. The main reason for this relatively unimportant role of highways in the transport of freight is the lack of good major highways and a low priority until recently on the manufacture of heavy-duty trucks. The Soviet Union has no four-lane highways of any length and no road system in any way comparable with the interstate system of the United States. The total highway network in 1987 measured 1,000,344 miles (1,609,900 kilometers) of which 743,158 miles (1,960,000 kilometers), or 74 percent, were paved. Just over half of the hardtopped roads had concrete or asphalt surfaces. This compares with a total highway network in the United States of some 3.9 million miles (6.3 million kilometers) of which 39,000 miles (63,000 kilometers) consisted of four-lane interstate highways.

Most hauls by truck are short distance, and trucks do not compete with the railroads for moving freight over long distances. Although most of the truck traffic is in the European part of the country, highways in Siberia and the Far East provide vital links with remote areas such as the Kolyma gold fields and Yakutsk. The future of truck transportation is linked to the further construction of highways and the success of the new Kama truck plant. However, the decision must be made by the planners, who must balance the relative costs of railroads and highways in terms of construction and use. Up to the present, the railroad has obviously received the higher priority.

Air

Transportation of freight by air is relatively unimportant, except for valuable cargoes, such as precious metals, diamonds, or furs. It is, however, of increasing importance in moving passengers. In 1988, 19 percent of passengers went by air, compared with 5 percent in 1960. The first services by Aeroflot, the state airline, began in the 1920s, and the number and length of routes have expanded rapidly, from 90,700 miles (146,000 kilometers) in 1940 to 634,000 miles (1.02 million kilometers) in 1984, or double the 1965 network. In spite of this increase in services, the total number of passengers carried by air is less than half those carried on regularly scheduled flights in the United States.

The airline network is centered on Moscow and other major cities, such as St. Petersburg, Kiev, and Sverdlovsk. The major gains for air passengers are made on the long flights to Siberia and the Far East, where several days of travel by train may be accomplished in hours. Air links with such remote places as Yakutsk, Magadan, or Norilsk are vital for the existence of these places, and for many inhabitants of Siberia air is the only practical means of travel. The international services of Aeroflot, based on the airports of Moscow, cover the capitals not only of the communist nations, but also of most of the rest of the world.

Sea

Sea transportation plays a greater role in the total movement of freight than might be imagined. In 1988, 12 percent of freight went by sea compared with 6 percent in 1950. This growth is largely accounted for by the increase of the short-distance trade along the coasts between the ports of the individual seas as well as the growth of trade between the ports of different seas. As a cheap method of moving bulky cargoes, it is used whenever possible.

The most important sea in terms of shipping is the Black Sea, which remains open all year with the exception of parts of its northern coast. The busiest ports are Odessa, Novorossiysk, and Batumi. Odessa is the main port for overseas shipping and, along with nearby Ilyichevsk, is the busiest in the country. The new port of Yuzhnyy to the east of Odessa has modern facilities for handling a variety of cargoes, including chemicals, coal, and ores. The Black Sea ports account for about 46 percent of all Soviet seaborne freight.

The Baltic Sea is less favorable for shipping than the Black Sea due to icing conditions in its northern part and frequent fogs. The major port of St. Petersburg is hampered by icing of the Gulf of Finland for 180 days in the average year, while the port of Riga is frozen for 80–90 days. Nevertheless, the Baltic ports handle about 14 percent of Soviet shipping. Apart from St. Petersburg, the major ports are Riga and Tallinn. The latter is relatively ice-free and handles part of St. Petersburg's freight in winter. A large new grain port is being built at Munga near Tallinn to ease the burden of the ports of St. Petersburg and Riga in the winter months.

The Caspian Sea is much more important in terms of sea traffic than might be expected for a landlocked lake. Its importance is due to the fact that it provides a major water link between the Caucasus and Central Asia and also as connects with the Volga, an important water route to the heart of European Russia. The most important cargo shipped is oil and petroleum products from Baku to Krasnovodsk in Central Asia. Baku is the largest port, followed by Astrakhan. Because of sand bars and shallow water, large ships cannot reach Astrakhan but have to unload their cargoes at an artificial island port in the Caspian; from there the goods are transported up the Volga in smaller vessels. The Caspian Sea accounts for about 23 percent of Soviet shipping.

Because of its severe climate, the Far East region has limited value for the development of ports. The most important ports are Vladivostok and the more favorably located new ports of Nakhodka and Vostochnyy. Vladivostok is located on a bay that is frozen over for about 110 days in winter, but Nakhodka, 100 miles (161 kilometers) to the east, is ice-free. It handles about 60 percent of the Soviet Union's trade with Japan. Because of problems in expanding Nakhodka, construction of a new port to the east began in the mid-1970s. This port of Vostochnyy was equipped by Japanese firms and handles container cargoes from Japan for transport to Europe by the Trans-Siberian Railroad. It exports timber and coal to Japan and is rapidly becoming one of the most important ports in the country. The completion of the BAM line and the port of Vanino will provide another major port some 600 miles (970 kilometers) to the north. Its neighbor, the port of Sovetskaya Gavan, is usually closed for a considerable period in winter. Magadan, on the Sea of Okhotsk, serves the Kolyma gold fields and can be kept open by icebreakers in a moderate winter. The Far East seas account for about 11 percent of Soviet shipping.

The major port on the Arctic coast is Murmansk, which is ice-free and is thus of great strategic importance as a naval base. Its unhindered access to the Atlantic Ocean and its location at the western end of the Great Northern Sea Route add to its importance. Murmansk is also a port for export of goods abroad. Arkhangelsk on the White Sea handles primarily exports of timber to Western Europe. The navigation season on the Northern Sea Route has been extended in recent years by the use of four large nuclear-powered icebreakers, and it is claimed that the western section can be kept open the entire winter, using a new route along the coast. Only about 6 percent of Soviet seaborne freight uses the Northern Sea Route. The ports of Dudinka on the Yenisey, which serves the mining settlement of Norilsk, Igarka, a timber port on the Yenisey, and Tiksi at the mouth of the Lena have a very limited navigation season.

Although it possesses a long coastline, in its attempt to become a major maritime power, the Soviet Union has had to struggle with severe climatic conditions; seas that are either frozen over for much of the winter, are landlocked, or have restricted access to the outside world; and coasts that are often not suitable for port development. In spite of these handicaps, the Soviet Union has not only emerged as a great naval power, but has also developed a merchant marine during the last 15 years that in the number of vessels rivals that of the major shipping countries of the world.

SPATIAL CONNECTIVITY

Despite its emphasis on self-sufficiency, the Soviet Union has developed a number of linkages and interdependencies with the rest of the world. These include trade, foreign aid, and military assistance to various countries. In the past, Soviet foreign policy, in particular, the long-standing rivalry with the West, has been a key influence in establishing these linkages. At present, it is not entirely clear how the lessening of tensions between the Soviet Union and the West and current changes in Eastern Europe and the Soviet Union itself will affect these relationships, although some trends can be noted.

FOREIGN TRADE

For a major world power like the Soviet Union, foreign trade can be an important element of foreign policy (Table 6.1). This is especially true in the case of a country where the control and direction of foreign trade has been in the hands of the government, which could to a great extent choose its trading partners. Added to the role of foreign trade as an aid to diplomacy is its importance as a means of earning foreign currency, especially hard Western currency. These two aspects of trade are very evident in the policy of the Soviet Union since the Revolution and the recently initiated reforms.

Soviet foreign trade prior to the changes in Eastern Europe in 1989 and 1990 must be examined in two separate categories: trade with Soviet Bloc countries and trade with the rest of the world. In the case of the Soviet Bloc, the Soviet Union attempted to tie the trade of the individual countries together in a system of interreliance through the COMECON organization. COMECON was established in 1949 as an answer to the Organization for European Economic Cooperation (OEEC), founded in 1948 to administer U.S. Marshall Plan aid to Western Europe. The purpose of COMECON was to develop trade among members and foster general economic cooperation. The organization did not really become active until 1954, after which attempts were made to direct the different members in developing their own areas of agrarian economic specialization. This international "socialist division of labor" was resisted by some countries with an agrarian economy, notably Romania, which objected to being classified as suppliers of agricultural produce to the more industrialized countries, such as East Germany and Czechoslovakia. These agricultural countries were discovering a good market in Western Europe for their products and were unwilling to allow the Soviet Union to dictate their trade policies. Attempts to keep all COMECON trade solely within the organization were abandoned, and many of the Eastern European countries developed trade with the West and borrowed money from Western banks. Nevertheless, even in 1980, the Soviet Union has maintained a hold over most COMECON countries by being their major supplier of raw materials, especially energy, in the form of oil, coal, and natural gas. Several of the major iron and steel plants in the Eastern European countries still rely on Soviet supplies of coal and iron ore. In return, the Soviet Union

TABLE 6.1

Soviet Foreign Trade (Millions of Rubles), 1987

	TOTAL TRADE	EXPORTS	IMPORTS
Total	128,883	68,142	60,741
ALL COMMUNIST COUNTRIES	86,321	44,200	42,121
COMECON countries	79,552	40,696	38,856
East Germany	14,729	7,636	7,093
Czechoslovakia	13,684	6,777	6,907
Poland	12,872	6,542	6,330
Bulgaria	12,828	6,276	6,552
Hungary	9,680	4,600	5,080
Cuba	7,559	3,732	3,827
Romania	4,886	2,539	2,347
Vietnam	1,773	1,454	319
Mongolia	1,541	1,140	401
Other Communist Countries	6,769	3,504	3,265
Yugoslavia	3,974	1,901	2,073
People's Republic of China	1,475	724	751
North Korea	1,232	800	432
Laos	88	79	9
INDUSTRIALIZED CAPITALIST COUNTRIES	28,059	14,186	13,873
West Germany	4,957	2,327	2,630
Finland	3,743	1,707	2,036
Italy	3,491	1,804	1,687
France	2,608	1,518	1,090
Japan	2,601	973	1,628
Great Britain	2,110	1,586	524
United States	1,198	279	919
DEVELOPING COUNTRIES	14,503	9,756	4,747
India	2,178	1,105	1,073
Iraq	1,113	325	788
Afghanistan	772	537	235
Egypt	592	296	296
Argentina	457	40	417
Syria	441	250	191

SOURCE: *Narodnoe khozyaystvo S.S.S.R. v 1987 godu.*

has imported engineering and other technical products from former East Germany and Czechoslovakia. Former East Germany was the Soviet Union's main partner in COMECON, accounting for about 19 percent of the Soviet Union's total trade by value with COMECON countries in 1987. In the same year, 67 percent of total Soviet trade was with communist countries, and 62 percent with COMECON countries, including Cuba, Vietnam, and Mongolia (see Table 6.1).

In 1987, 22 percent of total trade by value was with industrialized capitalist countries. The most important by far was West Germany, which accounted for 18 percent of trade with this group. The United States was seventh, after Finland, Italy, Japan, France, and Great Britain, in that order. In the case of the United States and Canada, imports by the Soviet Union greatly exceeded exports to these countries. This imbalance reveals a weakness of Soviet foreign trade. In its trade with Western Europe, the main Soviet exports are petroleum and natural gas, which account for about 80 percent of total exports to the industrialized world. These products are in great demand in Western Europe, but less so in the United States and Canada. Exports of Soviet machinery and other engineering products have met little success in the West, except to a limited extent in the case of automobiles and tractors. Problems of quality and the availability of spare parts have been the major causes for this lack of interest. The Soviet Union desperately needs Western technology in the areas of electronics, especially computers, oil-drilling equipment, and pipeline and other engineering products. It is willing to pay for this technology with gold exports if necessary. In the case of trade with the United States and Canada, the need to purchase large quantities of grain has also added to the trade imbalance. The eagerness shown by the Soviet Union to sell more of its oil and natural gas to the West is a reflection of its rather limited trade options. A country such as former West Germany, which needs the type of products supplied by the Soviet Union and in turn has technology to export, is the best trading partner from the Soviet point of view. With the unification of Germany in 1990, future Soviet trade opportunities with Germany will certainly be explored more thoroughly.

In 1987, 11 percent of Soviet trade by value was with the developing countries. Of these, India was the most important, followed by Iraq, Afghanistan, Egypt and Argentina. In the case of Argentina, Soviet imports greatly exceeded exports, due to large grain purchases. Trade with individual countries was small in volume, except

in the case of India, which rivaled that with several industrialized countries, including Japan. Soviet trade policy with the developing countries has been somewhat different than that with the industrialized countries. By underselling the West in these markets, the Soviet Union has managed to gain customers, not only for energy products but also for industrial goods. The Soviet Union in some cases has offered oil at below-world prices in order to wean away some countries from reliance on supplies from Western oil companies. Sometimes the Soviet Union has had to take commodities from the developing countries that are of little use to it. In the past, some of these goods would be reexported to other countries, often in the Soviet Bloc.

In 1987, 48 percent of Soviet exports consisted of fuel and energy, 18 percent of machines and equipment, and 10 percent of ores and metals. Timber products and pulp and paper made up about 3 percent. More specifically, exports included petroleum products, coal, natural gas, chrome ore, copper and manganese, gold and diamonds, and automobiles and tractors. Although exports of food products were relatively unimportant, tinned salmon and crab were sold in significant quantities to Western Europe. In the same year, 48 percent of imports consisted of machinery and equipment, 16 percent of food products, including grain, 16 percent of consumer goods, and 5 percent of ores and metals. In recent years, the Soviet Union has imported consumer goods from trading partners such as Italy and Japan, but quantities of such goods are small and are not always easily available to the Soviet public.

In 1986, the Soviet Union developed new rules to encourage foreign trade. A State Foreign Economics Commission was set up to supervise joint ventures with foreign companies. The Soviet enterprise would retain 51 percent interest in the joint venture, but the foreign company would be permitted to take its profits out of the Soviet Union and to control prices. It also would not be subject to Soviet planning constrictions. It should be noted that some other COMECON countries had already developed such policies. In 1991, COMECON was abolished, but it is significant to note that former trading links among these COMECON countries have yet to undergo dramatic changes.

Economic Organizations and Trade

The Soviet Union has aimed at self-sufficiency in all branches of the economy, especially with regard to industrial raw materials and food supplies. When domes-

A SOVIET-STYLE INTERNATIONAL COMMERCIAL SYSTEM: WILL PERESTROIKA SUCCEED?

R elations between the Soviet Union and other countries of the world have followed different patterns from those of the Western nations. The Soviet Union does not belong to the major international trade and financial organizations that are important instruments in the relations between Western countries and also between those nations and developing countries. Major Soviet trade and aid links have been conducted within the former COMECON organization or between the Soviet Union and individual countries. Traditionally, Soviet trade has had two basic aims: the acquisition of Western technology, such as electronic and oil-drilling equipment, and the penetration of markets in developing countries for political purposes. Today, however, more emphasis is being placed on providing goods and services for domestic consumption. The limited variety and quality of Soviet export goods have hampered trade with the West, which shows little demand for Soviet manufactured goods. For this reason, the Soviet Union has concentrated on the export of oil, natural gas, timber, gold, and diamonds. Manufactured goods, including armaments, have found a better market in some developing countries.

But what effects will perestroika have on the traditional Soviet policy of autarky (economic self-sufficiency)? Will increased cooperation with Western industrial nations expand trade and increase Western investment to aid in the development of new industries? Currently, the Soviet economy is struggling. It is experiencing a budget deficit estimated at 11 percent of the gross national product. Inflation has increased, and growth in agriculture and industry is falling short of required output.

To counteract some of these problems, restructuring of portions of the Soviet economy is being attempted. No longer do factories have to fulfill Moscow-dictated quotas as they now attempt to become more self-financing and profitable. It is hoped that cutbacks in military spending will shift a significant amount of resources to other segments of the economy. Joint economic ventures with Western companies such as Eastman Kodak, Chevron, and Johnson and Johnson are being encouraged in the hope of infusing technology and financial investment into Soviet industry. Nevertheless, Western business and international assistance agencies are hesitant to invest without major structural changes in the economic system being in place. No adequate business banking system is available, nor is there certainty about currency exchanges, loan regulations, and a host of other financial matters. Perhaps the greatest uncertainty is the legal clarification of ownership and control of natural resources, industrial facilities, and the other means of production in the various republics and newly independent nations. Without rapid "perestroika," investment and aid may not come within a time-frame acceptable to the people.

tic supplies have not been available, the Soviet Union has preferred to obtained its requirements from the COMECON countries. Apart from raw materials, the Soviet Union seeks supplies of foreign manufactured goods, especially products of the engineering and electronics industries. When these could not be obtained from such countries as former East Germany or Czechoslovakia, the Soviet Union has been forced to import them from the West. This inability to rely only on trade

linkages within the Soviet Bloc created problems for the Soviet Union and compelled it to develop a trade policy with the Western industrialized countries. As shown above, about 22 percent of total Soviet trade is with these countries. In order to conduct this trade, the Soviet Union has had to invest in the production of exportable items, such as natural gas, petroleum, timber, gold and diamonds, furs, and some manufactured goods, such as automobiles. At times when the price of oil has dropped on the world market, the Soviet Union has been forced to sell from its stocks of gold in order to pay for the import of grain and technology. Even during periods of rising oil prices the Soviet Union has found it difficult to increase production and thus exports due to inadequate technology. In spite of the development of a domestic electronics industry and other branches of technology, it is unlikely that the Soviet Union can significantly diminish its reliance on Western computer technology or oil equipment. The problem of paying the high price of the products of Western know-how was coupled in the past with the problem of the embargo on the export of certain types of Western strategic goods to the Soviet Union imposed by Western governments, especially the United States. This situation has somewhat eased, but the virtual cessation of the "cold war" does not mean that the West is willing to release all types of military and other strategic goods from export restrictions. Industrial espionage and the illegal export of goods to the Soviet Union by Western traders, often through a third country, are ways in which the Soviet Union has attempted to obtain Western technology at least cost.

FOREIGN AID

The Soviet Union has also maintained links with the developing countries, both communist and noncommunist. As well as conducting trade, the Soviet Union gives aid to many of these countries. Aid, like trade, has political as well as economic goals. The Soviet Union would like to see as many of the developing countries as possible come to rely on it for supplies of energy, goods, capital, and technical assistance and thus weaken such ties as they may have with the West. The former COMECON countries have also played a role in the extension of the Soviet trade and aid program.

The Soviet Union does not usually give economic aid in the form of large financial loans or grants for the country concerned to use in any way that its government may think fit. Instead, the Soviet Union generally agrees to provide aid for specific projects, often including research, planning, equipment, technical personnel, and training of local labor. The government involved pays the costs of local labor, raw materials, and transportation. About 80 percent of all Soviet economic aid is in the form of projects. The Soviet Union also gives credits to countries for the purchase of Soviet petroleum or manufactured goods, such as machinery or equip-

ment. These credits, either for development or trade, are usually repayable at a low interest rate in the products of the country, often over a 10-year period. In the case of developing countries, these products are usually raw materials, such as metallic ores, cotton, tea, coffee, rice, and citrus fruits. It should be remembered that repayment in products means that these countries cannot sell those goods on the world market and thus earn hard currency. Thus, Soviet aid and trade are combined in a pattern somewhat different from that of Western aid to the developing world.

The Soviet authorities do not give any detailed information on the direction or amount of their foreign aid. The Soviet Union is not a member of international economic organizations such as the International Monetary Fund (IMF), the World Bank, or the General Agreement on Tariffs and Trade (GATT) and few statistics on Soviet trade and aid are published except in Soviet statistical handbooks, which are very selective in their information. Trade figures are more easily obtainable than those on aid. Estimates of Soviet aid are published by Western agencies that obtain part of their information from the recipient countries.

It is clear that Cuba receives a large part of total Soviet aid, probably about the same amount as all the noncommunist developing countries together. Of the noncommunist countries, India over the years has been the greatest recipient of aid. At present, the Soviet Union is helping India to locate new oil deposits and construct power stations. Until the break with the Soviet Union in 1972, Egypt had been a major recipient of aid, including the billion-dollar Aswan High Dam project between 1960 and 1971. Among the countries that are receiving or have received aid are Ethiopia, Afghanistan, Guinea, Iraq, Vietnam, Nicaragua, and Angola, which in 1985 signed a billion-dollar aid agreement with the Soviet Union, which included aid to agriculture and the building of power stations. Most of these countries either have or had Marxist governments or some political links with the Soviet Union. However, countries such as Nigeria, Tanzania, and Turkey have also accepted Soviet aid, although they are not necessarily sympathetic to Soviet political aims. Libya receives considerable aid at present. The Soviets have recently completed a major oil pipeline and have equipped a nuclear research center, as well as assisting with oil exploration and the construction of an electric power grid.

In spite of the considerable amount of Soviet aid given to foreign countries, not all of it has been of high quality. Recipient countries have complained of deliveries of inferior equipment, lack of spare parts, late deliveries, and the inadequate technical skills of advisers.

Special mention should be made of the Soviet Union's relationship with Japan. Japan's requirements for raw materials for industry have led to an interest in the relatively close supplies of coal, oil, natural gas, and timber in eastern Siberia and the Soviet Far East. In return, the Soviet Union would like access to Japanese technology and Japanese help in the exploitation of its Siberian raw materials. The completion of the BAM railroad and the port of Vostochnyy near Vladivostok have strengthened the development of these links with Japan. Vostochnyy was built with Japanese assistance and is now the major port for the export of coal and timber to Japan; it also handles container traffic from Japan, which goes to Western Europe via the Trans-Siberian Railroad. At present, Japan is aiding the development of the South Yakut coalfields in eastern Siberia and the construction of a branch railroad to the mines from the BAM railroad. Japan has given the Soviet Union a credit of $450 million to purchase construction machinery in Japan. Japan also gets oil, probably from Sakhalin Island, as well as quantities of timber.

MILITARY AID

Although the Soviet Union offers scientific, educational, and cultural aid as well as economic aid to the developing world, these types of aid are relatively innocuous compared to the amounts of military aid delivered to a number of countries. Both military equipment and training are offered on credit terms. A wide variety of arms, equipment, vehicles, and ships are involved, but no nuclear weapons. The former COMECON countries have been involved in a coordinated Soviet policy of military aid, and countries such as Czechoslovakia, former East Germany, and Poland have specialized in the production of certain types of weapons. Major recipients of military aid include or have included India, Jordan, Libya, Nicaragua, Afghanistan, North Korea, Sudan, Syria, and Vietnam. Several countries, such as Egypt and Pakistan, which received Soviet military equipment in the past, including aircraft, have problems in obtaining spare parts and, as in the case of Egypt, may still have to approach the Soviet Union for help. Soviet weapons often find their way into the hands of various terrorist groups such as the Palestine Liberation Organization

(PLO), the Irish Republican Army (IRA), and various groups in Lebanon and southern Africa.

Most of the countries that have received Soviet military aid have either been allies, such as the Warsaw Pact countries (Bulgaria, Czechoslovakia, former East Germany, Hungary, Poland, and Romania), Cuba, and Mongolia or Marxist countries, such as Nicaragua under the Sandinistas, Vietnam, or Guinea. In addition, the Soviet Union has other links with many of these countries in the form of military, air, or naval bases on their territory. Soviet access to the outside world, and in particular to the oceans, has been restricted, not only because of the ice-fringed coasts of most of the northern and eastern regions of the country, but also because of foreign control of the entrances to the Baltic and Black seas. For want of better warm water ports on Soviet-controlled territory or on those of adjacent allies, the Soviet Union has looked elsewhere, often far afield. Since World War II, the Soviet navy has sought bases in the Mediterranean, the Indian Ocean, and elsewhere on the territory of friendly countries. This search has not always been easy. Soviet naval facilities in Egypt were closed after the Sadat government expelled Soviet advisers and military personnel in 1972. On the east coast of Africa, Soviet bases in Somalia were abandoned due to a change in government, but other facilities were acquired in Ethiopia after a Marxist government was established there. The existence of Marxist governments in some other African countries has offered opportunities. Madagascar offers port facilities to the Soviet navy, and Mozambique has been visited by Soviet warships. A military base at Pointe Noire in the People's Republic of the Congo has been used by Soviet and Cuban military advisers and troops as a jumping-off point for activities in Angola, where Soviet interests are represented by a large Cuban military presence. The Soviets use Angolan airfields and ports. In the sensitive area of the Horn of Africa, the Soviet Union has an ally on the Asian shore in the shape of the People's Democratic Republic of Yemen (PDR of Yemen). In 1990, the PDR of Yemen and Yemen joined to form the Republic of Yemen. Yemen controls the island of Socotra at the mouth of the Gulf of Aden, giving it the potential of controlling shipping entering or leaving the Red Sea. The Suez Canal is of importance to the Soviet Union since it gives the Soviet navy a shorter route from the Black Sea via the Red Sea to the Indian Ocean than the long sail through the Mediterranean and round the Cape of Good Hope. The use of the Suez Canal has, however, been rendered less attractive because of the friendship of the canal's owner, Egypt, with the United States.

The Soviet naval presence in the Indian Ocean and on the African coast is important to the West because of the routes used by Western oil tankers from the Persian Gulf through the Arabian Sea and around the Cape of Good Hope. In the event of hostilities, the Soviet Union is well situated to threaten the vital oil lifeline of the Western powers. In order to reduce this perceived Soviet menace, the United States has obtained bases in the Indian Ocean, notably at Berbera in Somalia (formerly a Soviet base), in Oman and Kenya, and on Diego Garcia, an island in the Chagos Archipelago. Thus, a situation of naval confrontation could exist in this region between the two world powers. Clearly, the Soviet Union would like to have control of territories or bases closer to the source of Western oil supplies. It is for this reason that the West looks at the political situations in Iran, Iraq, and other countries of the Gulf region with some nervousness. In the recent Iraq/Kuwait crisis, however, the Soviet Union supported the United States' position.

The greatest success achieved by the Soviet Union in obtaining a foothold in an area of great strategic importance is the case of Cuba. After the consolidation of control by Castro, the Soviet Union began to build up its military and naval presence on the island, leading to the missile crisis with the United States in 1962. Since that date, the Soviet Union has been more cautious in its actions on Cuba, but still maintains military and naval facilities there. The proximity of Cuba to the United States is too tempting for the Soviet Union to abandon its advantage, but too dangerous for the United States to permit a major Soviet military and naval buildup there. The establishment of a pro-Cuban and pro-Soviet government in Nicaragua in the 1980s and the struggle for control of El Salvador and Guatemala by leftist rebels increased the concerns of the United States about further Soviet penetration of territories close to the United States. Due to recent charges in the political situation in Nicaragua, however, the United States' concerns over Soviet penetration have abated somewhat.

The Soviet Union has also taken measures to improve its position in the Pacific Ocean, not only with regard to the United States, but also to the People's Republic of China. The political influence now exercised by the Soviet Union in Vietnam poses a potential threat to China's southern flank. The Soviet fleet uses the old U.S. base at Camranh Bay, which has become the largest Soviet overseas naval installation. It also has an operational airfield. In this way, the Soviet Union could threaten China's southern shores as well as U.S. bases on the Phillippines and on Guam. In recent years, however, the Soviet Union has shown a greater tendency toward cooperation rather than conflict in this area of the Pacific. If the present trends in relations between West and the Soviet Union continue it is possible that many of the geopolitical factors discussed above may cease to have so much significance. However, it should be remembered that the Soviet Union is still a major military and naval power with an arsenal of nuclear and other weapons.

PROBLEMS AND PROSPECTS

In view of the current changes taking place in the Soviet Union and the uncertainty of their outcome, it is difficult to discuss future prospects. It is clear that the Soviet economy is in disarray and needs perestroika, or restructuring, to use the current Soviet catchphrase. In particular, the entire farming system needs a radical overhaul. The tight control of prices and allocations of industrial materials by the state has also hampered economic growth. Other problems are increasing costs in the extraction and development of sources of energy, an overburdened transportation system, inadequate urban housing, and an increasing restlessness among the ethnic groups in the various republics for more autonomy.

On the international scene, prospects are somewhat more encouraging. The Soviet Union has opened up relations with the West, and the United States and the Soviet Union seem more ready to discuss mutual problems than at any time during the last few decades. The Soviets have terminated their unfortunate adventure in Afghanistan, are maintaining a low profile in the Middle East, and have modified relations with such allies as Cuba with regard to economic and military aid.

In Eastern Europe, sweeping reforms have occurred as the Soviet Union allowed its satellite countries to institute free elections and overthrow Communist party leadership in the region. From August to December of 1989, six countries in Eastern Europe broke the monopoly of the Communist party, and the Berlin Wall came

Soldiers near the Press Building in Vilnius, Lithuania, three days after the prime minister of Lithuania and her cabinet announced their resignations on January 8, 1991. Subsequent events eventually led to the political independence of Lithuania.

tumbling down. Restlessness, anxiety, and hope prevail over the region as some of Gorbachev's reforms are set into motion.

Several paths can be followed by the Soviet Union in its attempt to reorganize its society and its political, economic, and military structure, but it is still too early to predict how far the Soviet Union will go to embrace a Western style of democracy and free-market economy. The failed coup by hardliners of the Communist party in 1991 has in effect accelerated the pace of reforms and has drastically altered the political and economic structure of the Soviet Union. The Communist party is being dismantled, the national leadership is struggling to maintain the Union as reformers like Boris Yeltsin push for more change. The Union of Soviet Socialist Republics with its fifteen republics no longer exists. The Baltic Republics of Estonia, Latvia, and Lithuania have declared their independence and have been recognized by several foreign governments. Other republics have also declared independence. It is difficult to predict whether the Soviet Union will dissolve completely or what type

of new Union will emerge as the central government works on new political and economic policies regarding the relationships between the republics and the central government.

Suggested Readings

Adams, A. E., I. M. Matley, and W. O. McCagg. *An Atlas of Russian and East European History.* New York: Praeger, 1966.

Allworth E., ed. *Soviet Nationality Problems.* New York: Columbia University Press, 1977.

Bater, J. M. *The Soviet Scene: A Geophysical Perspective.* London: Edward Arnold, 1989.

Bogarko, S. *Power Industry. The Soviet Union Today and Tomorrow.* Moscow: Novosti, 1981.

Borisov, A. A. *Climates of the U.S.S.R.* Edinburgh: Oliver and Boyd, 1965.

Cole. J. P. *Geography of the Soviet Union.* London: Butterworths, 1984.

Dewdney, J. C. *A Geography of the Soviet Union,* 3d ed. Oxford: Pergamon, 1979.

——— . *The U.S.S.R. Studies in Industrial Geography.* London: Hutchinson, 1978.

Feshbach, M. *The Soviet Union: Population Trends and Dilemmas.* Population Reference Bureau, Vol. 37, No. 3, August 1982.

Goldman, M. *The Soviet Union and Eastern Europe.* Guilford: Dushken Publishing Group, 1988.

Lydolph, P. E. *Geography of the U.S.S.R.* New York: John Wiley & Sons, 1990.

Matley, I. M. *"Cities of the Soviet Union and Eastern Europe,"* in S. D. Brunn and J. F. Williams, *Cities of the World: World Regional Urban Development,* pp. 123–61. New York: Harper and Row, 1983.

Mellor, R. E. H. *The Soviet Union and its Geographic Problems.* London: MacMillan, 1982.

Soviet Census 1979. GeoJournal Supplementary Issue, 1, 1980.

Symons, L., J. C. Dewdney, J. M. Hooson, W. W. Newey, and R. E. H. Mellor. *The Soviet Union: A Systematic Geography.* London: Hodder and Stoughton, 1982.

CHAPTER 7

Eastern Europe

IMPORTANT TERMS

Hercynian

Karst

Buffer zone

Solidarity

Workers' management

CONCEPTS AND ISSUES

■ Diversity of the physical landscape.

■ Cultural differences and ethnic diversity within and between countries.

■ The role of the Soviet Union in Eastern and Western European affairs.

■ The future of economic development under democratic socialism.

■ The expansion of global links beyond the Warsaw Pact and COMECON.

CHAPTER OUTLINE

PHYSICAL ENVIRONMENT
Climate
Landforms
Vegetation and Soils
Rivers and Water Bodies

HUMAN ENVIRONMENT
Historical Background before World
    War II
World War II and Its Aftermath
1989: Sweeping Reforms
Population
A Possible Revival of the Religious
    Legacy
Agriculture and Forestry
Industry

SPATIAL CONNECTIVITY
Foreign Trade, Connectivity, and
    Interdependencies

PROBLEMS AND PROSPECTS

Although for the past four decades the countries of Eastern Europe have had planned economies modeled, in general, on that of the Soviet Union, their individual economic, social, and political structures have never been identical. Two of them, in fact—Yugoslavia and Albania—broke with the Soviet Union many years ago, although they have maintained certain features of planned economies. Now, with the recent end of Communist party rule in Eastern Europe, social democracies have begun to emerge in all of the former Soviet Bloc countries in the region. As these democracies develop, and the influence of the Soviet Union lessens, the countries will continue to diverge as the various governments experiment with their existing economic, social, and political structures. Nevertheless, despite these differences, it is possible to consider the countries as a coherent group and to examine Eastern Europe as a region with many common characteristics (Figure 7.1).

The correct title for this region has been the subject of some discussion among geographers. Such names as Eastern Europe, Central Europe, and East-Central Europe have been suggested, while some geographers detach the southern part of the region under the title of Southeastern Europe or the Balkans. This lack of agreement is due primarily to the diversity of natural features, climates, peoples, and cultures in an area that stretches from the Baltic Sea in the north to the Adriatic Sea in the south. "Eastern Europe" is the term most popularly accepted, reflecting post–World War II political and economic structures, and it will be used here, keeping in mind, however, that part of the region has southern European characteristics.

PHYSICAL ENVIRONMENT

CLIMATE

Eastern Europe encompasses three climatic zones. In the north, former East Germany, Poland, and Czechoslovakia have a humid continental climate, with cool summers and cold winters. The western part of former East Germany falls within the region of west coast marine climate, which has milder winters and, in general, more rainfall. South of the Carpathians, a type of humid continental climate with warm summers and cold winters is found. In this region the range of annual temperatures is greater than further north. The Adriatic

coasts of Yugoslavia and Albania have a Mediterranean climate, with warm, dry summers and mild, rainy winters. Southern Yugoslavia (Macedonia) and parts of southern Bulgaria have continental climates that show some Mediterranean influences. Table 7.1 gives examples of the climatic conditions for typical locations in these regions.

Much of the rainfall in Dubrovnik (on the Adriatic coast of Yugoslavia) falls in the winter. Some mountain locations along the Yugoslav coast experience more than 157 inches (400 centimeters) of precipitation in the average year. In general, mountain regions throughout Eastern Europe have higher levels of precipitation and lower temperatures than do the plains. The complex distributional pattern of landforms in specific geographic areas leads to considerable differences in local climates within the same major climatic region.

LANDFORMS

The main division in the topography of the region is between the generally flat countries of Poland, Germany, and Hungary, and the mainly mountainous and hilly countries of Czechoslovakia, Yugoslavia, Bulgaria, and Albania. Romania has both plains and mountains in about equal measure (Figure 7.2).

Beginning in the northern part of the region, the North European Plain extends from the North Sea coast eastward through the Netherlands, Germany, and Poland, where plains occupy almost the entire area of these countries. The northern edge of these plains is formed by the coast of the Baltic Sea. This coastline is marked by several small lagoons and bays. Some distance inland from the coast, a ridge of low hills of glacial origin extends from west to east. In the east, small lakes have been formed where these glacial deposits have disturbed the drainage. South of the hills are areas of sandy soils laid down by melting water from glaciers. This zone is crossed by shallow valleys, extending from east to west, formed by rivers that flowed along the southern edge of the retreating glaciers. These ancient valleys have poor drainage. The southern part of the plain is covered by deposits of loess. This zone has soils of great fertility and is the best area of the North European Plain for farming.

South of the great plain is a zone of hills and mountains. In the west a ring of mountains surrounds the western part of Czechoslovakia, known as Bohemia. These mountains are not high, rarely reaching more than 4500 feet (1370 meters). They were formed during

FIGURE 7.1

Eastern Europe

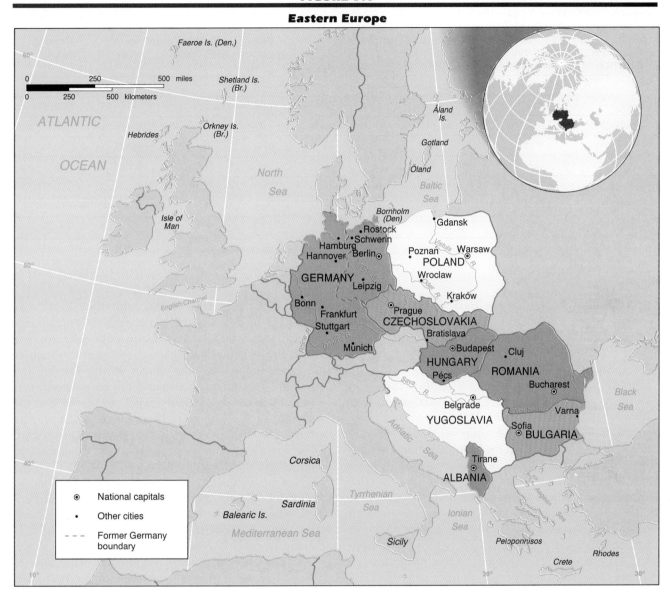

the **Hercynian** mountain-building period and are older than the Carpathians and the Alps. Eroded over long periods and then uplifted, they have a complex geology and contain a variety of metallic ores. Hence the name Erzgebirge (Ore Mountains)—Krušné Hory in Czech—for the range that forms the boundary between former East Germany and Czechoslovakia. To the east, the ranges of the Sudeten Mountains form part of the boundary between Czechoslovakia and Poland. They contain the highest mountain of the Hercynian group,

Sněžka (5255 feet; 1602 meters). The southern edge of Bohemia is flanked on the west by the mountains of the Šumava (Bohemian Forest), which form a large forested range. Farther east the hills of the Bohemian-Moravian Uplands cross Czechoslovakia, from south-west to north-east. Much of central Bohemia consists of rolling country, with areas of uplands.

Former East Germany contains two areas of mountains that are outliers of the complex patterns of hills and mountains found over the border in former West

STATISTICAL PROFILE: EASTERN EUROPE

REGION OR COUNTRY	Population Estimate mid-1990 (Millions)	Natural Increase (Annual)	Population Projected to 2000 (Millions)	Infant Mortality Rate	Life Expectancy at Birth (Years)	Urban Population (%)	Per Capita GNP, 1988 (US$)	Area, Thousands of Square Miles (Km²)	Population Density, No./mi² (No./km²)
Albania	3.3	2.0	3.8	2.8	71	35	—	11.1(28.8)	295(114)
Bulgaria	8.9	0.1	9.0	13.5	72	67	—	42.8(111.9)	209(81)
Czechoslovakia	15.7	0.2	16.3	11.9	71	75	—	49.4(127.9)	318(123)
Germany, East*	16.3	0.0	15.5	8.1	73	77	—	41.8(108.3)	390(151)
Hungary	10.6	−0.2	10.6	15.8	70	60	2460	35.9(93)	294(114)
Poland	37.8	0.6	38.9	16.2	71	61	1850	120.7(312.7)	313(121)
Romania	23.6	0.5	24.5	25.6	70	54	—	91.7(237.5)	254(98)
Yugoslavia	23.8	0.6	25.1	24.5	71	46	2680	98.8(255.8)	241(93)

* East and West Germany unified in 1990. The total population of Germany in 1990 was estimated to be 79.5 million.

Germany. They are part of the Harz Mountains, which formerly contained metallic minerals, now mostly exhausted, and the Thüringerwald (Thuringian Forest), which extends along the southern border with former West Germany.

Farther east the western ranges of the Carpathian Mountains occupy a large part of Slovakia, the eastern province of Czechoslovakia. The Carpathians were formed during the period of Alpine mountain building and are thus younger than the Hercynian ranges; consequently, they are higher and less eroded, with more rugged outlines. The Carpathians extend from Czechoslovakia and southern Poland through the Soviet Union to the great mountain arc that occupies much of central Romania. In Slovakia and Poland, the highest range of the Carpathians is the High Tatras, which reach 8700 feet (2655 meters). To the south of this main range, a number of smaller ranges form complex patterns. They include the Little Carpathians, White Carpathians, Low Tatras, Great and Little Tatras, and the Slovak Ore Mountains. The last, as the name suggests, contain metallic ores of which copper, manganese, and iron ore are the most important. Although not quite so high as the High Tatras, the long ranges of the Eastern and Southern Carpathians in Romania are more spectacular due to their variety of rugged relief. To the west of the great arc of the Romanian Carpathians is the smaller group of mountains known as the Western Carpathians. Their rounded summits permit human settlements as high as 4000 feet (1220 meters).

The Carpathians are bordered by foothills. In Poland an area of uplands covers much of central southern Poland, while in Romania the Sub-Carpathian foothills extend along the eastern and southern flanks of the Carpathian ranges.

South of the Carpathians is a second major area of plains. The Mid-Danube Plain (Pannonian Plain) extends from the borders of Austria to the Western Car-

TABLE 7.1

Climatic Conditions for Typical Eastern European Locations

		AVERAGE TEMPERATURES		AVERAGE ANNUAL PRECIPITATION
		January	July	
Humid continental (cool summers)	Warsaw, Poland	27°F (−3°C)	66°F (19°C)	21 inches (53 cm)
Humid continental (warm summers)	Belgrade, Yugoslavia	32°F (0°C)	73°F (23°C)	27 inches (69 cm)
West coast marine	Schwerin, Former East Germany	32°F (0°C)	64°F (18°C)	25 inches (63 cm)
Mediterranean	Dubrovnik, Yugoslavia	48°F (9°C)	77°F (25°C)	50 inches (127 cm)

pathians of Romania and from the Slovak Carpathians in the north to the mountains of Yugoslavia in the south. This plain is of great importance, due both to its location along the River Danube and to its fertile soils and favorable climate. The plain forms the core area of the country of Hungary and a large part of northern Yugoslavia. On it are located the capital cities of Hungary and Yugoslavia, Budapest and Belgrade, respectively. The city of Vienna occupies a strategic location in the northwestern corner of the plain where a belt of lowland, known as the Moravian Lowlands, cuts across the

"waist" of Czechoslovakia to join the Mid-Danube Plain with the North European Plain in southern Poland. The southwestern corner of the Mid-Danube Plain is linked to the Adriatic Sea by a gap in the coastal mountain ranges crossed by a railroad from Zagreb to the port of Rijeka in Yugoslavia. In the southeast, the Danube flows through the gorges of the Iron Gates at the southern end of the Carpathians to cross the third major area of plains, the Danubian (Wallachian) Plain.

The Danubian Plain lies between the foothills of the Romanian Carpathians and the Stara Planina Mountains

FIGURE 7.2

Major Physiographic Features of Eastern Europe

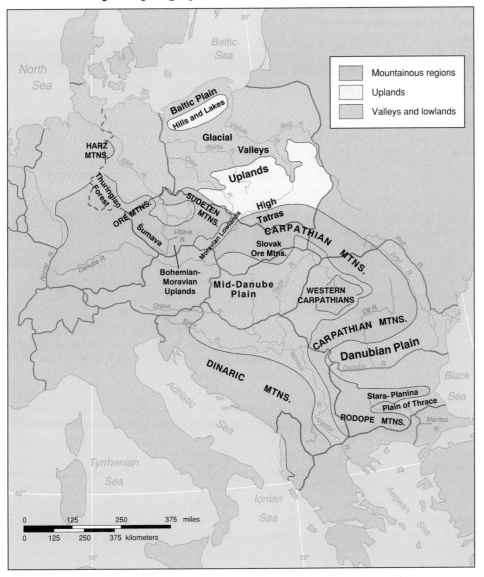

and their foothills in Bulgaria. Its southerly location and good soils make it one of the important agricultural areas in Eastern Europe. Along the Black Sea coast, the plain is connected with the plains of Soviet Moldavia and the southern Ukraine.

South of the Danubian Plain, a complex series of mountain ranges extend from the Austrian and Italian borders of Yugoslavia eastward to the Black Sea coast and southward through Yugoslavia and Albania to the southern tip of Greece. This mountainous region is often referred to as the Balkan Peninsula. In northwestern Yugoslavia, part of the eastern Alps extends over much of the territory of the republic of Slovenia. To the southeast, a number of ranges of the Dinaric Mountains line the coast of the Adriatic Sea. Most of these ranges contain large areas of limestone rocks full of fissures and sinkholes, which permit the drainage of water to underground caverns and rivers. This type of terrain is known as **karst.** In areas of karst, there are few rivers and lakes and surface water is scarce. Further inland the Dinaric ranges have few areas of karst and support streams and rivers. To the south the mountains of the republic of Macedonia belong partly to the Rodopi system of western Bulgaria and the Pindus system of northern Greece. The mountains of Albania form a northern extension of the Pindus group. In Bulgaria the Rodopi Mountains follow the same northwest-southeast trend of the Dinaric ranges, but in the north of the country, the Stara Planina range is, in fact, an extension of the Southern Carpathians. The Bulgarian mountains are separated from the Dinaric Mountains by the long valleys of the Morava and Vardar rivers, which form a major route between the Mid-Danube Plain and the plains along the north coast of the Aegean Sea. The Stara Planina and Rodopi ranges in Bulgaria are separated by the Plain of Thrace, through which the Maritsa River flows. Yugoslavia has no major coastal plains, but is fortunate in sharing a large part of the Mid-Danube Plain. Albania has little lowland except for a small coastal plain.

VEGETATION AND SOILS

Most of the original natural vegetation of Eastern Europe has been removed over the centuries. The ancient forests of the northern plains have largely vanished, but reforestation has replaced them with areas of coniferous and mixed forest. The mountain ranges that extend from western Czechoslovakia to Romania are mainly covered with coniferous forests on the upper slopes and mixed forests in the valleys. In some areas the forests have been removed to provide mountain pastures for grazing livestock. Farther south the forests became more sparse and are mainly deciduous, while much of the Yugoslav coast and Albania have predominantly Mediterranean-type vegetation, consisting of shrubs and sparse forests of oak and beech.

Areas of grassland are found on the Mid-Danube Plain and on the plains of eastern and southern Romania. Most of these grasslands have been plowed for agriculture, and little natural grass remains.

The soils of Eastern Europe vary considerably among the cool northern plains, the mountain areas, the Mediterranean region, and the grasslands. In the north grey-brown (alfisol) and brown forest soils of moderate fertility are found. Mountain soils are sparse and are of poor quality, but in areas of limestone rocks in the mountains of Yugoslavia, basins have developed with floors of red clay that form a fertile soil. Grassland soils consist mainly of brown forest soils with areas of fertile mollisols. Alluvial soils found along the banks of major rivers are of considerable value for agriculture. Areas of wind-deposited fertile loess soils are found in southern Poland and the Mid-Danube and Romanian plains.

RIVERS AND WATER BODIES

The rivers of Eastern Europe flow in three directions; northward to the North and Baltic seas, eastward to the Black Sea, and southward to the Mediterranean. The major river of former East Germany is the Elbe, which flows from its source in Czechoslovakia across the North European Plain through German territory to the North Sea. Its tributary, the Vltava, is the major river of western Czechoslovakia. Farther east, the Oder River forms part of the boundary between Germany and Poland. It enters the Baltic Sea through a lagoon, at the south end of which stands the Polish port of Szczecin. The Oder, which rises in Czechoslovakia, has been called the "Rhine of the COMECON," but its importance as an inland waterway route linking Poland, former East Germany, and Czechoslovakia should not be overestimated as it in no way carries the volume of traffic of the Rhine River. The Vistula, with its tributary the Bug, is a purely Polish river, crossing the country from its southern border to the Baltic Sea. The ports of Gdańsk and Gdynia are located near its mouth.

The major river of the southern part of the Eastern European region is the Danube. From its source in the

A view of the Danube River in Romania. The Danube, a major transportation waterway in Eastern Europe, passes through several countries before it empties into the Black Sea.

Black Forest of former West Germany, it flows for some 1800 miles (2900 kilometers) to the Black Sea. It enters Eastern Europe from Austria at the city of Bratislava in southern Czechoslovakia and crosses the territories of Hungary, Yugoslavia, and Romania. It is the most important waterway of Eastern Europe, linking the countries along its length. Its use as an outlet to the oceans of the world is limited by the fact that it flows into the Black Sea, from which access to the Atlantic Ocean involves crossing the length of the Mediterranean. The Soviet Union controls the north bank of the Danube at its mouth, which gives it considerable control over traffic entering or leaving the river. This situation has been somewhat modified by the construction of the Danube–Black Sea canal by the Romanians, which permits shipping to reach the Black Sea south of the Danube mouth.

The Danube has a number of major tributaries. The Tisa crosses the Mid-Danube Plain from north to south to join the Danube in Yugoslavia. The Sava and Drava flow parallel to one another across the width of northern Yugoslovia, while a number of rivers, such as the Olt, Jiu, and Siret, rise in the Carpathians and flow southward to the Danube.

The Morava and Vardar rivers arise near each other in southern Yugoslavia. The Morava, however, flows north to the Danube, while the Vardar flows in the opposite direction to reach the Aegean Sea near Thessaloniki in Greece. The valleys of these two rivers form the important Morava-Vardar gap. In Bulgaria the Maritsa River crosses the southern part of the country, finally forming the border between Greece and Turkey.

HUMAN ENVIRONMENT

HISTORICAL BACKGROUND BEFORE WORLD WAR II

The history of Eastern Europe is extremely complicated, because the region encompasses a large number of peoples, all of whom have their own culture, origins, and experiences. Until the nineteenth century, few of these peoples had a national state of their own, and they belonged instead to one or another of the empires that dominated the region.

The largest group consists of the peoples speaking Slavic languages. The Slavs originally came from an area located near the present Polish-Soviet border. The Western Slavs moved into the area occupied by the present-day countries of Poland and Czechoslovakia, and in the tenth century the first Polish and Czech states appeared (Figure 7.3). The Czech rulers encouraged an influx of Germans to exploit their forests and mines, leading to the early settlement of Germans in Bohemia. The Poles, Czechs, and Slovaks came under the influence of the Roman Catholic church. During the Reformation, many Czechs became Protestants, but their armies were defeated by the Catholic Austrians, who absorbed the Czechs into their empire.

Other groups of Slavs moved south to the Balkan Peninsula by crossing the Carpathians and the Mid-Danube Plain. These Slavic peoples have retained their own individual languages and cultural characteristics. The Slovenes settled in the Alps in the north of present-day Yugoslavia in close proximity to the Austrians. Farther south the Croats occupied much of the Adriatic coast and part of the interior. The Croats and the Slovenes became Roman Catholics due to Austrian and Italian influence. To the south and east, a group speaking much the same language as the Croats came under the influence of the Eastern Orthodox church. These were the Serbs, the largest ethnic group of Yugoslavia, who by the fourteenth century controlled much of the Bal-

FIGURE 7.3

Countries and Subregions of Eastern Europe

kans. A group of Serbs, known as Montenegrins, established a small state in the mountains of the southwest. To the east the Bulgarians had established a state of their own by the tenth century. All these early South Slav states and empires vanished when the Turks invaded the Balkans in the fourteenth century. Much of the region came under Turkish control, which lasted until the late nineteenth century. The rule of the Turks left an economic and cultural legacy that is still apparent in the region.

Two non-Slavic peoples are also found in this region: the Romanians and the Hungarians (Magyars). The Romanians are the descendants of Latin-speaking settlers who established themselves on their present territory in Roman times, mixing with the local Dacian population. Their language absorbed many words from neighboring Slavic languages, but it still remains a Romance language. The Hungarians probably came from east of the Volga River, crossing the Carpathian passes and settling in the Mid-Danube Plain in the late ninth century. The

Hungarians speak a Ural-Altaic language, quite unlike the Slavic languages. They formed a wedge of alien peoples who separated the South Slavs from the West Slavs. The Hungarians eventually rose to a position of power in the region. During the sixteenth century, the Austrian Habsburg family became heirs to large areas of Eastern Europe and along with the Hungarians established the Austro-Hungarian Empire. The Austrians controlled the Czech lands (Bohemia and Moravia), Slovenia, and part of southern Poland (Galicia), while the Hungarians ruled Slovakia, Croatia, and Transylvania (the western province of Romania). Austrian rule was more benevolent than Hungarian. Regions like Bohemia and Moravia saw considerable industrialization under the Austrians, but the Hungarians did little to develop the economies of their part of the empire and were hated by the Slovaks, Romanians, and others, who felt that they were oppressed. After World War I, the breakup of the Austro-Hungarian Empire resulted in the formation of Czechoslovakia and Yugoslavia as new states and the enlargement of Poland and Romania.

In the north the Poles had expanded their control over the lands between the Oder River in the west and the Bug River in the east by the beginning of the eleventh century. Over the next centuries, the Poles occupied territories to the east without much resistance. By the sixteenth century, Poland, in alliance with Lithuania, had annexed much of the area of the present-day Ukraine and Belorussia. However, the increasing power of the Germans and the Russians resulted in three partitions of the Polish state—in 1772, 1793, and 1795—among Russia, Prussia, and Austria. Poland ceased to exist as a state until 1918, when the collapse of the German and Russian empires after the World War I enabled the Poles to reestablish their country. The weakness of Russia also gave the Poles the opportunity to regain some of their old territories in the east, even though many of the inhabitants were of non-Polish origin. At the peace conference after the war, the Allied powers gave Poland access to the Baltic Sea by means of a corridor of territory that divided Germany proper from its province of East Prussia.

WORLD WAR II AND ITS AFTERMATH

The 1930s saw new threats developing to the stability of Eastern Europe. Nazi Germany used the German minority in western Czechoslovakia as an excuse to demand the annexation of the lands fringing Bohemia.

The German occupation of Austria in 1938 further weakened the Czech position, and later that year the Germans entered Czechoslovakia, resulting in the dissolution of the country. Hungary received part of the Slovak territory. In 1939 German expansionism was directed toward Poland, and the invasion of that country brought about the beginning of World War II. The Soviet Army also marched into Poland from the east and seized the eastern territories with their predominantly Ukrainian and Belorussian populations.

In southeastern Europe, the state of Yugoslavia had been in an unstable condition since its establishment due to strife between the two major groups, the Serbs and the Croats. Many Croats wished for an independent state of their own. The German invasion of 1941 led to the dismemberment of the country, with Germany, Italy, Hungary, and Bulgaria all taking a share. A bitter civil war was fought in the mountains between Serbian guerrilla groups and the troops of the puppet state of Croatia, which was established and supported by the Germans. A group of communist guerrillas known as the Partisans, under their leader Tito, appealed to all Yugoslavs to bury their differences and join in the struggle against the Germans. The Partisans were successful in liberating large areas of the country from German control and at the end of the war established a communist government.

After World War II, the boundaries of most of the Eastern European countries were restored to their prewar condition with some exceptions. Czechoslovakia lost a considerable amount of territory to the Soviet Union, which also retained the lands annexed from Poland in 1939 and part of Moldavia taken from Romania in 1940. Yugoslavia gained some Italian territory, including the port of Rijeka (Fiume). Poland gained extensive former German lands in the west and established its boundaries on the line of the Oder River and its tributary the Neisse River. As a consequence, the western boundary of Poland shifted westward, which resulted in Poland gaining a long Baltic coast, including the ports of Gdańsk (Danzig) and Szczecin (Stettin), as well as most of the important Silesian coalfield with its iron and steel industry. The partition of Germany into zones controlled by the United States, Great Britain, France, and the Soviet Union led in 1949 to the creation of the German Democratic Republic on the territory of the Soviet zone. The isolation of the Western-controlled areas of the old German capital of Berlin within the boundaries of this new communist state created periodic problems of access.

BUFFER ZONE

 astern Europe has frequently been referred to as a **buffer zone.** The term "buffer zone" is frequently used to refer to an area that is caught between competing power blocs from which it is physically and politically separate. In the case of Eastern Europe, it has been caught between the competing powers of the West and the Soviet Union.

The term **balkanization,** which refers to the process of creating many new states from a larger political unit, has also been used to describe the political changes that have occurred in the region as a result of the breakup of the Ottoman Empire and the Austro-Hungarian Empire.

Before World War I, most of Eastern Europe had formed part of the Austro-Hungarian and Ottoman (Turkish) Empires. The dissolution of these empires left the newly independent nations of the region open to pressures from Germany and the Soviet Union. During World War II, many countries were in-

vaded by the Germans and absorbed into the German sphere of influence. After the war, however, the region came firmly under Soviet control. The decline of Germany as a major military power and its division into two states left the Soviet Union without a major rival in Eastern Europe. Only Yugoslavia and Albania managed to leave the bloc of Soviet-controlled countries. Their geographical location on the western rim of the region made any

military intervention by the Soviet Union difficult. Thus, the region through time can clearly be viewed as an area that has served as a buffer zone.

The absorption of Poland, Czechoslovakia, former East Germany, Hungary, Bulgaria, and Romania into the Soviet Bloc after World War II partially eliminated the possibility of the area serving as a buffer zone between Western Europe and the Soviet Union. With the dissolution of the Berlin Wall, and the beginning of democratization in Eastern Europe in 1989, however, Eastern Europe could once again become a buffer between the Soviet Union and the West. Only the future will tell if Eastern Europe will continue to function as a buffer zone between the power blocs. Should Eastern Europe integrate itself more with the West or fall back under Soviet control, its function as a buffer zone would diminish considerably.

By 1949, the countries of Eastern Europe were firmly under Soviet control, with the exception of Yugoslavia, which had been expelled from the Soviet Bloc in 1948 because of resistance to Soviet dominance. Thereafter, Yugoslavia followed a somewhat different path from the other countries in terms of economic policies and political orientation. Albania remained closely linked to the Soviet Union until 1961 when Albania sided with the People's Republic of China in its dispute with the Soviet Union. In 1978, Albania broke with China due to disapproval of changes in the Chinese political and economic system, and at present it is somewhat isolated from the rest of the world.

As a result of Soviet domination, all the Eastern European countries adopted economic planning based on the Soviet model. Over the years, however, economic and political reforms in the various countries resulted in con-

siderable differences among the countries. In Poland religion was tolerated to a greater extent than in most other communist states. In Hungary a type of market economy with considerable free enterprise emerged. Bulgaria began a series of reforms intended to change its political, economic, and social structure. In Yugoslavia economic control by the state was greatly reduced by allowing the workers to run factories and other economic enterprises. Romania, on the other hand, maintained a rigid Soviet-type planned economy, but showed considerable independence in its foreign policy. The Czechs and Slovaks attempted basic and far-reaching political reforms in 1968, only to be crushed by a Soviet invasion; consequently, their government kept them on a tight rein for the next two decades. As a result, Czechoslovakia was the most conservative of the Soviet Bloc countries and introduced few reforms until 1989.

Standards of living also vary considerably between the various countries and do not always reflect the degree of state control of the economy. In general, the industrialized countries such as former East Germany and Czechoslovakia have the highest standards of living, along with Hungary, which has developed an efficient agricultural economy. Poland has suffered from political and economic upheavals in recent years, resulting in decreased industrial and agricultural output. Romania's economic problems are the result of bad planning and a highly rigid system. Albania has the lowest standard of living of all and is one of the poorest countries of Europe.

1989: SWEEPING REFORMS

The year 1989 will be remembered as the time of sweeping reforms in Eastern Europe. Although this "collapse of communism," as it was often called in the West, occurred quickly, it should be noted that this was not the first time Eastern Europeans had protested against communism and Soviet domination. The Hungarian Revolt of 1956, the reforms in Czechoslovakia in 1968 that led to an invasion by Soviet tanks, and antigovernment strikes in Poland in 1956, 1968, 1970, 1976, and in 1980–1981 by **Solidarity,** the first independent trade union in

the Soviet Bloc, all preceded Mikail Gorbachev, glasnost, and perestroika. Nevertheless, Gorbachev played an important role in 1989 with his willingness to let change occur.

Most of the sweeping changes occurred in the last six months of 1989 when Communist leaders were ousted in Poland, Hungary, former East Germany, Bulgaria, Czechoslovakia, and Romania. Probably one of the most dramatic events was the fall of the Berlin Wall at midnight on November 9. Built in 1961, the Wall had been an imposing symbol dividing the East from the West. Its collapse has been equally symbolic in signaling a new era in East-West relations. Prior to the opening of the Wall in November, an estimated 225,000 East Germans fled to West Germany through Hungary and Czechoslovakia. Many more Germans in Poland and the Soviet Union also fled to West Germany. Obviously, this mass migration placed a strain on West Germany, but the West Germans attempted to be as accommodating as possible.

What does the future hold for Eastern Europe? Will democracy succeed? Many experts believe that the countries of Eastern Europe could become neutral and nuclear-free and that emphasis will shift from military might to economics and technology. Nevertheless, many economic problems need to be addressed, most

Lech Walesa, a leader of Solidarity, who became president of Poland, speaks to a group in front of the shipyards in Gdansk, Poland.

East German refugees in
Prague, Czechoslovakia, wave
while holding their travel
documents as they prepare to
board buses that will take them
to special trains bound for West
Germany. This photograph was
taken on October 4, 1989,
about a month before the fall of
the Berlin Wall and a year
before German unification.

importantly, how to make the transition from predominantly state-controlled systems to increasingly free market economies. Several countries have already begun the process of abandoning their Soviet-style planned economies. Poland and Hungary have moved furthest toward adopting free market economies, but others are following. The East German economy was becoming integrated with that of West Germany even before unification took place. It is too soon, however, to predict how successful these changes will be. Nevertheless, as investments flow in from the West and the economies are transformed from within, it seems possible that Eastern Europe will become more economically integrated with the European Community.

Yet one must ask whether the forces of nationalism and ethnic rivalries can be contained during the transition period to allow long-term peace and prosperity to be established. German unification occurred on October 1, 1990, arousing some concern that the new Germany will be an economic and military power with aspirations to dominating Europe. Although other boundaries between countries and between the East and the West will remain unchanged, the character of the borders will probably change, and movement between countries will

become easier. Eventually, many ideological, cultural, and economic barriers may be lowered as the Eastern European nations free themselves from Soviet domination and begin to participate more fully in the larger global community.

The role of the Communist party varies from country to country. In some countries, such as Hungary and Poland, it has changed its name and claims to be a Western-style socialist party. In other countries, it has agreed to compete with other political parties in future elections. In general, the Communist party has surrendered its position of political monopoly.

POPULATION

A wide range of geographical conditions, together with differing economic conditions and historical influences, has resulted in an uneven distribution of population over the area of Eastern Europe.

The densest concentrations of rural population are found in the areas of fertile soils and favorable climate. The regions of loess soils in the south of the North European Plain, especially in former southern East Germany and Poland, have a particularly dense population.

To the south the Moravian Lowland also supports a relatively dense population. The Mid-Danube Plain has its greatest concentration of rural population in the south, while other parts of the plain, especially in the drier east, have fewer people. Local soil factors are also important in explaining the uneven distribution in this region. To the south the Danubian Plain of Romania supports a sparser population than might be expected. Although in general the soils are good, the low rainfall inhibited farm settlement before the days of large-scale irrigation. In fact, the foothills of the Carpathians carry a denser population than does the plain. The basin of Transylvania, contained by the great arc of the Carpathians, is relatively well peopled, as is the Valley of Thrace in Bulgaria.

The sparsest rural populations are found in the areas of sands and clays in the northern part of the North European Plain and the dry areas or lands likely to flood in the Mid-Danube and Danubian plains. The interior mountain areas of Yugoslavia are thinly populated due to the lack of cultivable land and karst conditions. The valleys and lower slopes of the Carpathians carry a somewhat denser population as do basins and valleys in the mountain regions of Bulgaria.

The urban population of Eastern Europe is characterized by concentrations in a number of large cities, often the capitals, with the rest of the urban population living in medium- and smaller-sized towns (Figure 7.4). This is especially true in the case of Albania, Hungary, Bulgaria, and Romania, where the capital cities, Tirana, Budapest, Sofia, and Bucharest, respectively, are considerably larger than any other town in the country. In Yugoslavia, Belgrade, the capital, is rivaled by Zagreb, while the more industrialized countries, such as Czechoslovakia and Poland, have a number of cities of considerable size apart from their capitals.

Industrialization during the period of communist rule accelerated the growth of some cities and resulted in the establishment of new "socialist cities," often iron and steel centers, such as Nowa Huta in Poland, Dunaújváros in Hungary, Košice in Czechoslovakia, and Nikšić in Yugoslavia. These centers, which were built to introduce industry to underdeveloped regions, exhibit features found in all communist countries, such as large standardized apartment blocks, a downtown square, often with the Party headquarters and a state-run department store and other shops, and community services, such as children's playgrounds and cultural centers. This concept of the socialist city was also applied to the new residential areas of the older cities where rows of huge apartment buildings take the place of the individual family homes found in the suburbs of Western cities. Many of the older cities have had their centers rebuilt to restore their attractive prewar appearance. The Old Town of Warsaw in Poland was completely reconstructed in the old style after suffering total destruction during World War II. The Czechoslovak government has financed the restoration of many old towns; even the old city center of Prague, undamaged by World War II, has experienced much recent refurbishing. The Bulgarians have preserved the old Roman parts of downtown Sofia, which can be visited by a tunnel under the level of the present streets. To Eastern Europeans, the artistic and architectural heritage of their past is as important as the present.

The most urbanized countries in Eastern Europe are former East Germany (77 percent) and Czechoslovakia (75 percent). Bulgaria has seen a large movement of people to the cities in recent years. The least urbanized countries are Albania (35 percent), Yugoslavia (46 percent), and Romania (54 percent), with Poland (61 percent), and Hungary (60 percent) in an intermediate position (see the Statistical Profile). These figures also reveal that the rural component of the population is, in general, large in Eastern Europe compared with most Western European countries.

Population trends in Eastern Europe, however, are very similar to those in the West. As the Statistical Profile indicates, birth rates have dropped to the point that the annual rate of natural increase of the population is approaching or has reached zero. Poland and Romania have higher birth rates than most other Eastern European countries; still, their annual natural increase is under 1 percent. Albania has the highest birth rate by far (26 per 1000), and with a death rate of 6, its annual rate of natural increase is 2 percent. Because of these low rates of growth, several countries have instituted measures to increase their birth rates. Former East Germany offered financial incentives to mothers to have more than two children, while Romania restricted access to methods of birth control.

Eastern Europe has seen many movements of population in its history. Since the arrival of the Slavs and the Hungarians, there has been a constant competition for territorial control between these groups and the Germans to their west. The boundary shifts in the region after both World Wars reflect the ascendancy of the Russians as the major power in Eastern Europe. These

FIGURE 7.4

Selected Cities of Eastern Europe

Cities with more than 1,000,000 persons

Cities with 250,001–1,000,000 persons

Cities with 100,001–250,000 persons

Cities with 25,000–100,000 persons

Sofia   National capitals

- - -   Former German boundary

changes in frontiers and political control were accompanied by major migrations of population. The annexation of German territory by Poland after World War II resulted in about 6.5 million Germans leaving Poland for West Germany. In the east, the Soviet occupation of Poland's eastern territories with their population of Belorussians and Ukrainians led to the migration of about 1.5 million Poles westward to Poland. Most of these Poles were settled in the lands taken from Germany. The Russians divided the former German province of East Prussia with Poland and expelled about 1.7 million Germans, replacing them with Russian settlers.

Czechoslovakia forced about 2.7 million Germans to leave their homes in the mountain fringes of Bohemia and replaced them with Czech settlers. Exchanges of population occurred between Hungary and Yugoslavia and between both those countries and Slovakia. However, the greatest movement of people after World War II was set in motion by the establishment of communist rule in former East Germany. Over 3 million people left that country for the West between the end of the war and 1989, with about 2 million moving during the period between 1955 and 1961. Since the dismantling of the Berlin Wall, however, thousands more have moved to former West Germany. Now that the two Germanies are unified, future population movements across the old border are somewhat difficult to predict. The pace of the movement to some extent will depend upon how rapidly the economy in former East Germany can be revitalized.

The population moves after World War II took place at a time when the whole of Eastern Europe was in ferment. Estimates of population changes at this time are complicated by war-related deaths. For example, about 6 million people are believed to have died in Poland during World War II, of whom about 3 million were Jews. A large number of Germans have never been accounted for following the period of population transfers.

These migrations have reduced the minority problems of several countries. Poland, having lost most of its former population of Germans, Belorussians, and Ukrainians, now has a relatively homogeneous population. Czechoslovakia no longer has a large German minority, and the small group of Ukrainians who lived in the eastern tip of the country were annexed by the Soviet Union. Some Hungarians also moved to Hungary. Yugoslavia also expelled its German population, and some Hungarians left. The only countries that saw little change in the ethnic composition of their populations after World War II were Albania, Bulgaria, and Romania. Romania did not expel its German population, although most have left in recent years for former West Germany; it still has a population of some 1.7 million Hungarians. The treatment of this Hungarian minority by the Romanian authorities has led to friction between the Romanian and Hungarian governments. Bulgaria has a Turkish minority of about 800,000 people (9 percent of the total population). Information on this group is difficult to obtain as the Bulgarian authorities did not recognize their existence. The Turks have a higher birth rate than the Bulgarians, which causes concern to the government. Attempts have been made to eliminate their identity by forcing them to take Bulgarian names and by banning the use of the Turkish language. About 300,000 Turks have left the country and moved to Turkey.

Yugoslavia is the major example in the region of a country with a large number of ethnic groups. Tensions between the Serbs and the Croats, as well as the existence of other Slavic peoples, such as the Slovenes, the Macedonians, and the Muslim Slavs, plus non-Slavic Albanians, Hungarians, and Romanians, have led to the establishment of a federal state that permits a great deal of autonomy to its constituent republics. Each major group has its own republic with its own government. The existence of a large Albanian minority within the

territory of the Serbian republic poses a major problem, however. Although the Albanians have their own autonomous region, there has been some agitation for either a full republic or secession from Yugoslavia and union with Albania. The Albanian problem has resulted in an increase of Serbian nationalism. The Croats and the Slovenes are worried by the development, and the latter have indicated that they might secede from the Yugoslav federation. Czechoslovakia has also adopted a federal solution and has established separate republics for the Czechs and the Slovaks.

A POSSIBLE REVIVAL OF THE RELIGIOUS LEGACY

Although the Communist party does not promote religious freedom, various religions continued to exist in Eastern Europe even under communist rule, and the developments since 1989 have encouraged Christians to rebuild some of their institutions. Numerically, Roman Catholicism has the largest number of followers. It is predominant in Poland, where it is estimated that 60 percent of the population are practicing Catholics, as well as in Czechoslovakia and Hungary and among the Croats in Yugoslavia. The Greek or Eastern Orthodox church is found in Romania and Bulgaria and among the Serbs and Macedonians in Yugoslavia. Protestantism (mainly Lutheranism) is dominant in former East Germany but is in the minority in the other countries throughout the region. Thus, all three branches of Christianity are found in Eastern Europe.

Islam is prevalent in Albania (although many Albanians claim to be atheists) and parts of Yugoslavia and Bulgaria. Judaism, which once had a large number of followers, is now much smaller than it was before World War II. Although religious statistics are difficult to obtain for Eastern Europe, it is believed that religious activities have been on the increase during the past 20 years with more persons now professing to be believers.

AGRICULTURE AND FORESTRY

As the population data indicate, the proportion of the population engaged in agriculture is relatively high in some Eastern European countries. The importance of the agricultural sector is a legacy of the past, when farming was the chief occupation of the population. In Bulgaria 82 percent of the labor force was involved in ag-

An Eastern Orthodox church in Bucharest, Romania. The Eastern Orthodox branch of Christianity has long had many adherents in southeastern Europe, particularly in Romania, Bulgaria, Yugoslavia, and Greece.

riculture in 1945, compared with 22 percent in 1984. Over three-quarters of the populations of Romania and Yugoslavia were involved in farming in 1945. Even in Czechoslovakia, a relatively industrialized country, 40 percent of the working population lived on farms 1945, compared with 14 percent in 1984.

Even if the size of the farming population still reflects the agrarian character of the prewar economy, the organization of agriculture does not. However, reforms have been taking place since 1989. In the 1930s, land holdings in countries such as Hungary and Poland consisted of large estates owned by relatively few families, mainly of aristocratic origin, and a large number of very

small farms, many of which could only be operated at a subsistence level. Rural overpopulation in some regions added to the problems of local food supplies. Family farms were usually fragmented into a number of scattered strips that made it difficult to run the farms efficiently. After the communist control of Eastern Europe was firmly established, the collectivization of agriculture on the Soviet model was introduced. This move was preceded by land reforms that broke up the big estates into smaller farms for the peasants. This stage did not last long, however, and was intended to placate the peasants temporarily. By the 1950s the collectivization drive was underway, resulting in the establishment of collective farms and state farms on the Soviet model. Only Poland and Yugoslavia did not collectivize, and the old system of small peasant farms persisted. State farms are, however, found in both countries.

Each country gave its collective farms a different name, such as "land production cooperatives" in former East Germany, "labor-cooperative farms" in Bulgaria, and "unified agricultural cooperatives" in Czechoslovakia, but in essence they were Soviet-style collective farms. Over the years, however, countries such as Hungary and Bulgaria reorganized their farming along more efficient lines, resulting in a deviation from the Soviet model. In Hungary the planning of agriculture was drastically reformed, and an arrangement was made between the collective farms and the owners of the private plots, which every member family is entitled to work for its own benefit. The collective gives land to its members, generally about 1.5 acres (6000 square meters), along with seed, fertilizer, fodder, and technical advice in exchange for a contractual agreement in which the members agree to deliver produce to the collective. Livestock is also given to the members of the collective on contract. The result has been very efficient agriculture, with high productivity particularly in corn, wheat, and pork. The collectives specialize in grain production, while the private sector concentrates on the production of vegetables, fruit, cattle, pigs, and poultry. In Bulgaria economic reforms introduced in 1979 resulted in changes in the system of fixing agricultural prices and the taxation of collective farms so that the most efficient were favored. Bulgaria and Hungary are the only East European countries that can maintain a significant export of agricultural products.

In other countries such as former East Germany, the huge collective farms established on the Soviet model have been dismantled and reorganized in smaller units.

In general, the Eastern Europeans have moved further than the Soviets in attempting to reduce the problems inherent in the Soviet system, but agriculture still remains hampered by the necessity to maintain at least the semblance of adherence to communist ideology. However, with the changes that occurred in 1989, most of these countries will feel less compelled to follow the communist ideology.

The pattern of farming activities and production varies greatly from one country to another due to environmental differences, historical developments, and variations in the system of agricultural organization. In former East Germany and Poland, rye and potatoes are the major crops. Some barley is also grown for fodder and brewing. These crops are well suited to the cool summers and relatively poor soils of the North European Plain. In Poland these crops are grown on the private farms of the peasants and on the few state farms, while in former East Germany they are found on collective and state farms. Further south in Poland, the loess soils and warmer summers permit the cultivation of wheat and some maize. In both countries sugar beets are an important crop, while flax and rapeseed are

grown for vegetable oil. Tobacco is also produced in southern Poland. Livestock is important, especially pigs, which are fed mainly on potatoes. Poland exports ham, bacon, and eggs, but must import much of its wheat requirements, as did former East Germany.

The pattern of farming in Czechoslovakia is somewhat different from that of Poland and former East Germany, due mainly to regional differences within the country. In the west (Bohemia), rye, potatoes, and flax predominate, but there is also an area of better soils where wheat, sugar beets, and barley are grown. The latter is used in brewing beer, one of Czechoslovakia's traditional industries. Further east, in southern Moravia and Slovakia, the summers are warm enough to permit the growing of corn, wheat, grapes for wine, tobacco, and sunflowers for oil. The rim of Bohemia and most of Slovakia are too mountainous for cultivation, and much is still forested, although some livestock grazing occurs on upland pastures.

South of the Carpathians, climatic conditions are more suitable for the growing of corn and wheat, and these crops predominate in countries such as Romania, Yugoslavia, Bulgaria, Hungary, and Albania. Sugar beets

A cooperative farm in Romania at harvest time. More mechanized equipment and modernized agricultural practices are needed in Romania and other Eastern European countries.

and sunflowers are also important in these countries. There are local variations in agricultural production. In Hungary, Romania, Bulgaria, and Yugoslavia, grapes are grown for wine, part of which is exported. Rice is grown in some parts of the Mid-Danube and Danubian plains, and in less favored areas, rye and potatoes are grown. Albania has a climate suitable for olives, figs, almonds, and cotton. Hungary has a large number of pigs, but in mountainous Yugoslavia and Romania, sheep outnumber other livestock. The mountain regions of Eastern Europe, especially the Carpathians and the Dinaric Mountains, have been used since ancient times as grazing areas for sheep, and in some areas herds are still annually moved to the high mountain pastures.

The forest reserves of Eastern Europe are considerable. In particular, the Carpathian Mountains of Romania contain some of the largest areas of forest in all Europe. This forest belt extends into Slovakia and southern Poland. The mountain rim of Bohemia is, in general, densely forested and supports an important timber and woodworking industry.

INDUSTRY

The communist governments of the Eastern European countries gave the highest priority to industrialization. However, before 1945 most countries had a basically agrarian economy with a predominantly peasant population and therefore required large investments of capital in order to achieve even a basic level of industry. In particular, the regions that had been under Turkish control before World War I had experienced little industrial development even during the interwar period and were still underdeveloped in 1945. These areas included southern Yugoslavia, Bulgaria, Albania, and parts of Romania. To the north, former East Germany, Bohemia, and Moravia, the southwestern portion of present-day Poland, Silesia, and Slovenia in northern Yugoslavia had benefited from German and Austrian economic policies and had developed manufacturing industries. The Transylvanian district of western Romania saw some industrialization under Hungarian control, but Slovakia received little attention from the Hungarians. The part of Poland under Russian domination also saw little development. The 1920s and 1930s saw some attempts by governments to develop their countries' industrial base, but with little notable success. Capital was lacking, and in some cases the raw materials were either unavailable or not yet exploited.

Because of the regional differences in the level of industrial development and the uneven distribution of raw materials (Figure 7.5), each country must be examined individually. First, however, it is helpful to divide the Eastern European countries into two major classes: those with a developed industrial economy and those with a less developed industrial base. In the first category, former East Germany and Czechoslovakia occupy the lead, followed by Poland and Hungary. The second category includes Bulgaria, Romania, Yugoslavia, and Albania.

Former East Germany

After World War II, the Soviets ran their German occupation zone to suit their own purposes. Whole factories were dismantled and shipped to the Soviet Union, and what remained of industry was geared to producing goods to pay for war reparations. The seizure of farmland by the authorities also took place at this time. These events along with fear of Soviet domination led large numbers of people to flee to West Germany, creating a labor shortage that is still felt today. After the 1960s, however, the economy stabilized and became one of the most developed in the Soviet Bloc. Although many problems are expected initially in the aftermath of 1989, West German aid and the integration of the two economies should serve to strengthen the East German economy in the future. However, during the transition period (perhaps six to eight years) necessitated by unification, unemployment in former East Germany could be higher than it is now as some of the present outdated industries are dismantled and newer, more modern ones are developed.

The present pattern of industrial development was laid down before World War II, when it was decided that former East Germany would concentrate on the chemical and power industries, based on large reserves of lignite (brown coal) that form the country's major energy resource. At present, a number of large electric power stations are located on the lignite fields, along with three nuclear power stations. A unified electric power grid joins the East German network with those of the other former COMECON countries, including the Soviet Union. The chemical industry is the second largest industry after engineering and is based on electric power, mineral salts, and petroleum, which comes from the Soviet Union by pipeline. Electric power is also used in the aluminum industry, which processes bauxite from Hungary.

The iron and steel industry operates on the basis of scrap metal and imported iron ore and pig iron from the Soviet Union and coal from Poland. It specializes in the production of high-quality steel. A number of centers produce steel, while pig iron comes mainly from the new "socialist city" of Eisenhüttenstadt.

The engineering industry employs about one-quarter of the labor force and produces about one-half of former East Germany's exports by value. Engineering plants are found in most cities, but the largest concentrations are in Magdeburg, Leipzig, Chemnitz (Karl-Marx-Stadt), and Berlin. Products include heavy indus-

FIGURE 7.5

Mineral Resources and Industrial Districts of Eastern Europe

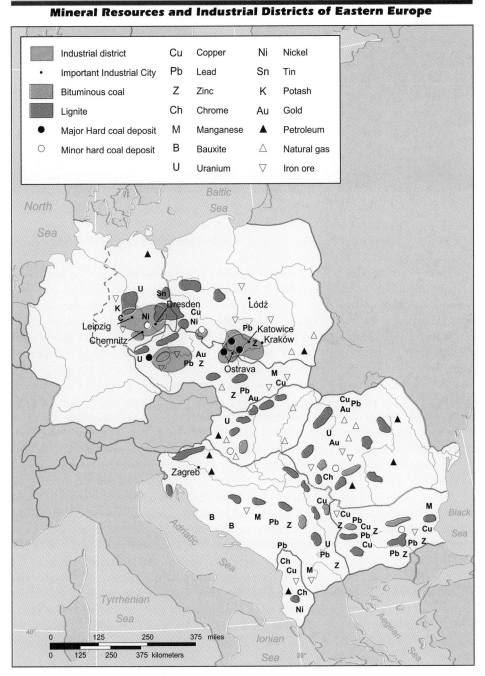

Although former East Germany had one of the leading industrial economies in Eastern Europe, many of its facilities were outdated compared to those in Western Europe. Now that Germany is unified, many East German plants must be modernized, and steps must be taken to reduce air pollution such as that visible in this photograph taken at Riesa, Germany.

trial equipment, machine tools, railroad equipment, and farm machinery. Automobiles and trucks are also manufactured and exported. Ships are built at Rostock. Dresden is the major center for electronics equipment, including radios and televisions. High-quality cameras and optical instruments are also produced and exported.

The textile industry is also important. Major centers are Zittau, Leipzig, Cottbus, and Chemnitz.

The transportation system is based mainly on the railroads, which carry more than twice the weight of freight that goes by highway. Former East Germany has part of the old Autobahn system built before World War II and has added to it since 1945. The country now has the best highway network of any of the Eastern European countries. The Elbe River carries considerable inland waterway traffic, which can use a system of canals linking the Elbe with the Oder. Magdeburg and Dresden are the major river ports. Sea traffic is not very large and is centered on the ports of Rostock and Wismar.

Czechoslovakia

The Czechoslovak industrial scene is one of great diversity, although certain branches of industry predominate. About three-quarters of the industrial labor force are employed in industry and metallurgy, but the man-

ufacturing of textiles, chemicals, glass, and footwear is also important.

The major energy resource is bituminous coal of good quality. It is found in the north of the province of Moravia, where part of the Silesian coalfield extends south of the Czechoslovak-Polish border. Deposits of lignite occur mainly in Bohemia and are used for electric power production. They are mostly mined by open-pit methods. Limited amounts of oil and natural gas are found in southern Moravia, but most of Czechoslovakia's supplies of oil and gas comes by pipeline from the Soviet Union. Hydroelectric power comes from dams on the Vltava River and from the rivers of Slovakia. There are three nuclear power stations, one in Slovakia and two in Moravia.

Iron and steel are produced at Ostrava on the Moravian coalfield and at Košice in eastern Slovakia. Much of the ore used at Ostrava comes from the Soviet Union. The Košice plant is the biggest in the country and was built to bring industry to an underdeveloped part of Slovakia. It uses coal from Moravia and iron ore from the Soviet Union and from local deposits. There are also iron and steel plants in Bohemia, close to the major centers of the engineering industry, which are concentrated in the Prague-Plzeň area. One of the largest is the Škoda plant in Plzeň, which produces heavy engineering products and armaments. Škoda automobiles are made near Prague and are exported to Western Europe. Prague factories manufacture a large variety of products, such as locomotives, aircraft, trucks, buses, motorcycles, and electrical equipment. Many of these goods are exported both to former Soviet Bloc countries and to the West. Brno in Moravia is also an important center for engineering, including the production of tractors and ball bearings. Bratislava, the capital of Slovakia, has an important chemicals industry. Engineering products and chemicals are also manufactured in a number of smaller cities distributed throughout the country.

Specialized industries are located in certain specific regions of Bohemia and Moravia. The textile industry was originally located in the Sudeten Mountains where water power was available, but at present textile plants are also found in Moravia and Slovakia. The Sudeten region is also the major producer of crystal and ornamental glass for export. The forests of the Šumava in southwestern Bohemia are the basis for a woodworking and furniture industry. The former Bata shoe factory in Gottwaldov in Moravia produces a variety of shoes and boots.

In the past the mining of metallic minerals was an important branch of the economy, but most of the ore deposits are now depleted. Except for coal and some iron ore, Czechoslovakia must import most of its industrial raw materials from abroad, chiefly from the Soviet Union.

As in most of the Eastern European countries, the railroad is the main form of transportation and carries about two-thirds of the freight. An international main line runs across the Moravian Lowland from Vienna to Warsaw, but at present the most important line is the one linking Prague with the Soviet Union via Slovakia. The highway network is less important than the railroad. A four-lane highway links Prague with Brno and Bratislava, but the rest of the network is less developed than that of former East Germany.

Poland

Poland entered the scene as a major industrial country only after World War II. Before the war, Poland's industrial development consisted mainly of a small iron and steel industry on its part of the Silesian coalfield, which was then largely owned by Germany, and a large textile industry, mainly located in Lódź. This industry had been developed by the Russians before World

War I, when they controlled much of Poland, and it was too large for the Polish domestic market. By the late 1930s, Poland had also developed armaments and aircraft industries, as well as some electrical engineering.

The situation changed dramatically after 1945, when Poland received the largest part of the Silesian coalfield due to the westward shift of its boundaries at the expense of Germany. The coal is of good quality for the iron and steel industry. Poland also obtained several major plants that had been built by the Germans, the largest being at Chorzów (former Königshütte). By the 1950s, most of these plants had been modernized, and Poland became an important producer of iron and steel in Eastern Europe. A large new iron and steel center was built at Nowa Huta near Kraków, and another new plant came into operation in 1976 in the older steel city of Katowice.

After the Soviet Union, Poland is by far the largest producer of bituminous coal in the former Soviet Bloc. Poland is less fortunate with its supplies of iron ore. There is a little ore of poor quality in southern Poland, but most of the country's requirements come from the Soviet Union. Poland lost an oilfield on the territory annexed by the Soviet Union in 1939 and must import most of its petroleum and natural gas from the latter country. There are some lignite fields in central Poland, which supply large electric power stations. A nuclear power station is under construction.

Deposits of copper in Silesia have given Poland the basis for a major polymetallic industry. The same ores also yield silver, nickel, lead, cobalt, and vanadium. An aluminum industry is based on bauxite imported from Hungary.

The metallurgical industries supply materials for the engineering industry. This industry has been built up on the basis of plants obtained by the annexation of German territory in 1945, as well as industry that had been developed in cities such as Warsaw, Poznań, Katowice, and Lódź. Major products include machine tools, mining equipment, farm machinery, construction machinery, automobiles, railroad equipment, and locomotives. Shipbuilding is an important industry and is centered in the ports of Gdańsk and Szczecin. Most of its products are exported. Among the largest Polish factories are the "Ursus" tractor plant and the "Polski Fiat" plant, which produces Fiat cars under license from the Italian company. Both of these plants are in Warsaw.

Poland also has a substantial chemicals industry, based on deposits of salts and sulfur, as well as on coal and on natural gas and oil from the Soviet Union. Textile manufacture has always been important and still remains so. It is concentrated mainly in the old center of Łódź.

The transportation system relies mainly on the railroads, which also carry much of the freight traffic in transit between the West and the Soviet Union. Coal is an important form of freight. The highway network consists mainly of two-lane highways, with short sections of expressway, the main one running from Berlin to Wrocław and Silesia. The port of Szczecin handles COMECON traffic along the Oder River, while Gdańsk is the major port for Polish traffic.

Polish industry has grown considerably since 1945 and has achieved considerable diversity. However, bad management and political unrest have plagued the Polish economy in recent years, culminating in the events surrounding the rise of the Solidarity movement and the state of martial law declared in 1983. The United States, which had earlier given Poland favored nation treatment in trade matters, cut back on trade and suspended economic assistance. In 1987, however, relations between Poland and the United States improved considerably. The events of 1989 could strengthen the ties between Poland and the United States even further. Revitalization of the Polish economy presents a difficult challenge to the government, however, particularly in light of the sweeping political reforms that have occurred in Poland and the attempts to develop a free market economy. If the new government cannot revitalize the economy, the Poles may become disenchanted with economic reform, and the government may drift back to a planned and regulated economy.

Hungary

Before the communist period, much of Hungary's industry was concentrated in Budapest, the capital. The development of heavy industry and the introduction of new industries have led to a greater spread of industrial activities, but Budapest still remains the single largest industrial center.

Hungary is not rich in energy resources. Most coal comes from the Soviet Union and Poland. Some oil and natural gas are available, but most supplies come by pipeline from the Soviet Union. There are three small hydroelectric power stations and some larger thermal plants using lignite. A nuclear power plant is located at Packs on the Danube, 60 miles (100 kilometers) south of Budapest. However, electric power has to be imported from the Soviet Union.

The iron and steel industry is concentrated mainly in the new city of Dunaújváros on the Danube; for the most part, it depends on imported raw materials, but also uses some Hungarian manganese and a little local coal. There are also a few small plants in scattered locations.

Hungary's most important industrial raw material is bauxite. Part of it is exported to the Soviet Union and other Eastern European countries. A national aluminum industry has been established and is located in a number of centers. Lack of electric power, so important in the production of aluminum, has hampered the development of the industry.

Much of the engineering industry is located in the Budapest area. Major products are machine tools, motor vehicles, especially buses, railroad equipment, farm machinery, and electrical equipment. There is a growing chemicals industry based mainly on imported raw materials. Production of pharmaceuticals is particularly important. There is also a textiles industry, based mainly in Budapest.

Although Hungarian industry is small compared with those of former East Germany and Czechoslovakia, its quality and productivity are high, due to economic reforms that permit a considerable degree of private and cooperative enterprise and initiative. Some factories make their equipment and facilities available to their employees after working hours to produce goods on their own time and for their own profit. This "secondary" economy is encouraged by the government. State enterprises must pay their own way and be able to compete in world markets. Consequently, the amount and quality of Hungarian consumer goods are higher than those of most other Eastern European countries. With a healthy agriculture to provide a variety of foodstuffs, the Hungarians lead a better material life than most of their neighbors. In return for these economic freedoms, prior to 1989 the government maintained a policy of strict allegiance to the Soviet Union, and dissent was less tolerated than in Poland.

As in most other East European countries, the railroad is the most important mover of freight. The highway system is not very well developed. A four-lane highway runs from Budapest to Lake Balaton, an important vacation region. The main internal waterway route is the River Danube, with Budapest as the major port.

In the late 1980s, Hungary began to encounter major economic problems, such as declining exports and too high levels of government spending. The government introduced an income tax (the first in any communist country) and attempted to close inefficient enterprises. Here too, challenges remain as Hungary moves toward a more free market economy.

Romania

Romania is fortunate in its variety of industrial raw materials, including important energy resources. Oil has been exploited in the foothills of the Carpathians since the 1850s, with Ploiești as the main center. Self-sufficiency in oil gave Romania a great deal of independence in its relations with the Soviet Union in the 1960s and enabled it to develop a foreign policy that was often at odds with that of its powerful neighbor. In recent years, however, the situation has changed. Romania's oil reserves have depleted rapidly and are not expected to last beyond the mid-1990s. Romania has thus become an importer of oil. The situation with other energy resources is more favorable. There are large deposits of natural gas in the center of the country. Some of this gas is exported to Hungary, but most is used for the production of electric power and in the chemicals industry. Bituminous coal is found mainly in the southwest, but quantities are limited. Lignite occurs in scattered locations and is used for electric power production. With many rivers flowing from the Carpathian Mountains, the potential for hydroelectric power development is good, and a number of dams have been built. A major hydropower installation has been built with Yugoslav cooperation on the River Danube at the gorges called the Iron Gates. It is the largest hydroelectric plant in Europe, excluding the Soviet Union (2.1 million kilowatt capacity). In spite of these developments, production of electric power per unit of population is the lowest of the European COMECON countries next to Hungary and is only half that of former East Germany. In 1985 a power crisis developed, leading to power cutbacks and serious shortages. To ensure adequate supplies of power, the army was put in charge of the power stations. During subsequent winters, further cuts and rationing of fuel took place.

The iron and steel industry dates back to the days of the Austro-Hungarian Empire when local coal and iron ore were used to establish plants at Reșița and Hunedoara in the province of Transylvania. When Romania acquired Transylvania after World War I, it received an important basis for the further development of its industry. Now these expanded plants require supplies of iron ore and coal from the Soviet Union, due to the inadequacy of local resources. A large plant was built at the Danube port city of Galați in the 1960s in spite of strong criticism by the Soviet Union, which thought that Romania's role in the former COMECON organization should be that of a supplier of agricultural products rather than an industrialized country. This plant uses Romanian coal and iron ore from the Soviet Union, Brazil, and India. The location of the plant makes it possible to use Soviet or other foreign ores brought by sea. Some smaller steel plants are found at other locations.

Romania has a variety of other metallic minerals, including manganese, chrome, and molybdenum. These minerals are used in the iron and steel industry. Non-ferrous metals are found mainly in northern Transylvania. Many of the ores are polymetallic and contain combinations of copper, lead, and zinc. Refineries are located near the ore deposits. Bauxite from the Western Carpathians is used in producing aluminum. Gold and silver are also mined in Transylvania, and although statistics are not available, Romania is probably the second largest gold producer in Europe after the Soviet Union.

Engineering has received great emphasis in recent plans, and a number of products, including equipment for mining and the oil industry, tractors, farm machinery, machine tools, automobiles, trucks, and railroad equipment, are produced. About one-third of Romania's exports consists of machinery. The engineering industry is concentrated mainly in the south of the country, in the region surrounding the capital, Bucharest, and in Brașov.

The chemical industry saw much investment in the 1960s when the new city of Gheorghe Gheorghiu-Dej was founded to manufacture a variety of chemicals from petroleum and local salts. Natural gas is also used in the production of fertilizer.

A cotton and wool textiles industry is located mainly in Bucharest and Brașov. Other minor industries, such as pulp and paper, cement, and shipbuilding have been developed since 1945.

Romanian industry has a good raw materials base compared with some of the other countries of Eastern Europe. However, the promise of the period of expansion in the 1960s has not been met, and the Romanian economy has deteriorated badly in the last decade. The standard of living of the people is probably among the lowest in the region with the exception of Albania. This sad situation is due mainly to bad management and planning on the part of the Romanian authorities. Over-investment in industry led to neglect of agriculture and subsequent food shortages. Unrealistic goals were set, little initiative was permitted to factory managers and collective farm committees. Romania is a good example of a country that has a good base for its economy but has been unable to utilize it effectively. However, with the recent ouster and execution of former leader Nicolae Ceausescu, who ruled Romania for 24 years, the new government must move to bring about economic reform in order to improve the country's economy.

Transportation in Romania is based on a railroad network that had been developed in two separate parts and was only united when Transylvania was added to Romania in 1918. Several new lines had to be built to join and improve the two networks. The highway system is not very modern. The Danube is the main internal waterway and is connected to the Black Sea by a new canal, which ends at the port of Constanța and removes the necessity of using the Soviet-controlled channel at the river's mouth. The main Danube ports are Galați and Brăila. The Black Sea coast has been developed for tourism by the construction of a number of resorts, the largest of which is Mamaia.

Bulgaria

In 1945 Bulgaria was a relatively backward country with little industry and an agrarian economy. Since that time, it has industrialized rapidly and in this sense is one of the most successful of the Eastern European countries. In 1988, 48 percent of the labor force was employed in industry, the highest proportion next to East Germany and Czechoslovakia. Like Yugoslavia, Bulgaria experienced several centuries of Turkish control, and even during the period of independence from 1878 until 1945, little industry developed.

Bulgaria possesses little in the way of energy resources and imports most of its requirements of coal, oil, and natural gas from the Soviet Union. Its dependency on outside supplies of energy is greater than that of any other Eastern European country. There are some deposits of lignite and a little bituminous coal. Small deposits of oil and natural gas also occur. Electric power production is based on lignite, hydropower, and nuclear power. The nuclear power station at Kozloduy on the Danube is the largest power station in the country. A hydroelectric power station has been built on the Danube with Romanian cooperation.

The country is more fortunate in its supplies of metallic ores, especially lead, zinc, copper, manganese, and tin. Tin is in short supply in the former COMECON countries, and Bulgaria is one of the few producers. Most of these metals are produced in sufficient quantities to permit exports.

Local iron ore and Soviet coal form the basis for the development of an iron and steel industry at Pernik and at Kremikovtsi near Sofia. The steel is used in the engineering industry, which has become the most important branch of Bulgarian industry. Railroad equipment, tractors, trucks, farm machinery, construction equipment, and ships are major products and form an important part of Bulgaria's exports. The chemical industry has also seen some development with Soviet aid. Before 1945, textile manufacture and food processing were the basic industries and are still important although they have been eclipsed by the growth of other branches of industry. Food processing, in particular, is more significant than in most other Eastern European countries due to Bulgaria's strong agricultural base. So-called agro-industrial complexes have been established that process a variety of farm products, mainly for export. The fruit and vegetable canning industry is particularly important, and much of its production is exported to Western Europe. Tobacco products and wine are also produced; Bulgaria is the largest exporter of wines among the COMECON countries. There are several large sugar refineries, and part of Cuba's exports of sugar to the former Soviet Bloc countries is processed in Bulgaria.

Bulgaria has developed a successful tourist industry, based mainly on the beaches of the Black Sea coast. It is the largest tourist industry of the former Soviet Bloc countries and employs about one million people. Bulgaria has followed a somewhat different direction than Romania in planning its resorts and has concentrated on achieving a variety of architectural styles and attractions that contrast favorably with the rows of large monotonous hotel blocks found in Romanian resorts such as Mamaia. Old Black Sea towns, such as Nesebar, have been well preserved as tourist attractions.

Due to neglect in the past, the transport network is less developed than in the other Eastern European countries. In recent years, the highway system has been improved by resurfacing. Three parallel transportation routes originate at Sofia in the west: the northern route leads to the Black Sea port of Varna, the middle route to the port of Burgas, and the southern to the Turkish border.

In addition to developing a tourist industry around the beaches of its Black Sea Coast, Bulgaria is promoting its ski resorts.

Bulgaria does not lack problems, but within the type of planned system imposed on the country in 1945, it has industrialized successfully and has given its citizens a rising standard of living over the years. The case of Bulgaria raises the question of whether a socialist planned economy is perhaps most successful in the early stages of a country's development, when it is moving from an agrarian to an industrialized society. At that stage some centralized authority may be needed to ensure the best directions to take for industrial development, obtain the necessary capital, and stimulate the extraction of raw materials. At a later stage of development, such as that reached by the Soviet Union and the other industrialized former COMECON countries, the planning of the economy down to its last detail and the centralization of economic control leads to stagnation, bureaucracy, and low productivity. Events in 1989 may serve to improve the Bulgarian economy at the time when it is becoming increasingly apparent that planned economies may have outlived their usefulness.

Yugoslavia

Like Bulgaria, much of Yugoslavia saw little development during several centuries of Turkish rule. However, the northern area, which had been part of the Austro-Hungarian Empire until the end of World War I and the foundation of the Yugoslav state, had experienced some industrialization. This historical fact still has relevance for understanding the present distribution of Yugoslav industry.

Energy resources are limited mainly to lignite and water power. There are some small bituminous coalfields, and petroleum and natural gas are found in the north of the country, but much has to be imported. About one-third of the country's electric power comes from hydroelectric power stations, mostly located on rivers flowing from the Dinaric Mountains. Yugoslavia shares power from the Danube power plant at the Iron Gates with Romania. In spite of having the largest hydropower potential in Europe next to Norway, however, Yugoslavia has invested relatively little in hydropower installations. In order to increase electric power production, a number of thermal plants have been built near lignite fields. In 1981, a nuclear power plant went into operation near Ljubljana in the north.

Yugoslavia is particularly rich in metallic minerals. It is the major producer of copper, lead, and antimony in Europe and has important supplies of bauxite, chrome, zinc, mercury, manganese, and iron ore. Many of these metals are exported.

Local iron ore is used as the basis for an iron and steel industry. The oldest plant, established during the period of the Austro-Hungarian Empire, is at Jesenice in the northern republic of Slovania. Other small plants at Zenica and Vareš were expanded in the 1930s, and new plants were built at Sisak and at Smederovo, near Belgrade. After 1945 the Zenica plant was enlarged and modernized into the largest in the country. New plants were built at Nikšić in Montenegro and Skopje in Macedonia. These iron and steel centers were built as much for political as for strictly economic reasons, the aim being to give these two underdeveloped southern regions an industrial base. In practice, however, the prestige of having a steel plant has outweighed the economic benefits. The Skopje plant is particularly high cost in its operations, and neither plant has a large market for its products. Although iron ore supplies are adequate, most Yugoslav plants must rely on imported coal.

Much of the engineering industry is located in the northern regions of Slovenia and Croatia, with secondary clusters around Zenica in Bosnia and Belgrade (the capital) in Serbia. Products include machine tools, farm machinery, and railroad equipment. The automobile industry has been developed with foreign aid and Fiat, Citroen, Daimler-Benz, and Volkswagen all have factories in Yugoslavia. A small cheap car, the Yugo, based on an old Fiat model, is being manufactured mainly for export to the United States. Shipbuilding is important and is located in the Adriatic ports of Rijeka, Pula, and Split.

The processing of nonferrous metals provides considerable employment in scattered locations. The chemicals industry has seen considerable development in recent years, but production cannot meet demand for mineral fertilizers, rubber, synthetic fibers, and other chemical products, which must be imported. The textile industry also relies largely on imported raw materials. Food processing is important for the export trade, with specialties in canned fruit and fish and tobacco products.

The development of industry in Yugoslavia has been complicated and at times hindered by its political structure. The practice of establishing a republic for each of the major ethnic groups and giving the governments of the republics a great deal of control over their own affairs has created a situation in which the republics compete for central government funds and investment. The less developed republics, such as Macedonia, Montenegro, and Bosnia, have demanded a higher level of federal investment in their industries than goes to the more industrialized republics of Slovenia, Croatia, and Serbia. These republics, on the other hand, point to their greater efficiency and to the fact that they produce most of the country's wealth and ask why they should be penalized to reward the less efficient southerners. The building of uneconomic "political plants" in the several republics has not helped the situation. The problem of uneven development in Yugoslavia has not been solved and remains a major political and economic issue. In the 1960s and 1970s, the problem of unemployment in the less developed areas of the country was partially solved by permitting large numbers of workers to emigrate to West Germany and other Western countries to find work. In 1984, about 600,000 Yugoslavs worked abroad. However, with economic recession in Western Europe in recent years, many Yugoslavs have returned home.

The problem of industrial employment is less easily handled in Yugoslavia than in most other communist countries. Yugoslavia has developed a system of **workers' management** by which individual enterprises, from factories down to retail stores, are run by the employees themselves with the aid of "workers' councils" and hired managers. These enterprises are thus not run by the state and receive no direct state investments, although they can borrow money from the state banks. They compete with one another like Western firms. An enterprise that does not make a profit may be closed down. In the case of essential industries, the state may come to the rescue, but in most cases an unsuccessful enterprise has to dismiss its workers. It should be noted that this system is not considered capitalist, as the enterprises in theory belong to the people, and the workers, who are "caretakers" for the people, share the profits and pay the manager. If they wish, they can replace the manager. Yugoslav newspapers contain advertisements for managers, while television carries commercials advertising the products of various enterprises. In Eastern European countries with typical Soviet-style economies, enterprises did not compete for customers, although advertisements for products from the state-run factories appeared in the press and on television. In most cases this advertising was meant to draw attention to a new product, to serve as a public service, or to encourage the purchase of goods that had been overproduced. Unemployment in these countries was not a major problem as inefficient or uneconomic enterprises still continued to function with state subsidies, and a considerable part of the labor force was, in fact, underemployed on tasks in factories, offices, and farms that would have been eliminated under the competitive conditions of capitalism. However, as economic reforms take hold, and capitalism and competition are intro-

duced, unemployment could be a problem as factories and offices streamline and modernize their operations.

Like other Eastern European countries that were divided in the past between the empires of Europe, Yugoslavia inherited an uneven transportation network at its independence. Little was done to improve it in the 1920s and 1930s, and it suffered considerable damage during World War II. The railroad is an important form of transportation, but its recent development reflects the competition between the various republics. The major rail route to the Adriatic coast runs from Zagreb to the port of Rijeka, with another branch to Split. Because these lines run on the territory of the Croat republic, the Serbs decided to build their own railroad to the coast. Money was even collected from the general public in Serbia, and the new line was built from Belgrade to Bar on the coast of Montenegro. Nationalism rather then economics was behind this difficult and costly project. The problems of rail construction in a mountainous country such as Yugoslavia have favored the development of the highway network, which parallels the railroad routes in many areas, but provides access to places that the railroad does not reach. More freight and passengers are moved by road than by rail. There are two major highways, one linking the Austrian border via Ljubljana, Zagreb, Belgrade, and Skopje with Greece, the

Adriatic coastal cities have traditionally attracted a large number of tourists from Western Europe. Dubrovnik is a popular resort city in Yugoslavia.

other running along the Adriatic coast. The latter is used by the large number of tourists, mainly from Western Europe, who visit the coastal cities and resorts, such as Dubrovnik, Split, and the Dalmatian islands.

The major ports are Rijeka, Split, Sibenik, Ploče, Koper, and Bar. Again, rivalry between the republics has caused excessive development of competing ports.

Albania

Albania has the lowest level of industrial development in Europe. This fact is due to the lack of interest of Albania's former rulers, the Turks and the Italians, rather than to a lack of resources. In fact, the country is quite well endowed with both energy resources and metallic ores. Oil production is sufficient to cover domestic needs and permit some export. There are several refineries. Electric power production is based largely on a number of hydropower stations, of which the largest is on the Drin River in the north. Some lignite is mined near the capital, Tirana.

The mining of metallic ores is of great economic importance. Albania is the largest producer of chrome in Europe. Iron, nickel, and copper ores are also mined. Much of these ores is exported, but some iron ore is used in a small iron and steel plant at Elbasan. Coal has to be imported. Much of Albania's industry was started in the 1950s with Soviet aid, and between 1961 and 1978 the People's Republic of China provided assistance. Textile production is the most important. There is only a rudimentary engineering industry, and a small chemicals industry produces some fertilizer and salt. Some factories process farm products, such as leather, tobacco, and cotton. Some of the industrial plants constructed with Soviet and Chinese aid are now falling into disrepair.

Due to the lack of a developed railroad network, most goods and passengers move by highway. The construction of railroads only began in 1947. Lines join the major port, Durres, with Tirana and Elbasan, with a continuation to the Yugoslav border. A coastal line joining the ports of Shkoder, Durres, and Vlora is under construction. A line from Shkoder to Titograd in Yugoslavia has been opened for freight traffic.

By its self-imposed isolation, due to disapproval of the paths followed by both the Soviet Union and the People's Republic of China and its disputes with Yugoslavia and Greece over borders and ethnic minorities, Albania has limited its economic development. With for-

eign aid, more could be done to develop its considerable resources. Recently, the government has shown some signs of a greater willingness to resume contacts with the outside world, which could only be to Albania's advantage.

SPATIAL CONNECTIVITY

FOREIGN TRADE, CONNECTIVITY, AND INTERDEPENDENCIES

In the preceding chapter the relationship of the members of the COMECON countries with the Soviet Union was discussed. It should be noted that neither Albania nor Yugoslavia was a member of COMECON, although in 1978 Yugoslavia agreed to develop a program of closer cooperation with the group. Table 7.2 shows the proportion of total trade of individual Eastern European COMECON countries by groups of countries. It reveals that most of the former COMECON countries traded predominantly with one another (including the Soviet Union, Cuba, and Mongolia), with the developed capitalist countries as their next major trading partners. Trade with the developing countries is more limited, except in the case of Romania.

Although there is a general pattern of trade for the whole group, there are individual differences that should be noted. Bulgaria is particularly dependent on inter-COMECON trade. Over 40 percent of its imports consist of fuel and raw materials, mainly from the Soviet Union. Its main export is machinery, but processed agricultural products such as canned fruit and vegetables, wine, and tobacco, which go to the West as well as the former Soviet Bloc countries, are also important. The Soviet Union is the largest trade partner, while former West Germany is the largest Western supplier of goods. Hungary follows much the same pattern of reliance on imports of fuels and metallic ores from the Soviet Union, but has a much larger trade with the West, especially former West Germany, Italy, and Austria. Hungary also exports food products and has built up an export of bauxite, buses, pharmaceuticals, clothing, and footwear. Hungary's imports of consumer goods from the West have been on a larger scale than in most other former COMECON countries.

Former East Germany is almost totally reliant on Soviet supplies of oil and iron ore, and its industrial prod-

TABLE 7.2

Trading Partners of COMECON Countries by Groups of Countries as a Percentage of Total Trade, 1987

	SOCIALIST COUNTRIES	COMECON COUNTRIES	DEVELOPED CAPITALIST COUNTRIES	DEVELOPING COUNTRIES
Bulgaria	81.4%	80.2%	11.0%	7.5%
Hungary	53.7	49.5	38.4	7.8
East Germany	68.8	66.6	27.2	4.0
Poland	76.4	72.7	19.5	4.1
Romania	56.7*	51.0*	Not available	Not available
Czechoslovakia	79.1	75.4	16.6	4.3

* 1985 data.

SOURCE: *Statisticheskiy yezhegodnik stran-chlenov Soveta Ekonomicheskoy Vzaimopomoshchi 1988.*

ucts, such as machinery, electrical equipment, and chemicals, are particularly sought by the Soviet Union. Nevertheless, former East Germany maintained trade connections with the West, with West Germany as its most important capitalist trading partner. Further moves toward integration of the East and West German economies after unification will cause considerable changes in former East Germany's trade pattern.

Poland relies on Soviet fuels to a lesser extent than the preceding countries because of its large coal reserves, but nevertheless has to import most of its oil and natural gas and much of its iron ore. Coal is a major export. Some copper and zinc are also exported. Shipbuilding and repair for foreign customers is important. Poland also earns foreign currency through its shipping services and the bunkering of foreign ships. It also specializes in the export of complete equipment for chemical and woodworking plants and sugar refineries. Excessive imports of Western goods, including machinery and equipment in the 1970s, followed by the social and political unrest in the early 1980s, which resulted in a drop in exports, have seriously affected Poland's trade balance.

Czechoslovakia has the best developed export trade of all the former COMECON countries; most of its industries that produce finished goods send part of their output abroad. About 20 percent of all industrial products, in particular, about one-third of its machinery, are exported. Czechoslovakia ranks high among European countries in its exports of industrial equipment, including machine tools, automobiles, motorcycles, and electric locomotives. Coking coal and steel are also exported. About 40 percent of all imports consist of raw

materials, such as oil, iron ore, and manganese, which come mainly from the Soviet Union. Czechoslovakia cooperates with the Soviet Union in projects in the latter country involving mineral extraction. Its main trading partners are the Soviet Union, former East Germany, and Poland. It also obtains metallic ores from Bulgaria, Romania, and Yugoslavia. Apart from exports of machinery to the former Soviet Bloc countries, Czechoslovakia has developed an export trade in industrial products with the developing countries.

Romania in the past relied less on imports of fuels and raw materials than the other COMECON countries, but at present must import ever-increasing amounts as its reserves become depleted. Over half of its imports consist of these items. Romania is an important exporter of chemicals. Exports of food products are less important than formerly, but the export of machinery is increasing, although it is still proportionally the lowest in the former Soviet Bloc countries.

Yugoslavia's trade relations have developed differently from those of the COMECON countries due to its independence from the former Soviet Bloc and from the European Community. It thus has had a greater variety of options in choosing its trading partners. Apart from cooperation with COMECON countries, Yugoslavia also has links with Western firms such as Fiat, Renault, Volkswagen, and Siemens. These companies and others from former West Germany, France, Italy, Great Britain, and the United States have entered the Yugoslav market and invested in or established enterprises, mainly in the automobile and chemicals industries. These countries are also Yugoslavia's major Western trading partners, with former West Germany and Italy in the lead. Most of the

TRADE LINKAGES

The major trading links of the Eastern European countries since World War II have been with one another and with the Soviet Union. In their relations with the rest of the world, they followed the lead of the Soviet Union. Of the former Soviet Bloc countries, only Romania developed a somewhat more independent foreign policy, while both Yugoslavia and Albania followed policies of nonalignment. The events of 1989 have now made it possible for all these countries to realign the directions of both their foreign policies and their trade links. They are seeking membership in various international financial and trade organizations and are attempting to solve the problem of the large debts that they incurred by borrowing from Western banks. Many countries are seeking further aid from the West, including the United States, to help rebuild their economies.

Eastern European countries have few products that are eagerly sought by the West, a fact that has hindered the development of their trade. For example, they lack the oil, natural gas, gold, and diamonds that have enabled the Soviet Union to maintain a more active trade with the West. Poland, however, has coal, and Yugoslavia and Albania have nonferrous metals that are in some demand in the industrial-

ized countries. Polish, Czechoslovak, and Yugoslav cars are exported to Western Europe and North America, but in general the countries of the developing world are the only markets for Eastern European manufactured goods, which, though cheaper, are of poorer quality than those of the West.

Economic aid follows somewhat the same lines as trade, but in the past military aid followed guidelines laid down by the Soviet Union. The Soviet Union used the armaments industries of such countries as Czechoslovakia and former East Germany to supplement its deliveries to its allies such as Cuba, Nicaragua under the Sandinistas, and Vietnam and to other customers such as India and Nigeria. However, Czechoslovakia recently announced that it would cease arms shipments abroad, and it is possible that other former Soviet Bloc countries may follow this example.

---

imports from the West are industrial goods. Yugoslavia imports fuels, iron ore, and cotton from its most important trading partner, the Soviet Union, and also trades with Poland, former East Germany and Czechoslovakia. Its exports include electrical and electronic goods, machinery, ships, furniture, clothing, and shoes.

The Yugoslav balance of payments is helped by a considerable transit trade, as well as by remittances sent by Yugoslavs working abroad and by the tourist industry. The considerable powers granted by the constitution to the individual republics permit them to conduct trade with foreign countries and to make their own trade agreements with them. However, Yugoslavia's total trade suffers from an imbalance of imports over exports, a situation that adds to the already considerable economic problems faced by a country that in some ways can be said to have been living beyond its means. In recent years, the standard of living has been falling due to the need for greater austerity and reduced imports of consumer goods from abroad.

Albania's relations with the outside world are very limited. The constitution forbids the acceptance of foreign credits or aid from capitalist countries or the establishment of joint companies. Foreign trade is conducted on a barter basis or for cash sale. Until 1978, almost half of Albania's trade was with the People's Republic of China, but at present trade is mainly with Eastern European countries, North Korea, Vietnam, and also with Italy, France, and India. This trade is based mainly on exports of oil, chrome, nickel, copper, and agricultural products, such as citrus fruit and tobacco. There are signs that Albania would like to expand this trade and at the same time improve its relations with the rest of the world. The resumption of rail traffic with Yugoslavia and orders placed with Western firms for new factory and transport equipment suggest that Albania's isolation may be coming to an end.

The organization of foreign trade and the establishment of relations with foreign firms varies by country. Yugoslavia is no longer the only country that permits cooperation with Western firms. Hungary has a number of specialized agencies to handle foreign trade and authorizes individual enterprises to conduct their own foreign trade relations and set up companies abroad. Participation in foreign firms is also permitted. Western firms are assisted in conducting trade within Hungary and can obtain help with market research. Czechoslovakia also has a number of trade agencies that act as independent companies with their own capital and managers appointed by the government. Bulgaria has a number of joint ventures with Western firms as do Poland and Romania. In most cases, the foreign firm is permitted to own up to 49 percent of shares in the Eastern European firm and is entitled to a share in the profits; since 1989, many restrictions on foreign firms have been lifted. Romania established trade links with the European Community and has agreements with it on trade in industrial products. In 1974, the first branch of a U.S. bank to be established in a COMECON country opened in Bucharest. Other countries are now permitting Western banks to open branches or even to buy a share in an Eastern European bank.

The pattern of aid by former COMECON countries to one another and to the developing countries is still quite similar to that of the Soviet Union. Most of the Eastern European countries have also given help in the form of projects to a number of developing countries such as Cuba, India, Algeria, Iraq, and Nicaragua. They have also cooperated with the Soviet Union in providing military aid in various forms. The members of the Warsaw Pact allowed the stationing of Soviet troops on their territory, although Romania resisted this and took part in Warsaw Pact maneuvers on a selective basis. Now most Eastern European countries are demanding the removal of Soviet troops from their territories. Neither Yugoslavia nor Albania has permitted the presence of foreign troops on its soil.

Like the Soviet Union, Yugoslavia provides foreign aid mainly in the form of projects and has cooperated in the development of industrial and agricultural projects in COMECON countries, including the building of hotels and other tourist facilities. In the developing world, Yugoslavia has assisted with the construction of factories, ports, electrical power transmission systems, and roads.

Some countries are members of international financial and trade organizations. Romania, Hungary, and Yugoslavia are members of the International Monetary Fund (IMF), as was former East Germany, and Hungary, Romania, and Yugoslavia are members of the International Bank for Reconstruction and Development (World Bank). Most countries belong to the General Agreement on Tariffs and Trade (GATT). Yugoslavia has a special relationship with the Organization of Economic Cooperation and Development (OECD). As members of these organizations or as independent agents, the Eastern European countries have borrowed large sums from Western banks. Poland, with a debt of some $31 billion, is the largest borrower, but Hungary, Romania, and to a lesser extent Czechoslovakia, former East Germany, and Bulgaria have all incurred debts that they are having difficulty repaying. Hungary has the largest per capita debt to the hard-currency countries. Poland's inability to meet the interest payments on its debts led to a major financial crisis in 1985, and a rescheduling agreement was made with Western banks. Since 1989, most Eastern European countries have begun negotiations for membership in most of the international financial and trade organizations and negotiated further aid from the West, including the United States.

The COMECON countries had economic cooperation treaties with the Soviet Union, which in effect spelled out their roles as trading partners with that country and still quite committed them to particular specializations in trade and economic development. Bulgaria even discussed the coordination of future Soviet and Bulgarian five-year plans. In the past, although each country individually might have developed certain trade and economic relations with countries of the capitalist West, allegiance to the Soviet Union and its policies took precedence and could not be superseded by other interests. Since 1989, however, these trading relations have begun to change.

Before 1989, political relations between the countries of Eastern Europe and the Soviet Union were reflected to some extent by their trade links and by their membership in COMECON and the Warsaw Pact. In other words, the Soviet Union provided the guidelines and directives that the other countries were expected to follow. Some deviations from the Soviet norm were permitted or, at least, tolerated, however. For example, religious freedom and a certain level of dissent were allowed to develop in Poland to a greater extent than in other Soviet Bloc countries, while Hungary was able to go further than the others in experimenting with economic reforms. Romania followed a flexible foreign policy, but maintained a rigid Soviet-type economy and society. Bulgaria followed the Soviet model in all respects, but is now radically reforming its political, economic, and social systems in response to Soviet leader Gorbachev's call for "restructuring" and "glasnost." Romania, to some extent, resisted these pressures from Moscow in line with its attempts to remain semi-independent, while Czechoslovakia has been instituting changes but is fully aware of the various problems associated with reform.

Political relations with the West and the developing countries vary from country to country, although, in general, they follow the pattern of trade and aid. Relations between the United States and Poland have improved since the visit of the vice president to Warsaw in 1987 and the overthrow of the Communist party leadership in 1989. Because of the large population of Polish origin in the United States, links between the two countries have historically been important. Although problems remain such as the disparity in living standards between former East and West Germany, unification ended the occasional tensions that had existed between the two Germanies since World War II. Political links with countries of the developing world have historically been less important than with the West and in the past usually followed the pattern established by the Soviet Union.

Tensions between the Eastern European countries exist in spite of their political, economic, and military ties and are likely to increase as these countries develop more independent foreign policies. Hungary and Romania are still at loggerheads over the disputed province of Transylvania, while Romania still resents the annexation of its province of Bessarabia by the Soviet Union in 1940. Bulgaria claims to have legitimate rights to control Macedonia, at present part of the Yugoslav state. Albania has only begun to patch up its long-standing disputes with Yugoslavia and Greece over boundaries and minority groups. Rivalry between the Czechs and Slovaks resulted in the division of Czechoslovakia into two federated republics. In spite of the former superimposition of Soviet control, and now the resurgence of democracy, old historical rivalries, nationalist tendencies, and the desire for independence from Soviet domination have produced a variety of attitudes on the part of the Eastern European countries, which reflect a lack of uniformity within the region.

PROBLEMS AND PROSPECTS

Since the end of World War II in 1945, the countries of Eastern Europe have been faced with a variety of polit-

ical, economic, and social problems. These have been largely caused by the imposition of Soviet control over the region and the introduction of an economic and social system that does not reflect the needs and aspirations of the people. In such countries as Poland and Hungary, the communist regimes have met with continuous opposition and lack of support from the population. Economic mismanagement and the slow growth in the standard of living compared with the Western countries are widely resented. The delineation of boundaries after World War II by the Soviet Union and the other Allied countries did not recognize the claims of some countries and rewarded others disproportionately. As a result, many Eastern Europeans have been dissatisfied with both the internal conditions and the external relations of their countries and have consequently been alienated from the governments, which have not represented the people.

Recent upheavals in Eastern Europe in which rule by the Communist party and its leaders has been overturned are the result of these past frustrations. Admittedly, many political and economic problems will not automatically be resolved by the installation of democratic institutions. Nevertheless, it is hoped that this newfound freedom will foster more social cohesiveness and inspire new solutions to political and economic problems. Will these countries assume more market-oriented economies? Will they become more fully integrated with the European Community and encourage investments from the West? Soviet interest in the political stability of its neighbors on its western border is still a factor despite the sweeping reforms recently set in motion. Obviously, the 1990s offer new opportunities for the region that it has not enjoyed since World War II.

Suggested Readings

Adams, A. E., I. M. Matley, and W. O. McCagg. *An Atlas of Russian and East European History.* New York: Praeger, 1967.

Alisor, N. V., and E. B. Valev, eds. *Economic Geography of the Socialist Countries of Europe.* Moscow: Progress Publishers, 1985.

Demek, J. *Geography of Czechoslovakia.* Prague, 1971.

Demko, G. J., ed. *Regional Development Problems and Policies in Eastern and Western Europe.* New York: St. Martin's Press, 1984.

Fischer-Galati, S., ed. *Eastern Europe in the 1980s.* Boulder, Colo.: Westview Press, 1981.

Gianaris, N. V. *The Economics of the Balkan Countries: Albania, Bulgaria, Greece, Romania, Turkey and Yugoslavia.* New York: Praeger, 1982.

Hoffman, G. W., ed. *Eastern Europe: Essays in Geographical Problems.* London: Methuen, 1971.

Matley, I. M. *Romania: A Profile.* New York: Praeger, 1970.

Mellor, R. E. H. *Eastern Europe.* New York: University of Columbia Press, 1975.

Rugg, D. S. *Eastern Europe.* White Plains, N.Y.: Longman, 1986.

Turnock, D. *An Economic Geography of Romania.* London: G. Bell and Sons, 1974.

U.S. Government. *Area Handbook for Albania.* Washington, D.C.: U.S. Government Printing Office, 1971.

———, *Area Handbook for Poland.* Washington, D.C.: U.S. Government Printing Office, 1973.

CHAPTER 8

Japan

IMPORTANT TERMS

Archipelago

Meiji Restoration

Shinto

Emulation

Animism

Kami

Balance of payments

CONCEPTS AND ISSUES

■ Export-based economy balances between resource dependency and strategic interdependency.

■ Natural hazards such as earthquakes, tsunamis, and typhoons as an accepted part of Japanese life.

■ Contributions of urbanization and industrialization to severe land supply and environmental problems.

■ Effects of feudalism and colonialism/imperialism on modern Japan.

■ Influences of cultural evolution and the adoption of innovations on the economy and culture.

CHAPTER OUTLINE

PHYSICAL ENVIRONMENT
Geology and Natural Hazards
Climate, Vegetation, and Soils
Mineral and Energy Resources

HUMAN ENVIRONMENT
Cultural History
Population
Cultural Environment
Economic Landscape

SPATIAL CONNECTIVITY
Trade
Strategic Interdependence

PROBLEMS AND PROSPECTS

Japan stands as a special case in East Asia because it has emerged as the only truly industrialized power outside Europe and North America. Its success serves as an example for the other nations of East Asia. As we approach the end of the twentieth century, Taiwan and South Korea seem to be developing in the same manner as Japan a decade or two decades ago. Other nations are learning from the Japanese experience and are seeking to stimulate their domestic economies by becoming more integrated into the world community.

Within a hundred years (between the 1870s and the 1970s), Japan has twice risen to become a world power—first, as a result of the Meiji Restoration (1868–1912), and again, with U.S. aid and support after World War II. The emergence of Japan as a modern, industrial power is all the more fascinating because it has overcome adverse environmental conditions and a poor natural resource base.

The physical geography of Japan is characterized by environmental hazards, a lack of natural resources, and insularity from the Asian mainland, all of which have played important roles in the evolution of Japanese society. Volcanoes, *tsunamis* (massive waves), typhoons, and earthquakes have plagued Japan throughout its history. Additionally, a relatively poor natural resource base has forced Japan to seek external sources of raw materials to fuel its economic development. This resource deficiency helps to explain Japan's expansionist policies since the mid-nineteenth century and its continued efforts to secure sources of raw materials—first, by war and, now, by trade.

The human landscape has emerged from a unique cultural history. The Japanese people have adopted, adapted, and modified beneficial characteristics from other cultures. The recent impact of Western culture on Japan, however, is changing many of the traditions of Japanese society, a trend that may have far-reaching consequences in the future.

As Japan has modernized, it has become increasingly involved in the world community, both economically and strategically. Japan depends on foreign trade to an extent unparalleled in the rest of the world. This interdependence is paralleled by Japan's strategic location vis-à-vis the People's Republic of China, the Soviet Union, and the United States, which necessitates that Japan perform a precarious balancing act among these superpowers so as to benefit from interdependence while not becoming overly dependent on any one country.

PHYSICAL ENVIRONMENT

The physical environment of Japan and its relative geographic location have played important roles in the evolution of Japanese society. At some points, the shallow Sea of Japan, which separates the islands of the Japanese **archipelago** from the east coast of Eurasia, is only 120 miles (193 kilometers) wide. Yet, despite Japan's proximity to the mainland, its physical detachment has provided periods of isolation. This isolation has contributed to the development of a very distinct Japanese culture.

The islands that comprise Japan (Figure 8.1) stretch for approximately 1800 miles (3000 kilometers) northeast to southwest, from 46° to 24° north latitude and from 123° to 149° east longitude. This geography provides a variety of environmental regions within a land area slightly smaller than California.

Japan consists of four main islands—Hokkaido, Honshu, Shikoku, and Kyushu—and hundreds of smaller islands. The four large islands account for 97 percent of the total landmass of Japan. The best-known group of smaller islands are the Ryukyu (Nansei) Islands. The Kurile Islands off the northern coast of Hokkaido have been occupied by the Soviet Union since the defeat of the Japanese in World War II. Japan has demanded their return, but without success.

GEOLOGY AND NATURAL HAZARDS

The Japanese archipelago is a relatively young geologic feature and consists of the exposed tops of a mountain range. As a result, over 85 percent of the nation is mountainous, and approximately 75 percent has a slope that exceeds 15 percent, making the land untillable. The highest elevation in Japan is the summit of Mt. Fuji at 12,388 feet (3776 meters). There are few relatively large plains. The most important are Kanto and Nobi on Honshu, Tsukushi on Kyushu, and Ishikari and Tokachi on Hokkaido, in addition to narrow coastal plains and intermontane basins (Figure 8.2).

The Japanese archipelago is also part of the geologically active Pacific Basin "ring of fire." Movement of the earth's crust subjects the Japanese to rather frequent natural hazards, some of which have been devastating. Of primary importance are earthquakes, tsunamis, and volcanoes. The archipelago also lies in the path of many Pacific typhoons.

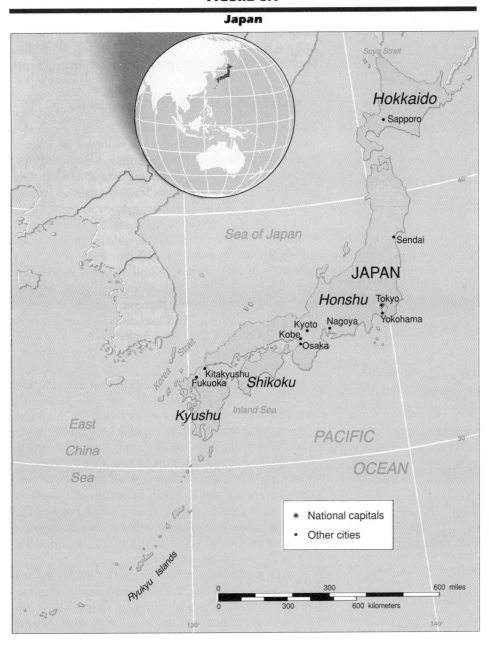

FIGURE 8.1

Japan

Over 1000 noticeable earthquakes occur each year, and about 200 of them are relatively strong. Most of these are centered along the Pacific side, where the islands lie near a deep ocean trench. Seventeen earthquakes have killed 1000 or more people during recorded history. The most devastating was the Tokyo earthquake of 1923, which resulted in over 100,000 deaths and billions of dollars in damage, mostly due to the ensuing fires. When the epicenter of an earthquake is located beneath the sea, a tsunami is created. The Sanriku tsunami of 1896 caused over 23,000 deaths.

After earthquakes, *tropical cyclones,* called typhoons, are the second deadliest natural hazard. Japan experiences 15 to 30 typhoons each year, usually in the late summer and early autumn. A major hazard is the possible flooding afterward. The most recent devastat-

FIGURE 8.2

Physical Features of Japan

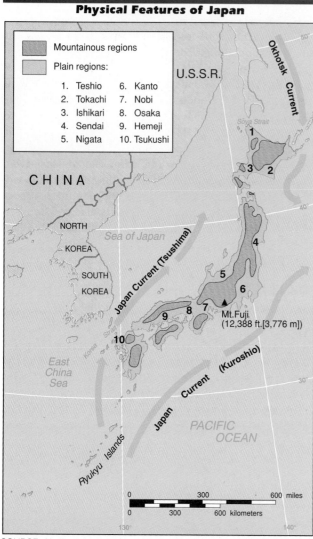

SOURCE: *National Atlas of Japan* (Tokyo: Geographical Survey Institute, Ministry of Construction, 1977).

located near the sea. Earthquakes, typhoons, and volcanoes have generally been most deadly when they caused tidal waves that killed people living near the sea. Although these natural hazards cannot be stopped, Japan is working hard to minimize their damage.

CLIMATE, VEGETATION, AND SOILS

The primary climatic influence on Japan is the monsoon winds that pick up moisture from the Sea of Japan on their northwesterly journey from the Asian continent in the winter and produce a wet southeasterly wind in the summer. Three major climatic regions are produced: humid continental, with cool summers, over much of Hokkaido; humid continental, with warm summers, on southern Hokkaido and northern Honshu; and humid subtropical throughout the rest of Japan (Figure 8.3). Climate is further affected by two major ocean currents (see Figure 8.2). The cold Okhotsk current from the north causes cool summers and fog along the east coasts of Hokkaido and northern Honshu. The stronger of the two, the equatorial Japan current from the south, warms the Pacific coasts of Kyushu, Shikoku, and Honshu as far north as Tokyo.

Average annual rainfall varies from less than 40 inches (102 centimeters) in the north to more than 120 inches (305 centimeters) in the higher elevations of the middle and southern regions. Peak precipitation along the Pacific coast occurs in late summer and autumn. Along the central Sea of Japan coast, peak precipitation occurs in the winter. Along the western and northern Sea of Japan coast, maximum precipitation occurs in the winter as well as in the late summer and autumn. January temperatures range from below 19°F (−7°C) in the north to 61°F (16°C) in the south, and July temperatures from 50°F (10°C) to above 90°F (32°C).

A large portion of Japan is covered by forests, ranging from deciduous forests in the north to temperate rainforests in the south. Acidic mountain soils predominate. Most important, however, are the fertile ***alluvial soils*** of the Japanese plains, which comprise only 15 percent of the total area of the country. These areas are important not only because of their utility for agriculture, but also because they provide the only flat land suitable for urban expansion.

MINERAL AND ENERGY RESOURCES

One of the amazing aspects of Japan's rise to the status of world industrial power is its paucity of natural re-

ing typhoon occurred in 1959, when over 5000 people died.

Volcanoes are also of great concern in Japan. Since 1850, more than 25 major volcanic eruptions have occurred. They are especially deadly when they cause landslides. The eruption of the Unzen volcano in 1792 produced a major landslide into the sea and resulted in more than 1500 deaths.

Many of the deaths that result from natural hazards occur because of the crowded settlements that occupy the limited amount of flat land, virtually all of which is

STATISTICAL PROFILE: JAPAN

REGION OR COUNTRY	Population Estimate mid-1990 (Millions)	Natural Increase (Annual)	Population Projected to 2000 (Millions)	Infant Mortality Rate	Life Expectancy at Birth (Years)	Urban Population (%)	Per Capita GNP, 1988 (US$)	Area, Thousands of Square Miles (Km²)	Population Density, No./mi² (No./km²)
Japan	123.6	0.4	127.5	4.8	79	77	21,040	143.8(372)	860(332)
East Asia average	1336	1.3	1510	35	69	29	2,460	3,946.6(10,222)	343(131)
China	1120	1.4	1280	37	68	21	330	3,705.4(9,597)	303(117)

sources. Japan must import food, wood products, minerals, and energy resources to survive. Since the continued well-being and growth of the Japanese economy are heavily dependent on trade to secure resources, such trade has been an important factor in Japan's foreign policy since the late 1800s.

Although Japan has deposits of several minerals, most deposits are of limited quantity. In terms of meeting demand, the most abundant mineral resources are zinc, manganese, lead, and sulfur. Small amounts of iron ore, copper, tin, mercury, tungsten, molybdenum, chromium, antimony, and magnesite also exist. The limited deposits of these crucial raw materials relative to Japan's large industrial demand for these materials makes Japan very dependent on the importation of these resources from other countries.

The lack of energy resources is even more critical. Coal has been the most abundant energy resource of Japan. It has provided the energy that fueled two industrialization drives since the 1860s. Japanese coal occurs in thin seams unsuited for large, cost-efficient mining equipment. This makes imported coal less expensive than domestic coal and has led to a reduction in Japanese coal production.

The most glaring resource deficiency facing Japan is the lack of petroleum. Domestic production yields about 2 percent of the petroleum that Japan consumes. Japan is also heavily dependent upon other nations for natural gas, an energy resource often used by industry. Ninety-two percent of the natural gas consumed in Japan is imported.

Hydroelectric power has become important since

The rugged Pacific coastline of Honshu, the major island of Japan, near Cape Irō.

FIGURE 8.3

Climate of Japan

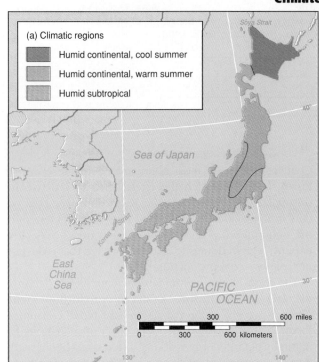

(a) Climatic regions

- Humid continental, cool summer
- Humid continental, warm summer
- Humid subtropical

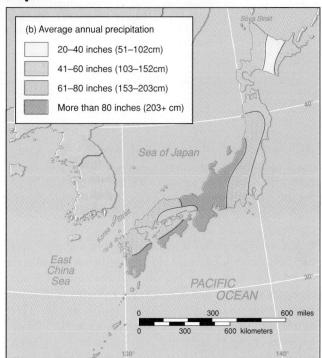

(b) Average annual precipitation

- 20–40 inches (51–102cm)
- 41–60 inches (103–152cm)
- 61–80 inches (153–203cm)
- More than 80 inches (203+ cm)

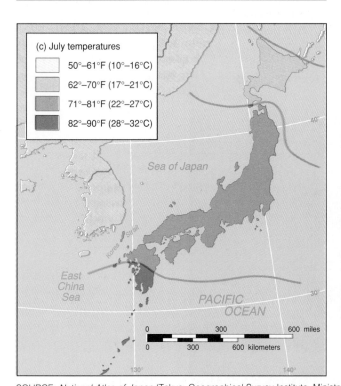

(c) July temperatures

- 50°–61°F (10°–16°C)
- 62°–70°F (17°–21°C)
- 71°–81°F (22°–27°C)
- 82°–90°F (28°–32°C)

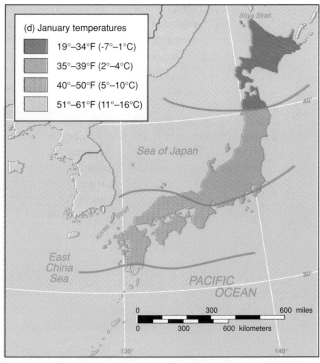

(d) January temperatures

- 19°–34°F (-7°–1°C)
- 35°–39°F (2°–4°C)
- 40°–50°F (5°–10°C)
- 51°–61°F (11°–16°C)

SOURCE: *National Atlas of Japan* (Tokyo: Geographical Survey Institute, Ministry of Construction, 1977).

1900. The physiography of Japan—mountains laced with small but swiftly flowing streams and rivers—is particularly well suited for hydroelectric power. Currently, it provides about 15 percent of Japanese electricity.

Since the Arab oil embargo in 1973, nuclear power has become more important. Today, it provides almost 34 percent of all Japanese electricity. Yet, nuclear power in Japan faces many of the same obstacles that it faces in the United States—high costs, waste disposal problems, and public opposition.

Thus, the continued growth and health of the Japanese economy are closely intertwined with its ability to continue to obtain resources from other nations. This is especially true of energy resources.

HUMAN ENVIRONMENT

The distinctive nature of Japanese culture and society has evolved from contacts first with China and later with the West, punctuated by long periods of isolation. At the same time, the nature of the Japanese people has had a major impact on the physical environment, enabling them to utilize their limited natural resources to the fullest. Through it all, the most important resource has been the Japanese people themselves.

CULTURAL HISTORY

One explanation for the emergence of Japan as the only non-Western, industrial superpower lies in the character of the Japanese. Only by examining the cultural history of Japan and the way in which Japanese society has evolved can one begin to understand and appreciate why Japanese development has traveled upon a path so different from its Asian neighbors, especially China.

Early Japanese Civilization

The earliest inhabitants of Japan were the Ainu, a Caucasian people who still exist as a community of about 20,000 on Hokkaido. The ancestors of most modern-day Japanese first arrived at the islands in the pre-Christian era. These people of Mongoloid stock came from northeast Asia via Korea and have the same ancestry as Koreans. Initially settling around the Inland Sea, these early Japanese lived an isolated existence until the fifth century A.D.

Under the influence of Chinese merchants who visited the islands, Japan became more civilized. Impressed by Chinese military power, wealth, and culture, the Japanese imported certain elements of Chinese culture. They adopted Chinese character writing, which they adapted to their own tongue. They emulated the centralized government of China by establishing an imperial court at Nara, although it did not carry the power or authority of its Chinese counterpart. Eventually, the Yamato clan emerged as the strongest of several contending groups; its leaders came to be known as emperors after the Chinese example. The divine right of the imperial family to the throne became accepted even though it did not have absolute control over the other clans.

By the twelfth century, the Minamoto clan had become the most powerful clan in the islands. The emperor appointed the leader of the Minamoto as the first shogun, or military dictator, and founded the Kamakura Shogunate. A system of dual central government was created with a military government **(shogunate)** under the shogun and a civilian government under the emperor. The shogun was the real power while the emperor became a revered figurehead. During a period of feudal warfare in the 1300s, the **samurai,** an aristocratic warrior class, emerged as a powerful force.

In the middle of the sixteenth century, Portuguese and Jesuit missionaries started arriving in Japan and proceeded to convert large numbers of people. Japanese leaders became convinced, however, that the missionaries were preparing Japan for a European invasion and began persecuting Christians. Additionally, all ports were closed to foreigners, except for Nagasaki, which was allowed to receive only 10 Chinese and 1 Dutch ship per year.

Yet, this was also a time of growth in Japan as many of the institutions that would be important later in its development emerged. The clans built ornate castles, around which towns grew. Agriculture prospered, domestic trade flourished, and an influential merchant class emerged, primarily as a result of making loans to both the **daimyos** (feudal warlords) and samurai.

Attempts by foreign powers to open Japan were unsuccessful until 1853, when Commodore Matthew Perry and a flotilla from the United States sailed into Tokyo Bay. The British, French, and Dutch soon followed. The entry into Japan of the Western powers marked the beginning of internal turmoil that eventually led to the overthrow of the shogunate government in 1867, after almost 700 years of rule.

COLONIALISM— JAPANESE STYLE

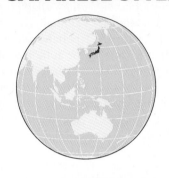

Japan was the only non-Western power that undertook colonialism in the European mode. Japanese colonialism began in the latter stages of the Meiji Restoration and ended with Japan's defeat in World War II. For half a century, Japan was the dominant imperial force in Asia. Like the European colonial powers, Japan undertook this direction in foreign policy because of intense nationalism and the need for resource hinterlands. The need for an expanded resource base, cheap labor, and captive markets motivated small, resource-poor Japan to engage in military and economic imperialism.

Japan first expanded during the Sino-Japanese War of 1894–1895, when the Japanese challenged China for hegemony over Korea. The war resulted in a striking military victory for Japan, which annexed Formosa (Taiwan) and the Liaodong Peninsula (in southern Manchuria). Russia, Germany, and France subsequently forced Japan to return Liaodong to China, which later leased it to Russia.

The Russo-Japanese War (1904–1905) established Japan as a modern world power. Its defeat of Russia returned Liaodong to Japan's protection under lease. In 1910, Japan annexed Korea and, during World War I, extended its empire to include the Shandong area of China, which had previously been under German control.

In 1931, Japan took advantage of the internal struggle between the Chinese Nationalists and the communists by annexing all of Manchuria (Northeast China), which became the puppet state of Manchukkuo. When the League of Nations condemned Japanese aggression in China, Japan left the league and began expanding its empire throughout Asia, first to Hong Kong, then to Malaya and Singapore.

In an effort to curtail Japanese military expansion, other nations, especially the United States, began to limit their exports of raw materials to Japan. Japan responded by seeking new

---

With the overthrow of the shogunate, the emperor was reestablished as the ruler of Japan, although he was controlled by a group of young samurai officers. Under the direction of these samurai, Japan set upon a policy of borrowing and adapting from the West to stimulate modernization. This period of modernization, lasting from 1868 to 1912, is called the **Meiji Restoration.** Several developments during this period were critical to Japan's emergence as a world superpower.

In the 1880s, there were movements toward democratic institutions. The emperor created a *Diet* (parliament) and accepted a constitution in 1889 that politically left him a figurehead even though it proclaimed him a virtual god in keeping with **Shinto** (the Japanese religion) tradition. During the early years of the twentieth century, a strong central government under the guidance of the samurai emerged with the emperor as a divinely sanctioned figurehead. Meiji leaders also abolished feudalism by eliminating the clan system and dissolving the samurai as a separate, elite class. The cen-

tral government established a modern communication system and a strong, centrally controlled military. In the economic sector, merchant families, called *zaibatsu,* grew in importance. These entrepreneurs provided the financial support needed for modernization.

As Japan became a modern nation, an intense period of nationalism ensued. This new sense of national power was first tested against the decaying Manchu Empire in China in 1894. The resulting Japanese victory eventually led to the annexation of Taiwan and Korea. A victorious campaign against Russia in 1904–1905 firmly established Japan as the leading power in Asia. The rise of Japan to power in a period of about 40 years was an amazing feat and truly unique in Asia. Why were the Japanese able to deflect Western incursions, when others, especially the Chinese, were not? How were they able to use their relationship with the West to their own advantage?

Although Japan had been isolated for a long period of time, the Japanese were well aware of the value of

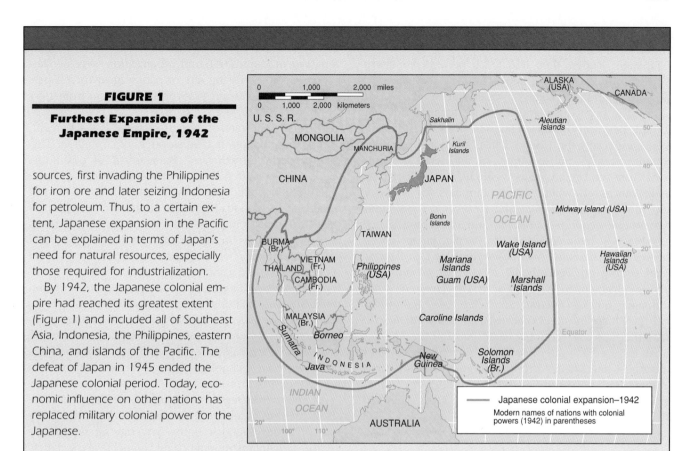

FIGURE 1

Furthest Expansion of the Japanese Empire, 1942

sources, first invading the Philippines for iron ore and later seizing Indonesia for petroleum. Thus, to a certain extent, Japanese expansion in the Pacific can be explained in terms of Japan's need for natural resources, especially those required for industrialization.

By 1942, the Japanese colonial empire had reached its greatest extent (Figure 1) and included all of Southeast Asia, Indonesia, the Philippines, eastern China, and islands of the Pacific. The defeat of Japan in 1945 ended the Japanese colonial period. Today, economic influence on other nations has replaced military colonial power for the Japanese.

borrowing, as they had been doing from the Chinese for centuries. Unlike the Chinese, who perceived China to be the center of the universe and the source of all knowledge and wisdom, the Japanese were willing to accept the idea of learning from another culture—in this case, the West.

In addition to the cultural traditions of adoption and adaptation, the nature of Japan's leaders allowed them to become proponents of modernization. Unlike China where the scholar-gentry class dominated the bureaucracy, Japan was ruled by feudal military men who were more realistic in their assessment of Western military superiority and the potential benefit of that technology to Japan.

Although foreign technology and capital were important components in Japan's drive to modernization, one should not underestimate the value of astute government leadership during the Meiji period. The Japanese did not want to be indebted to any foreign creditors so that no nation would have an "excuse" to invade or colonize Japan, as had happened in other Asian countries. Accordingly, the Japanese placed controls on foreign capital and commerce. The government reorganized the currency and banking, provided the capital for expanding and improving transportation and communications, and financial government mines and factories. Capital was raised internally by oppressive taxes on farmers. At the same time, the government stimulated industrial development in the private sector by providing access to necessary capital. The textile, paper and pulp, iron and steel, and chemical industries all developed quite rapidly under government encouragement.

World War II

Japan continued to grow and develop in the early twentieth century. The period after World War I was a period of transition. The military, with the emperor still as a divine figurehead, eradicated the democratic institutions and the autonomy of the zaibatsu and gained full control of the nation.

Japanese leaders then initiated a period of military expansionism. Japan seized Manchuria (Manchukuo) in 1931–1932 and eventually invaded much of eastern China. The occupation and harsh treatment of the occupied peoples brought criticism from the outside world, especially the United States, a major supplier of natural resources. Subsequently, oil and scrap steel imports from the United States were cut, forcing Japan to seek sources elsewhere—the Philippines for iron ore and Indonesia for oil. Reacting to what it considered provocative economic policies, and concerned about possible U.S. intervention in the Pacific, Japan forced the issue by attacking and destroying the U.S. fleet at Pearl Harbor on December 7, 1941, thus precipitating U.S. involvement in World War II. The Japanese army and navy quickly overran Southeast Asia, Indonesia, and the Philippines and other islands of the Pacific.

As the U.S. war economy came into full production, Japan incurred loss after loss. Japan surrendered in 1945 after the United States dropped atomic bombs on Hiroshima (August 6) and Nagasaki (August 9). The formal surrender occurred on the deck of the U.S.S. *Missouri* in Tokyo Bay on September 2, 1945.

Postwar Period

Since World War II, with U.S. aid and protection, Japan has emerged once again as a major world power. During this "second economic miracle," Japan has remained a close ally of the United States. Yet, Japan increasingly sees the necessity to play a balancing act between dependence on the United States and interdependence. Because of its proximity to the Soviet Union and the People's Republic of China, Japanese leaders have pursued a flexible foreign policy. Japan's relations with these two nations are the source of some of the problems and prospects that will be discussed later in this chapter.

POPULATION

In 1880, the Japanese population was estimated to be 38 million. By 1940, this number had increased to 71 million; the population grew at a 1.5 percent rate of natural increase during this period. Approximately 3 million Japanese (including 700,000 civilians) were killed during World War II. After the war, an upsurge in marriages that had been delayed by the war briefly caused the rate of natural increase to climb to 2.9 percent, but modernization and urbanization were accompanied by a decreasing rate of natural growth: 1.4 percent between 1950 and 1955, and 0.9 percent from 1955 to 1960.

Currently, the rate of natural increase is 0.4 percent, and the population in 1990 stood at 123.6 million. This relatively low growth rate, characteristic of postindustrial societies, portends future problems similar to those faced by the United States and Western Europe, namely an aging population. Figure 8.4 shows the sex-age pyramid for Japan compared to those for the United States and a typical developing country. In the near future, a greater proportion of the Japanese population will be older than 65, and a smaller proportion will be of working age. Already 11 percent of the population is 65 and over. Not only will this reduce the available labor force, it will also increase the number of people requiring social services while decreasing the number of people who will be available to pay for them. Continued improvements in technology will be required to prevent adverse effects from this trend.

As the demographic data in the Statistical Profile show, the basic demographic profile of Japan is clearly that of a developed country: low rate of natural increase, low infant mortality rate, high life expectancy, and a large percentage of the population living in urban areas. This profile is quite similar to the demographic profile of the United States and stands in stark contrast to the rest of East Asia, especially China.

These data represent a relatively recent urban transformation. In 1920 Japan had a total population of under 60 million, of which less than 25 percent lived in cities. By 1990 the population had increased to over 123 million, of which 77 percent are urban. Currently, growth rates have slowed to less than 1 percent, and the flow of migrants from rural areas to the cities has been replaced by migration from the central cities to the suburbs and exurbs, giving rise to the doughnut effect found in many postindustrial cities.

At the same time, the density of population has increased greatly. The average density of 860 people per square mile (332 per square kilometer) belies the true nature of population distribution. One must remember that only 15 percent of Japan is relatively level. Approximately 60 percent of the population inhabits just over 20 percent of the land. Population densities in major urban areas are some of the highest in the world, ex-

FIGURE 8.4

Japanese Population by Sex-Age Cohorts

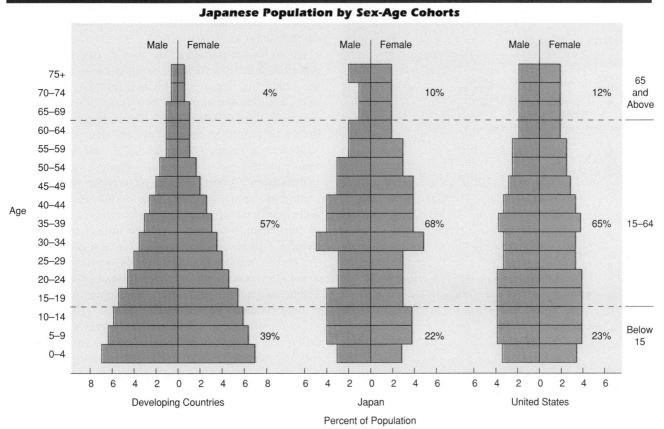

SOURCES: *U.N. Demographic Yearbook* (New York: United Nations, 1983); Population Reference Bureau, *World Population Data Sheet* (Washington, D.C.: Population Reference Bureau, 1986).

ceeding 15,500 per square mile (6000 per square kilometer) in Tokyo and other large metropolitan areas (Figure 8.5). The greatest concentration is focused on large cities in a corridor of urban expansion, called the ***Japanese Core.*** The core is comprised of only 17 percent of Japan's land area, but contains 40 percent of the population and produces 70 percent of the industrial output. Although Japanese planning agencies have identified the problem of overconcentration of population and industry in major urban centers and underconcentration elsewhere, as of yet they have been unable to deal with it effectively.

CULTURAL ENVIRONMENT

Although **emulation** may be an important ingredient in the success of Japan, this should not be taken to mean

that Japanese culture is merely an adaptation of characteristics from other peoples. The Japanese language and Shinto religion are unique to Japan, although they have been modified over the years by the influence of other societies. And, although the basic structure of the political system was borrowed from the West, it too has a distinctive Japanese flavor.

Language and Religion

Although Japanese and Chinese are not related languages, the Japanese adopted character writing from China as early as the sixth century A.D. There is some debate on the origins of Japanese. Some scholars say that it is not related to any other language. Others suggest that it is semantically related to Korean and may be related to the Altaic language family, with an influ-

FIGURE 8.5

Metropolitan Japan

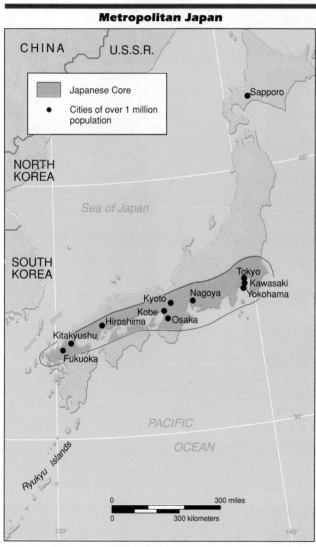

under the Yamato clan. Since the sixth century A.D., it has intermixed with other religions, especially Buddhism, Daoism, and Confucianism, producing many local variations and cults.

Shinto was important to the rise of nationalism in Japan during the nineteenth century because it supported the belief that Japan was created and favored by the gods, and that the emperor, as a direct descendant of a god, was divine. After World War II, the emperor was forced to renounce his divinity, and the popularity of Shinto as a religion waned, but it still has cultural significance.

Buddhism was first introduced to Japan with Chinese culture in the sixth century A.D. Buddhism both mixed and coexisted with Shinto to produce a unique variation of Buddhism, which had a major impact on Japanese society. Japanese Buddhism emphasized practical reality, human relations, family morals, and ancestor worship. The basic tenets of Buddhism were adapted to Shinto, and a large proportion of Japanese are members of both religions.

Confucianism, which is a code of ethics and social order, and Daoism, which stresses the continual search for "the way," were both compatible with Buddhism and were also introduced to Japan and quickly adapted to Japanese culture. They also influenced Japanese society, especially with respect to education, ethics, and political thought and conduct. Because of the inclusive nature of these belief systems, it was not unusual for people to follow two or three of them at the same time, especially after they had evolved into their Japanese variants.

Even Christianity found a niche within Japanese culture. The influence of Christianity on Japan has been much greater than one might expect given the small number of actual converts. Christianity is associated with the West, and the West represents modernization. Thus, Christianity acts as a conduit to introduce many aspects of Western culture into Japanese society, especially in the areas of education and social work. The overall impact of Christianity has, in fact, been cultural rather than religious.

Political System

The political structure of contemporary Japan was dictated by the U.S. occupation (1945–1952) after World War II, an occupation that had a positive impact on Japan. Expecting a harsh and cruel occupation, the

ence from the Malayo-Polynesian family. In either case, it has been modified by the adoption of words from Chinese and, more recently, English and other European languages.

Shinto is the indigenous **animistic** religion of Japan and is closely associated with Japanese culture. It is ritualistic rather than dogmatic, focusing on this world rather than the world after death. It stresses a close relationship with nature and with the many gods, or **kami,** residing in particular natural and human-made features, such as waterfalls, mountains, and trees. Shinto evolved in Japan before the unification of the country

Japanese received instead the benefits of a benevolent overlord, one who realized the importance of having a strong ally in the Far East. As a result, the Japanese transformed their sense of nationalism into a desire for cooperation and focused their energies on rebuilding their economy.

Several features of the U.S. occupation have affected the contemporary political system in Japan: (1) military demobilization, (2) security agreements with the United States, and (3) the 1947 constitution. As a result of the occupation, Japan's military force is limited to a lightly armed national militia while security agreements provide for the United States to defend Japan. These agreements have two major effects. First, they provide for Japanese dependence on the United States and alliance with the United States in East Asian matters. Second, since the United States pays for much of the defense of Japan, the Japanese government is relieved of a heavy financial burden.

The 1947 constitution is the cornerstone of the post war Japanese political structure. It provided for an elected Diet (parliament), a cabinet along British lines, universal suffrage, and guaranteed human rights. Imperial divinity was renounced. Land reforms for peasant farmers were instituted, and the zaibatsu monopolies were broken up. This form of government was not entirely new to Japan, however. The basic institutions of a parliamentary democracy had been established in the late nineteenth century and continued into the 1920s. The U.S. occupation ensured that postwar Japan would build upon these democratic traditions, rather than

A massive statue of Buddha at a shrine in Kamakura near Tokyo symbolizes the inclusion of Buddhism into Japanese culture.

The Heian Shinto Shrine is located in Kyoto. It was built in 1895 as a smaller replica of the first Imperial Palace.

upon the totalitarian traditions that had led it into World War II.

Several major parties have emerged since World War II. They include a small Communist party, a moderate Social Democratic party, and a more conservative Liberal Democratic party (LDP). Since its inception in 1955, the LDP has been in power without interruption. Despite internal turmoil, scandals, and domestic crises, the LDP remains firmly in control, to a great extent because of factionalism among the other parties.

ECONOMIC LANDSCAPE

Not only is Japan the only industrial superpower outside Europe and North America, but it also has one of the world's healthiest economies. As Figure 8.6 shows, 34 percent of the Japanese labor force is employed in the industrial sector of the economy, and 55 percent in the service sector. Uncharacteristic of advanced economies, however, is that 11 percent of the Japanese labor force is employed in agriculture and fishing, a relatively high percentage for a developed country. In 1987, the total gross national product (GNP) came from the following sources: agriculture, 3 percent; industry, 41 percent; and services, 57 percent. Japan compares favorably with the Western industrialized world and is strikingly different from China and the developing countries as a whole. Note that although an above-average percentage of labor is employed in agriculture, its contribution to the GNP is much lower, an indication of the labor-intensive nature of Japanese agriculture and the relative lack of mechanization.

Westerners generally associate Japan's economic success with government involvement in and support of private business—a sort of central government protectionism. The government does fund and coordinate activities that will enhance Japan's economy. In recent years, these activities have focused on computers, super conductivity, artificial intelligence, and materials science, all research areas that will lead Japan's industry into the twenty-first century. Competition between companies is keen in Japan, however, and government expenditures as a percentage of GNP are relatively low. In Japan, government expenditures account for 18.9 percent of GNP, compared to 25 percent in the United States and 30.1 percent for developed countries as a whole. In making this comparison, it is only fair to point out that part of this difference results from Japan having to pay relatively little for defense, not to mention small expenditures for social and welfare services. This point notwithstanding, overall growth of the Japanese economy in recent years has been healthier than for many other industrialized states. Average annual GNP growth between 1965 and 1983 was 4.8 percent, compared to 1.7 percent for the United States and 2.5 for industrial market economies on average.

Fishing, Agriculture, and Forestry

Although Japan still has the largest fish catch per year in the world and exhibits one of the highest agricultural yields per acre, these two areas of economic activity present problems that may hinder future development. Currently, Japan supplements domestic production by importing over 15 percent of the fish and approximately 30 percent of the agricultural products consumed by its people. This has been necessitated by increased demand from an increasingly larger population. Other causes include the loss of traditional fishing waters to the Soviet Union after World War II, a deficiency of **arable land,** a predominance of small-scale farms, and competition for land between agriculture and urban expansion.

Historically, fish have provided the Japanese with their main source of animal protein. Increased personal wealth has caused the per capita demand for fish and other desired foods to double in the last 40 years. Japan's fish catches are the largest in the world and its fishing grounds are an important resource. Recently, the nature of Japanese commercial fishing has changed. Many of the traditional fishing grounds have been overfished, become polluted, or have been declared off limits to the Japanese by other countries. As a result, the scale of Japanese fishing has become larger, shifting away from small coastal villages that fish mostly in local waters to large corporations with vessels that fish in waters all around the world. **Aquiculture** supplements this catch with sweetfish, carp, trout, and oysters from "sea farms" as well as green laver, a seaweed regarded as a delicacy by the Japanese.

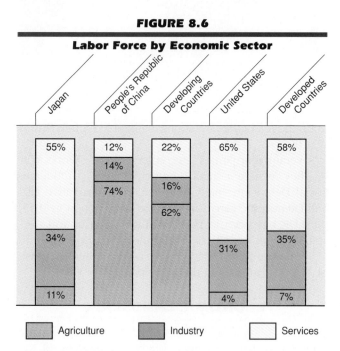

FIGURE 8.6

Labor Force by Economic Sector

SOURCE: *World Development Report, 1988.*

Ocean aquiculture in Kagoshima Bay off of Kyushu. Limited land and high prices for seafood encourage farming of the ocean.

Because of the terrain and the small size of Japanese farms, agriculture remains more labor intensive than in other industrialized countries. Of the 144,000 square miles (337,000 square kilometers) that comprise Japan, only 11.9 million acres (4.8 million hectares), or approximately 13 percent of the total land area, are arable. This constitutes only 0.15 acres (0.06 hectares) of arable land per capita, compared to 2.85 acres (1.15 hectares) per capita for the United States.

Urban growth has been reducing the land available for agricultural uses because the land most suitable for agriculture is also most desirable for urban expansion. Since the mid-1960s, arable land in Japan has decreased from 14.6 to 11.9 million acres (5.9 to 4.8 million hectares). This is a problem for all urbanizing societies, but the effects are more significant in Japan where good agricultural land is already at a premium.

The mainstay of Japanese agriculture is rice, which accounts for about half the value of total agricultural production. Wheat, barley, and potatoes are increasingly important. Fruits, flowers, and vegetables are also grown, depending on the local climatic conditions. The contribution of animal husbandry to Japanese agriculture is small, although it has been increasing in recent years as the Japanese demand for meat has increased. Chicken and pig raising are the primary kinds of animal husbandry.

The agricultural sector is subsidized and protected from foreign imports by the government. The results of these policies are a perpetuation of a relatively large number of small farmers, higher prices for Japanese consumers, and friction with trading partners, such as the United States, which would like to export meats, citrus, rice, and vegetables to Japan.

Timber is one natural resource that is abundant in Japan. Because Japan crosses several climatic regimes, it has a great variety of trees. These include both evergreen and deciduous, broadleaf forests as well as coniferous evergreen forests. Cedars, cypress, and pines are commercially most productive. Bamboo, palms, and other subtropical trees grow in the south. Most of mountainous Japan is devoted to commercial forests. Yet, even with the abundance of trees, the Japanese timber industry is still unable to meet domestic demand. Because of this, Japan is also a major importer of timber and timber products.

Transportation

A major contribution to Japan's economic growth has been the development of one of the world's most efficient transportation networks. In 1945 the national transportation system was in ruins. Since then, it has been rebuilt and improved to provide the linkages needed in an industrial society, especially one in which the physiography is so inhospitable to national unity. The basic nature of Japanese transportation also underwent a transformation from rail orientation to auto and air dependence. Today, rail and auto have an almost equal share of domestic passenger transport with approximately 40 percent each. Bus (14 percent), air (3 percent), and ship (less than 1 percent) transportation move fewer passengers, but air transportation is increasingly important. The Japanese continue to take advantage of low-cost sea transportation. In terms of ton-miles, ocean shipping accounts for half of freight movement, with trucking (39 percent) and rail (10 percent) making up the difference.

Even though the rail industry has suffered a relative decline, passenger-miles have tripled and ton-miles have increased by a third since 1950. The rail industry has also produced some major achievements, most notably the "bullet train," which carries passengers at speeds up to 130 miles per hour (210 kilometers per hour) along the Tokaido corridor. The full extent of the line now runs 664 miles (1069 kilometers) from Tokyo to Fukuoka on Kyushu.

Industry

Japan's emergence as an industrial power may seem to be a miracle because the islands have so few of the necessary resources for industrialization. The transformation of Japan into a world industrial giant can be explained by a combination of U.S. support, the basic ability of the Japanese to emulate, and the structure of the Japanese economy, in particular, intense competition at home and government-business cooperation for development of the foreign export market.

The key to recent Japanese success lies in the production of manufactured goods that are competitive on the world market. Japanese industrial products include telecommunications equipment, computers, televisions, radios and stereos, electrical machinery, office machines, automobiles, ships and boats, precision instruments, and photo equipment. At one time, Japanese goods had a reputation for being inexpensive and shoddy. Today, they are known for high quality and competitive pricing throughout the world.

One of the major stimuli to rapid economic growth and development has been the basic cooperation between government and business. Throughout the postwar period, Japan has been committed to the free mar-

Japanese motorcycles control the North American market due in part to rigorous quality testing and control. The only exception is in the market for large cruising motorcycles where the American Harley-Davidson retains prominence due to import restrictions, quality improvement, and a devoted clientele for Harley "hogs".

ket system. Yet, through various kinds of incentives and restrictions, the government has been able to direct investment toward high-growth sectors of the economy, thereby producing an economic atmosphere that strongly promotes big business. It also enables Japanese industries to be highly competitive internationally. Close ties between the developing nations for resources, the bureaucratic elite, big business, and export markets have been at the heart of this positive economic environment.

There are four major industrial zones and two lesser

FIGURE 8.7

Industrial Zones of Japan

ones in Japan (Figure 8.7). The Keihin, Hanshin, Chukyo, Kitakyushu, Tokai, and Setouchi industrial zones account for approximately half the total manufacturing output of Japan. These zones developed initially near ports during the Meiji Restoration. The geographical advantages of these regions have continued to be enhanced by the export-based nature of the contemporary Japanese economy.

The Keihin industrial zone, centered on Tokyo, Kawasaki, and Yokohama, is the largest of the zones. Steel, petroleum products, petrochemicals, ships, foodstuffs, automobiles, machinery, and high-tech products are some of the major industries in this diverse region that is advantageously located near major Japanese consumer and labor markets, as well as the ports of Tokyo and Yokohama.

The second largest industrial zone is Hanshin, centered on Osaka and Kobe. Major industries here include metals, iron and steel, machinery, textiles, foodstuffs, and chemicals. The Chukyo industrial zone, centered on Nagoya, produces textiles, ceramics, chemicals, automobiles, steel, and machinery, Kitakyushu, on northern Kyushu, is the smallest of the major zones; its primary industries include steel, chemicals, machinery, ceramics, and electrical appliances.

Two secondary industrial zones have emerged in Tokai, between Nagoya and Tokyo, and in Setouchi, on the Inland Sea. Major industries in these zones include chemicals, petrochemicals, steel, petroleum products, automobiles, and ships.

Urbanization

As Japan industrialized, it also urbanized. Today, Japan has truly become an urban society. Not only do 77 percent of the people live in cities, one of the highest percentages in the world, but the Tokyo urban agglomeration with over 11 million people (by some definitions, as high as 17 million) is the second largest urban area in the world. One-fifth of all urban Japanese live in the Tokyo urban complex, compared to 12 percent of U.S. urbanites in New York City and only 6 percent of the Chinese urban population in Shanghai.

Only 15 percent of the population lived in cities at the turn of the century. By 1920, this share had increased to 25 percent, of which 8 percent lived in the four cities with populations of over 500,000. By 1960, Japan was an urban society, with 63 percent of the population living in cities, 20 percent in the nine cities of half a million or more. Urbanization continued until, by the late 1980s, 77 percent of Japan's population lived in cities. Eleven cities had populations of over one million (Table 8.1), and an additional 10 cities had populations between 500,000 and one million. These 21 cities account for 42 percent of Japan's urbanites and one-fourth of the entire population. Over 55 percent of the population live in cities of 100,000 and above. The trend is for Japanese to live in large cities.

Along with urbanization and industrialization, the Japanese have achieved one of the highest standards of living in the world. Table 8.2 presents additional data

TABLE 8.1

Japanese Cities with Populations over One Million, 1985

	POPULATION (IN THOUSANDS)		POPULATION (IN THOUSANDS)
Tokyo	11,828	Kobe	1,411
Yokohama	2,993	Fukuoka	1,160
Osaka	2,636	Kawasaki	1,089
Nagoya	2,116	Kitakyushu	1,056
Sapporo	1,543	Hiroshima	1,044
Kyoto	1,479		

SOURCE: *The Far East and Australia 1987* (London: Europa, 1986).

that compare the Japanese quality of life with that of the People's Republic of China, a developing country, and that of the United States, another developed country. In each case, Japan exhibits the profile of a modern industrial nation with fewer persons per physician and higher levels of food and energy consumption than developing countries. The most frequently used measure of the wealth of a nation is the GNP per capita. The Japanese have a high per capita GNP of over $21,000.

Industrialization and urbanization have been accompanied, as in all societies, by increased levels of pollution. By the late 1960s, the problem was so grave that a nationwide effort was undertaken to deal with serious air and water pollution. The pollution of coastal waters by mercury resulted in the poisoning of many fishing villagers. Air pollution was so bad in Tokyo that oxygen stations were set up for pedestrians. Concern peaked in the 1970s and resulted in new laws to protect the environment and society. The levels of many pollutants were reduced, and by the 1980s the sense of immediacy had abated. Pollution, however, remains a problem that the Japanese must deal with if they are to enjoy continued improvement in the quality of their lives.

SPATIAL CONNECTIVITY

Japan occupies a unique and precarious niche in world economic and strategic affairs. Because of its dearth of natural resources, Japan is heavily dependent on imports to maintain its economic health. At the same time, it is a major supplier of high-level manufactured goods to both the developed and developing worlds, thus making many of these countries dependent on Japan for their economic well-being, as well as making Japan dependent on the willingness of these countries to accept Japanese products. Strategically, Japan attempts to steer a middle course between dependence on the United States and interdependence in its dealing with its East Asian neighbors, especially the People's Republic of China and the Soviet Union.

TRADE

Characteristically, Japanese trade involves the importation of primary products and the exportation of manufactured products. Primary goods comprise 79 percent

TABLE 8.2

Selected Characteristics Related to the Quality of Life

	GNP PER CAPITA (U.S. $)	CALORIC INTAKE PER DAY	ADULT LITERACY (%)	POPULATION PER PHYSICIAN	ENERGY CONSUMPTION PER CAPITA (KG)
Japan	$21,040	2864	99%	660	3232
Developing Countries	710	2384	50	5410	297
People's Republic of China	330	2630	66	1000	525
United States	19,780	3645	99	470	7265

SOURCE: World Bank, *World Development Report* (New York: World Bank, 1989).

of Japan's imports: fuels, 50 percent; other primary goods, 16 percent; and food, 13 percent. On the other hand, manufactured products account for 92 percent of all Japanese exports, with machinery and transport equipment comprising 56 percent of the total. Japan's **balance of payments** has been enormously favorable. In 1988 Japan enjoyed a trade surplus of over 83 billion U.S. dollars, with exports of $211 billion versus imports of $128 billion.

At the regional scale, Japanese exports and imports to other countries of Asia are almost balanced, while the value of imports from the Middle East, primarily oil, greatly exceeds exports. Japan uses its favorable export balance with the industrialized world to pay for the oil (Figure 8.8). Ten countries account for nearly two-thirds of all Japan's imports. The largest supplier of imports to Japan is the United States, which provides 23 percent of Japan's imports. Other major exporters of goods to Japan include the oil nations of Saudi Arabia, Indonesia, and the United Arab Emirates. Also included are Australia, the People's Republic of China, Taiwan, Canada, Korea, and Germany. Ten countries also receive over two-thirds of all Japanese exports (Table 8.3). Again, the

A view of the Tokyo megalopolis, which encompasses Yokohama and much of the coastal area of Tokyo Bay.

United States is the leader with 34 percent, followed by Germany, Korea, and Taiwan.

Primary goods dominate Japanese imports, with mineral fuels alone accounting for 26.6 percent of all imports (Table 8.4). Petroleum and petroleum products ac-

FIGURE 8.8

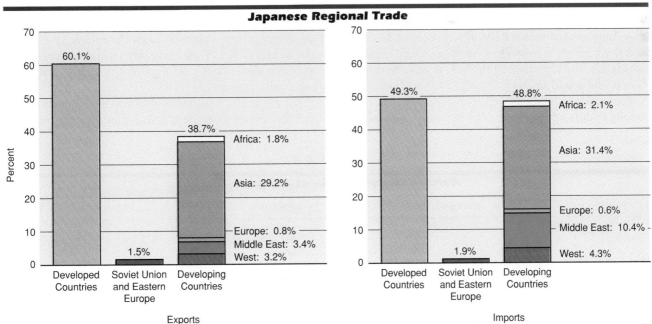

Japanese Regional Trade

SOURCE: International Monetary Fund, *Directions of Trade Statistics Yearbook* (Washington, D.C.: International Monetary Fund, 1988).

TABLE 8.3

Japan's Major Trading Partners

EXPORTERS TO JAPAN		IMPORTERS OF JAPANESE GOODS	
Country	Share of Trade (%)	Country	Share of Trade (%)
United States	23%	United States	34%
Korea	6	Germany	6
Australia	5	Korea	6
People's Republic of China	5	Taiwan	5
Indonesia	5	Hong Kong	4
Taiwan	5	United Kingdom	4
Canada	4	People's Republic of China	4
Germany	4	Singapore	3
Saudi Arabia	3	Australia	3
United Arab Emirates	3	Canada	2
Total share	63	Total share	71

SOURCE: *Direction of Trade Statistics Yearbook* (International Monetary Fund, 1989).

count for 18 percent of the total, while natural gas ranks second at 5 percent. Japanese exports, of course, are dominated by manufactured goods, which account for over 99 percent of the total. These products include transportation vehicles, with passenger cars alone accounting for 13 percent of all exports. Japan also exports consumer audiovisual equipment, e.g., cameras, radios, televisions, and stereos, as well as iron and steel, ships, and electrical machinery. In many of these areas, Japan has replaced the United States as the world's major exporter. Japanese steel, automobiles, and telecommunications equipment have all cut deeply into traditional American markets, as have high-tech products, which are already challenging U.S. goods for domination of the world market.

Four decades after its defeat in World War II, Japan is fully integrated into the world economy. It is a society that is dependent on the world economy for its economic survival and at the same time a society that the world depends on for its well-being. The export-based economy of Japan has several potential weaknesses, however. The Japanese must be concerned about growing competition from Hong Kong and the developing economies of Taiwan and South Korea, all of which are also export based. The possibility of protectionism in reaction to Japan's enormous trade surplus with other industrialized countries, especially the United States, presents another threat to the health of the Japanese economy. These concerns notwithstanding, the potential Achilles heel of the economy remains Japan's longstanding dependence on imports for natural resources, especially energy.

By the mid-1980s, Japan was the largest importer of energy resources in the world. Japan depends upon in-

TABLE 8.4

General Characteristics of Japanese Trade

IMPORTS	PERCENTAGE OF TOTAL	EXPORTS	PERCENTAGE OF TOTAL
Industrial supplies	34.9	Manufactured goods	99.5
Fuels	26.6	Agricultural products	0.3
Food	14.2	Mining products	0.2
Consumer goods	10.5		
Machinery	8.2		
Transport equipment	4.1		
Other	0.5		

SOURCE: United Nations, *International Trade Yearbook* (New York: United Nations, 1987).

ternational trade for 95 percent of its energy needs (Figure 8.9). It imports 93 percent of the coal it consumes, 96 percent of the natural gas, and 99 percent of the petroleum. Of these, petroleum may be the most important and the most vulnerable. Major suppliers of petroleum to Japan include Saudi Arabia, the United Arab Emirates, Indonesia, and Iran. These four countries account for over 70 percent of Japanese petroleum imports. Japan's dependency on imported oil places it in a precarious position if the price rises dramatically as occurred during the 1970s and in 1990.

STRATEGIC INTERDEPENDENCE

Japan's strategic position in East Asia is tenuous because it must ensure its security and independence in a region where the world's military superpowers have confronted one another. Accordingly, Japanese foreign policy has been guided by several needs: to maintain security agreements with the United States, to be a leading force in Asia, and to maintain good relations with the Soviet Union and the People's Republic of China.

Japan is still bound under provisions of the peace agreement that ended World War II and subsequent security agreements with the United States. Under these agreements, Japan is prevented from having its own military force; in return, its security is guaranteed by the United States. Yet, Japan realizes that it cannot depend forever on the United States, nor can it continue to serve merely as an extension of U.S. policy in East Asia.

There is currently debate within both countries concerning the extent to which Japan should be allowed to remilitarize and assume greater responsibility for its own national security. Factions in both countries favor Japan having its own military forces. In Japan, these sentiments are closely allied with renewed feelings of nationalism. In the United States, some individuals think that Japan is receiving an unfair economic advantage by not having to pay for its own defense. Six percent of U.S. GNP goes toward national security compared with approximately 1 percent in Japan. There are also groups in both countries that are afraid of the potential of a remilitarized Japan. The Japanese certainly have the capability to remilitarize and become a major military power in a short period of time and also to become a nuclear power very quickly.

Japan's tenuous situation is exacerbated by several noneconomic concerns related to its geographical position. It has served as the United States' strongest ally

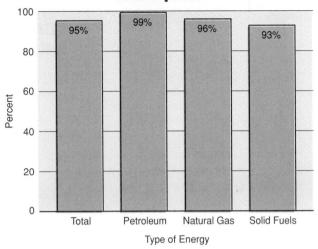

FIGURE 8.9

Imports as a Percentage of Japanese Energy Consumption

SOURCE: United Nations, *Energy Statistics Yearbook 1988* (New York: United Nations, 1990).

in this region since World War II and has acted as a surrogate for the United States in East Asian affairs. Yet, the Japanese have not unquestioningly adhered to U.S. foreign policy. On several occasions, they have followed an independent road in East Asian affairs, especially when dealing with the Soviet Union and the People's Republic of China. This independence is necessary because of Japan's geographical proximity to these two nations and the complementary nature of the Japanese economy with both the Soviet and Chinese economies.

Because of its geographical proximity to the People's Republic of China and the Soviet Union, Japan has pursued, and will continue to pursue, multilateral political and economic policies. To the Japanese, the improvement of U.S. relations with China and the Soviet Union in the 1970s seemed to undercut U.S. commitment to Japan's interests in East Asia and clarified Japan's need to develop other sources of security by dealing directly with the Soviets and the Chinese. Today, Japan hopes that maintaining relations with each of the superpowers will prevent it from becoming overly dependent upon a single nation while providing it with military security.

Another reason for improved relations with the Soviet Union and China stems from the complementary nature of their economies. Both are technologically backward and resource rich, while the Japanese are resource poor

INVESTMENT CAPITAL: THE NEW JAPANESE EXPORT

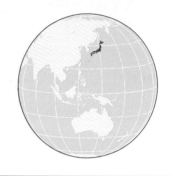

The financial system of banks and investment firms provides the apparatus for an interconnected global economy. Investment capital is the lifeblood of economic development and expansion. Japan dominates the purveyance of both of these. In the 1980s, a new kind of Japanese export emerged—investment capital. Corporate profits from its export-based economy and high levels of private savings (on average, 20 percent of disposable income) have produced a large capital surplus, allowing Japan to become a major overseas investor. By the mid-1980s, net exports of long-term capital exceeded $50 billion.

At first, these investments were concentrated in overseas firms. Now, approximately two-thirds of the Japanese income from foreign investments is derived from interest and dividends. With much of this investment directed at the

United States, Japan has become a major investor in U.S. stocks, bonds, and certificates of deposit. Direct Japanese investment in the United States now exceeds $5 billion, compared to only $1.5 billion in 1980. Japan is about to replace Great Britain as the

largest foreign investor in U.S. real estate. In addition, an estimated two-thirds of Japan's investment profits return to the United States to help finance the U.S. national debt.

It is in the banking industry that the full extent of Japan's strength can be seen. Japanese banks have now replaced American banks as the largest in the world. Whereas 5 of the world's largest banks were American in 1956, today only 2 American banks are in the top 25, while 7 of the 10 largest and 14 of the 25 largest banks are Japanese. In California, for example, 20 percent of total bank assets are in Japanese banks.

Between 1980 and 1984, Japanese profits from overseas investments quadrupled, from under $1 to over $4 billion. Japan has replaced the United States as the world's largest creditor nation.

and technologically advanced. Coupled with the geographical proximity, this sets the stage for greatly increased economic interaction, with Japan providing the capital and technology to develop Soviet and Chinese resources in return for natural resources. After an initial flurry of activity, which produced several joint ventures for the production of Siberian resources during the 1970s, this relationship with the Soviets has waned. The costs of developing the resources proved to be higher than first estimated, and the quality was inferior to less expensive resources available on the world market. On the other hand, trade with China, often used by the Japanese as a counterbalance against the Soviets, has steadily increased, especially during the post-Mao economic revival in China. Yet, Japanese-Chinese trade relations have also seen a recent downturn, a trend attributed to Chinese attempts to maintain a positive balance of trade and to Chinese memories of Japan's World War II atrocities against the Chinese.

Japan's security and independence cannot be viewed separately from its relations with the superpowers. Japan will continue to rely heavily upon U.S. military support and economic cooperation, while cautiously pursuing increased economic relations with the Soviet Union and China. Although Japan will use natural resources from these countries to help diversify and supplement its supply, it must be careful not to become too dependent on a potential adversary to supply essential items.

PROBLEMS AND PROSPECTS

Japan must resolve several problems to sustain its current level of economic development and continue improving its standard of living. Even with industrialization and modernization, Japanese society faces acute problems. There is a shortage of adequate housing, and social and welfare services have been poorly financed. These sectors of the economy have suffered because of the emphasis on supporting highly competitive, export-based industries.

The fact that the Japanese population is aging will further exacerbate some of these problems. This could have an adverse effect on the supply of labor—i.e., those available to pay for social programs—and thus on the ability of the nation to support, much less improve, social and welfare services.

In Japan's haste to reindustrialize, environmental concerns were often deferred. The high concentrations of industry and people in the core regions of the country have strained the environment, although steps are now being taken to lessen the impact and clean up the air, land, and water.

Economically, Japan must focus attention upon agriculture. The small scale of the farms and their competition with urban areas for land have caused agriculture to grow more slowly than other economic sectors.

Japan's most important international problem is its almost complete dependence on foreign sources for natural resources, especially energy. Additionally, it must also deal with growing resentment in the United States over the increasing bilateral trade imbalance, which produces the surpluses that Japan uses to pay for those resources. Finally, Japan must resolve its role in its own national defense as it continues to perform a diplomatic and economic balancing act among the People's Republic of China, Russia, and the United States.

These problems notwithstanding, future prospects for Japan are good. Overall, the Japanese economy exhibits a great deal of strength and vitality, and the political situation is relatively stable. Yet, it is the basic character of the Japanese people that holds the greatest promise for the future. Historically, they have proven their ability to adapt to the dramatic changes that have already taken place and will continue to occur in Japanese society. This is the strength of the nation and the ultimate explanation for Japan's success.

Suggested Readings

Association of Japanese Geographers, ed. *Geography of Japan.* Special Publication No. 4. Tokyo: Tekoku-Shoin, 1980.

Christopher, R. *The Japanese Mind: The Goliath Explained.* New York: Simon & Schuster, 1983.

Curtis, G. *The Japanese Way of Politics.* New York: Columbia University Press, 1988.

Fairbank, J., E. Reischauer, and A. Craig. *East Asia: Tradition and Transformation.* Boston: Houghton Mifflin, 1973.

Hall, J. *Japan: From Prehistory to Modern Times.* New York: Delacorte, 1970.

Hall, R. *Japan: Industrial Power of Asia,* 2d ed. New York: Van Nostrand, 1976.

Hane, M. *Modern Japan: A Historical Survey.* Boulder, Colo.: Westview Press, 1986.

Huddle, N., and M. Reich. *Islands of Dreams: Environmental Crisis in Japan.* New York: Autumn Press, 1975.

Ishida, H. *An Historical-Cultural Geography of Japan.* Hiroshima: University of Hiroshima Press, 1981.

Japan Statistical Yearbook. Tokyo: Statistics Bureau, annual.

Kirby, S. *Japan's Role in the 1980s.* London: Economic Intelligence Unit, 1980.

Pannell, C., ed. *East Asia: Geographical and Historical Approaches to Foreign Area Studies.* Dubuque: Kendall-Hunt, 1983.

Reischauer, E. *The Japanese Today: Change and Continuity.* Cambridge: Belknap, 1988.

Scheiner, E. *Modern Japan: An Interpretive Anthology.* New York: Macmillan, 1974.

Trewartha, G. *Japan: A Geography.* Madison, Wis.: University of Wisconsin Press, 1965.

CHAPTER 9

Australia and New Zealand

IMPORTANT TERMS

Migration

Replicative society

Federal state

Remoteness

CONCEPTS AND ISSUES

■ Development of the Outback.

■ Balance of industrial activities and agricultural output and exports.

■ Foreign capital investment and exploitation of resources.

■ Immigration policy and labor force needs.

■ Effects of isolation and remote location.

CHAPTER OUTLINE

PHYSICAL ENVIRONMENT
Climate
Vegetation
Soils
Landforms
Water Resources
Mineral Resources

HUMAN ENVIRONMENT
Historical Background
Population
Political Characteristics and Culture
Agriculture
Industry
Urbanization

SPATIAL CONNECTIVITY
Australia
New Zealand

PROBLEMS AND PROSPECTS

U nlike the western hemisphere and the European-African realm, Asia along the equator breaks up into a chain of islands, the largest being Australia and its companion state, New Zealand (Figure 9.1). Although geographically they belong to the greater Asian realm, culturally these two countries differ markedly from Asia. The almost empty spaces of the lands "down under" stand in sharp contrast to their crowded neighbors in Southeast Asia. Culturally and economically, the inhabitants of developed Australia and New Zealand are distinct from the millions of Southeast Asians with their intensive subsistence agriculture. In fact, Australia and New Zealand are separated by vast oceans from the places with which they have the strongest cultural and economic ties. Both countries were colonized by Europeans, most of them British, who brought their political, economic, and social institutions as well as their language. English remains the official language of both countries and is spoken by 90 percent of their respective populations. The two countries have always been closely allied, and their troops have fought side by side in the same wars. Together with the United States, to whom they both look for protection, they have formed the ANZUS alliance (a treaty of mutual protection among the three nations). Thus, by virtue of their culture, ethnic origins, and political orientation as well as their size and relative location, Australia and New Zealand constitute a region, although they still differ in some important respects.

Australia is the only continent that is entirely occupied by one country. It is a landmass of 2,969,000 square miles (7,686,861 square kilometers) and contains a population of 17.1 million, of whom 90 percent are descendants of Western Europeans. It has the lowest average elevation and lowest overall relief of all the continental landmasses. The continent is marked by a vast arid and semiarid interior, extensive plains, and broken peripheral moisture zones. The overall population is very small relative to the size of the landmass, less than six persons per square mile (two per square kilometer). Because of the continent's overall aridity, the population is very unevenly distributed. A very high percentage of the population is concentrated in five major urban centers along the east and southeast coasts and in one cluster on the southwest coast.

The Australian people enjoy a high standard of living. The continent produces a surplus of agricultural products, and the farms and ranches (called stations in Aus-

tralia) are highly mechanized. The country has vast amounts of mineral resources, some of which are just being tapped. Australia, which was once a British colony and is now a Commonwealth partner, depended on Great Britain for most of its imports for many years. Since World War II, however, it has greatly increased its manufacturing output, and the label "Made in Australia" is becoming increasingly common and popular.

The landmass of New Zealand, which lies 1200 miles (1930 kilometers) to the southeast of Australia, is small compared to that of Australia. It contains about 103,740 square miles (268,868 square kilometers), about the size of the state of Colorado. Like Australia, it has a small population, about 3.3 million inhabitants. New Zealand has an important pastoral and agricultural economy and relies on exports for its economic well-being. Its internal market is small, and it is too far removed from foreign markets to enjoy an industrial boom.

Despite their many similarities, the outlook of the two countries is somewhat different. Australia has a long Pacific coastline, but being a continental country, it looks inward and not essentially toward the Pacific. New Zealand, however, is the western base of the "Polynesian realm," of which Hawaii is the apex and Easter Island the eastern extremity. In a very real sense, New Zealand is committed to the Pacific. Auckland, its largest city, is the most important city of Polynesia. The Maori, a Polynesian people, first settled the islands that make up New Zealand, and today more than 300,000 still live in the country. There are also colonies of Samoans, Tongans, and Cook Islanders, whose later arrival constituted a second wave of immigration. The royal family of Tonga has a residence in Auckland as well as in the islands that they rule.

PHYSICAL ENVIRONMENT

Although New Zealand and Australia lie within the same general latitude (though New Zealand extends farther south and the larger nation extends across more degrees of latitude), the physical environment of the two countries differs considerably in several respects, most notably in their available water and their topography. The aridity of much of Australia contrasts sharply with the generally abundant moisture of New Zealand. Similarly, the mountainous terrain of much of New Zealand is

FIGURE 9.1

Australia and New Zealand

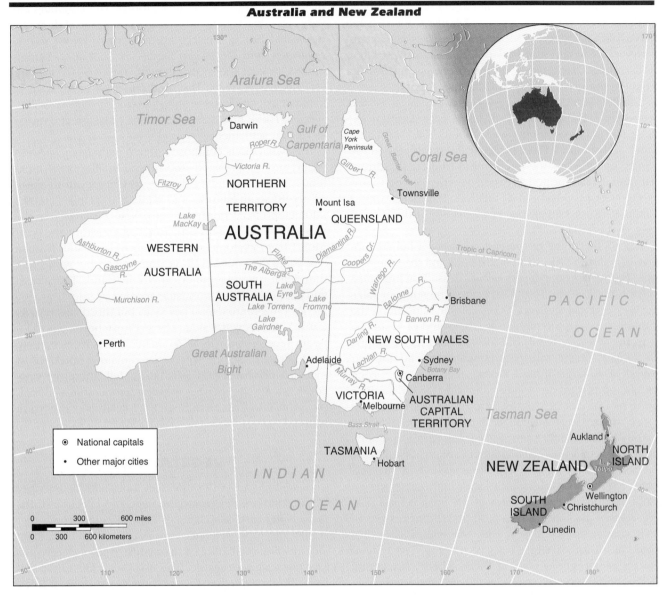

markedly different from the generally low elevations of much of Australia.

CLIMATE

The climate of Australia is partially controlled by three factors: its location between 10° and 40° south latitude, the generally low elevation of the land, and the Eastern Highlands (Great Dividing Range) that block the east winds blowing in from the Pacific (Figure 9.2). The continent is divided into desert, steppe, savanna, rainy tropical, humid subtropical, Mediterranean, and marine climates. The west central third of the continent is desert, one of the largest of the trade wind deserts, and receives less than 10 inches (25.4 centimeters) of rain per year. The arid lands extend to the coast in both the west and the south along the Great Australian Bight (the bay off the south coast). During the southern hemisphere summer—December to March—the intertropical front moves about 10 degrees south of the equator and brings heavy rains to the north coast of Australia. When the front moves back north during the winter—June to August—the dry, hot season is intense.

The southeast trade winds affect the east coast the entire year. The northern half has a rainy tropical climate and receives rainfall each month, with the maximum in summer. From Townsville north, the trade winds, enforced by the intertropical front, bring torrential rains. Tully, located between Townsville and Cairns in the state of Queensland, receives 180 inches (457.2 centimeters) of rainfall per year, making it the wettest spot in Australia.

The southern half of the east coast has a humid subtropical climate. The winters are mild and the summers hot. The winter rainfall is cyclonic, caused by low pressures drifting north from the Pacific. The summer rains are convectional showers or orographic rains along the eastern slopes of the highlands. The island of Tasmania is in the westerly wind drift all year and has a marine west coast climate with cool summers; it is humid year round. The large Indian Ocean high pressure zone moves northward in winter, and the westerly winds bring moisture to the southwest coast around Perth. Because of the Great Australian Bight, winter rains come to the region around Adelaide in South Australia. Rainy, mild winters and hot, dry summers are characteristic of the southwest coast—an area dominated by a Mediterranean type climate. Tropical and subtropical steppe lands are found on three sides of the great central desert. Rainfall averages between 10 and 20 inches (25.4–50.8 centimeters) per year and sometimes does not come at all for several weeks or months. The rainfall comes in the summer along the northern part of the steppe and during the winter in the east and south. This climatic zone supports grasslands but sometimes is plagued by drought and dust storms. The summers are very hot and usually dry. Marble Bar, located in the tropical steppe in Western Australia, rivals Death Valley, California, and the Libyan Desert as a hot spot. During one of its heat spells, it averaged 100°F (37°C) or more for 100 straight days. The northern coast of Australia is sometimes struck by tropical cyclones, called Willie-Willies. These hurricane-like storms cause great damage to coastal settlements.

New Zealand is composed of two main islands, North Island and South Island, and several smaller ones. It extends through about 14 degrees of latitude, from about 34° to 48° south latitude, and is about 1000 miles (1609.3 kilometers) from north to south. New Zealand is in the westerly wind system and receives moisture the year round from cyclonic storms originating in the Tasman Sea. Rainfall is both cyclonic and orographic with some thunderstorms in summer on North Island. The mountain ranges on the west side receive as much as 300 inches (762 centimeters) per year, but this drops rapidly toward the leeward side where Christchurch and Dunedin receive only 25 to 30 inches (63.5–76.2 centimeters) per year. Marine west coast climate is noted for its mild temperatures and reliable rainfall; therefore only South Island experiences freezing temperatures in winter. Snow and frost are rare on the lowlands of the North Island.

VEGETATION

Climate, together with slope, elevation, and the intensity of evaporation, determine the vegetation regime of the region. Because of Australia's long isolation from other continents, most of its trees are peculiar to the continent and are predominantly of two types: gum or eucalyptus and the acacia or mattle. The acacia is akin to the mimosa in the United States and bursts into brilliant flow-

STATISTICAL PROFILE: AUSTRALIA AND NEW ZEALAND								
REGION OR COUNTRY	Population Estimate mid-1990 (Millions)	Natural Increase (Annual)	Population Projected to 2000 (Millions)	Infant Mortality Rate	Life Expectancy at Birth (Years)	Urban Population (%)	Per Capita GNP, 1988 (US$)	Area, Thousands of Square Miles (Km²)	Population Density, No./mi² (No./km²)
Australia	17.1	0.8	19.1	8.7	76	86	12,390	2967.9(7686.9)	6(2)
New Zealand	3.3	0.8	3.5	10.0	74	84	9,620	103.7(268.7)	37(12)

FIGURE 9.2

Physiography of Australia and New Zealand

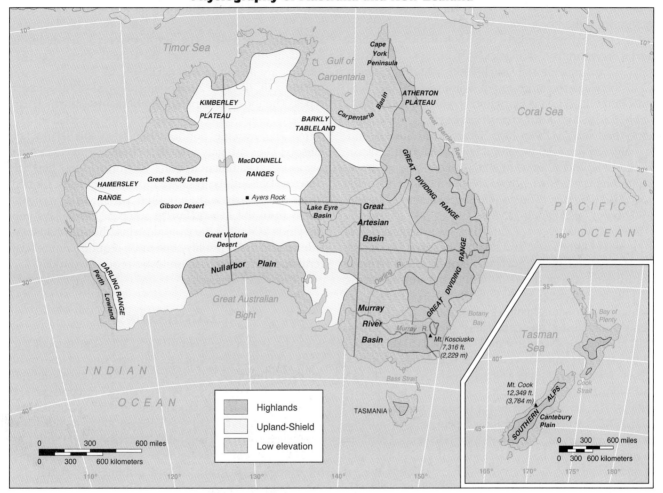

ers in the spring. More characteristic, however, is the eucalyptus, whose 600 varieties range from stunted shrubs in the subhumid areas to majestic mountain species that grow to 300 feet (91 meters) where moisture abounds. Scattered patches of rainforest occur at wetter sites along the east coast, particularly in Queensland. More widespread over the continent is a savanna-type open growth of eucalyptus woodland, which is the dominant natural association of the Eastern Highlands and much of the north coast.

Most of Australia, however, is treeless. The desert heart of the continent is largely barren, but there is sizable expanses of steppe grasses and shrubs around its margins. Notable among the shrub vegetation of inland Australia is spenifex, a tall porcupine grass that grows in circular clumps. Other plants include saltbush and bluebush, vigorous shrubs that provide excellent livestock forage; mulga, a low-growing variety of acacia that is widespread in semiarid areas; and malee, a low tree that looks much like mulga but is a member of the eucalyptus family.

The fauna of Australia are highly peculiar and justly reknowned. Due to the isolation of the landmass in prehistoric eras, the continent has a near-monopoly on marsupials, animals that carry their young in a pouch. The opossum and "opossum rats" of the Americas are the only marsupials found outside Australia. The largest marsupial is the kangaroo, which, with the ostrichlike emu is found on Australia's coat of arms. Much beloved by visitors is the koala bear, which lives exclusively on eucalyptus leaves, normally does not drink water, and carries its young on its back. Most bizarre of all Australia's creatures is the duckbilled platypus, which stands between the more recent mammals and the creatures of prehistory. It is amphibious with a ducklike bill, webbed feet, and a fur coat. It hatches its young from eggs, then suckles them. It is a very timid animal, and few persons see it in its natural environment. It is heavily protected by the government. Another nonmarsupial is the dingo, a yellow hound dog brought to Australia from Asia long ago. The country is also famous for the kookaburra or laughing jackass.

Like Australia, New Zealand has long been isolated from other landmasses, and this has had a similar effect on the types of flora and fauna found on the islands. North Island has its pines and tree ferns, and South Island its beech forests, especially on the west side. Tussock and other natural grasses occur on the drier eastern slopes, notably on the Canterbury Plain. Native mammals are lacking except for marine animals and the bat. Flightless birds (moas) were once common. Some varieties of the moa family were 12 feet (3.7 meters) tall and were hunted to extinction by the first Polynesian settlers. The kiwi and the kea, a predaceous parrot, are unique New Zealand fauna. Red deer from Scotland, the rabbit from Britain, and the opossum from Australia have been introduced to the islands and have become significant pests.

SOILS

All three of the soil-forming processes, laterization, ***podzolization,*** and calcification, are prevalent in Australia. Parent material and climate, together with vegetation, are the most important factors in the processing of the ultimate soil. Based upon these factors, the soils of Australia comprise a rather simple pattern. Since two-thirds of the continent is arid or semiarid, the aridisols and entisols constitute about the same percentage of the land. In the southwestern and southeastern areas of the country, as well as the savanna region of Arnhem Land, the podzolization process has been active, and the alfisols predominate. Along the eastern slopes of the Eastern highlands are found a sizable variety of soils, which are classified as mountain soils, except on the Cape

York Peninsula, where laterization is prevalent and a rather large area of ultisols is to be found. The semiarid grassland and brush country of the Great Artesian Basin, with its dark brown soils and high content of clay, constitute the vertisols. The great expanses of mollisols found in the Soviet Union, the United States, and Argentina are absent in Australia.

The soils of New Zealand include only three major types. The most widespread is mountain soil, which can be very complex because of slope and vegetation patterns. The extreme northern part of North Island has alfisols. They are gray-brown in color and have a rather high clay accumulation. Parts of North Island and the east side of South Island have mostly inceptisols, formed on volcanic ash and deposited on material washed down from the mountains. In some of the valleys and lowlands as well as the coastal plains, the soils had considerable natural fertility, but most require commercial fertilizers and crop rotation to be productive today.

LANDFORMS

Australia is a young nation historically, but in geologic terms it is an ancient land, perhaps one of the most ancient. The great "alpine upheaval" that threw up the great mountain ranges of all the other continents passed Australia by. The tallest mountain on the continent, Kosciusko, is a mere 7316 feet (185.83 meters) high. The long period of erosion and thus the relatively low elevation of the landmass—a mere 1000 feet (304.5 meters) on average, the lowest of any continent—account for the small number of physiographic regions.

Australia may be divided into three major physiographic regions: the Western Plateau, the Central Lowlands, and the Eastern Highlands (Figure 9.2). The Western Plateau occupies about three-fifths of the continent and consists of very old rocks that have been worn down to relative flatness. The monotonous surface is broken by a few knobs and ranges that rise to elevations of 1200 or more feet (365.76 meters). The most spectacular are Mount Olga and the monolith of red sandstone, Ayers Rock, located west of Alice Springs. On the north face of Ayers Rock is some grotesque erosion that Australians call "The Brain," because of its resemblance to a human skull.

The Central Lowlands, also called the Great Artesian Basin, extend from the Gulf of Carpentaria to the Murray River Basin and the Great Australian Bight. From north

An aerial view of Ayers Rock, the world's largest monolith. This mammoth foundation of redstone rises 1,143 feet (348 meters) out of the central desert of Australia.

to south, the Carpentaria, Lake Eyre, and Murray basins together include 800,000 square miles (2,072,000 square kilometers) or over one-fourth of the country's area. These basins are underlain by dipping layers of sedimentary rocks, chiefly shale and sandstone.

The Lake Eyre Basin is much like the Salton Sea area of southern California. The surface streams are intermittent, and the lake bed, usually dry, is 39 feet (11.89 meters) below sea level. The southern portion of the Central Lowlands is drained by the Murray River and its tributaries, the Murrumbidgee and the Darling.

The Eastern Highlands cover about one-sixth of the continent. The region consists of plateaus, tilted crustal blocks, and folded zones of varied rock types, along with small disconnected areas of lowland and coastal plains. Elevations of the highlands are moderate and vary from under 1000 to over 3000 feet (304.5–914.4 meters), with summits above 7000 feet (2133.60 meters) in the southeast. The highlands continue across the Bass Strait into Tasmania, a rugged island composed of uplands penetrated by flattish river valleys. Another remarkable feature of the east coast is the Great Barrier Reef, which parallels the Queensland coast for over 1000 miles (1609 kilometers). In stark contrast to the small, but well-watered coastal plains along the east coast is the Nullarbor Plain, which extends for hundreds of miles across the south coast of South and Western Australia. Nullarbor Plain, whose name means "no tree," is covered with small pebbles; this type of desert is referred to as "gibber." It bears only the marks of the human march across the continent, telephone and telegraph poles and a single-track, standard gauge railroad that runs arrow-straight for nearly 300 miles (483 kilometers)—the longest such stretch of railroad in the world.

New Zealand is much younger geologically than Australia and was a part of the great alpine uplift. A mountain chain extends the entire length of the islands and reaches elevations of over two miles (three kilometers). The highest peak, Mt. Cook, is 12,349 feet (3764 meters) above sea level. The southwest coast of South Island is called Fiordland and is indented by numerous fiords gorged out by prehistoric glaciers. The most spectacular is Milford Sound whose sea canyon is flanked for 10 miles (16.09 kilometers) by mountains rising as high as 6000 feet (1828.8 meters). Much of the Southern Alps, as the mountains are called, are snowcapped, and glaciers are still present. The South Island has many alpine lakes that are being harnessed for hydroelectric power. North Island has a greater variety of scenic beauty, including several volcanic cones, lakes, and geysers. The area once supported a magnificent forest of Kauri pines, one of the first resources of New Zealand to be exploited.

The Great Barrier Reef on the Queensland Coast is a maze of nooks and coral gardens in window-clear water. Built over eons by trillions of tiny limestone-secreting polyps, the reef covers 80,000 square miles; it is the largest structure on earth built by living organisms.

WATER RESOURCES

The most essential resource for human existence upon earth, water, is lacking in two-thirds of Australia. Only about half of the remaining land area receives ample and reliable moisture. The Eastern Highlands and the east coast, as well as Tasmania, get sufficient rainfall to support agriculture, forestry, industry, and grazing and provide domestic water for large urban centers. The southwest and the north coast receive seasonal rains that must be stored in natural or human-made reservoirs

for the dry season. Southwest Australia receives its rainfall during the winter months, and the summers are hot and dry. Evaporation is very high in the summer. The north coast gets heavy monsoon-type rainfall during the summer and the winters are dry. Evaporation is not as great in the north coast winters as in the summers, but if rainfall falls below normal, the water supply can become critical. Both underground and surface water are plentiful along the east and north coasts. The streams and rivers carry an abundance of water.

The Great Artesian Basin between the Western Plateau and the Eastern Highlands is one of the largest aquifers in the world. Wells (bores) can be sunk into the aquifer, and static pressure is ample to create a free-flowing artesian well. Unfortunately in many cases, the water is too hot or too salty to use for human or livestock consumption. However, it will support bush grasses for sheep or cattle grazing. Australia's longest and largest river system is located at the southern end of the Great Artesian Basin. The Murray River with its tributaries, the Murrumbidgee and Darling, has ample water to support considerable irrigated farming.

Unlike Australia, New Zealand has abundant water. Annual rainfall ranges from 25 to 300 inches (63.5–762 centimeters). No part is subhumid and even on the east side where rainfall is light, evaporation is also light. Irrigation is unnecessary, and ground water and surface water are ample for all needs. The islands have numerous alpine lakes, some more than a 1000 feet (304.5 meters) deep. The rivers are short, and some are swift. The lakes and rivers are kept full by orographic rainfall and melting snow.

MINERAL RESOURCES

The United States and the Soviet Union are the only countries in the world with a larger and more varied mineral supply than Australia (Figure 9.3). However, they also have much larger populations and are more highly industrialized than the country down under. Therefore, the two industrial giants are using their resources at a much faster pace. The extraction and processing of minerals were a major factor in the economic development of Australia almost from the earliest days of European settlement and remain important today. The extraction of Australia's mineral wealth started in 1851 when E. H. Hargraves returned from the gold fields of California and discovered gold in New South Wales. Soon a big strike was made near the town of Ballarat in

FIGURE 9.3

Mineral and Energy Resources of Australia and New Zealand

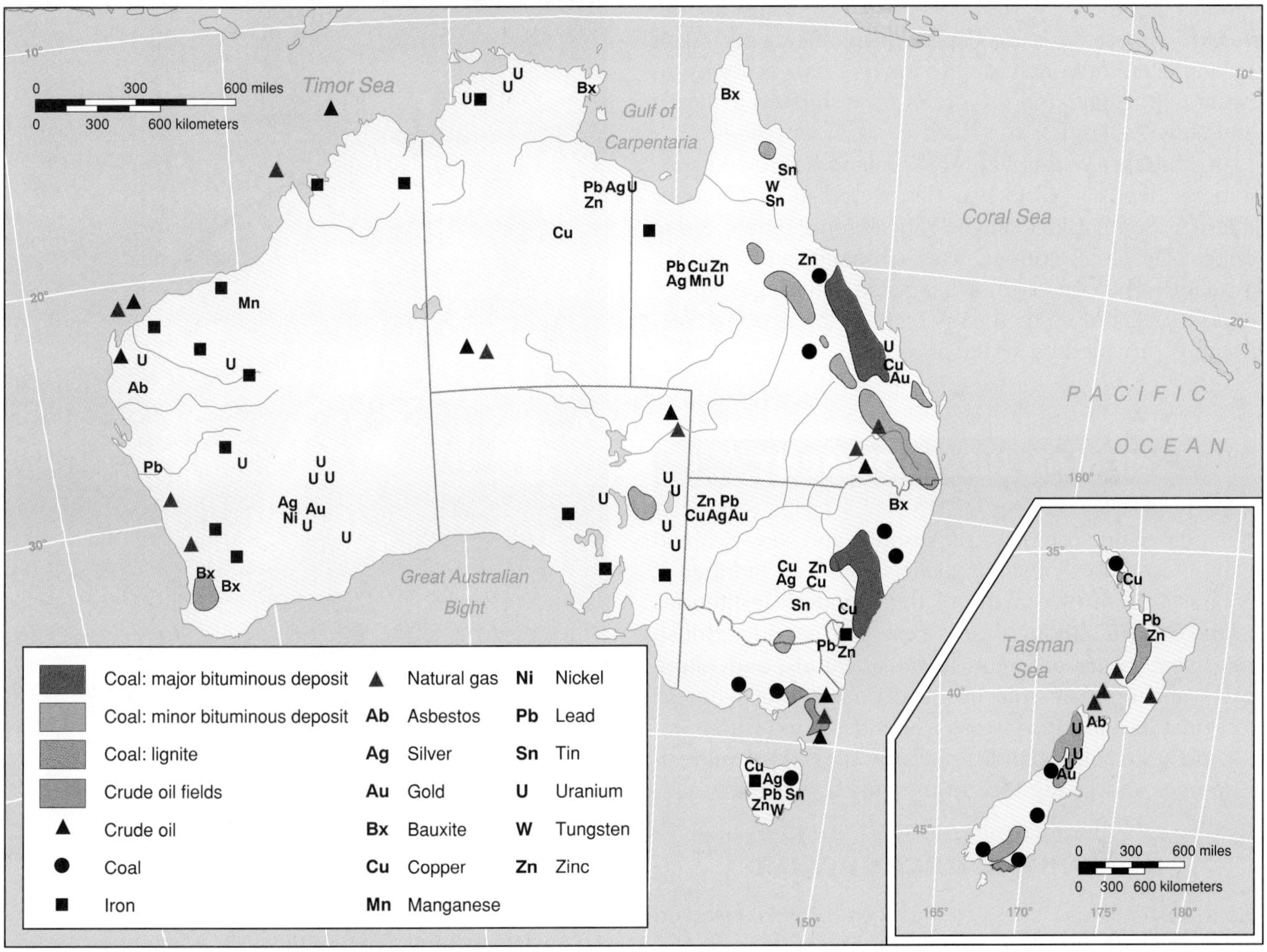

the state of Victoria. Later a big deposit was discovered at Kalgoorlie in Western Australia and a lesser one in Tasmania. During the decade of the 1850s, Australia supplies 40 percent of the world's gold. Today, it ranks fifth among world producers.

Mineral fuels are essential to industrialization, and Australia is well supplied with bituminous coal and also with large reserves of lower grades (subbituminous and lignite). The major coalfields are located along the Eastern Highlands close to the large urban centers, especially in New South Wales and Queensland. The discovery of oil and gas reserves in Queensland and offshore in the Bass Strait in the mid-1960s is doing

much to cut down on the large and costly imports of both fuels.

Australia is also well endowed with metallic minerals, both ferrous and nonferrous, as well as ferro-alloys. Iron ore is abundant, especially in Western Australia, and adequate supplies exist in South Australia. The export of minerals has been an asset to Australia's economy, and Broken Hill, New South Wales, and Mount Isa, Queensland, with their vast reserves of lead, zinc, copper, and by-products of gold, silver, manganese, and uranium have helped to keep Australia a creditor nation.

Ferro-alloys, a must in steel making, are found in several locations in the country. The deposits of tungsten in Tasmania, manganese in Queensland, and recently discovered nickel at Kambalda, south of the gold region around Kalgoorlie, are believed to be some of the world's largest reserves. Cape York, long suspected of having aluminum-rich ore (bauxite), has become an economic reality, and the ore is now shipped to Japan and New Zealand.

In contrast to Australia, New Zealand has few known mineral deposits. Gold was discovered in small quantities on South Island in the nineteenth century. Some deposits of coal, copper, lead, uranium, and zinc are available. The country also has some natural gas, silver, and tungsten. More than 85 percent of the nation's electricity is provided by water power.

HUMAN ENVIRONMENT

The human environments of Australia and New Zealand exhibit a number of similarities as well as a number of contrasts. Both owe many of their modern institutions to the British, although their experiences under British rule were rather different. In both nations, agriculture plays an important role in the economy, but whereas Australia has also developed a significant industrial sector, New Zealand, handicapped by its lack of mineral resources and its distance from markets, has not.

HISTORICAL BACKGROUND

Both Australia and New Zealand were originally settled by peoples who originated elsewhere—Australia by Melanesian peoples from Southeast Asia, New Zealand by Polynesian Maoris. Both countries also experienced a long period of British rule, which has had a lasting effect on their cultures. Nevertheless, despite these similarities, their historical experiences offer some interesting differences.

Australia

The human occupancy of Australia began about 25,000 years ago when ocean levels were lowered by continental glaciers and a land bridge by way of Indonesia and New Guinea connected Asia to the continent. Melanesian people moved in three separate waves into Australia, which at the time was more moist and more fertile. Rivers and lakes that existed at that time have

Australian Aborigines engage in a Night Dance, an organized corroboree to teach song and dance to the young.

since disappeared. The settlers, or Aborigines as they are called, developed a highly complex indigenous culture. The return of the sea to the north cut them off for millennia from their original homelands and thus from their former culture area. They developed a traditional hunting and gathering economy, but had no knowledge of seed-bearing plants and possessed no domesticated animals except the dingo (dog). Those who have escaped the influence of European culture, or have not been absorbed into a higher economic life, remain living examples of Stone Age culture. There were perhaps 300,000 Aborigines when the Europeans first settled in Australia, and today they number less than 225,000. They are located mostly in the northern part of the country, Arnhem Land, and the adjacent islands. Many of them now live on government and mission stations, supported by the state or religious missionaries. Others work as foresters or as stockpersons on the cattle stations, or on the railroads, in factories, and in domestic service. It is estimated that several thousand persons of mixed Aborigine and European ancestry now reside in Sydney. Several factors have contributed to the decline of the Aborigine population. By 1806 soldiers were driving them out of the colonized regions, and they were hunted like kangaroos, fed arsenic-laced food, introduced to liquor, and encouraged to fight among themselves.

The European settlement of Australia began in January 1788 with the landing of 750 convicts at Botany Bay under the care of Arthur Phillip, the first governor of Australia. The newfound continent was suggested as a penal colony by Sir Joseph Banks, a botanist, who had accompanied Captain James Cook on his historic voyage of 1770 when Cook landed on the east coast of Australia and claimed it for Great Britain. The British House of Commons accepted Bank's suggestion readily because, since the American Revolution, the American colonies were no longer available as a dumping ground for convicts. The area around Botany Bay was arid and waterless so Phillip set out to the north in a small boat and sailed into Sydney Harbor, one of the finest in the world, and modern Australian history began to unfold.

Before the exportation of criminals was abolished in 1868, over 168,000 convicts were sent to Australia. In 1803, to forestall the possibility of a French occupation, a party was sent out to Port Phillip Bay, and the settlement developed into Australia's second city, Melbourne, and the state of Victoria. By 1815 the Blue Mountains west of Sydney had been penetrated, and the pastoral country and the well-watered river basins of the southern Central Basin were discovered. Convicts as well as free persons were also arriving in Tasmania. Settlements in the states now known as Western Australia, South Australia, and the youngest of all, Queensland, were soon developed. Beginning in the 1830s, free immigrants started to come in large numbers, and in the next two decades over 200,000 Britons migrated to the country.

The first population explosion came with the discovery of large-scale gold deposits in 1851. The population increased mainly by **migration** from a few hundred thousand to over a million in 1870. Almost all of the immigrants were English-speaking Britons and Irish with a few Chinese who had been brought in by the landed gentry. The Chinese were later barred from entry, however. Eventually, a "White Australia Policy" was established that is still somewhat prevalent today.

After the gold fever died down, the country settled into an agricultural-pastoral economy exporting animal and agricultural products. None of this led to rapid population growth except in the urban centers where the

The Sydney skyline crowds the shores of one of the world's most beautiful harbors and forms a backdrop for two of its famous landmarks, the Sydney Opera House and Sydney Harbor Bridge.

processing and shipping took place. By the mid-1920s, Australia had a population of 7 million, about the same as New York City.

New Zealand

The settlement of New Zealand began about 1000 years ago with Polynesians called Maoris. Europeans reached the islands in 1642, when the Dutch navigator, Abel Tasman, discovered New Zealand and gave it its name. The British officer and navigator, Captain Cook, reached New Zealand in 1769. Soon settlers from Britain arrived but not in large numbers until 1840, when the islands were formally annexed by Great Britain. In 1840, the Maoris ceded sovereignty to the British Crown, and the 200,000 Polynesian inhabitants soon began to decline because of the white settlers' diseases and wars against them. Unlike Australia, New Zealand was never a penal colony, and many of the first European settlers were religious groups. At the end of the Maori Wars (1860–1872), the Maori population numbered less than 50,000; today they number more than 300,000 and control about 4 million acres (1.62 million hectares) of land. New Zealand has favored European settlers, especially the British, but some Dutch and Germans have also settled there. Today, over 90 percent of the population are descended from Western Europeans, and English is the official language.

North Island was settled first, but during the nineteenth century, the discovery of gold and the favorable conditions for sheep raising pushed South Island ahead in population. The warmer temperatures of North Island and the possibility of more intensive agriculture, especially dairying, have encouraged more people to settle in the north in recent years.

POPULATION

The population of Australia today is estimated at 17.1 million, but it is unevenly distributed and highly clustered. Along the east and southeast coasts, which are well watered, cities, towns, and farms have developed; these areas account for more than 80 percent of the population (Figure 9.4). Australia has a crude birth rate of 16 and a crude death rate of 8; therefore, its crude natural increase is a relatively low 0.8 percent. At that rate, it will take Australia about 90 years to double its population. Life expectancy at birth is a relatively high 76 years.

Just two cities, Sydney and Melbourne, account for slightly less than 40 percent of the country's population; overall, 86 percent of the population live in cities and towns. The country has a population density of six persons per square mile, (two per square kilometer). With an annual natural rate of increase of 0.8 percent (compares to 0.6 percent for the United States), the population is not increasing very rapidly even though immigration laws were relaxed after World War II, when a large number of mainland Europeans entered the country. After 1973, a few boat people from Southeast Asia made it into Australia.

In many respects, Australia is a **replicative society** of Great Britain; that is, it has adopted many aspects of British culture. Over 90 percent of the inhabitants speak English, and a vast majority attend the Church of England, or Anglican church. About one-quarter of the population are Roman Catholics. Catholicism was introduced by Irish immigrants in the early 1800s, and its following increased after World War II with an influx of Spaniards, Italians, Yugoslavs, and Greeks. Adherents of the Greek Orthodox, Lutheran, and Oriental religions constitute small minorities.

The present population of New Zealand is 3.3 million, with a population density of 32 persons per square mile (12 per square kilometer; see Figure 9.4). Its annual natural rate of increase is 0.8 percent, the same as Australia's. About 70 percent of New Zealanders live on North Island. Life expectancy for New Zealanders at birth is 74 years. The majority (over 85 percent) of New Zealanders are Protestants; Roman Catholics constitute about 14 percent of the population; and the Church of Latter-day Saints (Mormons) has made some converts among the Maori.

POLITICAL CHARACTERISTICS AND CULTURE

Australia is a **federal state;** that is, it has both a national government and governments for the constituent states. Today it has a prime minister, a parliament, a high court, and a federal capital. It has not always had this type of political structure. In 1859 when Queensland separated from New South Wales and became the youngest colony, the political division of the country was established, except for Canberra, the federal capital. There were now six states and one territory. Each state was autonomous, a situation that led to much political rivalry

FIGURE 9.4

Population Density of Australia and New Zealand

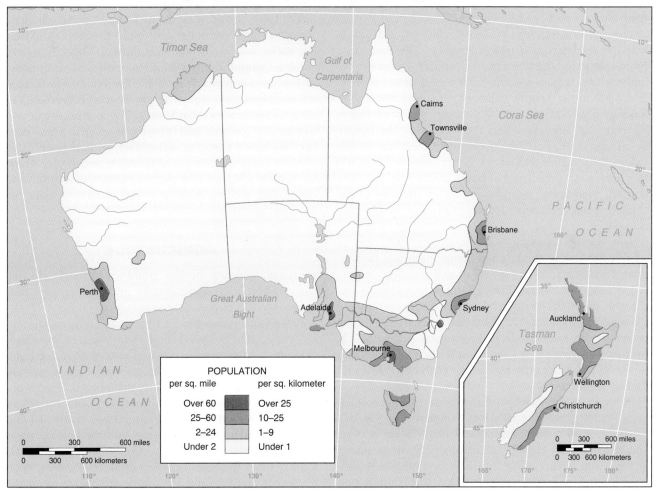

and economic feuding that hindered development for many years. Not only did each state build its own railroad system with a different gauge, but argued over trade policies and competed for hinterlands. In addition, there was no common defense system. Each state had to rely on Great Britain thousands of miles away. The colonies had many worries, including the Asian masses to the north who might migrate south, European powers that might establish a claim, and especially the Germans who were expanding in Africa and the Pacific during the 1880s. But eventually Australians recognized that cooperation was necessary among the several states if they

were going to have a place in international trade and develop world markets.

During 1900, lengthy conferences were held, and finally a constitution was created and approved by the voters of the colonies and confirmed by the British Parliament. On January 1, 1901, the colonies became the Commonwealth of Australia. Under this constitution, the federal government has power over many facets of government such as defense, customs, and external tariffs; it also has "concurrent" powers, which it shares with the governments of the six states, but in case of dispute the federal law prevails. The federal High Court, like the

American Supreme Court, is the final arbiter in disputes, although under a curious survival of British rule an appeal may still be made to the Privy Council of the United Kingdom, far away in London.

The one problem remaining after the 1901 federation was the selection of a national capital. The two most populous states, New South Wales, and Victoria, also had the two largest and most influential cities, Sydney and Melbourne. Shortly before the country became a federation, the premier of New South Wales managed to secure an agreement that the federal capital would lie in that state. For this, two concessions had to be made: first, Sydney would not be the capital, but a totally new city would be built at a jointly selected site; second, Melbourne would function as the temporary capital until the new federal site was ready for occupation.

The site of Canberra was chosen in 1908, and a worldwide architectural design competition was held in order to acquire the best city plan and building design. A U.S. architect, Walter Burley Griffin, was chosen for the task, and Canberra is now a spacious, modern, well-planned city of almost 250,000 residents. It has functioned as the seat of federal government since 1927. It occupies an area of 940 square miles (2500 square kilometers) and is the only major Australian city that is not located on the coast.

In terms of the standard of living and the per capita gross national product ($12,390, compared to $19,780 for the United States), Australia in many ways is very comparable to the developed nations of North America, Europe, and Japan. However, there are rather stark differences in the quality of life within the country between the urban and rural areas. While conveniences in the urban centers are similar to other Western nations, life in the Outback, the isolated rural areas, is not so well developed. However, the typical rural station homestead is substantial, even though the water is collected in a cistern or pumped by a windmill. The residents keep in touch with the world by shortwave radio. They use it to help educate their children or to summon a doctor or a flying ambulance should they need to go to the hospital. They dress formally every evening for dinner and travel to the cities for shopping and visiting at least two times each year.

Like Australia, New Zealand was a British colony, and its political institutions still reflect British influence. When Great Britain gave the colony of New Zealand a constitution in 1852, it established a statute for the country that in many ways functions as a constitution. Its clauses can be amended or replaced by the New Zealand Parliament through the normal legislative process. Since 1852, the New Zealand legislation has changed almost all its provisions, and for all practical purposes, the nation has no written constitution. The Colony became a Dominion in 1907 and an independent member of the Commonwealth in 1947. The country is divided into 63 counties. There is an 80-seat Parliament, of which 4 are reserved for Maoris. The main political parties are the National and Labour parties. The prime minister is elected from the ruling party in Parliament.

New Zealand enjoys a very stable social life. There are few very rich or very poor persons, thanks to a well-developed welfare state, a homogeneous population, and few minorities.

AGRICULTURE

Australia is one of the few countries of the world that can support itself agriculturally and still supply great quantities for export. The agricultural activity of the country can be divided into pastoral, commercial, and crop specialities. The Central Plain, west of the Great Dividing Range, was recognized as a potential grazing area in the early 1830s. Merino sheep were brought from Spain, and today many millions of sheep are grazed in an area extending from the southern part of Queensland southward into Victoria and westward into South Australia. Australia is the world's leading wool producer and exporter. This same area, especially the Murray River Basin, is a major producer of wheat. Winter wheat, barley, and rice are grown on many of the large sheep stations, making Australia a major exporter of wheat. There is another small grain region in Western Australia east and south of Perth where winter rainfall supports a commercial grain crop. On the southeast coast where moisture is available all year, thousands of sheep, used primarily for lamb and mutton, are grazed on improved pasture (Figure 9.5).

When refrigerated ships were developed in the 1880s, large-scale cattle ranching came to Australia. Cattle are raised mostly in Queensland and in the northern part of the Northern Territory and Western Australia. Cattle in the country number about 15 million and graze on the savanna lands on stations that measure in thousands of square miles because of the low carrying capacity of the land. Cattle must at times be driven long distances to railheads or moved in giant tandem freight trucks over

NEW ZEALAND: THE MAKING OF THE WELFARE STATE

The reason New Zealand has been a pioneer in social-welfare legislation can be attributed, perhaps, to several causes. Remoteness looms large because few wished to go so far from their native Britain. In 1840 when British rule was established, the European population numbered a mere 1000. The discovery of gold in 1861 brought an influx of settlers, and many remained to settle on the land after the gold rush ended. Outstanding leaders of British descent, wishing to see the country developed and populated, were responsible for many of the social programs. Sir Julius Vogel, prime minister from 1873 to 1875, initiated a policy of borrowing funds in London to provide state aid to more than 100,000 British immigrants to build railroads, bridges, and roads to foster trade. Sir George Gray, who came to power in 1877, introduced manhood suffrage and made education free, compulsory, and secular.

World depressions no doubt influenced social reform. In the early 1880s, a fall in world price levels of selected commodities brought financial depression to New Zealand. The first successful shipment of frozen meat to London in 1882 and the establishment of cooperative dairies gave promise of a brighter future for farmers.

During the decade of the 1890s, trade unions began to organize, and when the maritime strike of 1890 failed, the union leaders determined to make New Zealand's Parliament their future battleground. In 1891, there began a series of Liberal-Labour ministries whose bold political and social legislation evoked worldwide interest. Women received the right to vote in 1893. During that decade, the government passed numerous labor relations acts that protected workers and unions and set in motion the machinery to establish minimum wages. The government subdivided large estates, provided financial assistance to small farms, and passed the Old-Age Pension Act—the first of its kind in the English-speaking world. Numerous public works projects were also established. The government of New Zealand was becoming a foster parent to the nation's people and a partner of many of its farmers and business executives. Many of the settlers had come to this remote land at considerable government expense, and the country did not wish to lose them.

In 1936, while the world was experiencing another crisis in commodity prices, New Zealand had a second great period of social-welfare legislation when the government set basic wages and established a 40-hour work week for industry. In 1938, the Social Security Act was passed, which consolidated pension measures and established New Zealand's present extensive health and medical benefit plans. As a consequence, there are very few wealthy people in New Zealand; neither are there many poor. Everyone pays into the system, and although taxes are high by U.S. standards, everyone benefits.

rough and dusty roads. Two particularly interesting cattle operations are (1) the Kimberley Mountains–Ord River enterprise in Western Australia and (2) the King Ranch, Inc., of Texas cattle station on the east coast of Queensland near Tully. The Ord River has been developed into a series of catchments to store water for irrigation. The cattle are slaughtered and quick-frozen on the station and airlifted to the port of Wyndham or Derby. From there the ships go to Perth or to overseas markets. The King Ranch has cleared 51,000 acres of rainforest on the Queensland coast near Tully and sowed it in African grass that grows six feet (1.8 meters) in six months. The ranch is small as stations go in Australia, but it is expected to carry 25,000 head of cattle.

The country's dairy industry is concentrated along the well-watered east and south coasts where pastures can be improved and a dairy herd can be carried on a relatively small farm. The industry is also close to the large urban centers where the products are processed and consumed.

FIGURE 9.5

Agricultural Patterns of Australia and New Zealand

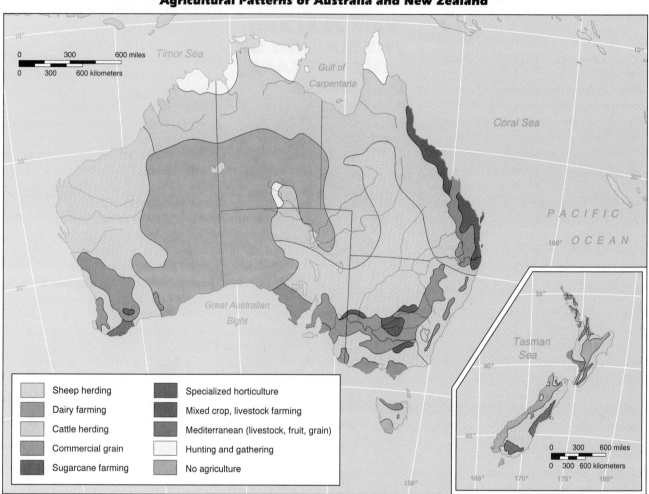

The speciality crops include sugarcane, cotton, and irrigated citrus, grapes and various fruits. From Brisbane northward, the coastal area separates into several river floodplains, each of which is planted in sugarcane. The cane belt stretches northward as far as Cairns. Away from the coast where it is slightly drier, cotton becomes the cash crop. In the Murray-Darling river basin, irrigation works make possible the production of citrus fruit, grapes, peaches, pears and apricots. Likewise, in the cool marine climate on the island of Tasmania, such fruits as apples, pears, cherries, and berries are grown.

In assessing the agriculture of Australia, three areas should be noted as special. They are the Barassa Valley of South Australia, Darling Downs of Queensland, and the Atherton Plateau of northwestern Queensland. The Barassa Valley northeast of Adelaide was settled by Germans in the 1840s. They planted vineyards, and today this valley is the wine capital of Australia. The Darling Downs, with a topsoil 40 feet (12.2 meters) thick, produces excellent crops of wheat, barley, potatoes, and hay without the use of commercial fertilizers. The Atherton Plateau is in the tropics and supports rainforest, but its elevation makes it cool enough to be much more comfortable than the Queensland coast. Here, tropical crops are grown: pineapples, bananas, mangoes, papaya, and a variety of vegetables.

The chief economic base of New Zealand is the livestock industry (Figure 9.5). Sheep grazing developed first and later was followed by dairy cattle. Due to the mild climate, especially on North Island, the cattle do not require elaborate housing and can graze outside the year round. Much scientific work has been devoted to improving livestock, pasture, and the processing of dairy products and meats. New Zealand cheese, butter, and dried and canned milk have a worldwide reputation. Wool, meat, and dairy products constitute over 90 percent of New Zealand's exports.

Field crops make up a small percentage of the occupied farmland, only 1.5 million acres (607,500 hectares). Wheat is grown mostly on the Canterbury Plain along with oats and potatoes. The country produces a variety of vegetables, fruits, berries, and nuts for home consumption. Only apples are exported in any quantity.

INDUSTRY

Industrial production in Australia was severely handicapped during the colonial period when the country was composed of six autonomous states. Each state was a small market, and each protected its home industry. Almost all manufactured goods were imported from Britain. After federation in 1901, some industries were developed, especially those that produced bulky commodities with high transport costs: furniture, building materials, lumber, and the ubiquitous food and dairy products. World War I emphasized the vulnerability of the country as Australia was cut off from most of its imports. During the 1920s and 1930s, the country once again relied upon imports, and its exports of wool, meat, wheat, fruit, and primary metals were more than ample to pay for the needs of a few million people. Britain went to war with Germany in 1939, and in early 1942 Australia was threatened by invasion from Japan. Once again, many of its imports were cut off. Since the end of World War II, the country has made giant strides in manufacturing for its domestic market. The country concluded that if it were to take its place among the developed nations, domestic industries were necessary.

The abundance of coal and its location close to the coasts helped to establish basic industry in Australia. Steel is produced at Newcastle in New South Wales and at Whyalla in South Australia. Iron ore is supplied from South Australia and Western Australia. The basic steel is used to manufacture cars, ships, machinery, and aircraft. There industries are located in and around the major

urban centers. Nonferrous metals are smelted and refined at Mount Isa in Queensland and Broken Hill in New South Wales. Bauxite from the York Peninsula is shipped to a smelter at Gladstone on the Queensland coast and refined into alumina. The smelter at Gladstone is owned by Australian, Canadian, American, and French interests. The alumina is shipped to Japan and New Zealand for further refining into aluminum, a process that consumes large amounts of electricity. Copper from Mount Isa is smelted at Townsville, Queensland.

Food processing is the most widely distributed industry in the country. Because Mackay, on the Queensland coast, is located in the heart of the sugarcane belt, it has several sugar refineries. The milling industries and processing plants for fruits, vegetables, meat, and dairy products are scattered throughout Perth, Adelaide, Melbourne, Sydney, and Brisbane. Almost all major cities and even some smaller ones produce spirits, beer, or wine, especially beer, the national drink of Australia.

A sheep herd in the highlands of New Zealand. Flocks like this help to make New Zealand the world's second leading wool exporter.

The construction industry is an important component of Australia's economy, and several cement, brick, and stone works, together with lumber mills, are scattered throughout the country. The timber industry is more localized, however. Beginning in the late 1800s, Australia imported seed and seedlings of foreign conifers and now has thousands of acres of these trees to supply lumber. One of the largest is the 200,000-acre (81,000-hectare) Monterey pine forest in extreme southeast South Australia north of Mount Gambier. The trees grow to 120 feet (36.5 meters) in 40 years. In addition, textiles and apparel industries are widespread and are found in all of the major cities.

In contrast to Australia, New Zealand is not noted for a rich variety of mineral resources and thus has not developed much heavy industry. As in Australia, industry increased during and after World War II, but still accounts for only a small percentage of the gainfully employed. Most of New Zealand's industry consists of processing and shipping animal products. It has sufficient energy from coal and hydroelectric power to support some industry, but its small domestic market and distance from foreign markets make expansion difficult. In-

dustries include textiles and clothing, wood products, fertilizers, building materials, metal products, and some machinery. A small steel mill has been opened near Auckland, and an aluminum plant that processes Australian bauxite is located on South Island. Both rely on New Zealand's ample hydroelectric power.

URBANIZATION

Australia is a land of contrasts between its densely settled urban areas and the **remoteness** of the sparsely settled Outback. Most Australians live on the periphery of the continent along the east, south, and extreme southwest coasts. Here the land is well watered, and the farms are small and much more concentrated. Five of the country's largest cities are located here. Australia is a land of urbanites. Eighty-six percent of the people live in cities, most of this urban population is concentrated in the five largest cities—Brisbane, Sydney, Melbourne, Adelaide, and Perth.

All of these cities are state capitals and were the first sites chosen for settlement in the early development of the country. Since each site was selected because of

An aerial view of the Canterbury Plain and coastline south of Timaru on New Zealand's South Island. The plain is a garden spot of the country, producing wheat, fruit, vegetables, lambs, and wool.

A view of the Outback, a term used by Australians to describe the vast and rugged inland region of their country.

access to water transportation, they have good harbors and conduct extensive commerce. Each built a rail and road communications network to its hinterland, which supports its industrial and commercial activities. The only Australian city of any size not located on the coast is Canberra, the federal capital. The city has a population of nearly 250,000. Hobart, Tasmania, is still relatively small, less than 200,000, but so is the overall population of the island, which is Australia's smallest state. In the early twentieth century, Australian cities developed much like the major cities of the developed world. They had a viable central business district, industrial

sectors along the waterfronts and railroad lines, and outlying residential sectors of varying price ranges. The central business districts of Australian cities are still quite viable because the cities have maintained good rapid transit systems. The cities have now grown to a size that makes going to the market daily too much of a burden. Consequently, malls and shopping centers along the major arterials away from the intercity have become common.

Despite their differences in size, climate, and physical makeup, Australia and New Zealand are very similar in their patterns of urbanization and peripheral development. New Zealand cities are located on the coast, are major ports, and contain 84 percent of the population. Auckland, with a population of 900,000, is located on North Island. Wellington (350,000), the capital, is located on the southern portion of North Island. Hamilton (152,000), the hub of an excellent grazing and mixed-crop livestock region, is on the west side of North Island.

Urbanization on South Island is dominated by Christchurch (325,000) and Dunedin (120,000). Christchurch, the hub of the Canterbury Plain, is called the garden city of New Zealand. The city was founded by members of the Anglican church, and the cathedral dominates the city square. Christchurch's port is Lyttelton, which is separated from the city by a narrow mountain range. Dunedin, on the southeast coast of South Island, serves as a fishing port and a starting point for numerous Antarctic expeditions. No other urban center has a population as large as 100,000, but several have 40,000 to 90,000 residents and serve as market and supply centers for grazing and agricultural regions. New Zealand has one of the highest per capita exports in the world, and the cities are service centers for rural areas and for the processing and shipping of animal products.

SPATIAL CONNECTIVITY

Due to their remoteness, both Australia and New Zealand face special challenges in establishing and maintaining trade and other relationships with the rest of the world. For Australia, the problem is compounded by the size of the country itself, which makes internal communication and transportation difficult.

AUSTRALIA

A very essential ingredient of economic growth and national development is accessibility. Due to Australia's size and its small overall population, goods must be transported great distances to market, and people must travel many miles between major urban centers. Going from Sydney to Perth is comparable to going from Los Angeles to Atlanta. Each Australian state has but one major city, and the states are large—several are larger than Alaska. Between 1860 and 1901, each capital city developed its own railroad network to tap the trade and resources of its hinterland. After federation in 1901, the lines were extended to connect the major cities. The country today has 28,000 miles (45,052 kilometers) of railroads and has made great progress in developing a standard gauge rail system. Several of the smaller coastal ports have lines connecting with the interior. There is only one east-west line, and it runs across the southern part of the continent. A north-south line extends as far north as Alice Springs.

The highway system has greatly improved since World War I, and it is now possible to drive between the major cities on all-weather paved highways, many of them with four lanes. Good hard-surfaced roads connect to interior towns as well. Australia today can count more automobiles per 100 persons and more road kilometers per person than the United States. The country is well served with airlines with several flights daily between major cities. Qantas, one of the most successful overseas airlines in the world, makes connections with every continent. Australia has well-developed coastal shipping and several international ports with freight and passenger connections to the world's major ports.

Communications are excellent (Table 9.1). Every city or town has a newspaper, radio stations are numerous,

TABLE 9.1

Communications in Australia, 1984

TELEPHONES PER 100 POPULATION	DAILY NEWSPAPER CIRCULATION COPIES PER 1000 POPULATION	TELEVISION RECEIVERS PER 1000 POPULATION	RADIO RECEIVERS PER 1000 POPULATION
53.6	334	423	1301

SOURCE: United Nations, *World Statistics in Brief* (New York: United Nations, 1986); *United Nations Yearbook: 1983–84* (New York: United Nations, 1986).

TABLE 9.2

**Value of Australia's Foreign Trade by Country of Origin
(in A$1,000), 1984–1985**

SELECTED COUNTRIES	EXPORTS	IMPORTS
Japan	$ 6,570,041	$ 5,366,190
United States	2,704,744	5,188,392
United Kingdom	1,134,210	1,740,161
New Zealand	1,400,821	921,748
Singapore	951,515	470,172
Hong Kong	612,004	552,176
Total	29,809,723	29,055,754

SOURCE: United Nations, *World Statistics in Brief* (New York: United Nations, 1986).

and television stations, including satellite reception, are found in all major cities and towns. Push-button telephones can connect with many places on the continent and overseas as well. Those Outback stations and settlements too remote for telephone lines have shortwave radios to stay in touch.

Australia has long had a favorable balance of trade (Table 9.2). The early exports of wool, meat, dairy products, and wheat went mostly to Great Britain. Exports today still include the products of farms and ranches, but exports of minerals and manufactured products have increased. Japan and the United States are major receivers of Australian exports. The country had sufficient trade income to pay for increased manufacturing until the decline in commodity prices in the past decade. The nation must rely on foreign investment to finance new industries. The home market is so small that increased exports are essential to ensure full employment and continued growth. Only 12 percent of the labor force produces the agricultural products, but nearly 50 percent are engaged in manufacturing and commerce. Increased exports are thus very essential to the nation's prosperity.

NEW ZEALAND

New Zealand is the most isolated of the Westernized-industrialized nations in the world. Its size, remoteness, and small population present many problems. However, the nation enjoys a high per capita income and standard of living. Education is free and medical service is excellent.

The nation has developed an excellent transportation system. All cities and major regions of the country can be reached by highway, railroad, or air service. The country is also served by international air service and has three international air terminals located at Auckland, Wellington, and Christchurch. New Zealand is moving into the twenty-first century in the field of technology. Everything present in Western scientific development can be seen or experienced in New Zealand.

The country has a large export trade in spite of its remoteness, which adds to shipping costs. Much of its foreign trade is with Great Britain as well as most of the foreign investment (Table 9.3). New Zealand is expanding its foreign trade, and Japan, Australia, and the United States are its leading trading partners. Since World War II, more foreigners have discovered the beauty of New Zealand, and the tourist trade is on the increase. On South Island, scenes comparable to parts of Scotland, Norway, Switzerland, and Austria can be experienced in a single day.

PROBLEMS AND PROSPECTS

World War II was the great watershed of change in the twentieth century. It brought about more independent nations and changes in trade, alliances, and defense treaties than any other event in modern history. Australia has been part of this reorientation. Before the war, as a Commonwealth member, the nation looked to Great Britain for trade and defense. At that time, Britain

For nations to survive and play an integral role in world affairs, they must be able to contact, communicate, and interact with other countries. Australia and New Zealand, although somewhat removed from the rest of the non-English-speaking world because of their British heritage and culture, are spatially connected with the world in many ways. Through their international ports, airports, satellites, and telecommunications, they keep in touch with the world community. Politically, they are members of the ANZUS military alliance and the British Commonwealth and play a unique role in the security of the Southwest Pacific.

Acculturation of Minorities and Urban-Rural Interconnectivity

The major culture problems Australia and New Zealand face are connectivity between the rural and urban societies and between the white and indigenous populations. These problems are more pronounced in Australia because of its size and its relative inability to assimilate the Aborigines. Conversely, New Zealanders have been better able to assimilate the Maoris into their culture. The Maoris are still landowners and include scholars, professionals, and eminent political leaders.

Australia tried to solve its minority problem by eliminating it. The Aborigines were driven out of the colonized areas and settled in Arnhem Land and on islands in the Gulf of Carpentaria. Today, it is estimated that fewer than 225,000 Aborigines remain, and many of those are of mixed ancestry. Few traditional communities exist today.

SPATIAL CONNECTIVITY ISSUES

Many Aborigines work on the cattle stations or live on reserves established by the government. Most live in poverty, are poorly educated, and lack health care. But external and internal pressures are forcing Australia to address the problem, and some progress is being made.

A very wide gap in communication exists between the rural and urban populations of Australia. The young people on the farms and stations have little in common with the surfers on Bondi Beach in Sydney. The great Outback is essential to the continental growth and prosperity of the country. The interior is responsible for most of Australia's trade and balance of payments. For continued growth, more people are required, but the urbanites are reluctant to leave the cities for the isolation of the interior. This is not altogether a new problem. Several decades ago, James McAuley, an Australian poet, addressed the problem when he so aptly stated:
". . . though they praise the inner spaces, when asked to go themselves, they'd rather not."

Immigration Pressure

The overwhelming immigration problem of the twenty-first century will be with the Asians. Australia and New Zealand belong to the Colombo Plan, which provides assistance for economic development to countries of South and Southeast Asia. Both countries have spent millions on foreign aid, capital investment, and subsidies for Asian students. In the past, Australia practiced a "eugenic" selective immigration policy favoring white immigrants from Europe. This policy began to change in the 1960s, and more Asians were admitted. The annual rate of nonwhite immigration is still small, and applicants applying for residence must have a needed skill. But Australia's close proximity and spatial access to Southeast Asia and its neighboring islands have made it vulnerable to accepting refugees. Many already have come from Vietnam and Cambodia. In the future, Australia can expect to receive illegal aliens when political and economic crises occur within neighboring countries. Australia's illegal alien problem, though small in scope, is similar to some extent to that of the United States and its rising tide of illegal Hispanic immigrants. Both the United States and Australia are perceived as countries offering economic opportunity and political freedom, and both have neighboring developing countries with burgeoning populations. Thus, one of Australia's challenges in the future will be how to handle the increasing number of Asians who wish to migrate legally or illegally to the land "down under."

TABLE 9.3

Value of New Zealand's Foreign Trade by Country of Origin (in NZ$1,000), 1984–1985

SELECTED COUNTRIES	EXPORTS	IMPORTS
Japan	$ 2,194,371	$ 2,287,303
Australia	1,835,672	2,108,863
United States	1,823,091	1,879,625
United Kingdom	985,521	989,714
West Germany	622,635	544,123
Total	10,571,700	10,468,300

SOURCE: United Nations, *World Statistics in Brief* (New York: United Nations, 1986).

still had a Far Eastern empire and bases from which to operate. The war changed this, and in May 1942 when a Japanese task force was turned back in the Coral Sea, the prime minister of Australia announced that henceforth Australia would look to the United States for protection. As britain lost its eastern empire, the United States has assumed the task of policing the Southwest Pacific. Australia, New Zealand, and the United States have formed a treaty of mutual protection (ANZUS). Australia has allowed the United States to establish space tracking stations as well as a monitoring station in the Northern Territory. Australia has supplied troops to all the wars the United States has fought in the Pacific.

Among Australia's advantages are its stable government, favorable balance of trade, and natural resources. At the same time, it faces pressure to allow more imports as well as more immigrants.

Despite Australia's stable government, size, and natural resources, it lacks the population and industrial base to be a major world power. Instead it will continue to play a major role in Southwest Pacific affairs, but mostly in conjunction with the United States.

Remoteness, small size, few people, and a poor resource base other than soil and water are the problems that New Zealanders must face daily. Many of the bright and well educated may possibly migrate because of the lack of opportunity within their own country. As a member of ANZUS, the country will remain a stabilizing force in the Southwest Pacific. At the present time, the United States and New Zealand disagree about atomic weapons. The country will not permit American ships with nuclear weapons to dock in New Zealand ports.

Selected Readings

Barrett, R. D., and R. A. Ford. *Patterns in the Human Geography of Australia.* South Melbourne, Australia: Macmillan of Australia, 1987.

Burnley, I. H. *Population, Society and Environment in Australia.* Melbourne, Australia: Shillington House, 1982.

Cochrane, P. *Industrialization and Dependence: Australia's Road to Economic Development, 1870–1939.* St. Lucia, Australia: Queensland, University Press, 1980.

Crabb, P. "A New Australian Railroad—Then North to Darwin." *Geographical Review* 72 (1982): 90–93.

Cumberland, K. B., and J. S. Whitelaw. *New Zealand.* London: Longmans Group, 1970.

Gale, F. "The Social Geography of Aboriginal Australia." In *Australia: A Geography,* edited by D. N. Jeans. New York: St. Martin's Press, 1978.

Gibson, K. D., and R. J. Horvath. "Global Capital and the Restructuring in Australian Manufacturing." *Economic Geography* 59 (1983): 178–94.

Heathcote, R. L. *Australia.* New York: Longmans, 1975.

Jeans, D. N., ed. *Australia: A Geography.* New York, St. Martin's Press, 1978.

Linge, G. J. R., and J. McKay eds. *Structural Change in Australia: Some Spatial and Organizational Responses.* Canberra, Australia: Australian National University, 1981.

McKnight, T. L. *Australia's Corner of the World.* Englewood Cliffs, N.J.: Prentice-Hall, 1970.

Nichols, J. L. "The Past and Present Extent of New Zealand's Indigenous Forests." *Environmental Conservation* 7 (1980): 309–10.

The Polar Regions

The regions that encompass the poles of the earth are Antarctica (Figure 1) at the South Pole and the Arctic Basin (Figure 2) at the North Pole. These regions are loosely defined by the Antarctic Circle at 66½° south latitude and the Arctic Circle at 66½° north latitude. Due to their high latitudinal positions, the polar regions receive minimal precipitation and limited sunlight and are cold and inhospitable to most creatures on earth including humans.

In the past, however, the land areas in both regions were exposed to a much milder environment. Fossils and fossil fuels are found in both regions, indicating that they were once in more temperate zones of the earth's surface. Antarctica, for example, has deposits of coal and petroleum. Likewise, fossil remains of warm-weather plants have been found in Greenland in the Arctic region. The large oil fields of Alaska are north of the Arctic Circle. Over millions of years, the drift of the tectonic plates underlying these regions has carried them from warmer environments to their current positions at the "ends of the earth," far removed from the moderate climates of the past.

Although both regions are located at the poles of the earth, they differ in many important respects. The Antarctic is a massive landmass of 5.5 million square miles (14.2 million square kilometers) covered almost entirely by an ice sheet averaging over one mile (1.6 kilometers) in thickness. This ice sheet is so massive that if it should melt as a result of climatic change, it is estimated that average sea level would rise 200 feet (61 meters). The average elevation of Antarctica is over 6000 feet (1800 meters). The combination of the large landmass and the high elevation makes Antarctica the coldest place on earth. The coldest temperature on record of −129°F

The Eskimos are a group of people who live in the Arctic. Their homeland stretches from the northeastern tip of the Soviet Union across Alaska and northern Canada to Greenland. The Eskimos live chiefly on seal and caribou meat and use two types of boats—kayaks and umiaks. This photograph shows an Eskimo mother with her two children and a umiak in the background.

(−89°C) was recorded at the Soviet Union's Vostok research station approximately 1000 miles (1600 kilometers) from the South Pole. The highest temperature ever recorded at the South Pole is 7.5°F (−14°C); no location on the Antarctic mainland has a month when the average temperature is above freezing.

Due to this harsh environment, Antarctica has only a limited number of mosses, lichens, and insects and a few flowering plants, all of which live on the coast of the continent. In fact, even bacteria are relatively absent compared to milder, less remote areas of the earth. The

FIGURE 1

Antarctic Region

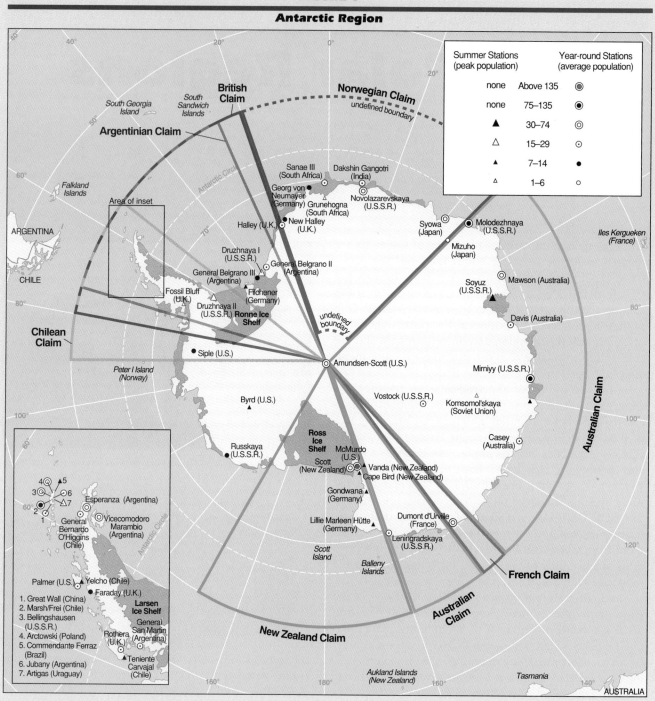

only animals that inhabit Antarctica are migrant penguins, seals, and scientists.

In contrast to the landmass of Antarctica, the Arctic is comprised mostly of the Arctic Ocean basin, which is covered, in part, by an ice sheet. The ocean is able to store more heat energy in summer and release it in winter than the ice-covered land of Antarctica. The result is a much more moderate polar climate in the north. For

FIGURE 2

Arctic Region

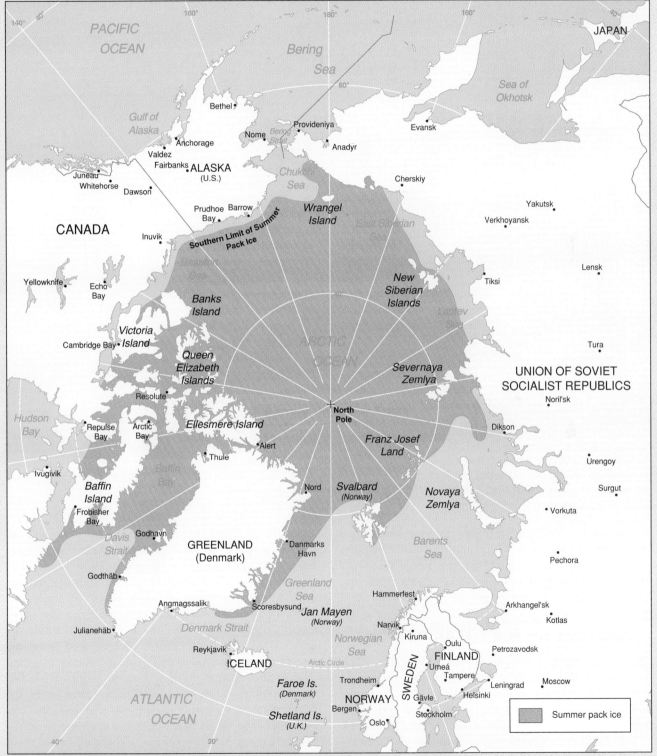

Although the whale looks much like a fish, it is a mammal. Whalers have greatly reduced
the number and species of whales by overhunting. The International Whaling Commission
has tried to regulate the whaling industry, but effective control of whaling is difficult
because regulations are hard to enforce.

this reason, the Arctic is the home of a great abundance
of plants, insects, and larger animals. Ironically, the
coldest temperatures in the Arctic are not at the North
pole, but hundreds of miles south in Siberia where the
effects of the moderating ocean are nonexistent.

CONTROL OF THE POLAR REGIONS

The Arctic region was first explored during the 1800s
by expeditions in search of the Northwest Passage linking Europe and Asia. In 1909 Robert Peary successfully
led an expedition across the ice sheet of the Arctic
Ocean to the North Pole after several others had failed.

Currently, the international laws and treaties governing the oceans and offshore territorial claims, such as

exclusive economic zones, guide the debate over control of the Arctic. As more natural resources are discovered and technology is developed to allow commercial
exploitation of these resources, more international disputes over boundaries are likely to occur among the
nations controlling land on the periphery of the Arctic
Ocean. These nations are Canada, the United States, the
Soviet Union, Norway, Finland, Sweden, and Denmark,
which claims ownership of Greenland. Equally important are the internal policies of these nations regarding
the resources, ecosystems, and native peoples of the
region. The regulation of fishing and sealing in the Arctic Ocean, the debate over resource exploitation in the
Arctic National Wildlife Refuge, and the effects of modern technology on the native Eskimo and Aleut popu-

lations are a few of the difficult questions that face the Arctic region.

Control of the Antarctic region is a very different situation. The land areas of the Antarctic region were initially discovered in the late 1700s and early 1800s when ships entered the southern oceans surrounding Antarctica in search of seals and whales. These oceans were the last haven of large herds of great whales, many of which were slaughtered to meet the demand for lubricants and fuel made from whale oil in pre-petroleum days. In the mid-1800s, British and American expeditions formally explored and mapped the region. During the late nineteenth and early twentieth centuries, other nations joined in Antarctic expeditions to justify their claims to fishing and whaling rights in the region. An expedition led by the Norwegian Roald Amundsen reached the South Pole in December 1911.

The scramble for Antarctica has intensified in recent years as new resources have been discovered and possible uses for the area have been identified. Large deposits of low-quality coal have been discovered along with deposits of gold, tin, silver, molybdenum, and platinum. Evidence suggests that petroleum may exist either on the continent or in the offshore areas. Tourism to one of the last frontiers for whale and penguin watching is becoming a profitable business. Due to its location far from centers of population, Antarctica is also being considered as a possible site for the storage of nuclear waste.

Seven nations including Argentina, Chile, Australia, New Zealand, Norway, France, and the United Kingdom have asserted territorial claims to parts of Antarctica. Several of these claims are overlapping (Figure 1). Several other nations have established zones of interest and research stations to bolster future claims to ownership and control. Currently, 36 countries are conducting scientific research in Antarctica. Under the Antarctic Treaty, which was ratified in 1961 by several of the nations involved, the region was declared a demilitarized zone, and all territorial claims were set aside for 30 years.

In 1988, the parties to the Antarctic Treaty proposed the Convention on the Regulation of Antarctic Minerals (CRAMRA), which addresses allowable mineral exploitation activities and environmental procedures. Two claimants, Australia and France, have proposed that the entire continent and the surrounding oceans be declared a nature reserve or world park. The Soviet Union and the United States have not made territorial claims to Antarctica, but still maintain their right to do so if they desire in the future. Given the level of commercial and environmental interest and the expiration of the 30-year limitation on claims in the Antarctic Treaty, the control of Antarctica is likely to be a major issue during the 1990s.

CHAPTER 10

Middle America

IMPORTANT TERMS

Vertical (altitudinal) zonation

Mestizo

Ladino

Haciendas

Ejido

Monroe Doctrine

LAFTA

CONCEPTS AND ISSUES

■ Dominance of the Spanish cultures on the region.

■ Instability and corruption in government.

■ Persistence of native peoples.

■ Interference by the United States in political and economic affairs.

■ Contrasts between plantation and traditional agriculture.

CHAPTER OUTLINE

PHYSICAL ENVIRONMENT
The Mainland
The Caribbean

HUMAN ENVIRONMENT
Columbus and the "New" World
Demographic Characteristics

Culture
The Political Climate
Technological Achievements
Agriculture
Industry
Urbanization

SPATIAL CONNECTIVITY
Dependence
Interdependence

PROBLEMS AND PROSPECTS

W here is Middle America? Is it in the midwestern portion of the United States? In Kansas? Nebraska? Maybe Iowa? Perhaps it is somewhere near the center of the nation, in the area of the Mississippi River? No, despite the somewhat confusing terminology, Middle America has nothing at all to do with the United States. It is a term that English-speaking geographers use when referring to Mexico, the Central American republics, and all the countries lying in the Caribbean Sea. Remember that although we mistakenly call the United States of America "America" for short, technically, the entire landmass of Canada, the United States, Mexico, Central America, the Caribbean, and South America are all, together, the **Americas.** Similarly, all lands south of the United States, from the northern limits of Mexico to the farthest reaches of Tierra del Fuego, comprise a region geographers call Latin America. It is Latin because the majority of the people speak a Latin-based language—generally Spanish, but also Portuguese and French—and because the many present-day ethnic groups of the region combine features related to both the original, pre-Columbian peoples and the conquering, post-Columbian peoples from Western Europe who brought their Latin-based cultures to the New World. Again, these were predominantly the Spanish, the Portuguese, and the French. Thus, the predominant (i.e., the most powerful) culture is Latin in origin; the physical setting is in the Americas.

Latin America is a huge area, and it can be broken down further into two parts: South America, which constitutes an entire continent and will be the subject of the following chapter; and Middle America, which, as we delimited it above, is the northern half of Latin America, the portion sandwiched between the United States and South America. For reasons that will become clear in this chapter, we shall later break Middle America into two major components: **Mainland Middle America,** which includes Mexico and the seven Central American states; and **Caribbean Middle America** (Figure 10.1). The Caribbean part of this region contains several large and medium-sized islands, such as Cuba, Puerto Rico, and Jamaica, and literally thousands of smaller islands, such as Martinique, Guadeloupe, and St. Lucia. Before we begin our exploration of Middle America, perhaps we should first ask why the region is important to study.

Situated adjacent to North America, and contiguous to the United States, Middle America's proximity to its powerful neighbor has been an important factor in the region's past. Likewise, its relative location will continue to be of vital importance to Middle America's future. For example, the majority of all Spanish-speaking immigrants to both Canada and the United States derive from this part of the world. This is also the part of Latin America that most Canadians and U.S. Americans who travel south of the U.S. border visit. Border crossings between the United States and Mexico—both legal and otherwise—increase yearly, and the number of Canadian and American tourists visiting the popular Caribbean resorts of Jamaica, Barbados, and the Bahamas climbs steadily. Prior to the 1959 revolution led by Fidel Castro, the island of Cuba—a mere 90 miles south of Key West, Florida—was a major beach and gambling oasis for American tourists. The history of the southwestern portion of the United States was integrally intertwined with the history of New Spain and, later, Mexico. Today, a significant percentage of the people living in Florida, Texas, New Mexico, Arizona, and California are of Latin American descent, and most trace their heritage to some portion of Middle America; place-names and city streets in the American Southwest attest to their abiding influence.

Similarly, sizable numbers of former Middle Americans are residing in Florida (Cubans), New York (Puerto Ricans), Ontario (Jamaicans), and many other urban centers across Anglo America, many of whom are recent arrivals from this neighboring part of the globe. Just a few years ago, many so-called boat people fled abominable conditions in Haiti. Many are now citizens of either Canada or the United States; some continue to arrive, as do many other boat people, or refugees, from the Dominican Republic and other parts of Middle America. As of 1986, Mexico and Honduras were also host to large numbers of refugees from Guatemala, Nicaragua, and other Central American states. For all these reasons, perhaps no other world region should be of such concern and interest to Anglo Americas as Middle America.

Nothing reveals the extreme importance of this region more than the daily news. Rarely does a week pass that the media does not remind U.S. citizens of the hostilities in Nicaragua. Panamá, too, has become a heated topic of political concern now that General Noriega has been ousted from power. Honduras is less than 1,000 miles (1600 kilometers) from Texas, far less from Miami. That we understand what is happening to our southern neighbors has never been as crucial as it is today. In 1987, virtually all forms of mass media—newspapers, magazines, radio, and television—were focused on the vast sums of money that were transferred from Iran, via

FIGURE 10.1

Middle America

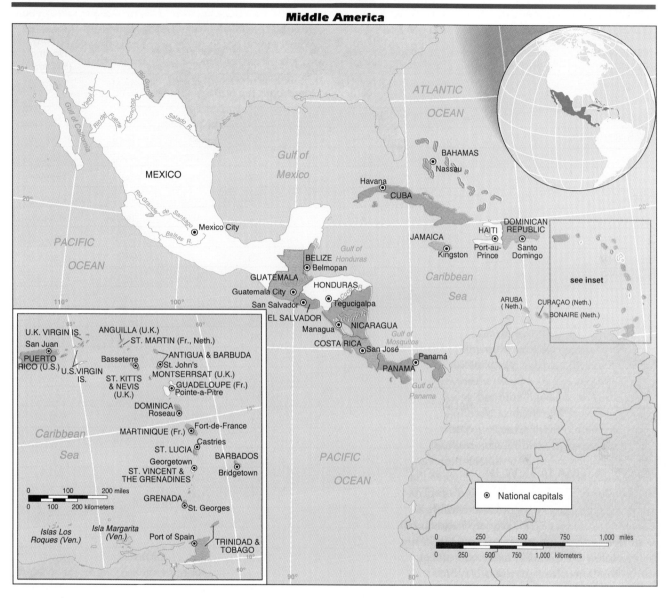

Switzerland, to the Contras in Central America who were fighting Sandinista "rebels." Likewise, in 1983, American Marines "invaded" the tiny Caribbean island of Grenada "to help restore democratic institutions in that country." In 1989, the United States invaded Panamá, charging that high Panamanian officials were funneling drugs into the United States.

Given the location and political concerns of the United States, and considering the ever-shrinking global economy that will undoubtedly develop in the 1990s, it is crucial that Americans appreciate these ongoing events. We must understand and familiarize ourselves with the peoples and nations of Middle America, learn a little about their past and the landscape they inhabit, gain an insight into their resources (both those they currently exploit and their potential sources), and try our best to appreciate our Middle American neighbors.

STATISTICAL PROFILE: MIDDLE AMERICA

REGION OR COUNTRY	Population Estimate mid-1990 (Millions)	Natural Increase (Annual)	Population Projected to 2000 (Millions)	Infant Mortality Rate	Life Expectancy at Birth (Years)	Urban Population (%)	Per Capita GNP, 1988 (US$)	Area, Thousands of Square Miles (Km²)	Population Density, No./mi² (No./km²)
Antigua and Barbuda	0.1	1.0	0.1	10	72	34	2,030	.2(.5)	588(227)
Bahamas	0.2	1.8	0.3	26.7	70	75	7,150	5(14)	37(14)
Barbados	0.3	0.9	0.4	13.1	73	32	4,680	.2(.5)	1523(588)
Belize	0.2	3.1	0.3	36	69	50	1,130	9(23)	34(13)
Costa Rica	2.8	2.7	3.7	18.9	74	48	1,290	20(51)	142(55)
Cuba	10.3	1.2	11.4	16.5	73	71	—	44(115)	238(92)
Dominica	0.1	1.7	0.1	13	75	—	1,160	.3(.8)	345(133)
Dominican Republic	6.5	2.5	8.4	70	63	52	810	19(49)	358(138)
El Salvador	5.3	2.6	7.2	65	66	43	710	8(21)	654(252)
Grenada	0.1	1.9	0.1	22	72	—	970	.1(.3)	752(290)
Guadeloupe	0.3	1.3	0.4	14	72	46	—	.7(2)	437(169)
Guatemala	8.4	3.2	12.2	71	60	39	1,240	42(109)	217(84)
Haiti	6.2	2.3	7.7	107	53	26	350	11(28)	579(224)
Honduras	4.7	3.1	7.0	69	63	40	730	43(112)	113(44)
Jamaica	2.5	2.0	3.0	20	73	54	940	4(11)	571(220)
Martinique	0.3	1.1	0.4	13	74	71	—	.4(1)	706(273)
Mexico	81.9	2.5	104.5	50	67	70	2,080	762(1973)	113(44)
Netherlands Antilles	0.2	1.4	0.2	10	75	50	6,110	.4(1)	522(202)
Nicaragua	3.5	3.4	5.1	69	61	53	850	50(130)	70(27)
Panamá	2.3	2.2	2.9	25	72	51	2,020	30(77)	81(31)
Puerto Rico	3.3	1.3	3.7	15.7	75	67	4,850	3(9)	999(386)
St. Kitts-Nevis	0.05	1.6	0.1	27.8	67	45	1,520	.1(.3)	481(186)
Saint Lučia	0.1	2.5	0.2	17.6	71	40	1,210	.2(.5)	420(162)
St. Vincent and the Grenadines	0.1	2.0	0.1	26.5	69	—	840	.2(.5)	667(257)
Trinidad and Tobago	1.3	2.0	1.6	20	70	34	6,010	2(5)	650(250)

Isolated for most of its history, Latin America remained a "closed society" even throughout most of the nineteenth century. Middle America, and indeed even Mexico located next to the United States, was a virtual mystery to most North American tourists until recently. Today, however, the recent discovery of large oil deposits in the Gulf of Mexico, the lingering tensions in Central America, the presence of American military personnel in Grenada in 1983 and in Panamá in the early 1990s, the 1977 agreement to return the Panama Canal to Panama by the century's end, and the United States' Caribbean basin Initiative—instituted to promote better trade, political, and economic relations between the peoples of the Caribbean and the United States—all point to the essential timeliness of Middle America in future world affairs.

PHYSICAL ENVIRONMENT

THE MAINLAND

With the exception of the Caribbean part of the region, Middle America can be viewed as a huge, north-south peninsular extension of North America. Baja California, separated from the mainland by the Gulf of California, is, of course, a true peninsula (800 miles; 1280 kilometers long), as is the Yucatán portion of Mexico, which extends into the Gulf of Mexico. Beginning as a relatively wide landmass at the U.S.-Mexican border, the mainland portion of the region gradually tapers to a thin band of land by the time it reaches Panamá, more than 3000 miles (4800 kilometers) away. Immediately prior to the point where Panamá (Central America) joins Colombia (South America), the region is at its narrowest, a mere 50 miles (80 kilometers) across. It is hardly surprising that this location was selected for the Panama Canal, connecting the Pacific Ocean with the Caribbean Sea. Some people believe that when the canal reverts back to Panamanian ownership at the end of the century, this isthmus may lose some of its strategic and economic importance, and an alternative canal may be proposed. Built long before today's supertankers existed, the canal is far too small to accommodate many ocean-going vessels, and it requires continual maintenance to keep it open. Much will depend on Panamá's ability to keep the canal operable and free of hostilities.

A large percentage of the mainland is composed of highlands, while the eastern coastlines and portions of the west coasts are flattened lowlands. For instance, the Yucatán Peninsula, most of the east coasts of both Mexico and Belize, the Mosquito Coast of Nicaragua and Honduras, and the west coasts of Panamá, Nicaragua, Costa Rica, and Guatemala are all lowland areas, lying at or near sea level. In contrast, Mexico City sits well above one mile (1.609 kilometers) high amid the Mexican Plateau, an uplifted and tilted range of mountains that forms one of the major physiographic regions within Middle America. Here one encounters perennially snow-covered volcanic mountains like Citlaltepetl (Orizaba) (18,701 feet; 5704 meters)—the highest point in all of Middle America—and the twin peaks of Iztaccihuatl (17,343 feet; 5286 meters) and the majestic Popocatepetl (17,887 feet; 5452 meters). The highland portion of the mainland is prone to earthquakes, some of severe magnitude. Northern Mexico, outside the Tropic of Cancer, is generally dry and less humid than the remainder of the region. Much of this area has been scarred by deep canyons, or *barrancas,* the most spectacular being *Barranca Del Cobre* (Copper Canyon) in the state of Chihuahua.

Most of the mainland lies south of the Tropic of Cancer, so the majority of Mexico and all of the Central

The Panama Canal viewed from the Pacific Ocean side. The canal was completed in 1914 after a formidable number of engineering and technological obstacles were overcome. The United States was to control the Canal Zone in perpetuity, but under a 1977 agreement with Panama, it is conducting a staged withdrawal from the Canal Zone and plans to have totally withdrawn from the Panama Canal itself by the year 2000.

Spatial connectivity usually implies the interaction or movement of goods, money, and services, but people (and their ideas and ethnic habits) are part of the concept as well. International migration is an old theme in geography, and the flow of numerous refugee groups within Middle America provides one example of how these countries are linked to one another. In this regard, it is especially useful to examine the refugee population of Mainland Middle America.

Without exception, all mainland states harbor refugees from all other mainland states. The situation was on the increase throughout the 1980s due, mostly, to the civil war in Nicaragua; the fighting, however, has now stopped. But before it did, Nicaraguan refugees had fled to El Salvador and Salvadoran refugees to Nicaragua. Indeed, the United Nations High Commissioner for Refugees (UNHCR) had targeted Central America as one of the world's four "hot spots" in terms of refugees in crisis. In 1987, the UNHCR spent nearly $38 million helping refugees in Latin America, the largest portion going to Central Americans. They assisted over 120,000 refugees in Central America alone, most of whom were from Nicaragua, El Salvador, and Guatemala. However, since many refugees were either reluctant or afraid to seek government aid (by officially registering with the UNHCR in their new country of residence), this figure represented only a small portion of the actual refugees in the area at any given time. Estimates at that time suggested that a more realistic figure might have been at least 300,000 in Central America, perhaps as many as 2 million in the entire region. The majority of those

MIDDLE AMERICAN REFUGEES AND SPATIAL CONNECTIVITY

Central Americans who did register, nearly 70,000, settled in Honduras, while another 44,000 (mostly Guatemalans) settled in Mexico. The figure of 44,000 represented known, registered Guatemalans in Mexico; Mexican government officials estimated at that time that the actual figure was closer to 100,000, and they made it clear that they would like all refugees who had not registered to leave. But, an increasing number of Guatemalan refugees in Mexico have had children, who, legally, are now entitled to Mexican citizenship. This, of course, complicates an already confused situation. Thus, even though the war has ceased,

many problems remain, particularly the aftermath of the refugee problem.

While a significant number of refugees resulted from the war in Central America, others have fled oppressive conditions and corrupt government practices. Some have opted to return voluntarily (to repatriate), whereas the majority are seeking asylum in their new (host) nations.

Typically, refugees are thought of as camp dwellers who stay put in one place until they either return home or obtain permanent citizenship status in their host country. This is not always the case. In fact, in Central America this is far from the truth. There, many refugees change their location on a frequent basis. Some choose to return home to aid other family members, and some, fearful of being caught, maintain a seminomadic life-style, regularly moving from camp to camp to avoid capture. Thus, the movement of "illegal immigrants" across the Middle American borders is great and continual. Mexico alone contains perhaps 200,000 registered refugees, ranking eleventh worldwide. There are far more "illegal" refugees in Mexico than those recognized as legal by the UNHCR.

Refugees in Middle America are not limited to the mainland, however. The Caribbean is burdened with refugees

TABLE 1

Refugees in Middle America, 1987

COUNTRY OF REFUGE	REFUGEES	COUNTRY OF REFUGE	REFUGEES
Belize	9,000	Honduras	68,000
Cuba	2,000	Mexico	175,000
Costa Rica	31,000	Nicaragua	8,200
Dominican Republic	6,000	Panamá	1,200
Guatemala	12,000		

Because of political turmoil in the region, refugees are found in every Middle American nation. These are Miskito Indian refugees at Tapamlaya Camp, Honduras, in May 1986.

from a great many countries as well. The largest number are from Haiti, which has one of the worst human rights records in the world. Thousands of Haitians have fled the country in search of more humane living conditions. Their situation is tragic at best, forcing scores to leave for literally any country willing to accept them. Estimates suggest that over one million Haitians—more than one-sixth of Haiti's total population—are in exile. Most have fled to the United States and Canada, while a sizable number are in the neighboring Dominican Republic. Other Haitians can be found throughout the region and beyond.

Statistics are particularly vague for Haiti, however, because the largest number left illegally—by boat or raft or as stowaways—so they are extremely reluctant to register when they arrive in a new home. Haitian "boat people" are also sensitive to the fact that, although it is prohibited by United Nations sanction, the U.S. Coast Guard has repeatedly sent Haitian refugees back to Port-au-Prince when they are discovered in U.S. waters. Since the departure of Jean-Claude "Baby Doc" Duvalier in 1986, a small percentage have repatriated, but the vast majority, fearful of reprisals (particularly by the hated *Tonton Macoutes*) and worsened conditions, have little desire to return in the near future. Meanwhile,

many continue to seek permanent residency in some country, necessitating that they remain "in transit," crossing numerous borders, in search of an eventual home. Leslie F. Manigat, one of Haiti's most recent presidents, was elected on his promise to normalize civil conditions and end the flow of refugees. But Manigat was ousted in mid-1988 and, as of this writing, the military was in power.

Table 1 lists the total number of "legal," UNHCR-assisted refugees in Middle America as of December 1987. Be aware that the total number of *all* refugees is much higher, and that the figure changes monthly; there are probably some in every Middle American country.

American republics are dominated by tropical climate and vegetation patterns. Exceptions to this pattern include the highlands, Baja California (desert conditions), and north central Mexico, where the land is wide enough to be affected by its own continentality (i.e., distance from the moderating effects of large water bodies), which results in the prevalence of large, semiarid steppe lands.

Mexico, by far the largest nation of Middle America (761,600 square miles; 1,971,782 square kilometers), is mostly a highland country where variations in elevation are more important indicators of weather and vegetation than changes in latitude or longitude. The relatively high elevations (above 6000 feet; 1830 meters) are referred to as the *tierra fria* (cold country), the middle elevations (between 2000 and 6000 feet; 610 and 1830 meters) are called the *tierra templada* (temperate country), and the lower, hotter parts of the country (generally, below 2000 feet; 610 meters) are known as the *tierra caliente* (hot country). The *tierra fria* extends up to the snow line and includes alpine grasslands and a variety of coniferous forests. The *tierra templada* is where much of the Mexican coffee crop is produced, and, taking advantage of the more moderate climate, this is also where the majority of the residents live (as did the Aztecs before them). The *tierra caliente* is mainly restricted to the arid and semiarid parts of northern Mexico and to the lowland, humid sections of eastern Mexico. Finally, there is the *tierra helada* (frozen country). This refers to those areas lying well above the *tierra fria,* a common rule being those places above 12,000 feet (3660 meters). In South America, these extremely cold areas are known as the *paramos* lands. As one might reasonably expect, the *tierra helada* includes only the highest elevations of the region, and, due to a lack of moisture and almost continually chilling temperatures, the variety of plant life is meager, especially on the shaded sides of mountains where there is little direct exposure the sunlight. Taken together, this elevation system, or division of the earth into a series of zones, is sometimes called **vertical** or **altitudinal zonation.**

The soils on the mainland vary considerably, mostly due to climatic, slope, and vegetation differences, as well as changes in elevation. In the wet and dry lowlands, for example, grayish brown alfisols are common, particularly on the gulf coast of Mexico along the Bay of Campeche. Farther east, however, in the predominantly flat, limestone portion of the Yucatán Peninsula, the soils turn to mollisols and become darker in color.

The soils in the dry, northern reaches of Mexico are typically aridisols, where some surface salt deposits (associated with this soil type) can be found. Most of the highlands of Mexico and Central America, on the other hand, contain so-called mountain soils, which tend to exhibit remarkable diversity from locale to locale. Mountain soils are less developed because they are relatively recent in origin. A thin strip of saturated inceptisols is found along the Mosquito Coast of Nicaragua, and farther south, around the Isthmus of Panamá, the soils turn reddish yellow in color and are mainly ultisols.

Mention of the tropics frequently brings to mind humid rainforests abounding with dense, luxuriant plant life; huge, glimmering, crystal-blue lakes; magnificent, snake-filled rivers carrying torrents of water; and numerous exotic forms of insects and wild animals, all living on or near the water's edge. In truth, although the mainland segment of Middle America lies almost entirely within the tropics, it is crucial to understand that there is rather little standing water throughout the eight contiguous states. There are few mighty rivers, only a modicum of the thickly clad forests often associated with the tropics, and far fewer plant, animal, and insect species than one typically encounters in the more vibrant rainforests of South America. The only significant bands of true tropical rainforest exist on the east coasts of Costa Rica and in Mexico, south of the city of Veracruz. From a global perspective, the region has only one world-class lake and no major rivers, though there are some modest rivers, and consequently a moderate amount of hydroelectric potential, in Costa Rica, El Salvador, Honduras, and Panamá. In the southwest corner of Nicaragua lies Lake Nicaragua, the largest body of inland water in all of Middle America and the twentieth largest natural lake in the world. It is nearly three times the size of the state of Rhode Island. The surface elevation of Lake Nicaragua is 105 feet (32 meters) above sea level.

The mineral resources of the mainland are diverse and scattered. Certain things are ubiquitous, such as limestone, especially in the Yucatán area and northern Guatemala; some minerals, however, are far less abundant, such as mercury, which is found principally in Honduras. Other exploitable minerals are fairly common in a number of the mainland republics. Copper, in particular, is found in worthwhile quantities in Mexico, Guatemala, Panamá, and Honduras. The Panamanian copper deposits are supposedly quite substantial, but, as yet, have not been mined to their anticipated poten-

Although more tropical rainforests exist in South America than in Middle America, deforestation is a problem in both areas as this photograph from Honduras indicates.

tial. Panamá is also endowed with manganese ore and iron ore, and smaller fields of iron ore exist in Honduras. Some bauxite is mined in Costa Rica, and there are several respectable nickel-producing areas in Guatemala. Gold, of varying quality, is mined and exported from Nicaragua, Honduras, Mexico, and Costa Rica. While both Honduras and Nicaragua have significant silver lodes, none compare to the large silver reserves found throughout Mexico. Almost immediately after the first Spaniards to arrive exhausted the rich gold mines they "discovered" upon conquering the Aztecs in Mexico, they exploited the even richer silver deposits of that nation—and some proved to be quite profitable indeed. Several Mexican cities are famous for their silver mines and silver craft items, the most notable being Durango, San Luis Potosi, Zacatecas, Guanajuato, Queretaro, and Pachuca. By some estimates, Mexico is the world's largest producer of silver, most of which is of fine, highgrade quality.

No discussion of mainland mineral reserves would be complete without mention of petroleum, especially the ever-burgeoning Mexican oil and natural gas industry. Mexico's only mainland "competitor" in the oil business is Guatemala, but the Mexican government is hardly worried about its so-called competition. Estimates change yearly, but, as of 1985, Mexico had proven reserves nearly 20 times those of neighboring Guatemala. Oil has been flowing from Mexican wells since early in this century, but it was only during the 1960s and 1970s that the country's probable potential was fully realized. Located mainly on the east coast and in the adjacent Gulf, Mexico, to everyone's surprise, continues to discover additional oil fields almost yearly. Some geologists suggest that the Gulf may in fact prove richer than the largest known fields yet discovered in the Middle East.

Lastly, it is important to note that two Central American states lack almost all valuable minerals. Neither Belize nor El Salvador has substantial quantities of any single mineral, and both lack commercial quality minerals that might otherwise justify exploitation. Both rely extensively on their forest reserves as a source of revenue, yet overcutting in Belize has greatly diminished its forestry potential, while unchecked soil erosion in El Salvador, due to poor (in some cases, nonexistent) management techniques, has nearly terminated its entire forest industry. At the opposite end of the spectrum, Guatemala is fortunate to have a number of rare, precious minerals, especially tungsten, which is coveted worldwide to harden steel, and antimony, which is used in metal alloys and in some medicines.

THE CARIBBEAN

As might be expected, there are substantial differences between the natural landscapes of the mainland and the Caribbean portions of Middle America. For example, unlike their mainland counterparts, the islands in the Caribbean are all fairly small in size, and, because the major ones all lie at the same general latitude, they experience similar weather conditions and climatic patterns. In short, the variety of weather types and natural vegetation found on the mainland is nearly absent in the Caribbean. The topography, however, varies considerably from island to island.

Cuba, the largest of all the Caribbean islands, is about the same size as the state of Pennsylvania and is the fifteenth largest island in the world. It is relatively flat with rolling hills and few notable mountainous areas. An exception is found along the extreme southeastern coast, where the Sierra Maestra Mountains parallel Guantanamo Bay and face neighboring Haiti. The Bahamas, comprising more than 700 islands of varying sizes in a chain that extends about 760 miles (1216 kilometers) from the northwest to southeast, are virtually featureless, with no significant highland areas over 200 feet (61 meters) above sea level. In contrast, Jamaica, Hispaniola (the island that consists of the two independent countries of Haiti and the Dominican Republic), and most of the Lesser Antilles (both the Leeward Islands and the Windward Islands) are studded with mountain peaks. In fact, the western part of the Dominican Republic and the whole of Haiti are highland areas, with little available flat land for agriculture. Jamaica's topography is similar, and several of the islands, including Puerto Rico, Dominica, and St. Vincent, for instance (the latter a mere 150 square miles; 388 square kilometers in area), are practically all mountains with only pockets of lowlands.

As islands surrounded entirely by the sea and with little landmass to protect them, the Caribbean nations are regularly susceptible to extreme, frequently harmful weather conditions. Hurricanes, all too often of severe magnitude, occur almost every summer (from June through October) in the Gulf of Mexico, and the Caribbean islands consequently suffer repeated, sometimes brutal, damage. The islands of Hispaniola and Cuba seem to be perennial targets, and their respective economies reflect this annual destruction of crops, homes, and public property; Hurricane Gilbert caused terrible damage to Jamaica in the summer of 1988. Likewise, due to their location and relatively small size, the islands tend to experience fairly uniform temperatures throughout the year. Thus, the variety of climatic types associated with Mainland Middle America is greatly reduced in the Caribbean; and none of the islands has enough landmass to result in even a moderate degree of continentality. One result of the stable climate type and uniform temperatures is less overall variety in the natural vegetation. There are, however, with some exceptions, detectable differences between the windward and leeward sides of the islands. Hence, most of the Caribbean can be considered tropical, with the exception of the extreme highland portions. There are no major inland water bodies or large rivers throughout the Caribbean, and fresh drinking water can consequently be a prob-

lem on some islands. This can become most acute following a hurricane.

The soils of the Caribbean are far less variable than on the mainland, mostly as a result of their origin. The islands are recent (geologically speaking), so the soils are typically mountainlike in nature, which indicate a relatively rapid rate of erosion. The vast majority of the islands have moderate to poor quality soils, generally a mixture of ultisols, alfisols, and entisols. One major exception, however, is Cuba. Throughout most of the island the soils are inceptisols, with a dull, brownish red hue. This long band of inceptisols results from a lack of surface relief on the island, which consequently produces a more uniform soil type.

Although one might assume that, due to their close proximity to one another, the Caribbean nations would have similar mineral deposits, this is not at all the case. The natural mineral resources of the Caribbean vary tremendously from island to island, with little or no apparent pattern. There are small reserves of oil and natural gas in Barbados and Trinidad and Tobago, and considerably less in Cuba and Jamaica. Gold is found in the Dominican Republic and Haiti, but not in significant amounts. Likewise, the Dominican Republic has some silver deposits, but they approach neither the quality nor the quantity of those found in Mexico. Cuba is a major producer of nickel, with lesser deposits located in the nearby Dominican Republic. Jamaica is one of the world's major sources of bauxite and alumina, and there are smaller but noteworthy amounts of bauxite in both Haiti and the Dominican Republic. Cuba contains moderate iron and manganese ore fields, and the Dominican Republic has the only exploitable mercury within the Caribbean. Copper is present in limited amounts in Haiti, Cuba, and the Dominican Republic, but the mines are nowhere near as large as those found in other parts of Latin America, most notably Chile.

The Dominican Republic has a considerable variety of valuable minerals, but most are either fully owned or controlled by multinational corporations, which are mainly based in the United States, Canada, and France. Therefore, little of the profit from these resources remains on the island. Finally, several additional precious minerals of high dollar value are found in the Caribbean, but most exist in limited known quantities. Chromium and cobalt, which can be found on Cuba, and aragonite, a special form of limestone mined in the Bahamas, are examples.

Some states, however, are unfortunately deficient in

almost all important minerals. Grenada, for example, has practically no minerals of sufficient quality or quantity to justify their exploitation. This makes it exceedingly difficult for such countries to cope, let alone compete, in a world where finite natural resources will come to mean more and more as global reserves continue to diminish year by year.

HUMAN ENVIRONMENT

Historical circumstances and physical geography have combined to create a rather unique blend of human activity in Middle America. Today, virtually all aspects of human initiative in the region, in the Caribbean and on the mainland alike, reflect a rich heritage of diverse cultures (both past and present) plus the variety of natural environments found within this part of the world. Yet, as we enter the 1990s, the human landscape of Middle America is undergoing a remarkable transformation. As Western technology and ideas diffuse throughout the region, change is the order of business, especially in the urbanized areas. The pace of life and the means to alter the region's standard of living have both increased dramatically, and, as a consequence, changes that were unexpected even 25 years ago have now become commonplace. Few parts of the region are unaffected by foreign influences, and this is perhaps most apparent in the agricultural and industrial sectors of the economy. It is also particularly visible in the largest cities of Middle America.

COLUMBUS AND THE "NEW" WORLD

The year 1992 will mark the quincentennial (500 years) of Christopher Columbus's supposed "discovery" of the New World. In truth, Columbus had no accurate sense of where he was on the globe. When he arrived in what came to be called the New World, he firmly believed that he had landed somewhere in Asia, probably India. In retrospect, it was therefore perfectly natural for him to call the native peoples he found "Indians." Of course, it was not India or any other part of Asia. In fact, it was nowhere near Asia. It was a part of the globe that few persons in Western Europe in the fifteenth century knew existed. He had landed in the Caribbean, but since he believed it to be India, several place-specific references were used, and many still remain; indeed, we

often refer to the Caribbean area today as the West Indies.

Columbus, and most of those who followed him, left a strong and lasting imprint on this "new" part of the known world. They gave a Spanish name to almost everything they encountered—the islands, the bodies of water, the mountains, the rivers, the plants, the animals, and the Indian villages. Today, those place-names and a lengthy list of Spanish words still exist on maps, in our vocabularly, and in the minds of the Middle American people. The Spanish also exported their ideas and worldviews to their colonies in the New World. They imposed their language and their religion on the native peoples, and the vast majority of all Middle Americans consequently speak Spanish and practice Catholicism today. Likewise, the cultural landscape of Middle America is still distinctively Spanish in flavor, representing the imposition of Spanish technology on the indigenous Middle American cultures. Spanish farming and mining techniques soon became the norm. Old World customs including dietary habits, styles of dress, musical preferences, religious celebrations, medical practices, and many forms of entertainment and children's games quickly became a part of the Indians' lives. Even the former Indian settlements began to assume a more Iberian character following the Spanish conquest. Streets were laid out in even, rectangular grids, and a central plaza was erected in nearly every Middle American town and city. The indigenous communities were either brutally destroyed by disease, warfare, or overwork, or buried beneath a new, Spanish-style settlement. In short, one might say the newcomers effectively transplanted a large portion of Spanish culture overseas, most of which persists to this day.

We will probably never know precisely how many native peoples occupied the realm of Middle America at the time of Columbus's arrival, but we do have fairly good estimates based on historical, archival, biological, and archaeological evidence. It is believed that in all of Latin America there were as many as 100 million people at that time of initial contact with the Spanish in 1492, and that **Mesoamerica,** a term that refers to the high-culture portion of Middle America, undoubtedly was one of the most densely populated portions of the entire New World. Recent findings suggest a figure of about 40 million people for all of Middle America, but that figure has changed dramatically over the years, and it tends to grow as research progresses and new evidence is uncovered. Without a doubt, however, the Mexican part of Middle America (exclusive of Baja California) contained the vast majority of people, a somewhat modest estimate being somewhere between 25 and 30 million Indians.

Most of the indigenous peoples of Middle America lived in the highlands of central Mexico, in the general vicinity of present-day Mexico City. Residing south of the Tropic of Cancer, mainly between 18 and 22 degrees north of the equator, the Aztecs understandably chose to settle in the cooler, high-altitude areas, where escape from the tropical, often torrid, climate of the lowlands was a welcome relief. The Aztecs were the most numerous of all Middle American native peoples and, by most accounts, were the most urbanized and the best organized as well. To the south of the Aztecs, in the humid lowlands of the Yucatán Peninsula and farther south, in what today constitutes the northernmost republics of Central America, were the Maya. At the time of European contact, the Mayan peoples do not seem to have been as efficiently organized nor as urban oriented as the Aztecs, but they may have been more advanced in terms of their technological achievements. Some argue that they were the most technologically advanced of all New World Indians and possessed a highly sophisticated system of scientific learning.

In the Caribbean portion of Middle America, the natives were far less numerous and apparently even less urbanized than both the Aztecs and the Maya. The two largest culture groups of the island Indians (at the time of European contact) were the Caribs, a comparatively "primitive" people—some claim they were cannibals—and the more densely settled Arawaks, a decidedly more sophisticated society, with an economy organized around agriculture and stable village life. The Caribs occupied what is commonly referred to as the **Lesser Antilles,** whereas the Arawaks were chiefly located in the **Greater Antilles,** primarily on the islands of Cuba, Hispaniola, and Puerto Rico. Together, the Arawaks, the Caribs, and several smaller groups probably numbered between two and three million people in 1492. Again, no one knows for certain how many there were, but these are the best estimates at present.

Irrespective of which group the Spanish encountered in Middle America, most were significantly reduced in size either due to warfare or from exposure to European diseases (to which they had no immunity), most notably smallpox. Former Indian hamlets and villages or, in the case of the Aztec Empire, former cities began to assume a Spanish form and character in very short order. Almost without exception, the newly named Spanish towns and

A portion of Palenque (Mexico), a ruin from the Classical Maya Period. At its height (A.D. 200s to the 800s), Mayan society may have included about 2 million persons. The Maya achieved outstanding success in astronomy and arithmetic. Today, more than 1.5 million descendants of the Maya live in the area where the Maya civilization flourished.

cities were built on either the dwelling or ceremonial sites previously occupied by the various Amerindian peoples. In fact, a map of current Middle American settlement patterns provides a good approximation of where the pre-Columbian Indians lived, the relative size of their respective influence on other Indians, and where and with whom they regularly traded.

Descendants of most mainland native groups still exist in Middle America, but the contemporary cultural landscape of this region, particularly in the urban areas, is decidedly Spanish-like in nature. Indeed, to the unwary tourist, there is little to indicate that the entire area was formerly under total Indian domination and once reflected Mesoamerican cultural values. To the untrained eye, the native influence is conspicuously absent except for numerous archaeological ruins, and in the more rural hamlets and villages.

In the islands, a different situation produced a different result. By and large, the island Indians were effectively eradicated due to disease and/or warfare. A small number escaped, but the vast majority died and little evidence of their influence, or even of their existence, remains in the Caribbean. To compensate for their rapid demise—those Indians who survived were used as slaves by the Spaniards throughout Latin America—a far greater proportion of African slaves were brought to the islands than were brought to the mainland, and they arrived in the islands long before their arrival on the mainland. Indeed, more Africans were transported, involuntarily, to the Caribbean than to any other single area of the western hemisphere; according to some estimates, Caribbean-bound slaves numbered more than 40 percent of the entire Atlantic slave trade. Transplanted Africans, mostly from the Ivory Coast region of West Africa, very quickly became a significant element of island society. Their legacy persists throughout the Caribbean today.

DEMOGRAPHIC CHARACTERISTICS

When looking at the population characteristics, or demographics, of Middle America, it is a good idea to separate Mexico from the remainder of the region. Mexico contains far more people than the other nations. Indeed, Mexico's population is significantly greater than all other Middle American countries combined and, due in part to its history and in part to its geography, it has a somewhat unique blend of peoples and ethnic groups. Mid-

dle America minus Mexico is sometimes regarded as the ***Caribbean Basin*** because all of the countries except Mexico and El Salvador touch, lie squarely within, or are economically oriented toward the Caribbean Sea. Due to historical precedent, and because of continuing economic relationships, the South American countries of Guyana and Suriname are also generally included in basin statistics. However, for the purposes of this chapter, Guyana and Suriname will not be included in the Caribbean Basin.

Caribbean Basin

A glance at the Statistical Profile will reveal some important characteristics about the Caribbean Basin. The total population is about 63 million people, but the anticipated population, by the year 2000, will be over 76 million, an increase of about 13 million. Fortunately, the majority of these nations have their population problems under partial control, and, by the latter years of this century, most of the population reproduction rates should fall correspondingly. If this prediction fails, however, then the Caribbean may grow substantially beyond its carrying capacity, at which point the population will exceed the ability of the land to provide adequate foodstuffs for the total number of inhabitants. Obviously, some countries will be able to curb their growth more effectively than others. For example, the countries of Costa Rica and Trinidad and Tobago are anticipating a mere 0.15 percent drop in their reproduction rates. One immediate answer to the population issue is, of course, birth control. In this regard, it is noteworthy that Costa Rica (the most European of all Middle American states) and Panamá (the nation that is perhaps most strongly influenced by the United States) have the greatest use of modern contraception techniques of all Latin American countries.

In strong contrast, however, Cuba and Barbados are predicting a 0.06 percent and a 0.02 percent increase in their respective rates of reproduction by the year 2000. Barbados is a small, densely populated country, so a larger population will pose a detriment in its ability to feed itself. Cuba, with a socialist, Soviet-dominated government, is less concerned about population control than the Western nations; communist ideology views human resources as the backbone of the communist cause: the more people the better. It is interesting to note, however, that Cuba, which was the most populated nation in the Caribbean Basin in 1990, will likely drop to number two by the end of the century, as Guatemala forges into the dubious lead. Note, also, that Dominica, Netherland Antilles, and Puerto Rico have the highest life expectancies in the entire Caribbean Basin. It is also significant to note that Nicaragua (3.4%) and Guatemala (3.2%) are presently experiencing the greatest rates of natural increase.

Although the developing world is often perceived to be overpopulated, it is important to note that many of the Middle American nations are nowhere near what might be considered heavily populated. For instance, note that Grenada, Belize, the Bahamas, Barbados, and seven other countries have less than half a million people each. Even by the century's end, they together will contain well under two million total inhabitants (about the same as Houston, Texas). Recall that the Bahamas contain over 700 islands. This means that the population density there is extremely low; thus, while the physiologic density (the ratio of people and animals to arable land) is high, the carrying capacity (the potential productivity of land, which varies according to technology) is probably not in any immediate danger. Hunger frequently results from a combination of technological, cultural, and outside (foreign) factors.

While the overall population of the Caribbean Basin will grow from 63 to 76 million for an increase of about 21 percent by the beginning of the next century, the urban areas will experience only modest growth. The switch from an agricultural society to an industrial economy normally necessitates a fairly large urban work force to accommodate the new, mainly urban job opportunities. This is a pattern found the world over, and it partially accounts for the massive rural-urban migration occurring in many developing nations. Although many Middle American nations experienced rapid urban growth during the 1960s and 1970s, the flow of people into the cities slowed, for the most part, during the late 1980s. This growth is expected to ease further during the 1990s, as word filters back to the countryside that urban jobs are far more scarce than many believed. The urban population throughout the basin is expected to account, on average, for approximately 58 percent of the total population by the year 2000. Consequently, Middle America's anticipated urban population (and associated problems), with exceptions, should be less severe than many regions of the world will experience by the next century.

Exceptions, as always, will occur. For example, Guyana, Haiti, Barbados, and Trinidad and Tobago each had urban populations in 1990 that were relatively small when measured as a percentage of their respective total populations—26, 32, and 34 percent respectively. By 2000, this situation is not expected to change dramatically. This is especially true for Barbados, and Trinidad and Tobago, where these countries should still see about two-thirds of their respective populations living outside the urban areas. Haiti is a special case. As the poorest nation in the Caribbean, with the lowest life expectancy in all of Middle America, a high rate of natural increase, and a substantial increase in urban growth expected by 2000, the slums of Port-au-Prince, already miserable by world standards, will certainly grow worse.

At the upper end of the spectrum are countries like Jamaica, Honduras, Guatemala, and the Dominican Republic. These states anticipate significant urban growth by the end of the century, and their ability to cope with that growth will, in part, determine their future socioeconomic well-being. The state of Grenada, consisting of the main island and a chain of northern smaller islands approaching St. Vincent and the Grenadines, is expecting a massive urban influx exceeding the percentage increase in the region as a whole. Industry has come to Grenada only recently, and Grenadians are looking ahead to 2000 when their urban areas will nearly double in population. Clearly, urban planning is a necessity if this tiny Caribbean country is to survive such tremendous population shifts.

Mexico

As we noted earlier, Mexico is significantly different from the remainder of Middle America in a number of ways. It had a total population of about 81.9 million in 1990, more people than the forecasted sum for the entire Caribbean Basin by 2000. Current best estimates suggest that an additional 23 million (or more) people will be added to the Mexican census by the year 2000. However, the average annual natural increase of the nation has been steadily decreasing since about 1975, and it should decrease even more sharply between 2000 and 2005. Rapid growth was a dire problem in Mexico during the latter part of the 1960s and the early and mid-1970s, but constantly improving educational programs, a greater number of family planning clinics, and federally subsidized birth control measures have successfully

Foreign influence can be seen throughout Middle America. This is a planned barrio on the southern edge of Tegucigalpa, Honduras.

contributed to bring an end to a potential crisis. It is now hoped that other Middle American countries will follow Mexico's lead and achieve substantially lower growth population rates by the new century.

Another difference between Mexico and its Middle American neighbors is the location of the Mexican population. Previously, we observed that present-day Middle American population clusters generally correspond to pre-Columbian settlement patterns. This is especially true for Mexico, but less so for the Caribbean where the indigenous peoples were ostensibly eliminated shortly after European contact. Much of Mexico's population is concentrated in and around Mexico City, and that area experienced a large increase in population during the past couple of decades. It is estimated that the current population of the metropolitan area of Mexico City is about 18 million. Mexico City is located on the site of Tenochtitlan, the ancient Aztec ceremonial center situated in a mountain valley on the Mesa Central and home to nearly one-quarter of a million native peoples during the fifteenth century. Today, that same approximate part of Mexico—the adjoining states of Puebla, Mexico, and the Federal District (Districto Federal)—is home to nearly 25 million Mexicans, close to one-third of the national population, and is growing.

Traditionally, Mexico has been a rural nation, but in recent decades the urban centers have grown out of proportion to the rest of the country. Peasants abandoned their traditional, farming life-styles in tremendous numbers (due to their swiftly growing population base); many left their villages in a desperate search for perceived employment opportunities elsewhere. Tens of thousands of migrants flocked to the cities in search of work—any type of work—and improved educational prospects. Greater Mexico City, greater Guadalajara, and greater Monterrey, together, recorded over one-fourth of Mexico's population in the 1970 census. But, as word filtered back to the countryside that the urban prospects were far less attractive than anticipated, if not worse, this rapid rural-urban migration pattern seemed to be on the wane in the late 1980s. Although the tide has slowed, the cities are nevertheless overcrowded and they continue to grow. Mexico City, by some estimates, has already reached its upper limits. A true *primate city,* with about seven times the population of the second largest urban center (Guadalajara with an estimated 3.3 million in mid-1990), Mexico City is now generally regarded as the world's most populated city with a metropolitan population of over 19 million in 1990. It is easy to see why urban analysts are currently predicting

Much of Middle America is dominated by primate cities that also function as capital cities. Mexico City is a typical example of this pattern.

a turn-of-the-century population of over 20 million or more. However, other researchers monitoring Mexico's population consider this a gross underestimate. Other densely settled cities include Monterrey, Ciudad Juarez, Tijuana, and Léon, all located north of the Mesa Central.

CULTURE

The ethnic make up of Middle America is complex, the region is intricate and vast, and it is therefore extremely difficult to form universally acceptable statements about the culture. Another geographer writing on the complexity of the area's culture had this to say:

> Culturally, Middle America includes a variety of ethnic, social, economic, and political patterns rivaled by virtually no other area of comparable size in the world. Political processes range from the functional democracy of Costa Rica to rule by presidential whim in Haiti and the "sugar communism" of Cuba. Despite the trend of political independence among Caribbean dependencies, some territories continue to have varying degrees of political association or partnership with outside powers such as Britain, the Netherlands, France, and the United States. In addition to the

Papiamento tongue of some of the Netherlands Antilles, there are at least three major linguistic communities—Spanish, English, and French—plus a proliferation of local dialects in Middle America.*

Thus, Middle America might be considered one of the most culturally diverse regions on the planet. Accordingly, to simplify matters, we will necessarily have to generalize about the people and their distinctive ways of doing things.

Language and Religion

As in most of Latin America, the Spanish language is commonly identified with Middle America. Spanish is the first and major language of education, and it serves as the lingua franca of the citizenry. Moreover, it is the "official" language of a majority of Mesoamerican states as well; as an official language, it is required in all government matters, as defined by local authorities. As Table 10.1 demonstrates, however, Spanish is not the official language of every country; it is a second language in some; it is not even spoken in others. Of the 24 states listed in Table 10.1 (some of these "states" are actually dependencies, possessing various levels of autonomy and independence), an equal number (9) are officially either English- or Spanish-speaking, 3 are constitutionally Francophone countries, while 1, the Netherlands Antilles (consisting of two groups of three islands each, but organized into four self-governing communities, namely, Aruba, Bonaire, Curaçao, and the Leeward Islands) uses Dutch. The officially recognized language in any given country is almost entirely a reflection of that state's colonial history, its current dependency status, and/or its relative (or perceived) position in international trade. In time, a lingua franca can occasionally become an official language.

Language is sometimes considered a good barometer of diffusion, which means the spread of an idea or concept from point A to point B. Spanish, of course, diffused to the New World from Spain in the fifteenth century. Today, Spanish is one of the major international languages, spoken by nearly 300 million people worldwide; in fact, there are more Spanish speakers in Mexico alone than in Spain. But Spanish varies considerably

*Robert C. West and John P. Augelli, *Middle America: Its Land and Peoples,* 2d ed. (Englewood Cliffs, N.J.: Prentice-Hall, 1976), p. 2.

from place to place, and the version of Spanish spoken in any particular part of Middle America is a fairly good clue to where the original Spanish landholders who settled in that vicinity came from originally—i.e., which part of Spain they emigrated from. Research based on the pronunciation of Spanish in the New World has shown, for example, that the first Europeans to arrive were primarily from the coastal, Mediterranean parts of Spain, while later conquistadors came from the northeastern and northwestern corners of Spain. Today, these respective language differences remain virtually extant, and the different dialects found in Middle America are in part a reflection of the diffusion of different forms of Spanish from different parts of Spain. Likewise, the mixture of Spanish with the local Indian languages also helps to account for regional variations.

Geography, in particular, elevation, is another factor that helps determine regional differences in dialect. Highland Spanish (often called Serrano) and lowland Spanish are significantly different throughout Middle America, and some linguists attribute the distinctions not to the origins of the first Spaniards alone, but also to the elevation and its effect on speech patterns. Most notably, their research has found that the oxygen content at higher elevations influences the pronunciation of certain consonants, thereby producing, over several generations, an overall sound quite distinct from the original lowland speech pattern. This helps explain why the highland peoples of Mexico and Central America sound relatively similar, whereas the lowland peoples of Middle America also seems to speak a related form of Spanish.

Amerindian languages were both varied and numerous. Throughout Latin America, as far as anyone can tell, some 2000 native languages existed at the time of first contact with Old World explorers, and perhaps 800 still exist intact today, the majority of them in South America. In Middle America, the figure is much smaller, but it should be remembered that it is practically impossible to give exact figures for the number of people who continue to use native languages as their sole mode of communication. Nahua or Nahuatl, the language of the Aztecs, is still used by about one million Indians in Mexico; perhaps a quarter of them continue to use it as their only language. The language of the Maya, spoken in southern Mexico and Guatemala, is even more common, with perhaps as many as two million people using one of the Mayan languages: mainly

TABLE 10.1

Major Middle American Languages

COUNTRY	OFFICIAL LANGUAGE	OTHER COMMON LANGUAGES
Antigua and Barbuda	English	Patois
Bahamas	English	Patois
Barbados	English	Patois
Belize	English	Spanish, Indian dialects
Costa Rica	Spanish	English, Indian dialects
Cuba	Spanish	English
Dominica	French	English
Dominican Republic	Spanish	French, English, Creole
El Salvador	Spanish	Indian dialects
Grenada	English	Patois
Guadeloupe	French	Creole, Patois
Guatemala	Spanish	Indian dialects
Haiti	French	Creole, Spanish, English
Honduras	Spanish	Indian dialects, English
Jamaica	English	Pidgin English
Martinique	French	Creole, patois
Mexico	Spanish	Indian dialects, English
Netherlands Antilles	Dutch	Spanish, English, and Papiamento
Nicaragua	Spanish	English, Indian dialects
Panamá	Spanish	English
St. Kitts-Nevis	English	Patois
St. Lucia	English	Patois
St. Vincent and the Grenadines	English	Patois
Trinidad and Tobago	English	Spanish, Hindustani, Hindi, French

Quiche, Yucatecan Maya, Cakchiquel, Mam, and Kekchi. Farther north, in the Mexican state of Veracruz, Huastec, another Maya language, is frequently spoken. Remnants of other pre-Columbian languages also survive in Mexico, including Zapotec, Otomi, Totonac, and Mixtec. In Central America, south of Guatemala, indigenous languages are spoken daily by a sizable number of people in El Salvador (Pipil and Lenca), Honduras (Sumo, Garifuna, Paya, Misquito, and Jicaque), Nicaragua (Misquito), Costa Rica (Bribri, Guaymi, Terraba, Cabecar, Boruca, and Guatuso), and Panamá (Guaymi, Choco, and Cuna).

In general, the Amerindian languages—whether unchanged from precontact times or, more commonly, a combination of native words and postcontact terms and phrases (sometimes called syncretism)—survived mainly in the hidden, most isolated portions of Middle America, where the mixing with Europeans was slight or nonexistent. Naturally, this tended to be the cool, remote highland areas and the densest, inaccessible jungles. On the islands and on the wide open bottomlands, however, where there was literally nowhere for the na-

tives fleeing European contact to escape or hide, the Indian languages have been almost entirely wiped out.

Besides speaking Spanish, the region's adherence to Catholicism is another major cultural trait of Middle America. Many other religions exist, however: evangelical Protestantism, which is generally associated with the American South, is finding a home in the mainland portion of the region, and there are even significant numbers of Mormons in northern Mexico. But Catholicism is, by far, the most common religion throughout the region. Western states, like the United States and Canada, are generally more secular than most developing countries, so religion tends to be less influential in the everyday lives of Western peoples. Many persons in the United States may attend church, some do so on a regular basis, and many even consider themselves religious in nature. Nevertheless, there is a difference in degree. Religion to most persons in Canada and the United States, for example, is a weekly event, something that we tend to ponder mainly on Sundays and on religious holidays. It is not a conscious part of our everyday lives. In the developing world, however, religious beliefs play a far more critical role in the life of the average person, and a large percentage of the populace practice some form of religion—whether animism, polytheism, or monotheism—on an almost daily basis. Middle America is no exception to this rule. The Spanish people brought more than their language to the New World; they brought their entire culture, and Catholicism was an integral part of fifteenth-century Spanish life. In consequence, Middle Americans are predominantly Catholic today, primarily Roman Catholic. Their Catholic heritage is evident in the landscape (an abundance of Catholic churches), the sanctified place-names, the numerous religious holidays, and, in some countries, even their currency.

A Catholic church in Atapec, Mexico. Roman Catholicism is the major religion in Mexico and the other countries in Latin America.

Indigenous religious practices do exist, but they are typically hidden, or partially buried, under the guise of Catholicism. To the trained observer, local ceremonies and icons reveal a litany of extant native influences camouflaged beneath a Catholic surface. In the islands, for example, especially in Haiti, Jamaica, the Dominican Republic, and Trinidad and Tobago, persistent remnants of African (and, in some areas, East Indian) religious practices are very common, often with a Catholic flavor. This is another form of syncretism, in which dual systems operate simultaneously, with both functioning as a single, unified system. What we said about language is also true for religion and, as might be expected, Catholicism is strongest where contact with indigenous peoples was greatest. Conversely, the remote locales in Middle America are still home to some native religious beliefs, and the Caribbean is a virtual mosaic of religions and religious sects.

The Ethnic *Mélange*

Ethnicity is very important in Middle America, less so than it was in historical times, but much more so than in the United States or Canada. Historically, a person's comparative status or standing in life was essentially determined by his or her birth, somewhat similar to the caste system, which is common throughout the Hindu portion of India. Purebred Spaniards held the highest esteem in old Middle America, followed by creoles, persons of Spanish heritage who had been born and raised in the New World. Next came the mestizos. **Mestizos** were individuals with one parent of European descent and the other parent a native Middle American. Pure Indians (Indios)* came next in this hierarchy, followed by Negroes at the very bottom of the social ladder. However, it was not as cut and dried as this generalized description might suggest, because Middle Americans also made important distinctions based on the precise ethnic and racial backgrounds of both the mother and the father. Obviously, in a society as heritage conscious as this, a multitude of permutations were possible. Moreover, years of intermarriage and interethnic interrelations led to widespread miscegenation throughout

---

*The term "Indian" denotes many things in Middle America. Depending on whether Indians are defined culturally, racially, or as a mixture of both, population totals will vary accordingly, and they will vary from country to country. In this chapter, the term is used in its broadest sense to refer to people of native, non-European descent.

the region. Table 10.2 shows the numerous categories used in Honduras to classify ethnicity. This Honduran example can be regarded as typical (with some exceptions) of Middle American society before the present century. A similar hierarchy with minor variations existed in the French colonies.

Today, the mestizo overwhelmingly dominates Middle American life, but the term needs to be used cautiously. Rather than maintaining its original definition, mestizo has come to stand for the "common man" (or half-breed, when used in a derogatory manner), since most Middle Americans possess some combination of Indian and European blood (if they trace their heritage back far enough). (In Central America, but especially in Guatemala, an Indian who has assimilated into the mainstream, Spanish-speaking culture is referred to as a **ladino.**) In terms of absolute numbers, Mexico has unquestionably the region's largest mestizo population. On the other hand, recall that it also contains the largest Indian population in Middle America (over 6 million). As a percentage of the total population, however, Honduras and El Salvador have the largest mestizo populations (92 percent and 89 percent, respectively) within the region; in contrast, about 75 percent of Mexicans are mestizos today.

Pure Europeans are no longer an important part of Middle American society, but this is not the case in every country. For example, approximately 90 percent of all Costa Ricans are of European heritage, a figure far greater than for any other single state. Cuba is second, with over 75 percent of its population of European de-

scent. In Costa Rica, this is partly due to the relatively small native population during the fifteenth century, and to the related fact that many early Spanish immigrants came as families and did not mix and marry as readily as their neighbors in nearby countries did. In Cuba, as is true of the entire Caribbean, it is a result of the inordinate number of Indians killed after European contact. As a consequence, African slaves were utilized considerably in the islands, and today blacks comprise a significant percentage of the population, while native peoples are effectively nonexistent.

Conversely, although pure Indians comprise only a minute percentage of the total population of the region, in some countries their numbers remain high. The best example is Guatemala, where good estimates suggest that essentially one-half of the present-day population is pure Indian. Throughout the Caribbean and all along the Caribbean coasts of the Middle American states, blacks of African heritage are found in significant numbers. In some states, such as Grenada, the Dominican Republic, St. Lucia, the Bahamas, Barbados, and Haiti, there is a preponderance of blacks, most notably in Haiti and St. Lucia where they number nearly 100 percent of the total population. Trinidad and Tobago, with an immense East Indian community and a sizable number of Muslims, is an ethnic kaleidoscope. In contrast, as mentioned earlier, very few Indians are found in the islands today.

In every case, whether current populations are predominantly European, Indian, African, mestizo, or some ethnic combination, it should be emphasized that they

TABLE 10.2

Ethnic (Social) Distinctions in Honduras (in Descending Order)

WHEN THE FATHER IS A:	AND THE MOTHER IS A:	THEN THE CHILD IS A:
Spaniard	Indian	Mestizo
Spaniard	Mestiza	Castiso
Castiso	Spaniard	Espanolo
Negro	Spaniard	Mulatto
Spaniard	Mulatta	Morisco
Morisco	Spaniard	Albino
Albino	Spaniard	Tornatras
Tornatras	Spaniard	Tent en el Aire
Negro	Indian	Lobo
Lobo	Indian	Caribujo
Lobo	Mulatta	Barsino
Lobo	Negro	Grifo
Indian	Mestiza	Chaniso

SOURCE: G. T. Kurian, *Encyclopedia of the Third World*, 3rd ed., (New York and Oxford: Facts on File, 1987).

are decidedly, and consciously, distinctive from their respective parent cultures in all ways, and indeed distinctive from one another. Costa Ricans, for example, though overwhelmingly European (Spanish) in origin, do not consider themselves Spanish; nor are they *quatemaltecos;* they are *costarricenses* first and foremost. Historical and geographical factors have come together to produce an array of peoples and nations, each with its own, characteristic culture complex. The fact that these similar yet distinct nations remain separate is somewhat a reflection of their peculiar ethnic identities and their sense of feeling unique from one another. Following in the footsteps of their North American neighbors, many persons in Caribbean and Central American countries are developing a strong and increasing awareness of their ethnic heritage. Mexicans are probably more conscious of their rich past than the others, but this is a rapidly growing trend in the region. Moreover, many Middle American states, even those with a strong European heritage, which meant so much in earlier decades, are today highly aware that they are no longer European in any sense of the word. Consequently, they are beginning to identify themselves by their nationality above all other considerations, past and present. A major challenge for the Middle American governments in the 1990s will therefore be to accommodate a rise in ethnic appreciation, while simultaneously trying to instill a broader sense of national pride.

THE POLITICAL CLIMATE

Generalizations about Middle American politics can be little more than gross approximations. There are no distinct patterns, and few countries have immediately comparable systems of government. Furthermore, some governments proclaim one system but practice something very different. Given their complex cultural makeup and their trend toward increasing heterogeneity, few states strictly conform to the Iberian political environment (which provided the basic model for most of Latin America). This is because many of these states have changed significantly since independence and continue to alter their political systems substantially. With the possible exception of Cuba, no government is similar to the Soviet political model, but even Cuba's system differs considerably from the Soviet model (which is also in a state of evolution at present). Several former British colonies possess parliamentary-like governments, but virtually no Middle American nation

strictly follows the American adherence to democratic, popularly, and freely elected officials. Many contend that the trend is toward democracy (throughout Latin America), and Costa Rica and Belize are usually held in high esteem as models of democratic-like governments to emulate. As we shall see, Costa Rica's democracy is most akin to the United States.

Perhaps the only consistency of the region is a tendency toward military strongholds, but here too we find exceptions. Most Middle American states exist because of a delicate balance between civil and military power, which is why coups are so common. A coup occurs when the military decides that the civil government is not functioning properly (meaning, usually, that it is not functioning the way the military would like it to function). The army tells the president to leave office (and the country) so that "peace and order" can be restored. In sharp contrast to coups in other developing states, however, those in Middle America are typically bloodless and are conducted with a minimum of violence; violence results only when the president puts up a fight or when the military is divided internally.

Thus, Middle American politics is a mosaic in which no single form of government prevails. It is a mixed bag: partly indigenous, partly inherited, partly borrowed, but always subject to change to accommodate unique circumstances. The states have generally worked against, rather than with, one another, so their political ties have been strained at best. Hence, it is quite difficult to paint a comprehensive picture of the average Middle American government. Yet, to leave it at that tells us very little; we will therefore need to make some broad generalizations and point out the extreme exceptions as well. No attempt is made here to describe each nation in the region, however: only the most important and those representative of the area will be discussed.

In one sense, these states are regarded from the outside as continually changing entities. Political leaders come and go, and we frequently hear about the military taking control in one country or another. As a result, we tend to think that the region is always in a state of change. Most recently, in addition to the somewhat chaotic conditions in some Central American states, we have read about the brutal killings in Haiti when that state tried to hold open elections. At present, the atmosphere in Panamá is similarly tense: President Eric Arturo Delvalle was deposed in February 1988 and General Manuel Antonio Noriega in 1989. But outside perception does not necessarily constitute reality. In fact, it

GEOPOLITICS: IDEOLOGICAL DIFFUSION IN MIDDLE AMERICA

Cultural diffusion refers to the spread of elements of culture such as ideas, innovations, and attitudes from the point of origin in one culture to other geographic areas within the same culture or to foreign cultures. The two important types of diffusion are *expansion diffusion* and *relocation diffusion*. In expansion diffusion, ideas and innovations are spread throughout a given geographic area in a contagious or snowballing fashion or in a hierarchical manner, where the idea or innovation leapfrogs from one important individual to another or from one organization to another. Relocation diffusion occurs when an individual or group physically moves to another area taking its idea or innovation along and spreading the innovation in a new geographic area. Ideology is an element of culture and can be diffused from one culture to another, just as other material and nonmaterial traits can be diffused.

Moreover, the diffusion of ideology can often have political consequences.

Soviet communism and American democracy have been quite energetic in their quests to win "converts" around the world, to their respective ideologies. Each country has tried to gain an effective foothold inside the other's sphere of influence and thus win the allegiance of the people in a given political entity. The United States, for instance, tries to make allies throughout the eastern hemisphere, while the Soviet Union tries its best at times to befriend as many countries as possible in the West. Both have the same goals, but the two have used different means to achieve their goals. Although this has happened in all parts of the world, Middle America remains a particularly good example of this geography of ideological diffusion despite the recently improved relations between the United States and the Soviet Union. In fact, in at least one respect—namely, the physical proximity of these two global, competing ideologies—Middle America is unique.

The Caribbean island of Cuba, sitting a mere 90 miles south of Key West, Florida, is something of an anomaly in world geopolitics. While both superpowers have had certain influences

is often just the opposite. These countries change politically far less than we might imagine.

In Canada and the United States, politics is virtually synonymous with stability. These two neighbors do not experience military coups, and a change of power results from an orderly process of voting (with distinct differences, to be sure). A stable government does not, however, mean that conditions rarely change. Canada and the United States are dynamic countries, where things appear to change all the time, seemingly on a daily basis. Middle America, however, is best viewed as a very slowly changing, male-dominated political environment. Stability is likewise a common feature of Middle America, but it is not the same sort of stability as in its North American neighbors. Political leaders remain in power because the military permits them that privilege, so they do not perceive change positively; rather, they strive to maintain the status quo. Change is not viewed as progress as is common in the Western world; instead, it typically is seen as a threat to authority. In consequence, conditions tend to change slowly in the region.

For many states in the region, stability might be better expressed as stagnation. Elections, where they exist (with the exception of Costa Rica), are not free and orderly, and the opposition party always cries "fraud" (the opposition is the party running someone against the chosen government candidate). Mexico is a good example. Although Mexico is the economic and political giant of the region, the nation is markedly conservative. The dominant political party, the Institutional Revolutionary party (PRI), literally controls politics in Mexico. Since its founding in 1929, the PRI has had complete and absolute control; it has won every single national election. The party is receptive to the people's needs, but only to the point where they opt for another party.

(often strong ones) in one another's territory, Cuba is, for all practical purposes, a virtual Soviet colony and, as such, is unique for the western hemisphere. Soviet influence in Cuba dates to 1959, when it was invited by Fidel Castro's revolutionary government to take an active role in Cuban affairs (both internally and abroad). But why has the United States worried about such a small island-state as Cuba?

The United States is not concerned because it fears that Cuba alone is a threat to Western democracy. It is not. Cuba is far too tiny, too underpopulated, and definitely too dependent on outside resources to do any (significant) harm to the United States; it is literally incapable of having a viable effect on U.S. affairs or security. The United States does fear, however, that Cuba will "export" its ideology to other countries within the region. Eventually, according to those who criticize this extension, or diffusion, of Soviet influence into the Caribbean, it will spread throughout Middle America and ultimately challenge the ideology of democracy in the West. On the other hand, recent evidence seems to suggest that the Soviets may be less intent on diffusing Soviet ideology to the region during the 1990s. The Soviet Union is presently facing severe domestic financial problems; perpetually debt-ridden countries like Cuba, thousands of miles from the Soviet Union, are simply too great a burden—despite Cuba's continued adherence to Soviet ideology.

Since its 1959 revolution, Cuba has made phenomenal strides in its education system; some have suggested that it may, in fact, possess one of the best teaching systems, and thus the most literate populace, in the world. It is Cuba's superb education system, together with its Soviet-backed, Marxist ideology, that Western analysts fear most. Some contend that this ideological diffusion from Cuba is already occurring, and they point to the former hostilities in Central America as evidence, if not proof, of this diffusion process. There is some evidence of Cuban involvement in Central American affairs, but the extent of their influence is unknown. Such claims are very difficult to verify, but, if true, it is an example of the diffusion process at work. Further, it would mean to suggest that political ideologies are spread not only by force or coercion, as is often believed, but also by example and invitation. This, of course, was how Cuba adopted its present, Soviet-backed political system. If indeed this is what is happening in the region, it raises two questions for further discussion: (1) Why were the Central American states seemingly a prime target for Soviet-backed aid? (2) Should the West (especially the United States) take an active role in Middle American geopolitics and try to curb this ideological diffusion?

The PRI is frequently accused of election fraud, and this is how stability is changed into stagnation: the options are limited by government decree.

Consequently, there is little political optimism among the peoples of Middle America. Change is seen as something distant, and one response is to move elsewhere. This is particularly true for oppressed, minority groups within a state. In Central America, where war raged throughout much of the 1980s, the political climate is even less positive, so many people have fled to less tumultuous areas.

One exception to the above scenario is Costa Rica. Costa Rica has not experienced military rule since 1949. Instead, it has a 5000-person Civil Guard and a 2000-person Rural Guard. Elections are perhaps freer than in any other Latin American country, and the pro-American government is generally Western in its outlook. Like the Democrats and Republicans in the United States, the liberal and conservative political parties of Costa Rica have tended to flip-flop in and out of power, an indication of a true democracy. Jamaica is another exception, where a relatively free people take part in the functioning of a socialist-type government. Because of its "calm" political environment, it has become a major tourist island for Americans and Canadians, and the Jamaican government is dong its utmost to maintain that image. In truth, however, there are distinct political factions in the country, and it is best viewed as a carefully guarded peace. Most of the other former British colonies in the region—Belize, Barbados, and the Bahamas, for example—are also exceptions to the above picture of Middle American politics. These are very stable countries in the Western sense of stability, and they provide tourists with vacations free from worry about political uprisings.

Puerto Rico is, of course, another major exception. Indeed, from a political perspective, it is difficult to consider Puerto Rico a part of Middle America since Puerto Ricans are U.S. citizens and the island is a self-governing Commonwealth and protected by the United States.

Panamá is almost two countries in one, a result of both history and continued outside political pressure. Its unusual aspect is, of course, the 10-mile (16-kilometer) wide, 40¼-mile (64½-kilometer) long Canal Zone, a strip of land that severs the country in two. Construction on the canal started in 1904, but shipping traffic did not ensue until mid-1914 when the *S.S. Ancon* became the first ship to travel the full length of the canal. Ever since, the canal has been of vital importance to international shipping. It has also been a point of heated debate and civil hostilities within Panamá. The Canal Zone was U.S. property until 1978, when preliminary treaties were signed between the United States and Panamá to provide for the transfer of the canal to Panamá at noon on December 31, 1999. Under the treaties, Panamá will receive more and more money as that date approaches. Yet, the zone remains a source of Panamanian animosity, and U.S.-Panamanian relations became increasingly strained during the 1980s. When General Manuel Noriega assumed control of Panamá in 1988, relations between the United States and Panamá became particularly tense until his ouster in 1989.

TECHNOLOGICAL ACHIEVEMENTS

It is extremely difficult to discuss technology without some personal bias. We all tend to evaluate other peoples' technologies in terms of their scientific achievements. By doing so, we compare the rest of the world to the nation or region that we perceive to be the most scientifically superior. But science is not a proper yardstick for measuring a region's technological level. Left alone, all nations and all regions have technology levels that are adequate for and suited to their specific needs. It is only when they try to borrow technology from other cultures that trouble arises, since they may not have the necessary cultural "mechanisms" to incorporate a foreign technology effectively into their own system. When the Spaniards arrived in the New World, for instance, the various Mesoamerican peoples had a sufficient technology to survive, in some cases with relative ease. Indeed, one way to prove this is to note that many of their indigenous ways of coping or surviving still persist; obviously, they worked then and they still work today. As we approach the twenty-first century, however, technology has become synonymous with the expression

"high tech." Because of this, it generally refers to the ability of a state or region to produce "sophisticated" material goods. In this regard, the technological level of the Middle American countries varies tremendously.

Mexico, without a doubt, is the most "advanced" nation within Middle America under this definition; indeed, it is one of the most advanced, industrialized nations in all of Latin America, irrespective of how we define technology. There is virtually nothing that Mexico cannot produce, and almost everything you can imagine is produced in Mexico. Mexico manufactures trucks, automobiles, heavy-duty farming machinery, delicate surgical equipment, and intricate electronic components and appliances and is also the region's major oil-producing country. In fact, Mexico is one of the world's major oil producers. Most of its modern industries have been imported from the West, as is true of the entire region, with the notable exception of Cuba. Today, Mexico has a huge, well-educated and well-trained work force capable of achieving and maintaining a standard of living comparable to the industrialized nations. Other factors, however, such as tradition, a large and growing population, political corruption, and a heavy degree of foreign investment, inhibit Mexico's realization of such a high standard of living in the immediate future.

Puerto Rico is another Middle American "state" with a very sophisticated level of technology. As a virtual economic and political appendage of the United States, it has been given tremendous financial capital, and today its level of technology is ostensibly equal to that of the United States. Indeed, due to the very favorable tax incentives for American corporations doing business in Puerto Rico, many companies have elected to take advantage of the situation and move their operations and, in some cases, their entire facility to the island. Several years ago, for example, most major American drug and chemical companies (which produce high-value, lightweight products) realized the financial benefits (cheap labor, low taxes) and the geographical advantages (centrally located to the Americas) of Puerto Rico's position in the Caribbean. When these factors were combined with the drastic decreases in air transport that had occurred over the years, there was no other rational choice but to move their headquarters south. Today, the island has become the pharmaceutical center for the western hemisphere.

All other Middle American countries are somewhere "below" Mexico and Puerto Rico in terms of their level

Many Middle American countries rely on the tourist industry to help support their economies. Jamaica has become a desirable tourist site for many Americans. Montego Bay (shown here) is a frequent destination.

of technological achievement or development. None, however, are what some might call "primitive," existing totally at a survival level and living removed from the affairs of other countries or regions. All Middle American states rely, to one degree or another, on outside support and manufactured goods. The range of their reliance on the outside world for their survival, however, is great. In other words, their respective levels of potential vary considerably. Haiti, for example, the poorest nation in the western hemisphere, can no longer survive without foreign aid. Panamá, on the other hand, although strongly reliant on others to maintain its standard of living, is decidedly more advanced than Haiti in its technological developments. The most remote, rural villagers in Panamá have a standard of living comparable to that of many urban dwellers in Haiti's capital, Port-au-Prince.

Determining a region's technological stage or level automatically involves some sort of comparison or value judgment: Middle America's technological accomplishments can only be described relative to some other region's accomplishments or potential accomplishments. In truth, however, the degree of technology a nation or a region has attained is in part due to its desire to be-

come like those at the "top"; in part, it is also due to those at the "top" permitting that nation or region to develop or achieve a new level. No nation becomes "developed" overnight, and in today's complex world it can do so only (1) if other more developed nations allow it to develop, and (2) at the expense of all developed and developing states and planned economies as well. In short, given the finite resources on this planet, as one country rises up the technological ladder, it does so at the expense of all other countries. Some countries have a vested interest in keeping a given state down; others have an equally vested interest in seeing that the same state prospers and grows. Those who are stronger will ultimately win. Finally, standards of technology are based, among other things, on cultural, social, economic, religious, linguistic, and other distinctly human factors. In today's interconnected global economy, however, a region's ability to become more "sophisticated" in any one or more of these factors is more often than not a direct consequence of political alliances, trade agreements, resource allocations, and other socioeconomic factors.

AGRICULTURE

In all of Latin America, agriculture has traditionally been the mainstay of the economy, but variations do exist. Figure 10.2 shows the amount of arable land (in thousands of hectares and thousands of acres) in each country. Although there is a rather strong relationship between the size of the country and the amount of arable land, it is still important to gain some insight as to the total amount of arable land in each country. Figure 10.3 shows the average annual rate of growth or decline in food production. Belize, Dominica, and Honduras experienced average annual rates of growth of more than 4 percent in 1987. In contrast, many countries such as Barbados and Cuba declined considerably in the same year. Since population growth rates are rising in Middle America, increases in food production are important as countries try to keep food production and population increases somewhat balanced. Middle American agriculture is best examined within a three-part regional framework: Mexico, Central America, and the West Indies.

In Mexico, a kaleidoscope of natural environments has created a mixture of land-use patterns. Commercial plantations so common in many colonies (including the Caribbean), which were principally oriented toward

FIGURE 10.2

Arable Land, 1987

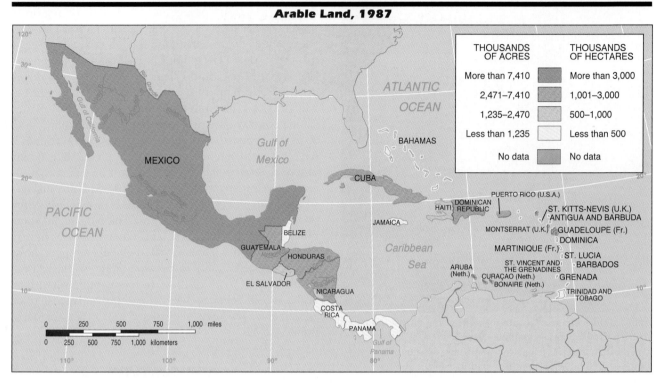

SOURCE: Data are from United Nations, *Statistical Yearbook for Latin America and the Caribbean* (New York: United Nations, 1988).

overseas trade, never developed fully in Mexico. Large land holdings, or **haciendas,** were created instead, which rapidly became the center of Mexican life. Haciendas were self-sufficient units: the Indians provided the labor, while the *hacendado,* or owner, was typically an absentee landlord who often resided off the hacienda (frequently outside Mexico). Communication between haciendas was minimal, to the extent that workers came to regard them as their entire world. Debt peonage was the norm, and breaking free of the hacienda's hold was extremely difficult.

During the early part of this century, agrarian reform measures resulted in the ejido movement. **Ejidos** are rural farming communities (or cooperatives) in which parcels of land are distributed to peasants, but ownership is retained by the ejido. Although ejido lands produce both a great variety of crops and the majority of Mexico's agricultural output—like haciendas, many ejidos occupy huge tracts of land—their per capita production is relatively low due to meager technological adaptations. In contrast, private landholdings are more productive, but less self-sufficient. Land reform, which is at the heart of Mexican politics, has been instrumental in carving up the Mexican countryside into smaller and smaller parcels. The trend is thus toward greater individual ownership, which results in smaller landholdings. The irony of this reform, however, has been the lower productivity of these smaller plots of land and a corresponding reliance on government-subsidized agricultural products. In the north of Mexico, where the terrain is dry and productivity relatively low, large landholdings still persist, whereas the remainder of the state's land continues to be divided systematically.

The Central American agricultural picture is notably different. Rural life continues to dominate Central America, so agriculture is of supreme importance to these countries. Many families are landholders, but few own significant amounts of land. Moreover, most of the land is mountainous, so traditional agricultural techniques do not apply as readily. Furthermore, due to a combination of the rugged topography and the inheritance system found throughout Latin America, many Central Ameri-

FIGURE 10.3

Growth of Food Production, 1987

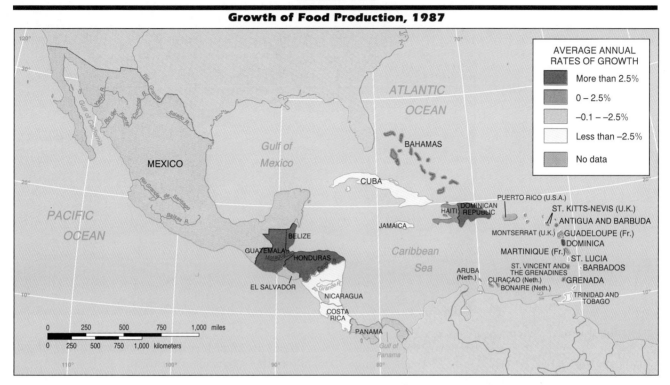

SOURCE: Data are from United Nations, *Statistical Yearbook for Latin America and the Caribbean* (New York: United Nations, 1988).

cans own discontinuous parcels of land, which makes it all the more difficult to produce substantial quantities of a given crop. Highland agriculture in this area is therefore primarily self-sufficient, and few commercial enterprises exist. Due to the overuse of the steep slopes to grow the same perennial crops and to an overall lack of conservation efforts, major parts of the region have experienced extensive erosion problems. A rising population has forced many to migrate to the lowland portions of Central America, where shifting agriculture is more widespread. This migration, however, has resulted in greater competition for arable land than existed in the past, which, in turn, has created additional problems.

Coffee, the first commercial export to be developed in the region, continues to be a major source of income throughout highland Central America. Costa Rica, Honduras, Nicaragua, Guatemala, and even tiny El Salvador produce significant coffee harvests that are sold worldwide. Most of the plantations are owned by a handful of individuals, especially in Guatemala and El Salvador.

Central American coffees are among the best in the world, and great care is taken in coffee production because it accounts for a tremendous percentage of the region's livelihood.

Bananas represent the second most important crop of Central America. Originally established in Costa Rica, today bananas are grown in all Central American states. Unlike coffee, banana plantations were dominated by two companies until the 1960s. The United Fruit Company and the Standard Fruit and Steamship Company literally monopolized production and sales in the region from the late nineteenth century until 1967. Due to growing foreign competition, however, banana production is less important to the local economy today than in the past, but it nevertheless provides millions of needed dollars to each country's economy.

Concentrated in the humid lowlands facing the Caribbean, banana production differs considerably from coffee. Bananas are more bulky than coffee, and both their perishable nature and their susceptibility to tropical diseases necessitate a different sort of marketing system.

Banana fields surround workers' homes in Costa Rica. Bananas are an important cash crop in Costa Rica.

In addition, since they are grown in the traditionally less inhabited (lowland) portions of Central America, huge investments in infrastructure were needed before production could begin. Coffee, on the other hand, is grown in the densely populated highlands where roads and marketing facilities were already in place. One further problem associated with the lowland location of bananas is that they are prone to insect and hurricane damage, which coffee, at much higher altitudes, is not.

Agriculture in the Caribbean is decidedly distinct and differs from that of both Central America and Mexico in important ways. For several centuries following European settlement, the islands served as commercial outposts of European trade. The Caribbean was the scene of initial Spanish incursion into Middle America and, in many ways, was considered the most crucial locale to defend against European competitors. The islands, by their very nature and number, were difficult to protect. Indeed, they were to change "owners" many times over the centuries, as pirates (Spanish, French, English, Dutch, and Portuguese) raided supply ships transporting goods destined for Old World ports. Following the deaths of the native peoples, vast quantities of African slaves were imported to work what were soon to become the busiest commercial plantations in the New World. By the mid-seventeenth century, tiny Barbados became the most populated, most congested, and rich-est colony in English-speaking America, due predominantly to the sugarcane industry.

Regardless of which European power held sway at any given moment, and irrespective of the particular island in question, a system of mercantilism persisted for many years, supported by slavery and dominated by the sugar trade. Four periods of sugar monoculture have been identified: the early phase (1638–1763) when the industry began on St. Kitts-Nevis, Jamaica, Martinique, Barbados, and Guadeloupe; the middle period (1763–1870) when sugar was extended to Tobago, Grenada, St. Vincent, and Dominica; the industrial period (1870–1948) when Cuba, Trinidad, Puerto Rico, the Dominican Republic, and Jamaica controlled the industry; and the modern period (1948–) when the sugar fields were nationalized and mechanized by the respective states. In addition to sugar, early island exports included large quantities of wood, salt, beef hides and tallow, salted beef, cacao, indigo, pimento, tobacco, coffee, rum, cotton, and ginger, most of which were shipped to Europe.

Today, crops vary from island to island, although no island produces enough of any single item to survive economically. Sugar remains important, of course, but often less so than in the past. Haiti exports coffee, and Cuba and Puerto Rico are major tobacco growers; in particular, they raise leaf tobacco for cigar production. Many tropical fruits, such as coconuts, bananas, pa-

payas, and mangoes, are grown for export, and a limited cotton industry persists. West African staples like millet, sorghum, okra, breadfruit, and yams were introduced during the slave trade, and they continue to flourish throughout the Caribbean, especially in Haiti, Jamaica, the Dominican Republic, Trinidad and Tobago, and Cuba, which today have large black African populations. Rum, which is distilled from sugarcane, continues to be a major export of several island countries.

INDUSTRY

The pattern of industrial location in Middle America mirrors the distribution of minerals and other natural resources noted earlier. As was true for minerals, industry can be seen in terms of two subregions: the mainland and the Caribbean.

As expected, Mexico dominates mainland mineral and oil production. Silver and gold mining (and, to a lesser extent, copper, lead, and zinc) used to be the mainstay of the Mexican mineral industry, but oil has now assumed ascendancy; lead (and more recently sulfur) is still an important source of income. Although Mexico lacks substantial amounts of coal and iron ore, steel production is important, especially in the Monterrey area. In fact, much of northern Mexico, where most of the natural gas and iron ore are found, is industrial

in nature. The largest manufacturing region, however, as one might expect, is the densely settled metropolitan Mexico City area, which contains the single largest concentration of consumers. Major producers of all wholesale and retail consumer goods are found in this area, even though few raw materials are located nearby. As the primary market of the country, it is cheaper to bring the raw materials to Mexico City for manufacture than to transport bulky manufactured goods from the north where the natural resources are found. Only the most technologically sophisticated manufactured items that need to be imported are not produced in this metropolitan area. Guadalajara and Tijuana are also manufacturing cities of regional significance.

Elsewhere on the mainland, industrial output is of far less importance than agriculture, the economic staple. The importance of industrial output varies from a high relative percentage of the gross national product (GNP) in Nicaragua to a low relative percentage in Panamá. In most states, small-scale manufacturing (of traditional goods for a local market) predominates. In terms of industrial location, though, each country basically parallels the conditions found in Mexico: sporadic manufacturing is found nationwide (particularly in Guatemala), primarily focused in and around the mineral-rich areas, but most is located near the national capitals, which are usually primate cities. Nearly 70 percent of all factory

Two workers harness a drill on a Pemex rig in Mexico. Income from oil has become an important part of the Mexican economy.

employment occurs in the 10 largest urban areas. In addition, since Panamá serves as a major port of entry for the entire region, a vast array of inexpensive manufactured goods from around the world finds its way into Central America, thereby lessening the need for further industrial development.

In the islands, the situation is considerably different. Due to their geology, the islands are virtually mineral-free. A few countries, most notably those in the Greater Antilles, have vast quantities of a single marketable resource, while most have virtually none. Jamaica, for instance, has large reserves of high-grade bauxite, which is shipped worldwide for aluminum production. Similarly, Trinidad and Tobago now has an important petrochemical industry. The remaining islands, however, have only minor manufacturing economies (with the exception of Puerto Rico's American-owned and financed pharmaceutical industry). Cuba reportedly has the potential to market quality manufactured goods, but little is known of that island's mineral reserves, and its lingering ties to the Soviet Union greatly hinder full exploitation of those reserves. Generally speaking, the Ca-

ribbean lacks the necessary raw materials to develop viable manufacturing industries beyond local needs.

URBANIZATION

A common pattern in the developing world is the presence of primate cities, and in this regard Middle America appears to conform to the trend. Assuming increased economic opportunities in the urban areas, many less fortunate Middle Americans migrated to the cities during the 1960s and 1970s; today that flow has slowed somewhat. Only 5 countries are more than 60 percent urban, however, 11 are less than 50 percent urban, and 6 are between 50 and 60 percent urban (Figure 10.4). Table 10.3, which lists each major Middle American state and its most populated city, reveals a number of important traits about urbanization in the region. At first glance, the most striking characteristic seems to be the dominance of the largest city vis-à-vis the entire population. El Salvador's capital, San Salvador, contains the smallest percentage of any Middle American primate-city popu-

FIGURE 10.4

Urban Population, 1990

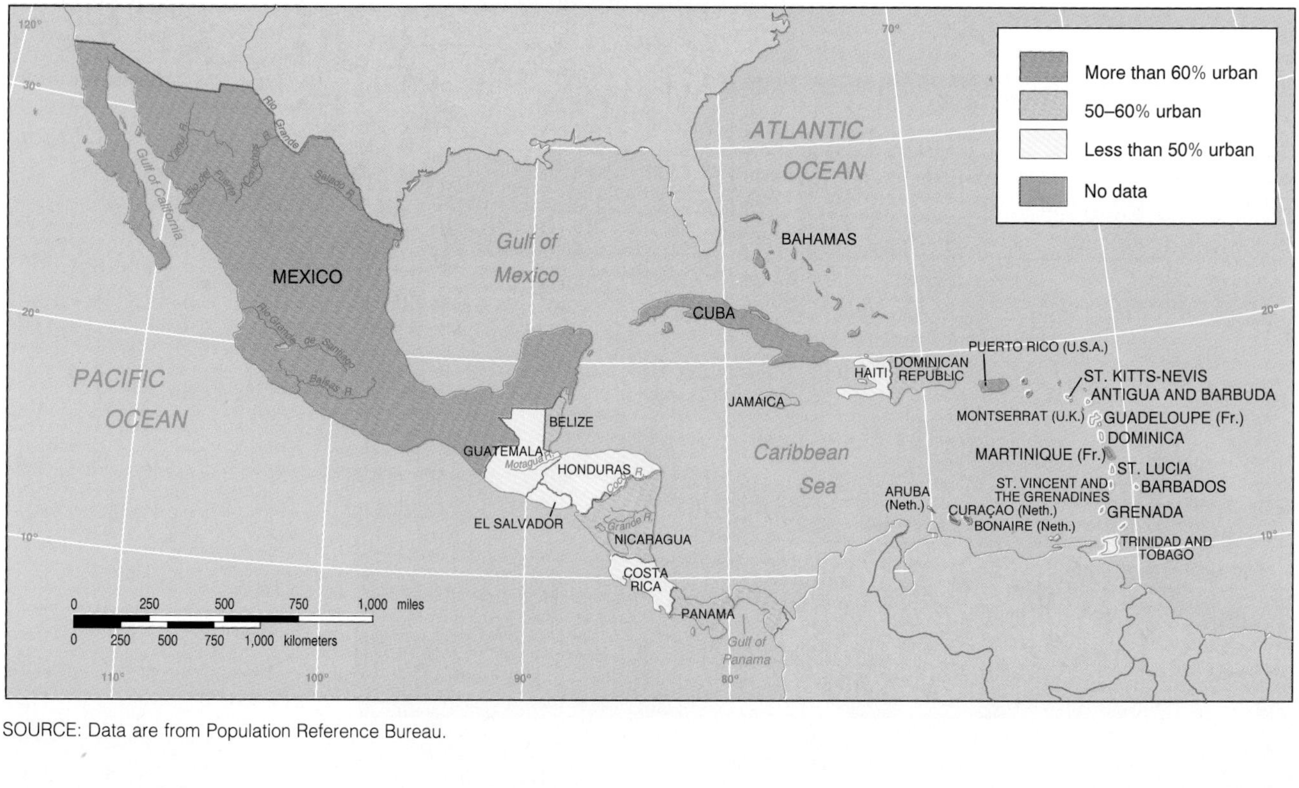

SOURCE: Data are from Population Reference Bureau.

TABLE 10.3

Middle American Countries: Largest Cities and Their Percentage of the Total Population

COUNTRY AND LARGEST CITY	POPULATION ESTIMATE, 1986	PERCENTAGE OF TOTAL POPULATION
Bahamas	235,000	
Nassau	140,000	59.6%
Barbados	253,000	
Bridgetown	105,000	41.5
Belize	168,000	
Belize City	45,000	26.8
Costa Rica	2,714,000	
San Jose	269,000	9.9
Cuba	10,200,000	
Havana	2,000,000	19.6
Dominican Republic	6,700,000	
Santo Domingo	1,500,000	22.4
El Salvador	5,100,000	
San Salvador	360,000	7.1
Grenada	90,000	
St. George's	30,100	33.4
Guatemala	8,600,000	
Guatemala City	754,000	8.8
Haiti	5,900,000	
Port-au-Prince	900,000	15.3
Honduras	4,648,000	
Tegucigalpa	509,000	11.0
Jamaica	2,290,000	
Kingston	494,000	21.6
Martinique	328,000	
Fort-de-France	116,000	35.4
Mexico	81,700,000	
Mexico City	18,000,000	22.0
Netherland Antilles	200,000	
Willemstad	149,000	74.5
Nicaragua	3,400,000	
Managua	1,000,000	29.4
Panamá	2,000,000	
Panama City	450,000	22.5
Trinidad and Tobago	1,204,000	
Port-of-Spain	300,000	24.9

NOTE: All cities are capital cities except Belize City, which was the capital of Belize until 1970. The population estimate for the Netherland Antilles is from January 1987 and does not include Aruba.

lation (7.1 percent), whereas Willemstad, the capital of the Netherlands Antilles, is at the other extreme, containing a staggering 74.5 percent of that country's total population.

In virtually every instance, the largest city in each country also functions as the capital city. The sole exception is Belize City, which was the capital until it was moved to Belmopan in the interior in 1970. Complicating this comparative analysis of cities, however, is the fact that all countries define city boundaries in a slightly different manner. Moreover, such comparisons ignore geographic factors, so perhaps it makes greater sense to look at the countries and cities with regard to their size, populations, and both absolute and relative location.

A second glance at Table 10.3 indicates that it is mainly the Caribbean states that have true primate cities. On the mainland, the largest city in four of the eight countries contains only 11 percent or less of the total population; the average is slightly more than 17 percent. In the islands, however, the situation is decidedly different. Each of the 10 states listed in the table contains a city that supports at least 15 percent of the national population, and the average is 34.8 percent, or double the rate found on the mainland. Clearly, the size of the country is also important. The relatively small Caribbean islands have little need for a large number of cities, as administrative functions operate adequately over a small area. On the mainland, though, where the landmass is generally greater and people are thus able to live less compactly, cities serve an administrative purpose in addition to the typical urban services they provide.

The population of a country is also a factor in determining the growth of potential primate cities. Looking a third time at Table 10.3, note the relationship between the total population of a given country and the relative importance of that country's largest city. Of those countries with populations below two million—in increasing order, they are Grenada, Belize, the Netherlands Antilles, the Bahamas, Barbados, Martinque, and Trinidad and Tobago—all have cities that house at least one-quarter of the total national population; two contain cities with well over half the total; the average is an unbelievable 42.3 percent. In other words, states with small populations tend to have one dominant city. At the opposite end of the scale, in those countries with populations above five million—in decreasing order, they are Mexico, Cuba, Guatemala, the Dominican Republic, Haiti, and El Salvador—the primate cities aver-

All large cities in Middle America are crowded. This is a bustling street scene near the main urban market in Tegucigalpa, Honduras.

age less than 16 percent of the national total, and none contains one-fourth or more of the national population.

Moreover, except for those mainland states where the densely populated native settlements were predominantly in the highlands, such as Mexico, Guatemala, Honduras, El Salvador, and Costa Rica (on top of which the Spaniards later decided to create their capital cities), the major Middle American capitals are located on the coasts, where they serve as important port cities. None of the islands deviates from this rule. On the mainland, Nicaragua, Belize, and Panamá have capital cities in lowland areas, but, again, this reflects indigenous settlement patterns and lack of a suitable highland area. In sum, mainland cities of importance are generally found in the highlands, while comparable settlements in the Caribbean are port cities situated on the coastal lowlands. Most are capital cities, and most functioned as important sites for native peoples too.

SPATIAL CONNECTIVITY

At one time, authors of geography textbooks commonly discussed places, states, and even regions as discrete, autonomous entities. Region A was always discussed by itself; no other regions were introduced in a text devoted to region A. The text assumed or implied that knowing the basic facts and details germane to some place was all that was necessary to understand how that place functioned or survived. Eventually, all the regions would be discussed in this discrete fashion, and the student would be left with the task of figuring out how all the pieces of the global puzzle fit together. This approach is obviously very simplistic, however, and, in some respects very naive.

In today's world, it is almost irresponsible to talk about a place, especially one as large as Middle America, without considering the outside factors dependent upon or responsible for its continued existence. Social scientists talk about a "global economy," a "one-world culture," and an "interconnected global network," and that has indeed become the norm. For better or worse, many developing nations and states are looking either to the West or to the model offered by the planned, socialist economies for guidance. However, with the economic and political restructuring currently evident in the Soviet Union and Eastern Europe, the Soviet Union may no longer be the planned/socialist model for other

countries to emulate. Consequently, while each state attempts to maintain its uniqueness, places are rapidly becoming more and more alike, particularly with regard to economic considerations. Long gone are the nations seeking to isolate themselves from the whole of humanity, and countries are finding it increasingly difficult (if not impossible) to extricate themselves from this global economic web. The superpowers are wholly and intricately linked to all major economies worldwide, and some would argue that their power is, in large measure, a by-product of their ties to all other parts of the planet and results from their ability to function globally. Middle America is tied not only to the rest of Latin America, but, as we shall see, to every corner of the globe.

DEPENDENCE

One indicator of dependency is the ability of states to work together for a common cause or belief. In this regard, Middle America is best viewed on three distinct levels: as a component of the western hemisphere idea, as a collection of Latin America states, and as a separate, semiautonomous entity with its own special interests. The latter is particularly true of the Caribbean countries (which sometimes see themselves as a single unit) and the Central American republics (which also sometimes regard themselves as a cohesive unit). Mexico, on the other hand, is often viewed by other states in the region as an "outsider" due to its vast size, its tremendous resource potential, and its relative political strength. At one or more of these three levels, the region (or the smaller units within the region) makes decisions based on a multitude of factors.

In December 1823, the U.S. president James Monroe delivered a message to the U.S. Congress warning European states not to meddle in the affairs of the West. Non-Western states were told, in no uncertain terms, that the West was "free" territory, and that the United States strongly opposed Europe reasserting its hold over former colonies. In part, the message stated that:

> In the discussions to which this interest has given rise, and in the arrangements by which they may terminate, the occasion has been deemed proper for asserting as a principle in which rights and interests of the United States are involved, that the American continents, by free and independent condition which they have assumed and maintain, are henceforth not to be considered as subjects for future colonization by any European power. . . . We owe it, therefore, to candor and to the amicable relations existing between

the United States and those powers to declare that we should consider any attempt on their part to extend their system to any portion of this hemisphere as dangerous to our peace and safety.

Eventually called the **Monroe Doctrine,** the proclamation was soon realized for what it was, namely, a declaration that the western hemisphere was a separate entity, that it wished to remain separate, and that the United States was willing and indeed ready to defend its sovereignty. From 1823 onward, the United States has been regarded sometimes as a benevolent "big brother" (and, sometimes as a more oppressive one) to the rest of the western hemisphere, especially to Latin America and particularly to Middle America. The latter has been especially vulnerable to U.S. influence because of its proximity to the United States. One major result of the doctrine has been to motivate the Latin American countries and the United States and Canada to act as a unit to create political and economic ties that would benefit the western hemisphere as a whole. In effect, since the initiation of the Monroe Doctrine, the countries of North and South America have tried to cooperate with one another, and a number of important western hemispheric organizations have emerged to serve that end. Of course, times and political circumstances have changed since 1823, and today the United States is regarded favorably by only a small number of its neighbors, while some openly fear its continuing influence in their affairs. Depending on the issue, some Middle American countries are friendly to the United States whereas others are openly hostile.

Table 10.4 lists the nine major organizations and associations designed either to improve regional trade relations or protect the western hemisphere as a unit. The table indicates the states that are members of these organizations and those that have decided not to join (or, in some cases, have been ousted). Other smaller associations also exist, some of which were founded prior to these nine, but no significant association was established before the end of World War II, and none is of great importance today. The earliest of the nine was the Inter-American Treaty of Reciprocal Assistance (IATRA), sometimes called the Rio Pact. Originally signed by 19 states on September 2, 1947 (in Rio de Janeiro), the pact provides for "peaceful settlement of disputes in the Western Hemisphere and common action against aggressors, whether American or an outside nation, within a defense zone encircling North and

South America and including Greenland and Antarctica." Canada, Nicaragua, and Ecuador refused to sign initially, but later both Nicaragua and Ecuador signed and were admitted along with Trinidad and Tobago; Cuba was suspended from IATRA in 1964. The following year, two similar but more effective associations were established: the Economic Commission for Latin America (ECLA) and the Organization of American States (OAS).

ECLA's function is to strengthen Latin American countries, both individually and as a group. It meets once every two years and, among other initiatives, tries to concern itself with issues of foreign trade, regional imbalances (within Latin America), and the regulation of foreign investment. The OAS resulted from the Rio Pact; thus, its purposes are clearly stated. It is a larger, more powerful body than ECLA, designed primarily to foster peace and solidarity throughout the western hemisphere. It is the most important organization of its kind in this part of the world and acts, in many respects, as an umbrella organization for a number of other international and/or interregional bodies. Originally, when its policies were dominated by U.S. ideology, the OAS concerned itself exclusively with political affairs, but social, economic, scientific, and cultural issues are now discussed regularly. (The OAS contains numerous subagencies, including the Pan-American Institute of Geography and History.) Several years after the formation of ECLA and the OAS, the Organization of Central American States (OCAS) came into being. Established to foster mutual understanding, promote joint assistance, and maintain peaceful relations among Central American states, the OCAS has been plagued by constant problems.

Two other organizations, the Latin America Free Trade Association or **LAFTA** (the name was changed to the Latin American Integration Association or ***LAIA*** in 1980) and the Inter-American Development Bank (IADB), were both founded in 1960. LAFTA, as its name implies, is a trade association similar to the European Economic Community (EEC) and was established to regulate, restrict, and expedite the transportation of goods between participating countries. Within LAFTA, there are "free trade zones" and internal, subregional agreements based on the notion of cooperation, the safe movement of goods and services, and tariff reductions for preferred states. The underlying concept of the IADB—that is, to finance large economic development projects throughout the Americas—dates from at least

TABLE 10.4

Western Hemisphere Organizations and Membership

	CARICOM	CDB	IADB	LAES	LAFTA	OAS	OCAS	ECLA	IATRA
Antigua and Barbuda	Yes								
Argentina			Yes	Yes	Yes	Yes		Yes	Yes
Bahamas		Yes	Yes					Yes	
Barbados	Yes	Yes	Yes	Yes		Yes		Yes	
Belize	Yes								
Bolivia			Yes	Yes	Yes			Yes	Yes
Brazil			Yes	Yes	Yes	Yes		Yes	Yes
Canada		Yes	Yes					Yes	
Chile			Yes	Yes	Yes	Yes		Yes*	Yes
Colombia		Yes	Yes	Yes	Yes	Yes		Yes	Yes
Costa Rica			Yes	Yes		Yes	Yes	Yes	Yes
Cuba				Yes		Yes		Yes	
Dominica	Yes	Yes				Yes		Yes	
Dominican Republic			Yes	Yes		Yes		Yes	Yes
Ecuador			Yes	Yes	Yes	Yes		Yes	Yes
El Salvador			Yes	Yes		Yes	Yes*	Yes	Yes
Grenada	Yes	Yes		Yes		Yes		Yes	
Guatemala			Yes	Yes		Yes	Yes	Yes	Yes
Guyana	Yes	Yes*	Yes	Yes				Yes	
Haiti			Yes	Yes		Yes		Yes	Yes
Honduras			Yes	Yes		Yes	Yes	Yes	Yes
Jamaica	Yes	Yes*	Yes	Yes		Yes		Yes	
Mexico			Yes	Yes	Yes	Yes		Yes	Yes
Montserrat	Yes	Yes							
Nicaragua			Yes	Yes		Yes	Yes	Yes	Yes
Panamá			Yes	Yes		Yes		Yes	Yes
Paraguay			Yes	Yes	Yes	Yes		Yes	Yes
Peru			Yes	Yes	Yes	Yes		Yes	Yes
St. Kitts–Nevis	Yes	Yes							
St. Lucia	Yes	Yes				Yes			
St. Vincent and the Grenadines	Yes	Yes							
Suriname				Yes		Yes		Yes	
Trinidad and Tobago	Yes	Yes	Yes	Yes		Yes		Yes	Yes
Uruguay			Yes	Yes	Yes*	Yes		Yes	Yes
United States			Yes*			Yes*		Yes	Yes
Venezuela		Yes	Yes	Yes*	Yes	Yes		Yes	Yes

NOTE: The United Kingdom is also a member of the CDB, the IADB, and ECLA; the Netherlands is also a member of ECLA. The IADB has other members outside the western hemisphere. An asterisk indicates that the organization has its headquarters in that country.

SOURCE: G. T. Kurian, *Encyclopedia of the Third World*, 3rd ed., (New York and Oxford: Facts on File, 1987).

the turn of the century. The organization, which is heavily dominated by the wealthier member states, makes loans and grants to poorer member governments and frequently provides technical assistance when needed. More recent organizations include the Caribbean Development Bank (CDB), which was founded in 1969 to provide economic strength and cooperation within the Caribbean basin; the Caribbean Community and Common Market (CARICOM), which is an outgrowth of an earlier, now-defunct body—the Caribbean Free Trade Association—and was created in 1973 to monitor and regulate the regional flow of money and goods and to focus attention on the weaknesses of the basic infrastructure in member countries; and the Latin American Economic System (LAES), which is composed entirely of Latin American states and was instituted in 1975 to prevent foreign countries (including the United States) from intervening in their domestic affairs. LAES is protectionist, believing that non-Latin American assistance is necessary only as a last resort. In large measure, it tries to halt outside coercion—in particular, the mighty multinationals whose influence is often difficult, if not impossible, to regulate—*before* the coercion takes place. Its success rate is questionable.

Table 10.4 also reveals some important points about the makeup of the region and indeed the whole world. Note that in both Middle America and all of Latin America, politics and history have a great deal to do with the membership profile of each organization. For example, former British colonies tend to join as a unit; either all belong to a given association, or none belongs. Thus, the vestiges of colonial status are still an effective influence on many developing countries, although they are "free" in theory. The duration of a state's independence is also a significant factor, of course. Cuba, as an extension of Soviet influence in the western hemisphere, is obviously a special case, as is Puerto Rico, which is a self-governing Commonwealth protected by the United States and for all intents and purposes, is considered a part of the United States for this chapter. In most instances, however, it is advantageous for the poorest states to belong to as many organizations as possible, while the richest states (which typically keep these associations afloat by financing the bulk of the expenses) are sometimes hesitant to participate.

Table 10.5 gives trade (export and import) figures for selected Middle American states. It shows the degree to which each state trades with neighboring Latin Ameri-

TABLE 10.5

Exports and Imports of Selected Middle American Countries, 1982

	EXPORTS (MILLIONS OF U.S.$)		IMPORTS (MILLIONS of U.S.$)
COSTA RICA			
World total	$897.7	World total	$870.0
Latin America	257.8	Latin America	352.1
Major partner: Guatemala	65.7	Major partner: Venezuela	103.6
Middle America	212.3	Middle America	228.6
Major partner: Guatemala	65.7	Major partner: Guatemala	51.6
CUBA			
World total	1,207.1	World total	1,565.2
Latin America	102.6	Latin America	200.2
Major partner: Mexico	85.3	Major partner: Venezuela	69.0
Middle America	98.9	Middle America	62.7
Major partner: Mexico	85.3	Major partner: Dominican Republic	34.4
DOMINICAN REPUBLIC			
World total	809.0	World total	1,250.0
Latin America	86.3	Latin America	496.3
Major partner: Cuba	31.2	Major partner: Venezuela	221.0
Middle America	55.2	Middle America	249.9
Major partner: Cuba	31.2	Major partner: Mexico	170.7
EL SALVADOR			
World total	879.2	World total	929.0
Latin America	180.5	Latin America	373.9
Major partner: Guatemala	119.6	Major partner: Guatemala	210.3
Middle America	180.6	Middle America	323.2
Major partner: Guatemala	119.6	Major partner: Guatemala	210.3
GUATEMALA			
World total	1,244.6	World total	1,340.5
Latin America	437.0	Latin America	514.7
Major partner: El Salvador	196.3	Major partner: Mexico	112.9
Middle America	425.3	Middle America	399.8
Major partner: El Salvador	196.3	Major partner: Mexico	112.9

NOTE: Figures are rounded and may not equal 100 percent.

can countries in general and within Middle America itself; major trading partners for both Latin and Middle America are also listed, and a value of total world trade is given to put the intraregional and interregional trade into clearer perspective. Several things become evident almost at once. In terms of imports, most Middle American countries seem to derive a significant percentage of their goods from other Latin American countries, generally about 35 percent. Cuba and Haiti are notable exceptions. The former has received an immense percentage of its imports from countries within the former Soviet Bloc. The latter gets the overwhelming majority of its imports in the form of subsidies from the United States and other Western allies.

Major Latin American trading partners, not surprisingly, include Venezuela, Mexico, and Brazil, the first two producing respectable oil exports each year, while Brazil is an industrial giant in all respects. Also, note that in almost every case imports from Latin America actually originate in Middle America. This is due both to intraregional trade agreements and to the cost of shipping goods from more distant locales. El Salvador, for instance, receives as much as 86 percent of its Latin American imports from adjoining Middle American countries, mostly from neighboring Guatemala; El Salvador cannot afford to purchase many goods from points farther than Middle America. In contrast, Mexico imports only about 4 percent of its goods from Latin America. Its major Latin

TABLE 10.5 (continued)

Exports and Imports of Selected Middle American Countries, 1982

	EXPORTS (MILLIONS OF U.S.$)		IMPORTS (MILLIONS of U.S.$)
HAITI			
World total	380.0	World total	524.9
Latin America	8.0	Latin America	60.7
Major partner: Dominican Republic	2.4	Major partner: Trinidad and Tobago	25.1
Middle America	7.3	Middle America	44.7
Major partner: Dominican Republic	2.4	Major partner: Trinidad and Tobago	25.1
HONDURAS			
World total	744.6	World total	710.8
Latin America	91.7	Latin America	256.5
Major partner: Nicaragua	27.7	Major partner: Trinidad and Tobago	100.3
Middle America	84.6	Middle America	206.4
Major partner: Nicaragua	27.7	Major partner: Trinidad and Tobago	100.3
MEXICO			
World total	21,163.0	World total	15,372.0
Latin America	1,818.0	Latin America	622.0
Major partner: Brazil	697.0	Major partner: Brazil	253.0
Middle America	813.0	Middle America	197.0
Major partner: Panamá	164.0	Major partner: Guatemala	50.0
NICARAGUA			
World total	365.5	World total	686.3
Latin America	63.9	Latin America	399.0
Major partner: Costa Rica	20.1	Major partner: Mexico	127.9
Middle America	63.5	Middle America	310.0
Major partner: Costa Rica	20.1	Major partner: Mexico	127.9
PANAMÁ			
World total	308.1	World total	2,901.3
Latin America	55.9	Latin America	518.2
Major partner: Venezuela	12.5	Major partner: Mexico	164.6
Middle America	36.0	Middle America	253.6
Major partner: Costa Rica	10.1	Major partner: Mexico	164.6

SOURCE: J. W. Wilkie and D. Corey, eds., *Political Abstract of Latin America*, vol. 25, (Los Angeles: University of California UCLA Latin American Center Publication, 1986).

American trading partner is Brazil. Because Mexico has the capacity to produce most basic necessities internally, it must obtain only a few items from its Middle America neighbors. Note how Mexico dominates the region in the value of imported commodities, by a margin of 5 to 1 over Panamá, the next largest importing state.

On the export side, a similar scenario exists. Although exports involve a smaller dollar amount than imports, Middle America again appears to sell a sizable percentage of its goods to other Latin American countries. Again, Cuba has fewer choices and therefore must transport its export products to other centrally planned economies (primarily to the other side of the world). Due to trade restrictions, Mexico is the major country in the region to receive Cuba's exports, but another important trading partner is the Dominican Republic. However, both account for a small percentage of Cuba's regional trade. Similarly, debt-ridden Haiti has little to sell on the international market, so the industrialized states help support its bankrupt government by purchasing a majority of its exports, often at artificially inflated prices. As with imports, the vast majority of all goods shipped from Middle American countries to other Latin American countries wind up in neighboring Middle America. Nicaragua, Honduras, El Salvador, Haiti, Guatemala, and Cuba sell virtually all of their Latin America–bound items to other Middle American countries, all to countries in close proximity. In fact, El Salvador does not sell anything in Latin America outside Middle America; it would therefore seem that the goods El Salvador has to offer do not command a price high enough to justify export to more distant regions. Once again, Mexico is an exception to the norm. The majority of all Mexican goods shipped to other Latin American countries go to countries outside Middle America, a reflection of the quality of goods Mexico produces for export and, as a result, their significantly larger overseas market. Like its import figures, Mexico's export figures exceed those of all other states in the region combined.

Note that the import-export ratio of each country is generally equal. It would appear that most Middle American states are doing a good job of exporting the same dollar amount of goods as they import. This provides for a healthy economy; if imports exceed exports by too great a value, then a state's economy is in grave financial trouble. Panamá is the sole exception, but because it is the major port of entry for the region, we can assume that a large percentage of Panamá's imports ultimately wind up in other Middle American states.

From this discussion of trade organizations, hemispheric alliances, and the degree to which Middle American countries rely on one another for trade, it should be clear that they are not only leaning upon one another for a remarkable portion of their import-export income, but that they are genuinely supporting each other in today's complex international economy. What started in 1823 as a presidential decree directed at Europe has today become a continuing necessity based on current global economic and political circumstances. The Middle American states see distinct advantages in acting as a unit. The poorer republics (with the exception of Cuba) are more dependent upon one another than the richer ones are, but even Mexico, the richest country in the region, finds it advantageous to devote a fair amount of its import and export trade to its Caribbean and Central American neighbors.

One of the major problems in Middle America is the long-term debt of many countries. Several countries including Mexico, Nicaragua, Panamá, Guatemala, and Jamaica have relatively large amounts of outstanding debt. One important measurement of debt is to calculate it as a percentage of the gross national product (GNP). As Figure 10.5 shows, the debt of Nicaragua, Jamaica, Costa Rica, and Panamá exceeds 100 percent of their GNPs.

Another problem plaguing Middle America is inflation and the rise in the general level of the annual consumer price index (CPI). Using the base year of 1980 = 100, statistics for 1987 show that the CPIs of Nicaragua and Mexico exceed 1000. Panamá (118.3) and Haiti (141.3), have the lowest indexes in the region (Figure 10.6). It is readily apparent that the combination of long-term debt and ever increasing inflation rates act to prohibit investment in Middle America, while it also maintains the status quo; Middle American countries thus tend to remain at their same respective levels vis-à-vis one another.

INTERDEPENDENCE

There is a general misconception in the developed world that developing countries have made a conscious decision to exist separate and distinct from the remainder of the world, that many are self-sustaining entities trying to remain aloof from world political struggles, that they do not wish to participate in international trade alliances, and that they are attempting to develop totally from within. While some developing states may advance such rhetoric in public, in reality, nothing could be fur-

FIGURE 10.5

Long-Term Debt as a Percentage of GNP, 1986

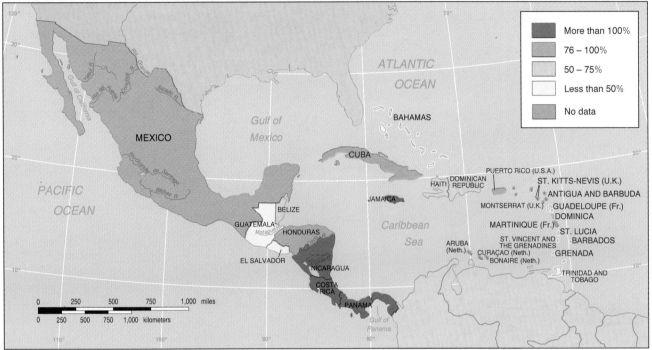

SOURCE: Data are from United Nations, *Handbook of International Trade and Development Statistics: 1988* (New York: United Nations, 1989).

ther from fact. This sort of "protectionist" attitude may have existed in the 1950s and the 1960s, but in the 1990s no country can function totally independent from the rest of the world. Moreover, the same can be said for regions as well. Like states, regions do not exist in a vacuum, and they must be examined in a global context. Middle America is merely one component within Latin America, and Latin America is no more than a single unit within a system of regions that together comprise a worldwide network of interrelated systems.

Middle American countries depend upon one another and upon their neighboring states throughout Latin America for their continued existence and basic needs. But these countries also rely heavily on economic, political, social, and religious contacts outside Latin America, literally in every corner of the world. According to the old saying, "When America sneezes, Canada catches a cold!" That expression captures, perhaps better than any other, the state of the world today and Middle America's place within it. In sum, the Middle American countries have increasingly become an integral part of

the capitalist world system, and they can hardly, nor would they necessarily want to, turn their backs on that system. Furthermore, it is decidedly to their advantage to remain within that web if they intend to perpetuate political and economic alliances. Alternatively, as in the case of Cuba, they may turn instead to the planned economies. (It is becoming increasingly evident, though, that even the planned, socialist states are resorting more and more to capitalism.) As one recent analyst noted of all of Latin America:

> With the exception of Cuba, the economies of Latin America are closely interlinked into the *world capitalist system*. With this one exception, the countries of Latin America have perceived their advantage in remaining and participating in the Western economic system. The relative stagnation of the Cuban economy since 1958 [following Castro's rise to power] and its heavy dependence on the Soviet government have not stimulated further resignations from the Western capitalist system. Indeed, Nicaragua [a good example because of Cuba's growing involvement in Nicaragua during the 1980s] maintains close trading and technological links

FIGURE 10.6

Annual Consumer Price Indexes, 1987

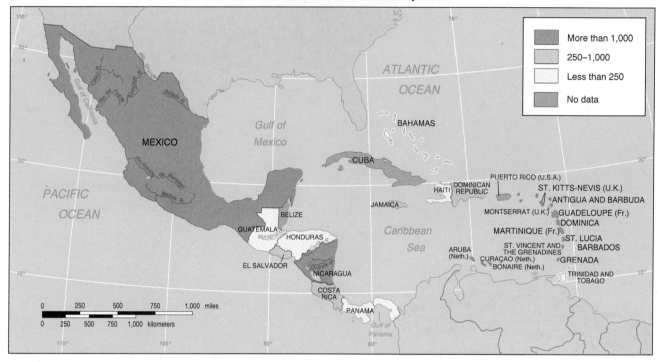

SOURCE: Data are from United Nations, *Statistical Yearbook for Latin America and the Caribbean* (New York: United Nations, 1988).

with West European countries. Generally speaking Latin American countries realize that they receive net benefits from interaction and trade with the industrialized countries of the West and those less developed countries that participate in the Western capitalist trading system. [Emphasis in original.]*

To varying degrees, the Middle American countries participate fully and willingly in this network because they derive enormous benefits from it. Cuba, as a Soviet satellite, has been "bound" to the former Soviet Bloc states. However, as is the case with Middle American countries tied to the West, Cuba too must look increasingly outward to survive.

Of the myriad international political and trade associations that exist today, Middle American countries belong to most and participate fully. It is impossible to name them all, but a partial list will help to illustrate the

*Robert N. Gwynne, *Industrialization and Urbanization in Latin America* (London and Sydney: Croom Helm, 1985), p. 3.

point. First is the Commonwealth. As an outgrowth of the British Empire, the Commonwealth's purpose is "to give expression to a continuing sense of affinity and to foster cooperation among states presently or formerly owing allegiance to the British Crown." Thus, all former British colonies worldwide, including 11 states in Middle America, are members. It is an economic organization, giving preferred trading status to member countries. The Commonwealth countries have also banned together to fight drug trafficking, international terrorism, nuclear and chemical weapon testing, and racial discrimination. In addition, they cooperate to improve health care facilities in member states. Except for Belize, all Middle American Commonwealth countries are in the Caribbean. (British royalty presides over the Commonwealth, so, as of 1991, Queen Elizabeth II is at its helm.)

The International Criminal Police Organization or INTERPOL, headquartered in France, is literally an international police force. Dating from the early 1920s, it is an information-based force, not a working police force in the traditional sense. The majority of the association's activities are concerned with drug trafficking, bank fraud, counterfeiting, international terrorism, and other criminal acts that transcend political borders. Eighteen Middle American states are members, and there is a subbureau in Bermuda. Another international organization that contains a large number of Middle American countries is the Nonaligned Movement. A somewhat unusual organization, the Nonaligned Movement has no headquarters and no politically powerful member states. It was created to ensure autonomous independence and sovereignty and to guard against racism, colonialism, foreign aggression, occupation, and domination. By definition, members are theoretically "neutral." In truth, however, no state or nation can afford to remain neutral in today's global economy, so the Nonaligned Movement is more of an ideal than a reality. Of the 101 member states, 10 are in Middle America; with the exceptions of Panamá, Nicaragua, and Belize, all are in the Caribbean. There are also "observer nations," which include Costa Rica, Antigua and Barbuda, El Salvador, Mexico, and Dominica; the Dominican Republic has "guest nation" status.

In an internationally focused world, international problems are certain to arise. To settle such problems, the Permanent Court of Arbitration was established at the turn of this century and is still headquartered in the Netherlands (The Hague). Today the association contains 75 members who try to resolve international disputes and conflicts. Members are also involved in the selection of candidates for the Nobel Peace Prize. Within Middle America, Cuba, Panamá, Guatemala, Haiti, Mexico, Honduras, Nicaragua, and the Dominican Republic are members. A very similar association is the International Court of Justice (ICJ). As a branch of the United Nations (UN), the ICJ (occasionally called the World Court) differs from the Permanent Court in that it is only supposed to treat issues that arise between or affect UN nations (see below). But since both courts have their headquarters in The Hague, there is often confusion about the purposes of the two bodies and their respective jurisdictions. Because the ICJ has more members, however, it is generally regarded as the more important and more influential of the two arbitrators.

Special mention should be made of two international, Soviet-based organizations because Cuba was a member of both. The former Council for Mutual Economic Assistance (COMECON) was the Soviet equivalent of the European Economic Community. Designed to foster cooperation, provide technical assistance, and simplify trade agreements between the planned economies of

the Soviet Union and kindred countries, COMECON had only 11 members, and Cuba was the only member in the western hemisphere. Mexico, however, was one of only three "cooperating countries," and Nicaragua was an "observer" to COMECON affairs. (Recent changes in Eastern Europe and the Soviet Union toward more free market-oriented economies have prompted the Soviets to disband COMECON. To date, however, it is too early to assess the impact of the break up of this organization and the economic future of the former Soviet Bloc countries.) A similar organization is the International Bank for Economic Cooperation (IBEC). The IBEC was established to be the equivalent of the International Monetary Fund (IMF) for the Soviet Bloc, and only states with planned economies were allowed to join. The IBEC was the financial heart of COMECON, supplying member countries with the capital to realize long-term, large-scale development projects. Cuba, once again, was the only member in this hemisphere and, so far as is known, had preferential status to borrow money as needed.

The most universally recognized international organization, one that links member countries to a global perspective, is, of course, the United Nations. Founded in 1945 and headquartered in New York, the UN is perhaps the most effective international association in the world. As of March 1987, there were 159 regular members in the UN General Assembly, including all Middle American states. The UN is a complex and massive institution, which is divided into a multitude of specialized agencies and organizations. Some of the major agencies include UNICEF, the United Nations International Children's Fund; UNDP, the United Nations Development Programme; UNDRCO, the United Nations Disaster Relief Coordinator's Office; UNHCR, the United Nations High Commissioner for Refugees; UNRWA, the United Nations Relief and Works Agency for Palestine Refugees in the Near East; and WFC, the World Food Council. Since these are all branches of the main body, all member states (and thus all Middle American countries) are members. In addition, the UN has virtually dozens of standing committees and hundreds of regular committees, and most countries in the region participate. More important to our discussion, however, are the special, peripheral agencies of the UN, to which members (and occasionally nonmembers) opt to affiliate. Table 10.6 lists the 18 most important agencies as well as the dates selected Middle American states were admitted to the UN. A brief analysis of the membership of these agencies will illustrate the region's strong ties to the international community.

TABLE 10.6

Major United Nations Agencies and Middle American Participation, March 1987

	DATE ADMITTED TO THE UN	FAO	GATT	IAEA	IBRD	ICAO	IDA	IFAD	IFC	ILO	IMF	IMO	ITU	UNESCO	UNIDO	UPU	WHO	WIPO	WMO	TOTAL PARTICIPATION IN THE 18 AGENCIES
Antigua and Barbuda	1981	Yes	Yes		Yes	Yes		Yes		Yes	Yes	Yes		Yes			Yes			10
Bahamas	1973	Yes	Yes		Yes	Yes				Yes	Yes	Yes	Yes	Yes		Yes	Yes	Yes	Yes	13
Barbados	1966	Yes	Yes		Yes	Yes		Yes	Yes	Yes	Yes	Yes	Yes	Yes	Yes	Yes	Yes	Yes	Yes	16
Belize	1981	Yes	Yes		Yes		Yes	Yes	Yes	Yes	Yes	Yes	Yes	Yes	Yes	Yes	Yes			14
Costa Rica	1945	Yes		Yes	Yes	Yes	Yes	Yes	Yes	Yes	Yes	Yes	Yes	Yes	Yes	Yes	Yes	Yes	Yes	16
Cuba	1945	Yes	Yes	Yes		Yes				Yes			Yes	Yes	Yes	Yes	Yes	Yes	Yes	14
Dominica	1978	Yes	Yes		Yes			Yes		Yes	Yes			Yes	Yes	Yes	Yes		Yes	14
Dominican Republic	1945	Yes	Yes	Yes	Yes	Yes	Yes	Yes	Yes	Yes	Yes	Yes	Yes	Yes	Yes	Yes	Yes		Yes	17
El Salvador	1945	Yes		Yes	Yes	Yes	Yes	Yes	Yes	Yes	Yes	Yes	Yes	Yes	Yes	Yes	Yes		Yes	16
Grenada	1974	Yes	Yes		Yes	Yes	Yes			Yes	Yes	Yes		Yes		Yes	Yes		Yes	14
Guatemala	1945	Yes		Yes	Yes	Yes	Yes	Yes	Yes	Yes	Yes	Yes	Yes	Yes	Yes	Yes	Yes	Yes	Yes	17
Haiti	1945	Yes	Yes	Yes	Yes	Yes	Yes	Yes	Yes	Yes	Yes	Yes	Yes	Yes	Yes	Yes	Yes	Yes	Yes	18
Honduras	1945	Yes		Yes	Yes	Yes	Yes	Yes	Yes	Yes	Yes	Yes	Yes	Yes	Yes	Yes	Yes	Yes	Yes	16
Jamaica	1962	Yes	Yes	Yes	Yes	Yes	Yes	Yes	Yes	Yes	Yes	Yes	Yes	Yes	Yes	Yes	Yes	Yes	Yes	17
Mexico	1945	Yes		Yes	Yes	Yes		Yes	Yes	Yes	Yes	Yes	Yes	Yes	Yes	Yes	Yes	Yes	Yes	18
Nicaragua	1945	Yes	Yes	Yes	Yes	Yes	Yes	Yes	Yes	Yes	Yes	Yes	Yes	Yes	Yes	Yes	Yes	Yes	Yes	18
Panamá	1945	Yes		Yes	Yes	Yes	Yes	Yes	Yes	Yes	Yes	Yes	Yes	Yes	Yes	Yes	Yes	Yes	Yes	17
St. Kitts-Nevis	1983	Yes			Yes	Yes	Yes	Yes			Yes			Yes	Yes					6
St. Lucia	1979	Yes	Yes		Yes	Yes	Yes		Yes	Yes	Yes	Yes		Yes	Yes	Yes	Yes		Yes	15
St. Vincent and the Grenadines	1980	Yes	Yes		Yes	Yes	Yes	Yes			Yes	Yes	Yes	Yes		Yes	Yes			12
Trinidad and Tobago	1962	Yes	Yes		Yes	Yes	Yes		Yes	Yes	Yes	Yes	Yes	Yes	Yes	Yes	Yes		Yes	15

NOTE: ICAO = International Civil Aviation Organization, headquartered in Montreal; IDA = International Development Association, headquartered in Washington, D.C.; IFAD = International Fund for Agricultural Development, headquartered in Rome; IFC = International Finance Corporation, headquartered in Washington, D.C.; ILO = International Labour Organization, headquartered in Geneva; IMO = International Maritime Organization, headquartered in London; ITU = International Telecommunication Union, headquartered in Geneva; UNIDO = UN Industrial Development Organization, headquartered in Vienna; UPU = Universal Postal Union, headquartered in Berne; WIPO = World Intellectual Property Organization, headquartered in Geneva; and WMO = World Meteorological Organization, headquartered in Geneva.

SOURCE: Arthur S. Banks, et al., ed., *Political Handbook of the World: 1987*, (Binghamton, N.Y.: CSA Publications, State University of New York, 1987).

First, 10 Middle American countries were among the original 51 countries that established the UN in 1945; all 10 are still members. Since obtaining independence, some as recently as the 1980s, all other independent states in the region have decided to join. Further, although most belong to the 18 specialized agencies in the table, some patterns are also evident. *All* Middle American member countries belong to only two of these agencies, the FAO (Food and Agriculture Organization) and UNESCO (United Nations Educational, Scientific and Cultural Organization), while all but one (St. Kitts–Nevis) belongs to WHO (World Health Organization). The FAO, which is headquarters in Rome, attempts to help countries raise their standards of living through better nutrition; it devotes particular attention to rural populations. The FAO works closely with WHO, a Geneva-based agency mandated to improve the standards of personal health around the world. The WHO acts primarily as a coordinating agency, in conjunction with numerous national and local health groups. In recent years, its main focus has been on the global eradication of the AIDS virus. Both associations work with UNESCO, which attempts to upgrade the world condition through scientific and cultural cooperation. With headquarters in Paris, UNESCO spends millions of dollars on educational programs, most aimed at raising literacy rates and improving cross-national scientific communication. In sum, all three are humanitarian in outlook, so a country might have fewer reasons to decide not to belong than in the case of some of the other, more political agencies.

With the exception of Cuba, all states in the region belong to the IMF and the IBRD, but Cuba has not joined for purely political reasons. The IMF (International Monetary Fund) and the IBRD (International Bank for Reconstruction and Development) are the Western equivalents of the IBEC. Both agencies are located in Washington, D.C., and both derive from the conference held at Bretton Woods, New Hampshire, on July 22, 1944. The IMF is a permanent international lending institution that fosters unencumbered exchange and foreign trade, helps states achieve a positive balance of payments, and tries to eliminate foreign debt. In short, it is a bank for countries needing money to offset low payments resulting from poor harvests, changing climatic conditions, poor market years, and the like. The IBRD, on the other hand, lends huge amounts of money to developing countries to finance major building projects and provides technical assistance unavailable in de-

veloping nations. Over the years, the differences between the two agencies have become largely a matter of scale. In the past, the Soviet Bloc states have not participated in the activities of these two agencies, choosing instead to utilize IBEC funds.

Looking at the membership pattern of the individual countries in the region reveals another story. Note that only three countries—Haiti, Mexico, and Nicaragua—belong to all of these 18 agencies. Haiti, of course, has little choice: given its poor relative status, it simply cannot afford not to join. Mexico, on the other hand, belongs for the opposite reason: as the most prosperous country in the region, it is virtually obligated to participate in the affairs of these agencies; in order to receive money from agencies like the IMF and the IBRD, Mexico must contribute significantly to other organizations, like the IAEA (International Atomic Energy Agency) and GATT (General Agreement on Tariffs and Trade). Nicaragua is another poor state with a rapidly growing population, so it makes sense for it to join as many associations as possible. The Dominican Republic and Guatemala, both of which belong to 17 UN agencies, are in a similar situation. Panamá and Jamaica are not as poor, but both belong to 17 agencies in an effort to maintain close ties with the United States (in the case of Panamá) and to foster a long-standing image of relative growth and prosperity (in the case of Jamaica).

Note that of the 21 countries listed in Table 10.6, 13 belong to 15 or more of the 18 agencies, a definite sign of global interdependence. There are hundreds of additional, less important international associations and subagencies, some of which are headquartered in Middle America.

Membership in international organizations is only one indicator of interdependency, however. Perhaps a better indicator is the global origin and distribution of Middle American imports and exports, respectively. Table 10.7 shows the 1982 values of such imports and exports for selected countries. Shown also are the worldwide totals (in millions of U.S. dollars) for each state, the portion of that total derived from both the industrialized countries (with a further breakdown for trade with the EEC) and the planned economies, and the major trading partner within both spheres. This is similar to the information in Table 10.5, and the two tables should be studied together.

Note, in every case, with the obvious exception of Cuba, the relatively small percentage of imports and exports that originates in or is sold to the planned econ-

TABLE 10.7

Exports and Imports of Selected Middle American Countries, (by Non-Latin American Region), 1982

	EXPORTS (MILLIONS OF U.S.$)		IMPORTS (MILLIONS of U.S.$)
COSTA RICA			
World total	897.7	World Total	870.0
Industrial nations	606.1	Industrial nations	503.9
EEC	231.0	EEC	87.0
Major partner: USA	307.3	Major partner: USA	331.1
Planned economies	19.6	Planned economies	1.5
Major partner: Albania	6.2	Major partner: Czechoslovakia	0.9
CUBA			
World total	1,207.1	World total	1,565.2
Industrial nations	521.3	Industrial nations	1,021.5
EEC	200.2	EEC	422.6
Major partner: Japan	103.5	Major partner: Canada	289.8
Planned economies	4,259.0*	Planned economies	5,081.0*
Major partner: USSR	2,455.0*	Major partner: USSR	323.0*
DOMINICAN REPUBLIC			
World total	809.0	World total	1,250.0
Industrial nations	623.4	Industrial nations	728.3
EEC	53.5	EEC	88.2
Major partner: USA	437.2	Major partner: USA	486.1
Planned economies	99.8	Planned economies	0.2
Major partner: USSR	68.5	Major partner: Former East Germany and Czechoslovakia	0.2
EL SALVADOR			
World total	879.2	World total	929.0
Industrial nations	536.7	Industrial nations	473.4
EEC	173.8	EEC	77.2
Major partner: USA	302.8	Major partner: USA	320.9
Planned economies	0.0	Planned economies	0.4
Major partner: None	0.0	Major partner: Czechoslovakia	0.3
GUATEMALA			
World total	1,244.6	World total	1,340.5
Industrial nations	676.2	Industrial Nations	777.5
EEC	201.1	EEC	136.3
Major partner: USA	331.5	Major partner: USA	428.9
Planned economies	4.3	Planned economies	3.5
Major partner: Czechoslovakia	3.1	Major partner: USSR	1.9

NOTE: Figures are rounded and may not equal 100 percent.
* Figures are from 1981 and are in millions of pesos.

omies. With minor exceptions, the greatest portion of each country's imports come from the United States, and the major percentage of its exports are sent to the United States; in other words, the United States is the major trading partner of these Middle American states. In fact, the United States apparently purchases more goods from and sells more goods to these countries than all EEC nations combined. Cuba, of course, has no trade relations with the United States, so it turns to the Soviet Union, Japan, Mexico, and Canada to compensate. (Canada, the United States' largest trading partner, flatly refused to cease trade relations with Cuba when the United States terminated diplomatic and trade agreements with Cuba in the early 1960s; consequently, Canada enjoys a substantial portion of Cuba's import market.) Panamá, sometimes called "the Hong Kong of the West," trades with a multitude of countries; the free trade zone around Colón has developed into the largest

TABLE 10.7 (continued)

Exports and Imports of Selected Middle American Countries, (by Non-Latin American Region), 1982

	EXPORTS (MILLIONS OF U.S.$)		IMPORTS (MILLIONS of U.S.$)
HAITI			
World total	380.0	World total	524.9
Industrial nations	369.1	Industrial Nations	445.2
EEC	59.3	EEC	57.4
Major partner: USA	296.0	Major partner: USA	328.5
Planned economies	0.0	Planned economies	1.7
Major partner: None	0.0	Major partner: Czechoslovakia	1.0
HONDURAS			
World total	744.6	World total	710.8
Industrial nations	645.6	Industrial Nations	433.9
EEC	121.2	EEC	61.1
Major partner: USA	386.9	Major partner: USA	302.1
Planned economies	0.0	Planned economies	3.4
Major partner: None	0.0	Major partner: Czechoslovakia	1.6
MEXICO			
World total	21,163.0	World total	15,372.0
Industrial nations	18,097.0	Industrial nations	12,375.0
EEC	2,385.0	EEC	1,717.0
Major partner: USA	11,887.0	Major partner: USA	8,921.0
Planned economies	40.0	Planned economies	127.0
Major partner: Cuba	22.0	Major partner: Cuba	94.0
NICARAGUA			
World total	365.5	World total	686.3
Industrial nations	270.1	Industrial nations	275.0
EEC	86.1	EEC	83.9
Major partner: USA	89.3	Major partner: USA	130.5
Planned economies	0.0	Planned economies	0.0
Major partner: None	0.0	Major partner: None	0.0
PANAMÁ			
World total	308.1	World total	2,901.3
Industrial nations	189.3	Industrial nations	1,894.5
EEC	42.3	EEC	223.3
Major partner: USA	127.9	Major partner: Japan	880.2
Planned economies	.5	Planned economies	1.8
Major partner: Cuba	.5	Major partner: Czechoslovakia	0.7

SOURCE: J. W. Wilkie and D. Corey, eds., *Statistical Abstract of Latin America*, vol. 25, (Los Angeles: University of California UCLA Latin American Center Publication, 1986).

free trade area in all of Latin America. Mexico exports more (in dollar value) than all the other Middle American countries combined; a comparable situation exists with regard to Mexican imports. Most countries in the region are forced to import more than they export, regardless of where they sell or buy their commodities.

Table 10.8 is similar to 10.7. Here the same Middle American states are shown in relation to their global trading patterns, further broken down by major, capi-

talist, industrialized partners. Again, except for Cuba, the United States dominates, but look closely at former West Germany and especially at Japan, both of which are solvent countries and rising dramatically within international financial circles; compare also their trade figures, and note the relative increase, as a percentage of total sales, over only a three-year period (1980–1982). When combined with the EEC total, the 1982 totals for the United States and Japan account for, in every case except Cuba's, greater than 32 percent of all imports to, and nearly 50 percent of all exports from, Middle America; in some cases, this figure is as high as almost 95 percent (Haiti). Clearly, the region is interdependent on the rest of the world and a long way from being isolated and inward looking. Figures are currently unavailable, but the total of Middle American foreign imports and exports seems to have risen steadily since 1982. The region exports a considerable amount of domestic products, but it also receives a very significant portion of its domestic needs from other world regions.

Middle America's isolationist era is apparently at an end. It has become an integral part of the global economy. In addition to what has already been mentioned, the region is interdependent with the remainder of the world in a variety of ways. For instance, migration—

TABLE 10.8

Selected Middle American Countries and Their Non-Latin American Trading Partners, as a Percentage of Total Trade

	UNITED STATES			UNITED KINGDOM			FORMER WEST GERMANY		
	1980	1981	1982	1980	1981	1982	1980	1981	1982
COSTA RICA									
Exports	33.8	32.5	33.5	0.3	1.0	2.3	11.3	12.2	14.3
Imports	34.3	33.3	35.6	1.8	1.5	2.7	4.8	4.6	3.9
CUBA									
Exports			0.1*	5.0*	2.2*	2.4*	4.7*	1.8*	2.4*
Imports			0.1*	4.6*	4.0*	7.7*	6.2*	5.0*	4.3*
DOMINICAN REPUBLIC									
Exports	46.3	67.0	54.0	0.2	0.4	0.2	0.2	0.1	0.2
Imports	44.8	43.0	38.9	1.3	1.3	1.1	2.6	2.4	2.2
EL SALVADOR									
Exports	39.7	17.1	34.6*	0.6	0.4	0.4*	19.4	21.8	15.7*
Imports	32.4	25.5	33.6*	1.3	0.9	1.0*	2.8	3.1	3.2*
GUATEMALA									
Exports	27.7	18.2	27.3	4.4	1.2	1.7	8.3	8.2	7.0
Imports	34.5	33.8	31.1	2.3	1.8	1.2	5.4	6.4	5.6
HAITI									
Exports	69.5	79.4*	78.6*	0.6	0.8*	1.1*	2.6	2.7*	3.2*
Imports	57.2	56.3*	61.0*	1.2	0.7*	1.3*	1.8	1.5*	2.3*
HONDURAS									
Exports	49.1	54.8	52.6	0.9	0.9	0.8	11.5	8.2	8.8
Imports	40.4	41.5	41.0	2.9	1.4	2.1	2.1	3.6	3.4
MEXICO									
Exports	63.2	55.3	52.0	0.5	1.2	4.2	1.7	4.6	4.3
Imports	65.6	63.8	59.9	2.2	1.8	1.8	5.2	4.9	6.1
NICARAGUA									
Exports	37.6	28.2*	24.4*	0.6	0.4*	1.4*	8.6	6.6*	9.3*
Imports	32.7	25.2*	18.9*	0.8	0.6*	1.4*	2.0	3.4*	2.2*
PANAMÁ									
Exports	49.3	52.7	41.5	0.1	0.1	0.2	5.1	7.6	5.9
Imports	33.8	34.8	35.0	1.1	1.1	1.2	1.8	1.8	1.8

* Estimated figure.
NOTE: Cuba's trade with the United States in 1980 and 1981 was zero or neglible.

both legal and otherwise—between Middle America and the rest of the world is at the highest level since the colonial period. Some Middle American cities are now home to many foreign ethnic groups who have settled there since World War II. Mexico City, for example, has become one of the most cosmopolitan cities in the western hemisphere, and several of the Caribbean islands are host to an ever-growing number of foreign visitors each year. Jamaica has become the center for writers from developing countries. It is unfortunate, but some Central American states have become major purchasers of foreign-made weapons, war-related technology, and other defense-related paraphernalia. People, food, goods, capital, and even ideas flow into Middle America from abroad. Interdependence, as much as dependence, has indeed become a way of life for the region.

TABLE 10.8 (continued)

Selected Middle American Countries and Their Non-Latin American Trading Partners, as a Percentage of Total Trade

	JAPAN			EEC NATIONS			UNITED STATES, JAPAN, AND THE EEC		
	1980	1981	1982	1980	1981	1982	1980	1981	1982
COSTA RICA									
Exports	0.8	0.5	0.7	23.2	22.5	25.9	57.8	55.5	60.1
Imports	11.6	9.8	4.2	11.1	11.8	10.9	57.0	54.9	50.7
CUBA									
Exports	15.4*	10.1	8.5	28.6	12.7	16.5	44.0*	28.8*	25.1*
Imports	13.6*	14.2	8.4	34.6	24.7	25.8	48.2*	38.9*	34.3*
DOMINICAN REPUBLIC									
Exports	0.9	0.8	0.8	8.4	5.5	6.6	55.6	73.3	61.4
Imports	8.0	6.2	5.1	8.8	3.8	4.1	61.6	53.0	48.1
EL SALVADOR									
Exports	3.3	4.5	2.3*	28.8	24.3	19.8	71.8	45.9	56.7
Imports	3.7	2.7	2.8*	10.0	13.7	8.1	46.1	41.9	44.5
GUATEMALA									
Exports	2.8	4.9	5.0	24.1	22.0	17.5	54.6	45.1	49.8
Imports	8.0	7.7	5.2	12.5	13.5	11.8	55.0	55.0	48.1
HAITI									
Exports	0.2	0.3*	0.7*	23.3	3.6	15.4	93.0	83.3	94.7
Imports	5.4	3.6*	5.2*	9.5	15.1	10.7	72.1	75.0	76.9
HONDURAS									
Exports	3.7	6.2	5.8	18.0	18.8	21.0	70.8	79.8	79.4
Imports	8.9	6.7	6.8	10.5	12.4	11.1	59.8	60.6	58.9
MEXICO									
Exports	3.7	6.0	6.7	6.6	8.5	12.0	73.5	69.8	70.7
Imports	5.3	5.0	5.7	13.8	12.6	14.8	84.7	81.4	80.4
NICARAGUA									
Exports	2.6	11.6*	12.0*	19.7	17.7	23.4	59.9	57.5	59.8
Imports	2.7	1.6*	1.4*	6.6	9.0	12.2	42.0	35.8	32.5
PANAMÁ									
Exports	0.4	0.1	0.1	12.9	17.6	13.7	62.6	70.4	55.3
Imports	6.1	6.1	7.6	6.4	5.6	7.1	46.3	46.5	49.7

SOURCE: J. W. Wilkie and D. Corey, eds., *Statistical Abstract of Latin America*, vol. 25, (Los Angeles: University of California UCLA Latin American Center Publication, 1986).

PROBLEMS AND PROSPECTS

To forecast the future of a region, or even of a single nation, is extremely difficult today. The planet is changing at a quickened pace, and rapidly changing events require immediate responses. Consequently, the world of tomorrow can really be no more than an "educated guess" at best. Predicting the future of an entire world region is, of course, much more difficult. No one in the 1970s could have accurately foretold what Middle America would be like today, and it is equally difficult to know what the turn of the century will bring from today's perspective. Therefore, we can only suggest broad approximations about conditions in the 1990s and beyond.

In some ways, it is almost necessary to know what other regions will do before we can suggest what might happen to this region. By definition, any state in the global economy affects the conditions within and the actions of any other state. Yet, some things are a relative certainty. For instance, unless the balance of politics in the world changes dramatically, we can be reasonably assured that Middle America will remain a close ally of the United States and the rest of the free world. Its proximity and long-standing trade relations with two economic giants, the United States and Canada, will certainly continue. Some countries in the region are lending support, both politically and economically, to the Soviet Union, but pressure from the democratic Latin America states and changes within the Soviet Union may help to encourage these countries to maintain ties with the major free nations of the western hemisphere. Their respective economies are far too reliant on this trade to make such a change profitable—at least not in the short run.

One thing is definite: Middle America will continue to develop at a rapid rate; and, by the early decades of the next century, unless something catastrophic befalls the region, most countries should be far more self-sufficient than they are today. Of course, this will depend on world conditions, especially the global economy. If all the major countries of the world spiral into a depression, then Middle America will have no choice but to be drawn into the depression. Some analysts also suggest that regions will follow the internal pattern of Western societies: as today's "rich" regions get richer, other regions will necessarily become poorer due to

dwindling resources. Some desperately poor countries, like Haiti, will probably become even poorer on a relative basis. Such countries are in such dire conditions already that their future holds little prospect of improvement—unless education reform and massive amounts of financial and humanitarian aid are forthcoming. But assuming the remainder of the 1990s are at least as prosperous as the 1980s (for the world in general), most of Middle America should experience progress and economic growth.

Population growth, however, will be the single greatest determinate of the region's potential. If Middle America loses sight of its potential population problem, then all the political stability in the world will mean nothing. Runaway population growth will prevent even the most benevolent governments from instituting beneficial social and economic reforms. Already destitute countries such as Haiti (expecting a threefold population increase by 2025) will suffer the most; states like Nicaragua and Honduras (with high reproduction rates that are not expected to ease much in the next 20 years) are likewise vulnerable. As the world increasingly moves toward a more capital-based, Western life-style, food consumption will become a sensitive issue. Given its geographic location (in the tropics and subtropics), the region is currently unable to produce a variety of foods common in the midlatitudes. If the Middle American states cannot begin to produce a significant proportion of what they consume, they will surely face considerable hardships in the coming decades. If there are too many mouths to feed, dramatic changes will necessarily follow.

Because of its geography, Mexico has probably the greatest potential of all. If it can reduce its currently high foreign debt, Mexico has all the resource potential to become self-sufficient. The other countries have less potential because they have fewer available resources. The natural resources of Mexico may, however, only be able to sustain Mexicans at a traditional or near-traditional standard of living, not in a Western manner. Nevertheless, like the remainder of the region, Mexico is moving more and more toward the Western model of development, and with the adoption of that model comes the adoption of a Western or a quasi-Western life-style. With a population expected to double by 2025, to over 154 million people, the time is rapidly approaching when Mexico needs to reconsider its long-term priorities. This question of priorities is a universal concern that all developing countries must address. The rapidly developing, multiresource states like Mexico will have to find solutions to the issue first; after all, in time they will become models of development for the other countries within the region.

Suggested Readings

Blakemore, H., and C. T. Smith, eds. *Latin America: Geographical Perspectives,* 2d ed. London and New York: Methuen, 1983.

Booth, J. A., and T. Walker. *Understanding Central America.* Boulder, Colo.: Westview Press, 1989.

Chin, H. E., ed. *The Caribbean Basin and the Changing World Economic Structure.* Groningen: Wolters-Noordhoff, 1986.

Davidson, W. V. "Geography of Minority Populations in Central America." In R. G. Boehm and S. Visser, eds., *Latin America: Case Studies,* pp. 31–37. Dubuque, Iowa: Kendall/Hunt, 1984.

Falcoff, M., and R. Royal, eds. *The Continuing Crisis: U.S. Policy in Central America and the Caribbean.* Washington, D.C.: Ethics and Public Policy Center, 1987.

Graham, N. A., and K. L. Edwards, et al. *The Caribbean Basin to the Year 2000: Demographic, Economic, and Resource-Use Trends in Seventeen Countries.* Boulder and London: Westview Press, 1984.

Gwynne, R. N. *Industrialization and Urbanization in Latin America.* London and Sydney: Croom Helm, 1985.

James, P. E., and C. W. Minkel, eds. *Latin America,* 5th ed. New York: John Wiley & Sons, 1986.

Merrick, T. W., et al. "Population Pressures in Latin America." *Population Bulletin* 41 (July 1986): 1–50.

REFUGEES (United Nations High Commissioner for Refugees) 48 (December 1987).

Sauer, C. O. *The Early Spanish Main.* Berkeley and Los Angeles: University of California Press, 1966.

West, R. C., and J. P. Augelli. *Middle America: Its Land and Peoples,* 2d ed. Englewood Cliffs, N.J.: Prentice-Hall, 1976.

Wilkie, J. W., and A. Perkal, eds. *Statistical Abstract of Latin America: Volume 24.* Los Angeles: University of California (UCLA Latin American Center Publication), 1985.

Wilkie, R. W. *Latin American Population and Urbanization Analysis: Maps and Statistics, 1950–1982.* Los Angeles: University of California (UCLA Latin American Center Publication), 1984.

World Development Report 1987. New York: Oxford University Press, for The World Bank, 1987.

CHAPTER 11

South America

IMPORTANT TERMS

El Niño

Slash and burn agriculture

Entrepôt

Plantation

Carrying capacity

Squatter settlements

Frontier migration

Greenhouse effect

CONCEPTS AND ISSUES

■ Pressure of frontier settlements in the Amazon lowland on tropical rainforests and native peoples.

■ Impact of cocaine production on local or regional economies and cultures.

■ Pan-Americanism and the dominance of the United States.

■ Rural migration to urban centers and frontier areas.

CHAPTER OUTLINE

PHYSICAL ENVIRONMENT
Physiography
Climate
Vegetation
Soils
Resources

HUMAN ENVIRONMENT
Aboriginal Inhabitants
Colonial Impacts

Modern Population
Culture
Human Landscape
Regions of South America

SPATIAL CONNECTIVITY
Economic Linkages
Trade
Transportation

Telecommunications
Tourism

PROBLEMS AND PROSPECTS
Population Growth
Environmental Degradation
Economic Development
Political Stability

South America is a continent of contrasts in terms of both physical and cultural variation. Glacier-capped mountains nearly 23,000 feet (7000 meters) high in the southern Andes stand in sharp contrast to the lowland pampas of Argentina and the coastal plains of the Guianas. The driest deserts on earth are found along the northern coast of Chile, yet across the Andes are the lush rainforests of the upper Amazon Basin. While some Amerindian inhabitants of the rainforest have only recently made contact with the outside world, the urbanites of Rio de Janeiro, São Paulo, and Buenos Aires look to Paris, Milan, and New York for the latest fashions in attire.

Inhabited on the eve of European discovery by perhaps 15 million aboriginal "Indians," whose cultures ranged from very traditional to highly developed, South America was quickly divided up between the Iberian countries of Spain and Portugal. Iberian colonists were sent to claim, settle, and exploit the new territories. Many demographic changes soon took place, including the decimation of much of the native population by introduced diseases. Following ill-fated attempts at enslaving the indigenous peoples, a labor supply for the mines and plantations was forcibly imported from Africa. Today a high proportion of the coastal population of the east and north coasts of South America is of African descent. Elsewhere, there was substantial mixing of the European and native populations, which resulted

in a large **mestizo** class that is characteristic of much of highland South America today. European immigration was important in the temperate southern South American countries in the late nineteenth and early twentieth centuries, and today portions of those countries remain quite European in terms of both population and cultural landscape.

South America's present population of nearly 300 million is distributed rather unevenly among 12 independent countries and the two small foreign "dependencies" (Figure 11.1). Over half the total population is in the Portuguese-speaking country of Brazil, which comprises slightly less than half of the total landmass. With the exceptions of the three sparsely populated Guianas—Guyana (formerly British Guiana), Suriname (formerly Dutch Guiana), and French Guiana (officially a French overseas *département*)—and the 2000-inhabitant offshore British colony of the Falkland Islands (or Islas Malvinas, as Argentina claims), the rest of the South American countries are Spanish speaking and trace their political roots to the disintegration of the Spanish colonial empire in the 1820s. Argentina, with slightly over 32 million inhabitants, is the most populous of the Spanish-speaking countries, and its northern neighbor Paraguay is the least populous with 4.3 million. Although the overall population density is relatively moderate (41 inhabitants per square mile; 16 inhabitants per square kilometer), South Americans are

Street Fair in Plaza, San Telma.
Buenas Aires, Brazil.

FIGURE 11.1

Political Units and Major Cities of South America

STATISTICAL PROFILE: SOUTH AMERICA

REGION OR COUNTRY	Population Estimate mid-1990 (Millions)	Natural Increase (Annual)	Population Projected to 2000 (Millions)	Infant Mortality Rate	Life Expectancy at Birth (Years)	Urban Population (%)	Per Capita GNP, 1988 (US$)	Area, Thousands of Square Miles (Km²)	Population Density, No./mi² (No./km²)
Argentina	32.3	1.3	36.0	32	71	85	2640	1068.3(2766.9)	30(12)
Bolivia	7.3	2.6	9.3	110	53	49	570	424.2(1098.6)	17(7)
Brazil	150.4	1.9	179.5	63	65	74	2280	3286.5(8512)	46(18)
Chile	13.2	1.7	15.3	18.5	71	84	1510	292.3(757)	45(17)
Colombia	31.8	2.0	38.0	46	66	68	1240	439.7(1139)	72(28)
Ecuador	10.7	2.5	13.6	63	65	54	1080	109.5(283.6)	98(38)
Falkland Islands	0.002	—	—	—	—	59	—	4.7(12.2)	0.4(0.2)
French Guiana	0.09	—	—	—	—	73	—	35.1(91)	2.6(1.0)
Guyana	0.8	1.9	0.8	30	67	33	410	83.0(2.5)	9(3)
Paraguay	4.3	2.8	5.5	42	67	43	1180	157.1(406.8)	27(11)
Peru	21.9	2.4	26.4	76	65	69	1440	496.2(1285.2)	44(17)
Suriname	0.4	2.0	0.5	40	68	66	2450	63.0(163.3)	6(2)
Uruguay	3.0	0.8	3.2	22.3	71	87	2470	68.0(176.2)	45(17)
Venezuela	19.6	2.3	24.1	33	70	83	3170	352.1(912)	56(22)

unevenly distributed. Many highland and coastal regions are so overcrowded that they are at or near their carrying capacities, at least under present levels of technology and economic development.

Population pressures have led both to the degradation of natural resources, such as agricultural and forested land, and to the proliferation of slums, or squatter settlements, in the numerous large cities scattered around the periphery of the continent. Rates of urbanization in several of the countries, especially those subject to high rates of European immigration during the last hundred years, equal or exceed those of most developed countries of the world. The traditionally more agrarian countries—with a higher percentage of Indian inhabitants—have a proportionally lower urban component and are, in regard to levels of urbanization, more in line with patterns of developing countries of Asia and Africa.

Although spared the nineteenth-century European colonial "scrambles" that characterized Asia and Africa, South America's economic dependency upon the developed world has resulted in numerous boom-and-bust cycles. Historically, the economy of South America has been oriented toward demand for its raw materials by

Europe and North America, and fluctuations in exports of sugar, gold, silver, tin, copper, rubber, bananas, and coffee have not allowed for the development of very stable economies. In the 1980s, new types of exports—illicit drugs, chiefly cocaine and marijuana—have fueled the economies of several South American countries. Controversies over this new trade have led to much antagonism between the beneficiary South American countries and the developed countries that are the destinations of the illicit exports. In recent decades, Common Market–style economic liaisons have stimulated internal economic growth, but the traditional trade linkages with Europe, North America, and now Japan remain as the strong spatial connectivity links today.

PHYSICAL ENVIRONMENT

PHYSIOGRAPHY

Stretching nearly 5000 miles (8000 kilometers) north to south (from about 12° north to 55° south latitude) and over 3000 miles (5000 kilometers) from east to west (from about 35° to 80° west longitude) not counting offshore islands, the familiar triangular-shaped continent of South America encompasses an area of nearly 7 million square miles (18 million square kilometers), almost twice the size of the United States.

Physiographically, the continent can be divided into three major units: a plateau region on the eastern side, lowland plains in the central portion, and a narrow range of high mountains on the western side (Figure 11.2). The two major plateaus—the Guiana and Brazilian highlands—are the oldest sections of the continent. They have the appearance of a highly weathered and eroded landscape, partly covered by stratified rocks and lava flows. Except for the lusher vegetation cover, the remote Guiana highlands resemble the eroded plateau landscape of Arizona's Monument Valley. Angel Falls, descending nearly 3300 feet (1000 meters) from a high plateau *(tepui)* in Venezuela, is the world's highest waterfall. The Brazilian highlands extend quite close to the Atlantic, forming a sharp escarpment, or cliff, at the edge of the narrow coastal plain upon which cities such as Rio de Janeiro and Salvador de Bahía are located. Although quite scenic in appearance, this escarpment historically has served as a barrier to easy colonization of Brazil's interior. The interior low plains of the continent have received many of the sediments eroded from the older highlands, and most of South America's major river systems—including the Orinoco, Amazon, and Paraná/Paraguay—are found here.

The third major physical unit is the geologically young Andean **cordillera.** South America is entirely embedded in a continental plate, and at its western edge, plate collision has forced the **subduction** of the Pacific continental plate under the South American plate. As a result, deep oceanic trenches lie immediately offshore, and extensive volcanic and earthquake activity has created a mountain chain with over 50 peaks exceeding 20,000 feet (6000 meters) in elevation. The Andean cordillera, although over 5000 miles (8000 kilometers) long, has an average width of only about 200 miles (320 kilometers), except near the Peru-Bolivia border. In this area, the Andes split into an eastern and western chain, and between them lies the **altiplano,** a basin 14,000 feet (4250 meters) high within which several lakes have formed from local snowmelt. Lake Titicaca, encompassing 3200 square miles (8300 square kilometers), is the largest lake on the continent.

CLIMATE

As much of South America lies within the tropics, tropical rainy climates prevail over a large portion of the

A vista of the Brazilian Highlands near Petrópolis. Petrópolis is located slightly north of Rio de Janeiro on the southern edge of the Brazilian Highlands. The highlands are important for agriculture, livestock ranching, and mining—particularly iron ore.

FIGURE 11.2

Physical Features of South America

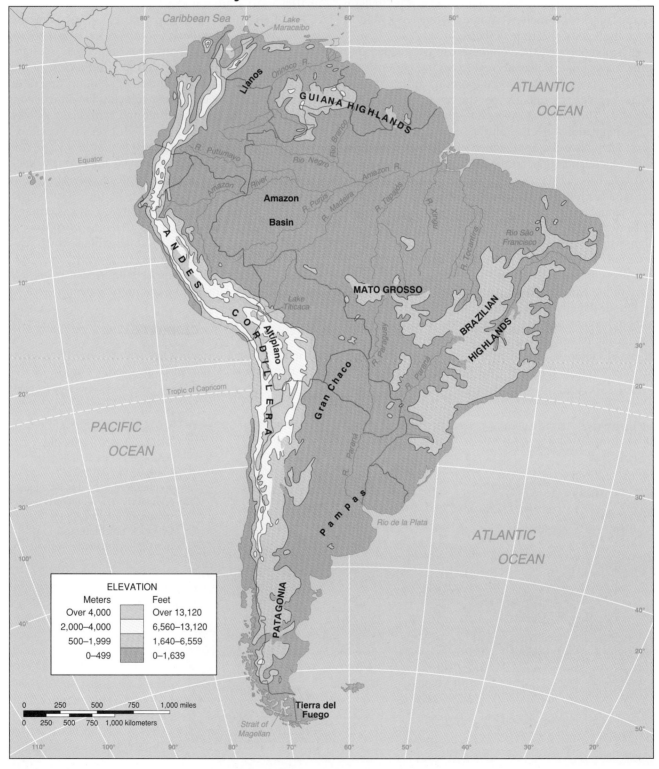

ELEVATION

Meters		Feet
Over 4,000		Over 13,120
2,000–4,000		6,560–13,120
500–1,999		1,640–6,559
0–499		0–1,639

continent (Figure 11.3). The Amazon basin, along the equator, is hot and wet year round with temperatures varying little from a daily average of 80°F (27°C) and annual rainfall of about 100 inches (250 centimeters). Flanking the Amazon Basin to north and south are tropical wet-and-dry (savanna) climates characterized by distinctive rainy seasons ("winter," or *invierno*) in their respective high-sun periods. The low-sun dry seasons are considered "summer," or *verano,* and shorter dry spells within the rainy season are locally referred to as a ***veranillo*** ("little summer"). The southeast trade winds and warm offshore currents bring moisture to a narrow coastal belt of Brazil, thereby extending a wet tropical climate to slightly south of Rio de Janeiro. Northeast Brazil is a dry region, however, receiving only between 10 and 20 inches (25–50 centimeters) of rainfall. Precipitation here is also quite variable, and several prolonged droughts have led to crop failures and out-migration of people.

The narrow southern portion of South America, although situated within temperate latitudes, is not subject to the extremes of heat and cold (continentality) that characterize the large temperate landmasses of Eurasia and North America. Much of southern Brazil, Paraguay, Uruguay, and northern Argentina is humid subtropical, quite similar to the southeastern United States. Central Chile is classed as Mediterranean, a climate characterized by a marked summer (high-sun) dry period.

The Andean cordillera has a very pronounced effect upon South America's climate. Because of extreme variations in elevation and latitude, many microclimates are found within the cordillera. In the tropical Andes, the climates are best described in terms of the *vertical zonation* concept introduced in the last chapter. As one climbs in elevation, hot and humid equatorial climates grade into temperate and even polar climates. The Andes also modify the distribution of precipitation by means of the orographic effect. Incoming moisture-laden air masses rise and cool as they reach the higher mountains, thus causing precipitation. Consequently, the windward sides of the mountains receive the bulk of the rainfall whereas the leeward sides, in the rain shadow of the mountains, remain dry. This pattern is best seen in extreme southern South America, where westerly winds have produced a wet marine west coast climate on the west side of the Andes while dry climates characterize Patagonia on the east side. Even within the Amazon Basin, the highest amounts of rainfall—and lushest vegetation cover—are found along the eastern slopes of the Andes, which in these latitudes represent the windward sides. The narrow band of west coast desert extending throughout Peru and northern Chile is produced both by the rain shadow effect and also by the cold, north-flowing offshore Peru (or Humboldt) Current. As the moist, cool ocean air above the cold current moves overland and is warmed, the relative humidity drops, thus precluding condensation (reaching of the dew point).

VEGETATION

The distribution of vegetation reflects the influence of the prevailing climate, the soils, and the availability of moisture (Figure 11.4). The Amazon Basin is covered with tropical rainforest, or ***selva,*** in which the continuous tree canopy precludes sunlight from reaching the forest floor. As the climate grades from "tropical wet" to "tropical wet-and-dry" with increasing distance from the equator, evergreen rainforests grade into more deciduous types of tropical forest as well as a variety of savanna (tropical wet-and-dry grassland) associations. Whether these savannas formed naturally or by periodic burning by Indians is debated, although extensive conversion of forest to savanna has been noted in historical times. Several large expanses of savanna, including the ***llanos*** of Colombia and Venezuela and the partially wooded ***campos*** of Brazil, are important cattle-ranching zones. The dry northeastern area of Brazil contains a thorny, drought-resistant woodland known as ***caatinga,*** and agricultural colonization in this region is periodically set back by extended droughts. Perhaps the most inhospitable terrain in South America is the Gran Chaco of western Paraguay. Containing a thorn forest somewhat similar to Brazil's caatinga, the Chaco is characterized by extensive desiccation of the topsoil during the annual dry season droughts and widespread flooding during the summer rainy season.

Biogeographically, South America exhibits a southern hemispheric equivalent of temperate North America. In Chile, for example, the vegetation grades from sparse xerophytic scrub in the northern Atacama Desert to evergreen woodlands in the central Mediterranean region to a southern temperate marine west coast forest. These distributions are quite similar to vegetative changes along the Pacific coast from Baja California to British Columbia. As with the Sierra Nevada and Cascade Mountains of the United States, the rain shadow zone east of the southern Andes (Patagonia) consists of steppe and desert vegetation. The humid subtropics of east central Argentina and Uruguay, similar to the Great

FIGURE 11.3

Climates of South America

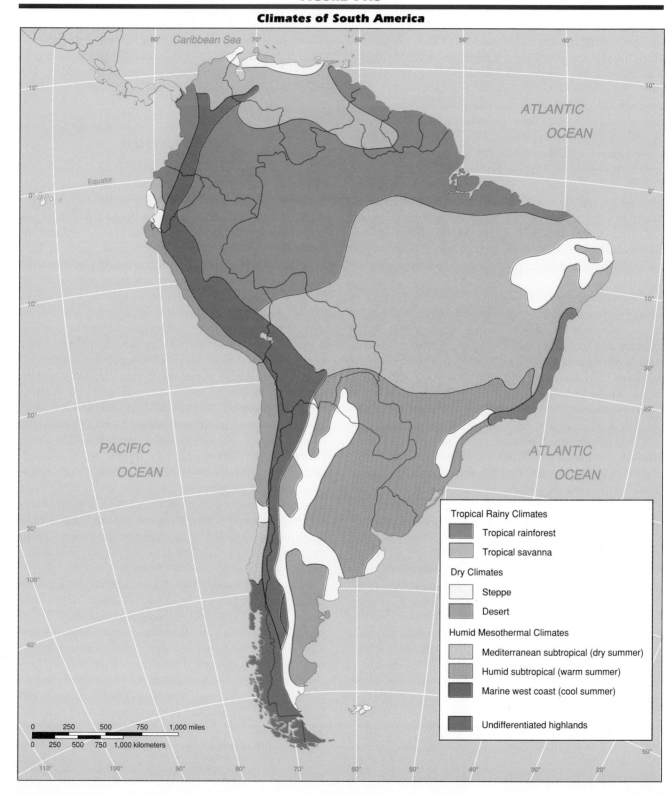

Tropical Rainy Climates
- Tropical rainforest
- Tropical savanna

Dry Climates
- Steppe
- Desert

Humid Mesothermal Climates
- Mediterranean subtropical (dry summer)
- Humid subtropical (warm summer)
- Marine west coast (cool summer)

- Undifferentiated highlands

FIGURE 11.4

Vegetation of South America

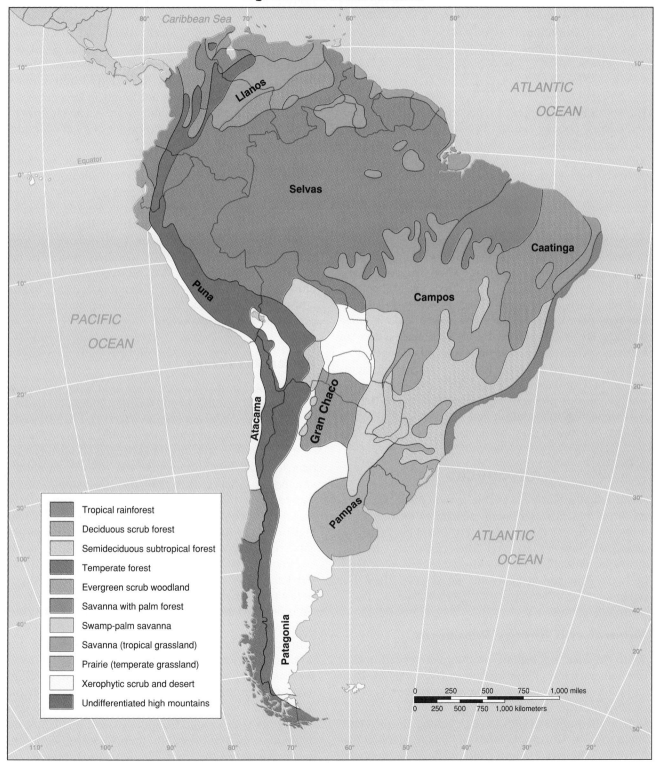

Legend:
- Tropical rainforest
- Deciduous scrub forest
- Semideciduous subtropical forest
- Temperate forest
- Evergreen scrub woodland
- Savanna with palm forest
- Swamp-palm savanna
- Savanna (tropical grassland)
- Prairie (temperate grassland)
- Xerophytic scrub and desert
- Undifferentiated high mountains

Plains, contain agriculturally fertile grasslands known as *pampas*. A small amount of tundra vegetation is found at the very southern tip of the continent and in the Falkland Islands, and similar cold-tolerant grasses—locally known as *páramos* and *puna*—occupy higher elevations within the Andes.

SOILS

The soils of South America reflect the influence of several factors, notably climate, vegetation cover, and parent material. From an agricultural standpoint, the richest soils are found in the pampas of Argentina and Uruguay, the narrow and flat coastal plain of Brazil and the Guianas, and intermontane Andean basins. River valleys, coastal plains, and highland basins contain a large amount of rich alluvial soils deposited by streams and rivers, and these are especially conducive to agricultural production. Even the dry river valleys of coastal Peru contain rich alluvial soils that, when irrigated, have proved to be agriculturally productive. The soils of the lush rainforests of the Amazon Basin—*oxisols*—are deceivingly poor for agriculture, however. The abundance of plant life is maintained by an elaborate system of nutrient cycling within the trees and other vegetation, and once the land is cleared for agriculture, the nutrients are leached, or dissolved, out of the upper layers of the soil. Often, this process of *laterization* leaves behind a hardpan, infertile reddish soil—rich in iron and aluminum, but low in plant nutrients.

RESOURCES

South America has considerable natural resources ranging from minerals and energy resources including petroleum to rivers that, if dammed, could generate hydroelectric power. Fish and forest products are also available, although these, like the other resources, must be well managed if they are to achieve their full potential.

Mineral and Energy Resources

South America is a leading producer of several important minerals and energy resources (Figure 11.5). Peru and Chile account for nearly 20 percent of world copper production. The agriculturally poor, yet mineral-rich oxisols of the humid tropics have benefited Brazil, Venezuela, and the Guianas in terms of bauxite (the raw material for aluminum), iron, and manganese deposits.

Other important mineral deposits include tin and tungsten (Bolivia), uranium (Brazil), and zinc (Argentina, Bolivia, and Peru). Historically, silver mining was important in the Peruvian and Bolivian highlands. Gold and diamond mining is locally significant in the older Brazilian and Guiana highlands. Colombia is also a small producer of gold (derived mainly from **placer mining**) in addition to being the world's leading source of emeralds. Coal deposits are found in several of the countries, and although volumes are relatively low by world production standards, they are of local economic importance.

Most of the continent's petroleum deposits are associated with the interior flanks of the Andean cordillera, although important coastal and offshore fields lie along the Venezuelan and Brazilian coasts. Petroleum exploration dates to the turn of the century when Venezuela's famous Lake Maracaibo oil fields were discovered. Venezuela and Ecuador are OPEC members, and Peru has also recently joined the list of oil exporters. More populous Argentina, Brazil, and Colombia are also important oil producers, but high internal demand has so far precluded any net exporting.

Water Resources

With ample annual precipitation, and a large interior basin, humid tropical South America is drained by several large river systems (Figure 11.2). The largest is the Amazon River and its tributaries, which drain nearly half of the continent. The 4000-mile (6400-kilometer) long river, only slightly shorter than the Nile yet carrying a volume 11 times greater than the Mississippi, cuts its way from the high Andes to the Atlantic Ocean approximately at the equator between the Guiana and Brazilian highlands. To the north, the Orinoco River drains the llanos (plains) of Colombia and Venezuela. The southern lowlands of South America are drained by the Paraná/Paraguay river system, which empties in the Río de la Plata estuary near Buenos Aires. The Río São Francisco flows to the Atlantic through the eastern Brazilian highlands.

In spite of the extensive network of rivers, their importance historically has been less than that of rivers such as the Mississippi or Nile. The Amazon, although navigable into Peru, flows through what historically has been an undeveloped wilderness. Because of the generally poor soils eroded from the eastern highlands, agriculture only flourished in narrow strips along the lower reaches of the river. The Río São Francisco con-

FIGURE 11.5

Mineral and Energy Resources of South America

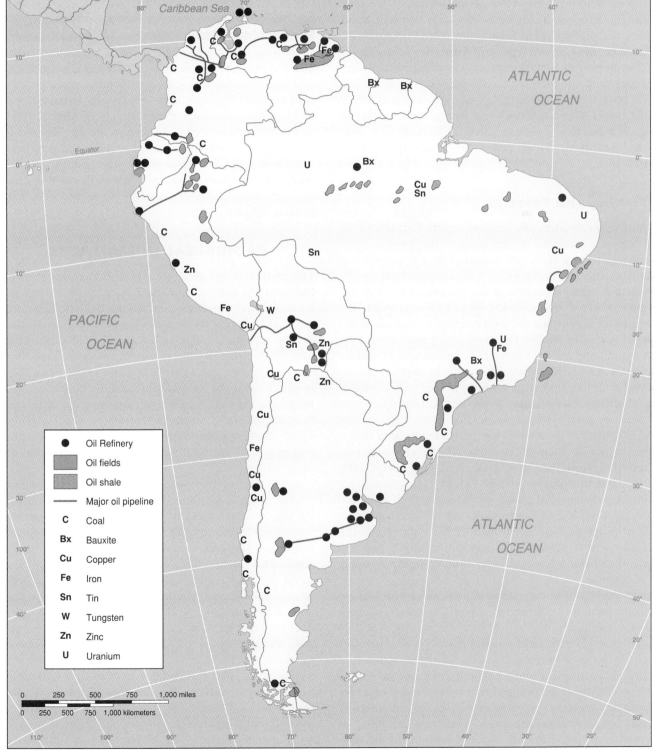

●	Oil Refinery
	Oil fields
	Oil shale
—	Major oil pipeline
C	Coal
Bx	Bauxite
Cu	Copper
Fe	Iron
Sn	Tin
W	Tungsten
Zn	Zinc
U	Uranium

SOURCE: Adapted from *International Petroleum Encyclopedia* (1980) and *Goode's World Atlas.*

tains rapids close to the coast, and this precluded easy access to the interior. The Paraná River served an important role as a transportation artery into the interior, but its shallow waters have severely limited its use in the modern era. The greatest resource value of South America's rivers lies in their hydroelectric potential, especially for populous developing nations such as Brazil. The upper Paraná contains 11 major hydroelectric plants, including the famous Itaipú Dam on the Brazil/Paraguay border, the largest power-generating facility in the world. On the western side of the Andes are many short rivers that, if dammed, would both alleviate periodic severe flooding and yield great hydroelectric potential.

Fishing and Forestry

Although subsistence fishing takes place throughout the continent, especially in the rivers of the interior, commercial fishing is largely restricted to the coastal and offshore areas. Shrimping is an important activity on the Atlantic coast of Brazil and the Guianas, and Peru has been a world leader in the harvest of anchovies. Because the west coast of South America is washed by the cold Peru (Humboldt) Current flowing northward from Antarctic waters, the upwelling (replacement of surface waters with colder water from greater depths) associated with this cold current brings rich bounty of plankton and anchovies close to the sea surface where the latter are easily harvested. In some years, Peru has attained the distinction of being the leading fishing nation in the world, although in other years freak warm currents (**El Niño,** the Christ Child) in winter or overfishing have led to collapses in the industry. A potential for greater commercialization of fishing exists in South America as long as the resources are well managed. Recent progress is apparent in both the commercial development of inland fisheries in the Amazon Basin and the expansion of aquaculture, mainly shrimp farming, in Ecuador.

Because of the high diversity of species and relative inaccessibility, commercial lumbering has been relatively minimal in the vast tropical rainforests. Selective cutting of valuable hardwoods is practiced, and attempts at tree farming have been made, but the economics of large-scale tropical forest lumbering are not favorable. (The rainforests are disappearing at alarming rates, however, but more as a result of conversion to low-intensity agriculture and grazing land.) Commercial forestry is more widespread in the temperate portions of the con-

tinent, especially in southern Brazil where extensive stands of softwoods are found.

HUMAN ENVIRONMENT

Like the physical environment, the human environment of South America exhibits considerable diversity. Native populations, voluntary immigrants from Europe, and involuntary immigrants (slaves) from Africa as well as various mixtures of these groups are all reflected in the population of South America today. Patterns of population density and distribution are similarly diverse; South America is the home of some of the world's largest cities as well as large sparsely populated areas and frontier regions.

ABORIGINAL INHABITANTS

Although archaeologists are still in disagreement as to the exact antiquity of humans in the New World, we do know that the Americas were among the last of the major world regions to become populated. Estimates of earliest settlement range as far back as 30,000 years before the present (B.P.), but recent revisions have lowered this figure to between 12,000 and 15,000 years B.P., or shortly after the last glacial maximum of the Ice Age. By 11,000 years B.P., however, descendants of the pioneering Asiatic forebearers had reached the southern tip of South America (Tierra del Fuego), and over the subsequent millennia, a variety of distinctive native cultures evolved.

At the time of European contact, perhaps 15 million natives of varying culture levels inhabited South America (Figure 11.6). Most of the continent was inhabited by tropical forest farming peoples who lived along the numerous lowland rivers and practiced shifting, or **slash and burn,** agriculture. Cassava (manioc) and sweet potatoes comprised the basic food staples, and protein was derived predominantly from fish but also from wild game and peanuts. Since these people were river oriented, they traveled considerable distances; notable examples were the Arawaks and Caribs who respectively had migrated, with their Amazon lowland culture traits, into the Antilles by the time Columbus "discovered" this New World. Other well-known lowland cultures include the Tupi, Gê, and Guaraní of eastern and southern Brazil and adjacent Paraguay. A temperate latitude slash and burn culture—the

FIGURE 11.6

Aboriginal Economies of South America

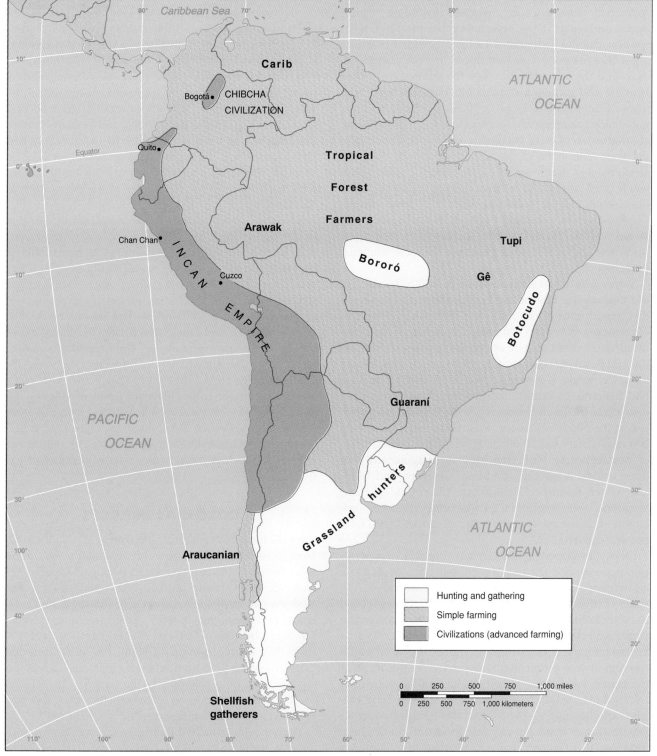

SOURCE: Adapted from Robert C. West, "Aboriginal and Colonial Geography of Latin America," in Brian W. Blouet and Olwyn M. Blouet, eds., *Latin America: An Introductory Survey* (New York: John Wiley & Sons, 1982), pp. 34–86.

Araucanians—occupied humid, forested south central Chile, and their agricultural basis was more complex. In addition to adopting food complexes prevalent in Mesoamerica and Andean America (including maize, beans, and squash), the Araucanians domesticated what we now call the Irish potato. Geographers and botanists argue that potatoes were introduced into the central Andes—and the rest of the world—from this source area. Hunting and gathering cultures inhabited several of the tropical grasslands (campos) of Brazil, the temperate grasslands (pampas) of Argentina and Uruguay, and the short-grass steppes of Patagonia. Fishing and shell fishing were important supplementary economic activities along most of South America's shoreline, and along the south coast of Chile and Tierra del Fuego, these activities constituted the dominant method of food supply. Charles Darwin, rounding Cape Horn on the research vessel H.M.S. *Beagle* in the mid-1800s, commented upon the impoverished appearance of these shellfish gatherers.

Two major civilizations existed in South America at the time of the Europeans' arrival: the Inca Empire of the central trans-Andes and the Chibcha civilization of contemporary Andean Colombia. Evidence of plant domestication dates to as early as 6000 years B.P. along the arid Peruvian coast, and by 3600 years B.P. elaborate irrigation works had been constructed, pottery making had become prevalent, and a temple had been built near Lima. Incipient coastal civilizations eventually gave rise to the Mochica Culture, which flourished from A.D. 200 until about A.D. 1000. Situated on the coastal plain of contemporary Peru, the Mochica fragmented into three coastal kingdoms of Chimu, Chancay, and Chincha, which flourished until about A.D. 1470. All three kingdoms were supported by elaborate irrigation agriculture, and the most famous city was Chan Chan, near present-day Trujillo.

Cities developed in the Andean highlands contemporaneously with the rise of a coastal plain civilization, and contact between the two areas was widespread in spite of the formidable topographic barriers. Several sites, including Kotosh in the north, Tiahuanaco in the south, and the Cuzco Valley between the two, became important centers of civilization. Political power within the region was eventually consolidated by the Incas of the Cuzco Valley during their empire-building push between 1438 and 1525. Under the family-run Inca Empire, an elaborate road network was established to facilitate transport of people and foodstuffs throughout the elongated territory that now reached from southern Colombia to central Chile. Unlike the extensive trade network of the enterprising Mesoamericans, the marketing infrastructure of the Incan Empire was the result of totalitarian policies. The Inca elite used forced labor in both agriculture and public works projects and instituted a socialistic food distribution system that ensured sufficient food supplies. **Quechua,** the Inca language, was also forced upon the conquered subjects, and its present distribution closely approximates the maximum extent of the empire. The Incas established a second capital at Quito in the early 1500s, and perhaps empire building would have continued had not the Spaniard Pizarro arrived in the region soon thereafter in 1531. At its peak, the Incan Empire contained up to 12 million inhabitants, or about half the size of the population of Mesoamerica.

A second, much smaller Andean civilization was that of the Chibchas of highland Colombia who numbered only 500,000 inhabitants. Although not architecturally as sophisticated as the central Andean cultures, the Chibchas formed a rudimentary political state, traded extensively with surrounding Indians of lower elevations, and crafted elaborate gold figurines. The famous myth of El Dorado (the Gilded One), which later led many expeditions to ill-fated endings in the tropical lowlands, stems from the Zipa (chief) of the Chibchas who annually sprinkled gold dust on his body and went for a swim in the cold Lake Guatavita north of Bogotá. Although the Spanish were well acquainted with this story before their arrival in the Colombian highlands, it had become so exaggerated that they did not recognize the source of the myth when they finally conquered the Chibcha in 1536.

COLONIAL IMPACTS

After Columbus's eventful discovery of October 12, 1492, territorial rights to world exploration and trade were divided between Spain and Portugal by papal decree. The 1494 Treaty of Tordesillas established a line of demarcation 370 leagues west of the Cape Verde Islands, or approximately $47\frac{1}{2}°$ west longitude, at what was thought to be a mid-ocean meridian (Figure 11.7). However, a large portion of South America protruded eastward of the line, and while accidentally sailing too far westward on a rounding of the Cape of Good Hope in 1500, the Portuguese discovered that they too had received a slice of the new world.

The Spanish launched their explorations in search of mineral wealth from their New World base in Santo

FIGURE 11.7

Loci and Routes of Diffusion of Colonial Settlements in South America

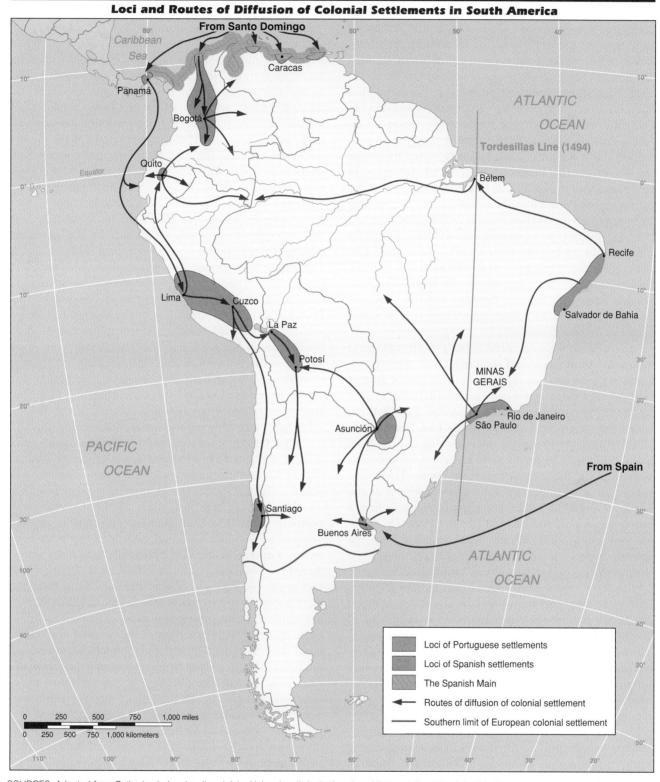

SOURCES: Adapted from Catherine L. Lombardi and John V. Lombardi, *Latin American History: A Teaching Atlas* (Madison: University of Wisconsin Press, 1983); and Robert C. West, "Aboriginal and Colonial Geography of Latin America," in Brian W. Blouet and Olwyn M. Blouet, eds., *Latin America: An Introductory Survey* (New York: John Wiley & Sons, 1982), pp. 34–86.

Domingo, following the failure of their first attempted settlement on the north coast of Hispaniola. An abundance of pearls along the north coast of South America (the Spanish Main) led to the establishment of small outposts from eastern Venezuela to Panama. Following Balboa's discovery of the Pacific Ocean (the South Sea) in 1513, the **entrepôt** city of Panama was soon established. From this link to the Caribbean, the Spanish proceeded down the west coast and discovered the Incan Empire. Following Pizarro's conquest of the Incas in 1531, Lima was established as the colonial capital of the viceroyalty of Peru. From this coastal base, the Spanish occupied Cuzco and other mineral-rich highland centers of Indian civilization, as well as more outlying territories within the Incan Empire. By the 1540s, the Spanish had discovered the immense silver deposits of Bolivia, known as Upper Peru because of the high elevations. Notable among the mineral deposits was the (still producing) silver mine of Potosí, near which several settlements were established.

Explorations were launched both northward and southward from Peru, and settlement loci were established at Quito and Santiago. From these nodes, further outward explorations and expansion took place, including down the Amazon River and across the southern Andes into Argentina. Buenos Aires had been established in the early 1530s by settlers arriving directly from Spain, but hostile Indians forced an inland relocation to Asunción, via the Paraná/Paraguay rivers. This Paraguayan locus of settlement subsequently served as the hearth of outward settlement expansion, including a later reoccupation of the site of Buenos Aires. The Chibcha civilization of Colombia was penetrated by the Spanish moving inland from the outposts along the Spanish Main, and Bogotá was established near the highland site of the Chibchan capital. Further explorations were launched from this base, mostly southward and northeastward, but also eastward in search of El Dorado. Spanish settlements were primarily concentrated in areas containing mineral wealth and a readily available supply of Indian labor to extract that wealth. Administrative centers were established either within those settlement loci or at coastal entrepôts. The mineral-poor Uruguayan grasslands were unimportant to the Spanish, at least until the Portuguese began expanding southward in the late 1700s. Likewise, the northern Amazonian lowlands, including the Guianas, had little to offer, and the Spanish subsequently avoided that part of the continent.

Like the Spanish, the Portuguese were interested in acquiring wealth, and the recently established trade with India and the Orient offered the easiest means to accomplish those aims. Early coastal explorations in the newly discovered territory yielded little evidence of mineral wealth, and the colony of Brazil (named after the dye-yielding brazilwood that constituted an early export) grew slowly. In 1580, an estimated 150,000 Spaniards had settled in the New World—approximately half of them in South America—yet only about 30,000 Portuguese inhabited Brazil. Settlement in Brazil was concentrated in two major zones: the older zone of São Paulo/Rio de Janeiro in the south and Salvador de Báhia/Recife in the north. Of these two, the northern became more important because of the establishment of a *plantation economy.* The **plantation,** an extensive feudal land tenure system dedicated to intensive production of an export crop, was first developed by the Portuguese in their African island colonies of Madeira and São Tomé and Principe. The tropical climate and fertile coastal plain of northeastern Brazil were well suited for the growing of sugarcane and tobacco, and a plantation landscape quickly developed. Although the search for mineral wealth and slave labor led to an inland penetration of Brazil well westward of the Tordesillas Line, including up the Amazon River and toward the Mato Grosso from São Paulo, most Portuguese settlement remained concentrated along a narrow coastal ribbon. Other Europeans, including Dutch, French, and British, adapted the plantation system to other portions of coastal South America, such as the Guianas and northern Brazil, that had been ignored by Iberian colonists. The Spanish, too, had experimented with sugarcane farming in the Canary Islands, and they introduced the plantation system to Hispaniola. By the early 1600s, the plantation economy had spread throughout the Caribbean and to the North American mainland.

Foremost among the population changes that took place in South America during the colonial era was a rapid decline in the native population, chiefly attributed to enslavement for work in the mines and on the plantations, introduction of diseases such as smallpox, measles, and typhus to which the Indians had no resistance, disruption of native economies, and psychological despair. The population declines were greatest in the areas of highest and densest population. In Peru, for example, an indigenous population estimated to be 12 million in 1530 rapidly declined to 1.5 million by 1570 and 600,000 by 1620. By this latter date, the population stabilized and a pattern of gradual increase began. Due to the virtual absence of accurate records for the less Europeanized segments of South America, we have little knowledge of either the aboriginal population size or the extent of the postcontact decimation. Orellana, the Spanish explorer who first descended the Amazon River, noted almost contiguous village settlement along the banks of the entire course of the river, a population density that still has not been regained today. Slave-raiding parties from coastal Brazil rapidly decimated the native Tupis, and the search for plantation labor drove the *Paulistas* (inhabitants of São Paulo) ever further into the interior. Most of the territorial expansion of Brazil—to approximately its modern boundaries—resulted from the search for slaves and gold by these *bandeirantes* (flag bearers).

The demand for labor could not be met by the enslavement of Indians alone, however, and the Portuguese began importing Negro slaves from their African territorial possessions, chiefly Portuguese Guinea and Angola. The demand for African labor in the New World increased rapidly, especially where the plantation economy was dominant. Estimates of the numbers of slaves imported from Africa to all of the Americas range from 10 to 15 million. Four to five million Africans were brought to South America, mostly to Brazil but also to the Guianas and the Spanish Main where the plantation economy had taken hold. The predominance of blacks is still quite evident in those areas today. The Spanish also used African slaves to augment the declining Indian labor supply in their gold and silver mines, but except for small enclaves in western Colombia and Ecuador, most evidence of Africans has disappeared through the generations because of intermarriage with Indians and Europeans in these areas.

The mixing of the races is another important result of the colonial impact on South America. Not only were African and Mediterranean European racial strains added to the aboriginal ethnic stock, but much intermarriage took place as well. Dozens of terms have come into local usage describing the various combinations of racial mixtures, most importantly mestizo (Indian and European), *mulatto* (African and European), and *zambo* (African and Indian). The mestizo population has become the dominant one in most of western South America, whereas blacks and mulattos are numerically important along the coastal lowlands of eastern and northern South America. In 1825, 50 percent of Brazil's population was black and 18 percent mulatto.

Whereas the Spanish settlement patterns did not significantly change throughout the entire colonial period, in Brazil the discovery of mineral wealth in 1693 triggered what has become known as the world's first gold rush. Centered in a region that became known as Minas Gerais, the gold rush attracted some 800,000 Portuguese immigrants during the eighteenth century. Not only did the balance of political power shift back to the south from Salvador, but the discovery of mineral wealth led to the first significant settlement inland of the Brazilian escarpment. Rio de Janeiro, the closest port to the interior mines, became the capital of Brazil in the 1750s.

MODERN POPULATION

The modern period in South America has seen several developments: another influx of European immigrants, dramatic population growth, and various forms of migration. The latter includes both rural-urban migration and movement to the frontier regions of the continent.

Immigration

Non-Hispanic European immigration did not become widespread until the nineteenth century. Under Spanish colonial rule, only Spaniards were allowed to immigrate into Spain's American colonies. Following independence from Spain in the early 1800s, many of the newly independent nations felt a need to improve their economies and to populate their sparsely settled territories. Immigrants from northern Europe were especially sought after because of their perceived strong work ethics. Brazil, too, opened its doors to immigrants as early as 1800, but the persistence of slavery until the end of the colonial period in the late 1880s precluded much demand for new sources of labor during most of the century. In the Guianas, abolition of slavery was followed by the importation of indentured plantation laborers, especially from the British and Dutch colonies of India and the East Indies, respectively. Their descendants comprise a significant percentage of the ethnic composition of Guyana and Suriname today. The few immigrants to French Guiana arrived by force, as France established several penal colonies—including the infamous Devil's Island—there.

Temperate, or southern, South America was the destination of the majority of European immigrants between 1870 and 1930. In Argentina, 6.5 million immigrants, 80 percent of them Italian and Spanish, arrived as the rich pampa grasslands were transformed from grazing lands into croplands. Most immigrants were non-landowning agricultural laborers, and they quickly flocked to the cities, thus stimulating the rapid growth of cities, including Buenos Aires. Today, a majority of Argentines are descended from these largely southern European immigrants, and even their pronunciations of Spanish words reflect Italian inflections. Uruguay received an ethnically similar influx of immigrants. Although smaller in number than in Argentina—perhaps one million total—the immigrant population was more significant because of the relatively low population prior to mass immigration. Brazil received five million immigrants, mostly from Portugal, Italy, and Spain, but also from Germany, Russia, and Turkey. Here immigration mostly followed the opening of coffee and other agricultural lands south and west of São Paulo, including the forested lands of Rio Grande do Sul where a small but locally important colony of Germans became established. European colonies, especially of Germans, became important in Paraguay following its defeat in the infamous War of the Triple Alliance in 1870. After a five-year battle against the combined forces of Brazil, Uruguay, and Argentina, Paraguay experienced a 90 percent decline in its male population. European immigration was numerically less important in other South American countries, although small German colonies were established in southern Chile and north central Venezuela. Oil-rich Venezuela continues to attract immigrants especially from Italy and the Canary Islands.

Population Density and Distribution

Statistics on population densities for South America indicate that the continent is relatively sparsely settled, yet these statistics mask the uneven distribution of the population (see the Statistical Profile). Averaged over the entire continent, the population density is slightly over half of the comparable figure for the United States and about one-third that of Mexico. On national levels, population densities are highest in Ecuador (over double the continental average) and lowest in French Guiana and the Falkland Islands. When one examines the distribution of the population, which is concentrated in the intermontane Andean valleys, along the coast, and in the cities, much higher values of localized population density emerge (Figure 11.8). The contemporary distribution of South Americans reflects patterns established in the precolonial, colonial, and modern eras.

Population Growth and Migration Patterns

The total population of South America is approximately 296 million (1990), of which slightly over half is in Brazil. Since 1950, about 135 million people have been added to the continent's population. This high rate of increase is largely explained by the demographic transition model, which forecasts a large decline in the death rate prior to any significant declines in the birth rate. Not including the effects of (international) immigration and emigration, which have been relatively unimportant since 1950, this signifies that birth rates have been substantially higher than death rates, and a population explosion has been the result.

Although present trends indicate a gradual drop in the birth rates, attributed largely to increasing practices of birth control, the pressures of population growth have led to extensive internal migrations of people. First and foremost, there have been extensive urban migrations. Many of South America's large cities are growing at rates of 7 or 8 percent per year, well above the national averages. Many traditional areas of settlement in the Andean valleys and along the coasts are at or near their **carrying capacities,** at least under present technological levels. In these regions, population growth is usually offset by out-migration, most often to the cities where the migrants seek jobs. Many urban problems, such as the difficulty in providing basic services, result from this rapid population influx, and **squatter settlements** are integral components of most large cities today. São Paulo, Buenos Aires, and Rio de Janeiro are among the most populated urban areas in the world with more than 10 million persons. Lima, Santiago, and Bogotá are also quite large with populations of more than 4 million each.

Other types of population migration include international migration, interregional migration, and **frontier migration.** Most international migration is restricted to the continent (intraregional, if the whole continent is considered to be one region) and usually entails the movement of citizens of a relatively poorer country to a richer neighbor. Consequently, there are substantial numbers (in the hundreds of thousands) of Colombians in Venezuela, and Bolivians and Paraguayans in Argentina. Because of distance factors, immigration to the United States is less important than for Mexico and Central America, although large Colombian populations may be found in Miami and New York City.

FIGURE 11.8

Distribution and Density of Population in South America

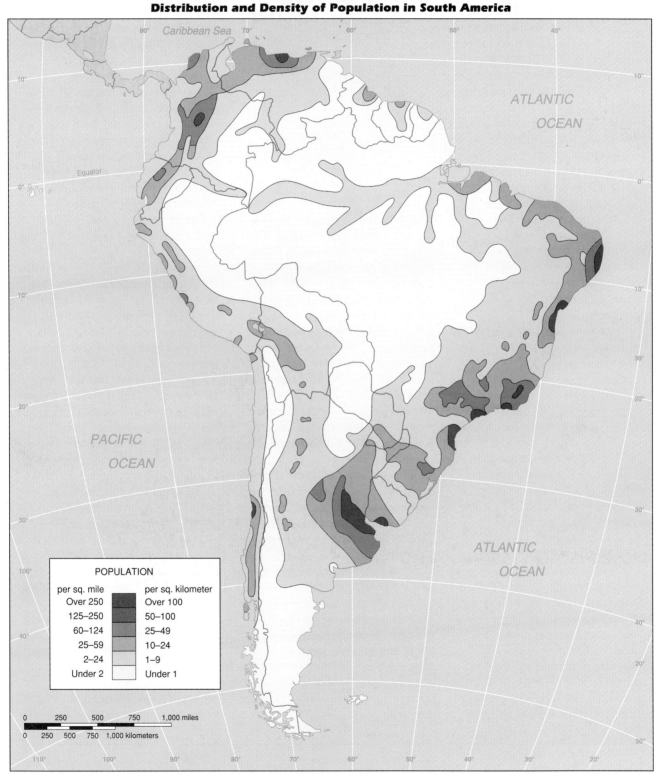

POPULATION

per sq. mile		per sq. kilometer
Over 250		Over 100
125–250		50–100
60–124		25–49
25–59		10–24
2–24		1–9
Under 2		Under 1

FRONTIER COLONIZATION

L and at the fringes of the main zones of settlement in a country comprises a *frontier* zone. In the United States, the earliest frontier was situated westward of the main settlement nodes along the Eastern Seaboard. As pioneering settlers colonized the often lawless frontier, and converted the woodlands and prairies to agriculture fields, the frontier zone was displaced westward. The historian Frederick Jackson Turner believed that the frontier not only offered ever-present opportunities for an expanding American population but that is also shaped American culture in general. With the possible exception of Alaska, which advertises itself as "The Last Frontier" on its license plates, the frontier in the United States has long been closed.

In South America, the frontier is very much alive although settlement zones are rapidly encroaching on the empty expanses of land nearby. A "cold-weather" frontier similar to Alaska exists in southern Chile and southern Argentina, and hardy colonists are continuing to migrate into the region. In Chile, the construction of a road to the southernmost extremes of the country is offering new settlement opportunities. Often frontier areas are only temporarily occupied, perhaps until a valuable natural resource can be sufficiently exploited to make a substantial profit. In that case, the frontier may still function as an empty territory beyond the margins of permanent settlement. Only the process of *permanent colonization* will change the status of the frontier.

The best example of a frontier in South America is the Amazon Basin. By the early twentieth century, pioneer colonists from the Andean countries had moved into the upper Amazon along the lower eastern slopes of the Andes (the **montaña**). Much of this movement took place prior to the opening of the road, and the direct stimulus for this *spontaneous* colonization was the increasing population pressure within the Andean basin settlement cores. The respective governments quickly recognized the value of frontier colonization, both in terms of occupying the national peripheral regions as well as relieving overcrowding in the traditional settlement areas. From Venezuela to Bolivia, quite a few *penetration roads* were built across the Andes to allow settlers to move in and to provide transportation routes for agricultural products to be transported to market. As the eastern slope frontier colonization zones grew and even coalesced, they became linked by crude highways. In 1957, former President Fernando Belaúnde of Peru, an architect by training, proposed linking all of the *frontier colonization zones* by a grandiose 3600-mile (5700-kilometer) long **Carreterra Marginal de la Selva,** or Marginal Forest Highway (Figure 1). Such a road, he felt, would not only open up new lands to settlers but would provide better transport cor-

ridors for marketing produce. From Venezuela to Bolivia, approximately two million settlers were expected to occupy this new frontier, and an overall population of nine million inhabitants was projected. Although the project as envisioned by Belaúnde was never built and colonists in the montaña number only in the hundreds of thousands, continued spontaneous colonization has led to increased linearity of settlement along the eastern Andean slopes. As the settlement zones become tied together by roads, one day the Carreterra Marginal will indeed become a reality.

The lower Amazon Basin is a frontier for Brazil. Although the Amazon lowlands periodically had been exploited for specific high-demand resources such as gold, slaves, and rubber, true frontier colonization has until recently been rather minimal. Politically, the Brazilian government has long perceived the Amazon as a vast untapped resource to be developed and populated. This optimistic view toward the interior was reinforced by the relocation of the capital from Rio de Janeiro to Brasília in 1960 and the extension of roads northward to Belém and westward through the Mato Grosso to Pôrto Velho. Significant settlement took place along these roads by both small-scale agriculturalists and corporate cattle ranchers (including the King Ranch of Texas). Development and settlement in the Amazon were also encouraged by the *growth-pole* concept, in which industrial or commercial development is initiated and transmitted to an area around it providing the basis for populating hitherto sparsely inhabited regions. The Amazon region was de-

FIGURE 1

Frontier Colonization in the Greater Amazon Basin

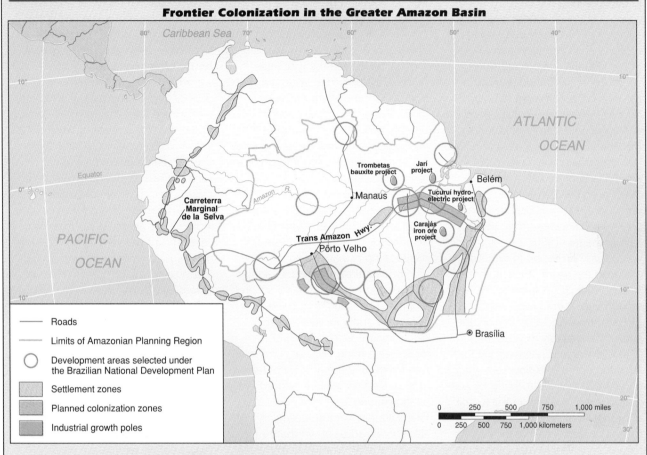

Roads
Limits of Amazonian Planning Region
Development areas selected under the Brazilian National Development Plan
Settlement zones
Planned colonization zones
Industrial growth poles

SOURCES: Adapted from Rosemary D. F. Bromley and Ray Bromley, *South American Development: A Geographical Introduction* (Cambridge, G. B.: Cambridge University Press, 1982), (repr. 1985); and Raymond E. Crist and Charles M. Nissly, *East from the Andes* (Gainesville: University of Florida Press, 1973).

clared a *duty-free* zone, and cities such as Manaus experienced minor booms in commercial and manufacturing activity. Also, billionaire Daniel Ludwig was able to acquire a Connecticut-sized piece of property along the Jarí River where, following an investment of $500 million, he converted thousands of acres to rice and *Gmelina arborea* trees, a fast-growing Australian pulpwood species. The gmelina trees were to supply his pulpwood processing plant, which was towed from Japan. Although Ludwig gave up in

1982, the growth pole of Jarí is now under Brazilian ownership, and 30,000 settlers live in the region.

In 1970, after a major drought in the populated Northeast, the Brazilian government embarked on an ambitious Amazonian development program, which was to include extensive road construction, infrastructure building for colonization zones, and industrial *growth-pole development* similar to Ludwig's Jarí. The first step in this program was the construction of the east-west **Trans-Amazon Highway,**

which was to connect the drought-stricken Northeast with the Peruvian border (Figure 1). Along this highway, as well as along the highway to Pôrto Velho, colonization zones were established and road-fronting land parcels of 250 acres (100 hectares) were made available to colonists. In addition, larger tracts of land—totaling 6.7 million acres (2.7 million hectares)—were set aside for private and corporate cattle ranches. It was estimated that one million colonists would be settled by 1980, most from the impoverished Northeast.

Although much spontaneous colonization has taken place along the newly opened highways and several of the industrial projects are under construction, the growth forecast by the Brazilian government has not been realized. Many problems have plagued the development process, including the lack of bridges, washed-out roads, lack of agricultural assistance, soil erosion, and depletion of nutrients. By 1980, only 8000 families had been settled in the colonization zones, and only 40 percent were from the Northeast. In the mid-1980s, out-migration was actually exceeding in-migration.

One of the major problems in the Amazon Basin is soil infertility. Although lush, species-rich rainforest is extensive throughout the region, the nutrients are rapidly recycled from the humus, or organic debris on the forest floor, through the shallow root systems of the trees. Clearing the land for agriculture allows nutrients in the topsoil to become rapidly leached deep into the soil by rainfall so that the root systems of commercial crops cannot tap them. With continued leaching, this process of laterization can lead to the formation of a mineral-rich, yet nutrient-poor, cementlike hardpan surface. One pattern has been for colonists to farm land productively for one or two seasons, then sell it to a cattle rancher as the soil fertility declines. The only "success stories" in the Brazilian Amazon have been the vast cattle ranches, established in large part to export beef for the North American fast-food industry and provide much-needed foreign revenues for Brazil. Beef production has increased in Brazil, but at the expense of widespread environmental destruction.

The environmental implications of inefficient frontier colonization in both Brazil and the Andean countries are tremendous. The tropical forests are disappearing at alarming rates (some worldwide estimates range as high as 40,000 square miles [100,000 square kilometers] per year), and scientists are concerned about local as well as global impacts. Locally, the lush rainforests are turning into nonproductive red deserts, and the amount of rainfall is decreasing because of reduced evapotranspiration from fewer trees. As forests are cut, we lose potentially valuable species, some of which may have pharmaceutical value (as does the bark of the *cinchona* tree in the treatment of malaria, for example). Also, not only is habitat for flora and fauna rapidly disappearing, but the habitat of aboriginal culture groups is disappearing as well. Globally, rainforests absorb carbon dioxide and produce half of the world's oxygen supply. Removal of the forests may well enhance the **greenhouse effect** and lead to global warming, thus stimulating the melting of polar icecaps and raising sea levels.

There are signs that rampant destruction of the Amazon is slowing. Official Brazilian policy in 1989 shifted toward greater concern for tropical rainforest resources, and a new democratic government—in place since 1990—has made environmental protection a key aspect of its Amazonian development program. Forest fires are now regularly monitored by satellites, and violators of anti-burning laws are increasingly being prosecuted.

Yes, there is a frontier in South America. But the humid tropics comprise a fragile frontier, quite different from the historic midlatitude frontier of the United States. With proper technology, such as modified versions of aboriginal methods of slash and burn farming and multicropping, frontier colonization can be a viable process. It should not be seen as the major method of solving the problems of population growth and economic stagnation that may characterize the settlement cores of the respective countries.

Interregional migrations consist of migrations from relatively impoverished regions to areas where opportunities are greater. Semiarid and highly populated Northeast Brazil is frequently affected by droughts, and the region has become a major zone of out-migration. Regions that attract migrants often result from the establishment of economic development centers, or *growth poles,* in previously undeveloped regions. Examples of growth-pole establishment include the inland relocation of a national capital such as Brasília, which has now grown to almost two million inhabitants, and the establishment of an industrial center such as Ciudad Guayana in southeastern Venezuela. In an effort to lower population pressures in the greater Buenos Aires area, Argentina has proposed moving its national capital south to Viedma, on the fringes of remote Patagonia.

Another type of migration is frontier migration, which refers primarily to pioneering agricultural colonization. In South America, this consists mainly of migration into the humid tropical lowlands, although Chileans and Argentines are also settling their cold southern frontiers. The last vast expanse of land in the continent is the

Amazon Basin, and pioneer settlement in Brazil and the Andean countries is gradually encroaching into this "last frontier," often at the expense of the native inhabitants who are forced out or forcibly acculturated.

CULTURE

Many aspects of South American culture including language and religion reflect the impact of the colonial period, although aboriginal languages persist in some areas and elements of native religions have been incorporated into Roman Catholicism. Similarly, the political history of the region has been shaped to a large degree by its colonial experience.

Language

With over one thousand aboriginal languages, South America exhibits great linguistic diversity. These languages collapse into 118 linguistic stocks falling into three major language groupings. All of the existing and extinct native languages ultimately can be traced to one protolanguage dating back at least 11,000 years. Therefore, it might be suggested that all South American Indians were descended from the same ethno-linguistic stock, and the great linguistic diversity is attributed to a high degree of isolation of the various individual tribes.

Several Indian languages became quite widespread, including Quechua, which was forced upon the subjects of the Incan Empire, but also several of the Brazilian highland and interior lowland languages, such as Tupi, Guaraní, Carib, and Arawak. Where Indians comprise significant proportions of the total population, native languages are still widely spoken. Notable examples are Quechua in Peru and Ecuador, Aymará in Bolivia, and Guaraní in Paraguay.

The colonial imprint on the South American continent has obscured the great diversity of aboriginal languages under an overlay of a few major European languages that have come to predominate. Portuguese is the official language of Brazil, and because of Brazil's large population, it is also the most widely spoken language on the continent. In the Andean nations and Uruguay, Spanish is the official language. Quechua shares official status with Spanish in Peru, however, as does Guaraní with Spanish in Paraguay. English is the official language in the Falkland Islands and Guyana, Dutch in Suriname, and French in French Guiana. On a micro-scale, linguistic diversity is greatest in the Guianas because of the great variety of cultures, including African, Amerindian, Hindi, Bengali, Indonesian, and Chinese. This diversity reaches its epitome in Suriname, where the official Dutch is less widely spoken than **_talkie-talkie_** (or taki-taki), a lingua franca derived from at least six or seven separate languages.

Religion

Most of South America is nominally Roman Catholic, the religion brought over by the Iberian colonialists. The aboriginal peoples were profoundly religious: most of the lowland groups were classified as being animist (imparting spiritual significance to animate or even inanimate objects), although among the Andean civilizations religion was highly organized and contained a pantheon of gods and a priesthood. With the forced introduction of Catholicism, many saints were substituted for indigenous gods, and the elaborate rituals of the new religion were blended with those of the past. In the plantation and slave economy of coastal Brazil, elements of animist African religions were incorporated into local versions of Catholicism.

In the non-Latin countries of South America, the earliest introduced Protestant religions (such as Anglican in the British colonies) have been augmented by numerous other religions. Indentured plantation laborers brought Hinduism and Islam, and these have become dominant in Guyana and Suriname. Missionizing efforts, mainly by Protestant sects including Baptists, Mormons, Seventh Day Adventists, and Jehovah's Witnesses, are ubiquitous throughout South America, and potential converts are usually those living at the margins of society (Indians and the socially and economically disadvantaged). Some religious groups, including the Mennonites, moved to South America not to make converts but to create a homeland for themselves, usually on the frontier. They have become quite dominant in Paraguay's Chaco region.

Political Characteristics

During the colonial period, settlement took place in several core areas deemed economically important by the colonial powers. These core areas were located in highlands proximate to sources of mineral wealth and/or a native labor pool or along coasts and/or rivers where plantation agriculture or entrepôt cities could be established. Politically, the Spanish colonies were closely administered from Spain, and colonial administrative capitals were established to control vast hinterlands known as *viceroyalties*. The three viceroyalties of South America were New Granada (comprising roughly modern-day Colombia, Panama, and Ecuador), Peru (Peru, parts of Bolivia, and Chile), and Río de la Plata (Bolivia, Argentina, Paraguay, Uruguay, and the Falkland Islands). The last two viceroyalties were further subdivided into 23 intendéncias and províncias. Venezuela was politically a captaincy-general administered from Santo Domingo.

Independence came in the 1810s to 1820s as the Spanish colonies, under the leadership of liberation fighters such as Simón Bolívar and José de San Martín, endured a long and bitter struggle to remove the yoke of colonialism. Brief attempts at postcolonial political alliances failed, but incipient forms of the nine present Spanish-speaking republics emerged—each centered on a node of settlement. Because the political geography of the new republics could be described easily by the *core-periphery* model, in which a relatively populated settlement core is surrounded by sparsely settled "peripheral" lands, territorial claims to these empty areas have led to numerous boundary disputes and redrawing of political boundaries. Bolivia is a good example of a victim of South American geopolitics, having lost its Pacific coastline to Peru and Chile in the War of the Pacific (1879–1884), its northernmost Amazonian lowlands to Brazil in 1867 and 1903, and portions of its Chaco to Paraguay in 1935 during the Chaco War (1928–1935). Boundary disputes are still quite prevalent in South America today, especially in areas suspected of being underlain by valuable mineral resources.

The viceroyalty of Brazil was politically divided into about 12 captaincies-general and subordinate captaincies, which had evolved from original large coastal land grants *(donatarios)*. As the armies of Napoleon swept through Europe in the early 1800s, the seat of the Portuguese monarchy was moved from Lisbon to Rio de Janeiro, and in 1815 Brazil was granted equal political status with the mother country within the kingdom of Portugal. After Prince Regent Dom João returned to Lisbon in 1821, his son Dom Pedro I stayed behind. The following year he established the independent monarchy of Brazil and had himself pronounced emperor. The Empire of Brazil, ruled only by Dom Pedro I and Dom Pedro II, lasted until 1889 at which time a republic was proclaimed. Unlike in the Spanish colonies, very little political fragmentation took place, and a geopol-

itically strong and expansionistic Brazil has managed to enlarge its territory since independence one century ago.

Although Brazil and the Spanish-speaking South American countries were proclaimed republics, their political histories have varied considerably. Political instability, stimulated by fighting between factions of oligarchical power brokers or between various socioeconomic classes, has characterized all of the countries to varying degrees, and military rule has prevailed throughout much of the postindependence period. Military dictatorships have ranged from conservative—even repressive—defenders of the oligarchical status quo (as in Paraguay, where martial law was in effect from 1954 until 1989 when General Stroessner was ousted in a coup d'etat) to progressive, socialistic governments (Peru during the early 1970s). Although Chile and Uruguay have had the longest histories of democracy in the twentieth century, both experienced military coups d'etat during 1973. Since the 1950s, only Colombia and Venezuela have consistently maintained civilian democratic governments. Although the future political status of the individual countries is difficult to forecast, the 1980s saw a trend to return government control to civilian—and democratic—hands; Peru (1980), Bolivia (1982), Brazil (1985), Uruguay (1985), and Chile (1989) have followed that route. Ecuador, a shaky democracy for a number of years, announced in 1988 that it would follow a more socialistic path of democracy and reduce its ties to the United States and Western Europe. Among the Latin nations of South America, only Paraguay retains a conservative military dictatorship, although elections are scheduled for 1992.

The colonial heritage of the Guianas and Falkland Islands is much more recent, and sparsely populated French Guiana and the Falklands remain European colonies. French Guiana is officially an overseas *département* of France, and Britain reasserted its sovereignty over the barren Falklands following the Falklands War of 1982. The Falklands, known as the Islas Malvinas to Argentines, have long been claimed by Argentina, and the 1982 invasion was an ill-fated attempt to assert sovereignty. Currently, 4000 British troops protect the 2000 inhabitants of these islands. Guyana received independence from Britain in 1966, and the development of local political factions along ethnic lines led to much political turmoil. Officially known as the Cooperative Republic of Guyana, the former British Guiana has followed a socialistic political path, and economic ties with the Soviet Union remain strong. Suriname was granted independence by the Netherlands in 1975, and democracy briefly reigned until a 1980 military coup d'etat. Suriname has headed down a similar socialistic path as Guyana, but political frictions remain strong, and by 1985 fully one-third of the entire population had emigrated to Holland.

Political unity among the Latin American nations exists primarily through the Organization of American States (OAS), which was formally inaugurated in 1948. Nineteenth-century efforts at establishing political unification, dating to the Panama Congress called by Simón Bolívar in 1826, were largely unsuccessful. *Pan-Americanism,* an idea promoted by the United States a century ago to establish a special relationship between North America and Latin America, led to an International Union of American Republics being formed in 1890. The OAS developed out of this organization as an effort to promote greater unity within Latin America, rather than unity vis-à-vis the United States, although the United States remains a member. Although the members agreed to the expulsion of Cuba in 1962, there remains much resentment against U.S. intervention in Latin America, including the 1962 Bay of Pigs invasion of Cuba, the 1965 military occupation of the Dominican Republic, and the invasion of Panama in 1989. The OAS has had limited success in settling minor disputes among Latin American countries as well as in establishing intraregional economic associations.

Breaking ground for planting in Valle Mantaro, Peru. Although plantation agriculture is practiced in areas along the coast, many Peruvians are engaged in subsistence agriculture growing crops for the domestic market.

Although efforts in the 1960s by the Cuban-backed guerrilla fighter Che Guevara to foment a Marxist revolution among the peasantry failed, left-wing guerrilla movements are active in many South American countries. Groups such as the M-19 in Colombia and the Sendero Luminoso (Shining Path) in Peru espouse Marxist/Leninist or Maoist philosophies and use violence to control vast portions of national territories. Outside arbitration has had little success in bringing peace to the guerrilla-held rural territories, and the national armies are largely ineffective in asserting control. The political role of drug cartels in these regions is still little understood, but evidence of "protection money" paid to guerrillas has surfaced.

HUMAN LANDSCAPE

Like the other features of South America, the human landscape exhibits a variety of patterns. Some have persisted from the precolonial period, others are a legacy of colonialism, and still others are a product of the modern era.

Agriculture

With the exceptions of southernmost South America and several expanses of Brazilian savanna, agricultural economies prevailed throughout the continent at the time of European contact. The lowlands east of the Andes supported primarily shifting cultivation, or slash and burn agriculture, and the Andean basins and valleys supported a variety of agricultural systems. These ranged from slash and burn subsistence to permanent subsistence farming and to elaborate market agriculture. The most elaborate agricultural technologies, including terracing of mountainsides, irrigation, and "sunken gardens" (to tap low water tables) developed in the Incan Empire. Although the intensive system of surplus food production of Peru was quickly abandoned following the decimation of the native population, many of the subsistence farming patterns survive to the present day.

The distribution of present agricultural land uses reflects the influences of aboriginal, colonial, and modern agricultural practices (Figure 11.9). The humid forested tropical lowlands still support primarily shifting cultivation, locally known as *roza* (*roça* in Brazil). During the respective dry season, a farmer—usually with members of his family or friends—selects a one or two acre plot of land, girdles or cuts several of the trees, and

FIGURE 11.9

Agricultural Land Uses in South America

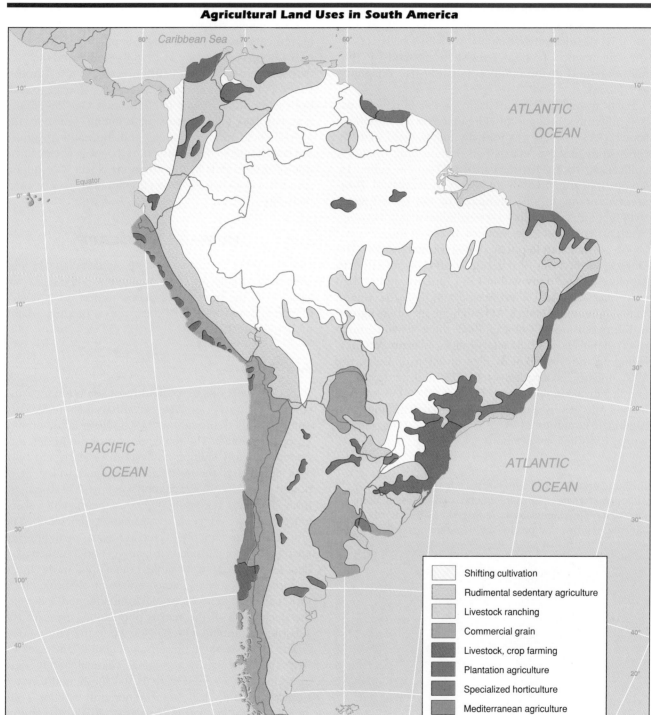

Legend:
- Shifting cultivation
- Rudimental sedentary agriculture
- Livestock ranching
- Commercial grain
- Livestock, crop farming
- Plantation agriculture
- Specialized horticulture
- Mediterranean agriculture
- Nonagricultural

burns the entire field. The ashes provide natural fertilizer, and at the onset of the rainy season, a locally specialized complex of crops is planted. The famous *Mesoamerican trilogy* of maize-beans-squash has been adopted along portions of the Andean montaña, but the crops of the Amazon lowlands are primarily root crops such as cassava and sweet potatoes. Because the fertility of the soil lies in the thin, easily leached upper humus layer, the agricultural plot is only productive for one or two seasons. As a new agricultural plot is selected elsewhere in the tropical forest, the old field is abandoned, and perhaps after 20 years or so it again reaches a stage of vegetative maturity and is ripe for renewed slashing and burning. Many of the humid tropical Andean slopes contain better soils, and the fallow time for shifting cultivation may be only 5 years or so.

The inhospitable Paraguayan Chaco and the colder and drier portions of the Andes, including much of the dry Pacific coastal plain, are not suitable for agriculture. Within the intermontane basins and valleys where fertile alluvial and volcanic soils have accumulated, a rudimental sedentary agriculture is practiced. This may be considered a form of subsistence agriculture, although slight surpluses of market crops maybe produced to provide a small amount of income for basic necessities. The staple food crops vary by region and elevation: maize is most important in the northern Andes at intermediate elevations (up to perhaps 8000 feet; 2440 me-

ters), and potatoes dominate in the southern Andes and in the highland basins of Peru and Bolivia (8000–12,000 feet; 2440–3660 meters). Wheat has become an important highland grain, especially in the northern Andes, and specialized grains such as quinoa (which can be grown up to 14,000 feet; 4270 meters) are found in the central Andes.

Commercial agriculture includes plantation agriculture, mechanized grain farming, and specialized crop farming. *Plantation agriculture* includes the traditional plantation crops such as sugarcane, cotton, and tobacco, as well as newer ones such as coffee (southern Brazil, Colombia), bananas (Ecuador, Colombia), and rice (the Guianas). Traditional sugar-growing regions along the east and north coasts of South America remain important for plantation agriculture, and new zones have been added in interior locations (e.g., sugarcane in Argentina and Colombia) and along the Pacific coast (e.g., bananas in Ecuador, sugarcane and cotton in the irrigated valleys of Peru). Mechanized grain farming is commercially most important in the pampas of Argentina and Uruguay, where crops such as wheat, sorghum, and alfalfa dominate. Soybeans are also becoming widespread, both in Argentina and in Brazil. Specialized crop farming includes truck farming around urban areas, fruit and vegetable growing, vineyards (especially for the Chilean and Argentine wine industries), and Mediterranean agriculture (dry grain farming, especially wheat).

Floodplain agriculture and relict terraces in Valle de Tarma, Peru. Usually, floodplains are quite fertile and, whenever possible, are used for agricultural purposes.

THE COCAINE CONNECTION

The concept of spatial connectivity may be applied to the flow of illicit drugs throughout the world. Increased demand for narcotics has been accompanied by an increased supply of narcotics, and since the major centers of demand and supply are usually widely separated, patterns of intercontinental drug flows have become established. The dominant world pattern in 1990 is one in which consumption of drugs is greatest in developed nations and drug cultivation and production is greatest in developing nations. A prime example of this pattern is cocaine, of which South America is the exclusive source.

Cocaine is derived from the coca plant (not to be confused with Cacao, from which cocoa and chocolate are made), which grows in the humid tropics. Two species of coca are cultivated in South America: *Erythroxylum coca,* which grows in the valleys of the eastern Andean slopes up to 10,000 feet (3000 meters) in elevation, and *Erythroxylum novogranatense,* which is a hardier species more prevalent in lower and often drier environments. When coca leaves are chewed in conjunction with an alkaline substance (such as legia in Bolivia), the narcotic is released, thus providing the user with a mildly euphoric effect. Traditionally, coca chewing was a culture trait of the highland Indians of Peru and Bolivia, and under the Incas only the elite classes were allowed to partake in the custom. After the Spanish crushed the Incan Empire and enslaved the Indians to work in the mines, coca chewing became quite widespread among the Indians, in part

as a means to escape the realities of their harsh existence. The tradition of **coqueo,** or coca chewing, has persisted to the present day, especially among poor male laborers, and only in Peru and Bolivia is the growing of coca legal.

The production of cocaine *hydrochloride (HCL)* from coca leaves is a laborious two-step process discovered in the 1850s. The first step entails the reduction—at a ratio of 250:1—of coca leaves into a coca paste by adding sodium carbonate and kerosene, agitating, and even stomping. In the second step, the paste is refined into cocaine via the addition of various chemicals; skilled chemists employ several variations in the production of the final product.

Initially promoted as a miracle drug in the late nineteenth century (even Coca-Cola began as a cocaine-based "brain tonic" in 1886), cocaine was classified as an illegal drug in the United States in the early 1900s following numerous deaths from overdoses. Its usage was confined largely to a small sector within the entertainment

industry until it became popular in the 1960s. By the 1980s, cocaine had become the "recreational drug" of the decade for all social strata in the United States, and estimates of regular users range from 6 to 20 million. At prices of $100 to $200 per gram of cocaine (diluted to only about 12 percent purity), upwards of $35 billion are spent on the illicit drug annually within the United States alone.

The increased demand for cocaine since the 1960s has not only significantly boosted production in the traditional coca-growing regions, but has also spawned many changes in the geographic distribution of coca cultivation, cocaine refining, and international marketing. Until the 1970s, virtually all cocaine production took place in Colombia—at refineries located in major cities such as Medellín, Cali, and Bogotá and in remote locations such as in the eastern llanos (Figure 2). Coca paste was imported into Colombia from the primary and traditional growing areas—Peru's Upper Huallaga Valley and Bolivia's Yungas and Chapare districts—that still today produce 80 percent of the world's coca on perhaps 250,000 acres (100,000 hectares) of land. Colombians, who became adept drug traffickers on the basis of the lucrative—and slightly older—marijuana trade, came to dominate much of the coca paste production process, especially in Peru. From the traditional primary and secondary source areas, where coca cultivation and coca-paste stomping provide lucrative employment opportunities, the paste was transported—usually by air—to the Colombian refineries, from which cocaine would be shipped to North America

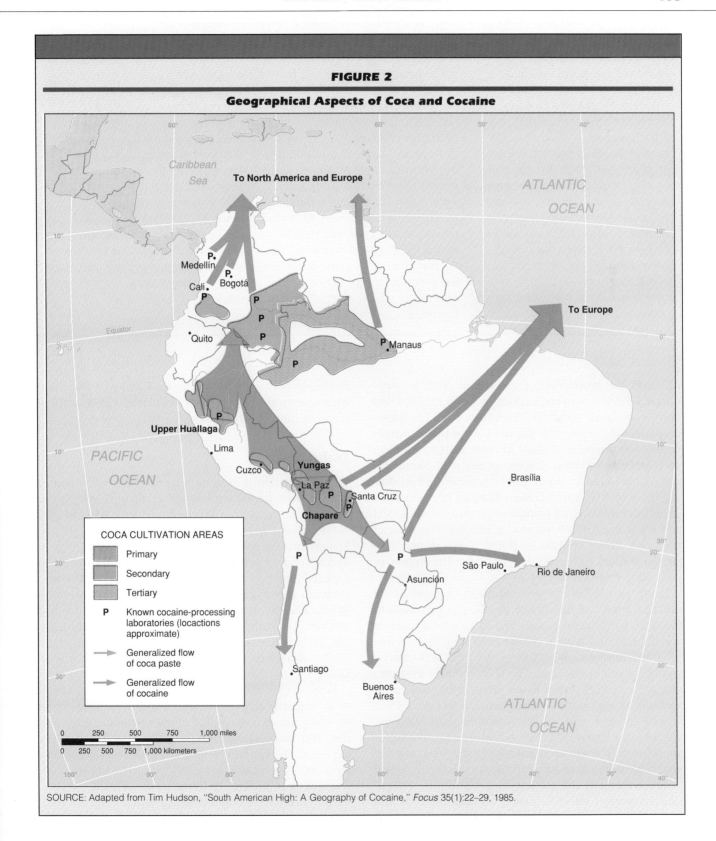

FIGURE 2

Geographical Aspects of Coca and Cocaine

SOURCE: Adapted from Tim Hudson, "South American High: A Geography of Cocaine," *Focus* 35(1):22–29, 1985.

and Europe by various carriers often via intermediate points in Central America, Mexico, and the Caribbean.

Because the potential profits from the cocaine trade are so high (totaling at least $4 billion in direct income), several new trends had become apparent by the late 1980s. First, the growing of coca has spread to new areas, especially in southern Colombia near the cocaine refineries and also along the Amazon River and its tributaries in adjacent Brazil. The highland-to-lowland shift of coca, attributed both to increased global demand for cocaine and to eradication efforts in traditional coca-growing zones, has led to

a substitution of *E. novogranatense* for the traditional *E. coca.* Second, cocaine refineries are being constructed in the new and traditional zones of coca cultivation as well as in locations optimal for distribution to the major market areas. Although Colombia still dominates the North American market, cocaine from Peru, Bolivia, Ecuador, and the Brazilian Amazon is capturing an increasing market share. Recently, cocaine refineries have been discovered in Paraguay and northern Chile, the market areas being the urban centers of Rio de Janeiro, São Paulo, Buenos Aires, and Santiago in addition to Europe.

The balance of supply and demand, coupled with the illicit nature of the cocaine industry, has led to changing patterns of spatial connectivity between the regions of cultivation and production and the regions of consumption. Economically, cocaine is very important to the source areas—in Bolivia, half of the government's foreign exchange earnings are from coca-based products and over 500,000 people are employed in the industry—and in response to both sustained demand and eradication efforts, the geography of the cocaine industry remains ever-changing.

Although the production of specialized crops is partly directed at local markets, exports to North America and Europe are quite significant, especially since the growing season is during the northern hemispheric winter when food prices are highest in those markets. Chilean and Argentine apples, grapes, pears, and plums are commonly found in North American supermarkets during the winter months. Other examples of specialized crops (sometimes even plantation grown) are marijuana and coca, from which cocaine is refined. The Andean countries, from Colombia to Bolivia, have gained substantial foreign exchange earnings from the export of these illegal narcotic crops.

Livestock ranching is widespread throughout South America, from Colombia's hot, dry Guajira peninsula to the cold, windswept steppes and tundras of Patagonia and the Falkland Islands. Where crop agriculture is feasible, the raising of beef and/or dairy cattle is often a secondary use of the land. The dairy industry in particular is concentrated near major urban areas, such as in southern Brazil, southern Chile, and the Río de la Plata region. In drier parts of the continent, including the wet-and-dry llanos and campos, livestock ranching is often the only major economic use of the land, and annual cattle drives across vast roadless landscapes are still common in many interior locations. As frontier colonization advances into the humid, tropical lowlands, the

raising of beef cattle is becoming a major activity, and extensive cattle pastures have replaced dense tropical forest in much of the Amazon.

Industry

The exploitation of mineral resources in South America is older than the earliest established Spanish mines; civilizations such as the Chibcha and Inca had held precious metals in high esteem. With the arrival of the Spanish in the sixteenth century, exploitation of gold and silver accelerated, and mining landscapes evolved especially in Peru and Bolivia. By the end of the seventeenth century, precious minerals were discovered in Brazil. The Minas Gerais gold rush, allegedly the first modern gold rush, triggered extensive searches throughout the Brazilian highlands, and several veins of high-quality diamonds were discovered in the ancient crystalline rocks.

Beginning in the mid-nineteenth century, deposits of other valuable economic mineral ores were discovered in South America, including copper, bauxite, coal, iron ore, and manganese. Like petroleum, which was first exploited in Venezuela in the 1890s, these minerals were mined primarily for export in raw form to North America and Europe. Until the mid-twentieth century, industrialization in South America remained quite rudimentary in spite of locally abundant raw materials. During the colonial era, the mineral resources were seen as the raw material for home-based industries. This pattern continued after independence, as foreign-owned companies were usually behind the major mining operations. It was in the interests of those companies to establish manufacturing facilities in their home countries near the major markets for the products.

Industrialization is normally considered the key to economic growth, as industries provide jobs and diversify the economy. North America, Europe, and Japan have achieved high standards of living, as measured by annual gross national product (GNP) per capita. Because the industrial revolution began in Europe and quickly spread to North America, those regions have had advantages in maintaining their early leads in industrialization. For developing countries, including those of South America, making the transition from being suppliers of raw materials for industrialized nations to being industrialized nations themselves requires considerable government investment in terms of capital outlay as well as infrastructure including highways, pipelines, electricity, and labor supply. Industrialization also requires a minimum threshold of production to be economically feasible, and countries with low populations and purchasing power do not provide a sufficient consumer base. One method of overcoming this problem is the formation of common market linkages with neighboring countries, but even arrangements such as this favor more populous countries. Although processes of industrialization are complex and not well understood, one simplistic means entails a simplified three-stage model of industrial development.

In the first stage of industrialization, low-technology industries are established. These include crude **reduction-in-bulk** type industries, such as the reduction of copper ore or tin ore to a bulk product with a higher concentration of minerals to ore, and also manufacturing industries aimed at local markets (such as textiles or soft drinks). Although the first example should not be considered true industrialization, but rather a convenience for avoiding exorbitant shipping costs, the second represents incipient industrialization. As local consumer classes became established in South America, especially in those countries populated by recent European immigrants and characterized by rapid urbanization, the production of consumer goods increased even more.

The second stage of industrialization consists of **import substitution,** by which items that were previously imported now are produced domestically. To enter this stage, government assistance is normally required, perhaps in the form of directly subsidizing a product, establishing protective taxation policies, or offering tax incentives for foreign companies to establish industries. Brazil may be considered the forerunner in South America in this regard, as an iron-and-steel industry was developed by the 1930s and 11 foreign motor vehicle assembly plants—using mostly domestically manufactured parts—were built by the 1950s. In spite of having a much smaller total consumer base, Argentina quickly followed Brazil's lead, and the remaining South American countries have also undergone import-substitution industrialization to some degree. In countries such as Peru and Colombia, however, where production of steel must be government subsidized to compete with imported steel, the economically beneficial effects of industrialization have not yet been realized.

The third stage of industrialization, or **general industrialization,** entails the expansion of all types of industries, including domestic and export-oriented consumer goods and even highly specialized products such as machinery parts. Brazil, with its domestic computer industry, is well into this stage, and Argentina is not far behind. In Brazil, major government support for industrial development has made this possible, and in view of the recently established industrial growth poles in Amazonia (see the box on Frontier Colonization), the industrializing trend is sure to continue. The government has embarked on an ambitious power-generation project, and in addition to the world's largest hydroelectric plant at Itaipú on the Paraná River, more plants are slated for several of the Amazon River tributaries. During the oil crisis of the 1970s, Brazil, a major oil producer yet a net oil importer, embarked on a program to distill ethanol from sugarcane, which could in turn be substituted for imported oil.

With the exception of industrial growth poles in frontier locations, most industrialization in South America has clustered in and around major urban areas (Figure 11.10). Although this trend makes economic sense, it also paints a national picture of apparent urban wealth and rural poverty that governments quickly like to dispel. And in view of the large numbers of rural inhabitants fleeing the countryside and moving to urban areas in search of jobs, many governments, notably Brazil and Venezuela, are establishing growth-pole cities closer to the raw materials than to the urban markets to foster industrial development in those peripheral zones.

FIGURE 11.10

Industrial Core Areas of South America

Although Brazil and Argentina have made great strides in manufacturing and industrialization, the remainder of the South American countries still have far to go (Table 11.1). Despite high levels of urbanization, economically many of the countries are still mostly agriculturally oriented, and overall income levels are still quite low. Economic trade associations were established in the hope of alleviating the low levels of industrialization in the countries with lower consumer bases, but so far this aim has not been accomplished. A large proportion of South America's population is engaged in economic activities that may be categorized as an "informal" economic sector, including sidewalk vending, lottery-ticket selling, cab driving, and various forms of legal and illegal hustling.

Urbanization

Unlike most developing nations in Asia and Africa, the South American countries are highly urbanized. Latin America as a whole is 69 percent urban, considerably more than Asia (29 percent) and Africa (31 percent) and

TABLE 11.1

Contribution of Manufacturing to GNP in South America, 1985

COUNTRY	VALUE ADDED TO GDP BY MANUFACTURING (MILLIONS OF U.S.$)	MANUFACTURING AS A PERCENTAGE OF GDP
Brazil	$58,089	28%
Argentina	17,954	31
Venezuela	6,157	23
Peru	3,426	20
Chile	no data	no data
Colombia	5,565	18
Uruguay	no data	no data
Ecuador	2,369	19
Bolivia	817	13
Paraguay	513	16

SOURCE: World Bank, *World Development Report 1988* (New York: Oxford University Press, 1988).

only slightly less than Europe (75 percent) and North America (74 percent). Within Latin America, tropical South America is 71 percent urban and temperate South America is 85 percent urban.

This relatively high urban percentage of population may be attributed to the immigrants, both the Iberian immigrants of the colonial era and the European immigrants of the post colonial period. Although several urban centers existed in aboriginal South America, most were restricted to the Central Andean region, and only a small proportion of the overall Indian population actually lived in cities. The arrival of Iberian colonists reoriented the pattern of settlement toward urban foci. Cities were established as administrative centers, mining centers, and port and entrepôt towns, and most Iberians lived in the cities. These cities also had a distinctive urban morphology, a legacy of the Roman city, in which the city was laid out on a grid orientation (the streets running north-south and east-west) around a nodal plaza. The plaza represented the central focus of the city, and on its sides were usually located the major government buildings, military barracks, the Catholic church, and perhaps the residence of the top administrator. Unlike in North America, where agricultural pioneers comprised much of the immigrant stock, the Iberian immigrants—especially the Spanish—were urban dwellers who came to exploit the natural resources or to administer. Even those colonists who were granted large pieces of rural real estate tended to maintain their permanent domiciles in urban areas. Censuses taken during the colonial period indicate very few nonurban Spanish settlers. In the Spanish colonies, the rural pop-

Farmers offer a variety of produce for sale at a periodic market in the central plaza in the village of Chincheros near Cuzco, Peru.

TABLE 11.2

The Millionaire Cities of South America
(Greater Urban Populations)

	POPULATION
1. São Paulo	16,832,000 (1989 est.)
2. Rio de Janeiro	11,141,000 (1989 est.)
3. Buenos Aires	10,750,000 (1985)
4. Lima	6,054,000 (1988)
5. Santiago	4,858,000 (1989 est.)
6. Bogotá	4,550,000 (1985)
7. Belo Horizonte	3,446,000 (1989 est.)
8. Caracas	3,373,000 (1989 est.)
9. Recife	2,945,000 (1989 est.)
10. Pôrto Alegre	2,924,000 (1989 est.)
11. Salvador de Bahía	2,362,000 (1989 est.)
12. Fortaleza	2,169,000 (1989 est.)
13. Medellín	2,095,000 (1987)
14. Curitiba	1,926,287 (1989 est.)
15. Brasília	1,803,000 (1989 est.)
16. Guayaquil	1,573,000 (1987)
17. Montevideo	1,550,000 (1987)
18. Cali	1,400,000 (1985)
19. Maracaibo	1,365,000 (1989 est.)
20. Belém	1,296,209 (1989 est.)
21. Valencia	1,227,000 (1989 est.)
22. Quito	1,138,000 (1987)
23. Córdoba	1,055,000 (1987)
24. La Paz	1,050,000 (1988 est.)
25. Rosario	1,045,000 (1985)

NOTE: est. = estimated. If no est. given, figures are from actual censuses.

SOURCES: Data compiled from Edward B. Espenshade Jr., and Joel L. Morrison, *Goode's World Atlas* (Chicago: Rand McNally, 1990); Europa Publications, *The Europa World Year Book 1990* (London: Europa Publications Ltd., 1990); John Paxton, ed., *The Statesman's Yearbook* (New York: St. Martin's Press, 1990); United Nations, *1988 Demographic Yearbook* (New York: United Nations, 1990).

ulations were mostly Indians, but in Brazil and the Guianas the majority of "rural" inhabitants were black slaves confined to the plantations. Following emancipation, a large rural exodus led to a corresponding increase in the populations of the coastal cities from Rio de Janeiro to the north coast of the continent.

Another major burst of urbanization took place during the period of European immigration, primarily between 1880 and 1930. Millions of laborers, mostly from Mediterranean countries, arrived in Brazil, Uruguay, and Argentina to work in the expanding agricultural sectors. They were not granted land, however, and most of those who did not return to their homelands drifted to cities such as São Paulo, Montevideo, Buenos Aires, and Rosario.

Although the largest urban agglomerations are in the southern part of the continent, large cities are found in all countries except Paraguay and the Guianas (Table 11.2). Half of the 20 largest cities are located in Brazil, most of them along the coast. Many of the rest are primate cities, which tend to dominate their respective countries economically and politically. Venezuela, Peru, Chile, Argentina, and Uruguay are prime examples of countries dominated by primate cities. All of the large urban centers are growing quite fast, several at rates of 7 to 8 percent per year, and the large influx of poor rural residents in search of jobs in the cities is causing severe strains on municipal services. All of South America's major cities contain extensive squatter settlements where conditions of housing, nutrition, and health are quite poor. To ease the burden of urban overcrowding, new growth in industry and even government is being encouraged in more remote locations. In addition to the remote industrial growth poles already discussed, the moving of Brazil's capital to Brasília in 1960 is a prime example of redirecting population growth to a relatively sparsely settled area. Argentina is planning a similar relocation of its capital from Buenos Aires to Viedma at the northern boundaries of Patagonia in the future.

REGIONS OF SOUTH AMERICA

The division of South America into subregions of relatively homogeneous cultural and physical traits provides a useful means for understanding the geographic variability of the continent. Traditional efforts at regionalizing South America usually place the major political units into four or five regional groupings and thereby mask the extensive physical and cultural variables found within even small areas. Peru, for example, contains landscapes of Amazonian rainforest, an eastern Andean agricultural frontier, barren glacier-capped Andes, densely populated intermontane basins, and dry and desolate coastal desert interrupted by irrigated floodplains. Rather than placing Peru in a broader region of "Andean America," for example, it would be more accurate to identify several distinctive regions that cut across the political borders of Peru. Accordingly, using a combination of physical (climatic, vegetative, and physiographic) and cultural (economic, ethnic, linguistic, and technological) characteristics, South America can be divided into two dozen subregions (Figure 11.11).

Perhaps the most prominent subregion is the Andean cordillera, which may be subdivided into northern, central, and southern sectors. The intermontane valleys and basins of the northern two-thirds of the Andes contained the major aboriginal centers of population, and these locations also attracted the Spaniards. Because the *Northern Andes* were flanked by insect-ridden humid tropical lowlands, the Spaniards established their settlements in numerous Andean basins from Cali to Caracas. This Northern Andean settlement core remains the most populated region in northern South America. The *Central Andes,* corresponding approximately to the maximum extent of the Inca Empire, likewise comprised a major locus of Spanish colonial settlement, especially following the establishment of mining centers. Yet high densities of Indian populations within the numerous valleys and basins, including the Altiplano, are reflected in a distinctive Indian cultural landscape in this region, which extends from southern Colombia to northern Argentina. The *Southern Andes,* by contrast, remain practically unpopulated because of their adverse physical geography. Extreme dryness in the north, the narrowness and high elevations of the cordillera, extensive glaciation (especially toward the south), and an absence of inhabitable basins and valleys have all rendered this region relatively unsuited for human habitation.

The continent's dry west coast comprises another subregion, one with a long record of aboriginal and colonial settlement. Although sparse in rainfall and natural vegetation, the *Desert Coast* of Peru and northern Chile was considered an improvement over the disease-ridden tropical lowlands of northern South America. In Peru much colonial settlement—including the primate city of Lima—focused on the aboriginally irrigated valley floodplains. Today, all of these valleys—mostly in Peru—are irrigated, and these "oases" produce a high proportion of the national agricultural output. The Chilean part of the Desert Coast is drier and contains very little coastal plain, yet several important port cities (notably Antofogasta) are located here, in part to export copper from the Andean hinterlands.

Chile's settlement core area, extending north and south from Santiago, may be considered *Mediterranean Chile* because of its climate and its cultural landscape that has come to resemble the Iberian homeland. The Central Valley of Chile, sometimes compared to California's Central Valley, is characterized by dry grain farming, livestock grazing, and vineyards (which produce the highest quality wines of the continent). This region is home to the vast majority of Chileans.

The temperate and heavily forested *Southern Chile,* including the large island of Chiloé, was the traditional homeland of the fiercely independent Araucanian (Mapuche) Indians, who still comprise a significant minority of the population. This region was still a frontier as recently as the late nineteenth century when several im-

A shantytown (Morro da Favela) in Rio de Janeiro, Brazil. Most major cities in the developing world have shantytowns inhabited by persons who have migrated to urban centers seeking economic opportunity, but for a host of reasons have been unable to find meaningful employment and are resigned to a cycle of poverty.

FIGURE 11.11

Subregions of South America

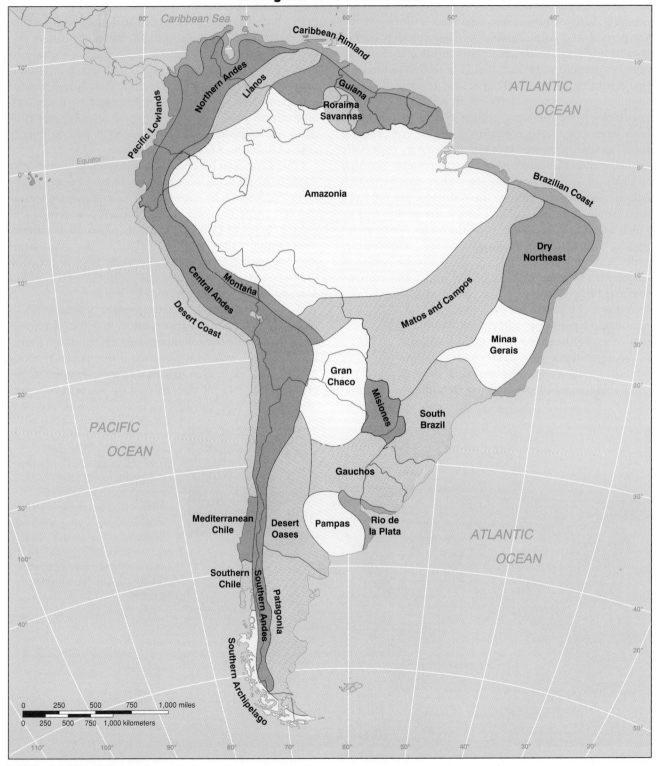

migrant groups, especially Germans, settled there. A very scenic region, similar to southern British Columbia, Southern Chile is the site of several lake resorts that attract domestic and international tourists.

Poleward of the region Chileans refer to as Southern Chile is the sparsely inhabited modern frontier zone of the *Southern Archipelago*. Consisting of heavily forested and fog-shrouded islands and Andean ridges separated by fjords, the archipelago wraps southeastward around the tip of the continent. Although lumber resources are extensive, the agricultural potential is limited, and the wet weather and lack of transportation render this region not very desirable for human habitation. The only notable city is the port of Punta Arenas, which lies on the southern margins of this region astride the Strait of Magellan.

East of the Andes, the two loci of early Spanish settlement were Paraguay and the rim of the Río de la Plata estuary. Although Paraguay has witnessed numerous boundary alterations throughout the course of its turbulent history, the portion east of the Paraguay River—along with the small adjacent province of Misiones in Argentina—comprises the subregion of *Misiones*. This was the heartland of the Guaraní Indians, many of whom were resettled in the numerous Jesuit missions that formerly were active. The mostly mestizo inhabitants share this agriculturally fertile region with Guaraní-speaking Indians and a small minority of European immigrants, mostly Germans. The land west of the Paraguay River is the agriculturally poor *Gran Chaco* subregion, a largely thorny, scrubby land subject to prolonged desiccation in the dry season and flooding in the rainy season. Extending into adjacent Bolivia and Argentina, this subregion is relatively devoid of human settlement, with the exception of several Mennonite colonies.

The *Río de la Plata* subregion, extending up the Paraná River to Rosario and including the cities of Buenos Aires and Montevideo, contains the majority of the inhabitants of Argentina and Uruguay. Control of the mouth of the river was important during the colonial period, but the great growth in this area resulted from massive immigration during the period of the opening of the agriculturally rich *Pampas*. Argentinians distinguish between a "Spanish pampa" and a "gringo pampa" on the basis of chronology of settlement and origins of the settlers. Most of the recent immigrants from Spain and Italy, however, regarded as "gringos" by the longer established Argentines of Spanish heritage, worked in the pampas but became urban dwellers. Not only do the major cities of Río de la Plata exhibit

an ambience reminiscent of Mediterranean Europe, but a cultural tradition of seaside resorts also blossomed. The northern and southern limits of this subregion are marked by Punta del Este and Mar del Plata, respectively, although scattered smaller resorts occur even further along the coasts of Uruguay and Argentina.

The twin subregions of Río de la Plata/Pampas, essentially representing an urban core and an agricultural hinterland, are encircled by three other subregions with limitations for human settlement. To the north lies an amorphous subregion we might call *Gauchos* in honor of the South American "cowboys" *(gauchos)*. Largely a grasslands environment similar to the Pampas but with slightly inferior soils, this subregion—encompassing much of northern Argentina, most of Uruguay, and southernmost Brazil—came to be dominated by a ranching economy that persists to this day. The rural gaucho culture, which prevailed in Argentina and Uruguay until the waves of mass immigration, is very romanticized in music and literature of the respective countries. Commercial grain agriculture is replacing the traditional ranching economy in parts of this area, especially in Entre Ríos province between the Paraná and Uruguay rivers (which is locally known as the Argentine Mesopotamia). To the west of Pampas and Gauchos lie discontinuous arid basins which extend along the inner flanks of the Southern Andes. The earliest settlers in this region crossed the Andes from Chile and established cities such as Mendoza and Tucumán in the numerous *Desert Oases*. Most of the oases occupy low basins within the eastern Andean piedmont, although the southern reaches—extending all the way from the Andes to the Atlantic Ocean—are characterized by ribbon-like oases within the floodplains of several large rivers. Today, the dozens of urban and agricultural oases are important centers of sugar, fruit, grain, vegetable, and wine production. Sometimes referred to as the "other Argentina," the Desert Oases resent domination by Buenos Aires within the national politico-economic framework. South of the Río Colorado was the home of the fierce Tehuelche Indians (*patagones,* or big feet), and the steppe lands of *Patagonia* were not settled until the Indians were forcibly removed in the 1850s. This colder, windswept, and oasisless (adjectives that also aptly describe the offshore Falkland Islands) southern extension of the Andean piedmont became important mostly for grazing livestock, especially sheep. Sparsely populated except for a few coastal ports and a recreational enclave at Bariloche at the base of the Andes, this region is still perceived as a frontier in Argentina.

The semiarid Andean foothills in northwest Argentina. Much of the western portion of Argentina lies on the leeward side of the Andes Mountains; consequently, it receives much less precipitation than southern Chile, which is located on the windward side of the Andes and receives more than 60 inches (150 centimeters) of precipitation annually.

The earliest impact on the New World landscape by the Portuguese was along the *Brazilian Coast,* where the plantation economy first developed and the first African slaves were imported. Although a labor-intensive plantation economy no longer prevails, blacks comprise a large proportion of the total population, and African influences remain strong in music and religion. The core of this subregion extends from about Recife southward to Rio de Janeiro, although Dutch and French influence was great along the north-facing coast during the early colonial era, and architectural vestiges of that era survive to this day. During the present century, this subregion has been extended to Belém, at the mouth of the Amazon River, and much tropical forest has been replaced by agriculture. This area, perhaps epitomized by Carnival in Rio de Janeiro or Salvador de Bahía, is often the Brazil envisioned by many outsiders.

Inland of the Brazilian Coast lie several distinctive subregions. Formerly these comprised the legendary *sertão,* an amorphous frontier region romanticized by Brazilians in literature and history. The sertão most often refers to the *Dry Northeast,* a semiarid, yet highly populated area that is periodically subject to prolonged droughts. This is the most impoverished region in Brazil, one from which out-migration and resettlement in new frontiers such as the Amazon are encouraged by the government.

Between the Northeast and São Paulo lies the highland subregion of *Minas Gerais,* site of the 1697 gold boom and the rich hinterland of Rio de Janeiro. This region, which includes an area slightly greater than the state of Minas Gerais, is characterized by its mines, including gold, diamonds, manganese, and iron ore. Although the original capital of Ouro Prêto is now but a scenic colonial tourist attraction, the newer capital of Belo Horizonte—since 1897—has become the seventh largest city in South America.

South Brazil, from São Paulo westward to the Río Paraná and southward to the land of the Gauchos, contains most of Brazil's temperate lands. The urban metropolis of São Paulo is the major industrial center of South America, but the region as a whole is defined on the basis of rich agricultural soils (largely volcanic soils on lava flow plateaus) and a high proportion of inhabitants descended from European immigrants. The northern part of the region comprises essentially the state of São Paulo, which is the prime coffee-growing area in the country. Toward the southern reaches of the region, the formerly heavily forested landscape supports a large lumber industry as well as a dairy industry. Many pockets of German immigrants, and preserved German culture, are found in Santa Catarina state.

Along the north coast of South America, the *Caribbean Rimland* is an extension of the same subregion that prevails throughout the Antilles and the eastern shores of Central America. In many ways quite similar to the Brazilian Coast, the region is marked by a plantation economy and great ethnic and linguistic diversity, and a high proportion of the population is descended from African slaves. Most of the population is concentrated in a narrow coastal plain belt. The Guianas, especially Guyana and Suriname, epitomize this subregion, which extends along the coast, in discontinuous fashion, westward to Central America. Although the north coasts of Colombia and Venezuela were historically less plantation oriented, coastal and Andean foothill plantations of bananas, sugar, coffee, and other products transformed the landscape—especially of coastal Colombia—during the present century. The *Pacific Lowlands* subregion of Colombia and Ecuador may be considered a west coast extension of the Caribbean Rimland. Although the extremely rainy west coast of Colombia is sparsely inhabited by Indians and blacks who placer mine for gold in the many streams, southernmost western Colombia and coastal Ecuador have witnessed the development of a banana plantation landscape in the twentieth century.

The remaining humid tropical lowlands of the continent, situated inland of the regions previously discussed, are frontier zones. Those areas subject to a pronounced seasonality of rainfall and naturally supporting

a savanna grassland have developed into major cattle-ranching areas where crop agriculture is a secondary economic activity at best. The *Llanos* of Colombia and Venezuela comprise one such area, which accounts for much of the beef production of those countries. Another small cattle-ranching area is the *Roraima Savannas* within the southern Guiana Highlands. A seasonally dry, upland tropical grassland environment, this sparsely settled area supports a small but important beef industry in northern Brazil and adjacent Guyana. Venezuela, with its extensive llanos, has not yet seen a need to develop its beef production potential in this region. A southern arc of savannas and savanna woodlands characterizes the *Matos and Campos* subregion of eastern Bolivia and what is considered the "interior" of Brazil. Typifying this region is Mato Grosso state in Brazil, one of the major beef production regions in the country. Although this area is an active frontier zone in Brazil bordering on the national capital of Brasília, its potential for extensive permanent agricultural settlement is limited by less-than-optimal soils. Nevertheless, the government of Brazil is actively promoting infrastructural development and encouraging colonization in this area.

The last of the interior subregions are the humid, forested tropics epitomized by the Amazon Basin lowland. *Amazonia* consists of a vast network of tributaries and associated floodplains separated by interfluves (areas

A savanna landscape in the Roraima territory in northern Brazil. Savanna-type climates have a prolonged dry season, and typically the landscape is one of grasses interspersed with trees.

Rice farming in the Rio Mayo Valley of northeastern Peru. Rice, together with vegetables, fruits, and wheat, is grown in Peru. Cotton and sugar are important export products.

between the streams) of poor soils yet luxuriant vegetation. Traditionally, settlement was along rivers, and shifting agriculture and fishing were—and for the most part still are—the major economic activities. Except for periodic boom-and-bust cycles of natural resource exploitation, human interest in this region was insignificant until the late 1960s. A rubber boom in the late 1800s and early 1900s brought a temporary wave of prosperity to the region, and Amazonian cities such as Manaus and Iquitos briefly flourished. Today, only Brazil is actively trying to develop the Amazon in terms of road building, natural resource exploitation, and agricultural colonization, but the environmental impacts and problems have been great. Fringing Amazonia along the foothills of the Andes is a narrow ribbon of naturally high luxuriant rainforest known as the *Montaña*. This area represents the zone of active frontier colonization for the Andean countries from Colombia to Bolivia. Although environmental and infrastructural problems are many, the soils are somewhat more conducive to agriculture than those of the lower Amazon *(hylea).* If present trends continue, one day the entire Montaña will be settled and linked by roads. A final humid tropical forest subregion is *Guiana,* which extends from eastern Venezuela almost to the mouth of the Amazon River in Brazil. Physiographically within the Guiana Highlands, vegetatively most of this region is dense tropical forest and consequently not very attractive for human settlement. The occurrence of gold and diamonds has attracted a variety of adventurers, and bauxite and iron ore mining are locally important at the northern fringes of this area. For the most part, however, Guiana remains among the least inhabited subregions.

SPATIAL CONNECTIVITY

ECONOMIC LINKAGES

Throughout the colonial period, the economic and trade linkages of Brazil and the Spanish-speaking territories of South America were almost exclusively with the respective mother countries. And although evidence shows a fair amount of trade with pirates and enterprising capitalists from northern European countries, these economic linkages were highly illicit and officially nonexistent.

Beginning in the nineteenth century, South America's economies became more closely tied to the more developed countries—notably the United States and Britain, but also Germany and France. These economic linkages were characterized largely by the export of raw materials and the import of manufactured products. The dependence of South America on the developed nations for both markets for its primary products and sources of manufactured consumer goods eventually led to local efforts at industrialization and economic integration.

Regional economic integration in Latin America in the postcolonial era has largely been dominated by the United States. The South American liberator Simón Bolívar eloquently proposed the formation of a Latin American confederation at the Panama Congress of 1826, but his plans did not materialize. Seeing itself as the "big brother" in the western hemisphere, the United States in the latter part of the century encouraged Pan-Americanism, which culminated in the 1890 formation of the International Union of American Republics (later known as the Pan American Union), which came to be based in Washington D.C. Recognized as being essentially an instrument of U.S. policy toward Latin America, this inter-American organization was rechartered as the Organization of American States (OAS) in 1948. Only Cuba has been formally barred from the organization, following its official conversion to communism in 1960.

It has been argued that movements for Latin American unity have developed in response to domination by the United States. Trade and economic development were spurred on by the 1948 establishment of the United Nations Economic Commission for Latin America (ECLA), which adopted a policy of industrialization. Also, during the 1950s, intraregional trade dropped significantly while imports from the rest of the world increased. In response, a Latin American common market known as the Latin American Free Trade Association (*LAFTA*) was created in 1960; its signatory members included all 10 South American republics as well as Mexico. As the more industrialized and economically more developed members of LAFTA—Argentina, Brazil, and Mexico—quickly began to dominate regional trade, an Andean subgroup comprised of Bolivia, Chile, Colombia, Ecuador, and Peru signed the *Andean Pact* in 1969. Venezuela joined in 1973. More ambitious than just a common market arrangement, this regional alliance promoted common foreign investment policies and planned industrialization, although with only limited success. Meanwhile, the United States tried once again to boost its role in the region through President Kennedy's 1961 Alliance for Progress program, which was stimulated by a desire to promote democracy and prevent Cuba-style conversions to communism. Although much aid and foreign investment resulted from this program, the domination of the United States in economic and political policies was much resented. In 1980, LAFTA was disbanded in favor of the similar, but supposedly more powerful Latin American Integration Association (*LAIA*). Among the changes in the new organization was a shift in emphasis from economic growth to defense of trade, i.e., protective barriers.

Guyana and Suriname, traditionally tied to the Caribbean region, belong to the Caribbean Common Market (*CARICOM*). Guyana is a full-fledged member, while Suriname maintains observer status.

TRADE

International trade in South America reflects both external (extraregional) and internal (intraregional) economic linkages (Table 11.3). By value, external markets are much more important than internal markets, in spite of the recent efforts at regional economic integration. As a generalization, *primary commodities* still comprise the bulk of extraregional exports, and manufactured goods make up most intraregional exports. In overall terms, the leading industrializing countries are slowly increasing their shares of exports of manufactured goods (Table 11.4). The only major increase in the export of primary commodities involves the export of oil (the leading export by value); Peru has joined OPEC members Ecuador and Venezuela as net exporters. Additional primary exports from South America include coffee, copper, sugar, beef, iron ore, bauxite, cotton, and bananas. In regard to the top 15 primary commodities, South America's share of the world market

TABLE 11.3

South American Imports and Exports (Millions of U.S.$)

	TOTAL INTERNATIONAL TRADE, 1987		
	Imports	Exports	Balance
Argentina	$ 5,819	$ 6,360	+ $ 541
Bolivia	769	569	− 200
Brazil	16,299	26,225	+ 9,926
Chile	4,023	5,102	+ 1,079
Colombia	3,907	4,642	+ 735
Ecuador	2,052	1,989	− 63
Guyana	255*	207*	− 48
Paraguay	509**	275**	− 234
Peru	2,886	2,577	− 309
Suriname	338*	63*	− 325
Uruguay	1,130	1,189	+ 59
Venezuela	7,951	8,402	+ 451
French Guiana	394	54	− 340

  * 1985 data.
** 1986 data.

SOURCE: International trade data are from United Nations, *International Trade Statistics Yearbook 1987* (New York: United Nations, 1989.

amounted to about 16 percent in 1983, a figure that has been steadily declining.

Several South American countries have been able to lower their imports of food and other primary commodities by boosting agricultural production (Table 11.5). Efforts at reducing imports of manufactured goods and machinery have been less successful. Furthermore, the push toward industrialization, coupled with world oil price hikes since the early 1970s, has greatly increased the countries' shares of fuel imports. Only the major oil exporters have benefited from increases in world petroleum prices, which rose until 1982, then resumed their upward climb after Iraq's invasion of Kuwait in 1990. In spite of the oil crises, annual trade balance figures ap-

TABLE 11.4

Structure of South American Merchandise Exports (in Percent), 1965 and 1986

	FUELS, MINERALS, AND METALS		OTHER PRIMARY COMMODITIES		MANUFACTURED GOODS*	
	1965	1986	1965	1986	1965	1986
Argentina	1%	4%	93%	73%	6%	22%
Bolivia	92	90	3	8	4	2
Brazil	9	19	83	41	9	41
Chile	89	66	7	25	5	9
Colombia	18	12	75	70	6	18
Ecuador	2	54	96	43	2	3
Paraguay	0	0	92	81	8	19
Peru	45	60	54	18	1	23
Uruguay	0	0	95	58	5	42
Venezuela	97	90	1	1	2	9

* Not including textiles and clothing.
NOTE: Totals may not equal 100 percent due to rounding.

SOURCE: World Bank, *World Development Report 1988* (New York: Oxford University Press, 1988).

TABLE 11.5

Structure of South American Merchandise Imports (in Percent), 1965 and 1986

	FUELS		FOOD AND OTHER PRIMARY COMMODITIES		MANUFACTURED GOODS AND MACHINERY	
	1965	1986	1965	1986	1965	1986
Argentina	10%	9%	27%	17%	63%	72%
Bolivia	1	1	22	12	77	78
Brazil	21	27	29	22	50	51
Chile	6	9	30	15	65	77
Colombia	1	6	18	14	80	81
Ecuador	9	2	14	8	77	90
Paraguay	14	18	16	16	70	76
Peru	3	1	22	16	75	84
Uruguay	17	21	23	11	60	67
Venezuela	1	0	17	18	83	81

NOTE: Totals may not equal 100 percent due to rounding.

SOURCE: World Bank, *World Development Report 1988* (New York: Oxford University Press) 1988.

pear favorable (Table 11.3). However, these figures mask the extensive borrowing incurred during the 1970s in the push toward industrialization. National debts and interest payments on money borrowed in the past have financially strapped some of the most economically promising South American countries. During the 1980s, even the oil exporters experienced economic hardships resulting from the drop in the world market price for oil after 1982. Ecuador has recently announced a suspension of foreign debt payments, and Venezuela has experienced rioting over belt-tightening austerity measures.

Examination of South America's trade linkages to the rest of the world reveals a continuation of the traditional strong ties to the developed nations of North America and Europe (Table 11.6). The United States remains the primary source of imports as well as the chief destination for exports. The United Kingdom, former West Germany, and the Netherlands comprise the main European trading partners, although Japan's share of the market has been steadily increasing. In addition, several Middle Eastern and African countries have become dominant fuel exporters to several oil-dependent South American countries. The relatively low importance (by value) of intraregional trade is seen by the paucity of regional trading partners. Exceptions include Trinidad and Tobago, which is a major trading partner of the Guianas (a reflection of the CARICOM connection), and several member nations of LAIA and the Andean Pact that trade extensively with their immediate neighbors.

TRANSPORTATION

Except for the Incan Empire, precolonial transportation patterns were rather rudimentary. In the lowlands, travel was primarily by canoe through the labyrinth of streams and rivers. In higher and drier lands, simple trails functioned as the conduits for regional connectivity. Within the Incan Empire, two major north-south arteries connected the far reaches of the empire. These roads, one highland and one coastal, were allegedly more elaborate than the roads of Rome. Joined at various points by east-west trans-Andean roads, an extensive Incan highway network was centered on Cuzco, the Incan capital.

During the colonial period, the Spanish expanded upon the Incan highway, extending a *camino real* (royal highway) from Lima to Caracas. Mountainflanking roads also ran north-south along the Andean chains in Colombia, and in temperate Argentina, a colonial highway connected the Andean colonies with the port of Buenos Aires. Although these roads were known as cart roads (the Spanish had introduced both the wheel and the mule), a high proportion of trade between the colonies was by ship. In Brazil, road networks also gradually evolved in the hinterlands of the major coastal ports. Roads penetrating the interior of the continent were virtually nonexistent, especially in the humid tropical lowlands and the barren steppe lands of Patagonia.

Railroads were the first significant form of mass transportation to penetrate interior portions of the continent.

TABLE 11.6

Major South American Trading Partners (Imports and Exports by Value)

	IMPORTS FROM (PERCENTAGE OF TOTAL)			EXPORTS TO (PERCENTAGE OF TOTAL)		
	1	2	3	1	2	3
Argentina	United States (22%)	Brazil (13%)	West Germany (9%)	Soviet Union (21%)	United States (13%)	Netherlands (8%)
Bolivia	United States (29)	Argentina (15)	Japan (11)	Argentina (52)	United States (26)	Netherlands (4)
Brazil	Iraq (19)	Saudi Arabia (15)	United States (15)	United States (21)	Japan (7)	West Germany (6)
Chile	United States (24)	Venezuela (7)	Brazil (7)	United States (21)	Japan (12)	West Germany (11)
Colombia	United States (35)	EEC (14)	Andean Pact (14)	EEC (38)	United States (29)	Andean Pact (6)
Ecuador	United States (34)	Japan (9)	West Germany (9)	United States (57)	Panama (10)	Colombia (6)
Guyana	Trinidad and Tobago (34)	United States (25)	United Kingdom (16)	United Kingdom (26)	United States (22)	Trinidad and Tobago (9)
Paraguay	Brazil (28)	Argentina (17)	Algeria (14)	Brazil (21)	Netherlands (14)	West Germany (12)
Peru	United States (33)	Japan (9)	West Germany (7)	United States (33)	Japan (15)	—
Suriname	United States (31)	Netherlands (19)	Trinidad and Tobago (15)	United States (35)	Netherlands (14)	Norway (13)
Uruguay	United States (12)	Nigeria (12)	Brazil (12)	Brazil (14)	Argentina (11)	West Germany (9)
Venezuela	United States (46)	Japan (8)	West Germany (7)	United States (37)	Netherlands Antilles (22)	Canada (10)
*French Guiana	France (52)	Trinidad and Tobago (19)	United States (9)	United States (48)	France (21)	Japan (13)

* European colony.

NOTE: EEC = European Economic Community.

SOURCE: Preston E. James and Clarence W. Minkel, *Latin America* (New York: John Wiley & Sons, 1986).

Although most of South America had achieved independence by the time the Railway Era began in 1850, the construction of rail networks by European and North American interests brought on a new form of "colonialism." Most rail lines in South America were built primarily to facilitate the export of raw materials (foodstuffs and mineral resources) to the rapidly industrializing nations of Europe and North America. In terms of spatial connectivity, the railroads were not very successful, as many were built exclusively to exploit a singular resource in the hinterland, such as copper and silver in Chile and Peru. Only in the flat and temperate areas of Argentina, Chile, Uruguay, and southeastern Brazil did extensive rail networks evolve. Here too, the main stimulus for rail development was the opening of agricultural lands such as the Argentine pampas and the coffee lands of the Paraná plateau in Brazil. Today, rail transportation has greatly declined in importance, especially in comparison to highways.

Highway development became an important concern of the South American nations as early as the 1920s. Expansionist countries like Brazil realized the value of providing a highway infrastructure for a rapidly growing population. Unlike the railroads, the highway networks were better planned and intranational spatial interconnectivity was vastly improved. The ***Pan-American Highway,*** the major north-south linkage in the Americas, has been essentially complete since the 1940s. Actually consisting of multiple routes in its southern reaches, the Pan-American Highway follows the Andean backbone through Colombia and Ecuador before dropping into the Peruvian coastal plain. The highway continues along the desert coast into Chile, although a branch cuts back into the Andes in southern Peru toward La Paz; from there one branch cuts eastward across Bolivia and Brazil toward São Paulo, and another branch cuts southeastward across Argentina to Buenos Aires. Although the proportion of paved highways is still relatively small, most are graded, all-weather roads. Oil-rich Venezuela boasts the highest percentage of paved roads—41 percent of its national total. The number of passenger cars on South American highways increased from 4 to 13 million between 1968 and 1980, a trend that will undoubtedly continue and stimulate both highway improvements and new road construction.

The greatest gaps in the highway network have been in the humid, tropical lowlands such as the vast Amazon Basin. In recent decades, however, road penetration into this territory has been accelerated to promote frontier colonization. A north-south highway built through the Amazon lowlands in 1976 allows one to drive from Caracas to Manaus on an all-weather road, take a ferry across the Amazon River, and continue on to Rio de Janeiro on paved highway. The east-west Trans-Amazon Highway, though unpaved and lacking bridges, will one day allow automobile traffic between Lima, Peru, and Recife, Brazil.

Air transportation has a long history in South America, a fact partly explained by the difficulty of overland transportation. The Brazilian Santos Dumont was a pioneer aviator before World War I, and commercial services date to soon thereafter. Avianca Airlines of Colombia lays claim to being the oldest commercial airline in the western hemisphere, having begun service in 1919. Prior to World War II, German, French, Italian, and American interests helped many of the individual countries establish national airlines, which came under domestic control by the end of the war. Air service has greatly improved spatial connectivity with remote national hinterlands, neighboring countries, and North America and Europe. Within the continent, the role of air transport has increased even more for freight than for passengers because of the difficulty of providing overland transportation in adverse terrain.

Overall, the development of integrated transportation networks is crucial to economic development, and South America has come far in providing infrastructure for rail, road, and air transport. Nevertheless, the transport system is still far from perfect, especially in overland linkages, and it remains to be seen to what degree new frontier highway infrastructure will stimulate development and spatial connectivity.

TELECOMMUNICATIONS

Although advances in telecommunications have greatly increased spatial connectivity both within countries and between countries in South America, the levels of participation in the media of mass communications still need improvement (Table 11.7). One measure of newspaper readership is the number of daily newspapers published, and Brazil and Argentina rank quite high using this method of measurement. The less Europeanized Andean and interior countries have the least accessibility to telecommunications, and rates of telephone and television ownership reflect this. Radios are ubiquitous throughout the continent, yet certain countries (Suriname, Guyana, Uruguay, Paraguay, and Ecuador) show

TABLE 11.7

Selected Telecommunications Data for South America

COUNTRY	RADIO RECEIVERS (THOUSANDS IN USE)	TV RECEIVERS (THOUSANDS IN USE)	TELEPHONES (THOUSANDS IN USE)	NUMBER OF DAILY NEWSPAPERS
Argentina	20,000[3]	6,500[3]	2,580[2]	188[2]
Bolivia	n.a.	n.a.	205[1]	10[2]
Brazil	75,000[6]	34,000[6]	9,082[6]	322[6]
Chile	4,000[3]	1,750[3]	629[1]	40[3]
Colombia	5,000[5]	3,250[1]	2,547[2]	31[2]
Ecuador	2,900[5]	800[5]	339[3]	22[2]
Falkland Islands	n.a.	n.a.	0.6[5]	n.a.
French Guiana	n.a.	n.a.	26[7]	1[7]
Guyana	307[5]	15[5]	33[5]	1[7]
Paraguay	645[5]	92[5]	93[4]	6[7]
Peru	5,000[5]	1,750[5]	n.a.	30[7]
Suriname	245[5]	35[5]	31[5]	2[7]
Uruguay	1,800[4]	520[4]	337[4]	14[7]
Venezuela	7,550[4]	2,500[4]	1,166[3]	25[7]

[1] 1983   [3] 1985   [5] 1987   [7] 1989
[2] 1984   [4] 1986   [6] 1988

SOURCES: John Paxton, ed., *The Statesman's Yearbook* (New York: St. Martin's Press, 1990); Europa Publications, *The Europa World Yearbook 1990* (London: Europa Publications Ltd., 1990).

relatively low levels of radio ownership. As a generalization, countries with strong historical or present ties to Europe have higher participation rates in telecommunications than the other countries.

TOURISM

The tourism industry is still not extensively developed in South America, and opportunities for greater interconnectivity as well as economic development exist. Historically, few scenic or archaeological attractions were developed for tourists, and travel agencies and airlines primarily promoted the major cities of South America as tourist destinations. Now, however, both to attract more international tourists and also to keep their own citizens from spending their vacations abroad, several governments are developing a more complete tourism infrastructure. Statistics from 1985 (Table 11.8) indicate that revenues generated from tourism are important to the respective national economies.

Various types of tourist attractions exist in South America, and some of the opportunities for development are only now slowly being exploited. The north coast of South America is also the southern shore of the Caribbean Sea ("America's Mediterranean"), and a beach resort/cruiseship port tourist landscape has developed in parts of Colombia and Venezuela. Distance to the United States, the major tourist market area, is not too great for continued "winter playground" tourism development.

A more recent trend in South American tourism has been the development of specialized tourism and ecotourism. These focus upon archaeological attractions, Indian and colonial cities, exotic natural settings (especially selvas), seasonal events such as Carnival in Rio or Salvador, seasonal sports activities such as skiing, or remote lands such as Antarctica. Package tours, often oriented around specific themes (birdwatching, penguin watching, Incan archaeology), are becoming more popular and are introducing tourists to portions of South America previously relatively untouched.

Domestic tourism is important in the more affluent countries of South America, notably Argentina, Brazil, Chile, Uruguay, and Venezuela. Beach resorts seasonally catering to national tourists line the shores of the countries. Some of these beach resorts also attract citizens of neighboring countries (such as Argentines vacationing in Viña del Mar, Chile), and the level of intraregional tourism is significant. Domestic tourism has also been promoted by the establishment of duty-free zones, where consumer items such as televisions, videocassette recorders, and automobiles can be purchased relatively tax-free. Although duty technically must be paid upon return to one's domicile, Colombians flock to offshore San Andrés Island, Venezuelans to Margarita Island, and Brazilians to Amazonia in search of bargains.

TABLE 11.8

Tourist Arrivals and Tourism Revenues in South America, 1985

	TOURIST ARRIVALS	TOURIST RECEIPTS (U.S.$)
Argentina	1,503,099	$ 673,000,000
Bolivia	127,027	36,000,000
Brazil	1,735,982	1,739,000,000
Chile	418,050	112,000,000
Colombia*	715,277	209,000,000
Ecuador	238,105	118,000,000
Paraguay	262,689	80,000,000
Peru*	278,783	258,000,000
Uruguay	702,100	129,000,000
Venezuela	239,030	367,000,000

* 1984 data.

SOURCE: James W. Wilkie and Enrique Ochoa, eds., *Statistical Abstract of Latin America,* Vol. 27 (Los Angeles: UCLA Latin American Center Publications, 1989).

As the savings often exceed the price of travel, this form of tourism may well increase.

PROBLEMS AND PROSPECTS

POPULATION GROWTH

Although advances in provision of education in rural areas and promotion of family planning have led to a general decline in the birth rate in South America, overall high rates of population growth still make it difficult for governments to ensure adequate standards of living. In tropical South America, the average rate of natural increase is 2.1 percent (meaning that the population will double in 33 years), and in countries such as Paraguay the rate is as high as 2.8 percent (doubling time of 25 years). These high rates of growth, although falling in comparison with Asia, Africa, or Middle America, often make it difficult for the respective governments to provide adequate medical care and education. The worst aspect of high population growth is lack of employment opportunities in rural areas (not to mention the threat of food shortages during droughts or other natural catastrophes), which in turn stimulates migration to already overcrowded urban areas. False hopes of rising standards of living in the urban "paradises" reduce people to living in unsanitary shantytowns and perhaps begging or stealing to make ends meet.

Rural attitudes toward large families paradoxically tend to persist for several generations, and this compounds the problems that governments face. Instead of perceiving children as financial liabilities (more mouths to feed and more clothes to buy), rural migrants to urban shantytowns often maintain that the more children a family has, the greater the chance that one or more will eventually land a well-paying job. If this unlikely scenario becomes reality, these children will then support the parents and less fortunate unemployed siblings. These types of rural attitudes toward family size help maintain high birth rates and high unemployment levels in South America's urban centers.

High population growth rates also preclude governments from making significant gains in increasing food production and stimulating overall economic growth. Extensive borrowing from Western banks in the 1970s has put several South American nations deeply in debt, and belt-tightening measures designed to put the countries back on course are causing drops in the per capita living standards, not to mention violent demonstrations.

Curtailment of high population growth would alleviate at least some of South America's environmental and economic problems. In view of the promising trends established within the last 10 to 15 years, it is quite likely that population growth rates will drop to levels characteristic of North American or European nations.

ENVIRONMENTAL DEGRADATION

Although environmental degradation in South America may be traced back to soil erosion in the days of pre-Incan civilizations, the greatest destruction in areal terms is occurring in the present day. Human movement into the tropical rainforest environment—in search of mineral wealth, lumber, farming opportunities, hydroelectric development, or military security—has been accompanied by rates of forest clearing unequaled in human history. Although the respective countries have undoubtedly gained certain economic advantages, the environmental damage in terms of soil laterization, species extinction, oxygen depletion, hydrologic alteration, and cultural survival has not yet been adequately measured. The promised frontier is witnessing substantial degradation, even in regard to its most essential activity—subsistence agriculture. Governments are providing access into this "forbidden jungle," but migrants unaccustomed to such environments are not taught proper agricultural and soil conservation practices. The degradation of the tropical forest environment is destroying not only the natural ecology, but also its potential for properly managed human exploitation. In recent years, the respective governments have shown a greater awareness of the problems, and one may be cautiously optimistic that better management practices will be instituted.

The movement into the frontiers, especially the Amazon lowlands, for purposes of growth-pole establishment and economic development should be carefully monitored by the respective governments to ensure a minimum of environmental destruction. Economic development in such marginal lands should be viewed as a long-term, not a short-term, endeavor. Short-term exploitation of natural resources is not compatible with long-term economic development, and although many problems characterize development efforts in the humid tropical environment, proper planning and management can minimize these problems and maximize future economic development.

Decline in agricultural fertility due to poor conservation practices or overgrazing is also a problem in other parts of South America, particularly densely settled ones. These regions are often quite vulnerable to enhanced degradation by natural catastrophes such as droughts or floods. The Brazilian Northeast has been subjected to several prolonged droughts that have led to desertification of the soil, starvation, and outmigration by the inhabitants. Continuing efforts by national and international scientists to provide technical assistance and education to rural agriculturists are making gains in conserving South America's agricultural lands.

ECONOMIC DEVELOPMENT

South America has made great strides in economic development over the past several decades. Among these are the development of manufacturing and industrial facilities, which have accounted for a reduction of the proportion of raw material exports to the developed nations of North America relative to manufactured consumer items. In spite of efforts at economic integration, many nations still maintain closer economic ties to major European or American powers than to their own neighbors. The promotion of stronger intraregional spatial connectivity, facilitated by improvements in avenues of overland transportation, should reduce the percentage of raw material exports even further and strengthen a continental consumer goods industry. Regional imbalances in economic productivity are currently responsible for inequitable distribution of revenues, and incentives must be offered to stimulate both manufacturing infrastructure and market demand.

POLITICAL STABILITY

Except for the Guianas, the countries of South America have been independent republics since the nineteenth century. This long period of independence has not yet created an atmosphere of political stability, however, in part because of a continuing struggle to achieve the perfect balance of free enterprise and social welfare. The legacy of the colonial era can still be seen as local oligarchies whose members are of European descent have concentrated wealth and political power, often at the expense of poor, uneducated Indian and mestizo masses. In the past, political stability was maintained by autocratic military dictatorships, but democracy coupled with social reforms appears to be the popularly desired political path. Socialism and communism have occasionally enjoyed brief periods of success (as in Chile between 1970 and 1973), but popular backlash has caused these regimes to quickly lose favor. Castro's 1960s attempts to export the "Cuban Revolution" to South America via his Argentine lieutenant Che Guevara were also unsuccessful. Colombia and Venezuela have had democratic governments for three decades, and most remaining South American nations are now following their examples. Guyana and Suriname, both recently emerging from under colonial rule and characterized by outdated plantation economies, have experienced much political instability and extensive flirtations with Marxism. Many residents of those countries have fled to England, Canada, and the Netherlands as socialism has come to dominate. The relatively unpopulated French Guiana and Falkland Islands will continue to remain European colonies, the latter especially in the wake of the 1982 Falklands War between Britain and Argentina. (The loss of that war, started by the then-ruling military junta partly to defuse civic discontent at home, led to the toppling of the military regime and the reinstatement of democracy in Argentina.)

The continued inequities in the social structure are perhaps the greatest threat to political stability in South America. Where Indians and peasant-level mestizos comprise the bulk of the population, there are great inequalities in wealth and education between the upper and lower classes. Economic benefits have not reached the masses in countries such as Colombia, Ecuador, Peru, and Bolivia, and now various guerrilla factions flying the banners of a multitude of ideologies are claiming to be taking up their cause. These guerrilla groups, espousing the philosophies of Marx, Lenin, and Mao in addition to those of local populist heroes, control vast territories within their respective countries. Some have strong ties to the illicit narcotics industry as well. Political violence in areas where these groups are strong is not only endangering the lives of innocent people but is threatening the economic and political stability of the region. Improvements in intracountry spatial linkages, especially highways, would better integrate these guerrilla-held territories with the rest of the country and bring them under national control. But even more importantly, raising the standards of living of the poorest classes might remove incentives to support guerrilla groups that threaten to totally offset the social, economic, and political gains and aspirations of South American nations.

Selected Readings

Blouet, B. W., and O. M. Blouet, eds. *Latin America: An introductory Survey.* New York: John Wiley & Sons, 1982.

Collier, S. H. Blakemore, and T. E. Skidmore, eds. *The Cambridge Encyclopedia of Latin America and the Caribbean.* Cambridge: Cambridge University Press, 1985.

Economic Commission for Latin America and the Caribbean. *Economic Survey of Latin America and the Caribbean,* 2 volumes. Santiago, Chile: United Nations 1989.

Europa Publications, *The Europa World Year Book 1990,* 2 volumes. London: Europa Publications Ltd., 1990.

Goodwin, Paul B., Jr., *Latin America,* 4th ed. Guilford, Conn.: The Dushkin Publishing Group, Inc., 1990.

Gwynne, R. N. *Industrialization and Urbanization in Latin America.* Baltimore: John Hopkins University Press, 1986.

James, P. E., and C. W. Minkel, *Latin America,* 5th ed. New York: John Wiley & Sons. 1986.

Lombardi, C. L., and J. V. Lombardi. *Latin American History: A Teaching Atlas.* Madison: University of Wisconsin Press, 1983.

Merrick, Thomas W., Population Pressures in Latin America, *Population Bulletin* 41, 3, July 1986.

Paxton, John, ed., *The Statesman's Yearbook.* New York: St. Martin's Press, 1990.

Skidmore, T. E., and P. H. Smith. *Modern Latin America.* New York: Oxford University Press, 1984.

United Nations, *1988 Demographic Yearbook.* (40th issue) New York: United Nations, 1990.

Wilkie, James W. and Enrique Ochoa, eds., *Satistical Abstract of Latin America,* Vol. 27. Los Angeles: UCLA Latin American Center Publications, 1989.

World Resources Institute, *World Resources 1990–1991.* New York: Oxford University Press, 1990.

CHAPTER 12

Southwest Asia–North Africa

IMPORTANT TERMS

Islam

Precipitation effectiveness

Dependency ratio

Sunni

Shia

Zionism

Fertile crescent

CONCEPTS AND ISSUES

■ Distribution and use of limited water resources.

■ Regional disparity in population distribution.

■ The pervasiveness of Islam.

■ Dependence of the economies and politics on petroleum resources.

CHAPTER OUTLINE

PHYSICAL ENVIRONMENT
Surface Features
Precipitation
Temperature
Rivers
Soils
Vegetation

HUMAN ENVIRONMENT
Population
Cultural Environment
Human Landscape
Regionalization

SPATIAL CONNECTIVITY
National Cohesion
Connectivity among Southwest
   Asian–North African Countries
Southwest Asia–North Africa and
   the World

PROBLEMS AND PROSPECTS

outhwest Asia–North Africa (SWANA), as used in this book, contains 21 countries (Figure 12.1). Together they cover an area of 4,963,500 square miles (12,855,465 square kilometers), which is about 8.6 percent of the earth's land surface. They have a combined population of more than 340 million, or about 6 percent of the world's population. The region has an east-west expanse of some 5600 miles (9010 kilometers), which is a little longer than the distance from New York to Honolulu. Morocco is closer to New York than it is to Afghanistan.

SWANA is distinguished by three geographic characteristics, two of which are cultural while one is physical. Culturally, the region is dominated by **Islam,** which is the major religion in 19 of the 21 countries. The excep-

tions are Cyprus and Israel. In addition to being a faith, Islam is a code of individual behavior, social interaction and responsibility, and justice. It is pervasive, so much so that it is part of the cultural heritage of Christian Arab communities in the region. The second major cultural characteristic is ***Arabic,*** the dominant language in 16 countries; the exceptions, in addition to Cyprus and Israel, are Afghanistan, Iran, and Turkey. Note that not all Muslims are Arab (Figure 12.2). In fact, most Muslims are not Arab, so that Arabs are a minority in the world of Islam. For example, the world's most populous Muslim countries are Indonesia, Bangladesh, and Pakistan, which have a combined population of about 420 million. All of them are located outside this region. Nor are all Arabs Muslim, though the great majority of them are.

FIGURE 12.1

Southwest Asia–North Africa

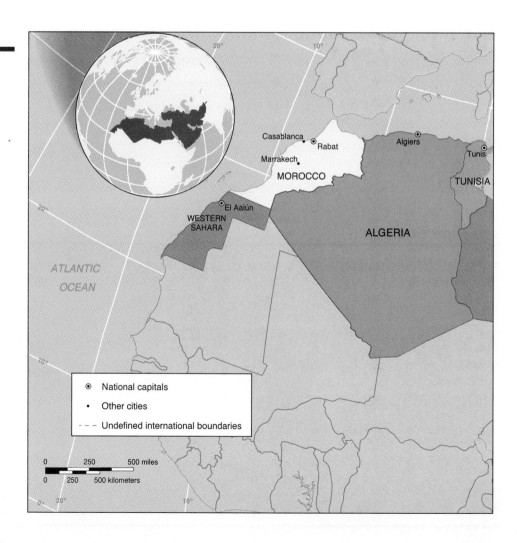

Those Arabs who are not Muslim are mostly Christian. At the same time, because Arabic is the language of the **Quran,** the Muslim religious book, the influence of the Arabic language reaches beyond those for whom it is a native language. Muslims learn to recite the Quran, or sections of it, in Arabic irrespective of their native language.

The prevailing characteristic of SWANA's physical geography is a precarious water supply. Arid or semiarid conditions are found throughout most of the region. The Sahara (which includes areas outside North Africa as used in this book) is larger than the contiguous United States. In Southwest Asia, dry conditions extend from the Arabian peninsula northeastward beyond Afghanistan into the Soviet Union and the People's Republic of China. In the nonarid areas, there is a sharp seasonality to the precipitation: wet winters and dry summers. This pattern, which is characteristic of the Mediterranean type of climate, is experienced in coastal Turkey, the eastern borderlands of the Mediterranean Sea, and the coastal area of northwest Africa. The Arabic word for rain (*shita*) is the same as the word for winter. An exception to this seasonality is found in the southwest of the Arabian peninsula, which is at the fringe of the monsoonal conditions that affect eastern Africa and South Asia: here summer is the rainy season. The great variation in the amount and timing of the precipitation that does fall makes water an especially precious resource.

Preoccupation with the water supply is part of the region's cultural and literary heritage. For instance, a

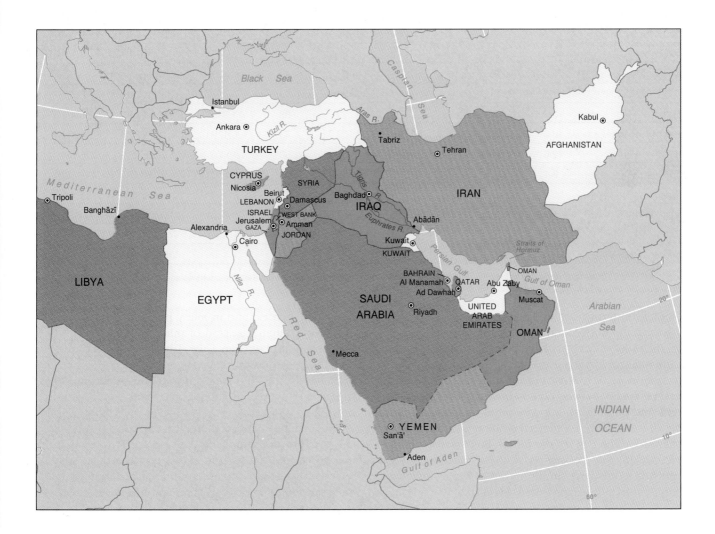

STATISTICAL PROFILE: SOUTHWEST ASIA–NORTH AFRICA

REGION OR COUNTRY	Population Estimate mid-1990 (Millions)	Natural Increase (Annual)	Population Projected to 2000 (Millions)	Infant Mortality Rate	Life Expectancy at Birth (Years)	Urban Population (%)	Per Capita GNP, 1988 (US$)	Area, Thousands of Square Miles (Km²)	Population Density, No./mi² (No./km²)
Afghanistan	15.9	2.6	25.4	182	41	18	—	250.0(647.5)	63(24)
Algeria	25.6	3.1	32.7	74	60	43	2,450	919.6(2381.8)	28(11)
Bahrain	0.5	2.3	0.7	24	67	81	6,610	0.2(0.6)	2172(839)
Cyprus	0.7	1.0	0.8	11	76	62	6,260	3.6(9.2)	197(76)
Egypt	54.7	2.9	69.0	90	60	45	650	386.7(1001.4)	141(54)
Iran	55.6	3.6	75.7	91	63	54	—	636.3(1648)	87(34)
Iraq	18.8	3.9	27.2	67	67	68	—	167.9(434.9)	112(43)
Israel	4.6	1.6	5.4	10	75	89	8,650	8.0(20.8)	572(221)
Jordan	4.1	3.5	5.7	54	69	64	1,500	37.7(97.7)	109(42)
Kuwait	2.1	2.5	2.9	15.6	73	94	13,680	6.9(17.8)	311(120)
Lebanon	3.3	2.1	4.1	49	68	80	—	4.0(10.4)	832(321)
Libya	4.2	3.1	5.6	69	66	76	5,410	679.4(1759.5)	6(2)
Morocco	25.6	2.6	31.4	2	61	43	750	172.4(446.5)	149(58)
Oman	1.5	3.3	2.1	100	55	9	5,070	82.0(212.5)	18(7)
Qatar	0.5	2.3	0.7	25	69	88	11,610	4.3(11)	116(45)
Saudi Arabia	15.0	3.4	22.0	71	63	73	6,170	830.0(2149.7)	18(7)
Syria	12.6	3.8	18.0	48	65	50	1,670	71.5(185.2)	176(68)
Tunisia	8.1	2.0	10.1	59	65	53	1,230	63.2(163.6)	129(50)
Turkey	56.7	2.1	69.0	74	64	53	1,280	301.4(780.6)	188(73)
United Arab Emirates	1.6	1.9	2.0	26	71	81	15,720	32.3(83.6)	49(19)
Yemen, North*	7.2	3.5	10.0	129	49	20	650	75.3(195)	95(37)
Yemen, South*	2.6	3.4	3.6	110	52	40	430	128.6(333)	20(8)

*North Yemen and South Yemen merged to form the Republic of Yemen in 1990, but statistics in this table list the countries separately. Also, Western Sahara (population 100,000) is not included in this list.

complex body of rules and regulations has developed to govern the apportionment of water. It is not surprising, therefore, that rivers have been of great importance in the region's history. Thus, we speak of a *Nile Valley* civilization or of a *Mesopotamian* civilization (Mesopotamia means between the two rivers, in this case the Tigris and Euphrates), but we do not speak in the same way of a Mississippi Valley civilization or a Rhine River civilization. The case of the Jordan River is also instructive. Given its association with significant biblical events, it looms large in the minds of most Christians. Many are greatly disappointed when they first see it; it

looks like a creek, and in many places it can be forded during the summer season of minimum discharge. The main channel, between Lake Tiberias (Sea of Galilee) and the Dead Sea, is about as long as one of the branches of the Nile Delta. But at the same time, the Jordan is a source of continuous water supply amidst prevailing aridity.

The juxtaposition of these three characteristics—Islam, Arabic, and a precarious water supply—is distinctive to SWANA and is true of none of the other world regions. At the same time, one must keep in mind that these attributes are not coextensive. None of them is true of all SWANA, and none of them is limited only to SWANA. More Muslims live outside the region than within it, and Christianity, Judaism, and Zoroastrianism all have important representations within the region. Arabic-speaking countries (defined in terms of membership in the League of Arab States) include non-SWANA Sudan, Somalia, Djibouti, and Mauritania, and non-Arab Iran, Turkey, Israel, and Cyprus are found within the region. Deserts, of course, are found elsewhere in the world, and within SWANA nondesert conditions are found in the mountains of Morocco, Algeria, Lebanon, Turkey, and Iran.

Finally, one must keep in mind the scope of the temporal and spatial scales at which SWANA is viewed. We speak, perhaps too readily, of millennia of land use and of history, a temporal sweep so grand that it can induce indifference to events spanning "only" a hundred years. A century becomes a fleeting moment, and therein lies a pitfall of hasty temporal generalization. In the United States, roadside "historical markers" are erected to immortalize happenings only a few score years old. By comparison, in 1970 Al-Azhar University in Cairo celebrated its one-thousandth anniversary as a continually functioning university. Spatially, also, SWANA contains great variations in the range of cultural and physical characteristics, variations that belie facile generalizations and are invisible to a distant viewer who sees the region only as a whole.

PHYSICAL ENVIRONMENT

The paucity and precariousness of the water supply are a matter of continuing preoccupation for the people of SWANA. The prevalence of aridity, especially exemplified by the great desert across northern Africa, the Sahara,* is related to the dominance of the subtropical high pressure cells found at approximately 20 degrees north and south latitude on the west coasts of continents. In order for precipitation to occur, air must be forced to rise, whether vertically or obliquely. High pressure is associated with descending drying and diverging air; thus the possibility of rain is nil. Exceptions to the prevailing dryness are rare showers resulting from considerable surface heating that causes intense con-

* "Sahara" comes from the Arabic word for desert. Thus, the phrase "Sahara Desert" is redundant just as "Rio Grande River" is redundant, since "rio" is the Spanish word for river.

FIGURE 12.2

Muslim and Arab Worlds

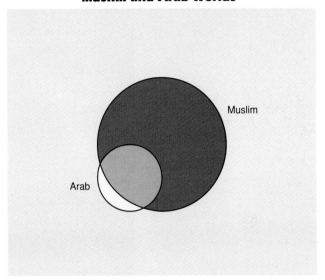

vection. These are like the "scattered showers" of mid-latitude weather forecasts, except that in the subtropics the showers are much more "scattered" in time as well as in space. Eastward from Africa into Southwest Asia, the drying effect of the subtropical high is extended by the enormous continental expanse of Asia, and it is accentuated by the blocking of sea influence by the virtually continuous highlands and mountain systems. The result is the world's largest dry region, stretching from Morocco's Atlantic coast to Mongolia and northern China.

Obviously not all of SWANA lacks water, otherwise how could the more than 340 million people live? Before looking at water supply and its distribution, it will be useful to examine the region's surface features. Mountains are directly relevant to the matter of water supply, both to its presence and to its absence. One side of a range may be in the path of moist air blowing from the sea, causing the air to rise, with the consequently enhanced possibility of precipitation. The other side of the range may well be dry, for here the air is descending. Such a situation is termed orographic from the Greek word for mountain (oros). The two sides are called the windward and leeward sides; the latter is also known as the rain shadow side. In an area where the moisture supply is marginal, a slight variation in altitude or in the orientation of a line of hills may result in locally significant variations in precipitation.

Much of North Africa is arid, but several oases dot the landscape.

SURFACE FEATURES

Moving from west to east across SWANA, one first encounters the Atlas Mountains, trending northeastward in most of Morocco, northern Algeria, and northern Tunisia. Geologically, these are related to the Alps, Pyrenees, and Apennines in Europe. They consist of several ranges, more or less parallel to the coast. There is almost no coastal plain, so that small low-lying areas become especially precious, both for agricultural and urban land uses. Likewise, interior valleys and plateaus have become foci of human activity. In Libya, the edge of the African plateau is close to the Mediterranean coast. Two small areas are locally of considerable human importance, one around Tripoli (the capital city) in the west and the other near Benghazi in the east. In both cases the coast reaches far enough north to benefit, barely, from winter cyclonic storms, and the possibility of rain is enhanced by the orographic effect of the plateau edge. The upland area behind Benghazi is called Jabal al-Akhdar (the Green Mountain). It is a small, narrow strip, with a rapid transition from a wooded landscape through grasslands to the desert. Between Tripoli and Benghazi, the desert reaches the sea. In Egypt, the entire country is desert. The highlands along the Red Sea are the eastern edge of the African plateau, and they are much too far south to benefit from cyclonic storms that move eastward along the Mediterranean.

In Southwest Asia, too, there is a shortage of coastal plains. Prominent along the eastern Mediterranean are the Lebanon Mountains. Turkey is dominated by the Anatolian Plateau whose surface averages 3000 feet (914 meters) above sea level. In the south the plateau is bounded by the high, though not entirely continuous, Taurus Mountains and associated ranges, and in the north, along the Black Sea, it is bounded by the more continuous Pontic Mountains. These two ranges come together in eastern Turkey in the complex Armenian Knot, dominated by Mount Ararat, which is associated with Noah's Ark. From here radiate two major mountain systems. To the east are the Elburz (Albruz) Mountains along the southern coast of the Caspian Sea, continuing after a brief interruption as the Koppeh (Kopet) Mountains along the border with the Soviet Union. Still farther east are the Hindu Kush of Afghanistan, which extend to the Pamir Knot from which the Himalayas and other mountain ranges radiate into central Asia. Extending southeastward from the Armenian Knot are the Zagros Mountains in Iran along the Iraqi border. These curve southeastward along the coast of the Arabian Sea. Interior Iran, like Turkey, consists of high basins and plateaus with elevations between 3000 and 5000 feet (914–1524 meters).

The Arabian peninsula is a massive plateau that abuts on the Red Sea in a prominent scarp whose elevation increases southward. The plateau dips northeastward toward the Gulf* (Arabian or Persian) and the Tigris-Euphrates lowland. In the southeast, along the Gulf of Oman, is the Jabal al-Akhdar range—the same name as the upland in Libya behind Benghazi, though here there is little tree growth. Because of the porous limestone, there are numerous springs along the foothills, and some scholars suggest that the name "green" is associated with the life-giving waters of the springs rather than with a vegetation cover.

PRECIPITATION

Precipitation distribution in SWANA is especially sensitive to surface features. Starting in northwest Africa, the Atlas Mountains are well reflected on a precipitation map. The local relief enhances the rain brought about by the winter cyclonic systems that are common in Western Europe all year. Thus, the north- and west-facing slopes of the Atlas Mountains are well watered, and intermontane valleys and plateaus experience rain shadow effects. The interior of Morocco, and especially that of Algeria, is dry. Tunisia has a well-marked north-south precipitation gradient. In Libya, the hinterlands of Tripoli and Benghazi are barely outside the arid category, and then only for a distance of less than 20 miles (32 kilometers). Along the eastern Mediterranean coast is a narrow zone of nonarid conditions, and precipitation amounts increase northward and upward. Turkey's periphery is well watered. The alignments of the Elburz and Zagros mountains, and to a lesser extent the Koppeh and Hindu Kush, are clearly reflected in the precipitation pattern, as is Jabal al-Akhdar in southeastern Arabia. Similarly, the western edge of the Arabian plateau stands out; here precipitation amounts increase southward as well as upward. Because this area is south of the dominance of the subtropical high pressure, summer is the rainy season, and there is the beginning of the monsoonal effect that is well developed over Ethiopia.

The correlation between altitude and precipitation can be illustrated by comparing elevations and amounts of annual rainfall. The Atlas Mountains peak at 13,665 feet (4165 meters) at Jabal Toubkal, in Morocco and approach 10,000 feet (3048 meters) in Algeria. Mount Ararat in eastern Turkey reaches 16,804 feet (5122 meters). The Lebanon Mountains peak at 10,131 feet (3088 me-

ters) and carry snow into early summer. The Zagros exceed 14,000 feet (4267 meters), and Mount Damavand in the Elburz reaches 18,386 feet (5604 meters) and is snowcapped all year. It even has a small glacier. In Afghanistan mountain ridges approach 20,000 feet (6096 meters), and in the Arabian peninsula the highest elevation is Hadur Shu'ayb at 12,336 feet (3760 meters) in the Republic of Yemen. For comparison, the highest peaks in the contiguous United States and North America are, respectively, Mount Whitney in California at 14,494 feet (4418 meters) and Mount McKinley in Alaska at 20,320 feet (6194 meters). As for precipitation, the following are illustrative of the amounts received in the higher elevations: 45–49 inches (114–124 centimeters) in Morocco and over 60 inches (752 centimeters) in Algeria; over 50 inches (127 centimeters) in Lebanon; about 40 inches (102 centimeters) in Yemen; 80 inches (203 centimeters) in northwestern Iran near the Caspian Sea; and over 100 inches (254 centimeters) in extreme eastern Turkey along the Black Sea. These values are upper limits and thus are by no means representative of the respective countries. But they do illustrate a considerable variety in environmental conditions, thus belying uninformed presumptions of almost universal aridity. One must keep in mind the marked seasonality of the precipitation. By comparison, Washington, D.C., has an annual precipitation of about 40 inches (102 centimeters) with no distinct dry season.

TEMPERATURE

Temperatures likewise vary greatly. Typically, the Mediterranean type of climate has hot dry summers and mild wet winters, a description that is satisfactory at an overall regional level. Given SWANA's highly varied topography, its interpenetration by large water bodies, its proximity to enormous continental influences, and its latitudinal location astride the subtropics and the southern borders of the midlatitudes, it is not surprising to find that the average conditions at the regional scale of analysis mask great variations at the local scale, the scale at which human societies make their living.

The average temperature for July, the warmest month, ranges from the low 70 degrees Fahrenheit (21°C) along the Atlantic coast of Morocco and the Black Sea coast of Turkey (and the low 60s [16°C] in the mountains of eastern Turkey) to the low 90s (32°C) around coastal Arabia from the Red Sea to the Gulf, and to the upper 90s (37°C) in the Algerian Sahara. For January, the coldest month, daily averages are in the teens (−10° to

*The gulf between the Arabian peninsula and Iran is known variously as the Persian Gulf or the Arabian Gulf. Because the nomenclature can arouse political controversy, many writers refer to it simply as the Gulf. This practice is used in this chapter.

A useful measure of precipitation effectiveness is the difference between the amount of precipitation on the one hand and evaporation (and transpiration) on the other, or the water balance. This measure is especially useful for agriculture and irrigation. In an equatorial climate, the water supply is abundant on a continuous basis, and there is a monthly surplus. This is far from being the case in a Mediterranean climate, where soil moisture is often temporarily used up. The year can be divided into four periods: (1) water surplus, (2) soil water use, (3) soil moisture shortage, and (4) soil water recharge. Water supply is measured in monthly time intervals. These stages are illustrated in a general way for Mediterranean and desert climates in Figure 1. Clearly, for SWANA there is a discordance between periods of maximum water supply and maximum loss. This divergence can be regarded as an advantage, in

WATER BALANCE

that during the period of maximum evaporation there is no rainfall to evaporate. On the other hand, the considerable soil moisture loss makes irrigation essential for cultivation—if the water is somehow available.

The moisture index* is calculated as follows:

$$\text{Moisture index} = \frac{P - PE}{PE} \times 100$$

$$\text{Moisture index} = \frac{S - D}{PE} \times 100$$

Where P = precipitation
PE = potential
        evapotranspiration
S = surplus
D = deficit

If precipitation (P) is greater than potential evapotranspiration (PE), the index will be positive. If it is less, the index will be negative. Note that the index may be positive for the year, but negative for certain months. This aspect is especially important in the Mediterranean climate, where the dry season is precisely when potential evapotranspiration is highest.

*These formulas were developed by C. Warren Thornthwaite in 1948. See Robert A. Muller and Theodore M. Oberlander, *Physical Geography Today* (New York: Random House, 1984), pp. 197–202; and Arthur W. Strahler and Alan H. Strahler, *Elements of Physical Geography* (New York: John Wiley & Sons, 1976), pp. 147–55. Most books in physical geography contain reference to Thornthwaite's work.

−7°C) in eastern Turkey and in the 70s (23°C) along Arabia's western and southern coasts.

Thus, the annual range of temperature, which is the difference between the highest and lowest average monthly temperatures, ranges from only 7 or 8 Fahrenheit degrees (4–5 Celsius degrees) along Oman's southern coast to 50 Fahrenheit degrees (28 Celsius de-

grees) in Tehran. Furthermore, there are considerable variations in the daily range of temperatures. These vary from about 10 degrees Fahrenheit (5.6 Celsius degrees) in Jiddah on Saudi Arabia's Red Sea coast (for either July or January) to 43 Fahrenheit degrees (24 Celsius degrees) in Bakhtaran in west central Iran (formerly Kermanshah). Note that these values are averages; thus they

FIGURE 1

Water Balance in Mediterranean and Desert Climates

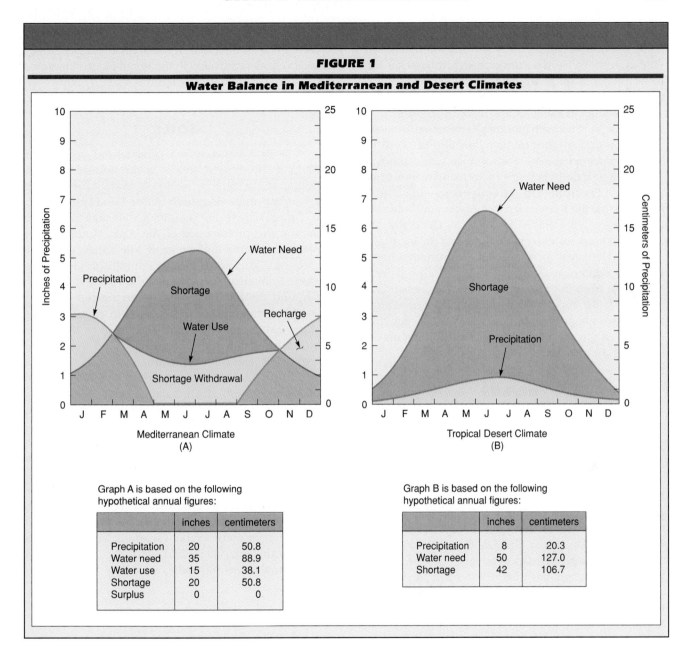

Mediterranean Climate
(A)

Tropical Desert Climate
(B)

Graph A is based on the following hypothetical annual figures:

	inches	centimeters
Precipitation	20	50.8
Water need	35	88.9
Water use	15	38.1
Shortage	20	50.8
Surplus	0	0

Graph B is based on the following hypothetical annual figures:

	inches	centimeters
Precipitation	8	20.3
Water need	50	127.0
Shortage	42	106.7

mask actual daily temperature ranges that frequently exceed 50 and 60 Fahrenheit degrees (23–34 Celsius degrees). The world record for a daily temperature range comes from the Libyan desert, where, at Al-Aziziyah (25 miles [40 kilometers] south of Tripoli) it is 100 Fahrenheit degrees (55.5 Celsius degrees), from 26°F to 126°F (−3.3° to 52.2°C).

Large daily ranges are common in dry areas because it is the moisture content of the air that holds atmospheric energy. When the air is dry, the earth's surface is heated directly by solar radiation; likewise it loses heat with little absorption by the overlying dry air. This is also the reason why in mountainous areas people can feel pleasantly warm when in the sun but quite cool as soon as they step into the shade; they are warmed directly by the sun while the dry air around them absorbs little energy. Those who travel in deserts quickly learn to be prepared for chilly night conditions even during the hottest time of the year.

Temperature conditions are directly relevant to the usefulness of whatever moisture is received from precipitation. Aridity, after all, is a function of these two sets of values combined. For instance, one cannot define a desert merely in terms of so many inches of rain per year. **Precipitation effectiveness**—the amount still available after losses by evaporation and transpiration—varies considerably in SWANA. Generally, it is higher with an increase in latitude and an increase in altitude. The areas best off, not surprisingly, are the highland regions in northwestern Africa and in Southwest Asia. Even in those areas where the climate is not a desert, water supply is a continuing concern. Despite the relative abundance of rain, there is a nagging uncertainty about its reliability for both the amount and timing of the precipitation vary considerably.

RIVERS

The importance of rivers such as the Nile, Tigris-Euphrates, and Jordan was mentioned earlier in this chapter. Several perennial streams flow from the Atlas Mountains to the Atlantic and to the Mediterranean. These are of great importance for local agriculture, even though their names are virtually unknown outside the respective countries: in Morocco, from south to north, there are the Sous, Tensift, Oum er-Rhibia, Sebou, and Moulouya; in Algeria the Chelif; and in Tunisia the Medjerda. Libya has no permanent streams; here seasonal or intermittent streams have been important for irrigation projects. Lebanon has the Litani River. Syria has the Orontes (which rises in Lebanon) and a stretch of the Euphrates. Turkey has numerous rivers that empty into the Mediterranean Sea to the south, especially into the Aegean Sea to the west, and into the Black Sea to the north. Iran has some streams that reach the Caspian Sea from the Elburz and the Koppeh Mountains, and others that flow westward from the Zagros Mountains, either to join the Tigris River or to empty directly into the Gulf. In Afghanistan several rivers flow out of the Hindu Kush, northwestward into the Soviet Union and southeastward into Pakistan; of promising potential for agricultural development in Afghanistan is the Helmand, which empties in the Sistan Basin along the border with

Iran. In none of the countries of the Arabian peninsula is there even an intermittent stream that reaches the sea. Some do flow year round in the Republic of Yemen, but dry out by the time they reach the coast.

SOILS

The soils of SWANA reflect climatic and topographic variations. The largest area is in the aridisol category, including stony surfaces and loose sand. In the northern parts of the three *Maghrib* states (Morocco, Algeria, and Tunisia), there are mountain soils and inceptisols, reflecting the prominence of the Atlas Mountains. Inceptisols are found in the narrow Nile Valley, especially the Nile Delta. Cyprus has mostly alfisols. Peripheral Turkey is dominated by mountain soils of the udic suborder, and the interior plateau is dominated by mollisols. Mountain soils are to be found in eastern Turkey and the Zagros range in Iran, and much of the rest of Iran has mountain soils belonging to the torric group and aridisols. Likewise, Afghanistan has torric mountain soils and undifferentiated rugged mountain scapes. Inceptisols are found in the southern half of Iraq in the Tigris-Euphrates Valley. Along the eastern Mediterranean borderlands, as elsewhere in SWANA, one finds small discontinuous areas of alfisols, a soil of great importance to agriculture because of its capàcity to hold moisture. A great disadvantage is its susceptibility to erosion if plant cover is removed, a feature that over the thousands of years of land use has resulted in severe erosion in many places. In the well-watered mountain areas, there are considerable variations within short distances, both horizontally and vertically, frequently including small areas of productive soil.

VEGETATION

Natural vegetation, too, reflects climatic and topographic variations. However, natural vegetation assumes that there has been no human interference. It refers to an abstract plant cover that would be the climax vegetation under "natural" conditions. Given SWANA's more than seven thousand years of settlement, it is difficult to imagine that any original natural vegetation can be found. The type associated with a Mediterranean climate is a broadleaf evergreen shrub, perhaps exceeding ten feet in height, though shorter than three feet in the drier areas. Wet mountain slopes were once heavily wooded, and in places some still are. The cedar of Lebanon, a majestic tree, once covered a continuous belt along the Lebanon Mountains, but only patches remain. The west-facing slopes of the Elburz Mountains are heavily wooded, as are the northern and eastern mountains of Turkey and the northern slopes of the Atlas Mountains. Most of eastern Iran and Afghanistan have grass and other herbaceous plants, growing singly or in patches.

HUMAN ENVIRONMENT

POPULATION

The combined population of the 21 countries that comprise SWANA is more than 340 million. The population distribution varies considerably, both from country to country and within countries. The three most populous countries, with about 55 million people each, are Turkey, Egypt, and Iran. In fact, these three together contain about half of the region's entire population. On the other extreme are Qatar (0.5 million), Bahrain (0.5 million), and Cyprus (0.7 million). Turkey's population is larger than the combined populations of the 14 least populous countries, which include Syria and Tunisia. Table 12.1 and Figure 12.3 show the 21 countries according to the cumulative percentage of population by country rank-size. Note that five countries contain two-thirds of the region's population, and six countries contain three-fourths.

The overall distribution is one of clusters, some with very large populations and some quite small. These clusters are quite discontinuous, akin to islands of settlement separated by desolate expanses. There is a strong association between population distribution and water availability. Whereas the presence of water does not necessarily mean the presence of human settlement, it is certainly true that human settlement cannot exist without water.

Moving eastward from northwestern Africa, in Morocco the heaviest concentrations are on the western and northern slopes of the Atlas Mountains. There are

TABLE 12.1

Rank Order of SWANA Countries, 1989 Estimates

RANK ORDER	POPULATION (MILLIONS)	PERCENTAGE OF TOTAL	CUMULATIVE PERCENTAGE
1. Turkey	55.4	17.63%	17.63%
2. Egypt	54.8	17.44	35.07
3. Iran	53.9	17.15	52.22
4. Morocco	25.6	8.14	60.36
5. Algeria	24.9	7.92	68.28
6. Iraq	18.1	5.76	74.04
7. Afghanistan	14.8	4.71	78.75
8. Saudi Arabia	14.7	4.68	83.44
9. Syria	12.1	3.85	87.28
10. Tunisia	7.9	2.51	89.79
11. North Yemen*	6.9	2.19	89.98
12. Israel	4.5	1.43	93.41
13. Libya	4.1	1.30	94.71
14. Jordan	4.0	1.27	95.98
15. Lebanon	3.3	1.05	97.03
16. South Yemen*	2.5	0.80	97.83
17. Kuwait	2.1	0.67	98.50
18. United Arab Emirates (UAE)	1.7	0.54	99.04
19. Oman	1.4	0.45	99.49
20. Cyprus	0.7	0.22	99.71
21. Bahrain	0.5	0.16	99.87
22. Qatar	0.4	0.13	100.00
	314.3	100.00	

*North Yemen and South Yemen merged to form the Republic of Yemen in 1990, but statistics in this table list the countries separately. Also, Western Sahara is not included in this list.

FIGURE 12.3

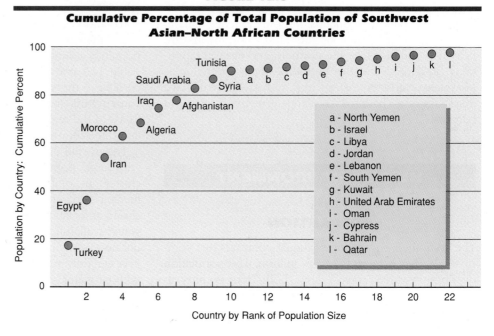

Cumulative Percentage of Total Population of Southwest Asian–North African Countries

a - North Yemen
b - Israel
c - Libya
d - Jordan
e - Lebanon
f - South Yemen
g - Kuwait
h - United Arab Emirates
i - Oman
j - Cypress
k - Bahrain
l - Qatar

also sizable populations in the interior valleys, including the important and historic settlements centered on Marrakesh, Fez, and Meknes. Eastern Morocco is on the leeward side of the mountains and on the periphery of the Sahara. In Algeria there is an even greater disparity between concentrations along and near the Mediterranean on the one hand and in the interior on the other. Most of Algeria is in the virtually uninhabited Sahara. In Tunisia the population density steadily decreases from north to south. In Libya, two small concentrations stand out, each associated with an important regional center: Tripoli in the west and Benghazi in the east. Otherwise the country is a sparsely inhabited desert.

In Egypt, population distribution is synonymous with the Nile, which is credited with supporting perhaps 97 percent of the country's population. Almost the entire population is within sight of the river and its delta. It is literally possible to stand with one foot in a cultivated field and the other in the desert. The entire country is climatically a desert, yet because of the Nile, Egyptians have been making the desert bloom for thousands of years.

In Southwest Asia, a narrow zone with a relatively high population concentration is found along the eastern Mediterranean borderland, namely, in northern Israel, Lebanon, and western Syria. In Turkey, concentrations are heaviest near the borderlands of the Aegean

and Black seas, and in Iran they are highest in the northwestern interior. Historically, concentrations in the Arabian peninsula have been greatest along the southwestern edge of the plateau, especially in Yemen. During the 1970s and 1980s, however, there was a burgeoning of settlement in the interior around the Saudi capital city of Riyadh, as well as in the east along the Gulf in connection with oil-related industrial activity.

The preceding description is quite generalized, and three observations need to be made. The first is that any subject can be studied at an ever-increasing level of detail. In this case, population distribution can be examined at a global level, at the SWANA regional level, at a country level such as Turkey, within one metropolitan area such as Tehran. Thus, the term "concentration" has a sliding meaning along a continuum of density. By its nature it is a comparative and not an absolute concept. Hence, and this is the second observation, concentration means something quite different in reference to the Nile Delta than it does with reference to the Arabian peninsula. Also, whereas Turkey has some 40,000 villages, Kuwait has barely a dozen settlements; thus, the two countries can hardly be compared in terms of population distribution.

Thirdly, the concept of **population density,** implied in the term concentration, must be used with caution. Population density is a ratio between individuals and units of area; for instance, so many people per square mile or square kilometer. As a ratio, it does not distinguish between peoples and areas in terms of productivity. One square mile in the Nile Delta is quite different from one square mile in the Zagros Mountains; further, one cannot assume that the former is more "productive" simply because it is agricultural. One square mile of barren desert in the oil fields of Saudi Arabia is not less valuable than a square mile in the agriculturally rich Tigris-Euphrates Valley. Likewise, we cannot equate two farmers simply because they are farmers; one may be a subsistence farmer barely able to produce enough to sustain himself and his family while the other is a commercial agricultural worker who produces enough to feed 30 people. Moreover, the population density of a country whose population is mostly urban (such as Saudi Arabia with over 70 percent urban) means something quite different from that of a largely rural society (84 percent and up in Afghanistan). Thus, population density figures are "unfairly democratic" in that they manipulate data in a mathematically uniform way, thereby treating as equal things that are not equal. How meaningful is it, for instance, to say that Egypt has a population density of 130 persons per square mile and that Bahrain's density is 2,172. Do we conclude that because Bahrain is 16 times more crowded that its socioeconomic well-being is 16 times worse off. Obviously not. By most measures of well-being, conditions are better in Bahrain. For example, Egypt's gross national product (GNP) per capita is $710 and Bahrain's is $6,610.

An alternative measure of density has been suggested, called physiological density. In this case only agriculturally productive land area is used in computing the ratio. But again similar questions need to be asked: land productive for what, and by whom? Any why exclude nonagricultural land, which may be underlain by oil as in the United Arab Emirates or by phosphates as in Morocco, which has two-thirds of the world's known deposits and is the world's leading exporter of this mineral fertilizer? Thus, simple population density figures are not very helpful, whether for description or interpretation. Worse, they can be misleading, unless contextually qualified in a very restrictive way. This is true of any region on earth. It is certainly true of SWANA.

The countries of SWANA exhibit a variety of population characteristics. In a global scheme of things, this region largely lies in the category called developing. Such a grouping masks variations that stand out at the regional level of analysis. Between 1950 and 1985, the region's overall population increased by 167 million, or

156 percent. During the same period, Cyprus and Lebanon are estimated to have increased, respectively, by 40 and 80 percent; this comparatively small increase is directly related to political concerns and conflict as well as to a low birth rate and emigration. On the other extreme are the small countries that experienced an oil boom, which, among other things, resulted in substantial in-migration: Kuwait increased by 1700 percent and the United Arab Emirates by 1300 percent. Less seemingly unusual and closer to the regional average are Algeria (133 percent), Oman (160 percent), and Jordan (164 percent). Of course, one can argue that each country is "unusual" in its own way. Jordan, for instance, has experienced a great deal of in-migration of Palestinians displaced by the establishment of Israel in 1948–1949 and by Israel's conquest of eastern Palestine (the West Bank) in 1967. Below-average percentage increases apply to three of the four most populous countries: Turkey 120 percent, Egypt 111, and Morocco 122. Other percentages are Tunisia 100, Iran 188, Syria 200, Iraq 202, and Israel 250 (reflecting substantial immigration especially during the early 1950s).

In SWANA as a whole the annual rates of natural increase range from 1.0 percent for Cyprus to 3.8 percent for Syria, a considerable spread indeed. Two comments need to be made about the rate of natural increase. Similar rates of natural increase can result from different combinations of birth rates and death rates. For instance, Afghanistan's estimated rate of natural increase is 2.6 percent, and Qatar's is 2.7, but the respective birth and death rates are 49 and 23 per 1000 and 29 and 2 per 1000. Birth rates of 49 and 29 imply important societal differences, as do death rates of 23 and 2. Figure 12.4 is a scatter diagram of the birth rates and death rates of all 21 SWANA countries. Also included for comparison are the values for the United States, the Soviet Union, and the People's Republic of China. The diagonal lines are along points that have the same rate of natural increase. Note that Egypt and Qatar have rates of increase approaching 3.0 percent, but the fact that they are far apart on the graph indicates a considerable difference in the component data, birth rates and death rates. Here again, death rates of 9 and 2, respectively, suggest a significant difference in health services and facilities. Other pairs are identical or quite similar in these demographically significant ways. Figure 12.4 also indicates clearly that there are important demographic differences within the Arab world (enclosed in the figure) as well as within SWANA. Again, one must be cautious before making generalizations, such as "Arabs are . . ."

FIGURE 12.4

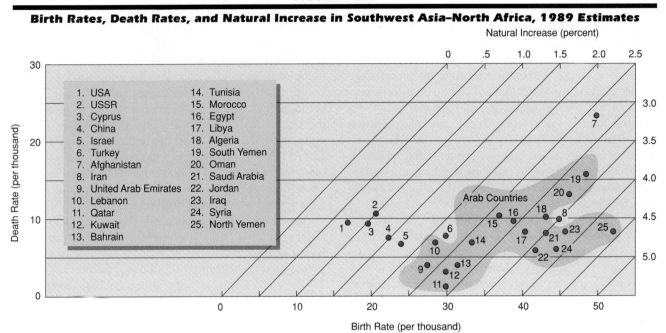

Birth Rates, Death Rates, and Natural Increase in Southwest Asia–North Africa, 1989 Estimates

1. USA
2. USSR
3. Cyprus
4. China
5. Israel
6. Turkey
7. Afghanistan
8. Iran
9. United Arab Emirates
10. Lebanon
11. Qatar
12. Kuwait
13. Bahrain
14. Tunisia
15. Morocco
16. Egypt
17. Libya
18. Algeria
19. South Yemen
20. Oman
21. Saudi Arabia
22. Jordan
23. Iraq
24. Syria
25. North Yemen

The scatter diagram in Figure 12.4 also indicates in an approximate way how SWANA countries compare with each other in terms of the demographic transition. Countries in the upper right are at an early stage of the demographic transition (both death rates and birth rates are high), and those in the lower left are at a later stage (both rates have declined).

The rate of natural increase is suggestive of future population numbers. It is suggestive, and not predictive because it assumes that birth rates and death rates will remain constant—an assumption that is not likely to be sustained. Since we cannot be certain how and how much these component rates will change, population projections often assume that current rates will remain constant; this assumption is used here. Figure 12.5 compares SWANA countries according to the number of years needed to double their 1989 populations, assuming current rates continue. The highest value is 71 years for Cyprus, which means that Cyprus has the slowest rate of population growth (1 percent). It is followed by Israel with 43 years. At the other extreme are Iraq and Syria, whose populations are projected to double in 18 years. Several countries have doubling times less than 22 years, but many countries are between 23 and 35 years. By comparison, equivalent values are 92 years for

the United States, 80 for the Soviet Union, and 49 for China.

An important aspect of any population is its composition by age. A population with a high proportion of young people is youthful, and one with a high proportion of aged persons is an old population. One implication of this variation is the burden that the working population has in order to support the rest (the very young and the very old). In making intercountry comparisons, it has become customary to use "under 15 years" and "65 years and over" as consumer/producer boundaries. One way to measure the "burden" is by means of a **dependency ratio,** computed as follows: the population younger than 15 plus the population 65 and older as a percentage of the population aged 15 through 64 (Figure 12.6).

To illustrate, in Algeria 46 percent of the population is under 15 years old and 4 percent is 65 years and older. The sum of 46 and 4 is 50, which means the remaining 50 percent are the producing population; 50 as a percentage of 50 is 100, which is Algeria's dependency ratio. This means that there are as many dependents as nondependents, so that each producer must support one other person. The higher the ratio the greater the burden, and among SWANA countries it is highest for Syria at 112.8. The minimum value is 40.8 for Qatar, in part reflecting a high labor in-migration. By comparison, dependency ratios are 49 for the United States, 51 for the Soviet Union, and 54 for the People's Republic of China.

Dependency ratios for SWANA countries are generally high (certainly in comparison with Western European countries) and reflect youthful populations, a situation that in turn implies a likelihood for rapid population growth as the young people reach the reproductive age. Therein lies one of the problems of developing countries: improved medical facilities lower the death rate (especially infant mortality), which increases the youthful population, thus increasing the demand on the producing population simply to stay at the same level of economic well-being. Developed countries, by contrast, are better able to produce a surplus to add to their already higher level of economic well-being. The gap between the haves and the have-nots thus increases.

CULTURAL ENVIRONMENT

SWANA has experienced a succession of cultures and civilizations for seven thousand years or more. It is the home of two major culture hearths, one in Mesopotamia (essentially today's Iraq) and one in the lower (northern) Nile Valley in Egypt. A succession of states and empires have dominated SWANA or parts of it (Table 12.2). None of these controlled the interior of the Arabian peninsula. Twentieth-century developments will be described later in the chapter.

The listing of states and empires in Table 12.2 may seem complicated, yet it is merely a very abbreviated summary of a complex history in a large area. It also suggests that a large number of varied cultural influences have affected SWANA. Given the large number of cultural superimpositions and attendant mixing of peoples, it is not possible to identify specific "ethnic" groups that have remained unchanged over the millennia. Over the years many groups have acquired a strong

FIGURE 12.5

Number of Years to Double the Population in Southwest Asia–North Africa

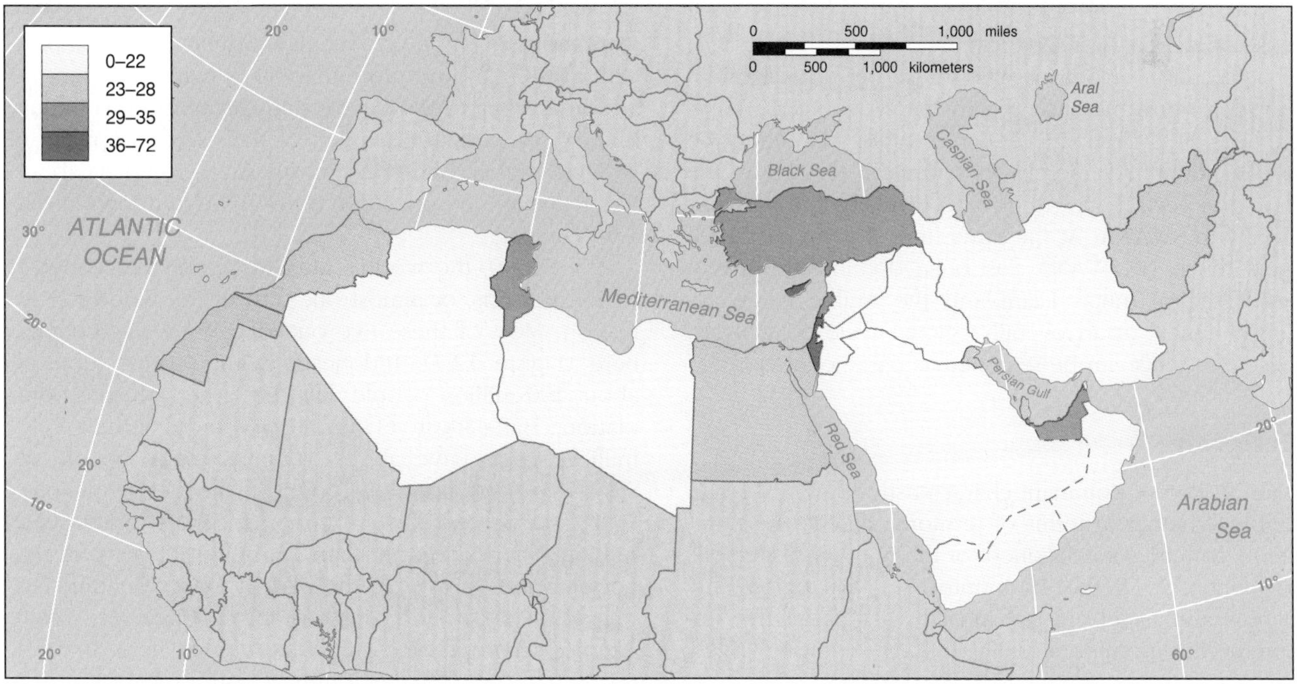

SOURCE: Data from Population Reference Bureau, *World Population Data Sheet* (Washington, D.C.: Population Reference Bureau, 1989).

FIGURE 12.6

Dependency Ratios in Southwest Asia–North Africa

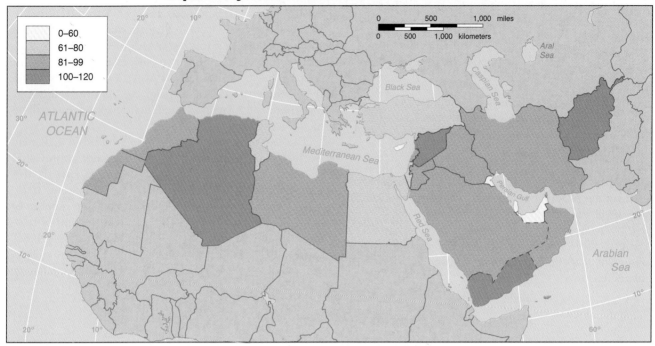

SOURCE: Data from Population Reference Bureau, *World Population Data Sheet* (Washington, D.C.: Population Reference Bureau, 1989).

sense of identity and separateness, whether in response to persecution or because they have developed an association with and an allegiance to identifiable communities. Still, members of such communities cannot be readily identified merely by physical features. One should expect, therefore, to find diversity and not uniformity in SWANA. At the same time, for the past 1300 years much of SWANA has been dominated by two basic cultural traits—Islam and the Arabic language. Consequently, at an overall regional scale of analysis, these two traits can be regarded as paramount.

Religion

Islam is a more dominant characteristic of SWANA than Arabic, whether in terms of territorial extent, the number of countries so classified, or the number of persons so identified. Of the 21 countries in this region only Cyprus and Israel are not Muslim, but both have important Muslim minorities. Cyprus and Israel are largely Greek Orthodox and Jewish, respectively. At one time Lebanon had a slight Christian Arab majority, but by 1990 the population was estimated to be at least 60 percent Muslim and no more than 40 percent Christian. In all other countries, the Muslim proportion is much higher, reaching 100 percent of the citizen population in Saudi Arabia. Christians, mostly Arab, comprise about 3 percent of the population of SWANA, and Jews about 1.5 percent.

As of 1990, the world's Muslims numbered about 1 billion persons, or almost one-fifth of the world's population. Most of these live outside SWANA as defined here (Figure 12.7). Indonesia, with a population of about 190 million people, has the largest Muslim population. Bangladesh (115 million) and Pakistan (115 million) each have about twice as many people as SWANA's most populous country, Egypt (55 million). India's 11 percent Muslim minority comes to about 92 million. There are more than 25 million Muslims in the Soviet Union, mostly in the part of Soviet Central Asia that borders Iran and Afghanistan. In Africa, too, Islam extends beyond North Africa as defined here. In 1971 an Organization of the Islamic Conference (OIC) was formed; it currently has 45 members (Figure 12.7).

It is important to understand the historic and geographic context of Islam and thus to appreciate its perception of itself and of Judaism and Christianity. Organized Islam started in A.D. 622, six hundred years after the time of Jesus, in the same general area as Christianity and Judaism. Moreover, its founder was familiar with both.

Thus, today's Christian Arabs are mostly descendants of Christians who go back to before the coming of Islam; many of them, therefore, do not fit into the denominational categories common in the West. In A.D. 313 the Emperor Constantine proclaimed toleration for Christianity and treated it as the favored religion of the Roman Empire (it became the official religion in 390). Shortly thereafter, in the year 330, Constantinople (today's Istanbul) became the imperial capital. The Greek Orthodox religion developed from Constantinople, as Roman Catholicism did from Rome. From Alexandria, which had become the center of a separate ecclesiastical administrative unit, developed the Coptic church, which endures as Egypt's largest Christian community. Antioch (today's Antakya in extreme southern Turkey near the Mediterranean Sea, in an area claimed by Syria) also was the center of an ecclesiastical unit; the Jacobite church (from Saint James) in Syria originated there. Other religious groups, such as the Armenians and Nestorians, had emerged by the fourth century. The Maronites go back at least to about A.D. 400 and are associated with a Syrian hermit, Saint Maron. They are mostly in what today is Lebanon. In 1054 a schism between Rome and Constantinople began, a split that still endures between the Roman Catholic and Greek Orthodox churches. For various political reasons, agreements were reached between some of these churches (or splits from them) and Rome, whereby the pope was acknowledged as the head of the local church while the latter maintained certain rites, such as a liturgical language or marriage for priests. Known as the Uniate churches, these are the Greek Catholics, Coptic Catholics, Armenian Catholics, Chaldean Catholics (from the Nestorians or Assyrians), and, after the twelfth century, the Maronites. Several small Protestant groups, started largely in the nineteenth and twentieth centuries by American, British, and German missionaries, are also present in SWANA. Sig-

TABLE 12.2

States and Empires of SWANA

	DATES
Akkadian (central Mesopotamia)	c. 2400–2000 B.C.
Babylonian (southern Mesopotamia)	c. 2100–1200 B.C.
Hittite (eastern Turkey)	c. 1700–500 B.C.
Assyrian (centered on northern Mesopotamia and reaching from the Gulf to central Egypt at its greatest extent)	c. 1500–612 B.C.
Persian (extending from the Indus River in modern Pakistan to central Egypt and eastern Libya, and beyond Istanbul into southeastern Europe)	c. 539–330 B.C.
Hellenistic (or Greek), established by Alexander the Great in 334–323 B.C. and including the Ptolemies in Egypt (until 312 B.C.) and the Seleucids (in Persia until c. 250 B.C.; in Mesopotamia until 141 B.C.; and in Syria and eastern Turkey until 64 B.C.)	4th century–1st century B.C.
Parthians (Mesopotamia to the Indus)	c. 250 B.C.–A.D. 200
Roman and Byzantine (Turkey, Egypt, and the eastern Mediterranean borderlands; in western Turkey until the 14th century A.D.)	c. 100 B.C.–7th century A.D.
Sassanid (centered in Iran and extending from Mesopotamia to the Indus)	c. A.D. 225–7th century A.D.
Arab dynasties (Umayyads, Abbasids, Fatimids, Mameluks)	7th century A.D.–15 century A.D.
Crusader states (in and around Jerusalem)	1099–1239
Ottoman (starting in west central Turkey and after 1500 extending from western Iran to Egypt and along the Mediterranean coast to Algeria)	1299–20th century

SOURCE: J. L. Bacharach, *A Middle East Handbook* (Seattle: University of Washington Press, 1984).

A Nubian in southern Egypt. Although his skin color is dark, his facial features are Caucasoid.

nificantly, converts to these new denominations came not from the Muslim population but largely from the older Christian churches.

But what of Islam? The word *Islam* means surrender (to the will of God). It comes from the same root as the word *peace;* thus, one is at peace with God's will. "Muslim" is the adjective; an adherent to Islam is a Muslim. Thus, in usage Islam is equivalent to Christianity, and Muslim to Christian. Occasionally, one sees references to Muhammedan and Muhammedanism. These are inappropriate and incorrect synonyms for Muslim and Islam. They are inappropriate because of an implied analogy with Christians and Christianity: Christ is divine and Christians worship Christ. Applying this analogy to Islam is wrong, because Muhammad is not divine and is not worshiped. He was a man—a very good man, but only a man. To think of him as divine is to commit heresy. He was the last in a long line of messenger "prophets" sent by God, going back to Abraham. In fact, according to Islam, Abraham was a Muslim so that the date A.D. 622 marks the institutionalization of the correct interpretation of the revelation rather than the beginning of the revelation itself. The Quran (spelled in various ways in English, such as Qur'an and Koran) is the religious book of Islam. The word *Quran* means "the reading" or "that which is read," and it refers to a command given

to Muhammad by the Angel Gabriel to read God's message. In addition to Abraham, the Quran refers to several Old Testament figures including Adam, Noah, Jonah, Jacob, Moses, Joseph, David, Solomon, and Job. It also mentions New Testament figures such as John the Baptist, Mary, and Jesus.

It is important to note that Muhammad was familiar with Judaism and Christianity and that Islam is part of the same tradition. The Quran refers to Jews and Christians as people of the Book, that is, people who have received God's revelation. But they erred. The error of the Jews was to regard the message as meant especially for them; hence, the mistaken concept of the chosen people. The error of the Christians was to worship the messenger as well as the message; hence, the mistaken regard of Jesus as divine. From the Muslim perspective, God's message was correctly transmitted and received through Muhammad, and God will send no more messengers. Muhammad is the last and greatest of the prophets and is, therefore, described as the seal of the prophets.

Like Christianity, Islam is a proselytizing or missionary religion (Judaism is not, for a missionary endeavor would be contrary to the concept of chosenness). But Islam does not engage in missionary work among Jews and Christians, since both are people of the Book. Both have received God's revelation; what they need is to

A Jewish man on his way to a synagogue in Zefat (Safed), Israel. Palestinians and Israelis live together and intermingle on the same landscape.

FIGURE 12.7

Member Countries of the Organization of the Islamic Conference

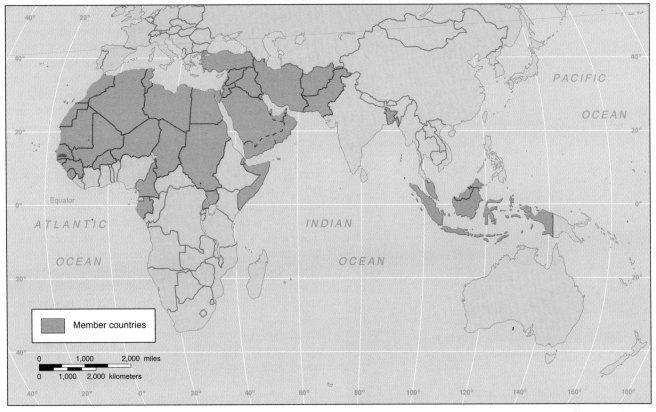

SOURCE: *The Middle East and North Africa 1983–84* (London: Europa Publications, 1985), p. 177.

make the correct interpretation. Christian missionary work among Muslims has been notably unsuccessful. It is difficult enough to convert those who know nothing about Christianity. But converting Muslims requires first persuading them to unbelieve what they regard as superior, specifically to Christianity, and then persuading them to believe what their initial religion specifically rejected as heretical. Whether or not Islam's theological perception is correct is not the concern here.

In the next section a comparison is made between languages spoken in SWANA, and the word God is used to illustrate linguistic and alphabetic comparisons. Allah is the Arabic word for God. It is not the Muslim word. It is not correct to define Allah as the supreme being of the Muslims any more than it is correct to say that Dieu is the supreme being of the French or Gott of the Germans. Christian Arabs say "I believe in Allah" when they recite the creed.

Like Christianity, Islam contains numerous subdivisions, or branches, or denominations. One fundamental division is worth noting here, that between **Sunni** and **Shia** (or Shi'a). Earlier in this chapter the point was made that Islam is a total human system, not just a religion. Muhammad was a political leader as well as a religious leader, a position taken for granted by the community of believers. Following his death in 632, an obvious and pressing question had to be addressed: Who would succeed him to authority? How was succession to be determined? The Arabic word for successor is *khalifa*, which in English is rendered *caliph*. One answer to the question was that the decision resided in the community's recognized leadership; that was how the first three caliphs were chosen. The first was Muhammad's father-in-law, and the next two were Muhammad's companions. The four caliph, Ali, was Muhammad's patrilineal cousin and son-in-law. His supporters

A mosque in Libya. The towers are called minarets and are used for calls to prayer five times a day. Nowadays, the call to prayer is often made by loudspeaker. In fact, in some mosques, the minarets are solid and serve purely decorative purposes.

Language

Arabic today is the dominant language in SWANA in terms of its spatial extent, the number of people whose native language it is, and the number of countries designated as Arab. It is spoken from the Atlantic Ocean to the Gulf by close to 200 million people. Of the 21 countries included in SWANA as defined in this book, 16 are Arab. The other 5 are Turkey (Turkish), Iran (predominantly Persian/Farsi), Afghanistan (largely Pushtu and Persian), Israel (Hebrew), and Cyprus (mostly Greek). Altogether there are 20 Arab countries, as defined by membership in the League of Arab States, or Arab League. Non-SWANA members of the Arab League are, from west and east, Mauritania, Sudan, Djibouti, and Somalia. The Arab League actually has one more member, Palestine, represented by the Palestine Liberation Organization. There are variations in the spoken Arabic, and generally the variation increases with distance. Thus, there is little difference in the Arabic spoken in Syria, Lebanon, and Jordan, though one can identify, for instance, a Lebanese dialect. Also distinct are Egyptian and Iraqi dialects. But a Syrian will have to struggle to understand clearly the Arabic spoken in parts of Algeria and Morocco. The written Arabic, on the other hand, is quite uniform across the Arab world. An Arabic newspaper from any country can be read in any other country. Radio and television Arabic also are standardized, though local accents may be used in certain features such as drama.

The standardization of the written Arabic is related to the importance given to it in the religion of Islam. The Muslim book of scriptures, the Quran, is believed to have been revealed in form as well as in content, so that the language in which the message is revealed is part of the revelation. The scriptures contain nine references to Arabic as the language of revelation. A Muslim, therefore, must read the Quran in the original Arabic. Some scholars have suggested that if it were not for this religious anchoring the Arabic language might have developed in a way similar to Latin: once a lingua franca, Latin has become a ritual language while spawning several offshoots such as Italian, French, Spanish, and Portuguese. Muslims who are not Arab, such as Turks, Persians, Pakistanis, Malaysians, and Indonesians, also must recite the Quran in Arabic. Often they memorize verses or even chapters without knowing the exact meaning of every word, only the general meaning of the particular passage. Still, one result is that many

argued that successors (caliphs) had to be descendants of Ali, and thus of Muhammad. These followers were called Ali's shia, which means Ali's partisans. Advocates of the first position are called Sunni, which means the established and correct path. The Sunni-Shia division has endured and hardened. Today, about 90 percent of the world's Muslims are Sunni, and 9 percent are Shia. Shia are especially dominant in Iran (88 percent of the population), followed by Iraq (57 percent) and Lebanon (30 percent). Numerically, the largest presence is in Iran (41 million), followed by Pakistan (15 million), India (14 million), and Iraq (9 million).

Arabic words have been incorporated into language of non-Arab-speaking Muslims, so that the importance of Arabic and its influence reach considerably farther than the territorial extent of its occurrence as the vernacular speech.

Of course, Arabic is not the only language spoken by a large number of people in SWANA. In a general way, it is spoken south and west of the Taurus-Zagros mountain systems. On the other side of this "divide" are non-Arab Turkey, Iran, and Afghanistan. Important exceptions to the south and west of the divide are Israel and Cyprus. Turkey, with an estimated 1990 population of 56.7 million people, has the largest population of any country in the region (Egypt has 54.7), and the national language here is Turkish. Iran, whose national language is Persian (or Farsi), with about 55.6 million people has the second largest population. Together these three countries alone have about 48 percent of SWANA's total population. But not everyone here speaks either Turkish or Persian. Kurdish is spoken in eastern Turkey, northwestern Iran, and northern Iraq by perhaps 15 million people. Arabic is spoken in western Iran by perhaps 3 million. In fact, Persian is spoken by fewer than half the country's population, about 45 percent. Afghanistan is also a country of minorities. About half of its people are Pushtu (or Pathans) and speak an Iranian language. In Israel Hebrew is the main language, though about 17 percent of the population is Arab. Cyprus is sharply divided between Greeks (82 percent) and Turks (18 percent). In western North Africa there is a large Berber population that preceded the coming of Arabic. The Berber language continues to be spoken by 40 percent of Morocco's population and by 25 percent of Algeria's. In addition to those noted above, numerous other language groups are present in various parts of SWANA, such as Armenian in most of the larger cities in the eastern Mediterranean borderlands.

Several language families are also represented. Arabic and Hebrew are Semitic. Turkish is Altaic. Persian, Greek, and Armenian are Indo-European. Berber is apparently Hamitic. The use of a certain alphabet in the written forms of a given language is not necessarily reflective of a particular language family. Because of the historic and religious importance of Arabic, it has influenced other languages. For instance, the Arabic alphabet is used in Persian and Berber. If you can read neither Arabic nor Persian, you cannot distinguish between an Arabic and a Persian newspaper; they look quite the same. Yet learning Persian is not an easy task for an Arab because of the linguistic difference. For several centuries the Arabic alphabet was used for writing Turkish, but in 1928 it was replaced by a modified Latin script. Thus, before 1928 Arabic and Turkish looked the same but were quite different linguistically. Since 1928, Turkish cannot be distinguished, say, from Swedish by a person who reads neither language. In fact, alphabetically Turkish now is akin to English, though it is by no means linguistically similar. Arabic and Hebrew are linguistically quite similar, but they use different alphabets. It is much easier for an Arab to learn Hebrew than any of the other SWANA languages mentioned here, even though the two languages do not look alike. These points are summarized in Table 12.3.

The distinction between alphabetic and linguistic comparisons can be illustrated by using Arabic and Hebrew. The Arabic word for god, or deity, is *elah;* in Hebrew it is *eloh.* In Semitic languages vowels are linguistically not important, so that the two words are in fact the same "elh." In Arabic God is designated by adding the definite article al, which is a prefix and not a separate word: al-elah. Elision shortens this word to allah, which in English is capitalized, Allah. Semitic languages do not have uppercase and lowercase letters. In Hebrew the change from god to God is accomplished by adding the masculine plural suffix (-im), resulting in elohim, which in English is capitalized, Elohim. Thus, linguistically Allah and Elohim stem from the same root, but alphabetically they are not at all similar. These points are summarized in Table 12.4.

In addition to Arabic's impact on Middle Eastern languages, it has also made contributions to English. There must be hundreds of words of Arabic origin; and the list in Table 12.5 is suggestive.

TABLE 12.3

Language Comparison: Linguistic and Alphabetic Similarity

	LINGUISTICALLY SIMILAR?	ALPHABETICALLY SIMILAR?
Arabic and Turkish	No	Yes, pre-1928 No, post-1928
Arabic and Persian	No	Yes
Arabic and Hebrew	Yes	No
English and Turkish	No	Yes

TABLE 12.4

Language Comparison: English, Arabic, and Hebrew

ENGLISH	ARABIC	HEBREW
god, deity	elah	eloh
God	al-elah	elohim
	allah, Allah	Elohim
	الله	אלהים

The Political Map

When does the present start? One can argue that the present does not exist, only the past and the future. One moment ago is already history, and one moment hence is still in the future. Yet the present is a convenient concept: it is a period within the continuity of time. What boundaries should be used for this period, and what is its duration? The answers depend on the given context and on given needs and objectives. The present can be defined to be this hour, today, this week, or this century, and so on as appropriate. In SWANA, where known history is as old as anywhere on earth, the question "when does the present start?" is as arbitrary as anywhere else.

In the context of the development of the contemporary political map, it is helpful to start at about the beginning of the twentieth century. First, however, one must note the cultural base over which the present boundaries were superimposed. The Arab empire, which reached its height about A.D. 750 (Figure 12.8), had several enduring consequences, most importantly, the widespread diffusion of both the Arabic language and the religion of Islam. At the start of the twentieth century, a prominent political presence in this region was the Ottoman* (or Turkish) Empire, which had

*The name Ottoman is derived from Osman (1259–1326), the founder of the dynasty.

FIGURE 12.8

The Arab Empire at Its Maximum Extent, A.D. 750

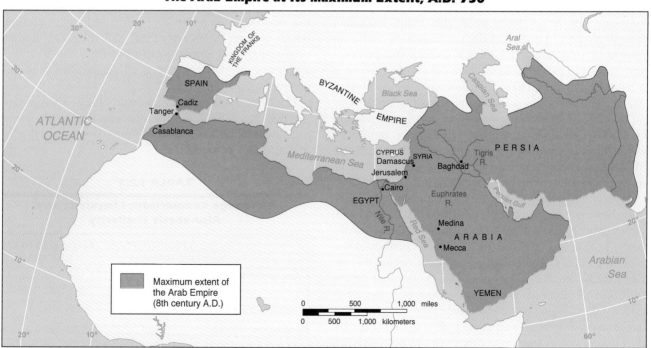

SOURCE: Adapted from Don Peretz, *The Middle East Today*, 4th ed. (New York: Holt, Rinehart and Winston, 1983), p. 26.

TABLE 12.5

Arabic Contributions to English

ENGLISH WORD	ARABIC ROOT	MEANING OF ARABIC ROOT
admiral	al-amir	commander of
albatross, Alcatraz	al ghattas	white-tailed sea eagle (literally, the diver)
alcohol	al-kuhl	powder of antimony (for staining the eyelids)
alcove	al-qubbah	the vault
algebra	al-jabr	the science of reuniting (for solving equations)
algorism	Al-Khawarizmi	name of Arab mathematician
apricot	al-birquq	the plum
arsenal	al-sina'ah	manufacturing
average	'awariyah	damaged goods (the loss is shared equally among the investors; hence the numerical average)
caliber	qalib	shoemaker's last (a block in the form of a foot)
candy	qand	sugarcane
carat	qirat	a measure of small weight
chemistry	al-kimya	the art of transmutation
coffee	qahwah	coffee
cotton	qoton (qutn)	cotton
damask	dimashq	Damascus
elixir	al-iksir	dry powder medicine
garble	ghirbal	sieve
gauze	Gaza	city associated with making the material
genie	jinnie	a spirit in human form
ghoul	ghul	evil spirit
guitar	gitar	guitar
jar	jarrah	large earthenware vase
lemon	limun	lemon
lute	al-'ud	lute
magazine	makhazin	storehouses
mattress	matrah	place (where something is thrown or placed)
mohair	mukayyar	choice (i.e., select)
muslin	muslin	Mosul, a city in Iraq
myrrh	murr	name of a tree
racket	rahet	palm of the hand
ream	rizmah	a bundle
safari	safra	trip
sesame	simsim	sesame seed
sherbet	sharbah	drink (noun)
soda	suda'	splitting headache (through medieval Latin, sodalum, a headache remedy)
sofa	suffa	carpet, divan
spinach	isfanakh	spinach
sugar	sukkar	sugar
sumac	summaq	sumac tree
talc	talq	talc
tambour, tambourine	tantur	drum
tariff	ta'rif	notification, declaration
zero	siphr	zero

SOURCE: Basheer K. Nijim, "English Words of Arabic Origin," *Middle East Outreach Council Newspaper*, vol. 7, no. 2, May 1985, pp. 8–13.

FIGURE 12.9

The Ottoman Empire at Its Maximum Extent, 1683

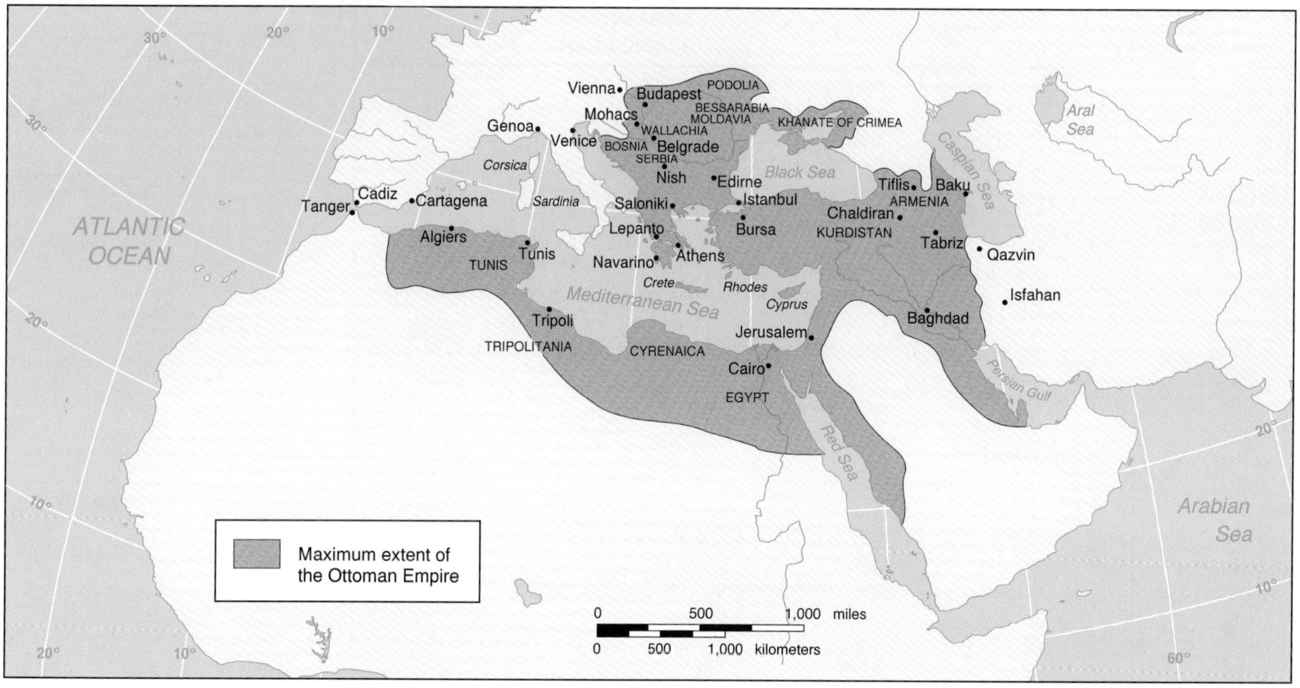

SOURCE: Adapted from Lois A. Aroian and Richard P. Mitchell, *The Modern Middle East and North Africa* (New York: Macmillan Publishing Co., 1984), p. 56.

reached its greatest extent in 1683 (Figure 12.9). During the nineteenth century, the Ottoman Empire, had steadily weakened and, in fact had come to be known as the "sick man of Europe." That happened to be a period of especially vigorous colonial expansion by the European powers, and their influence and control replaced that of the Ottomans in parts of SWANA. Figure 12.10 shows the political situation shortly before the start of World War I (1914–1918). By then Ottoman control was limited to Turkey, Iraq, and the eastern Mediterranean borderlands with nominal and tenuous control over Hijaz, a region in western Arabia. Egypt, too, was supposedly part of the Ottoman Empire, but it had been effectively under British control since 1882. When the war started in 1914, Britain unilaterally proclaimed Egypt to be a British protectorate. Two years earlier, in 1912, Libya had come under Italian occupation, and Italy had also established control over Eritrea on the African side of the southern entrance to the Red Sea. The French were already dominant in Tunisia, Algeria, and Morocco in western North Africa: Spain controlled the southern bor-

derland of the Strait of Gibraltar, plus two other areas on the Atlantic coast. In Sudan an Anglo-Egyptian condominium had been established in 1899, though Britain was very much the senior partner. The interior of Saudi Arabia had not been subdued, but Britain controlled the coastal periphery from Aden almost all the way to Kuwait in the Gulf. British and Russian "spheres of influence" had been superimposed on Persia (today's Iran), with a residual Persian state in between that was responsive to both powers (Table 12.6).

World War I and its wake had an enduring impact on the political landscape of SWANA, especially on the central part. During the war, the European powers had addressed the question of what should be done with the territories that comprised what was left of the Ottoman Empire. Britain and France, the primary actors in this regard, reached a secret arrangement in 1916, known as the Sykes-Picot Agreement after the diplomats representing the two countries. As shown in Figure 12.11 Britain would control much of Iraq, Jordan, and southern Palestine; France would control much of Syria,

Lebanon, and southeastern Turkey; and most of Palestine would come under international control. This agreement, to which Russia was a party, was made public by the Bolsheviks after the November 1917 revolution.

Meanwhile Britain had reached two other understandings, which were not only mutually exclusive but were both also contradictory to the terms of the Sykes-Picot Agreement. First, during 1915–1916, in an exchange of letters known as the Hussein-McMahon Correspondence, the British and an Arab leader (the great-grandfather of King Hussein of Jordan) agreed that the Arabs would revolt against Turkey in favor of Britain and that Britain would help the Arabs achieve independence after the war. Secondly, in November 1917,

Britain issued the Balfour Declaration whereby Britain supported the idea of "the establishment in Palestine of a national home for the Jewish people, . . . it being clearly understood that nothing shall be done which may prejudice the civil and religious rights of existing non-Jewish communities in Palestine. . . ." It so happens that the "non-Jewish communities"—that is, the Palestinian Arabs—comprised 90 percent of Palestine's population while Jews comprised about 9 percent.

After the war, at the 1919 Paris Peace Conference, the Allied powers established the League of Nations (neutral and enemy countries were at first excluded from membership, and the United States did not join). The league disposed of Turkish and German territories by establishing a mandate system whereby the Allied powers would

FIGURE 12.10

The Political Situation on the Eve of World War I

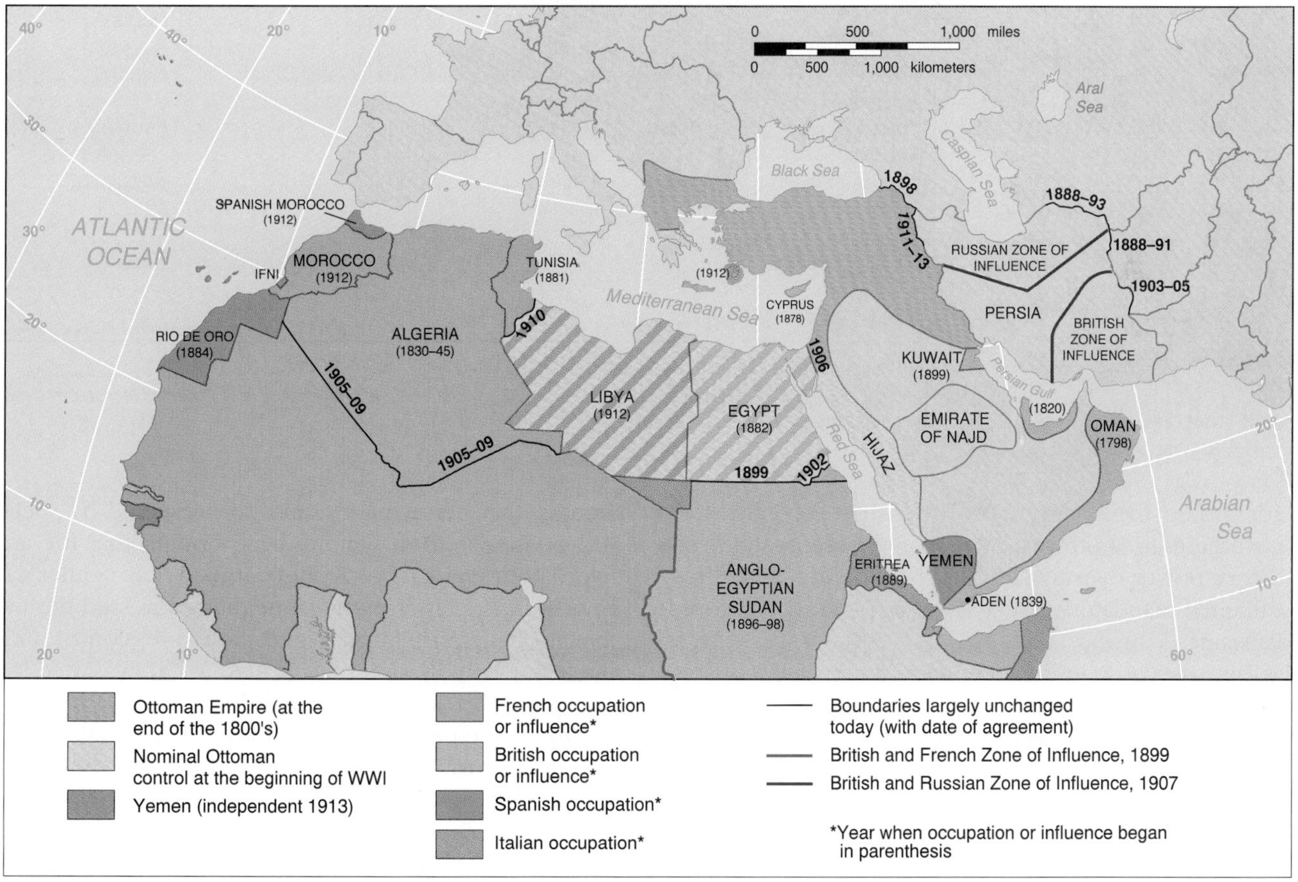

SOURCE: Adapted from Alasdair Drysdale and Gerald H. Blake, *The Middle East and Africa: A Political Geography* (New York: Oxford University Press, 1985), p. 43.

TABLE 12.6

Political Independence

COUNTRY	YEAR OF INDEPENDENCE	MOST RECENT COLONIAL POWER	TYPE OF GOVERNMENT
Afghanistan	1919	Britain, Russia	Republic
Algeria	1962	France	Republic
Bahrain	1971	Britain	Emirate
Cyprus	1960	Britain	Republic
Egypt	1922	Britain	Republic
Iran	Ancient	British and Russian influence	Islamic Republic
Iraq	1932	Britain	Republic
Israel	1948	Britain	Republic
Jordan	1946	Britain	Monarchy
Kuwait	1961	Britain	Emirate
Lebanon	1943	France	Republic
Libya	1951	Italy	Republic
Morocco	1956	France, Spain	Kingdom
Oman	1932	Britain	Sultanate
Qatar	1971	Britain	Emirate
Saudi Arabia	1932	British influence	Kingdom
Syria	1946	France	Republic
Tunisia	1956	France	Kingdom
Turkey	Empire defeated in World War I	Russian and British influence	Republic
United Arab Emirates	1971	British	Federation of emirates
Yemen Arab Republic (North Yemen)*	Local dynasties	None	Republic
Yemen, People's Democratic Republic of (South Yemen)*	1967	Britain	Republic

*North and South Yemen joined together to form the Republic of Yemen in 1990.

SOURCE: Based on A. Drysdale and G. H. Blake, *The Middle East and North Africa: A Political Georgraphy* (New York: Oxford University Press, 1985).

administer these territories until they became self-governing. In Southwest Asia, France became the mandatory power over Syria and Lebanon, and Britain became the mandatory power over Iraq, Transjordan, and Palestine.

The case of Palestine deserves further attention because of its involvement in international relations of regional and extraregional consequence. The Balfour Declaration, noted above, was a British response to **Zionism,** a two-decade-old movement that proclaimed itself the embodiment of Jewish nationalism. The World Zionist Organization was formed by European Jews in 1897 in Basel, Switzerland. Zionism was a Jewish answer to the problem of anti-Semitism. Some Jews

thought that persecution could be prevented by social and political reform while others thought that the answer lay in emigration. Zionists argued that neither avenue was a solution because neither addressed the essential condition that permitted the expression of anti-Semitism, namely, minority status. What was needed was a state for Jews, where any Jew could go and feel that he or she belonged. Zionist leaders negotiated with the powers of the day to secure territory for the establishment of such a state. Areas in Argentina, Uganda, Cyprus, and Sinai were considered or offered, but the chosen target was Palestine. The reason was simple. Immigrants were needed for the proposed Jewish state, and Jews had had a long religious association with the

FIGURE 12.11

The Sykes–Picot Agreement, 1916

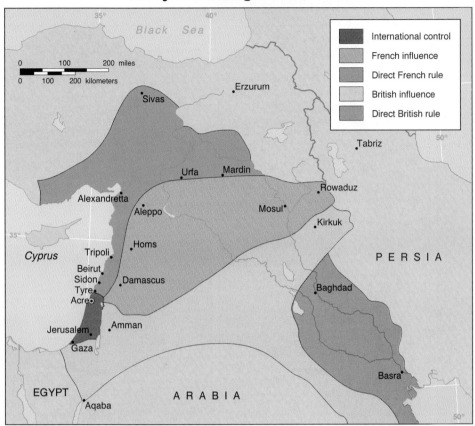

SOURCE: Adapted from Alasdair Drysdale and Gerald H. Blake, *The Middle East and Africa: A Political Geography* (New York: Oxford University Press, 1985), p. 64.

Holy Land. They might migrate to Palestine, but surely not to Uganda.

Migration thus was a crucial part of the Zionist objective, and the question of migration soon became a dominant issue in Palestine. Zionists wanted open Jewish immigration, and the Arab inhabitants opposed such immigration because of the professed intent of the immigrants to take over the country. The British mandatory government vacillated, permitting some immigrants and antagonizing both Zionists and Palestinians. Immigration increased, especially during the 1930s and early 1940s, because of Nazism in Europe.

By the time Israel was proclaimed a state in 1948, Palestine's Arab population comprised 66 percent of the total and Jews comprised about 33 percent, despite a higher Arab rate of growth. At this time Palestine's Jew-

ish population owned less than 7 percent of the land. In November 1947, the General Assembly of the United Nations approved a resolution recommending the partition of Palestine into three parts: an Arab state, a Jewish state, and an international zone to encompass Jerusalem and Bethlehem (Figure 12.12). As a result of the first Arab-Israeli war (1948–1949), Israel was established with an area larger than the partition resolution would have allotted. Two areas were held by the Arabs. What came to be known as the Gaza Strip in southwestern Palestine came under Egyptian administration. East central Palestine joined with Transjordan and came to be known as the West Bank when the country's name was changed to Jordan. Jerusalem was divided by the fighting between Israel and Jordan. In June 1967, Israel occupied the West Bank, the Gaza Strip, the Egyptian Sinai

FIGURE 12.12

Boundary Changes in the Palestine Area

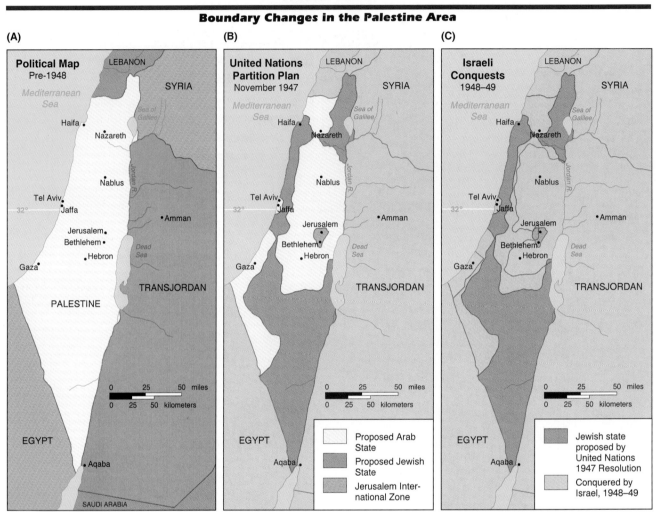

(A) Political Map Pre-1948

(B) United Nations Partition Plan November 1947

Proposed Arab State

Proposed Jewish State

Jerusalem International Zone

(C) Israeli Conquests 1948–49

Jewish state proposed by United Nations 1947 Resolution

Conquered by Israel, 1948–49

peninsula, and the Golan Heights in southwestern Syria. Within two weeks of the 1967 occupation, Israel announced the annexation of Arab Jerusalem plus adjacent areas, and in December 1981 it announced the annexation of the Golan Heights. Sinai was returned to Egypt in 1982 as part of an agreement reached in 1978 at Camp David outside Washington, D.C., with President Jimmy Carter as facilitator.

HUMAN LANDSCAPE

SWANA's human landscape is one of stark contrasts. Cities with millions of people are within hailing distance of subsistence agriculture and desert expanses. The cit-

ies themselves exhibit a myriad of varieties and gradations along the continuum between plush residences on the one hand and shantytowns or worse on the other. Methods of cultivation range from the seeming wizardry of growing vegetables in deserts, as in Kuwait and Libya, to practices that have changed little for centuries and perhaps millennia, such as basin irrigation in the Nile Valley and animal-drawn plows, for instance, in Iran. Fields may stretch to the horizon as in the wide plains of Syria and Iraq or climb laboriously up terraced mountain slopes as in Lebanon and Morocco. Some slopes have been overused and neglected and are environmental eyesores, while others are meticulously maintained and are obviously tended with caring hus-

I t seems self-evident that a country is a land area with clearly defined boundaries that is under effective control by a recognized government. No one questions that the United States is a country or where it is. Thus, it should be easy to know, for instance, how many countries there are in the world.

The answer is not quite that simple. Is membership in the United Nations an indication that an entity is a country (159 nations were members as of mid-1989)? If so, how should Switzerland, North Korea, and South Korea, none of which is a member of the United Nations, be regarded? What about microstates, such as Andorra, Liechtenstein, San Marino, and Monaco? If Bahrain with an area of 255 square miles (660 square kilometers)—a typical Midwestern county in the United States contains 476 square miles (1233 square kilometers)—were not wealthy, and the circumstances of its creation different, would it be regarded as a microstate? When did Lebanon come into being? In 1920, when France proclaimed the State of Greater Lebanon? In 1922, when the League of Nations established the mandate system and allotted Lebanon to France? In 1926, when it was proclaimed a republic? In 1941, when it was proclaimed independent by the Free French government? In 1943, when it was "granted" independence? In 1946, when the last French troops left? How should Lebanon be regarded in 1990, when it is partially occupied by Israel and Syria?

Israel itself raises other questions. It has never defined its boundaries except those with Egypt and that occurred 30 years after Israel was pro-

WHEN IS A COUNTRY A COUNTRY, AND WHEN DID IT COME INTO BEING?

claimed a state. Thus, we might ask, Where is Israel?

Egypt presents other problems. When did it become independent? Most reference books say 1922. But Britain, which ruled Egypt until 1922, retained control of Egypt's foreign relations, the status of foreign nationals within Egypt, Nile River regulation, and relations with what was ostensibly called the Anglo-Egyptian Sudan. The last British soldier did not leave what used to be called the Suez Canal Zone in Egypt until 1956.

Similarly, we might ask, When did Oman become a country? In the eighteenth century, when the ruling dynasty was established? In 1932, when it became a British protectorate? In 1951, when Britain signed a new Treaty of Friendship with the Sultanate of Muscat and Oman? In 1961, despite a United Nations resolution refusing to accept Muscat and Oman's independence? In 1970, when the

present regime was established? In 1971, when the country was accepted by the United Nations? Or has it always existed in some form?

The West Bank and Gaza are also cause for confusion. Palestine was an original nonvoting member (nonvoting because it was not independent) of the League of Arab States, founded in 1945. It declared itself a state on November 15, 1988, and it is recognized by over 50 countries. Yet Palestine does not exist as an independent country, and it has never existed as an independent state.

Finally, we might ask whether the Turkish Republic of Northern Cyprus, which was declared independent in November 1983, is a country, or what effect the Iraqi invasion of Kuwait in 1990 will have on the status of Kuwait.

The seemingly endemic conflicts in SWANA are related to this long, sometimes glorious and sometimes inglorious, history. Much of SWANA came under Muslim rule toward the end of the first millennium A.D. Thereafter it came under a succession of rulers, first, the Turks and then a variety of Europeans. These competing powers subdivided the area according to their strengths and interests. The boundaries they created and superimposed on the cultural landscapes are becoming entrenched and have acquired a life of their own.

Thus, the answer to the question of whether a country is really a country and when it came into being often depends on whom you ask. This is especially true of SWANA countries, whose existence as states has often been determined by forces beyond their control.

Arabs and Jews mingle on a Jerusalem street. Jews, Muslims, and Christians all regard Jerusalem as a holy city, and it is easy to see why these groups have such an intense interest in its future.

bandry. Nomadic pastoralists can still be found beyond the fringes of settled terrain, though even here one can see striking contradictions: in Jordan, for example, Toyota pickup trucks may be seen parked next to bedouin tents whose furnishings include transistor radios. Wide multilane highways span desolate and beautiful desert expanses in Saudi Arabia, and surefooted donkeys plod narrow and muddy mountainpaths in the Republic of Yemen. Solar emergency telephones dot highways within sight of communities with no electricity. Mules pull small kerosene/fuel oil tanks as their riders go from house to house selling fuel for cooking—and go to a gas station to fill up. Camels carry people and goods in Afghanistan, and they are trucked to market in Libya.

Urban Areas

It is appropriate to comment first on cities, for this is where at least half the people live and where, certainly, the major economic and political decisions are made. In terms of the extent of urbanization, the countries range from Kuwait's 94 percent to Oman's 9 percent urban. Urbanization rates in 15 of the 21 countries are 50 percent or greater and in 6 countries they are 80 percent or more (Figure 12.13). These percentages do not take into account absolute values. The countries of Kuwait (pre-August 1990) and Qatar rank in the top three in terms of percentage urban, but have a combined total population of only 2.6 million. Turkey's 53 percent urban means more than 24 million people live in cities,

as does 54 percent of Iran's 55.6 million and 45 percent of Egypt's 54.7 million. The combined *urban* populations of these three countries equal about one-third of the *total* population (urban and nonurban) of SWANA's 21 countries. Almost every country has an urban population of a million or more. Capital cities are generally the primate cities. Cairo with a metropolitan population exceeding 10 million is one of the world's largest cities. Cities are an ancient form in the Middle East, and, in fact, the world's oldest cities, whether archaeological or continually inhabited, are in this region.

Reasons for the establishment and development of cities can be varied and complex, but it is possible to identify certain factors that loom large, certain functions that are especially identified with some localities. Trade is one such function. In a region frequently described as the crossroads of three continents and the cradle of

An experimental farm in the Libyan desert. Wheat is growing out of pure sand within sight of an ancient volcanic cone. Heavy amounts of fertilizer must be used.

Open markets selling fruits, vegetables, and crafts are an integral part of SWANA. The production of various craft goods involves a sizable number of people who are quite skilled in their specific crafts such as this Iranian potter.

civilizations, active long-distance trade is as ancient as recorded history. A large number of locations are situated at abrupt changes in environmental conditions, a situation that peoples in different places and at different times have found advantageous. The boundary between the land and sea is certainly one such situation. SWANA's lands are interpenetrated by numerous water bodies: the Mediterranean Sea, the Black Sea, the Red Sea, the Caspian Sea, the Persian/Arabian Gulf, the Gulf of Aden, and the Gulf of Oman. The coastlines along these water bodies harbor the sites of such major cities as Casablanca, Algiers, Tunis, Tripoli, Benghazi, Alexandria, Haifa, Beirut, Izmir, Istanbul, Jiddah, Aden, Muscat, and Kuwait. In part because of this interpenetration of land and water, there are numerous straits and constrictions of maritime routes; these are guarded by such cities as Tanger (Tangier) in Morocco at the southern side of the Strait of Gibraltar, Port Said and Suez at the two entrances to the Suez canal in Egypt, and Aden near the entrance to the Red Sea in the Republic of Yemen.

Another type of spatial environmental change is climatic, in some cases a change from humid to subhumid conditions. Different products are likely to be associated with such change, enhancing the opportunity for commercial interaction. Damascus and Amman have such locations. Or cities may be clearly associated with rivers because of the relatively reliable water supply, such as Cairo and Baghdad. Topography may be relevant, such

as Jerusalem's location at the drainage divide and Mecca's near the western edge of the Arabian plateau. Many cities are associated with religious happenings and the sites of shrines. Examples, in addition to Mecca and Jerusalem, include Karbala (Iraq), Qum and Mashhad (Iran), and Kairouan (Tunis).

Most of the largest cities have served as administrative centers of one or more empires or dynasties, such as Tehran, Baghdad, Istanbul, Damascus, and Cairo. The colonial experience has enhanced the importance of coastal cities in particular, and some cities such as Tel Aviv are the result of particular historic circumstances. Ankara, an ancient city, grew rapidly in the twentieth century after it was designated in 1923 as Turkey's capital instead of Istanbul. During the second half of the twentieth century, several cities experienced extremely rapid growth because of the effect of oil-related activities. Examples include Abadan (Iran), Kuwait City, Abu Dhabi (United Arab Emirates), and Jubail and Yanbu (both in Saudi Arabia, on the Gulf and Red Sea coasts, respectively). Aqaba acquired prominence as Jordan's only access to the sea after that country lost its outlet to the Mediterranean with the establishment of Israel in 1948.

Urban growth is commonly associated with economic development, but it is not necessarily always a good thing, either for the individuals moving to the cities or for the cities themselves. In many of SWANA's countries, rural-urban migration has been a stream, not a trickle.

Ramses Square located in Cairo, Egypt. Many cities in the world, including London with its Trafalgar Square and Moscow with its Red Square, have squares that symbolize the city and its character.

FIGURE 12.13

Southwest Asia–North Africa Percent Urban, by Country (1989 Estimates)

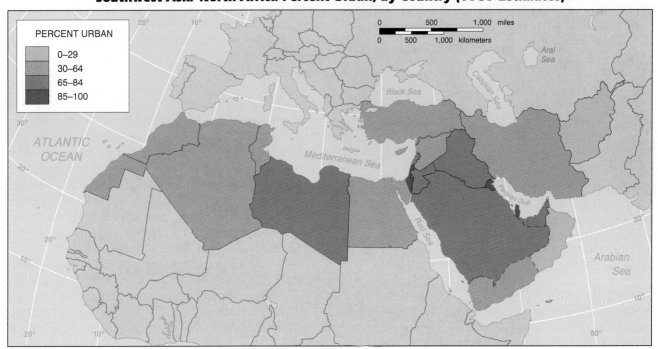

Shantytowns emerge almost overnight and rapidly acquire permanence, whether they are in Turkey, Iran, Egypt, or Morocco. Extreme poverty and high unemployment levels become endemic, and the services of the cities soon become inadequate in all sectors, such as housing, transportation, medicine, education, sewage, electricity, and water. Cairo is perhaps the primary example. During the 1970s, Libya offered incentives for people to stay in the rural areas or to return to farming, but without much success. Iraq tried to replace squatter settlements in Baghdad with subsidized housing, but the cost was still too high for those for whom the projects were intended. As the twentieth century comes to a close, the rapid rates of urban growth surely will not abate, and the concomitant challenges of management and administration will become even more demanding.

Agriculture

While at least half of SWANA's people live in urban areas, almost all the rest are villagers with agricultural pursuits as their mainstay.* The water factor is especially important, and agriculture is either rainfed (that is, of the dry farming type) or irrigated. In most of SWANA, the precipitation regime is Mediterranean, with mild wet winters and hot dry summers. Crops are sown in the fall and harvested in late spring.

Cereals have long been the dominant staple crops, and wheat and barley are of primary importance: both dot the landscape from Afghanistan to Morocco. Turkey produces about 3 or 4 percent of the world total for both crops. Maize, a summer crop, is a staple in Egypt because of the availability of year-round irrigation, and it is also grown along the Atlantic coast of Morocco. Rice is cultivated in the Nile Delta as well as in Turkey and Iraq. Egypt is also about the only country in the region with significant sugarcane cultivation because this crop needs both great heat and lots of water. Because both

* Material in this section on agriculture uses information in W. B. Fisher, *The Middle East*, 7th ed. (London: Methuen and Co., Ltd., 1978), pp. 204–231.

are found in Egypt, the country has a year-round growing season; in fact, two, three, or even four harvests of some crops can be grown on the same plot in one year. Millet is also grown in Egypt. In cooler elevations in Iran and Turkey, oats and rye are prominent. Sugar beets also are limited to the northern countries, and Turkey accounts for about 4 percent of world production.

Especially characteristic of the Mediterranean climate are olives, grapes, and figs. Olive products are a major export of Tunisia, and olive groves are also abundant in northwest Syria and coastal areas of western and southern Turkey. The olive needs dry summers and mild winters, but it can also be cultivated under irrigation in areas bordering desert conditions. Not only is the olive a source of vegetable oils, it is also an important component of Middle Eastern foods and is used in the manufacture of soap. In terms of world production, Turkey accounts for about 8 or 9 percent, Tunisia 6 percent, and Syria 4 percent. Another vegetable oil is cottonseed, and both Egypt and Turkey account for about 3 percent of world production.

Grapes are especially abundant in the coastal hill country of Algeria, Morocco, Tunisia, and Turkey, and in parts of western and northern Iran. They are used as a dessert and for making raisins, jams, and syruplike products and, especially in Algeria and Israel, for making wines. The distribution of the fig is similar to that of the olive, but more extensive because it tolerates greater amounts of moisture as well as quite dry conditions. Citrus trees, such as oranges, lemons, and grapefruits, are abundant in Turkey, Lebanon, Israel, Tunisia, Algeria, and Morocco, with Israel producing 3 percent of the world total. Other fruit trees are peaches, apricots, pomegranates, and increasingly apples. Nut trees, such as almonds, walnuts, and pistachios, are also abundant.

The date palm, popularly associated with the Middle East, thrives in hot and dry conditions. Temperatures must not fall below about 65°F (18.3°C) for extended periods. Although the flowers will not fertilize and thus not bear fruit if there is too much rain, the roots may be directly watered. Thus, date palms flourish in river valleys, such as the Nile, Tigris, and Euphrates, as well as in oases. The largest groves are on the banks of Shatt al-Arab, and Iraq for many years has accounted for more than 60 percent of the world's export of dates, followed by Saudi Arabia with about 8 percent. Other important Middle Eastern exporters are Tunisia and Kuwait.

Vegetables are a major part of the diet in SWANA.

With a rapid expansion in irrigation and a greatly improved connectivity in terms of transportation, vegetables have entered the regional international trade. Gulf countries such as Saudi Arabia, Qatar, and Kuwait are daily recipients of fresh vegetables from Jordan, Syria, and Lebanon.

Good quality cotton has long been associated with Egypt, and Turkey is now also an exporter. Each of these countries accounts for 3 or 4 percent of world production. Other producing areas are found in Iran, Iraq, Syria, and Israel. Tobacco is a cash crop grown in western Turkey; this country accounts for 3 to 4 percent of world production. Some tobacco is also grown on the coastal uplands of Syria and Lebanon. Fishing is a

The Blue Mosque in Istanbul, a very old and important city in Turkey. Most Turks are Sunni Moslems.

Modern housing in Hammanat, Tunisia. Adequate housing in urban areas is a major problem in many of SWANA's countries.

significant undertaking along the western shores of the Gulf and in the adjacent Arabian Sea and Indian Ocean. Tuna is caught in the western Mediterranean (Tunisia, Algeria, and Morocco), and sardines along the Atlantic coast of Morocco.

Livestock activity is dominated by sheep. Mutton continues to be the most common meat, and the animal is a source of milk and wool. Morocco, Algeria, Tunisia, Libya, Yemen, Turkey, Iran, and Afghanistan all have millions of sheep, with Turkey and Iran each accounting for about 3 or 4 percent of world production. Goats also are abundant and are likewise used for their meat, milk, and wool. Cattle are less common and are limited to the moister parts of the region such as western Turkey and coastal Algeria and Morocco. The water buffalo is used in Egypt and Turkey. Dairying is a recent development and is limited to some of the major urban areas. Camels, romantically associated with the Middle East, nowadays are rarely used for transportation. Like other livestock, they are a multipurpose product and are often bred for racing purposes. Swine are almost nonexistent because of the taboo on pork in Islam and Judaism.

Mineral Resources

In terms of industrial development, most of the countries of SWANA are in the early or intermediate stages.

A crucial prerequisite to industrialization is the availability of power resources, which can be human and animal labor, water power, coal, or oil. The first two have been the mainstay of handicraft and simple construction work from the earliest of times. Water power potential is almost limited to the Nile, Tigris, and Euphrates, though locally important projects have been undertaken in Morocco and Algeria. There are some coal deposits in northwestern Turkey near the Black Sea and in northeastern Morocco, but the region on the whole is poor in this mineral fuel.

Oil is, of course, the outstanding energy resource in the region. There are two major areas of substantial reserves: the Gulf basin and the central part of northern Africa. Oil deposits are associated with sedimentary rocks, and a major sedimentary basin lies between the plateaus of Arabia and Iran. The basin extends northwestward under the Tigris-Euphrates Valley. The countries that share this basin are the nations with substantial reserves: counterclockwise from the east, these are Iran, Iraq, Kuwait, Saudi Arabia, Bahrain, Qatar, the United Arab Emirates, and Oman. In North Africa, deposits are found mainly in east central Libya and east central Algeria.

Of an estimated late 1980s total world petroleum reserve of 670 billion tons, about 60 percent is in SWANA. A ranking by country appears in Table 12.7. Reserves are different from production, however. In the late 1980s total world production was about 20 billion barrels. The major producing countries are listed in Table 12.8. It is instructive to compare the countries in Tables 12.7 and 12.8. The association of Saudi Arabia with oil is indeed correct. This single country possesses one-fourth of the world's known reserves, and exploration has by no means exhausted its possibilities. Prospecting in the enormous sandy region in the southeast known as Rub' al-Khali (the Empty Quarter) has been going on since the 1960s, and as yet unexploited reserves are known to be there. In terms of production, Saudi Arabia accounts for about one-tenth of the world total; most of this production is for export markets. Each of the other SWANA countries is below 11 percent in both reserves and production, and in all cases most of the production is for export. The United States, by contrast, possesses some 4 percent of the world's reserves but produces 16 percent, and this production is barely enough to satisfy half of its demands.

Closely associated with oil is natural gas; Tables 12.9 and 12.10 rank countries in terms of reserves and production. In this case SWANA countries, except for Iran's reserves, are of minor importance in terms of both reserves and production.

SWANA's oil is primarily an export product, and the political geography of its international movement is worth noting. Oil is transported by pipeline and by tanker. Pipelines have the advantage of providing a constant flow of the product. Once constructed, the costs of their operation and maintenance are low. But a pipeline has a fixed location and volume, though there is some flexibility in the speed of flow. A pipeline usually lies above ground and is therefore vulnerable to sabotage, though notably such activity has been rare. More importantly, if the pipeline crosses an international boundary, it becomes subject to actions of another government. The latter, for instance, may increase the royalty price or may halt the flow of oil altogether in the pursuit of a particular foreign policy. Of course, such action also means that the host country loses income. Nevertheless, the potential for holding the pipeline "hostage" is there. An example is the record of the pipeline from central Iraq to the Mediterranean through Syria and Lebanon. The uncertainty of relations between Iraq and Syria has led Iraq to construct another pipeline northward to Turkey and then westward to the Mediterranean, thus bypassing Syria altogether.

Tankers, which carry the bulk of international oil, have the flexibility of changing destination in midvoy-

age. They are not labor intensive, so operating costs are low. They are unwieldy in terms of maneuverability, however, and are vulnerable to sabotage. They carry wartime insurance even in peacetime, and rates go even higher in areas of active conflict, such as in the Gulf during the 1980s. Tanker routes require negotiating several straits and narrows. For instance, a tanker leaving Kuwait for Western Europe must first exit the Gulf through the Strait of Hormuz between Iran and Oman. (Note that whereas most of the peninsula on the south-

TABLE 12.7

Petroleum Reserves, Late 1980s

COUNTRY	PERCENTAGE OF WORLD RESERVES
Saudi Arabia	25%
Kuwait	10
Soviet Union	9
Iran	8
Mexico	7
Iraq	6
United Arab Emirates	5
United States	4
Venezuela	4
Libya	3
People's Republic of China	3
Nigeria	3
United Kingdom	2

SOURCE: *International Petroleum Encyclopedia*, various years.

TABLE 12.8

Petroleum Production, Late 1980s

COUNTRY	PERCENTAGE OF WORLD PRODUCTION
Soviet Union	23%
United States	16
Saudi Arabia	10
Mexico	5
Iran	5
United Kingdom	4
People's Republic of China	4
Venezuela	3
Canada	3
Indonesia	3
Nigeria	2
United Arab Emirates	2
Kuwait	2
Libya	2

SOURCE: *International Petroleum Encyclopedia*, various years.

ern side of the strait is in the United Arab Emirates, the northern tip is an exclave of Oman.) The tanker then enters the funnel-shaped Gulf of Aden, which ends in the Strait of Bab el-Mandeb between the Arabian peninsula and Africa, specifically between the Republic of Yemen on the one hand and Djibouti and Ethiopia on the other. The importance of the ports of Aden (formerly under British rule) and Djibouti (formerly under French rule) is to a large extent related to this location. Bab el-Mandeb, an Arabic name, means Gate of Lamentations, or Tears; the name is related to the difficulty of navigation due to uncertain currents and winds. The Red Sea is almost a dead end, and, in fact, it was so before the opening of the Suez Canal in 1869. The tanker enters the Gulf of Suez before negotiating the 100-mile canal, both of which are bounded on both sides by Egyptian territory. The northern exit is to the Mediterranean Sea, which is almost a dead end. The Mediterranean has two distinct eastern and western basins, separated by the constriction between Sicily and Tunisia. The island of Malta is at the eastern approach to this passage and was of strategic importance to Britain during the prime of the British navy. The western exit from the Mediterranean is through the Strait of Gibraltar between Spain and Morocco, and Britain continues to hold a base at Gibraltar on the eastern side of the Spanish mainland. At long last the tanker is in the open Atlantic Ocean.

The route just described is what used to be called the Lifeline of the British Empire, essentially the route from Britain to India. It should be noted that most tankers now are supertankers, most of which are too large for the Suez Canal and thus go around southern Africa. Saudi Arabia has constructed a pipeline from Jubail on its eastern coast to Yanbu on its Red Sea coast, thus bypassing both Hormuz and Bab el-Mandeb and significantly shortening the distance. The pipeline is entirely in domestic territory. In Libya all pipelines to the Mediterranean are in domestic territory; likewise in Algeria, except for one that terminates on the southern Tunisian coast. Libya and Algeria have the advantage of being much closer to the European market than the Gulf countries.

Mention should be made of SWANA's other mineral resources, few and limited though they are. Turkey and Iran have chromite of significance on the world market. Also of world market significance are mineral fertilizers. Morocco produces about 15 percent of the world's phosphate rock, Tunisia between 4 and 5 percent, Jor-

TABLE 12.9

Natural Gas Reserves, Late 1980s

COUNTRY	PERCENTAGE OF WORLD RESERVES
Soviet Union	44%
Iran	15
United States	6
Saudi Arabia	4
Algeria	3
Canada	3
Mexico	2

SOURCE: *International Petroleum Encyclopedia*, various years.

dan between 3 and 4 percent, and Israel between 2 and 3 percent. Syria and Iraq have small amounts. Israel accounts for close to 4 percent of the world's production of potash, and Jordan produces a lesser amount. Turkey, Iran, and Morocco have some copper deposits, though in difficult-to-reach locations. Algeria produces two or three million tons of iron ore a year, and smaller amounts are produced by Turkey, Egypt, and Morocco. Morocco also produces manganese, nickel, and cobalt, and Turkey produces some manganese and tungsten. Lead is mined in Iran and Morocco, and zinc in Iran and Libya.

In terms of manufacturing activity, SWANA countries engage more in light than in heavy industry. Light industries include textiles, food processing (floor milling, soft drink production, and bottling), crafts, cement and other construction products, and assembly of imported consumer goods such as refrigerators and television sets. Petrochemical plants are found in oil-producing countries, and diamond polishing and armaments are especially associated with Israel. Turkey, Iran, Israel, Algeria, and Morocco have become the most diversified, and Afghanistan, Oman, and Yemen are the least.

REGIONALIZATION

At the beginning of this chapter, SWANA was described as being characterized by the dominance of the religion of Islam, the Arabic language, and the physical aspect of a precarious water supply. One can easily argue that the very delimitation of SWANA is arbitrary and is one of convenience, a statement that applies to other regions in this book. Within SWANA it is possible to identify several subregions, or areas that over time have exhibited a greater internal interaction than with other parts

TABLE 12.10

Natural Gas Production, Late 1980s

COUNTRY	PERCENTAGE OF WORLD PRODUCTION
Soviet Union	35%
United States	30
Netherlands	5
Canada	4
Romania	3
United Kingdom	2
Mexico	2
Algeria	2

SOURCE: *International Petroleum Encyclopedia*, various years.

of SWANA. Three subregions will be noted here: moving from west to east, the Maghrib, the Nile Valley, and the **Fertile Crescent.**

The *Maghrib (Maghreb)* refers to the three countries of Morocco, Algeria, and Tunisia. It is an Arabic word that means "west," and it refers to the western part of the Arab world. Some writers include Libya as part of the Maghrib. Properly, as the term was originally used by the Arabs, Maghrib referred to the region from the Tripoli area westward, so that it did not include Benghazi. Today, because of the consequences of colonial competition and compromise, Tripoli and Benghazi are within the same country, Libya; hence the lack of consensus as to whether Libya should be included in the Maghrib. This subregion has had frequent interaction with places on the north side of the Mediterranean, oftentimes more so than with the eastern Arab world (the Mashriq). In passing, it is worth comparing the Maghrib with the term "Middle East" and "Near East" commonly used in English; both these terms originated in Europe and reflect a European perspective of the region.

The *Nile Valley* is the next subregion. The importance of the river and its regime to the livelihood of the people along its banks has been credited with contributing to the development of centralized administration. It is almost the only source of water in Egypt. Without the Nile Egypt could not survive. A good deal of the country's history and foreign policy are related to the Nile River's water system. From Egypt one goes *up* to Africa, southward up the Nile Valley. Egyptians always have been sensitive to whatever affects the river's flow. In early times they were dependent on the annual flood, which starts late in summer and peaks in early September,

ber, going back to a minimum about the second week in May. Gravity-flow canals extended the time during which water was available as well as the area reached by water. Reservoir dams, known as "barrages," made it possible to store surplus water from the fall for use in spring and early summer, the time of minimum discharge. This is known as seasonal storage. The first such project was the Aswan Dam, built in 1902 and subsequently raised twice (in 1912 and 1934) to increase its capacity.

Increasing demands for the water made it necessary to address another problem, the great variation in the river's flow from year to year: in other words, how to achieve year-long storage, not just seasonal storage. One answer was to treat the entire Nile Valley as one system; in fact, such an undertaking was proposed in the late 1930s. The proposal envisioned a basin-wide coordination of 22 projects (some of which already existed) from Lake Victoria to the Mediterranean. The undertaking meant direct coordination among four countries—Uganda, Ethiopia, Sudan, and Egypt—however, and such coordination was not forthcoming. At the time Uganda was a British dependency, and Sudan was an Anglo-Egyptian condominium. Egypt had always been uneasy about not being in control of events affecting the Nile, and that is how the idea for a High Dam came about. It was first proposed in 1949. The intention was to construct a massive dam that would serve the over-year water storage need and yet would be within Egypt. The dam was started in 1960 and completed in 1970. It is located four miles from the old Aswan Dam with which it is sometimes confused: the correct name is the High Dam or the Aswan High Dam, not simply the Aswan Dam. In 1959 Egypt and Sudan reached a Nile Waters Agreement, but no comprehensive basin-wide arrangement has been reached by all the riparian countries, and such agreement is not foreseeable in the twentieth century. This raises the question of whether the Nile Valley is actually a region. It is a physical region, but the activities of peoples rarely correspond to physical regions, and political boundaries are of far greater consequence to the course of human history than physical boundaries. The High Dam is a striking example.

The third subregion is the *Fertile Crescent*. It encompasses the Tigris-Euphrates Valley (Mesopotamia) and the eastern Mediterranean borderlands; sometimes the Nile Valley in Egypt is included. The Fertile Crescent is the setting of the Middle East's early civilizations. It is a region where water is available and agriculture is pos-

A maintenance crew working on a pipeline near Damman, Saudi Arabia. Pipelines must be continually inspected and maintained, and in periods of political conflict, caution must be taken to prevent attacks or sabotage to the lines.

sible in stark contrast to the desert in between the ends of the Crescent and to the south. In the central part of the Fertile Crescent, along the eastern Mediterranean, the source of water is winter rain and snow, and at the two ends of the Crescent (if Egypt is included), water comes from the three great rivers, the Tigris, Euphrates, and Nile. In the river valleys, the land is level and rather easy to irrigate by means of gravity-flow canals. In between, the surface is hilly and even mountainous, so that, while precipitation is available, level land is at a premium. Artificial level land is created by terracing. Some of these terraces may be only 30 feet deep with a rise of 10 feet, and a terrace may carry only one row of a dozen trees and a ground crop.

The concept of the Fertile Crescent has been a useful mental construct for understanding patterns of settle-

ment and international relations. The empires that have developed in the great river valleys have come into competition and, not infrequently, into conflict. The path taken both by traders and warring armies has been along the "arch" of the Crescent and the Mediterranean borderlands. A look at a map suggests that the shortest distance between Mesopotamia and the Nile—for instance, along a straight line from Baghdad to Cairo— lies entirely south and east of the Mediterranean borderlands. But the terrain crossed by this line is not only a waterless and forbidding desert, it also contains an extensive field of broken and jagged basalt that has made movement even more difficult. This region has been an effective cultural boundary until the era of the automobile and the airplane. To people in the central part of the Fertile Crescent—for instance, in Jerusalem—

Mesopotamia was perceptually to their north, the abode of the enemy from the north, and not to the east as is correct geographically.

SPATIAL CONNECTIVITY

Every location is relative to other locations, and it is always interacting with other locations. This interaction, and its consequences, can be viewed at different scales of analysis, from the most local to the global. In this section this connectivity will be considered at the country level, at the regional level (that is, within SWANA and among its countries), and at the extraregional level (that is, between SWANA and its countries on the one hand and other regions and countries on the other).

NATIONAL COHESION

Political geographers talk of centripetal and centrifugal forces; that is, factors that contribute to the strengthening and to the weakening of a country. Perhaps the strongest of these is a sense of identification by the population with the respective country, its political system, and its government. Another factor is the accessibility of parts of the country to each other, of different segments of the population to each other, and of the populace to the decision makers; that is, the extent to which regionalisms are evident, a common centrifugal factor. Such connectivity can be measured, for instance, by the quantity and quality of the means of transportation and communication, by the exchange of goods and services, and by the reach and influence of different types of information media.

In the case of most SWANA countries, a major characteristic is the discordance between the political boundaries and the national groups over which the boundaries have been superimposed. These boundaries largely are nineteenth- and twentieth-century lines delimited by Europeans in a region with a complex cultural composition that reflects several thousand years of human movement and organization. Many of today's countries are mosaics of communities that do not share equally in the national *raison d'être,* the reason for being. Note the word "mosaic": it implies identifiable units that coexist, in contrast to the American concept of a melting pot. The communities are usually marked by language or religion, and often by both. Because of the preponderant importance of this ethno-linguistic factor in SWANA and its enduring consequence, it will be used to help summarize conditions in each of this region's countries, moving west to east.

In the Maghrib, the Berbers are a significant and enduring subnational group, accounting for about 40 percent of Morocco's population and 25 percent of Algeria's. Berber is more of a language than an ethnic identification. There is no Berber nationalism as such, nor a separatist or even divisive tendency. Morocco was divided in 1912 into French and Spanish protectorates, and it became independent in 1956 after many years of political conflict and guerrilla warfare. The recency of this date belies a long-standing continuity and sense of Moroccan identity: the ruling dynasty goes back to the year 1666. Moreover, the country has a sizable population base, a well-established agricultural sector, and an adequate transportation system.

In Algeria there is a particularly strong sense of nationalism because of the shared experience of a long and difficult struggle for liberation from France (1954–1962). It is a large country, about the size of the United States east of the Mississippi River. Most of its territory is part of the great Sahara and is virtually uninhabited. The rather substantial population is compactly concentrated in the north and is well integrated in terms of communication.

Tunisia has a considerable historic continuity and a separate identity, with no divisive regionalisms. It also is the home of ancient Carthage. Tunisia became independent in 1956 after 75 years of French rule. It has a diversified economic base, a well-developed commercial sector, and a good transportation network. The country sees itself as lying at the crossroads between the Arab West and the Arab East and between Africa and Europe.

Libya is at least 98 percent Arab and Muslim, but it has a distinct threefold regionalism: Tripolitania in the northwest, Cyrenaica in the east, and Fezzan in the southwest. Tripolitania and Cyrenaica have had a greater contact with regions across the Mediterranean than southward, whereas Fezzan's earliest contacts were southward into Africa. Libya was occupied by Italy following a 1911–1912 war with Turkey, and it became independent in 1951. It is an affluent country because of its oil resources; before the development of oil in the 1960s, it was one of the world's poorest countries. It has used its wealth to improve its infrastructure substantially, but because of its small population it has a small domestic pool of skilled labor.

In Egypt there is a clear and strong sense of national identity, perhaps more so than in any other SWANA country. It is more developed industrially than most SWANA countries, is a leader in education and publishing, and has a substantial population base. The latter has enabled Egypt to be a traditional exporter of labor, both skilled and unskilled, but at the same time the substantial population is a major constraint to economic development—many Egyptians live in conditions of chronic poverty. Whereas regionalism in the sense of a force compromising national integration is absent in Egypt, there is a well-defined religious minority, the Copts. The Monophysitic Christian Coptic church dates from the third century and thus precedes the coming of Islam by some four centuries. Today Copts comprise about 10 percent of Egypt's population, which in absolute numbers comes to more than five million people; they are mostly in Upper (southern) Egypt. In one sense the Copts are well integrated into Egyptian society: they are Arab, culturally they are Egyptian, and they think of themselves as Egyptian. On the other hand, they become conspicuous at times of Islamic revivals and can be at the receiving end of unwanted attention.

Moving to Southwest Asia, Israel is an example of a country with a strong sense of *raison d'être:* it came into being in 1948 as a Jewish state, it perceives Jewishness as a type of nationhood, and it seeks to bring about the immigration of Jews from all other countries. Reality is different from the ideal. On the one hand, most of the world's Jews have not elected to go to Israel (in fact, in several years in the 1980s more Jews emigrated out of Israel than immigrated to the country), and, on the other, 17 percent of Israel's population of 4.5 million is Arab, and the Arab population is projected to reach 20 percent of the total by the year 2000. Although Israeli Arabs are voting citizens, they can never feel that they fully belong in a country that defines itself as Jewish. A complicating factor is the Palestinian Arab population of 1.25 million in the Israeli-occupied territories of the West Bank and Gaza. The West Bank inhabitants were Jordanian until November 15, 1988. These two areas, which came under Israeli occupation in 1967, are not likely to be relinquished by Israel. Thus, in Israel and in areas occupied by Israel, there are 2 million Arabs and 3.5 million Jews, but the Arab numerical percentage is expected to increase. Israel's economic and military

well-being are expected to continue to be sustained by substantial assistance from the United States.

Jordan is almost always described as an artificial state (in a sense, one can ask, which country is not?); its boundaries were delineated in 1922 (as Transjordan) in the wake of arrangements made after World War I by the Western European powers. It became formally independent from Britain in 1946. Its population is Arab and mostly Muslim (about 12 percent are Christian Arabs), and perhaps more than 60 percent are of Palestinian origin. The country is well integrated with no disruptive regionalism.

Lebanon is a dramatic example of the dominance of communal over national allegiance. The civil war that started in 1975 and continued into the 1990s, though complex in nature, is related to the fact that in a country of three million inhabitants as many as 17 distinct communities, both Christian and Muslim, had been recognized by the political system.

Syria is Arab and mostly Muslim, with Christian Arabs comprising perhaps 15 percent of the population. The greatest cleavage is within the Muslim population because the different Muslim groups are regionally concentrated and have a strong sense of identity. About 65 percent of the total population are Sunni, 15 percent Alawi, and 3 percent Druze. The major enduring regionalism is between the capital Damascus and the city of Aleppo in the northwest. Other regionalisms center on Latakia, Homs, and Hama. Syria became independent in 1946 after an unhappy French mandate that had been established in 1922. It has a diversified and well-developed infrastructure and middle class.

Iraq has a sizable and skilled population, a well-developed agricultural base, and a reasonably well-developed transportation system. In Iraq the question of national integration is especially acute. The Kurds, a conspicuous minority, comprise about one-fifth of the country's population of some 18 million. Located in the northern part of the country, they have a strong affinity with the Kurds in adjacent areas of Syria, Turkey, and especially Iran. They are Kurdish nationalists, not Iraqi nationalists. The Kurds are Sunni Muslim. Arab Sunnis comprise about 25 percent of Iraq's population, and Arab Shi'ites about 55 percent, but the Sunnis have been more prominent in political and economic affairs. The Shi'ites are mostly in the center and south. The British mandate in Iraq, established in 1922, ended in 1932.

The largest of the eight countries that occupy the Arabian peninsula is Saudi Arabia, the Arab-Muslim or

Muslim-Arab state par excellence. Most Saudis are Sunni of the conservative Wahhabi group. There are some 300,000 Shi'ites (2 or 3 percent of the total population) in the eastern part of the country. Saudi Arabia is a large country (its area is about equal to that of the United States east of the Mississippi River) and is a composite of distinct regionalisms, including Hijaz in the west, Asir in the southwest, Najd in the interior (focused on the capital of Riyadh), and the areas bordering the Gulf in the east. Connectivity was greatly enhanced in the 1970s and 1980s with the implementation of ambitious and far-reaching development programs. Domestic air service is extensive and well integrated in the settled areas. Saudi Arabia has never come under foreign occupation or rule. The kingdom in its present form was established in 1932 by the founder of the Saudi dynasty; in fact, it is the only country in the world that is named after its ruling family.

The smaller countries bordering the Gulf (Kuwait, Bahrain, Qatar, and United Arab Emirates) are essentially city-states with highly concentrated populations, and all are prosperous. If some regionalism is to be sought, it might be found in the United Arab Emirates, which is a loose federation of seven principalities. Both Sunni and Shi'ite Muslims are present in the Gulf states, but internal tensions are more likely to be associated with the large number of expatriate workers who may number more than the indigenous populations. All countries had been under some form of British control before becoming independent: Kuwait in 1961 and the other three in 1971.

In Oman regionalism is an enduring domestic factor. Muscat on the coast has had an outward orientation and, in fact, was once the base of extensive commerce that reached eastern Africa and southern Asia. Fifty miles inland to the west and across the rugged coastal highlands is Nizwa, whose orientation historically has been to the interior. A third regionalism is associated with Dhofar in the southwest. Oman did not come under foreign rule during the twentieth century. Once poor, it is now wealthy because of its oil resources.

The Republic of Yemen is one of the poorest and least developed countries in SWANA. Regionalism continues to be strong in the former Yemen Arab Republic (North Yemen), with a Sunni sect in the agricultural and commercial southern highlands and coastal areas and a Shi'ite sect in the interior pastoral areas. Like Oman, the former Yemen Arab Republic has not experienced direct foreign rule. Regionalism is also of domestic importance

in the former People's Democratic Republic of Yemen (South Yemen). The rift is essentially between the coastal and peripherally located capital city of Aden and the vast interior, though the latter is a composite of numerous subregionalisms. The strategic port of Aden was captured by Britain in 1839, and by 1914 the interior had been organized as the Aden Protectorate. The country became independent in 1967. The two countries joined together in May 1990 to form the Republic of Yemen with San'a serving as the capital city.

In addition to Israel, there are four non-Arab countries: Cyprus, Turkey, Iran, and Afghanistan. The latter three are sometimes referred to as the northern tier. Cyprus became independent in 1960 from Britain, which had ruled it since 1878. Its population is clearly divided into a Greek-speaking Greek Orthodox majority (82 percent) and a Turkish-speaking Muslim minority (18 percent). The former have a sense of affiliation with Greece and the latter with Turkey. Since 1974 there has been a de facto partition, with a "Turkish Republic of Northern Cyprus" in the northern 40 percent of the island.

Turkey is the remnant of the Ottoman Empire that once dominated most of SWANA, and it has existed in its present territorial form since 1923. Though Muslim, it is a self-consciously secular state. Most Turks are Sunni, with perhaps a 20 percent Shi'ite minority mostly in the central and eastern parts of the country. The Kurds are a more conspicuous minority, amounting to about 10 percent of the population and concentrated in the rugged extreme eastern part of the country.

Iran is noted for both ethno-linguistic diversity and regionalism. Only about half the people speak Persian as the native language, though many learn it at school and as the language of business and government. Three of Iran's languages belong to the Altaic family (Turkoman, Qashqai, and Azerbaijani), five are Indo-European (Persian, Lur, Baluchi, Kurdish, and Bakhtiari), and one is Semitic (Arabic). Five are spoken across international boundaries in adjacent countries: Kurdish in the west and northwest, Azerbaijani in the northwest, Turkoman in the northeast, Arabic in the southwest, and Baluchi in the southeast. These distributions reflect well-established regionalisms that are separated by rugged mountain systems, notably the Zagros in the west and the Elburz in the north, and with the expansive desolation of deserts in the interior east and southeast. Iran, a Muslim country, is at least 90 percent Shi'ite. The sense of Iranian identity has its roots in the ancient Persian

Empire. In the twentieth century, Iran succumbed momentarily to the rivalries of Russia and Britain, which effectively ruled it between 1912 and 1919 and occupied it again in 1941 during World War II.

Afghanistan, too, is a Muslim country, mostly Sunni, and it has a considerable ethno-linguistic variety and a marked regionalism. The Pushtu located in the central, southern, and eastern areas of the country are perhaps the largest group. Others are the Uzbeks in the north, the Ghilzay in the east, and the Tadzhik in the west. Pushtu and Persian are recognized as official languages. Different parts of the country are poorly integrated because of a limited road system amidst a difficult physical geography. Afghanistan was recognized as independent in 1919; in 1979 it was invaded by the Soviet Union, but the Soviets have withdrawn their troops.

CONNECTIVITY AMONG SOUTHWEST ASIAN-NORTH AFRICAN COUNTRIES

The division of SWANA countries into Arab and non-Arab countries makes a good deal of sense in the context of spatial connectivity. Arab countries share, very importantly, the same language. True, there are dialectal variations from country to country, and these can be quite marked. An Iraqi, for instance, will have a difficult time conversing with a Moroccan if each uses the local dialect. However, standard "classical" Arabic is the same across the Arab world. It is the language of radio and television broadcasts, newspapers, magazines, and films, and all of these easily cut across political boundaries. It is also the Arabic studied in schools. Thus, the Iraqi and the Moroccan can readily converse in classical Arabic. One can generalize that a distance-decay function operates here: the farther individuals are from their home countries, the greater difficulty they have in communicating in the respective colloquial dialects. We see here the lasting impact of a Muslim belief, that the Quran was revealed in form as well as in content: Arabic was the language in which it was revealed.

In addition to language, Arabs have a shared history and a common pride in past days of glory and contributions to world culture and civilization, in such fields as medicine, astronomy, mathematics, geography, and architecture. Arabs also share a rich literature and a wealth of heroic and romantic poetry. And, for the most part, they share the faith of Islam, which is a way of life, not just a religion. Arabs also share a colonial experience, albeit under different powers and with varying conditions. They compare notes about their struggles for independence and exchange stories about their heroes.

But the varied colonial experience has also had a divisive and enduring impact. Boundaries that were nothing more than imperial compromises, that were superimposed on the human landscape as if the latter were a fresh mold, have acquired lives of their own. A Jordanian identity has evolved where no Jordan existed two or three decades earlier. A Libyan nationalism has had to be forged to encompass areas that had not co-existed separately from other areas. Moreover, the differing colonial experiences have led to diverging developments. Monarchic Britain introduced monarchies in Egypt, Iraq, and Jordan, though the first two were overthrown in 1952 and 1958, respectively. The association with republican France led to the formation of republics in Syria, Lebanon, and Algeria. English is the second language in the former countries, and French in the latter. Transportation systems are little integrated across international boundaries. There is much duplication of services within short distances, such as airports and harbors in the Gulf. Arab countries trade more with countries outside SWANA than with each other. Morocco, Algeria, and Tunisia are closer to Europe than to the eastern Arab world, functionally as well as geographically. Differences that can be attributed to the colonial experience are reflected in a host of other ways, such as in educational systems. One wonders, can we speak of an Arab world as a coherent entity? The answer is a qualified yes, because Arabs do speak of a common Arabness, their vast differences notwithstanding.

This Arabness has been expressed in a quest for Arab unity, elusive though the quest may be, and the quest has seen expression in surges of Arab nationalism. The roots of Arab nationalism trace back to the 1850s, with the stirrings of unrest under Ottoman rule. A history of Arab nationalism already was written in 1938 (George Antonius, *The Arab Awakening: The Story of the Arab National Movement* [Beirut, Lebanon: Khayats, 1938; also printed by Capricorn Books, New York, 1965]). During the first half of the twentieth century, there was much talk of a Fertile Crescent scheme that would encompass Iraq, Syria, Jordan, and Palestine. But political circumstances, such as dominance by Britain and France, militated against any such eventuality. There has also been talk of a union between Syria and Iraq, both of which are dominated by the same political party, the Baath ("renaissance"). In the 1950s Egypt and Syria actually united for a brief period to form the United Arab Republic (1958–1961), and in 1958 Jordan and Iraq formed the loose Arab Federation; this was almost still-born and ceased to exist when the monarchy in Iraq was overthrown the same year. The Yemens have al-ready joined together, and there is talk of a union be-tween the Maghrib countries, and between Egypt and Sudan (Sudan, an Arab country, is not included in SWANA as used in this book). One can say that these attempts and expectations are spasms of a dying past rather than harbingers of what is to come. One can also say that they did not succeed because of hasty rather than deliberate decisions.

More important than such dramatic efforts, and more enduring, are numerous inter-Arab organizations and ar-rangements that coordinate a host of mundane activities that touch the lives of individuals rather than mere pol-itics. Many of these are associated with the League of Arab States (the Arab League). Formed in 1945 in Cairo, the League has 21 members. By its own charter the League safeguards the sovereignty of its members. It has 16 standing committees—politics, culture, economics, communications, society, law, oil experts, information, health, human rights, administrative and financial af-fairs, meteorology, cooperation, Arab women, and youth welfare—with liaison officers to coordinate the trade and commercial activities of attachés in Arab em-bassies in other countries. Specialized agencies deal with such matters as maritime transport, civil aviation, study of dry regions, industrial development, research, literacy and adult education, labor, agricultural devel-opment, social measures to combat crime, standards and measures, postal service, satellite communication, broadcasting, and telecommunications. These agencies are based in different Arab cities, such as Rabat, Tunis, Cairo, Amman, Damascus, Baghdad, Riyadh, Kuwait, Dubai, Sharjah, and Khartoum. The fact of their exis-tence does not mean an active interaction, of course, but is symptomatic of dialogue and a desire for cooperation.

In addition, groupings of Arab countries have oc-curred outside the Arab League. In 1964 the Council of Arab Economic Unity was formed by Egypt, Jordan, Syria, Iraq, Kuwait, the United Arab Emirates, the two Yemens, Libya, Sudan, Somalia, Mauritania, and the Pal-estine Liberation Organization. Its intention is to move in the direction of open borders for the free movement of peoples, goods, and capital. The following year it established an Arab Common Market, patterned after

the European Economic Community, which several of its members joined while others are negotiating arrange-ments. A large number of inter-Arab business and professional organizations, such as trade unions, cham-bers of commerce, banks, and air carriers, exists, and there are even more numerous joint ventures in such varied enterprises as textiles, shipbuilding, pharmaceut-icals, mining, and livestock. Development funds have been established by Kuwait, Saudi Arabia, Abu Dhabi (United Arab Emirates), and Iraq for assisting projects in Arab and other (usually Muslim) countries. All 21 Arab League members are represented in the Arab Fund for Economic and Social Development, which has operated since 1974. These organizations and ventures have re-sulted in greater interaction than integration, for each country continues to be jealous of its sovereignty and protective of its particular foreign policy interests.

A nonorganizational development of substantial con-sequence has been international labor migration within the Arab countries. This movement started in the 1960s and increased rapidly in the 1970s. Opportunities have been associated especially with the Gulf countries and Libya, and Arab sources of labor notably have been Jor-dan, Syria, Egypt, the Yemen Arab Republic, Sudan, and Tunisia. Although the labor movements have not re-duced economic gaps between capital-rich and capital-poor countries (in fact, these gaps have increased), the two sets of countries have become more interdepen-dent, and social interaction across international bound-aries certainly has increased.

The one case of successful political integration is the United Arab Emirates, a loose federation of seven emir-ates ("princedoms" or "principalities") that has existed since 1971. Another promising case is the Gulf Coop-eration Council, formed in 1981 by Kuwait, Saudi Ara-bia, Bahrain, Qatar, the United Arab Emirates, and Oman with the objective of enhancing cooperation in all matters. It is modeled after the European Economic Community.

SWANA's non-Arab countries have much less inter-action with other SWANA countries. Iran, Turkey, and Afghanistan are members of the Organization of the Is-lamic Conference, which was established in 1971 and has 45 members (Figure 12.7). In 1980 Iran and Iraq started a war that lasted almost a decade; the conflict was triggered by a border dispute but was symptomatic of many points of contention. In 1990 Iraq invaded Ku-wait furthering conflict in the region. Afghanistan is pe-ripheral to this region functionally as well as geographically. Turkey has agreements with Iraq related to the Euphrates River, to an Iraqi pipeline, and to the control of minorities (namely, the Kurds) that straddle their common boundary. Israel and Egypt reached an agreement in 1979 at Camp David related to mutual recognition. Cyprus increasingly has been affected by the fallout from the conflict in Lebanon and elsewhere in the Middle East, with individuals, businesses, and organizations moving there from the war-torn country.

SOUTHWEST ASIA–NORTH AFRICA AND THE WORLD

The countries of SWANA already have wide-ranging interactions with other parts of the world, notwithstanding the recency of the independence of most of them. Over the years there has been a diversification in both the variety of exports and imports and the number of trading partners (Tables 12.11 and 12.12). In the late 1980s, only one country accounted for more than half of both imports and exports of a SWANA country; that was the Soviet Union as Afghanistan's trading partner.

About half of Oman's exports go to one country, Japan; in this case the commodity is oil, virtually Oman's only export. Oman's import partners are a bit more diversified: about 22 percent of its imports come from Japan and about 19 percent from the United Kingdom; these imports consist mostly of machinery and transport equipment and basic manufactured goods. Qatar's oil exports are also dominated by Japan (47 percent), followed by Italy, France, and Spain (6 or 7 percent each). Qatar's imports, which also consist mostly of machinery

TABLE 12.11

Destination of Exports from SWANA Countries, by Primary Recipients

EXPORTING COUNTRY	FIRST RECIPIENT	PERCENTAGE OF TOTAL EXPORTS	SECOND RECIPIENT	PERCENTAGE OF TOTAL EXPORTS	THIRD RECIPIENT	PERCENTAGE OF TOTAL EXPORTS
Afghanistan	USSR	59%	Pakistan	9%	India	6%
Algeria	France	32	USA	15	Italy	14
Cyprus	United Kingdom	19	Lebanon	15	Libya	12
Egypt	Italy	18	France	9	USA/USSR	7
Iran	USSR	20	West Germany	19	Italy	12
Iraq	Not available		Not available		Not available	
Israel	USA	33	United Kingdom	10	West Germany	9
Jordan	Saudi Arabia	35	Iraq	26	India	14
Kuwait	Japan	14	Taiwan	11	Italy/Iraq	8
Lebanon	Saudi Arabia	48	Syria	11	Kuwait	8
Libya	USA (pre-1986)	27	Italy	24	West Germany	10
Morocco	France	24	Spain/Italy	7	India/Netherlands	6
Oman	Japan	50	Singapore	14	USA	8
Qatar	Japan	47	Italy	7	France/Spain	6
Saudi Arabia	Japan	24	France	9	USA	8
Syria	Romania	30	Italy	16	USSR/France	11
Tunisia	France	23	Italy	16	West Germany	11
Turkey	West Germany	18	Iraq	13	Iran	11
United Arab Emirates (UAE)	Saudi Arabia	23	Japan	23	Qatar	11
Yemen Arab Republic (YAR)	PDRY	42	France	11	China/USA	9
People's Democratic Republic of Yemen (PDRY)	Japan	22	China	10	France/Italy	9

NOTE: Figures vary from year to year. The information given here is intended to suggest overall relationships.

SOURCE: Compiled from Europa Publications, *The Middle East and North Africa*, annual.

and transport equipment and basic manufactured goods, come from Japan (22 percent), the United Kingdom (17 percent), and the United States and former West Germany (8 or 9 percent each). By contrast, Saudi Arabia's oil exports go to more than 15 countries; about 24 percent goes to Japan, and 8 or 9 percent to both France and the United States. Its imports of machinery, base metals and metal products, transport equipment, and foodstuffs (about 12 percent) come from about 20 countries, the largest of which are the United States and Japan (about 20 percent each), former West Germany (10 percent), and Italy (7 percent). Oil products accounted for about 83 percent of Kuwait's exports before the August 1990 Iraqi occupation, and they went to about 15 countries; Japan, the number one destination, received only about 14 percent, followed by Taiwan (11

percent) and Italy (8 percent). Kuwait's imports of machinery and transport equipment, basic and miscellaneous manufactured articles, and foodstuffs (12 percent) came mostly from Japan (24 percent), former West Germany and the United States (about 14 percent each), and the United Kingdom and Italy (about 7 percent each).

Not surprisingly there is greater diversification in the trade of countries not dominated by oil. Eighteen percent of Egypt's exports of mostly crude petroleum and textiles go to one country, Italy, followed by France (9 percent) and the United States and the Soviet Union (about 7 percent each). Its imports of foodstuffs (25 percent), machinery and electrical apparatus, and transport equipment come from more than 20 countries, led by the United States (16 percent), former West Germany

TABLE 12.12

Sources of Imports to SWANA Countries, by Primary Countries of Origin

IMPORTING COUNTRY	FIRST ORIGIN	PERCENTAGE OF TOTAL IMPORTS	SECOND ORIGIN	PERCENTAGE OF TOTAL IMPORTS	THIRD ORIGIN	PERCENTAGE OF TOTAL IMPORTS
Afghanistan	USSR	59%	Japan	12%	Hong Kong	4%
Algeria	France	29	West Germany	14	Italy/Japan/ USA/Spain	8
Cyprus	United Kingdom	12	France	12	Italy	10
Egypt	USA	16	West Germany	11	Italy	8
Iran	West Germany	16	Japan	11	United Kingdom	6
Iraq	Japan	30	West Germany	16	Italy	8
Israel	USA	24	West Germany	15	United Kingdom	10
Jordan	Saudi Arabia	26	USA	16	Japan	13
Kuwait	Japan	24	West Germany	14	USA	13
Lebanon	Italy	15	France	10	USA/West Germany	9
Libya	Italy	25	West Germany	14	United Kingdom	8
Morocco	France	25	Saudi Arabia	13	Spain/USA	7
Oman	Japan	22	United Kingdom	19	UAE	17
Qatar	Japan	22	United Kingdom	17	USA/West Germany	9
Saudia Arabia	USA	20	Japan	19	West Germany	10
Syria	West Germany/France	8	Italy/Japan	7	USA	4
Tunisia	France	25	Italy	14	West Germany	11
Turkey	West Germany	14	France	11	USA/Iran	10
United Arab Emirates (UAE)	Japan	19	USA	13	United Kingdom	12
Yemen Arab Republic (YAR)	Saudi Arabia	19	Japan	13	France	8
People's Democratic Republic of Yemen (PDRY)	United Kingdom	13	Japan	10	Kuwait/Netherlands	9

NOTE: Figures vary from year to year. The information given here is intended to suggest overall relationships.

SOURCE: Compiled from Europa Publications, *The Middle East and North Africa*, annual.

(11 percent), and Italy and France (7 or 8 percent each). Turkey's exports of industrial products such as textile clothing and leather jackets (about 70 percent) and of agricultural products (20–25 percent) go to 15 countries, led by former West Germany (18 percent), Iraq (13 percent), Iran (11 percent), and Italy (7 percent). Its imports, dominated by liquid fuels (30–35 percent) and machinery (20 percent) come from about as many countries, led by former West Germany (14 percent), France (11 percent), and the United States and Iran (about 10 percent). In the case of Syria's imports, mostly mineral fuels and oils, but also manufacturing goods and chemical and pharmaceutical products, not one of its 20 or so trading partners accounts for as much as 10 percent.

It is still true, nevertheless, that SWANA's dominant trading partners are limited in number and geographic distribution (Tables 12.13 and 12.14; Figures 12.14 and 12.15). The primary export destinations are Japan, France, Saudi Arabia, the Soviet Union, the United States, the United Kingdom, Italy, and former West Germany. The primary countries of origin of SWANA's imports are Japan, France, the United States, former West Germany, the United Kingdom, Italy, Saudi Arabia, and the Soviet Union. Exports are primarily oil products, and imports are primarily manufacturing materials and manufactured goods. The trade is dominated by the developed countries of Japan and Western Europe. Conspicuously absent from both exports and imports are Latin American, African, and Asian countries.

Interaction, of course, goes beyond trade. There are organizations that bring together countries with common interests, whether general or specific, active or merely discursive. One group that is especially associated with the Middle East is the Organization of Petroleum Exporting Countries (OPEC). It has 13 members: *Algeria,* Ecuador, Gabon, Indonesia, *Iran, Iraq, Kuwait, Libya,* Nigeria, *Qatar, Saudi Arabia, United Arab Emirates,* and Venezuela. The nations in italic type are

TABLE 12.13

Countries Receiving SWANA Exports: Frequency of Primary, Secondary, and Tertiary Ranks

RECEIVING COUNTRY	TRADING PARTNER		
	Primary	Secondary	Tertiary
Japan	/////	/	
France	///	///	//
Saudi Arabia	///		
Soviet Union	//		//
Italy	/	/////	////
United States	/	/	////
United Kingdom	/	/	
West Germany	/	/	///
Romania	/		
Yemen, People's Democratic Republic			
Iraq		//	/
Spain		/	/
China		/	/
Singapore		/	
Taiwan		/	
Syria		/	
Pakistan		/	
Lebanon		/	
India			//
Libya			/
Kuwait			/
Iran			/
Qatar			/

NOTE: Rankings may vary from year to year. The information given here is intended to show overall relationships.

SOURCE: Data compiled by author from Europa Publications, *The Middle East and North Africa,* annual.

SWANA countries, so 5 are from other regions. Seven of the 13 are Arab countries, since Iran is Persian (see Figure 12.16). This organization received a great deal of media attention in the early 1970s for two reasons. First, after decades during which multinational oil companies determined the profit margin for the producing countries and thus kept the price of oil artificially low, the latter finally asserted their sovereignty and were able to establish the market price of their own product. As a result, the price increased dramatically over a short period. Secondly, in the context of the October 1973 war between Israel on the one hand and Egypt and Syria on the other, Arab oil producers were angered at the nature and extent of U.S. support of Israel and declared an oil embargo. Though only partially effective and of short duration, the embargo furthered the price increase and at the same time decreased the quantity of oil available. One consequence was that Western countries, especially the United States, initiated wide-ranging energy conservation measures and at the same time proceeded with vigor to explore and develop alternate sources of oil, such as in the North Sea and the Alaska north slope.

In the 1980s, oil prices came down and there was talk of a worldwide oil glut. This glut occurred despite the fact that two OPEC countries, Iran and Iraq, engaged in a debilitating war in the 1980s that not only greatly diminished their own oil production, but also interfered with the shipping of other countries, including the sinking of several tankers in the Gulf. Projections for the early 1990s suggest a rebound from the seeming surplus to perhaps some rather austere conditions. Recent conflict between Iraq and Kuwait has already affected oil prices worldwide.

Although OPEC members have met frequently and have attempted to coordinate their production and pricing policies, they have disagreed more often than they have agreed. Rather than a scheming cartel as some have characterized it, OPEC is really a loose organization whose deliberations are overshadowed by the national interests of the respective countries. Still, it does exist as a forum for the exchange of ideas, concerns, and intentions about a common resource that is likely to remain a major source of energy for developed and developing countries alike for many years.

Another grouping that most SWANA countries have joined is the Organization of the Islamic Conference (OIC). A noncountry member is the Palestine Liberation Organization. Nigeria, which has a substantial Muslim population especially in the north, has observer status, as does the "Turkish Republic of Northern Cyprus." OIC headquarters are in Jedda, Saudi Arabia, and the organization meets once every three years. It has acted to facilitate economic cooperation among its members, such as joint ventures in agriculture, industry, and com-

TABLE 12.14

Origins of Imports to SWANA Countries:
Frequency of Primary, Secondary, and Tertiary Ranks

COUNTRY OF ORIGIN	TRADING PARTNER		
	Primary	Secondary	Tertiary
Japan	/////	///////	//
France	////	///	/
United States	///	//	/////
West Germany	///	//////	//
United Kingdom	//	//	////
Italy	//	/	////
Saudi Arabia	//	/	
Spain			//
Hong Kong			/
United Arab Emirates			/
Kuwait			/
Netherlands			/

NOTE: Rankings may vary from year to year. The information given here is intended to show overall relationships.

SOURCE: Data compiled by the author from Europa Publications, *The Middle East and North Africa*, annual.

FIGURE 12.14

Countries Receiving SWANA Exports: Primary, Secondary, and Tertiary Recipients

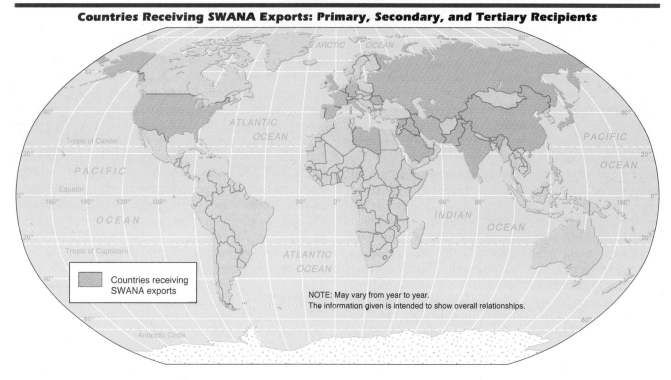

Countries receiving SWANA exports

NOTE: May vary from year to year.
The information given is intended to show overall relationships.

merce. It has supported activities and institutions furthering Muslim education, including the establishment of Muslim universities in Malaysia, Niger, and Uganda. It has provided humanitarian assistance to Muslim communities, such as those suffering from conditions of drought in the Sahel region along the southern fringes of the Sahara. It has called for the withdrawal of Israel from occupied territories, especially Arab Jerusalem. It also tried to mediate between Iraq and Iran during their Gulf war in the 1980s.

Regional organizations whose activities reach beyond SWANA include the Arab Bank for Economic Development in Africa. Established in 1975, it has approved loans and grants for more than a hundred projects. SWANA countries are intertwined with the range of United Nations (UN) agencies, such as the Economic Commission for Western Asia, the Economic Commission for Africa, UN Development Program, UN High Commissioner for Refugees, Food and Agricultural Organization, International Bank for Reconstruction and Development, International Development Association,

International Finance Association, International Fund for Agricultural Development, International Monetary Fund, the World Health Organization, and the UN Educational, Scientific, and Cultural Organization (UNESCO).

SWANA countries also have been involved with the European Community (EC) as part of the EC's program to engage in cooperative economic arrangements with Mediterranean countries that are not members of the EC, a program started in 1972. The EC has made cooperative agreements with Israel in 1975, with the Maghrib countries of Morocco, Algeria, and Tunisia in 1976, with the Mashriq countries of Egypt, Jordan, Lebanon, and Syria in 1977, and with the Yemen Arab Republic in 1984. Association agreements were made with Turkey (1963) and Cyprus (1972). A Euro-Arab dialogue was started in 1972 for the purpose of discussing economic and cultural issues, though it has been intermittently sidetracked by political considerations.

The regions into which this book is divided are a classification of convenience. In the final analysis, the regional boundaries used are arbitrary, and one can argue

FIGURE 12.15

Origins of Imports to SWANA Countries: Primary, Secondary, and Tertiary Countries

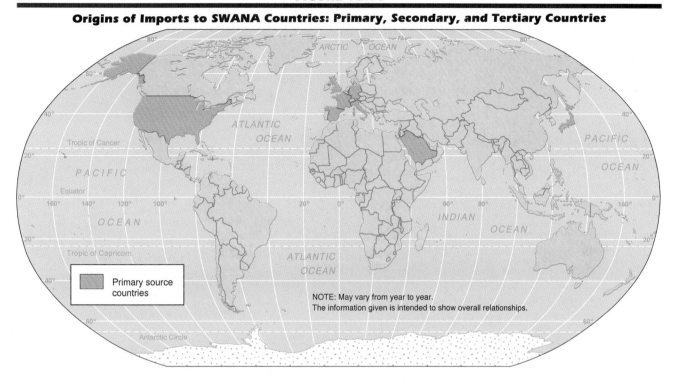

Primary source countries

NOTE: May vary from year to year.
The information given is intended to show overall relationships.

in favor of other ways of regionalizing the world. Boundaries have a way of acquiring a life of their own, and one should be careful not to become too enamored of them. Nevertheless, administrative boundaries certainly are of enormous import when it comes to the management of human affairs. It makes a great deal of difference if a person is on the Egyptian or Sudanese side of the 22nd parallel that separates the two countries, but at a regional nonnational scale Egypt and Sudan share a concern for the Nile River system that is more vital and more enduring than just about any concern that Egypt may share with Morocco, a co-SWANA country. Israel is in the Middle East, but in a way it is not of the Middle East. It was forged out of Arab land by European immigrants, and its closest ties are with the United States. It has received considerable and continuing U.S. economic, military, political, and diplomatic support. Turkey, by most definitions, is part of the Middle East or SWANA. Yet, within the administrative organization of the U.S. State Department, Turkey is in the category of "Europe" because of its membership in the

North Atlantic Treaty Organization (NATO). The Iraq-Iran war probably could not have lasted as long as it did without the tremendous infusion of weapons from all parts of the globe. The arms market knows no regional boundaries.

International organizations, which continue to proliferate, are not bound by regional boundaries, and they play a great role, witting or otherwise, in enhancing and making more intricate and more complex the interactions between all countries on earth. The annual volume *The Middle East and North Africa* (published in England by Europa Publications) has about three hundred entries in an Index of Regional Organizations. Most of these groupings do not make headlines, but they address concerns that affect the livelihood of ordinary human beings in dozens of countries. These concerns include such matters as education, health, labor, social welfare, music, technology, housing, rural development, literature, postal service, banking, civil aviation, livestock management, pest and disease control, transportation, adult education, satellite communication, accounting,

FIGURE 12.16

Members of the Organization of Petroleum Exporting Countries

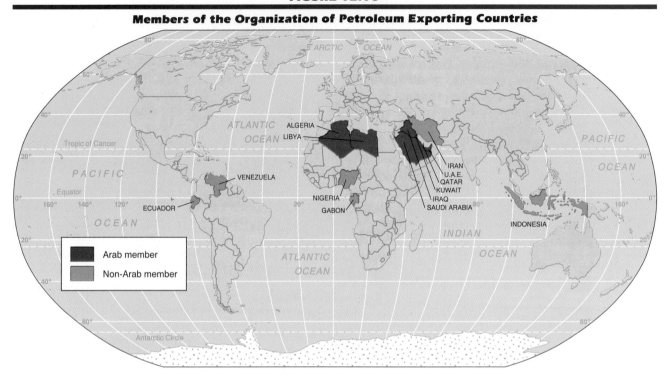

sports, broadcasting, publishing, law, displaced persons, religious schools, food, trade, fishing, international investment, cultural heritage, planned parenthood, women's groups, vocational training, chambers of commerce, commodity exchange, medical emergency assistance, defense, insurance, voluntary associations, forestry, disaster relief, solar energy research, Islamic art, environmental management, peacekeeping, exploration of natural resources, orphans, the handicapped, meteorology, and copyright and intellectual property, to name only a few.

PROBLEMS AND PROSPECTS

The centrality of SWANA to world affairs is evident from the frequency with which it commands media attention. It is strategically located astride three continents, Asia, Africa, and Europe. The Mediterranean Sea is more of a passageway than an impediment to movement. In fact, its very name means in the middle of the land (medi terra), and in the days of the Roman Empire it was

viewed as a Roman lake. Over the thousands of years of history, interaction has been in both directions—from SWANA outward and from other regions to SWANA. At times SWANA was the cradle of civilization and the home base of cultures and empires, and at other times it was at the receiving end of innovation and conquest. The last power to spread out from SWANA was the Turkish Ottoman Empire, and it was already waning by the year 1900. The twentieth century has been one of dominance by outsiders, subdivision into a score of countries, endemic weakness, and instability. The global competition between the United States and the Soviet Union has dominated the second half of the century, and competition and cooperation will continue to the next century. SWANA countries view both powers as having imperialistic intentions and have a relationship with the superpowers that is as much confrontational as cooperative. The superpowers are admired and feared, emulated and resented.

What will the twenty-first century bring? The basic physical geography will be the same, though not necessarily the unfolding human action. The region's location will still be strategic, a constant in international relations. The importance of a location, like that of a resource, is neutral until it is activated by people's needs and wants. Like many a resource, a strategic location will attract the active interest of outsiders. To those astride the location, this interest can be both a blessing and a curse. The United States and the Soviet Union will continue to assert their military presence from the Mediterranean Sea to the Indian Ocean. They will remain actively interested in the region's oil resources, though for different reasons and in different ways. Within the region many of the prevailing economic and social conditions of the 1980s are likely to continue at least through the 1990s. The active interaction of the Maghrib countries with Western Europe will remain strong. At the same time, inter-Arab economic and cultural activity is likely to increase. Egypt's economic difficulty is not

After Iraq invaded Kuwait in August 1990, a coalition of countries representing the United Nations forced the Iraqis to withdraw in February 1991. Pictured here is a Sheridan tank from the U.S. 82nd Airborne Division during maneuvers in Saudi Arabia prior to the drive into Iraq and Kuwait that successfully expelled the Iraqi forces from Kuwait.

likely to experience a dramatic amelioration, even though the decrease in the rate of population growth will probably be sustained. The need to be independent of outside controls and the desire for self-reliance and assertiveness will enhance the surge in introspective religious fundamentalism, whether of Judaism in Israel or of Islam elsewhere. Most of the existing political boundaries have the momentum of more than half a century of existence, and probably they will remain. But there is a certain fluidity to some border areas, such as around Israel, in Lebanon, between Iraq and Iran, and between Iraq and Kuwait. For instance, the West Bank of Jordan has experienced four regimes in 50 years: the Ottoman Turkish Empire, the British mandate in Palestine, the kingdom of Jordan, and Israeli military occupation. Why assume that the existing situation will endure? History does not stop with us, and change is virtually inevitable.

We seem accustomed to the expression that it is a small world, and to the presumption of a global village. In many ways these expressions are useful descriptions of the revolutionary developments in transportation and communications technologies. From the comfort of our living rooms, we have front row seats at a battle scene 10,000 miles away. But to most of us that is what it is: a scene, which conceals more than it reveals, for the scene does nothing to explain the setting, the people and their milieu, values, and beliefs. The communication may be instantaneous, but the physical distance has not changed. To speak of shrinking distances, a small world, and a global village is to speak in metaphors. They are useful metaphors, but only that. The moon looks small, so does the sun, and so do the stars. Anything looks small if you go far enough from it. You now know more about SWANA and its countries than you did when you started this chapter, and you can appreciate the considerable diversity and complexity of this region. An informed understanding is a safeguard against hasty generalization, against the false comfort of homogenizing significant heterogeneity.

Suggested Readings

Bacharach, J. L. *A Middle East Handbook.* Seattle: University of Washington Press, 1984.

Bates, D., and A. Rassam. *Peoples and Cultures of the Middle East.* Englewood Cliffs, N.J.: Prentice-Hall, 1983.

Beaumont, P., G. H. Blake, and J. M. Wagstaff. *The Middle East: A Geographical Study.* New York: Halsted Press, 1988.

Blake, G. H. and R. N. Schoefield, eds. *Boundaries and State Territory in the Middle East and North Africa.* Cambridgshire, England: Middle East and North Africa Studies Press, 1987.

Brawer, M., ed. *Atlas of the Middle East.* New York: Macmillan, 1988.

Drysdale, A., and G. H. Blake. *The Middle East and North Africa: A Political Geography.* New York: Oxford University Press, 1985.

Emery, J. J., N. A. Graham, and M. F. Oppenheimer. *Technology Trade with the Middle East.* Boulder, Colo.: Westview Press, 1986.

Faruqi, I. R., and L. Lamya al Faruqi. *The Cultural Atlas of Islam.* New York: Macmillan, 1986.

Fisher, W. *The Middle East: A Physical, Social and Regional Geography,* 7th ed. London: Methuen, 1978.

Haddad, H. S., and B. K. Nijim, eds. *The Arab World: A Handbook.* Wilmette, Ill.: Medina Press, 1978 (being revised 1991).

Held, C. C. *Middle East Patterns: Peoples, Places, and Politics.* Boulder, Colo.: Westview Press, 1989.

The Middle East and North Africa. London: Europa Publications, annual.

Naff, T., and R. C. Matson. *Water in the Middle East: Conflict or Cooperation?* Boulder, Colo.: Westview Press, 1984.

Rahman, M., ed. *Muslim World: Geography and Development.* Lanham, Md.: University Press of America, 1987.

Wagstaff, J. M. *The Evolution of Middle Eastern Landscapes: An Outline to A.D. 1840.* Totowa, N.J.: Barnes and Noble, 1985.

CHAPTER 13

Africa South of the Sahara

IMPORTANT TERMS

Gondwana

Neocolonialism

Pan-Africanism

Negritude

Subsistence cultivation

Irredentism

CONCEPTS AND ISSUES

■ Environmental degradation of forests and grazing lands.

■ Rapid population growth relative to the resource base.

■ Transference of ethnic identity and loyalty to national allegiance.

■ Ramifications of military versus civilian rule.

■ Infrastructure improvements to enhance accessibility, economic development, and interregional cooperation.

CHAPTER OUTLINE

PHYSICAL ENVIRONMENT
Climate
Vegetation
Soils
Physical Configuration
Resources

HUMAN ENVIRONMENT
Early History
Rise of Colonialism
Independence

Population
Culture
Agriculture
Urbanization and Modernization
Africa's Subregions

SPATIAL CONNECTIVITY
Economic Associations
Trade
Transportation

Telecommunications
Importance of Spatial Connectivity

PROBLEMS AND PROSPECTS
The Environment
Population Growth and Human
  Resources
Political Stability
Economic Development and
  Accessibility

Africa south of the Sahara has historically been rather isolated from other civilizations and world cultures, but the 51 independent states on the mainland and the islands have essentially resulted from colonial creations (Figure 13.1). During the "scramble for Africa" (1884–1914), many European powers carved out colonial possessions on the continent. Only Liberia and Ethiopia (except during World War II) escaped domination by a colonial power. Then after World War II, particularly in the late 1950s and early 1960s, most African countries gained their independence. Certainly, much has been written about the history and development of Africa before colonial contact, but the impact of colonialism and the rise of independent states have been the dominant themes of Sub-Saharan Africa during the past 100 years.

Because of its many countries and its cultural and environmental diversity, Africa is a complex area of study. The states vary in land size; some are large, some are small. Zaire, for example, is about three and one-half times the size of the state of Texas, and Swaziland is only slightly larger than the state of Connecticut.

More than 500 million people live in Sub-Saharan Africa. The population is not evenly distributed. Many areas are sparsely populated, while some parts of West Africa (especially Nigeria), East Africa (especially the highland zone), and South Africa are more densely populated. The population is growing rapidly; the annual rate of increase in most countries is more than 2.5 percent. Most of the people in Sub-Saharan Africa reside in the rural areas; fewer than 10 countries are more than 45 percent urban. Health problems continue to persist among the people, with high incidences of diseases such as malaria and sleeping sickness.

Economically, most of the people depend upon farming and the grazing of animals for their livelihood. Soil fertility is low in many areas, however, and precipitation is unevenly distributed across the continent. Much of the central portion is wet tropical rainforest, and deserts predominate in the north and south. Much of the southern fringe of the Sahara Desert, known as the Sahel, is drought-prone. Although Africa grows several cash crops such as coffee, tea, cocoa, and cotton, food shortages exist throughout the continent.

Significant mineral deposits are located in Africa, including gold, copper, tin, diamonds, iron ore, and some petroleum; in addition, cobalt, chromium, bauxite, manganese, and uranium are mined. Several countries rank among the world's 10 leading producers of certain minerals. Except for South Africa and Zimbabwe, however,

large deposits of coal are absent. International trade is important to Africa, but frequently intra- and interregional transportation connections are too poor to allow for the easy movement of goods and services.

Politically, the independent states of Sub-Saharan Africa are working diligently at the task of nation building. Most countries are faced with the legacy of colonialism, especially externally imposed boundaries, and must convince their various ethnic groups to form a greater political state rather than identifying exclusively with their respective ethnic units. In some instances, the transition from colonialism to independence has led to democracy and civilian rule; in others, it has led to dictatorships and military governments. While some countries have sought the path of socialism, others have developed market economies.

Sub-Saharan Africa remains part of the developing world and continues to have relationships with the superpowers—the United States and the Soviet Union—as well as various European countries, the People's Republic of China, Japan, and Latin America. Everyday life for Africans is a mixture of the traditional and contemporary society; most black Africans are a part of both worlds.

PHYSICAL ENVIRONMENT

CLIMATE

Because of Africa's equatorial position and the fact that the equator essentially bisects the continent, the African climate can be categorized into rather easily identifiable zones extending north and south from the equator. Beginning at the equator and going north or south, the climatic types follow a typical sequence: tropical wet equatorial rainforest, tropical wet-and-dry savanna, steppe, desert, steppe, and Mediterranean.

The continent of Africa is dominated by tropical rainy and wet-and-dry climates and arid and semiarid climates (Figure 13.2). Very little of Africa experiences temperate, midlatitude climate. Mediterranean climates are located along the northwest and southwest coastal areas of the continent, and some areas of marine west coast and humid subtropical climate are found in South Africa.

Seasonal temperature contrasts are minimal in the tropical areas, but rainfall amounts vary significantly. Many tropical areas receive over 50 inches (125 centi-

FIGURE 13.1

Political Units of Sub-Saharan Africa

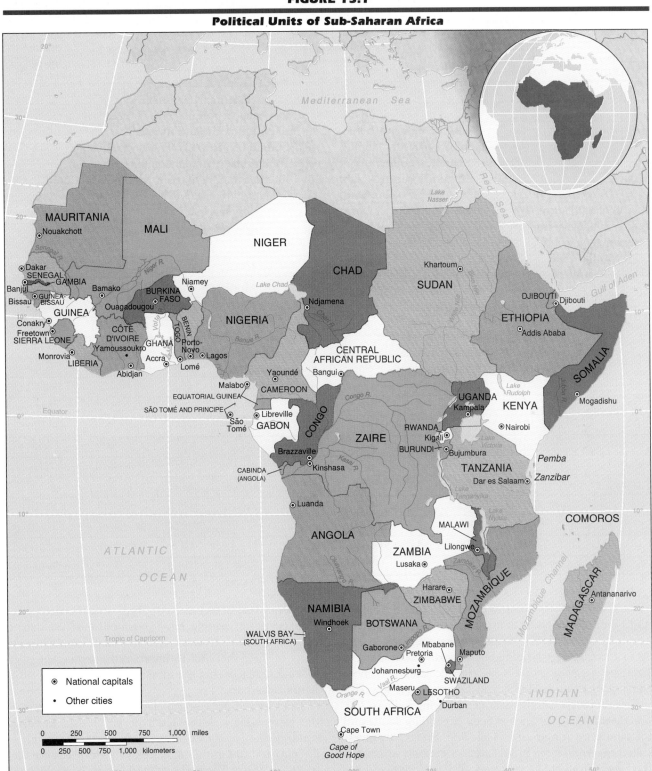

STATISTICAL PROFILE: AFRICA SOUTH OF THE SAHARA

REGION OR COUNTRY	Population Estimate mid-1990 (Millions)	Natural Increase (Annual)	Population Projected to 2000 (Millions)	Infant Mortality Rate	Life Expectancy at Birth (Years)	Urban Population (%)	Per Capita GNP, 1988 (US$)	Area, Thousands of Square Miles (Km²)	Population Density, No./mi² (No./km²)
Angola	8.5	2.7	11.1	137	45	25	—	481.4(1246.7)	18(7)
Benin	4.7	3.2	6.6	110	47	39	340	43.5(112.6)	109(42)
Botswana	1.2	2.9	1.6	64	59	22	1,050	231.8(600.4)	5(2)
Burkino Faso	9.1	3.2	12.5	126	51	8	230	105.9(274.2)	86(33)
Burundi	5.6	3.2	7.7	114	51	5	230	10.8(27.8)	525(203)
Cameroon	11.1	2.6	14.5	125	50	42	1,010	183.6(475.4)	60(23)
Cape Verde	0.4	2.8	0.5	66	61	27	—	1.6(4)	244(94)
Central African Republic	2.9	2.5	3.7	143	46	35	390	240.5(623)	12(5)
Chad	5.0	2.5	6.2	132	46	27	160	495.8(1284)	10(4)
Comoros	0.5	3.4	0.7	94	55	23	440	0.7(1.8)	663(256)
Congo	2.2	3.0	3.0	113	53	40	930	132.1(342)	17(7)
Côte d'Ivoire	12.6	3.7	18.5	96	53	43	740	124.5(322.5)	101(39)
Djibouti	0.4	3.0	0.6	122	47	78	—	8.5(22)	48(19)
Equatorial Guinea	0.4	2.6	0.5	120	50	60	350	10.8(28)	34(13)
Ethiopia	51.7	2.0	70.8	154	41	11	120	471.8(1222)	110(42)
Gabon	1.2	2.2	1.6	103	52	41	2,970	103.4(267.7)	11(4)
Gambia	0.9	2.6	1.1	143	43	21	220	4.4(11.3)	197(76)
Ghana	15.0	3.1	20.4	86	55	32	400	92.1(238.5)	163(63)
Guinea	7.3	2.5	9.2	147	42	22	350	94.9(245.9)	77(30)
Guinea-Bissau	1.0	2.1	1.2	132	45	27	160	14.0(36.1)	71(27)
Kenya	24.6	3.8	35.1	62	63	20	360	225(582.6)	110(42)
Lesotho	1.8	2.8	2.4	100	56	17	410	11.7(30.4)	151(58)
Liberia	2.6	3.2	3.7	83	56	43	450	43(111.4)	61(24)
Madagascar	12.0	3.2	16.6	120	54	22	180	226.7(587)	53(20)

meters) annually, while most deserts such as the Sahara and Namib receive less than 10 inches (25 centimeters) of rainfall annually. The semiarid steppe regions receive 10 to 20 inches (25–50 centimeters) yearly. Although the Kalahari is called a desert, its climatic pattern is more semiarid than arid. The Sahara Desert covers approximately 30 percent of the continent. During the past several decades, the Sahara has been expanding southward into moisture fringe areas (semiarid areas), a process called *desertification*. In this area, commonly known as the Sahel, drought has persisted since the 1960s and has impacted the life-styles and economy (Figure 13.3). As

the Sahel has become overgrazed, and the ecological balance upset, people have migrated and resettled in other areas. Some action is being taken to revegetate parts of the Sahel and to improve the management of grazing practices, but the long-range future of the Sahel is still uncertain.

The wet-and-dry tropical climate (tropical savanna) is the most widely distributed type of climate in Africa south of the Sahara. It occurs both north and south of the equator and in the higher elevations of eastern and southern Africa. The higher altitude areas normally have cooler temperatures. The wet-and-dry tropical savanna

REGION OR COUNTRY	Population Estimate mid-1990 (Millions)	Natural Increase (Annual)	Population Projected to 2000 (Millions)	Infant Mortality Rate	Life Expectancy at Birth (Years)	Urban Population (%)	Per Capita GNP, 1988 (US$)	Area, Thousands of Square Miles (Km²)	Population Density, No./mi² (No./km²)
Malawi	9.2	3.4	11.8	130	49	14	160	45.8(118.5)	200(77)
Mali	8.1	3.0	10.7	117	45	18	230	487.8(1240)	17(7)
Mauritania	2.0	2.7	2.7	127	46	35	480	398(1030.7)	5(2)
Mauritius	1.1	1.3	1.2	25.2	68	41	1,810	0.8(2)	1354(523)
Mozambique	15.7	2.7	20.4	141	47	19	100	309.5(801.6)	51(20)
Namibia	1.5	3.2	2.1	106	56	51	—	318.4(824.3)	5(2)
Niger	7.9	3.0	11.1	135	45	16	310	489.2(1267)	16(6)
Nigeria	118.8	2.9	160.8	121	48	31	290	356.7(923.8)	333(129)
Réunion	0.6	1.8	0.7	14	71	98	—	1(2.5)	614(237)
Rwanda	7.3	3.4	10.4	122	49	6	310	10.2(26.3)	715(276)
São Tomé and Principe	0.1	2.7	0.2	61.7	65	38	280	0.4(0.9)	337(130)
Senegal	7.4	2.7	9.7	128	46	36	630	75.8(196.2)	97(37)
Seychelles	0.1	1.7	0.1	17.0	70	52	3,800	0.1(0.3)	629(243)
Sierra Leone	4.2	2.5	5.4	154	41	28	210	27.7(71.7)	150(58)
Somalia	8.4	3.1	10.4	132	45	33	170	246.2(637.7)	34(13)
South Africa	39.6	2.7	51.5	55	63	56	2,290	471.4(1221)	84(32)
Sudan	25.2	2.9	33.6	108	50	20	340	967.5(2505.8)	26(10)
Swaziland	0.8	3.1	1.1	130	50	26	790	6.7(17.4)	116(45)
Tanzania	26.0	3.7	36.5	106	53	19	160	364.9(945.1)	71(27)
Togo	3.7	3.6	5.2	114	55	22	370	21.9(56.8)	168(65)
Uganda	18.0	3.6	25.1	107	49	9	280	91.1(236.1)	197(76)
Zaire	36.6	3.3	50.3	108	53	40	170	905.6(2345.4)	40(15)
Zambia	8.1	3.8	11.6	80	53	45	290	290.6(752.6)	28(11)
Zimbabwe	9.7	3.2	13.1	72	58	25	660	150.8(390.6)	64(25)

climates typically have summer wet and winter dry seasons, but those areas located closer to the equatory usually have higher amounts of precipitation than those located further away from the equator. Temperatures in all months in tropical climates average above 64.4°F (18°C). Daytime temperatures average between 70° and 80°F (21° and 27°C), with diurnal ranges seldom exceeding 15°F (8°C). In the steppe and desert areas, maximum temperatures range between 80° and 100°F (27° and 37°C), but the diurnal temperature range is far greater than in the tropics, and nighttime temperature lows can drop to 50° to 60°F (10°–15°C).

VEGETATION

There are five great biomes—forest, savanna, grassland, desert, and tundra—and all, even the tundra, which can be found in the higher elevations, are located in Africa. Vegetation patterns correspond rather closely to climatic patterns, though this correspondence is less pronounced in areas south of the Sahara. Broadleaf evergreen trees predominate in the tropical rainforests. Savanna vegetation, a matrix of grasses with interspersed trees (which tend to be shorter near the steppe and semiarid lands) can be found in the wet-and-dry tropical

FIGURE 13.2

Climates of Sub-Saharan Africa

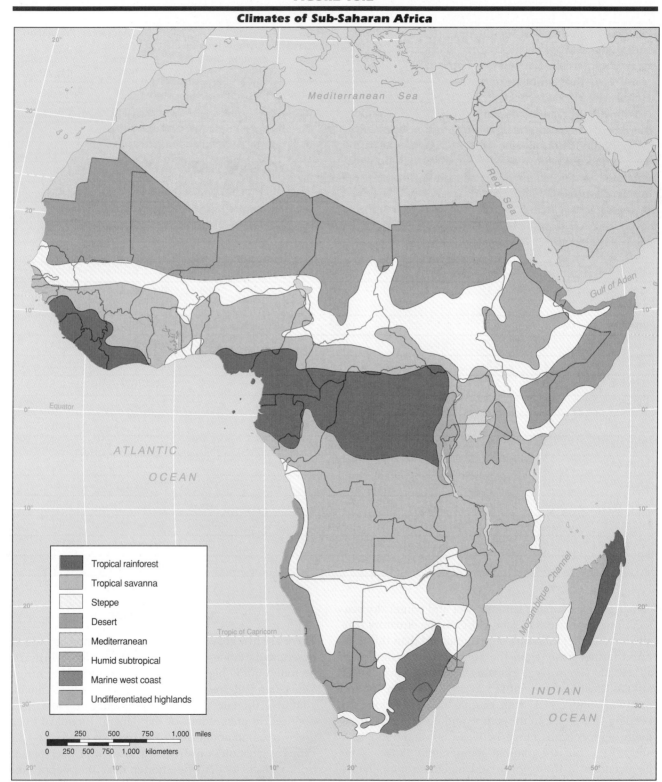

Tropical rainforest
Tropical savanna
Steppe
Desert
Mediterranean
Humid subtropical
Marine west coast
Undifferentiated highlands

FIGURE 13.3

Desertification of the Sahel

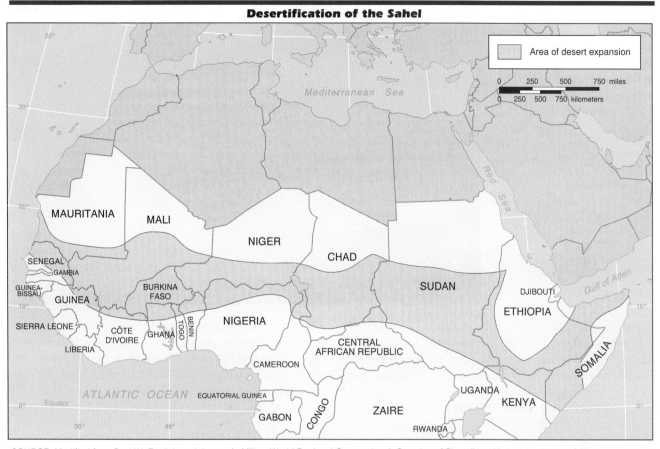

SOURCE: Modified from Paul W. English and James A. Miller, *World Regional Geography: A Question of Place* (New York: John Wiley & Sons, 1989), p. 511.

climatic areas. Acacia trees frequently predominate in the savanna. In wetter savanna areas, grasses are often more prevalent than trees. As the savanna merges into the steppe, grasses occur in clumps rather than as continuous cover, and short thorny shrubs and bushes growing singly or in groups or patches replace trees. However, some broadleaf deciduous trees are located in the steppe areas south of the savanna in such countries as Botswana and Zimbabwe. In the desert areas, vegetation is largely or entirely absent, although in some areas shrubs are present.

SOILS

Soils are affected by climate, vegetational cover, parent material, drainage, slope, and elevation. Soils in Africa vary quite widely in fertility, amount of organic matter (humus), and color.

In the wet tropical rainforests, soils are generally low in nutrients, heavily leached, rich in iron and aluminum oxides, and rather reddish or yellow-red in color; they are classified as oxisols. In areas with these soils, shifting cultivation is frequently practiced, and rubber trees and oil palms are grown on plantations. In the wet-and-dry climates of the subtropics, soils are not as severely leached and contain more calcium and humus than in the wet equatorial rainforests. Consequently, they are somewhat more fertile and, with adequate fertilization and good soil conservation practices such as crop rotation and fallowing, can be somewhat productive. However, these soils can also develop excessively thick crusts or hardpans (cemented soil particles) during the dry season.

Soils in portions of the highland areas of East Africa are rather fertile as are those in parts of South Africa, Zimbabwe, and Mozambique. Some of the fertile soils in East Africa and Cameroon are volcanic in origin and are capable of producing high yields.

A view of savanna-type vegetation in Kenya, with Masa Lodge in the background. In the wetter areas of the savanna, grasses grow much higher and trees are taller, but as the savanna begins to merge into the drier steppe areas, the grasses become much shorter and more widely dispersed, and the trees become shorter and fewer in number.

Desert soils, called aridisols, are basically infertile and lack organic content. In the steppe areas south of the Sahara, the horn area of Africa, and the dry areas of southwest Africa (parts of the Kalahari), ranching and nomadic herding are the leading economic activities. If water supplies were more reliable, more agricultural production could take place.

PHYSICAL CONFIGURATION

Africa straddles the equator and extends north to slightly more than 37° north latitude and south to about 35° south latitude. The north-south extent of the continent is about 4800 miles (7700 kilometers). At its widest point, it extends from about 17° west longitude to about 51° east longitude, a distance of about 4500 miles (7200 kilometers). The land area of the continent covers 11,708,000 square miles (30,325,000 square kilometers), approximately one-fifth of the land surface of the world. By comparison, the land area of the United States is only 3,618,000 square miles (9,370,000 square kilometers).

The outline of the coastline is relatively smooth, with few inlets and natural harbors. Of all the continents, Africa has the smallest amount of coastal plain. Much of the land surface south of the Sahara is plateau, with elevations ranging from 1000 to 5000 feet (305–1525 meters) above sea level. Some of the major plateaus and uplands include the Adamawa Plateau, Bihe Plateau,

and the Damaranama Upland. However, the general plateau surface is punctuated with basins and mountainous or highland areas. Some of the major basins include Djouf, Chad, Sudan, Zaire (Congo), and the Kalahari. Some of these basins, however, are still above 1000 feet (305 meters) in elevation. The Atlas Mountains predominate in the north, and the Drakensburg Mountains in the south. In places, these mountains rise above 5000 feet (1525 meters). Other major mountainous areas include the Ethiopian Highlands—volcanic mountains in east Africa that are dominated by Kilimanjaro (19,340 feet; 5895 meters), Mt. Kenya (17,058 feet; 5200 meters), and Mt. Elgon (14,178 feet; 4321 meters)—and the Cameroon Mountains in West Africa where the highest peak is Mt. Cameroon (13,354 feet; 4070 meters) (Figure 13.4).

One of the other dominant physical features of Africa is the complex Rift Valley system, which extends from the Jordan River Valley through the Dead Sea, Red Sea, central Ethiopia, and southward into South Africa. The Rift Valley in Africa varies from 20 to 50 miles (32–80 kilometers) in width, with steep walls sometimes 3000 feet (915 meters) in height. Associated with the Rift Valley system are the great lakes of East Africa, with Lake Victoria being the third largest lake in the world. Other rather large lakes include Lake Albert, Lake Tanganyika, and Lake Malawi.

Five great rivers drain much of the continent: the Nile, Zaire (Congo), Niger, Zambezi, and Orange. Because of

FIGURE 13.4

Landforms of Sub-Saharan Africa

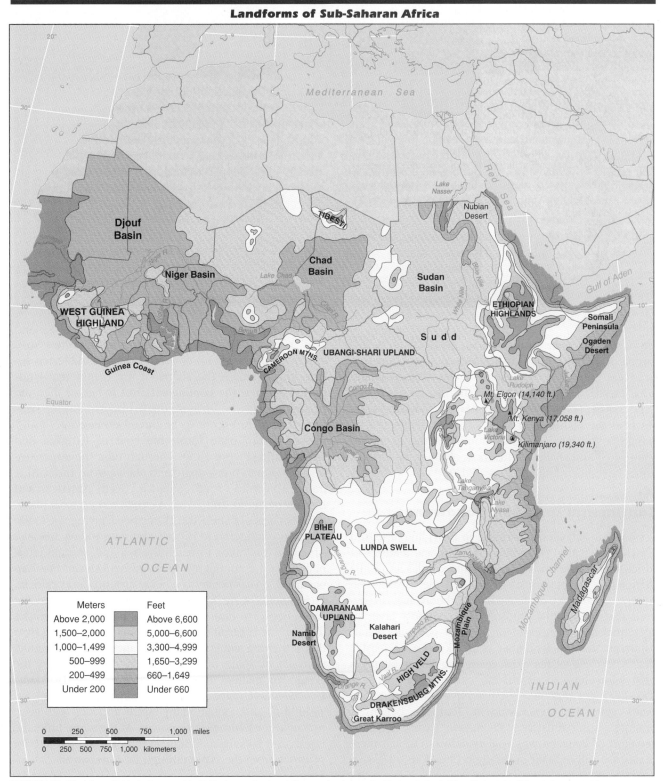

SOURCE: Modified from R. H. Jackson and L. E. Hudman, *Regional Geography: Issues for Today*, 3rd ed. (New York: John Wiley & Sons, 1990), p. 475.

A view of the Rift Valley. The Rift Valley, a significant landform feature of Eastern Africa, was formed as India, Australia, South America, Madagascar, and Antarctica began to drift away from Africa more than 200 million years ago.

the plateau structure of the African interior and the escarpment in the southern half of the continent, few major rivers are navigable for any great distance inland from the sea. Some of the better harbors occur between Cape Town, South Africa, and Maputo, Mozambique. Much of the coastline of West Africa is dominated by mangrove swamps, and the mouths of the rivers are often silted and shallow; thus most of the ports in West Africa have been artificially created.

Continental Drift and Gondwana

Continental drift theory is useful in providing at least a partial explanation of some of the physical features of Africa such as the Rift Valley, the folded Atlas and Cape mountains, the pattern of rivers, and the escarpment. Recent research in continental drift theory has discovered that the earth's lithosphere (outermost layer of solid rocks) is comprised of a set of geologic plates. Although the exact number is not known, at least a dozen tectonic plates exist, averaging about 60 miles (100 kilometers) in depth, but extending as deep as 200–250 miles (322–402 kilometers). About 250 million years ago, the supercontinent *Pangaea* began to emerge. Approximately 135 million years ago, the breakup of Pangaea was well underway. Today, these plates are still moving in different directions, at a very

slow rate. Where they collide, the earth's crust may crumple, producing much earthquake and volcanic activity that changes the shape of the continental landforms. Because the continental crust is lighter than the oceanic crust, when plates collide underwater, the denser oceanic crust tends to be forced downward.

The African plate forms the core of Pangaea and **Gondwana** (Figure 13.5). Through time the supercontinent began to break apart, and Madagascar, India, Australia, Antarctica, South America, North America, and Eurasia began to drift away. As the breakup took place, different types of stress forces occurred, giving rise to the physical configuration of the continent today. Initially, Africa's rivers failed to reach the sea and drained internally into the continent's great basins. As the breakup occurred, coasts began to emerge around the continent, escarpments were formed near the coastline, and the rivers began to drain to the coasts cutting deep gorges through the escarpments and eventually producing river deltas that are similar in geologic age. Thus, the middle portions of some of Africa's great rivers such as the Congo are "mature," flow slowly and can be used for transportation; in contrast, the lower reaches of the rivers are "youthful" and contain many rapids and falls with great potential for hydroelectric power. At the same time, the rapids and falls make it difficult to penetrate the interior of Africa from the coast through its waterways.

Because the drifting away of the continents put tensional stress on Africa, the creation of the Rift Valley is undoubtedly connected to continental drift. Many of these longitudinal trenches eventually filled with water resulting in a large number of lakes in East Africa.

RESOURCES

Africa produces more than two-thirds of the world's gold, diamonds, and cobalt, slightly more than 40 percent of the world's platinum, chromite, and manganese, and more than 10 percent of its phosphates, copper, uranium, bauxite, and iron ore. The Republic of South Africa accounts for much of the gold, diamonds, platinum, and chromite. Diamonds are also found in Zaire and Namibia, and cobalt is important in Zaire. Other metals of significance include vanadium, antimony, titanium, beryllium, and tin. Other leading nonmetal minerals include vermiculite, talc, and asbestos.

Much of the mining is controlled by multinational corporations, which tend to have the financing, technol-ogy, managerial skills, and marketing know-how to conduct business efficiently. Although some African states have attempted to nationalize portions of their mining operations, many African countries still do not fully share in the monetary gain accrued from their mineral deposits and mining activities. Frequently, the raw materials are shipped to other countries to be processed and used in making manufactured goods, and expensive mining equipment is often imported.

Africa possesses some energy resources but not enough to be self-sufficient. Coal deposits appear to be limited. According to some estimates, Africa has only about 3 percent of the world's coal reserves. Approximately 95 percent of the continent's coal production is in South Africa. Although Nigeria, Gabon, the Congo, Cameroon, and Angola produce oil for export, most African states are net importers of energy. Hydroelectricity is also generated along some of Africa's major rivers, but its production is limited because many African rivers have seasonal regimes, flooding in one season and dry in another. Dams can help in flood control, irrigation,

FIGURE 13.5

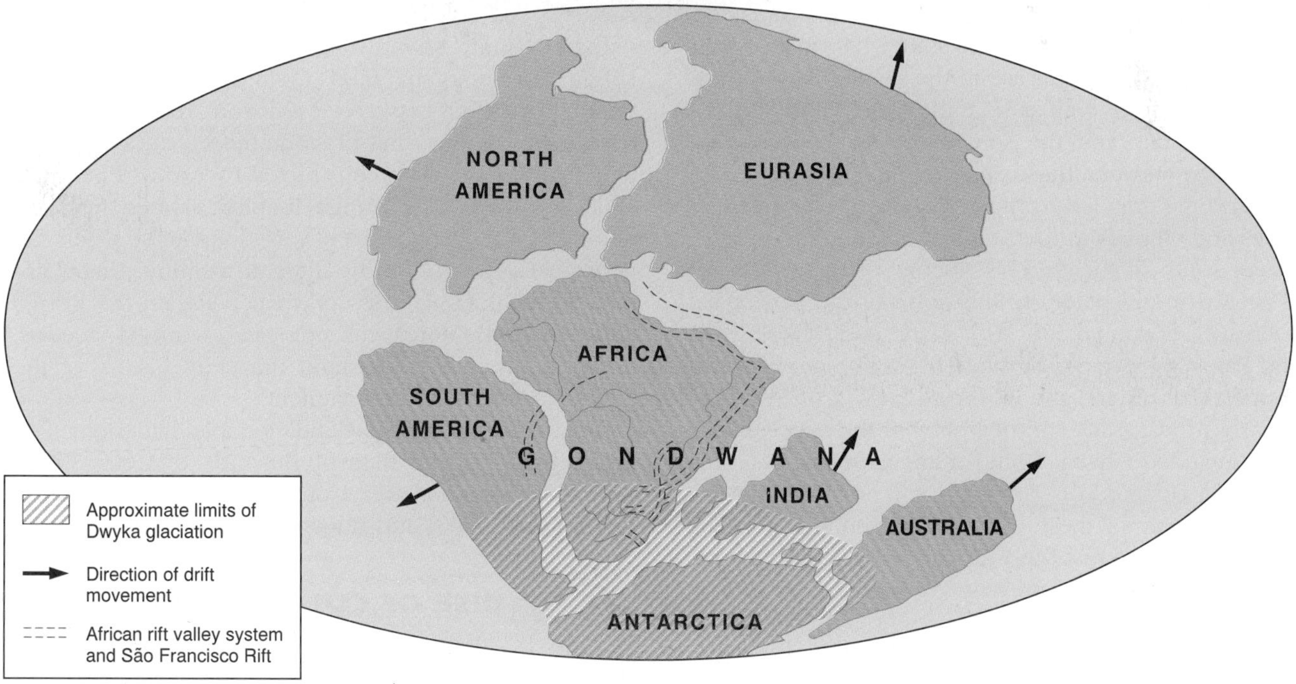

SOURCE: Modified from H. J. de Blij and P. Muller, *Geography: Regions and Concepts,* 5th ed. (New York: John Wiley & Sons, 1988), p. 425.

and energy production, but care must be exercised in their location even though the potential is great.

Fish and forest resources are also available in Africa, but neither has been fully developed. Much fishing is done in Sub-Saharan Africa, for example, but for the most part it is not well developed commercially except in South Africa and Angola. Fishing also takes place in the lakes of East Africa and in the many rivers, swamps, and lagoons across the continent, but much of the catch is used for local consumption and internal markets.

Although Africa has about 25 percent of the world's total forest area, it is still a net importer of forest products. A large amount of unrecorded cutting is used for domestic fuel consumption and for cooking. Some countries such as Ghana, Nigeria, Cameroon, Gabon, and the Congo export wood and wood products, but many other African states have not developed their forestry resources to their full potential.

HUMAN ENVIRONMENT

EARLY HISTORY

Prior to European contact, Bantu-speaking people had moved eastward and southward from West Africa. Although Sub-Saharan Africa was somewhat isolated from the early empires that arose in the Fertile Crescent and the Nile Valley, as well as from the Greeks and Romans, human contact and the spread of ideas nevertheless took place between these empires and Sub-Saharan Africa. In time, this led to the creation of several African states and empires in the Sahelian corridor and in other places south of the Sahel. Evidence of trade between West Africa and places to the north of the Sahara has been well documented. Gold, ivory, ebony, kola nuts, and slaves were traded to the north in exchange for salt, copper, dried fruits, and other products.

Early African states and empires in the Sahelian corridor included Ghana, Bornu, Kanem, and Hausa, and later Mali centered on Timbuktu and Songhai centered on Goa. Many of these existed between A.D. 700 and 1900. Ghana, the first West African state, was located in the western portion of present-day Mali and focused on the town of Ghana (Figure 13.6). Although not much is known about its origin, it is often dated at about A.D. 700, but it may have existed as early as the fourth century A.D. With the fall of Ghana, Mali arose in the area surrounding the middle portion of the Niger River. Mali

reached its zenith in the early 1400s. Its leaders then extended the territory of the empire further than they could control, and by 1500 the empire was in decline. Songhai, which succeeded Mali, included territory to the east of the former Mali Empire and at one time encompassed portions of the kingdom of Hausa. To the west, it included land formerly under the influence of Ghana. Moroccan invaders from the north and intrafamily rivalries for leadership eventually led to the downfall of Songhai.

African states also arose on the West African coast, including Oyo, Benin, Ashanti, and Dahomey. Although some of these states originated in the fifteenth and sixteenth centuries, their importance increased after European contact when seacoast trading posts were established. As contact and trade increased along the coast, the Sahelian states and towns began to decline, and land trade routes across the Sahara dwindled.

In East Africa, Greeks and Arabs are thought to have sailed down the Red Sea and into the Indian Ocean as far south as the Mozambique Channel to conduct trade. Contact between East African ports and the civilizations of Mesopotamia and the Mediterranean Basin existed as early as A.D. 120. Discoveries of Chinese porcelain at various locations along the East African coast suggest that trade with the Orient existed as early as the twelfth century.

Further south toward the interior, the Kongo, Luba, Lunda, Monomotapa, and Changamire empires and states flourished between 1400 and 1800. The Zimbabwe ruins, consisting of stone towers and walls, are believed to have been constructed by Africans between 1000 and 1400. The Kongo Empire, located near the mouth of the Congo River, was flourishing when the Portuguese found it in the fifteenth century. Later, diplomatic relations were established between the Kongo and Portugal. More archaeological work is needed, however, to fully understand the early history of the interior portions of the continent.

In East Africa, the BaGanda people had organized themselves into the state of Buganda by 1600. Other states in the general area around Lake Victoria included Kitara (1200–1500) and Bunyoro (1220–1500).

RISE OF COLONIALISM

Portugal was the first colonial power to contact Africa south of the Sahara. The Portuguese had explored much of the West African coast by the latter part of the fifteenth century and had rounded the Cape of Good

FIGURE 13.6

Early African States and Empires

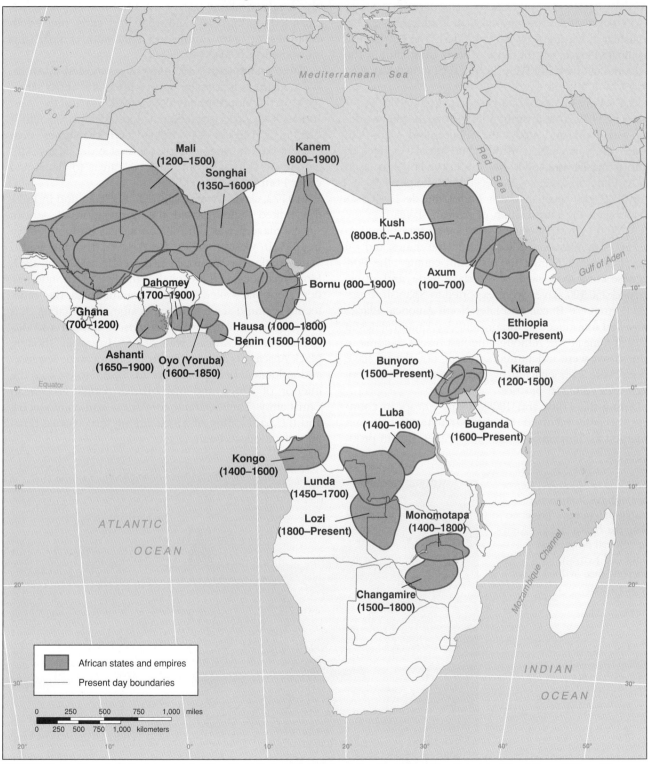

Mali (1200–1500)

Kanem (800–1900)

Songhai (1350–1600)

Kush (800B.C.–A.D.350)

Axum (100–700)

Bornu (800–1900)

Dahomey (1700–1900)

Ghana (700–1200)

Hausa (1000–1800)

Benin (1500–1800)

Ethiopia (1300-Present)

Ashanti (1650–1900)

Oyo (Yoruba) (1600–1850)

Bunyoro (1500–Present)

Kitara (1200-1500)

Luba (1400–1600)

Buganda (1600–Present)

Kongo (1400–1600)

Lunda (1450–1700)

Lozi (1800–Present)

Monomotapa (1400–1800)

Changamire (1500–1800)

African states and empires

Present day boundaries

0   250   500   750   1,000 miles

0   250 500 750 1,000 kilometers

Mediterranean Sea

Red Sea

Gulf of Aden

Equator

ATLANTIC OCEAN

Mozambique Channel

INDIAN OCEAN

SOURCE: Modified from P. W. English and J. A. Miller, *World Regional Geography: A Question of Place* (New York: John Wiley & Sons, 1989), p. 497.

Hope by 1500. Soon other European countries came to Sub-Saharan Africa, and coastal trading stations were established along the West African coast. The Portuguese dominated trade on the East African coast during the sixteenth century, but were expelled from much of East Africa by 1740.

The selling of slaves had been a major activity in Africa long before the arrival of the Europeans. Slave trading was conducted across the Sahara, and Arabs had raided the east coast of Africa shipping slaves back to Arabia, Persia, and India. But the Europeans increased the volume of slaves to unprecedented heights.

In 1517 the Spanish priest Bartolomé de Las Casas encouraged the introduction of African slaves into the New World. By 1530 slaves were being transported to the Spanish West Indies to work in the mines and on the plantations. During the sixteenth century, the Portuguese controlled much of the slave trade, and an estimated 250,000 slaves were exported to Spanish America, Portuguese Brazil, and later North America. Blacks were first introduced to the present-day United States in 1619 at Jamestown. Although the first blacks to come to Jamestown were indentured servants, not slaves, slave codes were later enacted by the colonies, and slavery became a common practice.

During the 1700s and 1800s, many European countries became involved in slave trading, but Britain emerged as the principal maritime power. Britain probably shipped about half of the total number of slaves exported during the eighteenth and early nineteenth centuries; at times during the eighteenth century, this amounted to 60,000 annually.

Three major source areas for slaves existed in Africa. West Africa was a major source area, and most slaves from there were shipped to North America and the West Indies. Two other source areas existed further south—one near present-day Angola and the other near present-day Mozambique on the east coast. Many Africans were shipped from these two areas to Brazil and other Portuguese possessions.

In 1772, slavery was abolished in Britain, and in 1808, the United States ceased the importation of African slaves. Slavery was officially abolished in South Africa in 1833. In western Africa, freed slaves from the United States settled in Monrovia in the early 1820s. In 1847, Monrovia became the capital of the first black independent republic called Liberia. Eventually in the nineteenth century, slave trading became unprofitable, thus bringing an end to three centuries of turmoil and disruption of the African way of life. Ironically though, black Africa was still not free from European penetration and influence, because Europeans began to explore the interior of Africa and missionaries sought to spread Christianity. The "scramble for Africa" had commenced, and the partitioning of Africa among the colonial powers was about to begin.

Fort Jesus in Mombasa, Kenya. Although the Portuguese established an early presence along the coastal area of Kenya, the colonial power that controlled Kenya prior to independence in 1963 was Great Britain.

FIGURE 13.7

Imperial Ambitions in Africa

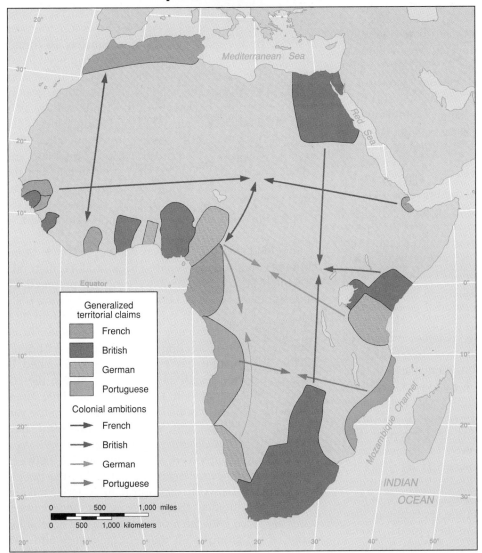

SOURCE: Modified from C. A. Stansfield and C. E. Zimolzak, *World Regions: Changing Interactions* (Columbus: Charles E. Merrill Pub. Co., 1982), p. 455.

Beginning with the Berlin Conference in 1884–1885, the colonial powers began to partition and divide Sub-Saharan Africa, a process that was completed in a mere 30 years from 1884 to 1914. International boundaries were drawn and spheres of influence were recognized. France wanted to establish links that would extend from Dakar on the west coast to the horn of Africa in the east, from Algeria south to the West African coast, and from the Congo centered on Brazzaville to the north (Figure 13.7). Britain wanted territories that would link

Africa from Cape Town to Cairo. Portugal wanted to connect Angola and Mozambique, and Germany sought to connect Tanganyika (Tanzania) with Cameroon and Southwest Africa (Namibia) to Cameroon. King Leopold of Belgium controlled the Congo Basin, and West Africa was divided up primarily among France, Britain, Germany, and Portugal. Italy, a latecomer to Africa, proceeded to occupy Eritrea and portions of Somaliland in the horn of Africa.

The political and social policies implemented by the

colonial powers in their African possessions varied according to the way the colonies were administered. France administered its colonies as overseas territories and encouraged the spread of the French languages and other French cultural traits. Africans who accepted French culture and became assimilated were entitled to French citizenship and supposedly enjoyed all the rights and privileges that native French citizens enjoyed. The Portuguese also had a policy of assimilation, but they viewed Africans as children and "savages" until they became Europeanized citizens. Portuguese possessions were considered to be provinces of Portugal and part of the mother country, not colonies that were to be steered toward eventual independence. On the whole, the Portuguese lagged far behind the British and French in permitting Africans to participate economically, politically, and socially in the administration of their provinces. At the same time, the Portuguese were extremely exploitative, and the economy of Portugal was greatly enhanced by its overseas provinces. Because of its attitude toward its provinces, Portugal was one of the last European powers to give up its colonial possessions in Africa.

Britain's administration of its colonies varied, although a policy of indirect rule was frequently employed. In some places, indigenous participation in government was encouraged, and local power structures were left intact. In contrast, Belgium took a paternalistic attitude towards its subjects and did little to encourage higher education for the Africans or to prepare their possession, the Congo (now Zaire), for independence and self-government. The Germans did much to improve the infrastructure in their possessions by building railroads and establishing plantations. Evidence of their former presence is still visible on the landscape in their former colonies. In characterizing the various colonial powers, some Africans today describe the British as "gentlemen," the French as "playboys," and the Germans as "brutal." Obviously, this is a rather simplified generalization, but it does reveal how some Africans view their treatment by the colonial administrators.

After World War II, colonialism began to wane and European influence weakened. Various countries in other parts of the world began to achieve independence. Starting with the Philippines in 1946, many countries in South and Southeast Asia became independent. Demands for freedom soon followed in Sub-Saharan Africa, and in 1957 Britain granted independence to Ghana (Figure 13.8).

INDEPENDENCE

After Ghana achieved its independence, other African countries gained their independence from the former colonial powers. Guinea, a French possession, followed in 1958. By the end of 1960, approximately 20 former French, British, and Belgian possessions had been freed. Much of the early independence occurred in western and central Africa. The wave of independence continued to sweep through Africa in the 1960s, and by 1970 only the Portuguese provinces of Portuguese Guinea (Guinea-Bissau), Angola, and Mozambique plus Rhodesia (Zimbabwe), Southwest Africa (Namibia), and French Somaliland (Djibouti) were still controlled by the colonial powers. The Portuguese withdrew in 1975, Djibouti became independent in 1977, Zimbabwe was established in 1980, and Namibia became independent in 1989.

Independence has not been without its price. Some countries were better prepared to go it alone than others. In those countries that were poorly prepared for independence, violence between factions seeking political control erupted. In other countries, the transition period after independence was much smoother. In all countries, much reordering of the political, economic, and social structure has taken place, necessitated in large part by the legacy of colonialism. When the colonial powers partitioned Africa, they paid little attention to the already existing boundaries between the various ethnic groups and simply superimposed new boundaries on preexisting cultures (Figure 13.9). As a result, ethnic groups were sometimes divided between two colonial powers. For example, the Yoruba people were divided between the British in Nigeria and the French in Dahomey (Benin). Now the independent states are faced with the monumental task of combining many ethnic groups with different languages and customs into a nation-state. Convincing the various ethnic groups to put aside their former boundaries and dealing with the problems posed by the separation of ethnic groups and jealousy among the groups are formidable undertakings. African states have made much progress in the short period of time they have been independent, but problems still persist.

POPULATION

Africa south of the Sahara has a population of more than 500 million persons, about 10 percent of the world's total population. But black Africa has a rapidly growing

FIGURE 13.8

Colonization and Independence in Sub-Saharan Africa

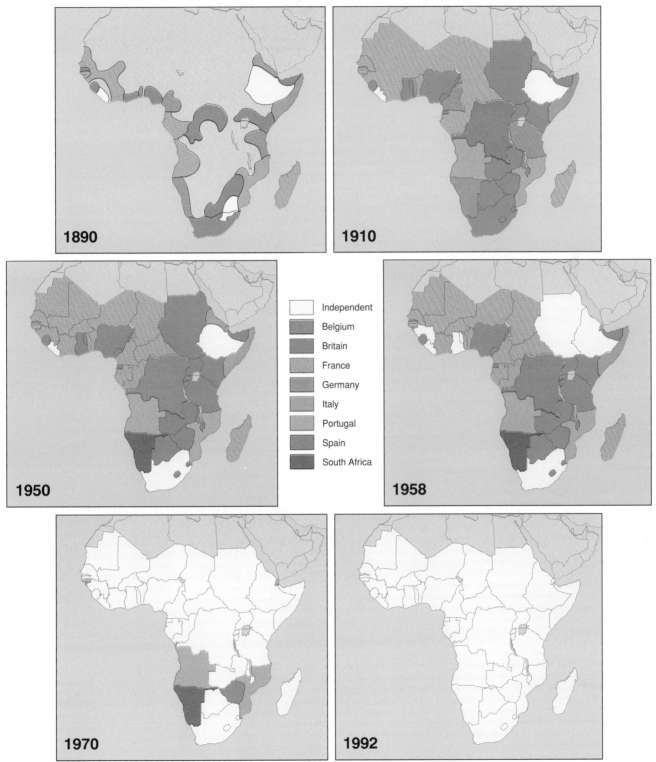

Independent
Belgium
Britain
France
Germany
Italy
Portugal
Spain
South Africa

1890

1910

1950

1958

1970

1992

SOURCE: Modified from H. J. de Blij and Peter Muller, *Geography: Regions and Concepts*, 6th ed. (New York: John Wiley & Sons, Inc., 1991), p. 415.

FIGURE 13.9

Ethnic Territories and Superimposed Boundaries in Sub-Saharan Africa

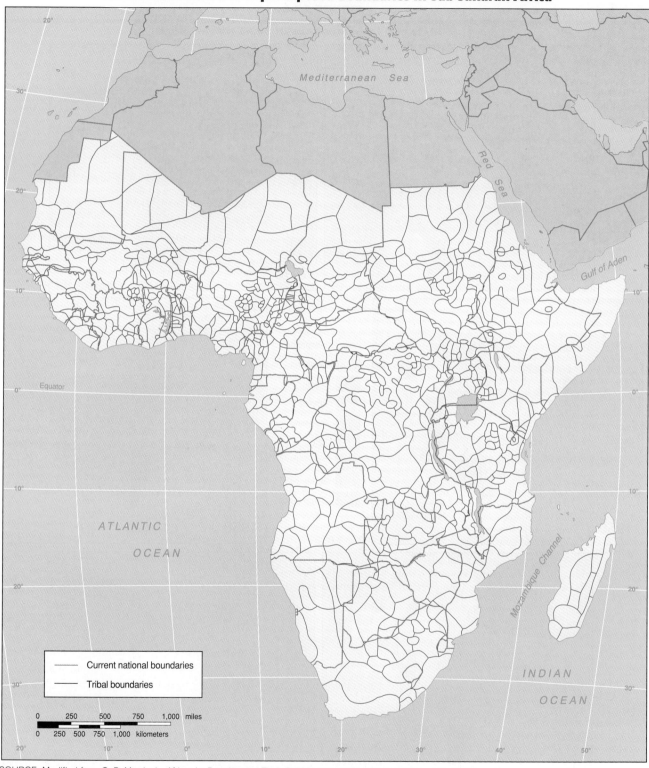

Mediterranean Sea

Red Sea

Gulf of Aden

Equator

ATLANTIC

OCEAN

Mozambique Channel

INDIAN

OCEAN

Current national boundaries

Tribal boundaries

| 0 | 250 | 500 | 750 | 1,000 | miles |

| 0 | 250 | 500 | 750 | 1,000 | kilometers |

SOURCE: Modified from G. P. Murdock, *Africa: Its Peoples and Their Cultural History* (New York: McGraw Hill, 1959).

population. Crude birth rates are very high in most countries and frequently exceed 45 births per 1000 persons. By comparison, the world average is only 27. The population of black Africa is growing at a rate of 2.7 to 3.0 percent annually, which is higher than the average for the developing countries in Asia or Latin America. By contrast, European countries, the Soviet Union, the United States, and Canada are all growing at rates of less than 1 percent. In 1976 the estimated population of Sub-Saharan Africa was about 330 million. By 1990 it had added more than 170 million persons. In another 15 years, Sub-Saharan Africa could have more than 700 million inhabitants, more than double the total in 1976.

In general, leaders in the various countries have not been overly concerned with the population explosion. African families are large, frequently averaging six children. Abortion is illegal, and birth control is not widely practiced. Continued indifference to population growth may cause future problems. Only in Cameroon, the Central African Republic, Côte d'Ivoire (Ivory Coast), Rwanda, and Sudan has food production outstripped population growth. Since most countries cannot produce as much food as they consume, commercial food imports have risen dramatically over the last several years. Future population growth policies are certainly a topic black Africa will need to address before the end of this century.

Nigeria is the only country with a population of more than 100 million. Ethiopia, South Africa, Tanzania, and Zaire have populations between 25 and 52 million. The remaining countries have fewer than 25 million inhabitants with several having fewer than 10 million.

Population Density and Distribution

The population of the continent is not evenly distributed. Islands of high population density are scattered about mainly in West Africa and in a north-south alignment in East Africa. Populations are concentrated near major urban centers, and in the rural areas clusters are found in locations where water is available and the soil is fertile (Figure 13.10). Most Africans are still engaged in agricultural or pastoral activities, and very few countries have urban populations exceeding 50 percent. Only 18 percent of the population is urban in eastern Africa, 30 percent in western Africa, and 37 percent in middle Africa, with southern Africa averaging 53 percent. South Africa, which is 56 percent urban and contains 88 percent of the total population of the countries

in southern Africa, tends to distort the percentage of urban dwellers for the region. Some of the remaining countries in southern Africa such as Botswana, Lesotho, and Swaziland have urban populations below 27 percent.

Much of the African population is still afflicted with disease. The spread of malaria, sleeping sickness, and river blindness has yet to be slowed to any great extent. The incidence of AIDS is also spreading rather rapidly. Infant mortality is high, malnutrition or protein deficiency affects about 20 percent of the population, and Africa's life expectancy at birth (about 50 years) is the lowest in the world. Food, health, education, and housing problems plague Africa, and as the gap in the quality of life between rural and urban Africa widens, many Africans are migrating to the urban centers in search of a better way of life. However, numerous urban centers are struggling with the influx of rural migrants. Many migrants cannot be absorbed into the urban economy and may be forced into a living style that may not be better than they experienced in the rural areas.

CULTURE

The culture of Sub-Saharan Africa combines indigenous elements with elements introduced during the colonial period. Thus, the European languages of the colonial powers coexist with indigenous languages, and traditional African religion survives alongside Christianity. Politically, too, the legacy of colonialism is apparent, as are new problems that have appeared since independence.

Language

Opinions differ as to the number of languages in Sub-Saharan Africa, but it is estimated that 800 to 1000 exist. Some scholars suggest, however, that there may be as many as 2000 different languages in Africa. Some of these "languages" may actually be dialects, but this argument of language versus dialect should not distract from the basic concept that Africa contains numerous languages and language groups. Most Africans can speak many languages—not only the language of their local area, but also that of their neighboring ethnic group(s), and often that of the former colonial power.

European languages tend to be the official languages of all Sub-Saharan independent African states. Some countries, like Cameroon, have two official languages.

FIGURE 13.10

Population Distribution in Sub-Saharan Africa

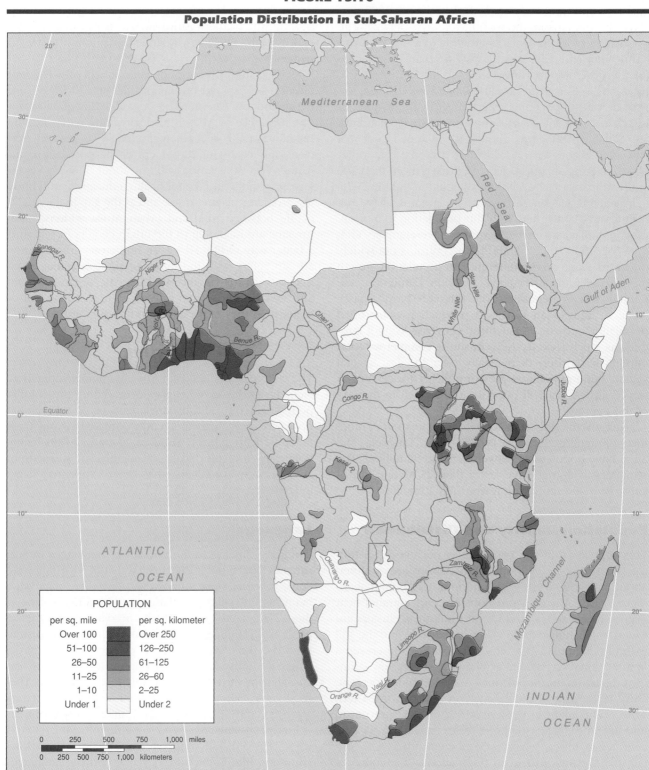

POPULATION

per sq. mile	per sq. kilometer
Over 100	Over 250
51–100	126–250
26–50	61–125
11–25	26–60
1–10	2–25
Under 1	Under 2

English is spoken in the western portion of the country and French in the remainder. European languages also serve as a convenient language for official communication within and between countries. They also tend to unite countries that have a multiplicity of indigenous languages. It is not uncommon for a country to have 20 or 30 indigenous languages. The various ethnic groups obviously want to maintain ties to their ethnic language, but in education, business transactions, radio, television, and newspaper communication, the need to unite under an official European language becomes obvious and is therefore promoted as a way to unify and communicate.

Political boundaries may be helpful in distinguishing the official language of the country, but are much less useful in analyzing the distribution of African languages because these languages tend to transcend modern political boundaries. Therefore, it is difficult to regionalize African languages. Nevertheless, four major linguistic families can be identified on the continent, plus one in Madagascar (Figure 13.11). Afroasiatic languages are found in north Africa and in the Sahelian corridor. They correspond rather closely to the distribution of Islam. Arabic, Berber, Amharic, Hausa, and Somali are representative examples of this group. The Congo-Kordofanian family, which predominates in western, central, and southern Africa, is the largest family in spatial extent and number of speakers. Examples in West Africa include Yoruba, Iko, Tiv, Ewe, and Fula. The predominant linguistic subgroup in central and southern Africa is Bantu. Swahili, Kikuyu, Bemba, Shona, Xhosa, Zulu, and Swazi are examples of related languages under the Bantu subgroup, which belongs to the larger Congo-Kordofanian family. The Nilo-Saharan linguistic family is located along parts of the Nile and Niger rivers. The fourth linguistic family in Africa is Khoisan, located mainly in the Kalahari Desert of southern Africa. The Khoi (Hottentots) and San (Bushmen) comprise the important ethnic groups speaking Khoisan, which is characterized by several sounds made with a click of the tongue or lips. Finally, Malayo-Polynesian, located in Madagascar, is related to Indonesian, thus giving credence to early trans-Indian Ocean contact between Madagascar and the islands lying off the coast of Southeast Asia.

Religion

A mixture of Christianity and traditional religion is found in most of Africa south of the Sahara, but Islam is a major influence in the Sahelian corridor and the horn of

east Africa. Religion including belief in a supreme being is an important part of daily life in Africa, but the traditional African concept of God, or the supreme being, differs markedly in many fundamental ways from Western or Christian religion. In traditional religion, which varies immensely across the continent, the supreme being is manifested in a variety of ways such as spirits, souls of ancestors, and totemic animal spirits. Traditional religion attempts to describe and explain the complexities of human conflict and relates to the values of the family, the community, and the cycle of agricultural tasks.

The introduction of Christianity, besides spreading the different beliefs from Catholicism to Protestant fundamentalism, also had an effect upon the diffusion of modern medicine and the establishment of schools. Throughout much of Sub-Saharan Africa, one finds a blend of traditional religion and Christianity. Many Africans practice both to some extent. In some instances, Africans have combined Christian teachings with traditional beliefs to form separatist churches; in other cases, they have attempted to Africanize certain practices and teachings of the Christian church to make them more relevant to the African way of life. Specific data on religious practices in Africa are difficult to obtain. Certainly, however, Christianity and Islam will continue to spread across the continent, and the interaction of these two faiths with the practice of traditional religion will continue with limited modification as Africa moves into the twenty-first century.

Political Characteristics

Black Africa has changed from European colonialism to independent self-governing countries. The surface area of the various states varies enormously. Approximately 9 states in Sub-Saharan Africa are larger than 386,100 square miles (1,000,000 square kilometers) in size. These 9 states encompass approximately 5,200,000 square miles (13,400,000 square kilometers), or more than 50 percent, of the slightly less than 9,650,000 square miles (25,000,000 square kilometers) in Sub-Saharan Africa. At the opposite end of the scale, 15 countries are less than 38,610 square miles (100,000 square kilometers) in size. The remaining countries range between 38,610 and 386,100 square miles (100,000–1,000,000 square kilometers).

Fourteen countries are landlocked, more than in any other continent. Some of these landlocked states were the least integrated into the former colonial system.

FIGURE 13.11

Major Language Groups in Sub-Saharan Africa

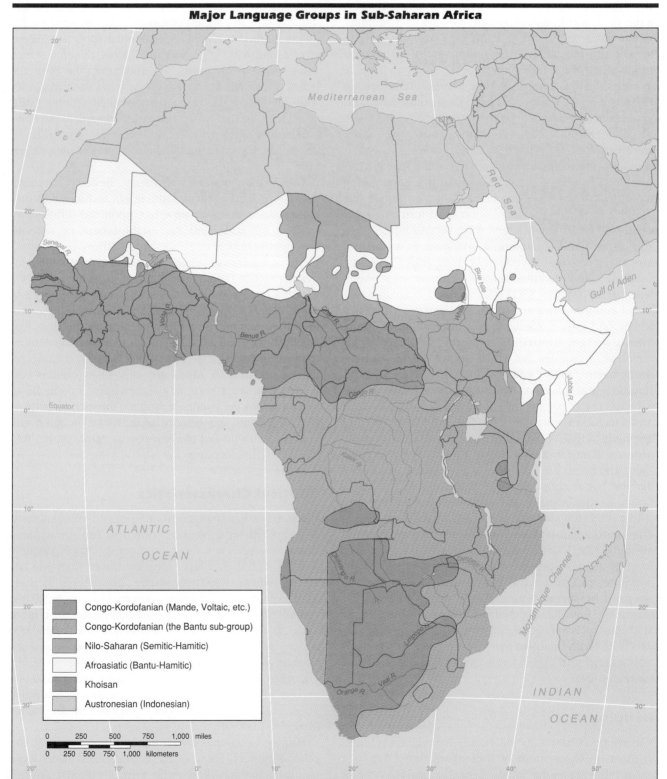

Congo-Kordofanian (Mande, Voltaic, etc.)

Congo-Kordofanian (the Bantu sub-group)

Nilo-Saharan (Semitic-Hamitic)

Afroasiatic (Bantu-Hamitic)

Khoisan

Austronesian (Indonesian)

SOURCE: Modified from R. H. Jackson and L. H. Hudman, *Regional Geography: Issues for Today*, 3rd ed. (New York: John Wiley & Sons, 1990), p. 492.

There are exceptions, but even today, the landlocked states are among the poorest and weakest, politically and economically, on the continent. Size (land area and population), wealth, and individual leadership have been important variables for states achieving political recognition and status in Sub-Saharan Africa. Nigeria and Zaire have attained important political status through the years and have a large number of diplomatic representatives in their respective countries. Ethiopia has attained a fair amount of political status because Addis Ababa serves as headquarters for several African international organizations such as the Organization of African Unity (OAU) and the Economic Commission for Africa (ECA).

After achieving independence, most African states attempted to resolve the problems associated with their colonial legacy while at the same time coping with the new problems resulting from independence. Disillusionment often occurred when the realities of independence began to surface with the realization that political independence did not ensure economic independence. The challenges of **neocolonialism** (dependence upon Western economic and financial institutions after independence) are a problem for most African leaders. Many manufacturing and industrial establishments, mining operations, plantations, and banking institutions were controlled by Europeans at the time of independence. The transfer of political and economic power from European to African control has been a long and arduous task. Some countries made the transition more quickly and were able to develop their mineral and agricultural resources and create new industries. Technical assistance and sources of capital may still be needed from the former colonial powers and from other nations in the world, but African countries must play a major role in their own economic development.

Maldistribution of wealth and problems associated with agriculture, transportation, industrialization, health, and education remained after independence. Some new governments were faced with trying to govern states that were barely economically viable. Some African leaders abused their political power, and corruption often existed at various levels within the government. Oftentimes the new governments could not achieve instant success, leading to frequent changes in political power and coups d'etat in many countries. Early secessionist movements in Zaire and Nigeria and civil war in Burundi did much to disrupt political unity in these countries. More than 50 coups d'etat have occurred in

postcolonial Sub-Saharan Africa. Benin, Ghana, and Burkino Faso have experienced several successful coups d'etat. More than 20 Sub-Saharan states currently have military rulers. Seldom has political power passed from one civilian government to another civilian government through the election process, and rarely do military leaders voluntarily return power to civilians. Even civilian governments must have a strong, supportive military. But despite the changes in leadership and turmoil experienced by several countries, the former colonial boundaries of the independent African states have basically remained intact. Boundaries have remained unchanged in part because the OAU continues to stress the preservation of the inherited colonial boundaries.

Most African states have just one political party because they fear that a multiparty system will invite disunity. In a few instances, the leader who came to power upon independence still remains in power.

Despite the various internal economic and political conflicts and struggles with European nations, various African states have joined international organizations and attempted to form regional economic and political unions. **Pan-Africanism,** a movement that began around the turn of the century, pressed for spiritual unity and equality of all black people in the world. Early leaders in the movement included W. E. B. DuBois, George Padmore, and Henry Sylvester-Williams. Between 1945 and 1955, the political aims of the movement changed slightly, as it began to focus on the liberation and independence of black people, particularly in Africa. As independence was achieved, the focus shifted again to African unity, self-sufficiency, and a concept of "oneness" as leaders promoted cooperation among persons of African descent. In Africa, continental nationalism was stressed; some leaders wanted to see a single currency, one army, and an African form of the common market. As the debate subsided, however, many leaders tended to prefer regional federations or units rather than a single continental federation.

The forces of Pan-Africanism were instrumental in the founding of the Organization of African Unity (OAU) in 1963. The OAU stressed self-government, concern for political and economic development, and territorial integrity and promoted international cooperation while endorsing nonalignment. Although the organization is a forum for discussing African issues, competing interests, regional differences, and ideological disagreements continue to be centrifugal forces that work against a strong unity.

Besides Pan-Africanism, the philosophy of **Negritude** was particularly important in the Francophone-speaking

Boundaries serve as a symbol of a state's territorial claim. They mark the limit of the state's jurisdiction and its control of various social, economic, and political functions. Sometimes boundaries function as impermeable barriers or separators, as did the Berlin Wall, but in most instances they are fluid and flexible, allowing for easy passage of people and goods. Political boundaries are human made, but they may be delimited by **physiographic** features such as rivers, crests of mountain ranges, or midpoints of lakes. Sometimes boundaries are **geometrically** delimited by using latitude and longitude or other mathematical measurements. Alternatively, **anthropogeographic boundaries,** which try to conform to the spatial extent of a specific culture, may be drawn.

Boundaries that are placed on the landscape prior to the full development of the cultural landscape are referred to as **antecedent;** those drawn after the cultural landscape becomes more developed are called **subsequent.** In some instances, boundaries are **superimposed** on societies as the result of colonialism or wars. Most of the present-day boundaries between Sub-Saharan African countries were superimposed by the colonial powers upon the existing cultural landscape with little understanding of the spatial extent of ethnic groups. Frequently, ethnic groups were divided between two colonial powers. Upon independence the new countries were faced with accepting the former colonial boundaries. With sovereign claims over the territory, the new states began the task of nation building and attempted to unify the numerous ethnic groups; this process entailed inducing their citizens to place less emphasis on ethnic allegiances and to think of themselves instead as members of the national unit.

SUPERIMPOSED BOUNDARIES, DECOLONIZATION, AND NATIONALISM

Cameroon is a good example of how superimposed boundaries, decolonization, and nationalism have affected a modern African state. In the "scramble for Africa," Germany took control of Kamerun (Cameroon) in July 1884 and governed it until 1914 (Figure 1). During World War I, the British and French forces were successful in expelling the Germans from Kamerun, and in 1916 the country was partitioned between France and Britain. This partition was later ratified by the Treaty of Versailles in 1919 under which the French assumed control of the eastern portion of the country, and the British took a narrow strip of land along the western side. Subsequent differences in French and British administrative practices led to distinctions in the commercial, economic, social, educational, and political development of the two areas.

Decolonization began in 1960 when France granted independence. People in the British sector were divided as to whether they wanted to join the new independent state of Cameroun or the Federation of Nigeria. The United Na-

tions held a plebiscite, and in 1961 the northern section of former British Cameroon became fully integrated into Nigeria, and southern Cameroon joined the Cameroun Republic. Upon unification, the new state was known as the Federal Republic of Cameroon. In 1972 the country adopted a unitary system of government, and the name changed to the United Republic of Cameroon.

Upon independence, the Cameroonian leaders began the process of reunification and nation building. Not only did the former French and British sectors need to be unified politically, but problems resulting from differences in language, education, and culture needed to be addressed. Having an English- and French-speaking country was difficult enough, but Cameroon is comprised of more than 200 ethnic groups, and 24 ethnic languages are recognized. Thus, besides regional cultural differences, the country faced the problem of unifying a diverse ethnic culture mix as well.

Cameroon is in the early stages of nationhood, its government has been reasonably stable since independence, its economy is fairly strong, and progress has been made in education, health care, housing, and transportation and communication. Ethnic rivalries between some of the larger groups such as the Bamileke, Tikar, Widekum, and Bamoun in the western highlands, the Fulani and Matakam in the north, and the Pahouin and Baloundau-Mbo in the south have been minimized as the groups strive to create a united Cameroon. Thus, the experience of Cameroon with its boundary superimposed by Germany, its decolonization from France and Great Britain, and its difficulties in forging a new nation exemplifies the problems many other African states have faced.

FIGURE 1

The Boundaries and Ethnic Groups of Cameroon

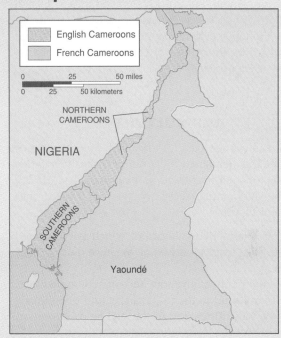

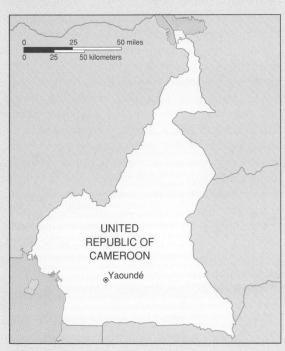

countries in West Africa. The philosophy of Negritude, which promotes black cultural pride, had its roots in the West Indies and French universities. Negritude began as a literary movement, but quickly evolved into a cultural and political philosophy. Today, it has expanded beyond the French-speaking countries of West Africa, and important works are being written by Nigerians, Ghanaians, and other Africans.

AGRICULTURE

Approximately 80 percent of Africans live as subsistence cultivators and herdsmen, depending upon their fields and herds for their livelihood and survival. The African traditional agricultural economy can be divided into three major types:

1. Shifting subsistence cultivation, which is conducted in the wet tropical areas of western and central Africa.
2. Sedentary grain cultivation and mixed agriculture, which are found in the tropical wet-and-dry climatic areas of West Africa and in the highlands in eastern Africa stretching from Ethiopia to South Africa.
3. Pastoralism in the Sahel and semiarid areas in East Africa and in Southwest Africa near the margins of the Kalahari and Namib deserts.

Subsistence cultivation refers to the production of crops such as sorghum, cassava, yams, corn, and vegetables for family or local consumption. Usually, the cultivated land is small in acreage with low yields. Specific agricultural methods vary depending upon such environmental factors as soil fertility, temperature, and rainfall. Historically, land in Africa is viewed as being communal and belonging to a specific ethnic group. Members of the group can use the land in egalitarian fashion. Private ownership of the land and its resources is not a widely accepted practice. However, modernization has made inroads on traditional agricultural practices in Africa, and more pressures are developing between traditional practices and modern commercial agriculture.

One type of subsistence cultivation is *shifting agriculture*, sometimes called *slash and burn agriculture*, which involves clearing the land and burning the vegetation. Crops are then grown on the cleared land for a period of one to three years with the length of time being dependent upon the fertility of the soil. Most families cultivate about two to three acres a year, and many

of the villages number fewer than 50 persons. Different crops are planted at different times of the year to ensure a year-round food supply. After the first year of production, soil fertility begins to decline. Eventually, the group abandons the fields and moves to a new location to begin the process all over again. The abandoned land will usually recover in about 20 years, and the same group or another group may move back onto the land and begin to practice agriculture again. This type of shifting cultivation is mainly practiced in the tropical forests of central Africa (i.e., Zaire, Congo, Gabon), selected coastal locations and savanna regions in West Africa, and the highlands in East Africa (Figure 13.12).

A second type of subsistence cultivation sometimes known as *sedentary subsistence* differs slightly from shifting cultivation in that the villages are usually permanent, and animal manure and other plant wastes may be used to help fertilize the soil. In this instance, villages may be larger, numbering as many as 2000 to 3000 persons. Individuals may plant crops and vegetables next to their homes and also cultivate an area close to the village itself. Usually, the soils are cropped for longer periods of time, such as three to five years, and the land lies fallow for shorter periods. Animals frequently graze on the land that is being fallowed. Diets of persons engaged in subsistence cultivation are frequently high in carbohydrates and starch and low in protein.

Most *sedentary grain cultivation* occurs in the transition areas between the tropical rainforests and the arid areas in the northern and southern portions of the continent. Some mixed farming, combining animal stock and crop raising, exists in the temperate highlands in East and South Africa. Rice, grown in the wetter regions, and sorghum, millet, and maize in the drier locations are the principal grain crops cultivated. In areas where grain cultivation is predominant, the people practice a sedentary subsistence form of life. In some areas, however, subsistence-type agriculture is undergoing changes as commercial marketing becomes more important. Some mechanization of the farmland is occurring, although the traditional hoe is still an important farming implement.

Commercial agriculture is growing in significance, and with the cash obtained from the sale of crops, Africans can begin to purchase many needed goods produced in their home countries or imported from abroad. However, many African states lack the infrastructure, particularly adequate transportation and communications networks, necessary to foster extensive commercial agriculture.

FIGURE 13.12

Agricultural Economies of Sub-Saharan Africa

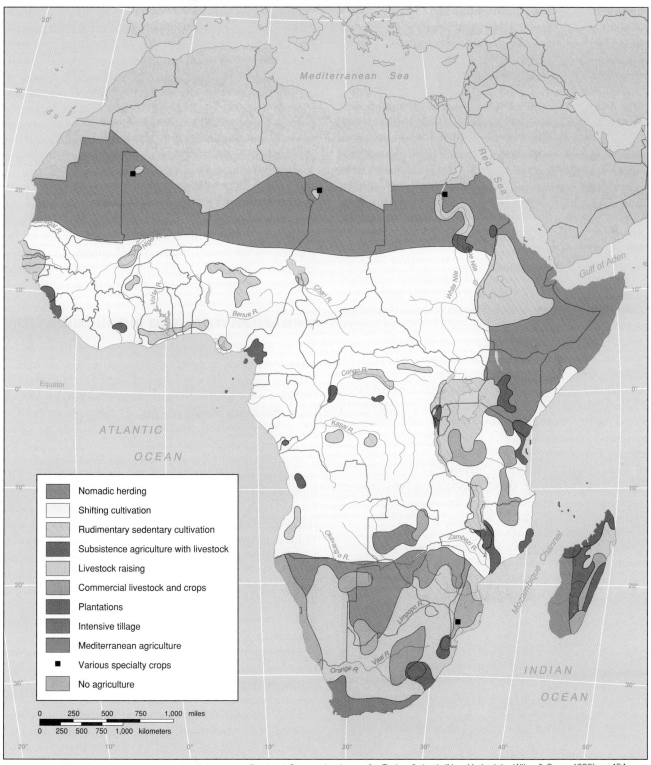

Nomadic herding

Shifting cultivation

Rudimentary sedentary cultivation

Subsistence agriculture with livestock

Livestock raising

Commercial livestock and crops

Plantations

Intensive tillage

Mediterranean agriculture

■ Various specialty crops

No agriculture

SOURCE: Modified from R. H. Jackson and L. E. Hudman, *Regional Geography: Issues for Today,* 3rd ed. (New York: John Wiley & Sons, 1990), p. 484.

Cultivation of tea in Kenya. Tea and coffee are important cash crops in Kenya.

Historically, many commercial crops grown for export were raised on plantations owned by foreign companies. Today, plantations are owned and operated by various African state governments as well as foreign companies. Small African-owned commercial farms also exist. In West Africa, cacao (cocoa), palm oil, bananas, rubber, coffee, and peanuts have been important agricultural export products. In East Africa, Kenya has been producing coffee, tea, cloves, and sisal for export. In South Africa, sugarcane, dairy products, and cotton are produced commercially, but tend to be consumed internally rather than exported.

Cattle grazing in Senegal. Most cattle owners in Africa are sedentary farmers rather than pastoralists. Many cattle owners sell their cattle for meat, but to some pastoralist groups in selected areas in eastern and southern Africa, cattle are more important as a measure of their owner's wealth and prestige than as a source of food. The tsetse fly is a dreaded threat to all cattle herders.

Pastoralism is practiced in the drier areas of Africa, where pastoralists herd sheep, goats, cattle, and camels. The predominance of one type of animal herding over another is frequently related to environmental conditions, income, and tradition. **Nomadism,** the migratory cyclic movement of animals and people, is on the decline in Africa as many nomadic people have begun to settle in villages and towns or change to sedentary agriculture. Drought and overgrazing have accelerated the decline of nomadism in some areas. Spatial movement of the pastoralists varies among the groups; some move with the seasons, others move several times a year. Population pressure and the expansion of other forms of agricultural economic activities into former grazing lands have reduced the amount of land available to the pastoralists for animal grazing. Still, among many African ethnic groups, wealth and prestige are associated with the number of cattle an individual possesses.

URBANIZATION AND MODERNIZATION

In precolonial Africa, most of the towns that existed were trading or commercial centers, served as administrative centers, or were agriculturally oriented. Industrialization as we perceive it today had not yet developed. Although many precolonial cities are not extinct, many have survived, and the majority of Africa's cities today were indigenously initiated.

Associated with the rise of colonialism were the introduction and development of seaport cities and administrative capitals; frequently, the same city performed both functions. In some places on the continent, the process of urbanization was tied to the development of mining. Quite often towns sprang up near the mining areas as people migrated, or were forced to migrate, to the mining districts seeking employment (Figure 13.13). Many of the colonial towns were planned and laid out in accordance with the style favored by the specific colonial power. Many towns had a French, English, Portuguese, or German imprint.

After independence, the process of urbanization continued at an accelerating pace as Africans in increasing numbers migrated from the rural areas to the urban centers and regions of economic development. Migration streams are particularly prominent in specific locations in West, East, and South Africa (Figure 13.14). Various governments have promoted and invested large sums of money in their urban centers; frequently, this has occurred at the expense of the agricultural and pastoral sectors of the economy, thus widening the development

FIGURE 13.13

Mineral Resources in Sub-Saharan Africa

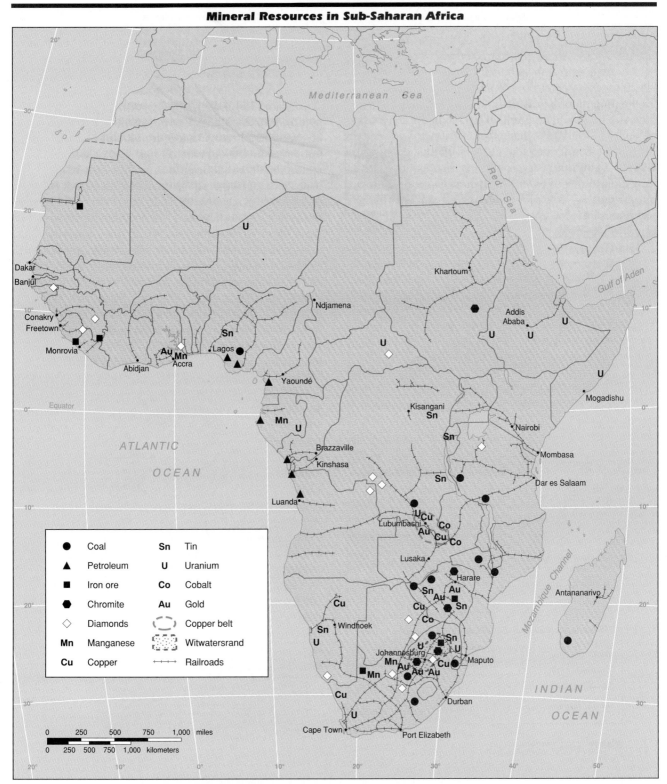

FIGURE 13.14

Migration and Economic Development in Sub-Saharan Africa

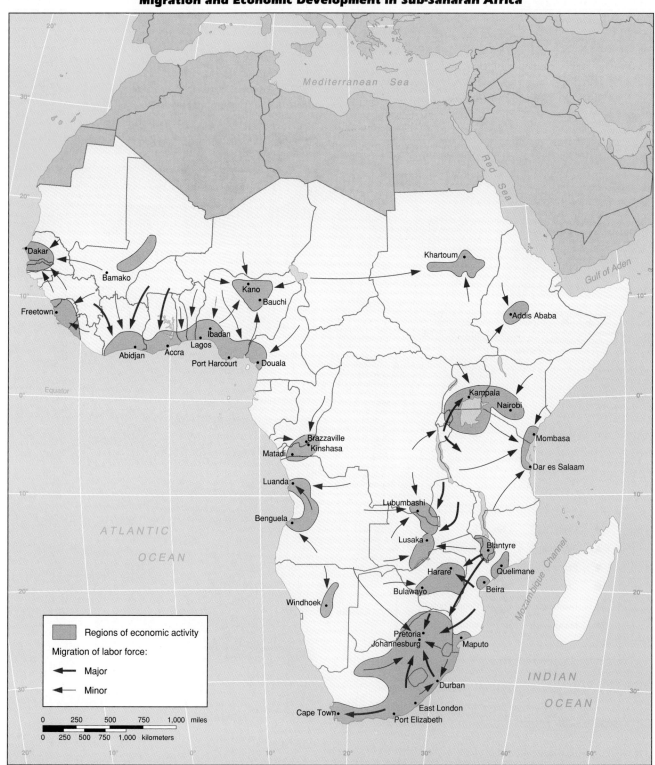

SOURCE: Modified from Paul W. English and James A. Muller, *World Regional Geography: A Question of Place* (New York: John Wiley & Sons, 1989), p. 520.

gap between cities and the villages and rural areas. But the huge influx of new urban dwellers has often occurred so rapidly that the cities have been unable to create enough jobs, provide adequate housing and medical facilities, or furnish sufficient water and sewer, electrical, educational, and transportation services. Shantytowns exist in virtually every African city. Family separation is a common problem as men migrate to the cities seeking employment, which may take months, while the women and children remain in the villages working in agriculture and other traditional activities. Oftentimes the men work in the cities but do not live there on a permanent basis, thus participating in both the modern urban world and the traditional rural life. More than 50 percent of the increase in urban growth is attributable to rural-urban migration. Thus, in most African cities, internal disparities exist between the educated urban elite, who maintain external linkages with the international community, and the poorer masses who promote internal linkages within the city and with the rural countryside. But as in most countries in the world, the political power, wealth, and future hopes and aspirations of the people are closely tied to the cities.

Africa is the least urbanized continent in the world, with less than 30 percent of the population in many countries living in urban areas. Frequently, the capital city is also the primate city of the country. Only a few exceptions exist, mainly in Benin, Cameroon, Malawi, and South Africa. Despite the fact that most capital cities are also primate cities, few (except Accra, Addis Ababa, Abidjan, Dakar, Lagos, Cape Town, Khartoum, Kinshasa, and Nairobi) have populations exceeding a million. Yet it is estimated that about one-third of Sub-Saharan Africa's urban population lives in the primate cities.

Not including the capitals of the landlocked states, most of the African capitals are seaports, often of colonial creation. Upon independence, most colonial capitals were transformed into national capitals. Some countries, however, like Botswana (Gaborone) and Mauritania (Nouakchott) created new capital cities. Since most of the coastal countries have maritime-oriented cities for capitals, these cities by the nature of their location are generally not geographically centered within the individual countries. Thus, they are in the margins of the country with external rather than internal orientations. Capitals with peripheral geographical locations have usually been unable to promote the kind of economic development and national unity that is needed in postcolonial Africa. As a result, some countries are

Nairobi, Kenya, is a modern African city with many high-rise buildings. Certain African cities initially possessed various distinctive characteristics such as architectural styles or city plans that were similar in character to those of the former colonial power. The impact of Britain on Nairobi is still discernible.

making plans to move their capitals inland to more desirable central locations.

Because so much of the rural-urban migration has been centered on the primate cities, they are being overtaxed to provide the necessary jobs and services. Thus there is some wisdom in attempting to develop new cities, or expand existing smaller ones, to help siphon off some of the potential migrants to the primate cities. Some countries like Tanzania, Nigeria, and Côte d'Ivoire (Ivory Coast) are developing new capital cities. Tanzania plans to move the capital from Dar es Salaam to Dadoma; Dadoma is not a new creation, but when improvements in transportation and other services and facilities are completed, the administrative functions at Dar es Salaam can be officially moved to the new capital. Yamoussoukro will replace Adidjan as the capital of the Côte d'Ivoire (Ivory Coast), and Akuja, a new city that opened in 1982, will eventually become the seat of government in Nigeria.

Much of the industrialization in Sub-Saharan Africa has taken place in the coastal cities and in the mineral-producing areas in South Africa and Zimbabwe and in the copper regions of Zambia and southern Zaire. Much of the industry and commerce in Africa is still closely tied to agriculture and the processing of raw materials and minerals. South Africa still remains the richest and most industrially advanced country in Sub-Saharan Africa.

AFRICA'S SUBREGIONS

Africa can be broadly divided into six major geographic subregions (Figure 13.15). First is *North Africa,* which includes those countries bordering the Mediterranean Sea—Egypt, Libya, Tunisia, Algeria, and Morocco (this area is included in the chapter on Southwest Asia). In addition, the northern portion of Sudan is culturally part of North Africa and the Arab world. This subregion contains several unifying factors such as Islam, the Arabic language, and the Arabic way of life. In addition, most of the countries face similar environmental and economic problems.

West Africa includes the interior northern tier of countries of Mauritania, Mali, Burkino Faso, and Niger, and the coastal states of Senegal, Gambia, Guinea-Bissau, Guinea, Sierra Leone, Liberia, Côte d'Ivoire (Ivory Coast), Ghana, Togo, Benin, and Nigeria. Sometimes Chad is included, but Chad is more properly regarded as a mixture of North, West, and Equatorial Africa. Most of West Africa was formerly controlled by France and Britain. Thus, French and English are the major unifying languages in the region. Historically, this area had a cultural vitality that was not extinguished by the colonial interlude, and persons living in the region today can identify with pride their historic ties to the former West African kingdoms and states like Mali, Songhai, Ghana, and Ashanti. Ecologically, the countries face similar geographical environmental problems as reflected in the parallel east-west extensions of specific climatic zones beginning with the tropical climates on the coast and extending north through the savanna, steppe, and desert.

Mauritania, Mali, and Niger and to some extent Burkina Faso (Upper Volta) receive little rainfall, and the majority of these countries lie in the Sahara Desert and the Sahel. These states are largely engaged in livestock grazing, but recurrent drought has wrought havoc with the carrying capacity of the land, causing widespread hunger and the loss of many animals.

The coastal states extend from Senegal to Cameroon. Because of increased precipitation, agricultural activities dominate. Cassava and other root crops, together with rice, fruit, and vegetables, are prevalent. Caloric intake for the region as a whole is still marginal in most countries. However, several economic development enclaves have developed that contribute substantially to the economy and serve as centers of migration for laborers who seek employment in these areas.

Some of the least developed nations in the world are located in West Africa. These include Benin, Burkino Faso, Gambia, Guinea-Bissau, Mali, Niger, Sierra Leone, and Togo. Several problems in these countries contribute to their underdevelopment: the environmental problems are extensive; fluctuations in weather conditions can be critical—particularly in single-ecosystem countries; diseases such as malaria, yellow fever, schistosomiasis (bilharzia), and onchocerciasis (river blindness) are prevalent; GNPs are low; levels of educational attainment are also low; and birth rates are high. The industrial sector in these countries is quite small and frequently focuses on the processing of mineral and agricultural products, such as beer, soft drinks, textiles, and leather goods, and small essential household items.

Some of the intermediate economically developed countries include Mauritania, Côte d'Ivoire (Ivory Coast), Ghana, Nigeria, Senegal, and Guinea. Some countries in this group are dominated by a single economic sector (i.e., minerals or agriculture), while others have a more balanced or diversified economy that combines mineral resources, forestry, and agriculture.

Liberia's economy has been built around rubber plantations and iron ore production. Liberia is Africa's leading exporter of iron ore, and approximately half of the labor force is employed in the rubber industry. Côte d'Ivoire (Ivory Coast) is one of the more prosperous West African countries and is a leading exporter of coffee, cocoa, and lumber. Ghana, as is typical of other West African countries, emphasizes the export of cash crops. Although it currently has economic problems, it is still a leading exporter of cocoa. Nigeria has a rather diverse economy. Cassava, cocoa, oil palm, and rubber are grown in the coastal regions, while millet, peanut, and cattle raising dominate in the northern portions of the country. The importance of petroleum has added

FIGURE 13.15

Regions of Sub-Saharan Africa

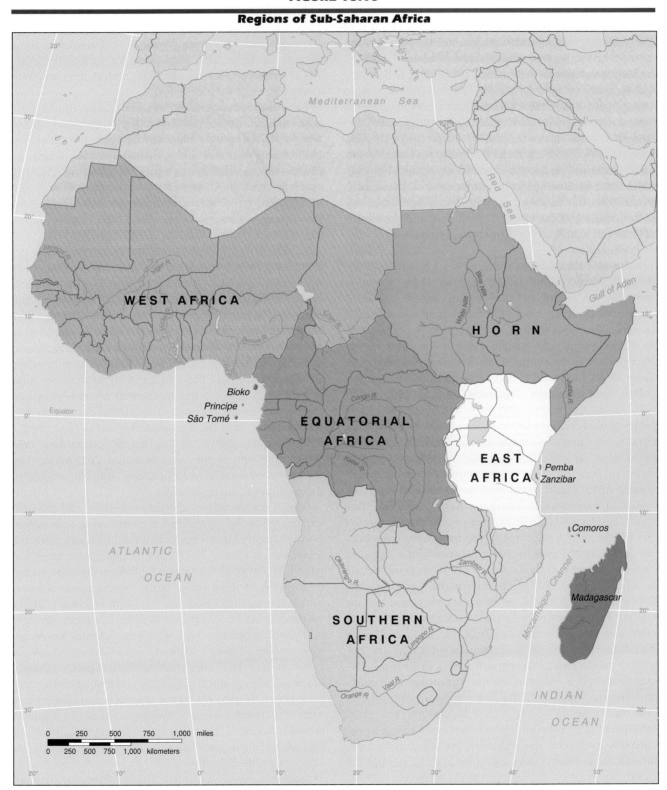

greatly to the economy of Nigeria, making the country one of the 10 leading exporters of petroleum and petroleum products. Nigeria is also the most populated of the African states with approximately 120 million inhabitants.

Equatorial Africa is comprised of Cameroon, the Central African Republic, Equatorial Guinea, Gabon, Congo (Brazzaville), and Zaire. Except for Zaire and Equatorial Guinea, this area was controlled by the French. Although France was never able to interconnect the various countries of the region effectively, it did leave various imprints upon the landscape, principally its language. A common geographical environment also ties these countries together. Subsistence agriculture, together with some commercial agriculture and the export of minerals and wood products, characterizes the region.

Zaire (905,063 square miles; 2,354,409 square kilometers) is the second largest state in Africa, excelled only by the Sudan (987,649 square miles; 2,505,813 square kilometers). Most of its inhabitants are engaged in agriculture, but mining also contributes significantly to the economy and enables Zaire to be classified as intermediate in economic status. On a worldwide scale, Zaire ranks high in the mining and export of copper, cobalt, and industrial diamonds. Gabon, particularly with its oil and timber resources, and Cameroon have fairly well developed economies. In contrast, Congo (Brazzaville) and the Central African Republic have limited resources, and their economies have lagged behind those of their neighboring equatorial states; they are among Africa's least developed countries.

East Africa includes Tanzania, Kenya, Uganda, Burundi, and Rwanda. Lake Victoria provides a central focus for the area since three of the major countries border it. Much of East Africa is a plateau with a savanna climate, but each of the countries has its own individual developmental problems. In Kenya the highlands are fertile and important for agriculture. Coffee and tea are leading export products. Tanzania, under the early leadership of Julius Nyerere, has been the leading African socialist state. Although Tanzania does not have the extensive moist highlands that are found in Kenya, agriculture is still a mainstay of the Tanzanian economy, and cotton, coffee and sisal are important cash crops. The economies of Tanzania and Kenya are classed as intermediate. In contrast, Uganda has been trying to reestablish itself after the chaotic reign of Idi Amin. Although the economy is in shambles and the people divided, the country is working toward recovery.

Burundi, Rwanda, and Uganda fall into the least developed countries category. Their resource bases are not large, production is low, and they are landlocked. Transportation routes are poor, and their reliance upon Kenya, Tanzania, and even Zaire for access to the sea for imports and exports places an added economic burden on these countries.

Ethiopia, Somalia, Djibouti, and Sudan form the *African Horn* subregion. The region sometimes suffers from a self-identity problem since it regards itself as part of the Muslim world and also part of black Africa. All of these countries fall into the least developed countries category.

Although Ethiopia is a large country with a varied environment, it has not been able to develop economically. Except for a brief interlude between 1935 and 1941 when the Italians controlled it, Ethiopia was not controlled by a colonial power, but it nevertheless has suffered and is still suffering from political turmoil. It has been plagued by a struggle between Christians and Muslims for political power. Although the province of Eritrea is part of Ethiopia, Eritrea entertains the idea of future independence, and this has also caused problems for Ethiopia. Recent drought and famine have ravished the countryside, but the country does have some good quality grazing land, and its mineral resources include copper, platinum, and potash. Industrial development is in its infancy, and the transportation network is poor.

Most of the people in Somalia are pastoralists; less than 1 percent of the land is under cultivation. The country is ethnically homogeneous, with 98 percent of the population being Somalis.

Somali **irredentism** (cultural extension and political expansion aimed at an ethnic group living in a neighboring country) is strong in Ethiopia, Djibouti, and to a lesser extent Eritrea and Kenya. Somalis have been engaged in a war in southeast Ethiopia's Ogaden region in an effort to incorporate many Somali nomads living in that territory into Somalia. Although Djibouti is about 89 percent desert wasteland with its economy strongly tied to trade services with Ethiopia, Somalia would ultimately like to annex Djibouti.

Sudan is a land of contrasts with a predominantly black populated south and an Arab north. Blacks comprise about 52 percent of the country's population and Arabs 39 percent. Most of the population (70 percent) is Sunni Muslim. The economy is based on agriculture and cotton is the leading crop along the irrigated lands adjacent to the Nile River.

Southern Africa includes Angola, Zambia, Malawi,

An aerial view of a village in Uganda. Most African countries are less than 50 percent urban. The majority of Africa's population remain dependent on agriculture for their livelihood and live in small towns and villages.

and Mozambique and all countries to the south of this northern tier of states including Namibia, Botswana, Zimbabwe, Swaziland, Lesotho and South Africa. It is an area where certain countries have rich mineral deposits, independence was late in coming for some countries, and the **apartheid** policy (now mostly dismantled) of South Africa has caused international concern.

After the Portuguese relinquished authority in Angola and Mozambique in 1975, political power struggles developed, and both countries came under the Soviet sphere of influence. In Angola, Soviet-backed MPLA (Movimento Popular de Libertacao de Angola) has tried to develop the country under difficult circumstances. But in central and southern Angola, rival UNITA (Uniao Nacional para a Independencia Total de Angola) forces, backed by South Africa and the United States, control most of the rural areas. Political stability and economic development are still somewhere in the future. In 1988, however, South Africa, Angola, and Cuba signed a cease-fire agreement providing for the withdrawal of South African and Cuban troops from Angola.

Angola has economic potential, but upon independence, many Portuguese technicians, professionals, and commercial farmers returned to Portugal. Cuban advisers moved in at that time to assist in the process of rebuilding, but production of coffee, cotton, sisal, and sugar and the mining of diamonds, iron ore, and manganese have dropped significantly since the civil strife

began in the mid-1970s. Petroleum production has been less affected by the civil war, however. Thus, the economic potential of Angola is high, but so far it has been hampered by political instability.

Mozambique derives much of its income from the sale of cash crops such as sugar, cotton, and cashew nuts, charges levied on neighboring countries for the use of rail and port facilities in Maputo, and revenues derived from South Africa for the use of its citizens who work in South Africa. Mozambique may be a little more stable than Angola, and race relations are possibly better than in South Africa. Portugal's early policy of *assimalado* (culturally assimilating blacks into Portuguese culture) did not prohibit intermarriage or integration of residential neighborhoods as South Africa's apartheid does. Nevertheless, many blacks did not choose to become assimilados.

The countries of Zambia, Zimbabwe, and Malawi were formerly under British control and were all unified under the name of the Federation of Rhodesia and Nyasaland (1953–1963). Eventually, Northern Rhodesia received independence and became Zambia (1964) and Nyasaland became Malawi (1964). Southern Rhodesia, later Rhodesia, and then finally Zimbabwe, became the "last stand" of European colonialism in southern Africa. In 1961 Southern Rhodesia drafted a constitution, and later Ian Smith issued a unilateral declaration of independence (UDI). Britain refused to recognize the UDI

The Nchanga Copper Mine in Kitwe, Zambia. Copper is Zambia's major export, but it is subject to fluctuations in world market prices. Because of its overdependence upon copper, the government has tried to diversify the Zambian economy by emphasizing agricultural development.

and imposed economic sanctions upon the country. Rhodesia worked at attaining self-sufficiency, but eventually white rule was broken, and Rhodesia became independent under black control and was renamed Zimbabwe in 1980.

Zimbabwe has an excellent resource base and is an important producer of chromite, coal, asbestos, copper, nickel, gold, and iron ore. Much of the mining and industrial development has taken place along the Great Dyke between Bulawayo and Harare. Kariba Dam on the Zambezi River supplies much of the country's energy needs, and tobacco and cotton are leading commercial crops.

Copper, mined in the north central portion of Zambia, accounts for more than 90 percent of the country's foreign exchange revenue. Zambia has also embarked on a program of economic diversification, however. Efforts

have been made to increase food production not only for internal consumption but also to increase exports of such crops as sugar, tobacco, and maize.

Malawi does not have the mineral resources found in Zambia or Zimbabwe; as a result, it continues to focus on agricultural exports such as sugar, tea, and tobacco. Industrial development and urbanization lag rather far behind neighboring countries. Population densities on the agricultural land are high since only about 14 percent of the population is urban.

Botswana, Lesotho, and Swaziland, formerly called the High Commission Territories, received their independence in 1966, 1966, and 1968, respectively. Botswana, formerly dominated by pastoral activities, has increased its GNP as a result of diamond and coal mining. Although part of its labor force is still tied to South Africa where many males migrate to work in the mines, the country is trying to develop more economic independence.

People in Lesotho and Swaziland have traditionally engaged in grazing activities. In addition, both countries are economically tied to South Africa and send several thousand workers there a year. Swaziland is more prosperous than Lesotho, has a somewhat larger resource base, and is developing small-scale consumer industries.

South Africa

In 1652 the Dutch East India Company established a trading station in Cape Town. The Dutch encountered the cattle-herding Khoi-Khoi (Hottentots) and within a short time had enslaved many of the men for use as laborers. In the early years, the enslavement of these people was based more on religious belief than on racial prejudice; in a short time, however, racial discrimination became an integral part of the Dutch policy.

The British arrived in South Africa near the turn of the nineteenth century, establishing themselves at the Cape in 1806. Soon friction developed between the British and the Boers (Dutch farmers). The British began to Anglicize the Cape and in 1833 outlawed slavery. The Boers believed their Afrikaans-speaking society was threatened and decided to move further inland. This Boer migration, which began in 1836, is known as the "Great Trek" and resulted in the eventual creation of Transvaal and the Orange Free State in 1852 and 1854, respectively. The conservative life-style of the Dutch was interrupted in 1867 with the discovery of diamonds near Kimberley and of gold near Johannesburg in 1886. The discovery of diamonds and gold caused the second and third treks as many people migrated into the Dutch territories to seek their fortunes. Friction continued between the Boers and the Anglos resulting in the Anglo-Boer War of 1880–1881, and the second and more important Boer War between 1899 and 1902. The British won, and a union of the republics (the British Cape and Natal and the Dutch Orange Free State and Transvaal) was formed in 1910. The British did attempt to be conciliatory toward the Dutch in some ways, however, and adopted the Boer view that only whites should participate in the political process. Thus, the Union of South Africa became a state ruled by an all-white electorate. Although blacks have not had a significant say about events that affected their welfare in southern Africa, blacks residing in Basutoland (Lesotho), Bechuanaland (Botswana), and Swaziland did not become part of the Union of South Africa, but became the "High Commission Territories" under the control of Britain. This status enabled these former British protectorates to achieve eventual independence in the 1960s, thus escaping political incorporation into South Africa.

In 1948, the National Party, led by Daniel Malan and dominated by Afrikaners, was elected to power in South Africa. Malan and his party had run on a platform that stressed racial separation (apartheid), economic prosperity, and a plan to make South Africa a republic and withdraw it from the British Commonwealth. The policy of apartheid ("apartness") called for the separation of the races (i.e., whites, Asians, Coloureds—persons of mixed race—usually black and white, and blacks) and for reserving separate lands for each group. Homelands (bantustans), proposed by the Tomlinson Commission, were established for Africans where their particular ethnic group was highly concentrated. Near urban centers, "townships" were established for blacks who were temporarily needed to work in mining, commercial, or industrial jobs. Thus, under the National Party, the policy of apartheid was implemented, the economy grew, and South Africa broke away from the Commonwealth and became a republic in 1961.

Under South Africa's governmental policy of separate development, 10 bantustans have been created, each of which is to be granted independence on becoming a viable political and economic state. South Africa's black population numbers more than 28 million, and the homelands occupy about 14 percent of the land area. More than 55 percent of the black population resides outside the homelands, however. Bantu-speaking Africans account for about 69 percent of the total population (over 40 million) in South Africa with whites amounting to about 7 million (18 percent) of which more than 50 percent are Afrikaners; Cape Coloureds account for 10 percent of the population and Asians (mostly Indians) 3 percent.

The first bantustan to be given independence was Transkei in 1976, Bophuthatswana followed in 1977, Venda in 1979, and Ciskei in 1981. Others scheduled for independence include Kwazulu, Lebowa, Gazankulu, KaNgwane, Ndebele, and Qwa Qwa (Figure 13.16). However, South Africa may not create any more new homelands since it recently repealed the Group Areas Act and the Population Registration Act—thus ending the classification of the population by race. As of 1991 most of the major apartheid laws have been removed, and the South African government, together with the

ANC (African National Congress), must agree on a new constitution which will give blacks the right to vote. Many problems remain, but the major "pillars" of apartheid have been dismantled. The world community continues to denounce South Africa, but the recent dismantling of apartheid may cause some countries to recognize South Africa and lift economic sanctions.

South Africa recently relinquished control of Namibia. Namibia, which was formerly controlled by Germany, became a League of Nations mandated territory entrusted to South Africa; in 1945 its mandate was transferred to the United Nations, which through the international courts challenged South Africa's control of Namibia. In October 1966, the United Nations terminated the South African mandate over Namibia, but South Africa did not accept the decision and continued to control Namibia as though it were a fifth province of the Republic. Recently, South Africa withdrew from Na-

mibia and it became an independent country. Although Namibia now has a pro-Marxist parliamentary majority, it is attempting to maintain a market-oriented economy.

Much of the industrial and urban growth in South Africa is concentrated in Johannesburg, the country's leading manufacturing center, Cape Town, Durban, Pretoria, and Port Elizabeth. The South African economy is built around black labor, and no major urban center has a white majority. South Africa has a strong economy, and it exports strategic minerals together with gold and diamonds. The Witwatersrand district near Johannesburg is important for gold. Much of the coal comes from Witbank and New Castle. Although many foreign companies are withdrawing their investments in South Africa (referred to as disinvestment), South Africa still has the largest GNP of any country south of the Sahara. Its major urban centers are linked by railroads, which bring blacks to work daily. Besides manufacturing and min-

FIGURE 13.16

South Africa and Its Homelands

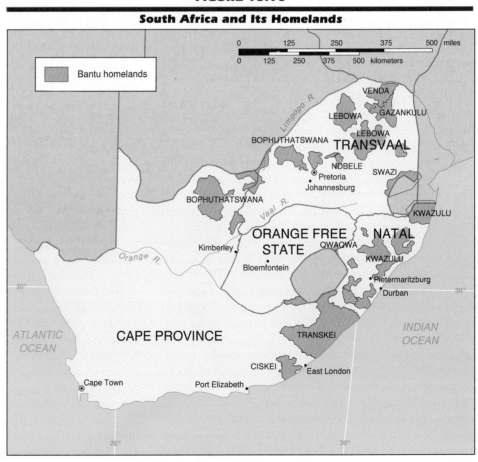

ing, the High Veld is a leading area for agricultural crops and livestock production. Corn, cotton, grapes, tobacco, and citrus fruits are some of the major crops.

The Islands

Some islands dot the west coast of Africa, but most are located in the Indian Ocean. Equatorial Guinea, located on the mainland, controls the island of Fernando Poo and several other small islands. St. Thomas and Prince Islands (São Tomé and Principe) received their independence from Portugal in 1975, as did the Cape Verde Islands.

The major islands in the Indian Ocean are Mauritius, Réunion, the Comoro Islands (Comoros), the Seychelles, and Madagascar. All are now independent from Britain or France except Réunion, which is still held by France. Madagascar's economy is based largely upon agriculture, with rice being the most important food crop and coffee the major export product.

SPATIAL CONNECTIVITY

ECONOMIC ASSOCIATIONS

Traditionally, individual countries in Sub-Saharan Africa have been intrinsically connected to their former colonial power, be it Britain, France, Belgium, Portugal, Germany, Spain, or Italy, by political, economic, and linguistic ties. Many Europeans migrated to the colonial possessions to become involved in agriculture, mining, and commercial activities. In turn, many Africans went to various European countries to study and then returned to assume leading roles in their home countries. Thus, much interaction occurred between Africa and Europe. Upon independence many countries modeled their new governments on European models, and many continued to be tied economically to Europe. As the newly independent states developed and progressed, however, many African countries tried to weaken their ties to Europe and attempted to develop their economies within their borders, with other neighboring African countries, and with the world community. Many African leaders recognized that continental economic and political integration must first be achieved if Africa was to extricate itself from the trappings of neocolonialism.

A street scene in Johannesburg, South Africa. Johannesburg, an important economic center and one of South Africa's largest cities, has a black African majority, a large white minority, and a small number of coloured and Asian inhabitants.

Following independence, African countries began to focus on African unity, self-sufficiency, and cooperation among persons of African descent. The ideals of Pan-Africanism and the early calls for Africa to unite were espoused by such leaders as Kwame Nkrumah, Sekou Toure, and Haile Selassie. Early in 1960, Nkrumah was instrumental in forming the "Union of African States" comprised of Ghana, Guinea, and Mali. He hoped that this experiment would lead to full continental unity, but the Union failed in 1963. During this time, African leaders were split on how to achieve unity. Some, like the Casablanca group, put political unity first, while others, like the Monrovia group, promoted economic associations. As the issues were debated, it became obvious that continental unity would be difficult, and smaller or regional federations were to be preferred. The Organization of African Unity (OAU), established in 1963, is now the leading continental organization promoting unity. All of Africa's black states belong except for South Africa. This organization helps to promote education, eliminate illiteracy, improve health standards, and promote economic regionalization; it is also committed to national liberation movements in those parts of Africa still struggling with colonialism.

Although African unity is an important rallying point and serves as a symbol of spiritual unity among black nation-states, practicality dictates that continental unity must first be fostered through regional organizations. Several examples of intraregional organizations exist in Sub-Saharan Africa.

Among the early regional economic organizations were the East African Community, the West African Economic Community, and the Customs and Economic Union of Central Africa (CEUCA). Some of these early organizations have collapsed, like the East African Community, while others have changed their names and modified their goals. However, the CEUCA, comprised of Cameroon, the Central African Republic, Congo, and Gabon, is still intact.

Currently, several regional groupings exist to enhance economic development. The Economic Community of West African States (ECOWAS) promotes regional cooperation in agriculture, industry, energy, trade, transportation, and telecommunications. Nigeria is the largest member in terms of population and per capita GNP. All of the members of CEAO (Communauté Economique de l'Afrique de l'Ouest; Economic Community of West Africa) are also members of ECOWAS. The major purpose of CEAO is to lower customs duties between members and promote economic development; however, since these goals are also shared by ECOWAS, CEAO may cease to exist in the future.

OCAM (Organization Commerce Africaine et Mauricienne), which promotes technical and cultural cooperation, used to be comprised of CEUCA members plus other Francophone-speaking countries in Central and West Africa. But since its founding in 1965, many countries have withdrawn their membership from this group.

In southern Africa, SADCC (South Africa Development Coordination Committee) was formed by Botswana, Zambia, Angola, Tanzania, and Mozambique in 1979. Subsequently, Lesotho, Malawi, Swaziland, and Zimbabwe joined. Now that Namibia is independent, it has been invited to join. Coexisting with this group is the South African Customs Union (SACU) to which South Africa and Namibia as well as Botswana, Lesotho, and Swaziland belong. Obviously, rival ideologies exist between these two organizations over economic development in southern Africa as South Africa attempts to keep the SACU members firmly under its control.

These regional groupings have as their objectives to promote unity, economic development, and independence and to withstand foreign domination. Because these regional groups are rather young and suffer from political immaturity and internal disunity, they have often been frustrated by the difficulties of achieving regional and continental integration. Geographical contiguity is frequently not enough to offset conflicting ideological differences, dependence upon former colonial ties, and economic inequalities between member states. Overcoming the long-established economic links to European powers to promote intercountry trade and development within Africa has not been easy. So even though these organizations exist to promote continental unity and economic independence, Sub-Saharan Africa is still a long way from having viable interconnections within and between countries and regional groupings.

TRADE

Petroleum has the largest percentage share of Sub-Saharan African exports. Other export items include metals and minerals, principally copper and iron ore, and coffee and cocoa are the leading items in the food and beverage category. With regard to imports, manufactured goods, machinery, and transport equipment are the leading items. Export and import trade figures reveal that the oil exporting states of Angola, Nigeria, and Gabon ranked high in export value (Table 13.1). Ghana, Côte d'Ivoire (Ivory Coast), and South Africa also ranked high, but most of the remaining states, particularly the low-income and semiarid states do not have high trade values, and most are experiencing trade deficits.

Sub-Saharan Africa's share of world trade is not large, and commodity price changes in petroleum and other export trade goods have affected the purchasing power and balance of trade for most countries. In some instances, there is little reason for optimism with regard to future expansion of exports and an increased role for Sub-Saharan Africa in world trade. Rapid population growth on the continent has increased consumption of food and goods, thereby decreasing the exportable surplus of some crops. Conversely, statistics show that Africa has been losing ground in its effort to feed itself, particularly since the mid-1960s (Figure 13.17). Increasingly, Sub-Saharan countries must import more food, and for the past 20 years food imports have risen about 1 percent per year. Trade restrictions imposed by various countries outside the region have affected its ability to penetrate certain markets. Moreover, African economies have been rather traditional and inflexible, making it difficult for them to diversify and produce products with rapidly growing markets.

How is Sub-Saharan Africa interconnected with the world markets? Table 13.2 shows the changes in export trade between Sub-Saharan countries and countries out-

TABLE 13.1

Trade: Imports and Exports (Millions of U.S. $), 1983

	IMPORTS	EXPORTS
LOW-INCOME SEMIARID		
Chad	$ 109[b]	$ 58[b]
Burkina Faso	288	57
Gambia	116	48[c]
Mali	344	167
Mauritania	227[c]	305[c]
Niger	442[b]	333[b]
Somalia	330[b]	199[b]
LOW-INCOME OTHER		
Benin	460[a]	40[a]
Burundi	187[c]	148[c]
Central African Republic	127[b]	109[b]
Ethiopia	875[c]	403[c]
Guinea	310[b]	410[b]
Guinea-Bissau	50[b]	12[b]
Lesotho	530[b]	40[b]
Malawi	279[c]	293[c]
Madagascar	387	296
Mozambique	635	132
Rwanda	279	80
Sierra Leone	171	119
Sudan	1,354	624
Tanzania	822	366
Togo	284	162
Uganda	270[b]	300[b]
Zaire	480[b]	569[b]
MIDDLE-INCOME OIL IMPORTERS		
Botswana	799[a]	413[a]
Cameroon	1,217	940
Ghana	2,534	2,029
Ivory Coast (Côte d'Ivoire)	1,340[c]	2,067
Kenya	1,361	980
Liberia	415	429
Senegal	1,039[c]	543[c]
Swaziland	550[a]	410[a]
Zambia	566[c]	1,648[c]
Zimbabwe	959[b]	1,008[c]
MIDDLE-INCOME OIL EXPORTERS		
Angola	636[c]	2,029[c]
Congo	807[b]	806
Gabon	853	1,975
Nigeria	13,440[c]	11,317[c]
MIDDLE-INCOME OTHER		
South Africa	14,528	9,671

SOURCE: United Nations, *World Statistics in Brief* (New York: United Nations, 1986).

NOTE: *a* = 1981; *b* = 1982; *c* = 1984.

TABLE 13.2

Destination of African Merchandise Exports (as a Percentage Share)

	INDUSTRIALIZED MARKET ECONOMIES		SUB-SAHARAN AFRICAN COUNTRIES		OTHER DEVELOPING COUNTRIES		CENTRALLY PLANNED ECONOMIES		CAPITAL-SURPLUS OIL EXPORTERS	
	1960	1979	1960	1979	1960	1979	1960	1979	1960	1979
LOW-INCOME COUNTRIES	76 w	64 w	6 w	8 w	16 w	22 w	1 w	3 w	1 w	4 w
Low-Income Semiarid	76 w	71 w	18 w	6 w	6 w	11 w	0 w	0 w	0 w	13 w
1. Chad	73	30	27	13	0	52	0	—	0	5
2. Somalia	85	18	0	1	15	2	0	(.)	(.)	80
3. Mali	93	68	7	15	0	17	0	(.)	(.)	0
4. Upper Volta (Burkina Faso)	4	75	96	9	0	16	0	(.)	0	0
5. Gambia	97	93	3	1	0	6	0	(.)	0	0
6. Niger	74	97	26	1	0	0	0	—	0	2
7. Mauritania	89	88	11	2	0	9	0	—	0	1
Low-Income Other	76 w	63 w	5 w	8 w	17 w	23 w	0 w	3 w	1 w	3 w
8. Ethiopia	69	72	4	(.)	20	11	1	7	6	10
9. Guinea-Bissau	—	29	32	22	—	38	—	1	—	0
10. Burundi	—	89	—	1	—	9	—	1	—	0
11. Malawi	—	84	—	12	—	4	—	1	—	—
12. Rwanda	—	80	—	4	—	16	—	1	—	(.)
13. Benin	90	89	8	2	0	8	2	1	0	(.)
14. Mozambique	29	43	5	4	66	45	(.)	1	(.)	7
15. Sierra Leone	99	98	1	1	0	1	0	—	0	(.)
16. Tanzania	74	57	4	4	21	36	1	2	0	1
17. Zaire	89	64	5	26	6	10	(.)	(.)	0	(.)
18. Guinea	63	69	10	3	9	26	18	—	0	2
19. Central African Republic	83	78	9	2	8	20	0	(.)	0	(.)
20. Madagascar	79	67	18	4	2	29	1	(.)	(.)	(.)
21. Uganda	62	67	7	3	31	27	0	1	0	2
22. Lesotho	—	—	—	—	—	—	—	—	—	—
23. Togo	74	67	26	8	0	17	0	8	0	—
24. Sudan	59	36	2	(.)	27	45	8	9	4	10

TABLE 13.2 (continued)

Destination of African Merchandise Exports (as a Percentage Share)

	INDUSTRIALIZED MARKET ECONOMIES		SUB-SAHARAN AFRICAN COUNTRIES		OTHER DEVELOPING COUNTRIES		CENTRALLY PLANNED ECONOMIES		CAPITAL-SURPLUS OIL EXPORTERS	
	1960	1979	1960	1979	1960	1979	1960	1979	1960	1979
MIDDLE-INCOME OIL IMPORTERS	88 w	75 w	3 w	9 w	7 w	13 w	3 w	3 w	0 w	0 w
25. Kenya	77	63	7	21	15	15	0	(.)	(.)	1
26. Ghana	88	70	2	2	3	15	7	13	(.)	(.)
27. Senegal	89	59	4	27	7	14	0	(.)	0	(.)
28. Zimbabwe	—	—	—	—	—	—	—	—	—	—
29. Liberia	100	86	0	(.)	(.)	14	0	(.)	0	(.)
30. Zambia	—	82	—	2	—	16	—	(.)	—	(.)
31. Cameroon	93	84	3	6	3	8	1	2	(.)	(.)
32. Swaziland	—	—	—	—	—	—	—	—	—	—
33. Botswana	—	—	—	—	—	—	—	—	—	—
34. Mauritius	97	95	2	4	1	1	0	0	0	0
35. Côte d'Ivoire	84	78	3	6	13	11	0	5	0	(.)
MIDDLE-INCOME OIL EXPORTERS	89 w	84 w	2 w	3 w	8 w	13 w	1 w	0 w	0 w	(.)w
36. Angola	64	33	7	(.)	27	66	2	0	0	1
37. Congo	93	72	(.)	1	7	27	0	(.)	0	(.)
38. Nigeria	95	87	1	2	3	11	1	(.)	0	(.)
39. Gabon	87	60	6	8	7	32	0	0	0	(.)
SUB-SAHARAN AFRICA	82 w	78 w	4 w	5 w	12 w	15 w	2 w	2 w	(.)w	1 w
All low-income countries	51 w	61 w			29 w*	29 w*	19 w	5 w	1 w	5 w
All middle-income countries	68 w	67 w			24 w*	26 w*	8 w	4 w	(.)w	3 w
Industrialized countries	67 w	69 w			30 w*	24 w*	3 w	3 w	(.)w	4 w

*Includes exports destined for Sub-Saharan Africa.

NOTE: *w* = weighted mean.

SOURCE: World Bank, *Accelerated Development in Sub-Saharan Africa* (Washington, D.C.: The World Bank, 1981).

FIGURE 13.17

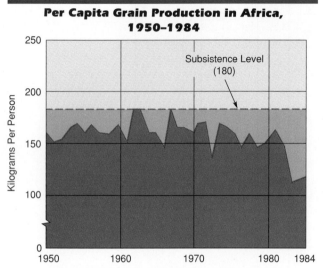

Per Capita Grain Production in Africa, 1950–1984

SOURCE: L. R. Brown and E. C. Wolf, *Reversing Africa's Decline,*
Worldwatch Paper No. 65 (Washington, D.C.: Worldwatch Institute, June
1985), p. 20.

side Africa between 1960 and 1979. Although a slight increase from 4 to 5 percent was experienced in merchandise exports among Sub-Saharan African countries between 1960 and 1979, exports to the industrialized market economies fell slightly from 82 to 78 percent. Exports to other developing countries rose from 12 to 15 percent, while trade to the centrally planned economies and capital-surplus oil exporting countries remained about the same.

Upon further analysis of the data in Table 13.2, it is evident that several countries such as Mali, Malawi, Zaire, Kenya, and Senegal increased their exports to Sub-Saharan countries rather dramatically, while Burkina Faso (Upper Volta), Niger, Guinea-Bissau, and Togo declined significantly. Most countries dropped in exports to industrialized market economies and increased their exports to other developing countries. Although these statistics do not reflect drastic changes, they do show a slow decline in the reliance of Sub-Saharan countries upon industrialized market countries for their exports.

The tourist industry is also an important variable in measuring interconnectivity. Some countries have built a stronger tourism trade than others. Kenya, Botswana, Senegal, Côte d'Ivoire (Ivory Coast), and Zimbabwe have averaged more than 200,000 tourist arrivals annually (Table 13.3). Kenya and parts of southern Africa have a number of national game parks that attract visi-

tors to see the wildlife. Most of the tourists visiting Africa are Europeans and North Americans, but there is also a major local tourist industry in southern Africa. Views differ on the contribution that tourism makes to the economy. Proponents stress that it brings in foreign exchange earnings, provides employment, improves the infrastructure, and promotes wildlife conservation. Critics argue that the earnings from tourism are small and benefit only a small sector of the population, and that the jobs created are menial. The reality of the situation is that both views have their merits, and while Africa will continue to exploit the tourist industry, it is only a partial contributor to the overall economic development strategy.

TRANSPORTATION

Besides overcoming the political obstacles to increased trade and economic development, the Sub-Saharan countries must improve their transportation infrastructure. Many trade routes were established in precolonial times. When the Europeans arrived, they frequently established trading posts along the coastal areas and connected them to precolonial sites in the interior or to newly created centers where resources were being developed. Thus intercountry connectivity remained weak throughout the colonial period. But as the model in Figure 13.18 reflects, in the modern period countries have attempted to improve their transportation networks with better connections in the hinterland.

The existence of several port cities along the coastal areas of Sub-Saharan Africa with rail connections into the interior is a legacy left from colonialism. These port cities provide important links in the export-import trade between Africa and the world market. Most of the coastal countries from Senegal to Cameroon in West Africa boast leading seaports such as Dakar, Freetown, Abidjan, Lagos, Port Harcourt, and Douala. Cape Town, Port Elizabeth, Durban, Maputo, Beira, Dar es Salaam, and Mombassa are important seaports in southern and East Africa (Figure 13.19). Some of these ports need to expand their berthing capacity and improve their container-handling facilities and shipping services for better regional distribution; the collection of containers also needs to be improved.

Railways

Southern African has the largest interconnected rail network in Africa. It not only connects the mineralized

TABLE 13.3

Tourists and Travel in Sub-Saharan Africa

	TOURIST ARRIVALS, 1983 (IN THOUSANDS)	MOTOR VEHICLES IN USE, 1982 (NUMBER PER 1000 INHABITANTS)
LOW-INCOME SEMIARID		
Chad	—	—
Burkina Faso	45[b]	3.0
Gambia	37	10
Mali	28[b]	—
Mauritania	—	—
Niger	21	—
Somalia	39	—
LOW-INCOME OTHER		
Benin	48	—
Burundi	22	1.6
Central African Republic	7	—
Ethiopia	56	1.3
Guinea	—	—
Guinea-Bissau	—	—
Lesotho	35	4[a]
Malawi	49	2.2
Madagascar	15	—
Mozambique	—	—
Rwanda	—	1
Sierra Leone	53	—
Sudan	22	8
Tanzania	100	3[d]
Togo	103	—
Uganda	13	0.8
Zaire	25	—
MIDDLE-INCOME OIL IMPORTERS		
Botswana	306	10
Cameroon	112	—
Côte d'Ivoire	202	29[c]
Ghana	44	9[a]
Kenya	366	13[c]
Liberia	—	—
Senegal	233	—
Swaziland	88	25[c]
Zambia	127	12[a]
Zimbabwe	200	31[c]
MIDDLE-INCOME OIL EXPORTERS		
Angola	—	23[c]
Congo	43	—
Gabon	16	14
Nigeria	103	14[c]
MIDDLE-INCOME OTHER		
South Africa	—	130[c]

SOURCES: United Nations, *World Statistics in Brief* (New York: United Nations, 1986); *United Nations Yearbook: 1983–84* (New York: United Nations, 1986).

NOTE: *a* = 1981; *b* = 1982 *c* = 1983; *d* = United Nations estimate.

FIGURE 13.18

Circulation Systems and Interconnectivity

The organization of space is a critical dimension of modernization and economic development. This simplified model illustrates the continuing imprint of spatial systems originally developed in the colonial period on the organization of life and the economy in modern Africa. In the precolonial period (1), local centers dotted the African coast, taking advantage of the abundant food supplies found where land meets sea. In the colonial period (2), selected local centers became trading posts for the export of gold, ivory, and slaves. In many colonies, railroads or roads were built into the interior to connect areas of agricultural and mineral production with port cities which were emerging as major centers of economic activity. In the late colonial period (3), transportation networks were extended deeper into the interior, and connections with other major centers along the coast were strengthened. After independence (4), the modern states of Africa made substantial efforts to develop fully articulated national economies with fully developed circulation systems. The persistence of the colonial patterns of transportation and communication, however, has been virtually indelible. (Adapted from H. C. Weinand, "A Spatio-Temporal Model of Economic Development," *Australian Geographical Studies*, vol. 10, 1972, pp. 95–100.)

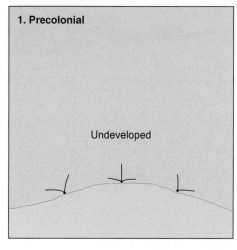

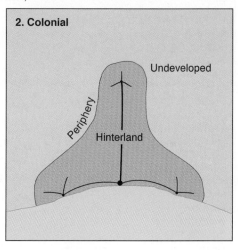

—— Major tranportation routes   ⊙ Major center

—— Minor transportation routes   ● Regional center

· Local center

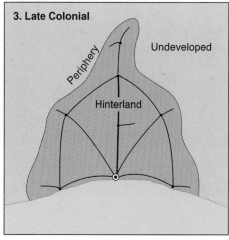

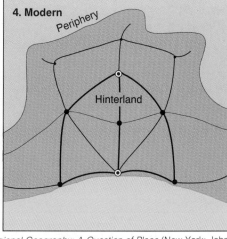

SOURCE: Paul W. English and James A. Miller, *World Regional Geography: A Question of Place* (New York: John Wiley & Sons, 1989), p. 531 (used by permission).

zones in the Durban-Johannesburg corridor, but also extends to the Harare-Bulawayo axis in Zimbabwe and to the copper belt to Zambia. As Zambia tried to reorient itself away from the South African sphere of influence, it was successful in getting the Chinese to construct a railroad (Tazara-Tanzania-Zambia Railway) from the copper belt region to Dar es Salaam. The railroad has not been particularly economically successful, however.

In East Africa, rail connections from Mombassa and Dar es Salaam into the interior have aided interconnectivity in the region. And in West Africa, rail extensions from the interior to coastal port cities exist throughout the region with Dakar, Abidjan, Port Harcourt, and Douala being examples. In Central Africa, rail and water routes extending from the minerally rich Shaba region in Zaire have also been important. Despite these important rail routes, examination of railways in Sub-Saharan Africa will show the basic lack of interconnectivity between countries (Figure 13.19).

Railways are important in areas where mineral production is significant. Not only do some new rail lines need to be built and old ones refurbished, but the operating efficiency of the government-owned railways needs improvement. More training of management and staff personnel is needed to improve operations in order to be more competitive with other modes of transport.

Roads

Besides rail and water networks, road transport has been important in African development. Both the construction of new roads and the paving of existing ones have been intensified during the past several years. Road transport is now the fastest growing means of transportation, and the number of passenger cars and commercial vehicles has risen steadily during the past two decades (Table 13.3). South Africa has the most motor vehicles in use and the highest number of vehicles (130) per 1000 persons. Zimbabwe and Côte d'Ivoire (Ivory Coast) were above 25 vehicles per 1000. For comparison, the United States has 700 motor vehicles per 1000 inhabitants.

Car ownership is increasing, and as countries work to improve intra- and intercountry connections, improvement in highway development has indeed been significant. But most roads run parallel with, and duplicate, routes already served by rail. This network helps to strengthen trade patterns outside the country but intra- and intercountry connections remain poor. The planned Trans-African Highway connecting Lagos, Nigeria, to Mombassa, Kenya, which would improve intercountry connectivity, is still far from completion.

More commercial vehicles are needed, and if agricultural production is to be increased, roads must be improved to enhance access to the rural areas. More feeder roads connecting to the main trunk network are needed. Locally available human resources, equipment, and supplies should be involved in the building and maintenance of these roads.

Air Transport

The growth and development of national airlines in Sub-Saharan Africa has been significant during the past 20 years. Most countries have tried to develop their own airline. Air networks for domestic and international flights have risen dramatically. Connections to Europe, North America, and the Soviet Union are good. Interconnectivity of air routes linking one African country with another is still poor, however, and few transcontinental east-west routes exist. Frequently, Africans trying to fly from east to west Africa have to go by way of Europe. Since traveling by air is still relatively expensive, most of the passengers are government officials, business executives, and diplomats. But small aircrafts have the potential for opening up isolated areas, providing them with skilled technicians and access to spare parts and other essentials. If more growth is to occur in domestic air transport, governmental controls on national airline tariffs may need to be loosened.

TELECOMMUNICATIONS

To further increase intra- and intercountry connectivity, greater improvements are needed in telecommunications. Interpersonal telecommunications such as telephone, postal, radio, newspaper, and television services are still quite underdeveloped. Some of the more moderately developed countries such as Zimbabwe, Swaziland, Botswana, Kenya, Zambia, Gabon, and the Congo average more than 10 telephones per 1000 persons. A similar pattern exists for the number of television receivers with Côte d'Ivoire (Ivory Coast), Gabon, Zambia, Liberia, and Zimbabwe averaging more than 10 receivers per 1000 inhabitants. Radios are more prevalent, and of the 39 countries listed in Table 13.4, 11 average more than 100 radios per 1000 inhabitants. South Africa, as expected, ranks high in telephone, television, and radio receivers per 1000 inhabitants.

With regard to daily newspaper circulation, nine

FIGURE 13.19

Transportation Networks in Sub-Saharan Africa

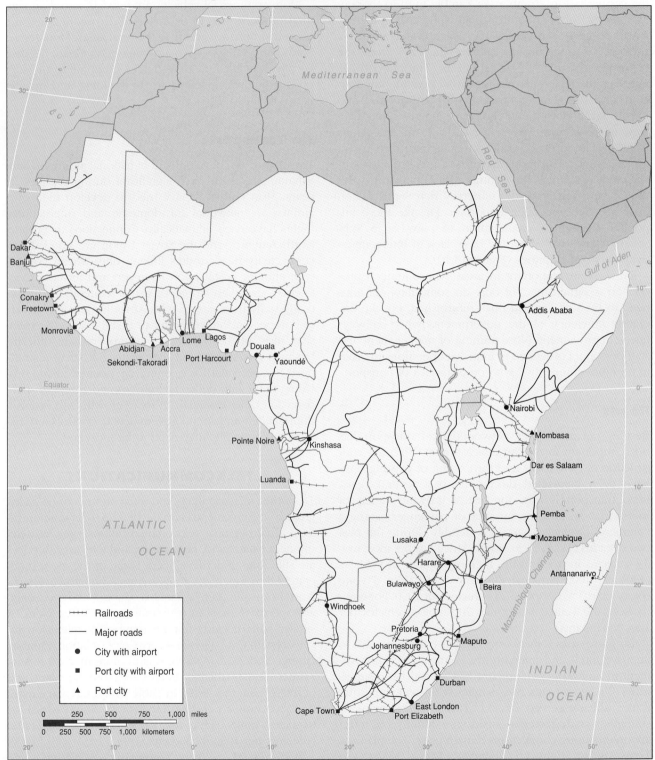

countries have a circulation of more than 10 newspapers per 1000 persons. Tanzania and Ghana rank high in readership for nondaily newspapers.

IMPORTANCE OF SPATIAL CONNECTIVITY

The importance of intracountry development of transportation and telecommunications systems, in addition to intercountry connections, cannot be overemphasized if Africa is to achieve any sophistication in economic development and interregional connectivity. The impact of colonialism has certainly imprinted the existing spatial structural patterns of transport and communication as early networks connected important agricultural or mineral resources. Usually, this meant connecting places in the hinterlands to important coastal port cities. This early pattern of development did little to promote intracountry lateral connections or intercountry movement. Thus various cores of economic development or growth poles occurred along coastal areas, producing regional inequalities within countries. Thus, the early transport network greatly influenced and promoted growth in selected areas.

This pattern of growth and development along the transport lines and in the growth centers should not necessarily be regarded entirely negatively since the initial colonial structure provided the impetus upon which countries today can build to promote further development. The task ahead for most countries is to expand the rail, road, air, and telecommunications networks within and between countries to promote better social and economic development while at the same time providing for better linkages to reduce the disparities between rural and urban Africa. Priority must be given to roads linking landlocked countries with the sea. Future economic development and interregional trade within Africa are clearly tied to improvements in transportation and communications.

PROBLEMS AND PROSPECTS

Sub-Saharan Africa's problems and prospects revolve around the recurring themes of the environment and natural resources, population growth and its consequences, political structure and stability of the various governments, economic growth and development, and increases in the quality of life for its inhabitants.

THE ENVIRONMENT

Much of Sub-Saharan Africa has a tropical climate and experiences irregular amounts of rainfall. Droughts and famine are frequently associated with the Sahelian countries. Even though deficiencies in rainfall are common, human pressures have been placed upon the land. Overgrazing, expansion of cleared land areas (which affects evaporation rates), and an overpopulation of animals have contributed to the drought problem. Government action to assist in reforestation and controlled utilization of pastures and fields may help to stem the spread of the desert, but desertification is a problem that must be dealt with, yet appears to have no short-term solutions.

Most of the soils are rather fragile, and soil conservation and more productive farming techniques need to be practiced. Soil erosion must be controlled, watersheds better managed, and new methods of crop rotation, seed production, and protection need to be implemented. The continuance of population pressure upon the land has shortened the fallow periods, which are needed to help regenerate soil fertility, thus diminishing certain soil capabilities. More use of manure and other fertilizers is also needed.

Associated with good conservation practices is reforestation. Wood supplies are being diminished at a rapid rate. An increase in reforestation practices will retard ecological deterioration and improve the fuelwood supply. Thus, rapid population growth combined with the expansion of agriculture onto marginal lands, declining agricultural productivity, severe deforestation and erosion, and misguided agricultural policies, as well as the inability to afford technological advances in farming, have all contributed to a degradation of the environment and few improvements in the living standard for most Africans.

POPULATION GROWTH AND HUMAN RESOURCES

That population is rapidly expanding in Sub-Saharan Africa is not an unknown fact. At its current rate of growth (2.7–3.0 percent), the population could double in about 25 years. Sub-Saharan Africa has the highest rate of growth of any continental area in the world. The magnitude of the consequences associated with this growth are rather enormous. Rural populations are predicted to swell despite migration to the urban areas. As rural-urban migration increases, normally there is a reduction in the number of people engaged in agricultural activi-

TABLE 13.4

Communications in Sub-Saharan Africa

	TELEPHONES IN USE, 1983 (NUMBER PER 1000 INHABITANTS)	TELEVISION RECEIVERS IN USE, 1982 (NUMBER PER 1000 INHABITANTS)	RADIOS, 1982 (NUMBER PER 1000 INHABITANTS)	DAILY NEWSPAPER, 1982 (CIRCULATION PER 1000 INHABITANTS)	NONDAILY NEWSPAPERS, 1982 (CIRCULATION PER 1000 INHABITANTS)
LOW-INCOME SEMIARID					
Chad	—	—	26	—	—
Burkina Faso	2[b]	3.0	19	—	—
Gambia	—	—	120	—	—
Mali	1[b]	—	16	—	—
Mauritania	3	—	98	—	—
Niger	2[b]	1.2	48	1	2
Somalia	—	—	25	—	—
LOW-INCOME OTHER					
Benin	—	2.8	75	0	—
Burundi	1	—	37	—	—
Central African Republic	—	0.5	56	—	—
Ethiopia	3	1.0[a]	92	1	0
Guinea	—	1.4	29	—	—
Guinea-Bissau	—	—	45	10	—
Lesotho	—	—	28	33	8
Malawi	5	—	46	5	—
Madagascar	—	8	206	6	0
Mozambique	4[a]	0.1[a]	25	4	—
Rwanda	—	—	30	0	—
Sierra Leone	—	6	177	3	—
Sudan	4[a]	6	75	—	5
Tanzania	5	0.4	28	—	36
Togo	4	3.4	209	—	—
Uganda	4	6	22	2	—
Zaire	1[b]	0.4	—	—	—
MIDDLE-INCOME OIL IMPORTERS					
Botswana	18	—	116	22	4
Cameroon	—	—	90	4	—
Côte d'Ivoire	—	41	128	8	—
Ghana	6[a]	6	172	—	24
Kenya	13	4.1	34	12	—
Liberia	—	11[c]	173	—	—
Senegal	—	0.8	62	8	—
Swaziland	23[a]	2.7	152	16	—
Zambia	12	12	26	18	—
Zimbabwe	33	11	40	21	—
MIDDLE-INCOME OIL EXPORTERS					
Angola	—	4.3	20	7	—
Congo	11	2.5	63	14	—
Gabon	11	27	176	27	—
Nigeria	7[a]	6	80	—	—
MIDDLE-INCOME OTHER					
South Africa	134	71	274	66	—

SOURCES: United Nations, *World Statistics in Brief* (New York: United Nations, 1986); *United Nations Yearbook: 1983–84* (New York: United Nations, 1986).

NOTE: *a* = 1981; *b* = 1982 *c* = 1983; *d* = United Nations estimate.

ties. Agricultural practices become more mechanized and productive allowing for a smaller work force, which in turn pushes persons out of the rural areas. Concordant with this are increased employment opportunities in the urban areas that pull rural inhabitants to the cities. However, this classic scenario may not be totally operational in Sub-Saharan Africa. Since much of the agriculture practiced is overwhelmingly subsistence-oriented farming, increases in crop yields through mechanization and fertilization may not fully materialize.

Because of the expected rapid population growth in the rural areas, increases in urban migration (although large) may not offset population growth in the rural areas. This, in turn, will apply more pressure to the land and natural resources to support the growing rural and urban population.

Similarly, the cities may not be able to generate enough jobs for the newly arrived migrants. Their demands for housing and the basic amenities of water, sanitation, and electricity cannot be met. Transportation facilities, waste disposal, health care, and communication services may not be able to keep pace with the burgeoning population. All of these factors do little to encourage the rising expectations of urban migrants or enhance the quality of life in urban areas.

Food self-sufficiency may be difficult to attain in the immediate future. It will be extremely difficult to achieve much improvement in the living standards for the majority of the population if current trends continue: that is, if the population continues to grow at about 2.8 percent; if women continue to average more than six live births; if per capita incomes grow between 1 and 2 percent; if demand for food stays at the current rate of about 4 percent; and if agricultural production grows at slightly less than 2 percent.

Local governments need to continue to devote their attention to the issue of population growth and the development of human resources. More countries need to begin to slow down their fertility rates and move along in the "demographic transition." Fertility rates can be slowed through child spacing, accomplished through prolonged breastfeeding and abstinence and increased use of contraceptive techniques. So far, African leaders have not seriously considered the full implications of rapid population growth. Unless this problem is seriously addressed, there will be continued population pressure upon the land, rapid urbanization, declining quality of life, and little improvement in basic services.

POLITICAL STABILITY

Despite instances of struggle, decolonization in many countries was reasonably peaceful, but in the wake of independence there have been outbursts of violence, internal power struggles, and conflicts. As a result, several states have experienced coups d'etat and now have military governments. Whether military or civilian governments rule, they are faced with the issue of nationalism and the promotion of national integration stemming from the pluralism of African societies within their borders. They must also attempt to reorganize the civil service, decentralize administration, and institute better planning systems. The allocation of resources must be executed properly in order to solve economic and social problems, but corruption and diversion of resources to military spending to accomplish political objectives occur far too often. Most African countries allocate an average of about 3 percent of their GNPs for military expenditures, but some countries like Somalia, Guinea-Bissau, Zimbabwe, and the Congo average more than 5 percent. For comparison, the United States averages about 6.6 percent, while Iran, Iraq, North Korea, Nicaragua, Oman, Saudi Arabia, the Soviet Union, and Syria average more than 10 percent.

ECONOMIC DEVELOPMENT AND ACCESSIBILITY

Promotion of economic development and trade, together with increased accessibility through improvements in transportation and communication, are needed if the African states are to make continued progress. More growth in exports is needed. While much of Sub-Saharan Africa is dependent upon the export of primary products, statistics show a decline in the region's share of nonfuel exports. Improvements must be made to reverse this trend. Incentives must be used to increase the export of primary products as well as manufactured goods.

Transportation and communication remain a problem for most countries. If Sub-Saharan Africa is to modernize its economic structure and improve delivery of important social and human services, transportation and communication services need to be updated. Better transportation would facilitate the movement of goods and people within the various countries and would promote interconnectivity between the various states. As it is now, interconnectivity between cities and regions

MIGRATION AND REFUGEES: GEOGRAPHY OF EXILE

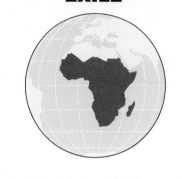

The movement of people is an integral component of the concept of spatial connectivity and interaction. Human beings have always tended to be rather mobile, and this is evidenced by the linguistic, religious, social, and racial mixing of a large portion of the world's population. The mobility of humans may be enhanced in some of the developed countries because of access to more and better transportation and communication facilities, but the basic fact remains that in all parts of the world "humanity is on the move."

Some movements are *cyclic;* they are journeys that begin and eventually terminate at the initial place of origin. For example, traveling to work, doing the daily or weekly shopping, going to church, or making social visits are all part of cyclic movements. Some types of movement are *periodic*. In this instance, the individual is temporarily away from his or her normal place of residence. Examples include students away at school, persons in the military service, and migrant laborers. *Migratory* movements are a third type; they involve a permanent change in residence for a substantial period of time in a new political unit. People may move from one political unit to another but remain within the same country, but if they legally move to another country, they are considered *immigrants*. If persons enter a country illegally, they may be referred to as *illegal aliens*. In other cases, persons may be *impelled* or *forced* to leave their homeland and enter another country because of racial, political, social, or religious persecution or because of some natural hazard such as a flood or drought. Most of the people moving under these circumstances are called *refugees.* Frequently, the host country will admit refugees conditionally with the hope that eventually they will be able to return to their homeland. Sometimes the return journey becomes possible, but in other instances the refugees remain and settle permanently into the larger population of the host country.

Political conflict, insecurity, economic conditions, and sporadic drought have been contributing factors prompting large numbers of persons to leave their home countries to seek comfort and security within another country in Sub-Saharan Africa. Many believe that Africa has more refugees and internally displaced persons than any other continent. Much of this refugee activity has occurred in southern Africa and in the northern portions of Sub-Saharan Africa, principally in the Horn and Sahel regions (Figure 2).

In southern Africa, political conflict in Mozambique and the war between the Angolan government forces and the South African–backed UNITA rebels have generated a significant movement of refugees from these two countries into neighboring states. Mozambicans have fled to Malawi, Tanzania, Zambia, Zimbabwe, Swaziland, and South Africa, with the largest number thought to be in Malawi. Associated with this movement are the problems the host country faces in providing protection and assistance for the refugees. Many refugee camps have become overcrowded, and security problems have prompted several countries like Tanzania and Zambia to encourage Mozambicans to settle further away from the areas that border Mozambique. Estimates vary, but it is believed that more than 600,000 Mozambicans are currently displaced in countries surrounding Mozambique. Similarly, many Angolans have fled to Zaire and Zambia, and in late 1987 Angolan refugees in these two countries were thought to number close to 400,000.

In the Horn of Africa, political turmoil and drought have caused many Ethiopians to flee from the famine-ridden areas to Somalia, Djibouti, and Sudan. Although some voluntary repatriation movements have occurred in Ethiopia as many international organizations attempt to transport food to the drought victims, many Ethiopians are still exiled in neighboring countries in the Horn.

Civil conflict in the southern part of Sudan has accelerated an exodus of refugees from this area to Ethiopia. Conflict in Uganda caused many Ugandans to move to Sudan, but a recent stabilization of conditions in Uganda has induced many refugees to return. Violence in the mid-1980s

FIGURE 2

Refugees in Sub-Saharan Africa

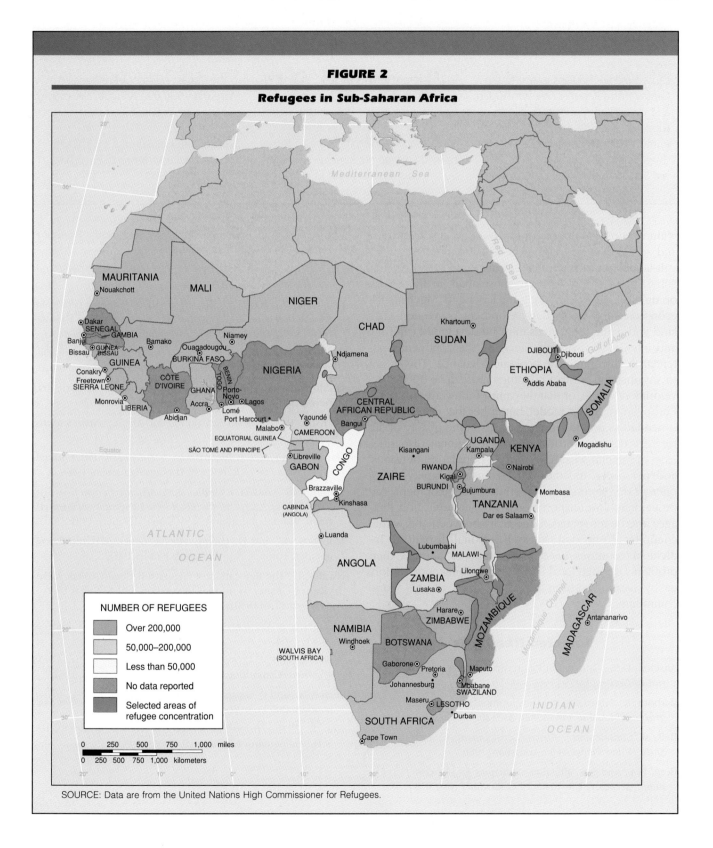

NUMBER OF REFUGEES

- Over 200,000
- 50,000–200,000
- Less than 50,000
- No data reported
- Selected areas of refugee concentration

SOURCE: Data are from the United Nations High Commissioner for Refugees.

caused many Chadians to flee to the Central Africa Republic, western Sudan, and Cameroon. But as the violence subsided, many Chadians returned from the Central Africa Republic. Many Chadian refugees still remain in western Sudan.

The refugee map of Sub-Saharan Africa will continue to change in the future as it has in the past. Internal conflicts, famines, and outbreaks of diseases will remain the moving forces behind refugee movements. Currently, Sudan and Somalia are host countries

to the largest number of refugees. More than 900,000 refugees are estimated to be in Sudan and over 700,000 in Somalia. More than 300,000 are estimated to be in Zaire, and Burundi and Tanzania each have more than 200,000.

within countries is poor, and without good connectivity and accessibility, development will continue to lag.

It is difficult to generalize about an area as diverse as Sub-Saharan Africa, but like most areas in the developing world, development is the key word. Much social, economic, and political development is needed. The land and natural resources must be developed carefully and intelligently without damaging the continent's potential or disrupting the ecological balance. This is not to say that the African countries must develop for development's sake, or that they must follow an American, European, or Soviet model, but rather that development is needed to provide improvements in the quality of life for Africans by Africans. But great regional disparities exist in mineral wealth, resources, population densities, and per capita incomes, which further amplify *development differentials,* or disparities in development. Improvements in transportation and communication linkages and better spatial connectivity between and within countries are needed to lessen regional disparities. Individual countries need to continue to address their problems through local initiatives, regional cooperation, or by assistance from other members of the world community.

Selected Readings

Altschul, D. R. "Transportation in African Development." *Journal of Geography* 79 (1980): 44–56.

Barbour, K. "Africa and the Development of Geography." *Geographical Journal* 148 (1982): 317–26.

Bell, M. *Contemporary Africa: Development, Culture and the State.* White Plains, N.Y.: Longman, 1986.

Berg, R. J., and J. S. Whitaker. *Strategies for African Development.* Berkeley: University of California Press, 1986.

Christopher, A. J. *Colonial Africa: An Historical Geography.* Totowa, N.J.: Barnes & Noble, 1984.

_____ . "Continuity and Change of African Capitals." *Geographical Review* 75 (1985): 44–57.

Franke, R. W., and B. H. Chasin. *Seeds of Famine: Ecological Destruction and the Development Dilemma in the West African Sahel.* Montclair, N.J.: Allanheld, Osmun, 1980.

Goliber, T. J. "Sub-Saharan Africa: Population Pressure on Development." *Population Bulletin* 40 (1985): 1–46.

Griffiths, I. L. L., ed. *An Atlas of African Affairs.* London and New York: Methuen, 1984.

Harrison, P. "A Green Revolution for Africa." *New Scientist* (May 7, 1987): 35–39.

Harrison Church, R. J. *West Africa: A Study of the Environment and Man's Use of It,* rev. ed. London: Longman, 1980.

Martin, E. B., and H. J. deBlij, eds. *African Perspectives: An Exchange of Essays on the Economic Geography of Nine African States.* London and New York: Methuen, 1981.

Mountjoy, A., and D. Hilling. *Africa: Geography and Development.* Totowa, N.J.: Barnes & Noble, 1987.

Nwafor, J. "The Relocation of Nigeria's Federal Capital: A Device for Greater Territorial Integration and National Unity." *GeoJournal* 4 (1980): 359–66.

O'Connor, A. M. *The African City.* New York: Holmes & Meier, 1983.

Oliver, R., and M. Crowder, eds. *The Cambridge Encyclopedia of Africa.* Cambridge: Cambridge University Press, 1981.

Senior, M., and P. Okunrotifa. *A Regional Geography of Africa.* London and New York: Longman, 1983.

Udo, R. K. *The Human Geography of Tropical Africa.* Exeter, N.H.: Heinemann Educational Books, 1982.

CHAPTER 14

China and East Asia

IMPORTANT TERMS

Dynasty

Confucianism

Unequal treaties

Four modernizations

Demilitarized zone (DMZ)

Double cropping

Chu'che

CONCEPTS AND ISSUES

■ Residual effects of dynastic rule and European colonialism on modern Chinese ideas and actions.

■ China's efforts to balance communist ideology with economic development and integration with the world economy.

■ China's relationships and its adjacent neighbors, especially the Soviet Union, Nationalist China, and a divided Korea.

CHAPTER OUTLINE

PHYSICAL ENVIRONMENT
Landforms
Climate, Soils, and Vegetation
Natural Resource Base

HUMAN ENVIRONMENT
Evolution of Chinese Society
Mongolia
Korea
Population and Culture
Political Situation
Economic Activity

SPATIAL CONNECTIVITY
Trade
Strategic Security

PROBLEMS AND PROSPECTS

O ver one-fifth of the world's population lives in China and the developing countries of East Asia (Figure 14.1). As the world's largest population cluster, with some of the most densely populated areas on earth, many of the problems of the developing world are intensified here. The emergence of Hong Kong, Taiwan, and South Korea as "latter-day Japans" and recent changes in the economic policies of the People's Republic of China appear to have put this region on the road to development. Especially in the case of China, which alone contains over one billion people, the twists and turns this new road takes will have a profound effect on relationships within the region as well as for the balance of world power.

In many ways, China is the regional antithesis of Japan. It is a large country with a large, mostly rural population. The resource base is vast, and the potential for exploitation of these resources is enormous. Yet, China lacks the capital and technology to develop its resources.

Recent developments in China, however, suggest major changes in its regional and global role. Since the death of Mao Zedong, China's leaders have taken a more balanced approach to economic development, as embodied in the policy of the "four modernizations," and placed greater emphasis on integration into the world economy. These changes have been coupled with an active policy of restricting population growth. At the same time, the repression of the democracy movement and the massacre in Tiananmen Square in June 1989 indicate that China's hard-line leadership will limit political freedom while continuing economic modernization and development. The future of East Asia will be closely tied to the development of China.

One should not forget the other countries of East Asia, each of which also contributes to the overall character of the region. For centuries these countries were part of or vassals to the Chinese empire. Today, they represent a wide range of political systems, nationalities, cultures, and levels of economic development.

Mongolia is the least developed, acting partially as a buffer between the Soviet Union and China. Here, a pastoral economy still prevails. Long tied to China, it has become a close ally of the Soviet Union in the twentieth century, and its economy is integrated into the Soviet economy.

Korea has historically been the bridge between China and Japan. Yet, it has its own unique character, distinct from the dominant Chinese and Japanese cultures. Korea is now divided into the communist north and capitalist south, providing a region of direct confrontation.

Taiwan has experienced economic growth that emulates that of Japan in many ways. Recently, official recognition of the People's Republic of China by the United States has politically isolated Taiwan and called into question its future as a sovereign and independent state.

Hong Kong has long been a symbol of economic wealth and is one of the last vestiges of European colonialism in Asia. Currently under British control, its return to China in 1997 raises serious concerns about its future. It will put to the test the new Chinese policy of "one country, two systems."

A portion of the Great Wall in northern China. The wall was built to keep foreign invaders out and maintain China's isolation.

STATISTICAL PROFILE: CHINA AND EAST ASIA

REGION OR COUNTRY	Population Estimate mid-1990 (Millions)	Natural Increase (Annual)	Population Projected to 2000 (Millions)	Infant Mortality Rate	Life Expectancy at Birth (Years)	Urban Population (%)	Per Capita GNP, 1988 (US$)	Area, Thousands of Square Miles (Km²)	Population Density, No./mi² (No./km²)
People's Republic of China	1119.9	1.4	1280.0	37	68	21	330	3705.4(9597)	302(117)
Hong Kong	5.8	0.8	6.3	7.4	77	93	9230	0.4(1)	14,534(5612)
Korea, North	21.3	2.1	24.9	33	70	64	1000*	46.5(120)	458(177)
Korea, South	42.8	1.0	46.0	30	68	70	3530	38.0(98)	1,125(434)
Mongolia	2.2	2.8	2.8	50	65	52	880*	604.3(157)	4(2)
Taiwan	20.2	1.2	22.1	17	74	71	5420*	12.5(32)	1,623(627)

* From *The Europa Year Book,* Vol. II (London: Europa, 1989); and *The World Fact Book 1990* (Washington, D.C.: CIA, 1990).

This chapter puts its greatest emphasis on the People's Republic of China, which is the world's third largest country in area with the world's largest population. North and South Korea and Taiwan are discussed in moderate detail with lesser attention given to Mongolia and Hong Kong.

PHYSICAL ENVIRONMENT

The physical features and patterns of climate, vegetation, and soils vary greatly throughout East Asia, as one might expect from a region that covers 4.6 million square miles (11.8 million square kilometers) or almost 9 percent of the earth's landmass. Of this, China contributes 3.7 million square miles (9.6 million square kilometers) or over 80 percent of East Asia, and thus will be addressed separately and in greater detail. The pattern of physical characteristics helps to explain the patterns of economic activity and population distribution.

LANDFORMS

China may be divided into two major geologic regions. These include the high plateaus and basins of the west and north, many of which drain to the interior, and the hills and plains of the east and south, which are drained by the east flowing rivers. Within these two regions, it is possible to identify six physiographic environmental

regions: the Tibetan Highlands, the Xinjiang-Mongolian* Uplands, the Northeast, the North, the South, and the Southwest (Figure 14.2).

Western China: High Plateaus and Basins

The Tibetan Highlands and the Xinjiang-Mongolian Uplands comprise the western and northern regions of high plateaus and basins. Together they account for approximately half the landmass of China.

The Tibetan Highlands, sometimes referred to as the "roof of the world," account for about one-fourth the landmass of China, although they contain less than 1 percent of the population. This region includes some of the great mountain ranges of the world: the Himalayas, with the highest mountain in the world—Mt. Everest at 29,028 feet (8850 meters); the Pamirs; the Karakorum Shan (mountains); Kunlun Shan; Nan Shan; and Qin Ling Shan. These mountains serve as the source region for some of the great rivers of the world. From the Tibetan Highlands, the Huang He and Chang Jiang (Yangtze) flow into China. The highlands are also the

*The Pinyin system for spelling place-names is based on the Mandarin Chinese pronunciation. It was adopted in 1958 to provide a uniform system for all China. According to the Pinyin system of transliteration x is pronounced hs, q is pronounced ch, and j is pronounced zh. For consistency, this chapter uses the transliterated Chinese words for river and mountains. In the north, "He" means river, e.g., Huang He; in the south, "Jiang" means river, e.g., Chang Jiang; and "Shan" means mountains, e.g., Qin Ling Shan.

FIGURE 14.1

China and East Asia

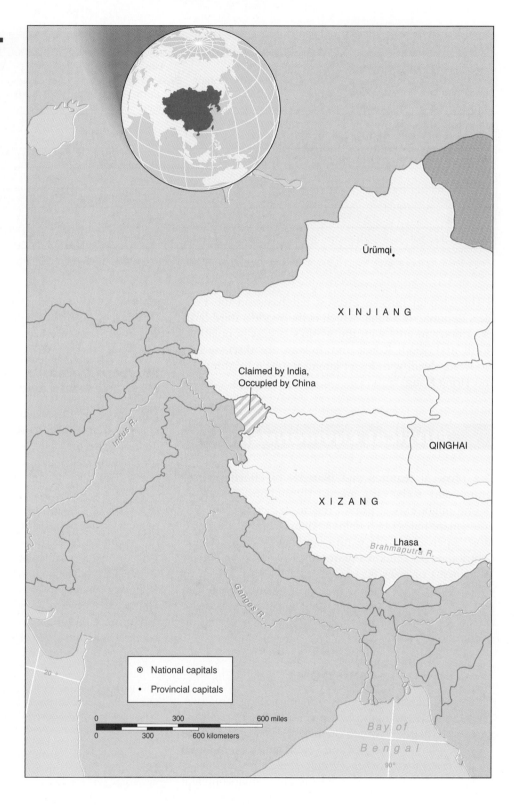

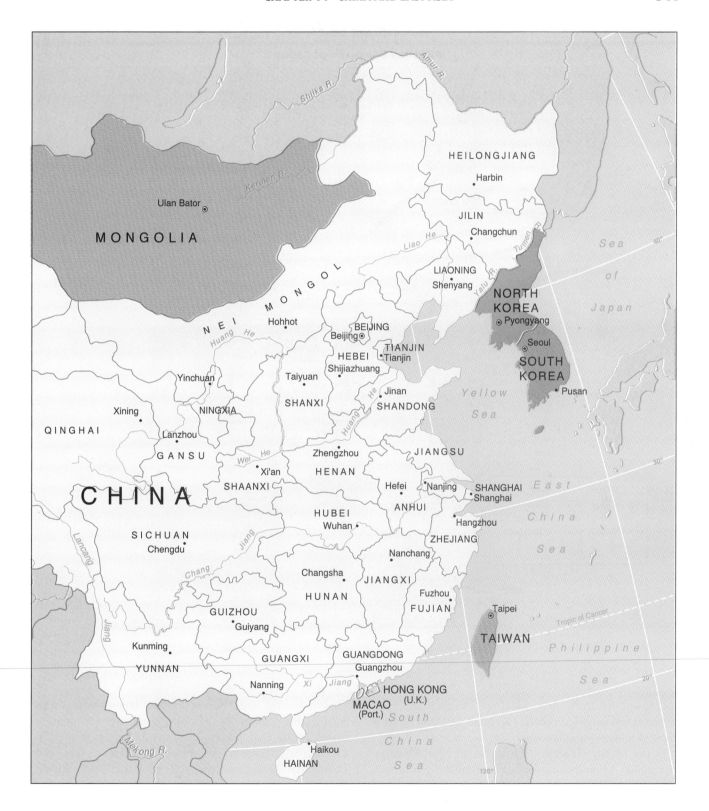

FIGURE 14.2

Physical Features of China

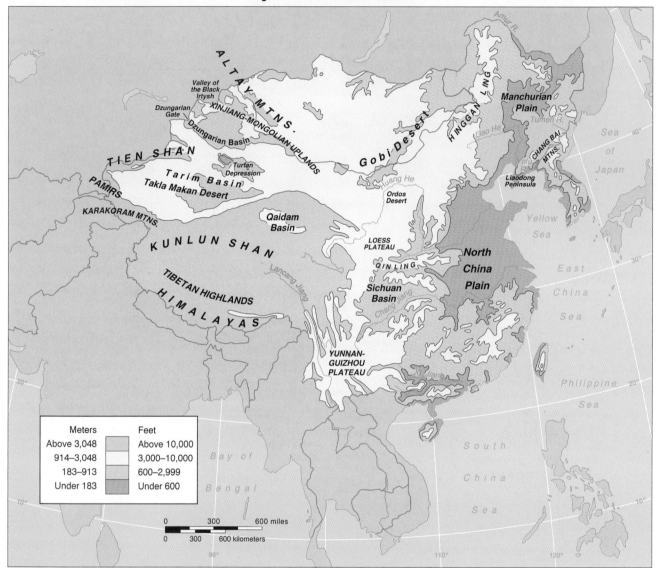

Meters		Feet
Above 3,048		Above 10,000
914–3,048		3,000–10,000
183–913		600–2,999
Under 183		Under 600

0   300   600 miles

0   300   600 kilometers

source of the Irrawaddy, Mekong, Brahmaputra, and Salween rivers. Additionally, the great Qaidam Basin is located in these highlands.

North of the Tibetan Highlands are the Xinjiang-Mongolian Uplands, another environmentally inhospitable region. Two great mountain ranges, the Tien Shan and the Altay, penetrate this region, which also encom-passes several large depressions or basins—the Tarim Basin, the Turfan Depression, and the Dzungarian Basin. In the Dzungarian Basin two gateways to the Soviet Union pass through these mountain ranges: the Dzungarian gates and the Valley of the Black Irtysh. Several vast desert areas are found in this upland region, including the Takla Makan, the Ordos, and the Gobi.

Eastern China: Hills and Plains

The hills and plains of eastern China comprise the cultural, agricultural, demographic, and industrial core of China. This section of the country can be subdivided into four environmental regions: the Northeast, the North, the South, and the Southwest.

The Northeast is dominated by the Manchurian Plain, a rich and productive agricultural region and source of many important resources. This region also includes the Hinggan mountains to the west and north, and the Chang Bai Shan along the border with Korea. Major rivers include the Amur along the border with the Soviet Union and the Liao, which empties near the Liaodong peninsula.

The North is dominated by the North China Plain. Also of importance are the Loess Plateau and the valleys of the Huang He and Chang Jiang, which are the main lifelines of the region. To the west lie the Qin Ling mountains, which may be the most significant geographic feature in China; they separate the northern, dry, wheat-growing areas from the southern, warm and humid, rice-growing areas.

The South is comprised primarily of the middle and lower drainage basins of the Chang Jiang, producing the fertile regions of the Chang Jiang Plain. Another particularly fertile area in this region is the delta of the Xi Jiang, near Hong Kong.

The Southwest is comprised of two major areas: the Sichuan Basin and the Yunnan-Guizhou Plateau. Sichuan is located in the drainage area for the Chang Jiang, while the southern hills and mountain regions benefit from the Xi Jiang, where one finds extremely productive agriculture.

The physical geography of Mongolia provides an inhospitable environment of contrasts. The west and north are dominated by mountains, while the southeast is dominated by the Gobi Desert, which extends into northern China.

Korean Peninsula and Taiwan

The Korean peninsula is dominated by mountains. Traveling south from the Yalu and Tumen rivers, which form the border between North Korea and China, one can identify six main land regions. These include the inhospitable northern and central mountains; the northwestern, southwestern, and southern plains; and the eastern coastal lowlands. Agriculture is most productive in the plains and lowlands, and the majority of the people live in these regions.

The island of Taiwan is also dominated by mountains, which form a north-south spine running the length of the country. On the east coast, steep slopes have pro-

Unusual landform features called karst towers near Guilin in south China.

hibited agriculture and settlement. Most of the productive areas and centers of population are located on alluvial plains and basins on the west coast.

CLIMATE, SOILS, AND VEGETATION

East Asia's climate is determined by the *monsoon*. The winter monsoon tends to be much stronger than the summer monsoon. In winter, the Siberian high over Mongolia produces strong, dry and cold northwesterly, northerly, and northeasterly winds. In summer, this high pressure system gives way to a low pressure system centered over South and Southwest Asia, providing warm, moist air to the continent by July and August. Typhoons and depressions (intense low pressure systems) bring hurricane-force winds, high seas, and heavy rain from July to September. Plateaus and mountains modify these influences to produce a basic climatic dichotomy between the arid west and the humid east.

In southern and southeastern China, Taiwan, and southern Korea, we find a warm and humid climate. This changes to temperate to cool and subhumid in northern and northeastern China and northern Korea.

Average annual precipitation generally decreases from southeast to northwest. South of the Chang Jiang it averages over 50 inches (127 centimeters) per year. North of the Huang He, it averages under 20 inches (51 centimeters). The maximum precipitation generally falls during the summer. Figure 14.3 shows the average annual precipitation and temperature ranges throughout East Asia.

There are two basic soil regions. To the north, one finds lime-rich, basic soils, and to the south, acidic soils dominate. Yet, the most important soils are the alluvial deposits of flood plains, deltas, and alluvial fans. Soil types have an important impact on the spatial distribution of agriculture. Rice and tea prefer acidic soils; maize and tobacco prefer neutral to slightly acidic soils; and wheat, sorghum, and cotton prefer basic soils.

Five major vegetation zones can be delimited in China. In eastern China, one finds tropical forests and savanna in the south, subtropical evergreen forests north to the Chang Jiang and west to the Tibetan Plateau, deciduous forests from the Chang Jiang north to the Amur Basin, and taiga in the Amur Basin. Western China is dominated by deserts and steppe.

Mountain soils dominate the soil regimes of Korea and Taiwan, covering over two-thirds of the landmass. Likewise, the primary kind of natural vegetation is the forest. As one moves north to south in Korea, the kinds of trees vary from coniferous to deciduous to mixed evergreen. Taiwan is dominated in the north and center by broadleafed evergreen forests and in the south by rainforests, with deciduous broadleafed and smallleafed trees and alpine shrubs at the higher elevations.

NATURAL RESOURCE BASE

Natural resources are unevenly distributed throughout East Asia. The fact that some countries have better access to resources than others is an important factor in economic development, although, as we have seen in the case of Japan, not the only one. To understand the development potential of the region, we shall examine the spatial distribution of arable land, minerals, and energy resources.

Arable Land

Eleven percent of the total land area of China is arable, only slightly below the world average of 12–13 percent. This compares to 20 percent for Korea and 25 percent for Taiwan. As we saw in Chapter 8, 16 percent of Japan is arable.

China's arable land is concentrated in four major plains: The northeastern plain, the North China Plain, the plains of the middle Chang Jiang, and Nan Ling. These regions account for the majority of China's 249 million acres (101 million hectares), which translate into 0.3 acres (0.12 hectares) of arable land per person. Taiwan's 2.3 million acres (923,000 hectares), which provide for 0.2 acres (0.08 hectares) per capita, are mostly concentrated in the plains and basins of western Taiwan. In Korea, arable land is found in the narrow valleys and small plains of the western and southern regions of the peninsula. These provide for 0.3 acres (0.12 hectares) per capita in North Korea and 0.1 (0.04 hectares) in South Korea.

With the exception of Mongolia, all of East Asia is a deficit region for food trade. Since World War II, China, Korea, and Taiwan have all experienced major famines. It is evident that lack of arable land, relative to the large population, is a major problem throughout all East Asia.

Mineral and Energy Resources

The mineral and energy resource base of China is enormous. To a great extent the potential wealth of this country is unknown. China has among the world's largest reserves of iron ore, lead, zinc, tin, bauxite and al-

FIGURE 14.3

Climatic Characteristics of East Asia

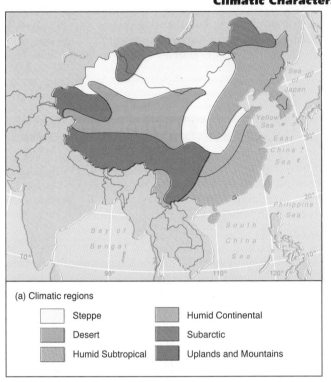

(a) Climatic regions

Steppe	Humid Continental
Desert	Subarctic
Humid Subtropical	Uplands and Mountains

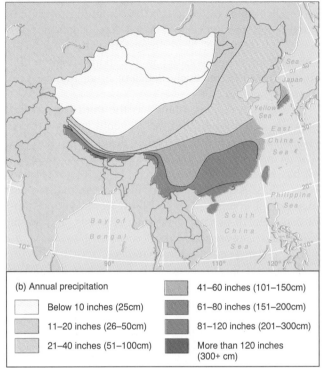

(b) Annual precipitation

Below 10 inches (25cm)	41–60 inches (101–150cm)
11–20 inches (26–50cm)	61–80 inches (151–200cm)
21–40 inches (51–100cm)	81–120 inches (201–300cm)
	More than 120 inches (300+ cm)

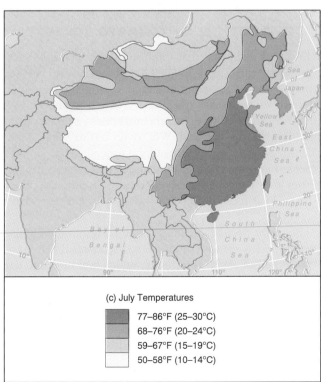

(c) July Temperatures

77–86°F (25–30°C)
68–76°F (20–24°C)
59–67°F (15–19°C)
50–58°F (10–14°C)

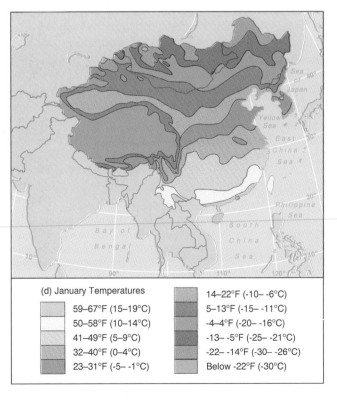

(d) January Temperatures

59–67°F (15–19°C)	14–22°F (-10– -6°C)
50–58°F (10–14°C)	5–13°F (-15– -11°C)
41–49°F (5–9°C)	-4–4°F (-20– -16°C)
32–40°F (0–4°C)	-13– -5°F (-25– -21°C)
23–31°F (-5– -1°C)	-22– -14°F (-30– -26°C)
	Below -22°F (-30°C)

umina shale, tungsten, antimony, mercury, manganese, magnesite, and molybdenum. China is apparently deficient, however, in copper, nickel, chromium, and platinum, all of which will be needed for continuing industrialization and modernization. Korea has significant reserves of iron ore, lead, tungsten, and magnesite, although more are located in the north than in the south. Taiwan is relatively deficient in all mineral and energy resources.

In energy resources, China has the third largest coal deposits in the world and the eleventh largest proved reserves of petroleum. The other developing countries of East Asia are deficient in these resources, although oil has recently been discovered off the South Korean coast. China is the only country in East Asia that is a net exporter of energy, whereas North Korea imports approximately 10 percent of its energy needs, and South Korea and Taiwan both import about 90 percent.

The production of China's mineral and energy resources is highly concentrated in the eastern half of the country (Figure 14.4). Iron ore production is dominated by Liaoning Province in the Northeast and the Nei Mongol autonomous region in the North. Although deposits are substantial, the inferior quality of the iron ore means that China must import iron ore and pig iron. Other nonferrous and ferro-alloy resources are found throughout eastern China.

Chinese energy is dominated by coal, which still accounts for over half of all energy consumption and approximately three-fourths of commercial energy use. Coal production is concentrated in the north, with one-third of all Chinese coal being mined in Shanxi Province. Petroleum has become an increasingly important source of energy. Production, however, is relatively backward in China, and future growth will depend on the discovery of new fields and the introduction of more advanced technology. Currently, over one-half of all petroleum production comes from Daqing in Heilongjiang Province, Shengli in Shandong, and Daqang in Hebei.

The use of hydroelectric power varies considerably throughout East Asia (Figure 14.5). In North Korea, where the topography is particularly conducive to the development of hydropower, this source accounts for 59 percent of the country's total electricity. In China, this proportion is still high at 20 percent, while it decreases to 15 percent in Taiwan and 5 percent in South Korea. The high proportions of hydroelectric power in North Korea and China are probably more suggestive of relatively less developed economies than they are of effi-

cient management of fossil fuels for the production of electricity.

Electrical energy produced from nuclear power is an important aspect of the resource base in Taiwan and South Korea. Approximately one-third of Taiwan's and about half of South Korea's electricity come from nuclear plants.

HUMAN ENVIRONMENT

China has evolved over the last four thousand years, making it one of the world's oldest civilizations. Therefore, an examination of the evolution of Chinese culture helps us to better understand contemporary China. Why is China still a relatively backward country while Japan is a major industrial power? Did the overthrow of the Manchu dynasty in 1911 and the Communist Revolution of 1949 cause an absolute break from the past, or is the twentieth century simply a short diversion from the mainstream of Chinese evolution? And, how do the other developing countries of East Asia—Mongolia, North and South Korea, the Republic of China (Taiwan), and the British colony of Hong Kong—fit into the picture?

EVOLUTION OF CHINESE SOCIETY

Historians estimate that prehistoric people inhabited China as early as 1,700,000 B.C. From these people, three agrarian societies emerged during the Neolithic period between 6000 and 5000 B.C. The best known was the Yangshao culture, which inhabited a region of the middle Huang He and Wei river basins near Xian. These people were sedentary farmers, who supplemented farming with hunting and gathering and fishing. From the Yangshao evolved the Longshanoid culture (3200–2000 B.C.), the main prototype for Chinese civilization.

Between 3000 and 2000 B.C., the Longshanoid culture developed a clearly stratified social system with specialized labor and a capital city, which in time led to the establishment of the first Chinese **dynasty.** The Xia dynasty (c. 1994–1766 B.C.) may, in fact, only be mythological; its existence has not yet been proven archaeologically. Whether it is mythological or real is not as important as the fact that during this period China developed an ordered society with a code of conduct en-

FIGURE 14.4

Natural Resources of China

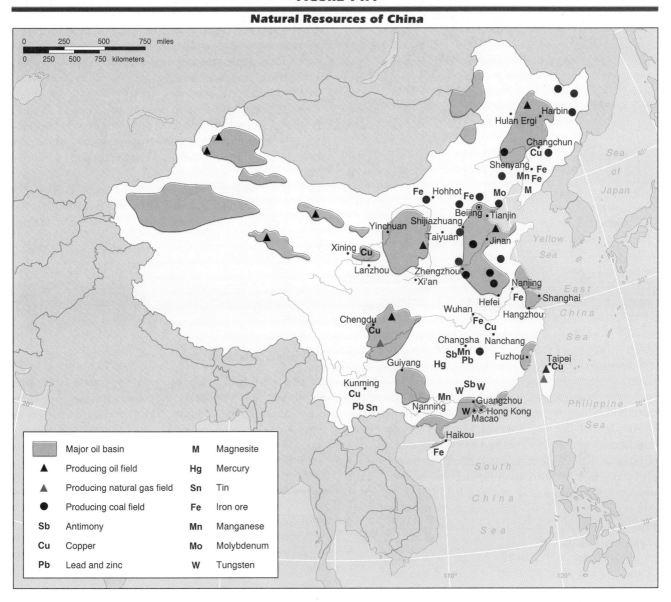

Major oil basin

▲ Producing oil field

▲ Producing natural gas field

● Producing coal field

Sb Antimony

Cu Copper

Pb Lead and zinc

M Magnesite

Hg Mercury

Sn Tin

Fe Iron ore

Mn Manganese

Mo Molybdenum

W Tungsten

forced by the military; this occurred within the dynastic system, which would endure for almost four thousand years.

Several great dynasties ruled China between 1766 B.C. and A.D. 1911 (Table 14.1), and each made important contributions to the evolution of Chinese culture. Unfortunately, time and space do not allow a description of each dynasty, but an overview of the basic character of Chinese society in the middle of the nineteenth cen-

tury, when the Chinese were confronted by the British will help explain why the Chinese reacted as they did to the British presence.

Characteristics of Dynastic China

Among the many facets of Chinese society as it evolved during the dynastic period, several characteristics, in

FIGURE 14.5

Production of Electricity by Type

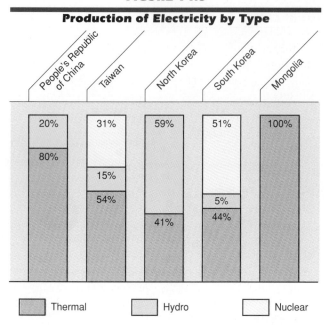

SOURCE: *Energy Statistics Yearbook 1988* (New York: United Nations, 1990); and *Republic of China 1989* (Taibei: Hilit, 1989).

particular, help explain the basic nature of Chinese society and why it responded as it did to the British.

Throughout the evolution of Chinese society, great importance was always attached to the arts, culture, and education. The Chinese believed that their own classics, especially **Confucianism,** contained the essence of all wisdom. The ideals and philosophy of Confucianism mingled with Buddhism and Doaism to produce a rich system of thought. Thus, the Chinese did not find it necessary to borrow from other cultures. The Chinese referred to China as the Middle Kingdom, a concept that defined China as the cultural center of the world; all other cultures were considered barbarian. This helps to explain why the Chinese were not especially interested in widening their contacts with the British or other Western nations.

Confucianism taught that society was strictly structured such that all relationships were between an inferior and a superior. There was little appreciation for the concept of relationships between equals. Because of this, the Chinese continued to demand a position of superiority in their relations with the British.

The Chinese bureaucracy was recruited from the scholar-gentry class and remained basically unchanged

for two millennia. Dynasties rose and fell, but the bureaucracy remained in control of the government and society. As one of the highest classes in China, the scholar-gentry leadership was relatively ignorant of the West and unable to fully appreciate the technological superiority, especially militarily, of the Western powers.

At the head of this society were the emperor and the imperial family. Given a mandate from heaven to rule, they represented the epitome of the highest moral values. The emperor was, in fact, the state, to whom all subjects owed their loyalty. There was no sense of loyalty to "China," which was regarded as equivalent to the entire civilized world.

When foreigners intervened in the nineteenth century, China was under the control of the Manchu (Qing) dynasty, which had conquered China in the seventeenth century. It is important to keep in mind that the Manchus were a foreign dynasty. Although they were assimilated into Chinese culture, the Manchus were still perceived as foreigners by the dominant Han Chinese (Hanren).

Foreign Incursion

Although the British had first come to China in the sixteenth century, until 1842 they were restricted to only one Chinese port—Canton. Based on Confucian ideas of society and the concept of China as the Middle King-

TABLE 14.1

Chinese Dynasties

DYNASTY	DATES
Xia (?)	c. 1994–1766 B.C.
Shang	1766–1122 B.C.
Zhou	1122–221 B.C.
Qin	221–207 B.C.
Han	207 B.C.–A.D. 220
Three kingdoms	A.D. 220–A.D. 280
Southern and northern dynasties	280–581
Sui	581–618
Tang	618–907
Five dynasties, ten kingdoms	907–960
Song	960–1279
Mongol	1279–1368
Ming	1368–1644
Manchu	1644–1911
Republic of China	1912–1949
People's Republic	1949–present

dom, the Chinese treated foreign traders as inferior and kept contacts to a minimum, since they believed there was little of value to be gained from other cultures.

The British, of course, wanted to be treated at least as equals, and they wanted greater access to Chinese markets. Tensions increased between the two cultures, and the British handed the Manchu dynasty a humiliating defeat in the Opium War (1839–1842). The war ended with the signing of the Treaty of Nanking, the first of the **"unequal treaties,"** in which all the benefits went to the foreigners, and none to the Chinese. Under the terms of these treaties, additional ports were opened. These **treaty ports** would total approximately 100 by the twentieth century. Westerners were also afforded special treatment in China through *extraterritoriality.*

The vulnerability of China was again illustrated by its defeat at the hands of the Japanese in 1894–1895. This was followed in 1900–1901 by a great internal rebellion—the Boxer Rebellion. This was an attempt to rid the country of all foreigners, although, in fact, it had the support of the imperial court. The rebellion failed, however. As a result of those defeats, by the twentieth century, China had come face to face with the fact that in economic and military affairs it was a weak and backward nation.

Republican China

The Manchu dynasty was finally overthrown in 1911, and a republic was established in 1912 by Dr. Sun Yat-sen. Almost four thousand years of dynastic rule were thus brought to an end. The years of the republic from 1912 to 1949 were years of turmoil and struggle for control. Chiang Kai-shek, leader of the Nationalist party, or **Guomindang,** took nominal control of China in 1928 and retained it until 1949. The Chinese Communist party (CCP), which was formed in 1921, was the main adversary of Chiang's Nationalist party. The CCP was initially an urban-based movement, following the example of the Soviet Union. This failed, however, and the CCP eventually turned to the countryside to seek peasant support. Mao Zedong emerged as the leader of the CCP, a position he would hold until his death in 1976.

From the beginning of the CCP, there were open conflicts with the Nationalist party. The CCP was almost completely eradicated by Chiang in 1934–1935. The communists, however, were able to escape from their bases in the south on the famous "Long March" to northern Shaanxi Province near Yan'an, where they recovered, rebuilt, and began to expand their support and control.

Support for the CCP greatly increased in the countryside during World War II. While the military arm of the CCP—the **People's Liberation Army (PLA)**—focused its attention on fighting the Japanese, the Nationalist army was more concerned with fighting the Communists. There were many instances of Nationalists collaborating with the hated Japanese enemy against the PLA. This collateration helped to solidify peasant support for the CCP.

Additionally, inflation, government corruption, and political divisions within the Guomindang helped to weaken the Nationalists' hold on the country. After World War II, the Nationalists were defeated and forced to flee to the island of Taiwan, which to this time had been a prefecture of China. It then became home to Chiang Kai-Shek's Nationalist "government in exile." On October 1, 1949, the People's Republic of China (PRC) was established, with Mao Zedong as the first chairman of the ruling Chinese Communist Party.

The People's Republic of China

With the help of the CCP apparatus and the support of the PLA, Mao became a virtual dictator. During this time, China began a program of heavy industrialization and collectivization of agriculture along the Soviet pattern. The PRC progressed from disorganization and backwardness toward a centralized, autocratic, and rationally administered state on the road, albeit a slow road, to modernization.

Culturally, the CCP attempted to forcibly change society by breaking traditional Chinese ties. Confucianism was replaced by the teachings of Chairman Mao. The state sought to replace the traditional bonds of family and religion with an allegiance, first and foremost, to the state.

In foreign policy, the 1950s saw China closely allied with the Soviet Union and other communist states. Many of the early programs and policies were based on the Soviet example. This reliance ended with the breaking of relations between the Soviet Union and China in 1960 due to differing interpretations of communist doctrines aggravated by tensions over boundary areas and intervention in each other's foreign initiatives.

By the mid-1950s, Mao was losing support within the CCP. In response, he initiated major policy changes to

An aerial view of the palaces, tombs, and gardens of the Forbidden City in Beijing. The Forbidden City has been maintained as a tourist attraction since the Communist Revolution.

strengthen his own position. These culminated in the "Great Leap Forward" (1958–1960), which emphasized the need for continuing revolution. After initial successes, especially in agriculture, this movement proved to be disastrous for the economy. The overall effect was probably a weakening of Mao's position.

Mao instituted a second major campaign in the mid-1960s. Known as the Great Proletarian Cultural Revolution, this was an attempt to silence his critics. The Cultural Revolution had an even more devastating effect on the Chinese economy than the Great Leap Forward. Relying on loyal youth groups, called the "Red Guard," and the PLA, Mao sought to purge all opponents, both real and imaginary, of his radical approach to growth and development, i.e., true Marxism-Leninism-Maoism.

Eventually, domestic stability was restored by more moderate factions within the CCP, headed by Premier Zhou Enlai. This initiated a period of better relations with the West. As a first step, the U.S. Ping-Pong team visited China in 1971, and President Richard Nixon visited China in 1972. Since then, Sino-American relations have continued to improve and expand and have been politically legitimized with U.S. recognition of the PRC in 1979.

Within China, the mid-1970s proved to be an important turning point on the road to development. After Mao died in September 1976, the leaders of the radical component of the CCP—the *"gang of four,"* led by Mao's widow Chiang Qing—were purged. By publically convicting these four as symbols of the misguided policies of the radicals within the CCP, the moderates were able to discredit the policies of Mao Zedong, while leaving his image intact.

In the 1980s, Deng Xiaoping became the most powerful man in China. His policies emphasized practical economic goals, centering on the **"four modernizations,"** or balanced growth and development in agriculture, industry, science and technology, and national defense.

Nationalist China (Taiwan)

Although initially of Fukiense and Hakka (Guangdong) descent, the people of Taiwan (then named Formosa) evolved as a culture distinct from the mainland Chinese. When one million followers of Chiang Kai-shek and his Guomindang party fled to the island in 1949, it became home to the Nationalist government of China in exile.

To the native Taiwanese, however, these mainland Chinese were foreigners.

The early years following the revolution were troubled. Possible attack from the mainland was a constant threat, although the United States stepped in to ensure Taiwan's security. There was also internal dissent caused by the fact that the mainlanders, who constituted about 20 percent of the population, took control of the government, finances, security, and the military. This friction began to dissipate in the 1960s as Mandarin was introduced as the official dialect of Taiwan, intermarriage broke down cultural differences, and the island experienced a period of economic growth. Since then, Taiwan has evolved from a rural society to an urban, industrializing country.

The future status of Taiwan, however, remains in doubt. It has been expelled from the United Nations, and its strongest ally, the United States, has withdrawn official recognition in favor of the PRC, although the United States still maintains close unofficial contacts with Taiwan. Taiwan refuses to negotiate with the PRC over a possible reunification with the mainland, although the PRC has offered to allow Taiwan to continue as a bastion of capitalism within the PRC, the same arrangement that has been made with Hong Kong. Taiwan's economic future is bright, and its political situation has stabilized; yet, its future vis-à-vis Communist China remains uncertain.

MONGOLIA

Prior to the sixteenth century, the nomadic Mongols had been a conquering people. Genghis Khan and his successors at one time extended the Mongol empire to include all of China, central Asia, much of Siberia, and southern European Russia. The empire weakened after the death of Kublai Khan however, and was conquered by its neighbors. In the seventeenth century, Mongolia became a vassal of the Manchu dynasty in China, a status it retained into the twentieth century. After the fall of the Manchus in 1911, Russian influence increased. The Communist Revolution of 1921 established the Mongolian People's Republic and ensured close links with the Soviet Union, a relationship that has persisted to the present. Today, for the most part, Mongolia's economy is simply an extension of the Soviet economy, and Mongolia serves as a *buffer zone* between the Soviet Union and China. Recent demands for greater in-

dependence within the Soviet Union have spread to Mangolia as well, and call into question the future relationship between the two countries.

KOREA

Korea has been inhabited since the third millennium B.C. These initial inhabitants, who were the ancestors of present-day Koreans, were similar in origin to the Turkic-Manchurian-Mongol people of Asia. By the second century B.C., they began to be influenced by the Chinese. Northern Korea became part of the Chinese empire, and in the first century B.C., Korea was divided into three kingdoms—Koguyro in the north, Silla in the southeast, and Paekche in the southwest—all of which were under the cultural influence of China, as reflected by their adoption of Buddhism, Confucianism, and character writing.

Korea became "independent" from the Manchu dynasty in 1876 as Western powers became interested in the region. It then signed a series of unequal treaties with Western powers and Japan. The Sino-Japanese War (1894–1895) extracted Korea from China, and the Russo-Japanese War (1904–1905) established Korea as a protectorate of Japan. Japan retained control of Korea until World War II.

The Japanese occupation of Korea had both positive and negative features. Japan introduced modern technology and advanced educational opportunities, both of which had long-term benefits for the Koreans. At the same time, the Koreans suffered greatly under the Japanese. The occupation was enforced by a harsh military rule, whereby Japan exploited Korean labor and agriculture for its own benefit. Japan tried to assimilate the Koreans into Japanese culture by making them adopt Japanese names and establishing Japanese as the official language of Korea. These efforts only served to embitter the Koreans against the Japanese.

At the close of World War II, it was decided that the Soviet Union would secure the northern half of Korea and the United States would secure the south. The boundary between Soviet and American troops was the 38th parallel. Negotiations between the north and south to reunify the country after the war were unsuccessful. In 1948, two separate countries were created: the Democratic People's Republic of Korea in the north and the Republic of Korea in the south.

In an effort to reunite the two by force, North Korea

invaded South Korea in 1950, initiating the Korean War (1950–1953). The North Koreans almost totally overran the south. Military intervention by the United States pushed the North Koreans out of the south all the way to the Yalu River, where the Chinese intervened against possible U.S. incursions into China. The Chinese pushed the U.S. forces back south to the 38th parallel. After two years of negotiation, the border between North and South Korea was relatively unchanged at the 38th parallel.

Since then, little interaction has occurred between the north and south. A communist government remains in control of North Korea, which is closely tied to the communist world. A one-party democracy, with strong military influence, controls South Korea, which is economically tied to the capitalist world. Recently, economic negotiations have begun between the two Koreas as a first step at increased interaction.

POPULATION AND CULTURE

East Asia represents the largest clustering of people in the world. One of every four people on earth lives in East Asia. In China alone, there are more than one billion people, or over 20 percent of the world's population. These people are unevenly distributed throughout the region (Figure 14.6). Most of China's one billion people are located in the eastern third of the country, especially in the North China Plain, the Chang Jiang basin and delta, the Sichuan basin, and the deltas of the Xi and Zhu rivers; Korea's 65 million people are primarily located along the west and south coasts; and Taiwan's 20 million are concentrated in the western plains and basins.

Size and Growth

As the Statistical Profile indicates, the demographic profile for the developing countries of East Asia is quite different than for Japan. Following World War II, population growth in East Asia exploded, reaching one of the highest rates of natural increase in the world. This has changed dramatically, primarily as a result of China's ability to reduce its rate of growth to slightly over 1 percent during the mid-1980s. Estimates put the 1950 population of East Asia at approximately 600 million, of which China accounted for 560 million, North Korea for 9 million, South Korea for 21 million, and Taiwan for 7 million. After a period of rapid growth in the 1950s and 1960s, the rate of natural increase dropped in the 1970s

and 1980s. With the exception of North Korea and Mongolia, the developing countries of East Asia have adopted family planning policies to bring the birth rates under control. The most notable case is China.

During most of the Mao years, official policies did not, for the most part, attempt to slow population growth rates. Not until after Mao's death in 1976 did the leadership of the PRC deal with this matter in an effective manner. Realizing that an ever-increasing population diminished the benefits of increased economic productivity, rigorous measures were imposed to slow the rate of natural increase. By all indications, these policies were highly effective. The estimated rate of increase of 1 percent in 1986 was totally uncharacteristic of a country with the level of economic development found in China. An easing of birth control policies has resulted in China's growth rate increasing back to 1.4 percent since 1986, however.

Nevertheless, one would expect China to have a rate of increase in excess of 2 percent (as it actually was well into the 1970s). Through widespread propaganda and education, penalties for having too many children, and improved and more readily available contraception, the government was able to reduce growth to a level more characteristic of Japan and the other developed countries of the world. China's current growth rate puts it in a class with the developing countries of Taiwan and South Korea.

Growth rates for Taiwan (1.2 percent) and South Korea (1 percent) are more characteristic of their levels of development, as newly emerging economic powers along the lines of the Japanese model. And growth rates for North Korea (2.1 percent) and Mongolia (2.8 percent) are characteristic of developing countries.

A final, and important, demographic characteristic of the region is the proportion of the population 15 years of age and younger. A large proportion of young people portend larger numbers of women coming into their childbearing years in the future, thus diminishing or negating potential effects of family planning programs. Nearly 30 percent of all the people living in East Asia are younger than 15 years of age. Japan and the United States, for example, have 22 percent of their populations in this category.

Ethnicity

There is a great deal of ethnic homogeneity in the countries of East Asia. Mongolia is dominated by Mongols, who comprise over 95 percent of the population. Ko-

FIGURE 14.6

Population Density of China

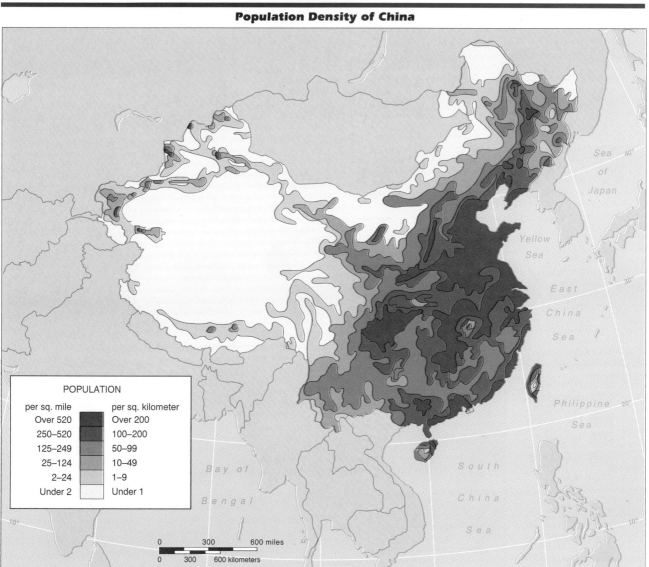

POPULATION

per sq. mile		per sq. kilometer
Over 520		Over 200
250–520		100–200
125–249		50–99
25–124		10–49
2–24		1–9
Under 2		Under 1

reans dominate the population of both the Koreas. Even in China, 94 percent of the population are Han Chinese. Yet, China is a state of many nationalities. In addition to the Han Chinese, there are about 55 minority nationalities. Although they comprise only 6 percent of the population, they occupy approximately 60 percent of China's territory (Figure 14.7).

Cultural assimilation is being accomplished in a number of ways. Millions of Chinese have been settled in minority regions. In only two autonomous regions does the minority nationality outnumber the Chinese. Minority students are sent to Chinese centers of higher learning and return to their native regions afterward. It appears that most of the minorities have accepted communist rule. The next step, from the point of view of the central government, will be full socioeconomic and political integration. For the time being, cultural autonomy will be allowed.

FIGURE 14.7

Major Ethnic Groups in China

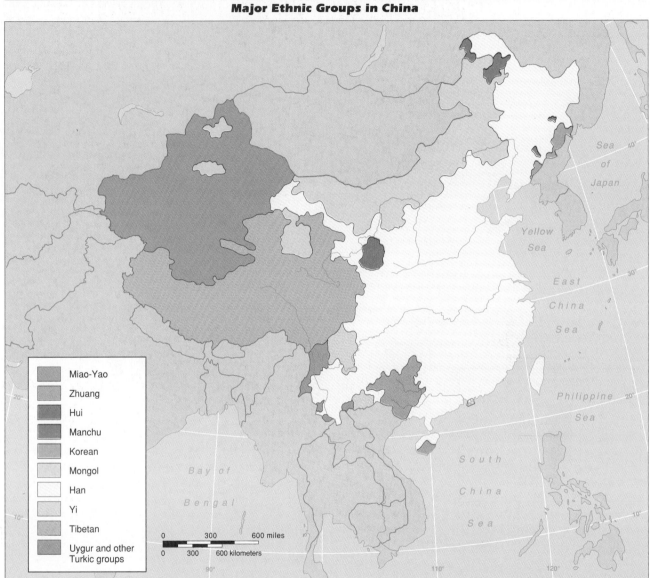

Religion and Ideology

The indigenous philosophies of Confucianism and Daoism and the assimilation of Buddhism from India were primarily responsible for a rich system of thought that structured Chinese society for over two thousand years. Even the Western religions of Islam and Christianity were introduced into China with some success. The most recent ideology to affect Chinese society has been Marxism-Leninism as interpreted by Mao Zedong and his successors.

In reaction to the social and political turmoil of his time, Confucius (Master Kong, 551–479 B.C.) did not seek to create a religion. Rather, Confucianism was a code of ethics that sought to define relationships between people, based on certain principles of social conduct, in order to produce harmony within an orderly system. Each relationship had clearly defined roles in which one person was superior to the other. That notwithstanding, each role carried with it privileges and responsibilities, and each relationship included obligations. In this respect, Confucianism promoted humane interaction nurtured by conscience and character as the basis for social harmony. The highest achievement in Confucianism was service to fellow humans. The closest Confucius came to religion was in his promulgation of ancestor worship and reverence for heaven.

In the second century B.C., Confucianism was accepted as state orthodoxy, or "religion," because it promoted order, harmony, and the status quo. It served as the basis for Chinese society and politics for two thousand years, affecting every aspect of Chinese life. Although Confucianism lost its official support after 1911, its heritage was preserved until the Communist Revolution in 1949. One must question whether the influence of Confucianism, however diminished, has been or can be eradicated entirely by communist rule.

The other indigenous philosophy of China was Daoism, which is most closely associated with the philosopher Laozi (c. fifth century B.C.). Although Daoism has provided an alternative, even an antithesis, to Confucianism for over two millennia, they are not mutually exclusive. It was not unusual for a Chinese to practice both.

Daoism developed as a metaphysical system of thought that found harmony in nonaction because action provoked conflict and reaction. It saw within nature two opposing forces, the yin and the yang, which created the Dao, or "the way." Because it was impossible to know the way, Daoism emphasized spontaneity and individual freedom in order to be in harmony with nature's rhythms and changes. Instead of Confucianism's moral principles, social structure, and government regulations, Daoism promoted a kind of primitivism.

The Buddha, or "Enlightened One," was the name given Prince Siddhartha Gautama who lived in northeastern India (c. 563–483 B.C.) and was a contemporary of Confucius. Buddhism, which derived in part from Hindu tradition, taught that all life was suffering and that salvation lay in the upward movement of a good person through reincarnation to higher levels of existence until one reached the final reward—Nirvana, or nonexistence. The good life disassociated one from worldly concerns and emphasized love, hospitality, and charity. Like Confucianism, Buddhism emphasized ethics as the basis of society. Like Daoism, it valued withdrawal from worldly concerns.

Buddhism spread to China in the second century B.C. Over the centuries, it has been modified and adapted to indigenous Chinese beliefs, just as it in turn has affected the evolution of Confucianism and Daoism. Confucianism was the dominant code of ethics for dynastic China, and Buddhism the dominant religion. Just as Confucianism has survived communist repression, so has Buddhism, which is very much intertwined with secular Chinese culture.

Islam was introduced to China in the eighth century A.D. by Muslim traders and conquerors. This was particularly true of northern China, where the Islamic culture is still found in several of China's largest minority groups. These include the Uygurs and other Turkic minorities in the northwest and the Hui in the north center. Because these peoples inhabit large frontier areas, Islam has been tolerated, even by the communist state.

Christianity has also persevered. Although it was apparently introduced to China as early as the eighth century A.D., its first real foothold dates from the sixteenth century. Christianity went through periods of great success and suppression, and its importance to China lay more in the transfer of Western knowledge than in religious dogma. In fact, the impact of Christianity on Chinese culture was slight. Its appeal was to the elite and it found little mass support. Both Islam and Christianity were severly repressed in the late 1950s through the 1970s, but have survived to enjoy greater, although limited, freedom in post-Mao China.

The 1949 Communist Revolution brought China a new ideology—Marxism-Leninism. Just as Chinese culture has assimilated most foreign influences, Mao Zedong's interpretation of Marxism-Leninism, not to mention post-Mao adaptations, has produced a uniquely Chinese brand of communism. The most obvious manifestation of this was when the Chinese based their revolution on the peasants, rather than on the urban proletariat.

Socialism did bring with it collectivization of agriculture and nationalization of industry as the road to development. Additionally, Chinese communism upholds the dictatorship of the proletariat and the preeminence of the Communist party. Yet, these principles of socialist development have been modified in the post-Mao era. Examples of a more pragmatic approach under the banner of the "four modernizations" include the dissolution of the commune and the reestablishment of the household farm as the basic agent of agriculture, private com-

ETHNIC MINORITIES IN CHINA

 lthough they comprise only 6 percent of the population of the People's Republic of China, ethnic minorities have been accorded special treatment by the central government through special administrative status. At the lower administrative levels, the government has established autonomous prefectures, counties, leagues, and banners, all based on ethnicity. At the highest level, it established autonomous regions (ARs), of which there are currently five. These five regions are inhabited by numerous groups, although each AR has a dominant nationality—Tibetans in Xizang, Uygurs in Xinjiang, Mongols in Nei Mongol, Hui (Muslims) in Ningxia, and Zhuang, China's largest ethnic minority, in Guangxi. This political accommodation is necessitated by the geographic dispersion and strategic locations of these minority groups.

The ARs are intended to guarantee political equality for minorities and to give special consideration to the development of minority areas. In fact, they are politically bound by the decisions of the central government. They have been allowed a certain degree of cultural autonomy, however. Recent intervention by Chinese troops in Xizang (Tibet) to deal with the followers of the Dalai Lama indicates the limits to this autonomy.

The underlying reasons for this special consideration are threefold. First, many of the minorities occupy border areas and are strategically important. The Xizang, Xinjiang, and Nei Mongol ARs occupy vast, sparsely inhabited areas that act as frontiers, or buffer zones, between the Han Chinese and their traditional enemies. In the southwest, the Xizang AR (Tibet) provides a buffer with India and other countries of South Asia. In the northwest, the Xinjiang AR provides a frontier with the Soviet Union. And in the north, the Nei Mongol AR serves as a buffer with the Soviet Union and Mongolia.

In addition to their strategic importance as frontiers, ARs occupy areas that possess important natural resources. These include petroleum in Xinjiang, coal in Ningxia, and iron ore in Nei Mongol.

Finally, the Chinese are concerned about the political image of socialism, which calls for the elimination of differences between all peoples and regions. Regional autonomy, as reflected in the ARs, helps to legitimize claims for the superiority of the socialist system. It also helps to assimilate these diverse ethnic groups into the greater Chinese polity, a process accelerated by the migration of large numbers of Han Chinese to these regions. Currently, only the Xinjiang and Xizang ARs have a greater number of minority people than Han Chinese.

---

merce and competition in agriculture and industry, and openness to the West.

The impact of the Revolution has been dramatic. On the other hand, communism has been in China for only four decades, which is "a drop in the bucket" within the context of Chinese history. It has already evolved into a philosophy strongly influenced by Chinese culture. This does not suggest that there will be a return to republicanism or dynastic rule. It does suggest, however, the permanence of a culture that has evolved over four thousand years and a basic social and political philosophy that evolved for over two thousand years.

POLITICAL SITUATION

East Asia is the scene of a major confrontation between two antagonistic ideologies. Here the communist countries of the People's Republic of China and North Korea come face to face with the capitalist countries of South Korea, Taiwan, and Japan.

It might be safe to say that the major political event of this century for the developing countries of East Asia was the Chinese Revolution and the installation of a communist government in China. Not only did this establish a communist foothold in East Asia, but it also intensified U.S. interest in and increased support for the capitalist countries of this region.

The People's Republic of China

We saw earlier in this chapter that dynastic China was greatly influenced by Confucian and Buddhist ideals, with inputs from other philosophies, especially Daoism. To a certain extent, many of the characteristics of dynastic China can be found in present-day China.

Yet, there can be no question that the philosophy of greatest importance to contemporary China is Marxism-Leninism-Maoism, which sought to break all ties with the past and refute the traditional values of four thousand years of Chinese society. As it evolved in China, Marxism-Leninism emerged as something quite different from the Soviet example. Mao criticized the Stalinist model, which he thought was lacking in dialectics and balanced development. In response, the Chinese model emphasized self-reliance and the need for continuing revolution within China to rid the Communist party and the country of all bourgeois elements. Many of the economic upheavals during the Mao period were justified by the Maoists within the context of continuing revolution. The Great Leap Forward and the Cultural Revolution were cases in point.

After the death of Mao and the fall of the radical "gang of four," the more moderate elements in the CCP have placed less emphasis on continuing revolution and self-reliance in economic development. The post-Mao period has seen a basic change in the nature of Chinese communism. In fact, some scholars have suggested that recent changes in the PRC are a reflection of the continuing influence of precommunist traditions. Chinese communism may well be more Chinese than it is communist. It will be quite some time before we will be able to assess the relative impact of communism on the overall evolution of Chinese society.

Administrative Hierarchy.   One of the major accomplishments of the PRC has been to extend control of the central government to the local level and to involve local people in urban neighborhoods and communes in the economy and politics. It has provided for a two-way line of communication extending from the central government to local residents and vice versa. The administrative hierarchy that has been implemented during the communist period can be divided into three basic levels.

First-order units are administered directly by the central government. These include the 21 provinces, five autonomous regions, and three special municipalities (Figure 14.1). The second level of the administrative hierarchy includes prefectures, autonomous prefectures, and provincial cities. Third-order units include counties, autonomous counties, and prefectural cities.

Until recently, the rural administration was based on the commune, which was created in 1958, during the Great Leap Forward. The average commune contained 15,000 people, 3300 households, and 5000 acres of farmland. These were further divided into production brigades, which approximated traditional villages, and production teams, which contained about 33 households each. The commune was modified by the economic reforms of 1979 when the primary responsibility for agricultural production was returned to the individual household farms; land is still held collectively, however.

Urban administration is centered on the urban neighborhood. Large and medium-size cities are divided into urban districts, roughly the equivalent of an urban commune, which are divided into urban neighborhoods, which usually contain tens of thousands of people. The neighborhood office is the basic organ of political power. Its functions include directing the work of residents' committees, implementing Party policies, transmitting the requests and opinions of the people to higher authorities, distributing ration coupons, mobilizing residents for housing and social work, and running neighborhood shops, restaurants, and factories.

In larger cities, the neighborhood is further subdivided into residential organizations. These include residents' areas, courtyards, and groups. These "grass roots" organizations (1) provide for the area's public welfare, (2) transmit views and opinions to higher officials, (3) mobilize residents, and (4) mediate disputes between residents.

Taiwan

Taiwan has functioned basically as a one-party democracy under the rule of the Nationalists. During the entire period, the island has been under martial law as the government still sees itself in a state of war with the mainland.

Under pressure from within as well as from the United States, there appears to be movement toward political reform and progress toward a greater degree of democracy. In the 1990s, opposition to the government has been allowed for the first time.

Korea

Since 1948, Korea has been divided into the pro-communist Democratic People's Republic of Korea (DPRK) in the north and the pro-capitalist Republic of Korea (ROK) in the south. The two countries are separated by a 2½-mile (4 kilometer) wide **demilitarized zone (DMZ),** which extends 150 miles across the peninsula. Although negotiations concerning possible reunification have occurred from time to time, there seems little chance that Korea will be united in the near future.

North Korea.   Since World War II, North Korea has had a communist government under the control of the Korean Worker's party. In fact, Kim-Il Sung, his family, and close associates have ruled North Korea autocratically throughout its entire history. After the Korean War, North Korea maintained close ties with both the Soviet Union and China. In recent years, it has favored China, although it still maintains relations with both.

South Korea.   Although nominally a democracy, the history of South Korea's government has been characterized by assassinations, military coups, and martial law. South Korean presidents maintain the support of the military, which is essential to remain in power. Recently, there has been some movement toward increased participation and liberalization of democratic processes. Martial law has been partially lifted and opposition parties are allowed.

Hong Kong

British Hong Kong represents one of the last vestiges of European colonialism in East Asia. It has long been a symbol of free market capitalism and economic prosperity. It already serves as an import transshipment port for mainland China and an important economic center for East Asia and the world. Leased from China under the unequal treaties, it will return to Chinese control in 1997. Under an agreement reached between Great Britain and the PRC, Hong Kong will remain a center of capitalism and will maintain its economic role after its return. Deng Xiaoping has referred to the policy of allowing Hong Kong to remain capitalist as a part of communist China as "one country, two systems."

ECONOMIC ACTIVITY

The economic landscape of East Asia exhibits some stark contrasts between the lesser developed, communist countries of China and North Korea and the developing, capitalist countries of South Korea and Taiwan, not to mention the industrial power, Japan, which was discussed in Chapter 8.

Table 14.2 presents economic data for the developing countries of East Asia, and for comparison, Japan. The PRC clearly has the profile of a developing country with almost 70 percent of the labor force employed in agriculture, one-fifth of the population living in cities, and

TABLE 14.2

Economic Data for East Asia

	GNP PER CAPITA (U.S.$)	LABOR FORCE		
		Agriculture (%)	Industry (%)	Services (%)
People's Republic of China	$ 330	69%	19%	12%
Taiwan	5420	20	41	39
Mongolia	880	40	21	39
North Korea	1000	45	30	27
South Korea	3530	36	27	37
Hong Kong	9230	2	51	47
Japan	21040	12	39	49

SOURCES: *World Development Report, 1988 and 1989* (New York: World Bank, 1988–89); and *Republic of China, 1986* (Taibei: Hilit, 1986).

a per capita gross national product (GNP) of only $330. The remaining developing countries of East Asia appear to be in transition, with over half of their labor forces in industry and services, over half the population living in cities, and per capita GNPs ranging from $880 (excluding Hong Kong). In this respect, South Korea and Taiwan appear to be the most developed economically.

For North Korea, with an estimated GNP per capita of $1000 in 1989, the economic profile may be somewhat misleading. During the Japanese occupation of Korea from 1910 to World War II, the northern part of the peninsula was developed as an industrial region, while the southern extension served as a source of agricultural goods for the Japanese empire. In this respect, the data for North Korea may reflect North Korea's historical advantage in industrial development under Japan, while representing an increasing rate of industrialization and growth for South Korea. Taiwan displays characteristics that are even more suggestive of a developing country in transition. Yet, South Korea and Taiwan are still well below the level of development of Japan and the developed countries of the world.

The People's Republic of China

China represents a vast economic potential, still hampered by a relative lack of capital and technology. Its economic backwardness can be explained in many ways. The traditional Chinese reluctance to borrow from other cultures, the more than 100 years of internal turmoil following the Opium War, and anti-Western attitudes following the Communist Revolution have all contributed. Yet, with the four modernizations development policies coupled with population control, China seems poised for possible development into a major economic power.

Fishing and Agriculture.   China accounts for one of the largest catches of fish in the world. Its marine fish catch is the fifth largest after Japan, the Soviet Union, the United States, and Chile. Over 60 percent of China's fish production, primarily carp, comes from fish farms in agricultural fields. The system of integrating freshwater ponds and agricultural land to produce an ecosystem supportive of both has been followed for thousands of years in China

The wide variety of climates and soils in China provide for a great diversity in agricultural production. The main division between agricultural zones is the Qin Ling

In Jiangsu Province, rice is harvested with machinery appropriate for intensive farming of small fields.

Mountains, which divide the temperate to cool and dry north from the warm and humid south. In the northeast, corn, wheat sorghum, and soybeans are the dominant crops.

The alluvial plains of two of the major rivers of China, the Huang He and Chang Jiang are major agricultural regions. The valley of the Huang He produces wheat, barley, corn (maize), millet, and cotton. The area also produces cool-climate fruits, such as apples, and is a major pork production area. The Chang Jiang Basin, being further south, has a hotter and more humid climate. Rice is the staple crop, although tea, cotton, millet, and sugar cane are also grown. **Double cropping** is possible in the extreme south where the growing season is long enough to permit two harvests per year. Typical crops in this area are rice, tea, citrus, and grains, Unfortunately, much of this southern area has shallow, infertile soils that limit agricultural production.

As throughout most of Asia, animal husbandry plays a relatively small role in agriculture. Only about 15 percent of the protein in the average Chinese diet comes from meat, 90 percent of which is pork. China has the largest swine population in the world, 40 percent of the total.

Recent policy changes have had a major impact on agricultural productivity to the point where China is on the brink of agricultural self-sufficiency. Reforms initiated in 1978 allowed local farmers to have greater latitude in determining the types of crops to be grown, to expand specialized services, and to sell food surpluses for profit. The land is still owned by the state, and quotas are set for basic products. As a result of the reforms, however, Chinese farmers now have a greater degree of control over their farms and the opportunity to improve their standard of living. Since 1978, agricultural production has increased by 8 percent annually, and

TABLE 14.3

Gross Value of Chinese Industrial Output by Province, 1952 and 1983 (Percent)

	1952	1983
NORTH	14.1%	15.6%
Beijing	2.4	4.1
Tianjin	5.3	3.7
Hebei	3.9	4.1
Shanxi	1.9	2.5
Nei Mongol	0.6	1.2
NORTHEAST	21.9	15.7
Liaoning	13.2	8.4
Jilin	3.2	2.7
Heilongjiang	5.5	4.7
EAST	40.5	37.1
Shanghai	19.0	11.0
Jiangsu	7.5	9.2
Zhejiang	3.2	4.3
Anhui	1.8	2.6
Fujian	1.2	1.6
Jiangxi	1.7	1.7
Shangdong	6.2	6.6
SOUTH	13.7	18.8
Henan	2.6	3.8
Hubei	2.8	5.1
Hunan	2.2	3.3
Guangdong	5.1	5.0
Guangxi	1.0	1.5
SOUTHWEST	6.6	8.0
Sichuan	4.8	5.5
Guizhou	0.8	1.0
Yunnan	1.0	1.4
Xizang	—	—
NORTHWEST	2.4	4.9
Shaanxi	1.1	2.1
Gansu	0.7	1.4
Qinghai	0.1	0.2
Ningxia	—	0.3
Xinjiang	0.5	0.9

SOURCE: Robert Field, "China: the Changing Structure of Industry," in *China's Economy Looks Toward the Year 2000*," Vol. 1 (Washington, D.C.: U.S. Government Printing Office, 1986), pp. 526–27.

rural incomes have increased by 250 percent. While the ultimate goal is agricultural self-sufficiency by the year 2000, greater central control was reestablished over certain aspects of agriculture in 1990.

Industry.   The level of industrialization in China remains relatively low, although it has greatly improved under the communist regime. Since 1950, industrial output has increased 23-fold. Still, less than one-fifth of the labor force is currently employed in industry, which accounts for over 40 percent of the country's GNP. Figure 14.8 shows the distribution of industrial centers in China.

Notwithstanding avowed policies since 1949 that are designed to locate and relocate industry away from the strategically vulnerable coastal cities, the PRC has taken advantage of the colonial structure to facilitate industrial growth. The spatial distribution of industrial production remains concentrated along the coastal regions where Western intervention in the treaty ports and the Japanese occupation of Manchuria have provided the basic *industrial infrastructure.*

The North, Northeast, and East accounted for nearly 80 percent of the gross value of industrial output in 1952 and still produced almost 70 percent in 1983 (Table 14.3). Today, Shanghai, Liaoning, Jiangsu, Heilongjiang and Tianjin alone account for one-half of the country's total industrial output.

Shanghai continues its traditional role as the industrial center of China, specializing in light industry, such as electronics, transportation equipment, textiles, chemicals, as well as in shipbuilding. The heavy industrial center of Anshan, located in Liaoning Province, specializes in steel and agricultural equipment. In Jiangsu Province, one finds the light industrial center of Nanjing. Heilongjiang Province (Manchuria) has been the traditional center of Chinese heavy industry, centered on Harbin. Tianjin, which serves as the port for Beijing, specializes in shipbuilding and light industry, such as textiles, chemicals, and engineering.

Postrevolutionary China saw a great deal of industrial growth between 1949 and 1960; during that period, annual growth rates averaged over 20 percent. This growth was stimulated by the importation of technology from the Soviet Union. The close economic ties between China and the Soviet Union were severed in 1960, however, China then set upon a period of "self-reliance" from 1960 to 1978. The rate of industrial growth dropped during this period due to lack of foreign tech-

FIGURE 14.8

Industrial Regions of China

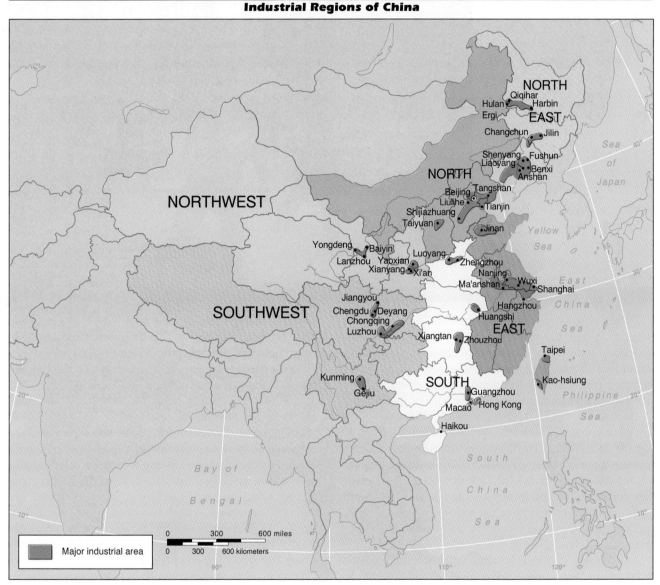

Major industrial area

0   300   600 miles
0   300   600 kilometers

nology and the adverse effects of the Great Leap Forward and the Cultural Revolution.

The most recent change in industrialization occurred in 1978 under the policy of the "four modernizations." Although this policy calls for a more balanced approach to economic development in which a greater portion of government investment is distributed to other sectors of the economy, industry has once again begun to benefit from the acceptance of foreign technology as a means

to industrialization and modernization. The overall goal is for China to become a major industrial power by the year 2000.

Currently, several kinds of products dominate Chinese industry. These include iron and steel, of which China is the fourth largest producer in the world. Textiles and clothing are also significant, followed by light industrial and consumer goods. The chemical and transportation equipment industries have recently become

Workers doing morning
exercises in Shanghai.

important. Yet, China still retains the flavor of prerevolutionary industry in that small-scale, cottage industry continues to play an important role. This, of course, is a major problem area that must be addressed if China is to reach its industrial goals by the turn of the century.

Recent changes in the basic structure of the economy may have a major impact on China's attempts to modernize its industry. The privatism introduced into agriculture has now been extended to the industrial sector. Hundreds of thousands of factories have been given greater flexibility and local autonomy. Levels of production are now set by individual industries and factories; prices are determined by supply and demand; and pay bonuses are based on worker productivity.

Guangdong Province is ideally located to take advantage of the economic reform in China. It lies on the southeastern coast bordering the economic mecca, Hong Kong. A number of rivers allow water access to Hong Kong and the world, thereby eliminating the need to rely on poor roads and railroads. Firms operate as free enterprises trading in hard currencies, exporting products, and importing raw materials from other countries, even Taiwan. The national government does not control firms in Guangdong, but does take a share of the profits.

Transportation. The degree to which the Chinese economy will be able to develop fully will depend, in part, on its ability to build a fully integrated transportation network. Linking this vast empire has been a major problem throughout Chinese history. To this day, the transportation system remains inadequate for the demands of an economy planning to become fully integrated by the turn of the century.

For most of China's history, waterways provided the primary means of transportation. The great river systems of eastern China, especially the Huang He, Chang Jiang, and Xi Jiang and their tributaries, were the backbone of this network, supplemented by an extensive canal system. The waterway system has been improved and expanded in the postrevolutionary period to the point that it now handles over 40 percent of all freight traffic in China (up from 20 percent in 1952).

During the colonial period, Western powers introduced steel rails to China, and the railroad became the dominant mode of transportation. Its role has also been enhanced since the Revolution. Rail lines have more than doubled, and railroads account for 50 percent of all freight traffic.

To the railroads and waterways have been added roads and air transportation. The primitive roads of dynastic China have been improved and expanded during the colonial and postcolonial eras. Air transportation still plays only a minor role in the overall transportation scheme.

The transportation infrastructure is a major hinderance to future economic expansion and integration. This fact suggests that future development will continue to be concentrated in the traditional industrial centers, served by a more integrated transportation system.

Urbanization. In most of East Asia, the population has become increasingly urban, as would be expected with increased economic development. The one exception is China, where only about 20 percent of the population lives in cities. That equals 200 million people, however, or almost as many as live in American cities. Projections estimate that by the end of the century China could have as many as 300 million urban dwellers!

Although cities have existed in China since the second millennium B.C., the traditional dynastic Chinese city dates from the Qin dynasty and has thus served as the center for Chinese urban society for two thousand years. In addition to their ceremonial functions, the earliest Chinese cities served primarily as military outposts and administrative centers for the growing Chinese empire.

These cities were characterized by a surrounding wall and a morphology that reflected the Chinese perception of the universe. The city wall and gates were oriented to cardinal directions, streets were set in a grid pattern, the city was divided into quadrants, and palaces and other government buildings were placed in a specific arrangement. In total, the dynastic Chinese city reflected the interrelationship between society and nature.

A major change occurred in the middle of the nineteenth century as a result of the unequal treaties with the West and the subsequent opening of nearly 100 treaty ports to Western influence. The development of these ports was strongly influenced by their interaction with foreigners. The combination of foreign technology and capital along with local entrepreneurs greatly stimulated the growth of these cities after the mid-nineteenth century.

Thus by 1949, there were two basic kinds of cities in China: a traditional system of interior cities, isolated from the treaty ports and not integrated into the transportation network, and a network of treaty ports that were in contact with the world economic system, yet had little impact on the rest of Chinese society. Upon this existing urban pattern, the communists attempted to build socialist cities for the new Chinese communist state. These cities were to reflect the benefits of a new kind of society. They would eliminate the contradictions and inequalities of capitalist cities, both traditional and colonial, and would produce a humane environment for all Chinese citizens.

The first goal for the new socialist city was to decentrate industry and disperse these activities into the countryside. This would not only produce more humane cities, but would also diminish differences between the cities and the countryside. The key to effecting this goal involved regulating the size of large cities and promoting the growth of small and medium-size cities. This would be accomplished through a system of passes and permits.

These policies notwithstanding, urbanization has proceeded at a rapid pace as a result of both natural increases and rural in-migration, and large cities have continued to grow and dominate Chinese society. This has been especially true for former treaty ports, where the infrastructure was already in place. Since the Revolution, for example, the population of Shanghai has doubled, and the population of Beijing and Tianjin have tripled. Approximately 15 percent of the urban population of China live in these three cities. Although still a predominantly rural society, China contains 9 of the world's 50 largest cities (10, if Hong Kong is counted). Table 14.4 and Figure 14.9 indicate the size and distribution of China's largest cities and urban agglomerations.

TABLE 14.4

China's 20 Largest Cities

CITY	POPULATION (MILLIONS)
Shanghai	6.3
Beijing	5.5
Tianjin	5.2
Shenyang	3.9
Wuhan	3.3
Guangzhou	3.2
Chongqing	2.7
Harbin	2.5
Chengdu	2.5
Zibo	2.2
Xián	2.2
Liupanshui	2.1
Nanjiang	2.1
Changchun	1.7
Taiyuan	1.7
Dalian	1.5
Kunming	1.4
Tangshan	1.4
Zhengzhou	1.4
Lanzhou	1.4

NOTE: Data are for administrative city only.

SOURCE: *1988 Demographic Yearbook* (New York: United Nations, 1990).

FIGURE 14.9

China's Major Urban Agglomerations

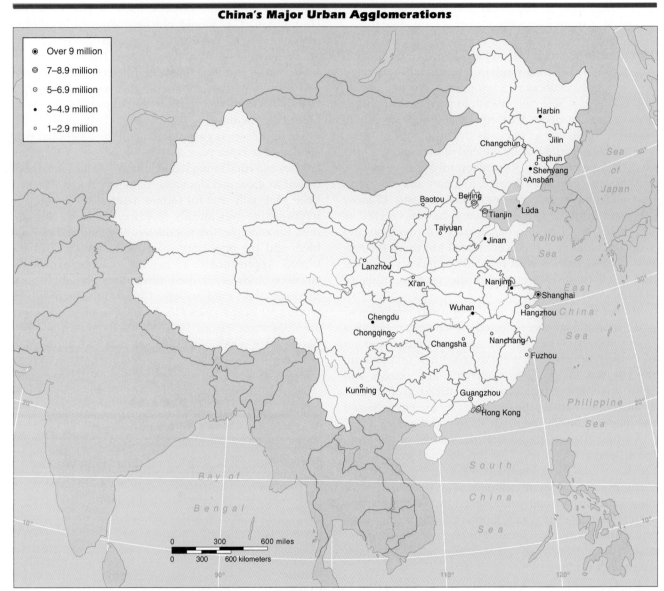

At the intraurban scale, Chinese communist policies seek to create a more positive urban environment. Urban planners are attempting to allocate land use in a rational manner, separating factories from residences, establishing adequate housing, public transportation, green spaces for recreation, and city centers for public and political functions, and providing the necessary goods and services. As of yet, these goals are far from being met. Many cities suffer from inadequate housing, slums, severe pollution, and shortages of goods and services. Once again, the needs of a society to modernize and industrialize appear to have 'overshadowed, for the time being, the desire to create ideal urban environments. The key to continued attempts to deconcentrate and to integrate more cities into the urban network will be related to improvements in the transportation system.

The Koreas and Taiwan

Korea and Taiwan, with economies somewhere between China and Japan, are more representative of developing countries. Between 1965 and 1980, the percentage of the North Korean labor force employed in industry went from 23 percent to 30 percent while South Korea's change was from 15 percent in 1965 to 27 percent in 1980. In fact, South Korea and Taiwan are often referred to as "latter-day Japans."

North Korea. Although the Japanese occupation of Korea from 1910 to 1945 was harsh and cruel, there were some benefits. One of these was the building of an industrial infrastructure in the north, where most of the peninsula's mineral resources were located. Although much of this infrastructure was destroyed during the Korean War, the infusion of Soviet and Chinese capital and equipment after the war helped to rebuild a relatively strong industrial base.

Because the economy is centrally controlled, heavy industry, especially iron and steel, machine building, and chemicals, has been emphasized during the communist period at the expense of light industry and consumer goods. In recent years, however, agriculture has received a greater investment priority.

Weaknesses in the economy have lessened North Korea's ability to develop into a truly industrialized society.

These include the lack of petroleum and other industrial resources, the inferiority of its goods on the world market, and a bad credit rating because of its inability to repay past loans. This bad credit rating limits North Korea's ability to purchase foreign technology.

Japan is North Korea's principal noncommunist trading partner, although problems involving the payment of debt persist. Otherwise, most North Korean trade is conducted with the socialist world, primarily the Soviet Union and China. Recurring debt problems have forced the North Koreans to turn increasingly to **chu'che,** a policy of self-reliance, to develop their economy.

Although current data concerning North Korea are difficult to obtain, best estimates put the per capita GNP at about $1000. Although better than China and many other developing countries of Asia, it is substantially below South Korea's $3530, not to mention Japan's $21,040. As in most developing, centrally planned economies, where investment priorities have emphasized the development of heavy industry, North Korea is able to provide the basic necessities of life—food, shelter, clothing, education, and health care. The major weakness is in the quality and quantity of goods and services available to the people.

South Korea and Taiwan. There are some striking similarities between the South Korean, Taiwanese, and Japanese economies. South Korea and Taiwan have

The modernization of South Korea is apparent in this view of downtown Seoul, the political and economic center of the country.

export-based economies, exporting many of the same items that Japan exports. Initially, these included textiles and clothing; then, electrical and transportation equipment; and now, high-tech goods.

Like Japan, South Korea and Taiwan are heavily dependent on imports of raw materials, especially energy resources. They also import large quantities of food. The health of the South Korean and Taiwanese economies depends on these sources of raw materials.

Another similar feature is the close cooperation between business and government, especially in South Korea, where close ties with the government and easy access to capital enable five conglomerates to dominate the economy.

South Korea receives economic and military support from the United States, which is also its major trading partner. In terms of value of goods, South Korea imports slightly more from Japan than from the United States, but exports more than twice as much to the United States. South Korea has made a significant penetration into the markets long controlled by the Japanese, especially in automobiles and consumer electronics. Thus, the Korean economy is closely tied to both the U.S. and Japanese economies. The same description is applicable to Taiwan, except that economic and military support are indirect.

Recent economic growth along the path of the Japanese example has given the South Koreans a per capita GNP of $3530 and the Taiwanese a per capita GNP of $5420. These data, however, are somewhat misleading. There are still major shortages in housing, education, and social welfare.

SPATIAL CONNECTIVITY

Recent changes in China's economic policies promise to have a profound effect on the world economy. Since 1978, China has abandoned its emphasis on self-reliance and has accepted the use of foreign technology as a means to achieve the goals of the "four modernizations" by the start of the next century. In the post-Mao period, China has already achieved a much greater degree of integration into the Asian and world economies.

In this regard, China once again appears as the antithesis of Japan. The Chinese possess a vast resource base waiting to be explored and developed. Yet, they

lack the necessary technology to develop these resources fully. In this respect, there is a seeming complementarity between China and North Korea, on the one hand, and Japan, South Korea, and Taiwan, on the other. The communist countries of East Asia are resource rich and technology poor, while the noncommunist countries are resource poor and technology rich.

TRADE

East Asia exhibits two basic profiles of trade. The communist countries of Mongolia, China, and North Korea, with relatively lower levels of development, depend heavily on exporting primary products and importing manufactured goods. The noncommunist countries of South Korea and Taiwan, as in the Japanese model, import primarily raw materials and export manufactured products. China, South Korea, and Taiwan have trade surpluses, while Mongolia has a small deficit. Only North Korea has serious foreign debt problems.

The basic pattern of foreign trade for the developing countries of East Asia is displayed in Table 14.5. With the exception of Mongolia, one recurring theme is the heavy involvement of the industrialized countries of the world, especially the United States and Japan. This has long been the case for South Korea and Taiwan, but has now become a recent trend for the PRC. Even North Korea, whose trade is heavily oriented to the socialist world, has increased trade relations with Japan.

Mongolia

For the most part, Mongolia can be treated as an extension of the Soviet economy. The level of integration is so great that some primary products are developed jointly solely for export to the Soviet Union. Eighty percent of all Mongolian trade is with the Soviet Union.

The benefits from Soviet involvement have been bilateral. The collectivization of agriculture has greatly increased the productivity of livestock herding, long the dominant economic activity of Mongolia. Livestock and livestock products still account for about half of all Mongolian exports. There has been, in addition, a dramatic increase in the production of other agricultural goods, primarily cereals.

Since the 1950s, the Mongolian economy has diversified, increasing the role of extractive and industrial activities. Exports of ores and minerals, especially copper

MODERNIZATION AND CHINA'S INTEGRATION INTO THE WORLD ECONOMY

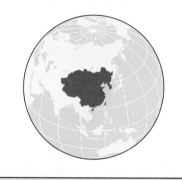

ne key element to the success of China's new policies of modernization is increasing integration into the world economy. This post-Mao "open door" policy is intended to attract capital and technology from other countries, especially from Japan and the industrialized West. Increased interaction has taken various forms: foreign investment, joint ventures, direct trade, and foreign tourism.

The 1978 economic reforms eliminated ideological opposition to foreign borrowing and opened China to foreign investment and trade. To encourage foreign investment, the government established "special economic zones" in Guangdong and Fujian provinces, where tax and investment incentives have been offered to foreign investors. These funds are invested in a wide variety of projects, primarily heavy and light industries and resource extraction. Similarly, joint ventures offer a wide range of tax and investment incentives. During the initial five-year period following the initiation of China's open door policy, total foreign investment in China approached $7 billion, and plans call for the Chinese to expand their efforts to increase the flow of foreign capital.

Chinese foreign trade has followed a similar pattern of increased interaction. This is especially true of trade with the developed countries. Total trade has increased from $4.2 billion in 1970, at the height of Mao's closed door policy,

to $18.8 billion in 1978 to $103 billion in 1988. China achieved a trade surplus in the early 1980s. Sixty-five percent of all Chinese imports in 1983 were from the developed countries; Japan and the United States were the two largest trading partners. Whereas exports are dominated by textiles and natural resources, imports are dominated by steel, machinery, and transportation equipment, all of which are important to modernizing the Chinese economy.

The final manifestation of China's new openness is found in the expanding tourist industry, aimed at attracting foreign capital. Between 1978 and 1984, the number of foreign tourists increased fivefold, and the amount of foreign exchange increased almost six-

fold. By the mid-1980s, well over one million tourists a year were visiting China to see attractions, such as the Great Wall and the Forbidden City. China is earning over $1 billion a year from its tourist industry.

Despite these successes, China has also suffered setbacks. The massacre in Tiananmen Square in 1989 retarded the integration of China into the world economy. Residents of Hong Kong have become more uncertain about life under Chinese authority. Many of China's economic partners placed joint ventures on hold. The World Bank reconsidered $700 million in loans, and hotel occupancy in tourist hotels averaged only 30 percent in the weeks after the tragedy. Over the long term, however, foreign investors will return in spite of the political repression. Most Chinese remember the chaos of the Cultural Revolution and are hesitant to return to similar disorder in the 1990s. It is likely the Chinese populace will sacrifice some individual rights for the better standard of living created by capitalism and a placid social order.

It is ironic that China's refusal to open itself to the West in the mid-nineteenth century, as Japan did, resulted in a century of political turmoil and economic instability. China has now apparently accepted an old Japanese cultural trait of borrowing from other cultures in order to further its economic development and modernization.

TABLE 14.5

Direction of Trade in East Asia

	IMPORTS FROM (PERCENTAGE OF TOTAL IMPORTS)		EXPORTS TO (PERCENTAGE OF TOTAL EXPORTS)	
Mongolia	USSR*	na	USSR*	na
People's Republic of China	Japan	29	Hong Kong	32
	Hong Kong	13	Japan	15
	USA	11	USA	9
Taiwan	Japan	34	USA	44
	USA	22	Japan	13
	Germany	5	Hong Kong	8
North Korea	USSR**	na	USSR	na
	Japan	na	PRC	na
	PRC	na	Japan	na
South Korea	Japan	34	USA	40
	USA	21	Japan	16
	Germany	4	Hong Kong	5
Hong Kong	PRC	31	USA	37
	Japan	19	PRC	14
	Taiwan	9	Germany	8
Reexports to			PRC	33
			USA	18
			Japan	5

* The USSR accounts for 80% of Mongolia's total trade turnover.

** The USSR accounts for 40% of North Korea's total trade turnover.

SOURCE: *The Europa Year Book,* 2 vols. (London: Europa, 1988 and 1989).

and molybdenum, and consumer goods, primarily of livestock products, now comprise a greater share of Mongolia's exports. Imports are dominated by machinery and equipment (35 percent), fuels and minerals (30 percent), and consumer goods (20 percent).

The People's Republic of China

The economy of China has undergone a major transformation, as described earlier, since the late 1970s. This has had an impact on foreign trade, whereby the leadership now seeks assistance from the industrialized, capitalist world in the development of the economy. This assistance takes the form of technology and high-level manufactured goods. These are used, in part, to develop China's vast natural resources; in part, to build the industrial and transportation infrastructure; and in part, to satisfy the needs of the Chinese people.

Primary goods and basic manufactured items account for over half of China's exports. Most valuable are pe-

troleum, textiles, and clothing. Agricultural products, especially fish, are also important export items.

Manufactured goods, industrial machinery, transportation equipment, chemicals, and food account for approximately 80 percent of all Chinese imports. Most valuable are machinery, transportation equipment, and cereals, especially wheat. Also important are chemicals, and iron and steel. In recent years, China has experienced a relatively balanced mix of imports and exports.

Currently, over 55 percent of all Chinese imports come from industrialized, capitalist countries. Most important are Japan, Hong Kong, and the United States, which alone account for over 50 percent of all Chinese imports. Conversely, over 55 percent of all Chinese exports go to industrialized, capitalist countries, most importantly to Hong Kong, Japan, and the United States.

In this regard, Hong Kong serves as the second most important port for the PRC. In addition to its own domestic exports, Hong Kong reexports goods. **Reexports** comprise 48 percent of all Hong Kong exports.

Of these, 33 percent are reexported to China. In this manner, countries like South Korea and Taiwan, which do not have relations with the PRC, trade with it indirectly. In Hong Kong, the level of interaction among the PRC, Japan, and the United States as well as Taiwan and South Korea is apparent. Under the current agreement with the British, Hong Kong will retain this role after its return to the mainland in 1997.

North Korea

Available trade data for North Korea are sparse, although basic characteristics can be identified. It exports magnesite, zinc, silver, lead, fish and fish products, steel, and military equipment, primarily to other communist countries. It is dependent on imports for petroleum, aluminum, wheat, chemicals, machinery and transportation equipment, and precision equipment.

Strong trade ties are maintained with Japan, which is Korea's third largest trading partner after the Soviet Union and China. In order to facilitate the transfer of capital and technology, North Korea is now encouraging joint ventures with Japan, whereby Japan helps to develop North Korean resources in return for the resources themselves. Similar arrangements have been employed by the Japanese in Soviet Siberia and China.

North Korea's need to import petroleum is reflected in the fact that Saudi Arabia is responsible for over 10 percent of all North Korean imports. This dependency on imports is also reflected in the overall trade deficit, which amounted to $6 billion in 1988, much of it with Japan.

South Korea and Taiwan

South Korea and Taiwan, as emerging economic powers in the region, have foreign trade patterns that are quite similar. Both countries have export-based economies, a dearth of natural resources, and a dependence on the United States. In both cases, Japan and the United States account for over half the country's total imports. Germany accounts for 5 percent of Taiwan's imports and 4 percent of South Korea's. Exports are dominated by the United States and Japan, followed by Hong Kong, which primarily reexports them to the PRC.

In the mid 1980s, South Korea made the transition from a trade-deficit to trade-surplus country. By 1988 the trade surplus reached nearly $9 billion. This trend follows in the footsteps of Taiwan's trade surplus, which amounted to over $14 billion in 1989.

South Korean exports are dominated by textiles and clothing, which account for over one-fourth of total ex-

Much of South Korea's economy depends on exports to other countries through ocean ports, such as Pusan on the east coast.

ports, transportation equipment (20 percent), and electrical machinery (10 percent). Manufactured goods also comprise 95 percent of all Taiwan's exports. These include textiles and clothing, plastic articles, industrial equipment, electrical equipment, and various kinds of consumer goods. South Korea and Taiwan are both moving into markets once dominated by Japan, especially with respect to steel, ships, automobiles, and other kinds of consumer goods. Concurrently, they are challenging Japanese markets, especially in Asia and the United States.

South Korean and Taiwanese imports are influenced by the need for mineral and energy resources and products to continue to fuel their economic growth. Petroleum, agricultural products, especially wheat and cotton, industrial machinery and supplies, and transportation equipment account for over 80 percent of these countries' imports.

Major trading partners include Japan and the United States, which together account for over half of all foreign trade for these two countries. Germany is a distant third as an exporter to South Korea and Taiwan. Hong Kong provides the third largest market for their exports, a portion of which are reexported to the PRC.

The communist and noncommunist countries in East Asia appear to complement each other, thereby increasing the level of economic interaction. Even though there are no political relations between North and South Korea, or between the PRC and Taiwan, the need for technology, high-level manufactured goods, and consumer goods in China and North Korea, and the need for mineral and energy resources in South Korea and Taiwan, not to mention Japan, suggest that this interaction will increase, especially if China continues on the development road begun in the late 1970s.

Energy Requirements

One area of particular vulnerability is energy dependency. Table 14.6 shows the relative dependency of East Asian countries on energy imports. China is the only East Asian country that is self-sufficient in energy, primarily as a result of petroleum production. China exports one-third of the oil it produces. North Korea imports only 9 percent of its energy needs, primarily in the form of petroleum, for which it is totally dependent on imports. South Korea and Taiwan, on the other hand, are heavily dependent on imports for their energy needs. Almost 90 percent of the energy consumed in Taiwan is imported, while South Korean energy imports account for three-fourths of its energy. Both are totally dependent on imports for oil.

STRATEGIC SECURITY

Although there is continuing economic interaction among the countries of East Asia, strategically it is still a region of direct confrontation between the communist and noncommunist worlds. We have already discussed the tenuous position of Japan. For the rest of East Asia, we see two countries divided: China into the People's Republic on the mainland and Nationalist China on Taiwan, and Korea into North and South. The DMZ in Korea and the 100-mile (65 kilometer) Strait of Taiwan serve as buffer zones between these ideological and economic antagonists. Yet, a siege mentality can still be found throughout the region, not without some justification.

Post–World War II Conflicts

Since World War II, there have been several notable military confrontations between these East Asian neighbors

TABLE 14.6

Energy Dependency

	CHINESE ENERGY RESOURCE EXPORTS (AS A PERCENTAGE OF PRODUCTION)	ENERGY RESOURCE IMPORTS (AS A PERCENTAGE OF CONSUMPTION)			
		Taiwan	North Korea	South Korea	Mongolia
Total	8%	89%	9%	76%	36%
Coal	1	—	1	46	—
Oil	33	—	100	100	100

SOURCES: *Energy Statistics Yearbook* (New York: United Nations, 1985); and *Republic of China, 1986* (Taibei: Hilit, 1986).

FIGURE 14.10

Post World-War II Conflicts

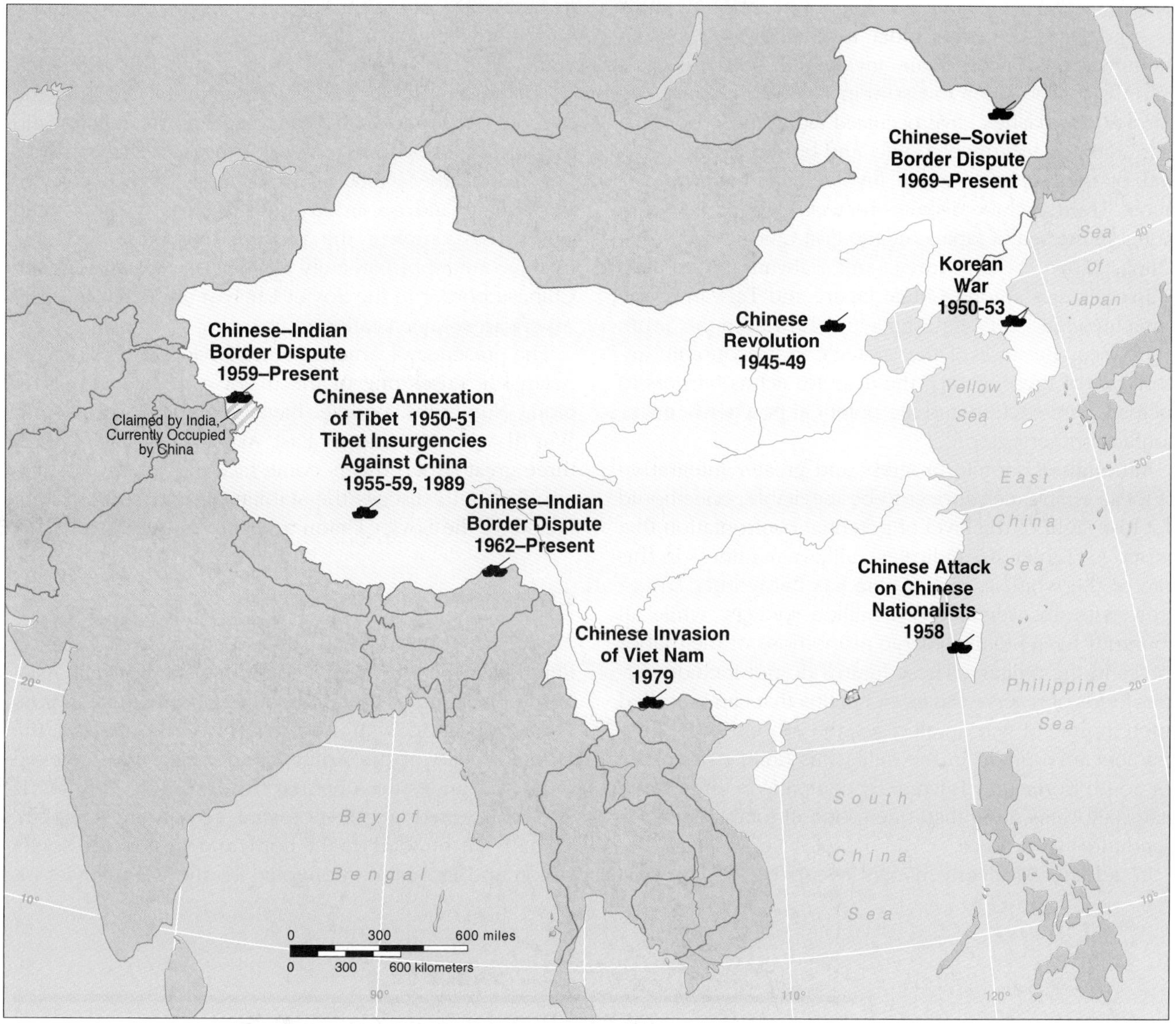

(Figure 14.10). The Korean War (1950–1953) brought North Korea and China into conflict with South Korea and the United Nations, primarily in the form of the United States. There have been continual skirmishes and incidents along the DMZ since its establishment. However, the two Koreas are showing signs of a rapprochement in recent years.

The situation between the PRC and Nationalist China is similar. Although the Chinese have not attempted to invade Taiwan, perhaps because of the long-time presence of the American 7th Fleet, China did bombard the Taiwanese islands of Quemoy and Matsu in 1958. Although most believe that the bombardment was intended to test the resolve of the Nationalist government, it elicited a response from the United States in the form of increased American commitment to Taiwan and East Asia.

During the 1950s, China also sought to expand its

realm in the southwest. In 1950–1951 China invaded and annexed Tibet, which eventually became a Chinese autonomous republic. China was also involved in border clashes with India in Ladakh in 1959. In the northeast, China became involved in border clashes with the Soviet Union in Ussuri in 1969. Continuing and intermittent clashes have occurred along these borders.

In addition to armed conflict and border clashes, several of these countries are involved in maritime disputes. These include disputes between Japan and South Korea in the Sea of Japan and the East China Sea; among China, Japan, South Korea, and Taiwan in the East China Sea and among China, Japan, and Taiwan in the same area; and between China and Taiwan in the South China Sea. Because of the strategic and economic importance of these waters, the disputes act as barriers to increased interaction and are potential powder-kegs for conflict in the region.

Although economic progress and greater integration of these economies appear to be inevitable, one should not lose sight of the level of potential confrontation that exists. As Table 14.7 indicates, military readiness in this part of the world is high. China has the world's largest army with an estimated 3–4 million soldiers, while all countries have relatively large proportions of their labor forces in the military. These figures do not include various kinds of reserve forces and militia that can be called upon in time of war. In all cases, the proportion of government investment in the military is high, particularly for South Korea and Taiwan. By comparison, the United States allocates more than one-fourth of total federal expenditures to defense.

In addition to conventional weapons, China possesses medium-range ballistic missiles, which are capable of hitting targets in the Asiatic Soviet Union and in its European core. Although no other East Asian country currently has nuclear capabilities, South Korea and Taiwan are on the verge of developing a nuclear capability.

The data in Table 14.7 do not take into account the five Soviet divisions in Mongolia nor the military aid the Soviet Union provides to Mongolia. They also do not include the 40,000 U.S. troops in South Korea nor the military aid accorded South Korea by the United States. To complete the picture, one must also take into account approximately 50 Soviet divisions along the Chinese border in the Soviet Far East and almost 50,000 American soldiers stationed in Japan.

The presence of American and Soviet soldiers on East Asian soil raises one of the major concerns of this region. Numerous people have suggested that World War III will originate in East Asia, where the world's three great superpowers come face to face. In fact, it is impossible to discuss the stability of East Asia without including the Soviet Union and the United States.

China and North Korea

As the heir of the great Chinese empire, the People's Republic of China is still the dominant force in Asia. Territorial disputes that originated with the Russian government continue with the Soviets over the status of the area northeast of the Amur River, which was annexed by Russia under the unequal treaties in 1853. China is also concerned about potential Soviet encirclement. China is surrounded on the north and west by the Soviet Union and its vassal, Mongolia. To the west, Soviet in-

TABLE 14.7

Military Data for East Asian Countries

	ARMED FORCES (THOUSANDS)	PERCENTAGE OF LABOR FORCE	DEFENSE AS A PERCENTAGE OF TOTAL EXPENDITURES
Mongolia	37	10%	13%
People's Republic of China	3160	< 1	8
Taiwan	406	5	37
North Korea	838	11	24*
South Korea	629	4	29

* Although official North Korean statistics show defense as only 14 percent of total expenditures, most experts agree that actual expenditures are closer to 23 percent.

SOURCE: *The Europa Year Book*, 2 vols. (London: Europa, 1988 and 1989).

fluence in Afghanistan touches Chinese territory. And to the south, pro-Soviet regimes now rule Vietnam, Laos, and Cambodia. Chinese mistrust of Soviet intentions runs high.

In response, China has greatly expanded its contacts with the United States. Partially in an effort to block a possible U.S.-Soviet alliance against China, and partially to facilitate the transfer of technology, Chinese relations with the United States have steadily improved since the early 1970s. China has also reached out to Japan for increased interaction. With respect to foreign trade, China represents an intervening opportunity for Japan, since China and the Soviet Union both have many of the resources that are needed by the Japanese. In this respect, increased trade between Japan and China means decreased trade between Japan and the Soviet Union.

China also seeks to regain lost territories. In this regard, Hong Kong and Taiwan are the most important. Under the system of unequal treaties, Hong Kong was leased to the British in 1898 for 99 years. The terms of its return have already been worked out. As we have seen, under the concept of "one country, two systems," Hong Kong will retain its status as a center of capitalism within China. Uncertainty over the future, however, has led to an outflow of funds from Hong Kong to more secure places and had a chilling effect on new commercial ventures. The PRC has offered to treat Taiwan in the same manner, but the Nationalist government refuses to negotiate with the communists, maintaining that it is the true government of China. For the time being, China is not forcing the issue and is content to wait.

It is important not to assume that North Korea is simply an extension of Chinese policy. Over the years, the North Koreans have proven to be independent. There is strong evidence to suggest that North Korea invaded South Korea without the prior knowledge of either China or the Soviet Union. The fact that North Korea has maintained ties with both the Soviet Union and China after the Sino-Soviet split in 1960 is indicative of North Korea's intent to act in its own best interest.

South Korea and Taiwan

South Korea and Taiwan, like Japan, have maintained close relations with the United States and have received U.S. economic and military aid in return. Although ties with Taiwan have been officially severed, indirect U.S. support still plays a major role in the health of the Tai-

wanese economy. Although South Korean and Taiwanese soldiers are among the best trained in the world, it is questionable whether they would be able to withstand a Chinese or North Korean attack without U.S. support.

U.S. Involvement

The Chinese Revolution, the Korean War, and wars of national liberation following World War II impressed upon the United States the vulnerability of East Asia and its importance to world peace. These events ensured a greatly increased American presence in the region. Although history has proven the fallacy of assuming that communism is monolithic and run solely from Moscow, as many believed in the 1950s, this misconception determined the United States' role in the region. The political and economic effects of this commitment have been far reaching. The economies of Japan, South Korea, and Taiwan have all been greatly helped by the American presence. Now, it appears that much of the dialectic of the postwar period has diminished and that increased interaction between the antagonists is possible.

Thus, in East Asia we can be cautiously optimistic. The criteria for confrontation and conflict remain, and many territorial disputes are still to be resolved. Economic considerations, however, have recently superseded these political considerations as the countries of this region, especially China and North Korea, see the importance of increased integration into the world economy to stimulate development. As these two communist nations become increasingly connected to their East Asian neighbors and the world, the chances of major conflicts decrease, and the opportunities for peaceful solutions increase.

PROBLEMS AND PROSPECTS

One can only imagine what the future impact of an economically developed and integrated East Asia would be. All the basic components for an economic superpower exist. Culturally, all these societies have felt the common influence of China. Economically, these countries have vast natural resources as well as technological and industrial expertise. They also represent a major market potential because of the number of people. An East

View of Hong Kong, an outpost of capitalism on the coast of China. The return of
Hong Kong to China has many citizens and businesspeople of Hong Kong concerned
about the future.

Asian "economic community" could be a formidable force on the world market.

Yet, the prospects for that happening in the near future are slight. Numerous obstacles stand in the way. First and foremost are historical antagonisms, including North Korea's status as a vassal of China and the not forgotten repressive nature of the Japanese occupations of China, Taiwan, and Korea. Current disputes over borders and seas are also hindrances. And, of course, the basic antagonisms between communist and capitalist societies are also present. As the various countries continue on their separate ways, each must address several issues to ensure continued economic growth and improvements in the standard of living.

In China, continued economic successes will depend on even greater improvements in family planning and the future integration of China into the world economy.

Apparently, the government is committed to reaching zero population growth by the end of the century. Yet, a 1 percent increase for a population of one billion translates into an additional 10 million people in one year. China's family planning efforts cannot stand on their laurels.

Additional stabilization of the economy and further integration into the capitalist world economy, as proposed in the "four modernizations," are key components. The more moderate approaches to development must be allowed to affect economic change. More radical elements, who wish to return to the policies of Mao, are waiting to return to power if current reforms fail. The ability of Deng Xiaoping's protégés to retain control will be crucial. Events in 1989 indicate that China will most likely not embrace democracy as Eastern Europe has and former Soviet Union have.

In Hong Kong, there is a tense calm over the future return of the territory to China in 1997. This situation is sensitive to economic and political conditions in the PRC, however. If more radical elements return to power in the PRC, this calm could shatter.

For both North and South Korea, a diminution of the military threat and tension between the two would greatly facilitate future development. Psychologically, it would reduce the siege mentality in the two countries, especially South Korea. Economically, it would allow the governments to expend a greater proportion of their budgets on nondefense sectors of the economy.

In North Korea, recent overtures to Japan are being expanded to include greater interaction with technologically advanced capitalist countries. South Korea needs to diversify its supplies of essential raw materials.

In Taiwan, domestic political tensions are being eased. Recent moves toward opposition parties and the succession of the first native Taiwanese president need to be extended. For the time being, tensions with the PRC have eased as the Chinese concentrate on economic development. Taiwan also needs to diversify its supply of raw materials and continue developing industries that help strengthen its export-based economy.

The prospects for peace and economic integration in East Asia are better now than they have ever been. The example of Japan and the promise of economic growth and an improvement in the quality of life overshadow rivalries between nations. Yet, the situation is delicate. Any incident could reignite historical, cultural, and political antagonisms.

Suggested Readings

Adshead, S. A. M. *China in World History*. New York: St. Martin's Press, 1988.

Cannon, T., and A. Jenkins, eds. *The Geography of Contemporary China*. New York: Routledge, Chapman, and Hall, 1988.

The Contemporary Atlas of China. Boston: Houghton Mifflin, 1988.

Fairbank, John, Reischauer, Edwin, and Craig Albert. *East Asia: Tradition and Transformation*. Boston: Houghton Mifflin, 1973.

Grinter, L., and Young Whan Kihl, eds. *East Asia Conflict Zones: Prospects for Regional Stability and Deescalation*. New York: St. Martin's Press, 1987.

Kolb, A. *East Asia: China, Japan, Korea, Vietnam: Geography of a Culture Region*. Trans. by C. A. M. Sym. London: Metheun, 1971.

Kraus, W., and W. Lükenhorst. *The Economic Development of the Pacific Basin*. New York: St. Martin's Press, 1986.

Leung, CK., and N. Ginsburg, eds. *China, Urbanization, and National Development*. Chicago: University of Chicago Press, 1980.

Nahm, A. *Korea: Tradition and Transformation*. Seoul: Hollym, 1988.

Pannell, C., and L. Ma. *China: The Geography of Development and Modernization*. New York: Halsted, 1982.

Shirk, S., ed. *The Challenge of China and Japan: Politics and Development in East Asia*. New York: Praeger, 1985.

Steinberg, D. *The Republic of Korea: Economic Transformation and Social Change*. Boulder, Colo.: Westview Press, 1989.

Tregear, T. R. *China: A Geographical Survey*. New York: Halsted, 1980.

World Bank. *China: Long-Term Development Issues and Options*. Baltimore: Johns Hopkins University Press, 1985.

Zhao Songqiao. *Physical Geography of China*. New York: John Wiley & Sons, 1986.

CHAPTER 15

South Asia

IMPORTANT TERMS

Caste system

Dharma

Centripetal force

Centrifugal force

Cottage industry

Import substitution

Green Revolution

Land reform

CONCEPTS AND ISSUES

■ Cultural diversity and the resolution of religious and ethnic conflicts.

■ Population growth exceeding carrying capacity.

■ The Green Revolution and rural industrialization as strategies for development.

■ Coping with massive urbanization.

CHAPTER OUTLINE

PHYSICAL ENVIRONMENT
Landforms
Climate and Natural Vegetation

HUMAN ENVIRONMENT
Historical Background
The Population Dilemma

Linguistic Diversity
Political Change
Industry
Agriculture
Urbanization

SPATIAL CONNECTIVITY
Spatial Interaction within South Asia
South Asia and the Global
    Economy

PROBLEMS AND PROSPECTS

With a population exceeding one billion, South Asia represents one of the world's great concentrations of human population. One-fifth of all human beings reside here. The region is dominated by India, whose population is expected to exceed that of its neighboring giant, the People's Republic of China, early in the next century. South Asia comprises six countries: three large countries, India (853 million), Pakistan (115 million), and Bangladesh (115 million); and three smaller ones, Sri Lanka (17 million), Nepal (19 million), and Bhutan (1.6 million) (Figure 15.1).

In the sixteenth and seventeenth centuries, the region was renowned for its fabulous wealth. Its spices, silks, and brassware attracted traders from Europe and inspired the voyages of discovery as Europeans searched for safe routes to the Indies. By the nineteenth century, the region, with the exception of Nepal and Bhutan, had become part of the vast British Empire. But the unity imposed by British control was tenuous. It was a veneer that masked the continuing dominant characteristic of the region, its diversity and complexity. The variations in religion—Hindu India and Nepal, Islamic Pakistan and Bangladesh, and Buddhist Sri Lanka, and Bhutan—are one example of this diversity that also extends to language, customs, economics, politics, and even topography.

Our images of South Asia today are very different from those of the seventeenth-century European traders. Each country in this region is struggling with poverty at the most intense level. Although the processes of economic development and modernization are under way, these countries still fall into the category of the fourth world, "managing poor" (India, Pakistan, Sri Lanka), and the fifth world, "extremely poor" (Bangladesh, Nepal, Bhutan). The depth of this struggle with poverty can be seen in the Statistical Profile. Population growth rates, though falling, continue to be above 2 percent annually, representing a doubling of population every 25 to 35 years. In India alone, 17 million people are added each year. Infant mortality and life expectancy at birth, among the most sensitive barometers for measuring the quality of life, present a depressing picture for all countries except Sri Lanka. Per capita income for 1988 was under $400 for almost the entire region. These statistics raise many questions about the future of this region. As the population passes the one billion mark, can the land base support the stress of more intensive cultivation? With the bulk of the population still engaged in agriculture, can these countries cope with the rural-urban migration that must follow as people move off the land to search for jobs in already overcrowded cities?

The statistics, though formidable and overwhelming, do not tell the whole story, however. There have been some remarkable achievements, though too often they have been overshadowed by the catalog of problems. The success stories are as significant as the problems and must not be ignored. They are a testimony to the resilience and strength of South Asian society. The examples can be local, such as the success of wheat farmers in the Punjab or reforestation projects in Nepal, or they can be national in scale. Sri Lanka's success in promoting rural health care and implementing an aggressive family planning program has made it a model to be emulated by other developing nations. India's achievements in industry have made it virtually self-sufficient in manufactured goods, thereby breaking the long-standing dependent relationship between the colonizer and its former colony. Finally, the survival of India as a sovereign state is a noteworthy achievement for a country that is as culturally complex, diverse, and varied as the entire continent of Europe.

The juxtaposition of poverty-ridden shantytowns, with modern buildings, as in this view of Bombay, is a common feature of South Asian cities.

PHYSICAL ENVIRONMENT

LANDFORMS

At first glance South Asia would seem to be surrounded by impenetrable barriers that protect the region from the continental extremes of Central Asia (Figure 15.2). To

FIGURE 15.1

South Asia: Major Political Divisions

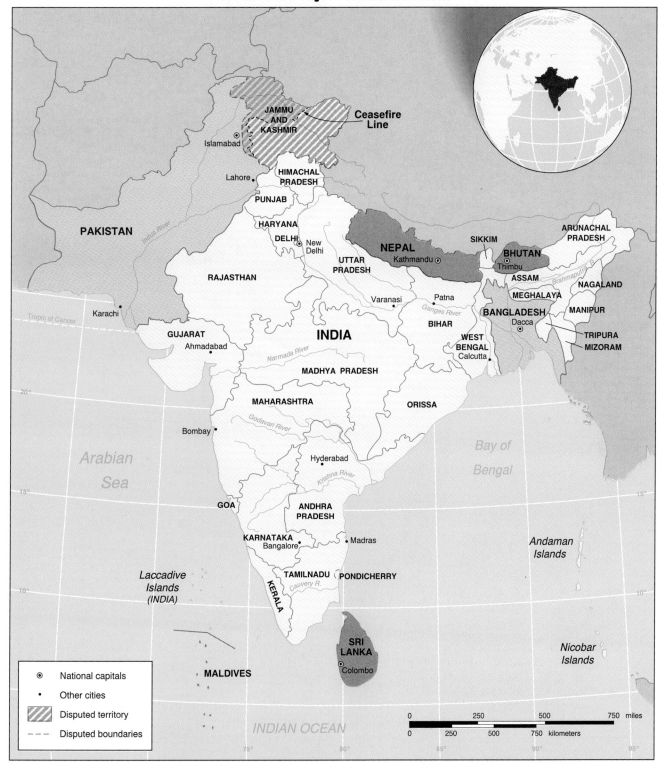

FIGURE 15.2

Major Landforms of South Asia

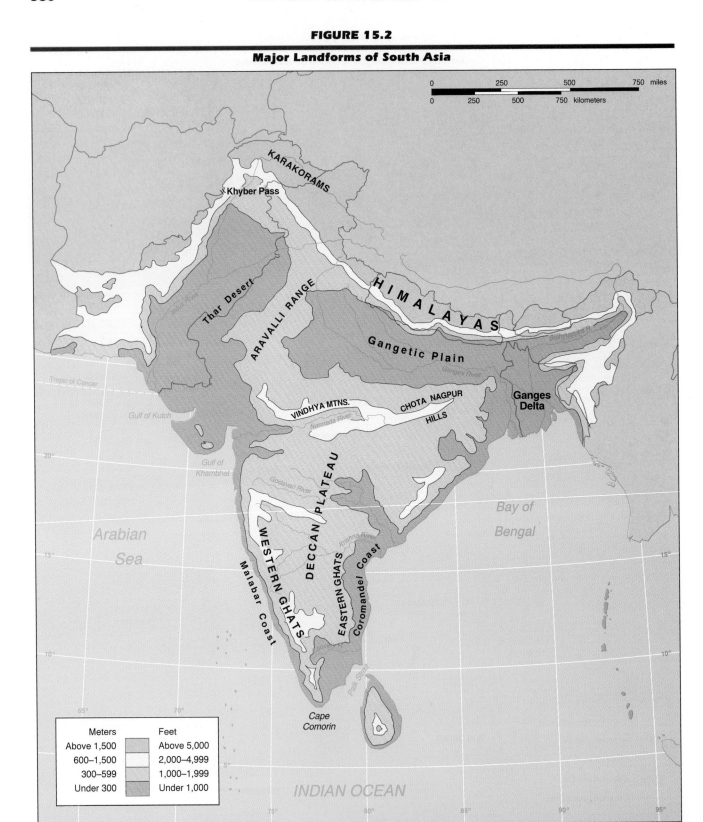

KARAKORAMS

Khyber Pass

H I M A L A Y A S

Thar Desert

ARAVALLI RANGE

Indus River

Gangetic Plain

Brahmaputra R.

Ganges River

Tropic of Cancer

Gulf of Kutch

VINDHYA MTNS.

CHOTA NAGPUR HILLS

Ganges Delta

Narmada River

Gulf of Khambhat

DECCAN PLATEAU

Godavari River

20°

Arabian Sea

WESTERN GHATS

EASTERN GHATS

Krishna River

Coromandel Coast

Bay of Bengal

15°

15°

Malabar Coast

10°

10°

Palk Strait

Cape Comorin

65°

70°

INDIAN OCEAN

75°

80°

85°

90°

95°

Meters	Feet
Above 1,500	Above 5,000
600–1,500	2,000–4,999
300–599	1,000–1,999
Under 300	Under 1,000

0   250   500   750   miles

0   250   500   750   kilometers

STATISTICAL PROFILE: SOUTH ASIA

REGION OR COUNTRY	Population Estimate mid-1990 (Millions)	Natural Increase (Annual)	Population Projected to 2000 (Millions)	Infant Mortality Rate	Life Expectancy at Birth (Years)	Urban Population (%)	Per Capita GNP, 1988 (US$)	Area, Thousands of Square Miles (Km²)	Population Density, No./mi² (No./km²)
Bangladesh	114.8	2.5	146.6	120	54	13	170	55.6(144)	2064(797)
Bhutan	1.6	2.1	1.9	128	48	5	150	18.2(47.1)	17(6)
India	853.4	2.1	1042.5	95	57	26	330	1269.3(3287)	672(259)
Maldives	0.2	3.7	0.3	76	61	26	410	0.120(.31)	1882(727)
Nepal	19.1	2.5	24.3	112	52	7	170	54.4(141)	352(134)
Pakistan	114.6	3.0	149.1	110	56	28	350	310.4(804)	369(142)
Sri Lanka	17.2	1.5	19.4	22.5	70	22	420	25.3(66)	679(262)

the north lie the massive Himalayas and beyond, the arid, cold Tibetan Plateau. In the west, the sharp ridges of the Hindu Kush lie adjacent to hundreds of miles of desert. To the south lies the Indian Ocean. Yet the region has maintained continuous and vital relations with other places by land and by sea. It was mainly from the west, through the mountain passes of the Khyber and the Bolan and through the narrow gateway along the Arabian Sea, that a long succession of conquerors, invaders, and other cultural influences came, creating the enormously complex civilization that today characterizes South Asia.

The physical geography of South Asia is complex with a wide variety of landforms and climate types. Landforms include the world's highest mountains and tropical coastal plains. Climates range from the extreme aridity of the Thar Desert in the northwest to the torrential monsoon rains of Cherrapunji in the Khasi Hills of Meghalaya in the northeast.

South Asia can be divided into four broad physiographic regions; the northern mountains, the plains and delta region, the Deccan Plateau, and the tropical coasts of Malabar and Coromandel.

The Northern Mountains

The northern mountains are dominated by the Himalayas, one of the world's most imposing barriers. Average elevations are above 20,000 feet (6096 meters), while at least 25 peaks exceed 25,000 feet (7600 meters). There are few passes and those that exist are at very high elevations. The mountains have a major effect on climate, protecting South Asia from the extreme cold of Central Asia and generating a tremendous orographic effect on

their southern slopes as moisture-laden monsoon winds blow from the ocean and are forced to rise. Three of the region's major rivers, the Indus, Ganges, and Brahmaputra, have their sources in these mountains. The Himalayas also form a cultural buffer zone between the Chinese and the Indian cultural regions. Nepal, Bhutan, and Sikkim (part of India since 1975) occupy the central portion of these mountains. The region is sparsely populated except in the more favorable sheltered valleys like the Vale of Kashmir or the valleys of Nepal.

Radiating southwest from the Pamirs are two desert ranges, the Hindu Kush of Afghanistan and the Sulemain Range of Pakistan, with their strategically located mountain passes. A frontier zone of mountains and hills dissected by the Brahmaputra River lies to the northeast. The hills of Assam and the Myanmar (Burma) border constitute difficult terrain with jungle-clad gorges and heavy rainfall.

The River Plains and Delta Region

The valleys of the Indus, Ganges, and Brahmaputra form the heartland of South Asia, the social, political, and economic cores of Pakistan, India, and Bangladesh. The Indus and Ganges plains, which were once important centers of historic civilizations, have extremely high concentrations of populations and a dense network of roads and railroads. These alluvial plains, 200 to 300 miles wide, are among the most fertile regions of South Asia. Agriculture occupies the bulk of the population.

In the arid west, the Indus flows across the plains of the Pakistani portion of the Punjab, a fertile region where wheat, cotton, and rice are grown under perennial irrigation, making this one of the two vital food-

The Himalaya Mountains have the highest peaks in the world.

producing areas of Pakistan. The region is centered on the ancient city of Lahore and the newly constructed capital of Islamabad. Further south, the river parallels the Thar Desert and broadens as it flows through the second of Pakistan's irrigated agricultural zones, the Sind, just north of the major port of Karachi. Though irrigation has expanded along the Indus since independence, major environmental problems have arisen from *salinization* (the concentration of salts in the soil) and water logging.

The combination of deep alluvial soils, sufficient rainfall, and a long growing season makes the Ganges Valley India's most productive growing area. Precipitation gradually increases from west to east. In the plains of Punjab on the western-most edge, precipitation is around 40 inches (100 centimeters), and wheat, cotton, and sugar cane are grown, Eastward, the precipitation

increases to around 100 inches (250 centimeters). Wheat fields become increasingly rare and irrigation less pronounced as rice and jute dominate. Population densities are exceedingly high throughout the plain, sometimes reaching densities of over 1035 per square mile (400 per square kilometer) in the more humid east. From Delhi in the west to Calcutta in the east, a line of India's leading urban centers, including Agra, Allahabad, and Patna, stretches out like beads on a necklace.

The Brahmaputra is much narrower and less navigable than either the Indus or Ganges and is somewhat isolated from the rest of the country. For much of its course, it flows through the densely forested frontier area of Arunchal Pradesh. The river widens along the densely settled Assam Valley where rice and jute are grown. On the higher slopes, tea gardens now cover some 40,000 acres (98,800 hectares). As pressure for land grows, there has been widespread deforestation on the steep valley slopes causing silting and flooding downstream.

The Ganges and Brahmaputra meet in the hot, wet delta region and empty into the Bay of Bengal through numerous channels. The delta area is extremely fertile, and practically every foot of soil is under rice or jute. Ownership of the land is highly fragmented, however, and population pressure is intense. With much of the land less than 13 feet (4 meters) above sea level, flooding and the occasional penetration of tropical cyclones, typhoon-type storms, are a major hazard.

The Deccan Plateau

The Deccan Plateau is an ancient massif composed of weathered crystalline rocks. The plateau is bounded by a series of mountain ranges and hills. To the west lie the steeply sloping Western Ghats with elevations exceeding 6000 feet (1829 meters). The heavily dissected plateau slopes gently eastward to the much lower, discontinuous hills of the Eastern Ghats. To the north lie the Satpura and Vindhya ranges and on the southern tip of the plateau, the Cardamon and Nilgiri hills. While soils vary in fertility from the very fertile black volcanic soils of the northern Deccan to less fertile soils elsewhere, it is the availability of rainfall that determines the nature of agriculture. Most of the Deccan is semiarid or has considerable variation in rainfall, partly because of the rain shadow cast by the Western Ghats. Cotton is grown on the rich volcanic soils while drought-tolerant crops such as millet and sorghums are grown elsewhere. Much of India's mineral wealth including copper, gold, manganese, iron, and coal is found in this plateau.

The Tropical Coasts

The Malabar and Coromandel coastal plains meet at the southern tip of the Indian peninsula. Of the two, the Malabar is more clearly defined, flanked by the steep ridges of the Western Ghats. The warm, moist tropical climate and fertile soils combine to make these plains highly productive agricultural zones. Rice is the major crop but spices, coconuts, and oil seeds are also important. Tea is grown on the adjacent hills. These plains have had a long history of trading with the Middle East, Africa, and Europe and with Southeast Asia. In the sixteenth century, the Portuguese established the port city of Goa. A century later, the British developed India's great port cities of Bombay and Madras.

The island of Sri Lanka is physically similar to southern India. The high mountains of the south are surrounded by dissected foothills. Beyond this lies a broad, moist, tropical coastal plain where rice is the dominant crop on the lowlands and tea, rubber, and coconuts are grown as plantation crops on the lower slopes of the mountains.

CLIMATE AND NATURAL VEGETATION

Although the climatic pattern of South Asia rivals that of its landforms in diversity, there is one dominant climatic force in the region: the monsoon. Derived from an Arabic word "mausim" meaning season, the term "monsoon" refers to a seasonal reversal of winds. During the summer, humid, unstable air moves from the ocean to the land bringing rain. In winter, a dry wind, having its origins over the continent, blows outward toward the ocean (Figure 15.3). The causes of the monsoon are complex and not yet fully understood. They are associated with the alternating heating and cooling of the ocean and continent and the corresponding pressure changes that occur as these bodies respond to the changing seasons. The huge landmass of the Asian continent intensifies contrasts arising from the usual heating and cooling difference between land and oceans.

Recent evidence suggests that winds in this region adjust to the shifting location of the jet stream north and south of the Himalayas. In winter, the jet stream lies south of the mountains, blocking the moist air from the ocean and preventing it from moving inland. In summer, the jet stream lies north of the Himalayas, allowing moist air from the ocean to penetrate inland. The contrast between the two seasons is striking. By the end of the dry season, the land is parched and desiccated, and average maximum temperatures in much of India are above 104°F (40°C). The monsoon rains start early in June bringing welcome relief and depressing average maximum temperatures by as much as 18°F (10°C). During the months of June to September, the summer monsoons bring 80 to 90 percent of India's total annual rainfall.

The date of the onset of the monsoon, the intensity of rain, and the degree of penetration inland vary greatly, however (Figure 15.4). Where moist air is forced over barriers like the Himalayas or Western Ghats, heavy precipitation occurs. Cherrapunji in the Khasi Hills of Meghalaya, which lie at right angles to the advancing winds, receives a yearly average precipitation

FIGURE 15.3

The South Asian Monsoon

The South Asian monsoon is characterized by a strong onshore flow in summer and somewhat less pronounced offshore flow in winter.

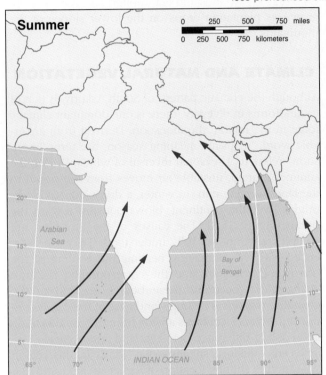
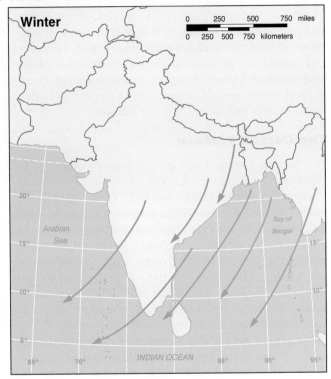

of 425 inches (1090 centimeters), but the record for this town is 1042 inches (2671 centimeters). Other areas receiving abundant rainfall include the Ganges Delta, the Indian Coastal Plains, and much of Sri Lanka. The Deccan Plateau, on the other hand, lying in the rain shadow of the Western Ghats is plagued by inadequate summer moisture. Pakistan is also generally arid, and only the coastal region receives the full impact of the monsoon.

Throughout most of South Asia, the rhythm of agricultural life and success of harvests are tied to the onset of the summer monsoon. The phrase that the monsoon is the "basis of life" is no exaggeration. When the rains are late or are inadequate, as has occurred two out of every five years in the last two decades, famine results. In 1973 and 1974, for example, the monsoons did not develop properly, and there were major food shortages. The heavy monsoon rains of 1975 were a welcome relief, and India produced a record crop of grain. Where

the monsoons strike with unusual intensity, however, rains may destroy crops and cause widespread flooding.

One important advantage to South Asian agriculture is that a year-round growing season is possible in all areas except the Himalayas. When water is available, two or three crops may be grown. Unfortunately, in many areas, facilities for storing water for the dry season simply do not exist. It is difficult to believe that Cherrapunji with its high average rainfall often experiences severe water shortages in April and May.

Like eastern China, much of South Asia has been occupied by a dense population for a very long time. It is, therefore, difficult to speak of "natural vegetation." While the lack of freezing temperatures and season or year-round rainfall in most of the region allow for abundant vegetative growth, much of the land has been cleared for intensive cultivation. Further destruction of the original cover has been caused by grazing animals,

FIGURE 15.4

Distribution of Precipitation in South Asia

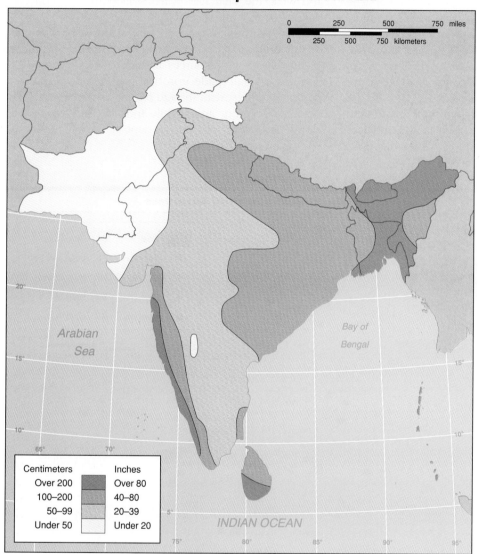

wood gathering for fuel, and repeated burning from slash and burn agriculture.

The original vegetation, in all but the wettest areas, was probably a tropical wet and dry forest with species adapted to the heavy rainfall of the monsoon followed by a period of drought. Much of the original forest has been reduced to scrub forest and bamboo. Today, the only significant stands of tropical forest are found in northeast India along the foothills of the Himalayas and the steep gorges of Assam and the Hill States.

The Resource Base: Land, Water, Soils, and Minerals

South Asia's resource base, particularly that of India, is impressive, especially when compared to the rest of Asia. Over half of India's land, an area one-third larger than that of China is arable. At present, only one-quarter of this land is under irrigation. Agricultural production could increase significantly if irrigation systems were expanded since most of the region has a year-round growing season. The Indus and Ganges plains, with their rich

alluvial deposits, are among the most productive lowlands in the world. In the Ganges Delta of Bangladesh,
the ratio of arable land to total area is even higher. The
rich black earth of the northwest Deccan is extremely
fertile, but elsewhere the red-yellow latosols need careful application of lime and fertilizer to be productive.
While much of the plateau can sustain agriculture, availability of water is a problem. The same is also true for
northwest India and much of Pakistan.

The areas least favorably endowed with arable land
include the mountains of western Pakistan and the king-

doms of Nepal and Bhutan. Here the rugged terrain and
harsh climate limit settlement to sheltered valleys. In the
arid west, the Thar Desert has soils with high concentrations of calcium and salt.

India has the strongest mineral resource base in the
region and one that can facilitate industrial development
(Figure 15.5). Its resources include one of the largest
deposits of high-grade iron ore in the world. Coal deposits are widespread though mining is concentrated in
the Chota Nagpur area. While India ranks sixth in the
world production of coal, its deposits are generally of

FIGURE 15.5

South Asia's Mineral Resources

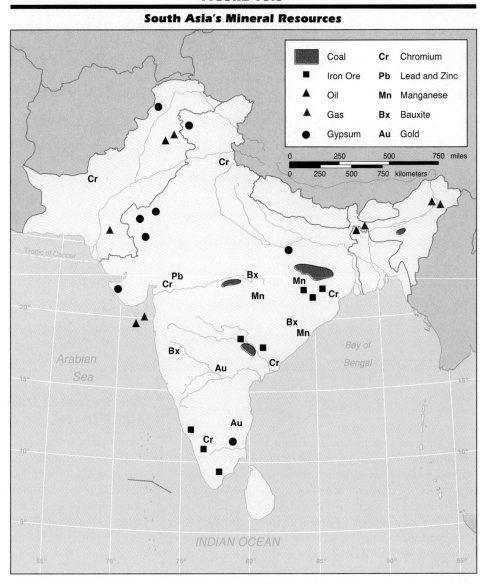

low quality, which handicaps the growth of the steel industry. Other minerals include lead, zinc, copper, asbestos, manganese, chrome, gold, and uranium. Pakistan possesses extensive reserves of coal but, as in India, the quality is poor.

The most serious deficiency for South Asia is oil. Imports of this commodity impose a major drain on foreign reserves. The outlook for India, however, has brightened considerably in recent years. Petroleum production in Assam and Gujarat, which supplied one-third of the country's needs, has been supplemented by offshore discoveries on the western continental shelf off Bombay. The first of these, the Bombay High, began commercial production in 1976 and is producing 10 million metric tons annually. Five other offshore oil fields are now in production, and it is expected that India will soon be able to meet 70 percent of its domestic needs. The region also possesses considerable hydroelectric potential, but as yet development has been fairly limited. In India, hydroelectric power is best developed along the Western Ghats, while Sri Lanka's Mahewali Project should significantly decrease dependence on imported oil.

One other serious resource deficiency should be mentioned. Throughout most of South Asia, fuel wood is becoming increasingly scarce. Firewood prices in Delhi were eight times higher in 1983 than in 1960. As firewood becomes scarce, cow manure becomes more important as a fuel, depriving soils of its use as a fertilizer. The rising demand for firewood leads to rapid deforestation in mountain areas with serious consequences for local communities and for the millions of farmers downstream in the flood plains. Loss of trees and watershed protection cause flooding and silting downstream while landslides endanger mountain communities. Nepal has mounted a vigorous campaign to replant its denuded hillsides. Tree nurseries and plantations have been established in 350 villages. Unfortunately, projects like these are few and the situation seems likely to deteriorate.

HUMAN ENVIRONMENT

HISTORICAL BACKGROUND

The cultural diversity that today characterizes South Asia evolved from a long succession of invaders, conquerors, and ideas that advanced into the region through the narrow mountain passes of the northwest. Forces that shaped the Indian civilization came from the west, not from the rest of monsoon Asia. There was very little contact between the Indian and Chinese civilizations.

Five thousand years ago, a river civilization had emerged in the Indus Valley, making it a contemporary of the civilizations of Mesopotamia and Egypt. Around 2000 B.C., waves of caucasoid Indo-Aryans invaded from the west, absorbing the tribes of the Indus Valley. They moved into the Ganges Valley, pushing many of the darker-skinned **Dravidians** into southern India. The lighter-skinned Aryans brought with them the Sanskrit language that today forms the basis of most of north India's languages. In time, the religious beliefs and practices of these groups were formalized into a religion, Hinduism, and an elaborate social structure—the **caste system.** These two institutions had a more profound influence on South Asia than any of the subsequent invaders including Alexander the Great, the Arabs, Persians, Afghans, and Europeans.

Hinduism, which literally means "Indianism," is difficult to define. It has no body of dogma and encompasses a wide variety of beliefs and practices. Yet the apparent lack of structure is deceptive, for Hinduism provides the spiritual foundations for one of the most clearly defined social organizations in the world. Hinduism is as much a way of life as it is a faith.

There emerged in village life a rigidly structured hierarchical system, based on occupation (the caste system). At the head of this hierarchy were the priests, scholars, and teachers (Brahmins). Other broad groups included the warriors (Kshatryas), the traders and farmers (Vaisyas), and the servant class (Sudras). At the very bottom were the **untouchables** (Harijans). It was believed that they could contaminate members of higher castes by their proximity. Within many villages, untouchables continue to live in separate communities and have their own wells. It has been suggested that the system originated as a color ban between the lighter-skinned Aryans and the darker-skinned Dravidians. Over time, the system became more complex and hundreds of subgroups emerged. Each caste formed an exclusive social group following the same occupation and ritual observances, eating together, sharing water facilities, and imposing a ban on marriage outside the caste. The system provided a way to meet everyone's needs in village life. Each village was virtually self-sufficient. Hinduism prescribed that every caste and every individual in that caste had a duty to perform

THE MONSOON

F or much of South Asia, the most critical event of the year is the arrival of the summer monsoon. The success of harvests, the well-being of villagers, and vital economic decisions are tied to this event. For rural areas, the monsoon determines the pace and rhythm of village life. For governments, the economic welfare of the nation, through decisions concerning flood relief, government aid, market value of crops, and purchase of expensive grains on the world market, depends on the presence or absence of the summer monsoons.

The basic physics of the monsoon was first described three centuries ago. From May to September, summer monsoon winds blow inland from the ocean bringing heavy rain. In winter, the winds are reversed as cool, dry air blows out from the land toward the ocean (Figure 15.3). One author has described it as Asia breathing in and then breathing out. The summer monsoon develops in two branches. In the western branch, winds blow across Sri Lanka and India's Malabar Coast buffeting the Western Ghats with heavy rain and reaching Bombay by early June. The second branch moves across the Bay of Bengal bringing violent storms and heavy rains to Bangladesh and northeast India, especially along the southern flanks of the Himalayas. The storms move westward up the Ganges Valley merging with the western

branch sometime in early July. For India, this summer monsoon produces nearly 90 percent of its total annual rainfall. Sri Lanka and the southern coast of India benefit from both summer and winter monsoons because the winter monsoon first blows offshore across warm water and then onshore again.

What is critical is the timing of the monsoon. Crops must be planted and established before the arrival of the summer monsoon. If it is late or fails to arrive, the soil will not have enough moisture to support the growing plants, which will wither and die. If the rains come too early, seedlings rot in waterlogged soil. Equally important is the amount of rainfall generated. Too much rain can wipe out a year's crop and flood entire communities in low-lying areas. Bangladesh is particularly vulnerable to such flooding. When the monsoon falters, and it can

skip entire regions in some years, drought and famine follow.

The summer monsoons exacerbate the problem of soil erosion. The rapid loss of forest cover to agriculture and fuel wood has become serious in mountain areas, especially along the slopes of the Himalayas. The exposed slopes cannot withstand the sustained downpour of the monsoon. In Nepal, where nearly a third of the forests have been destroyed in the last three decades, the situation is particularly critical. Each year more frequent and more powerful landslides sweep the hills. Fields, livestock, and entire villages have been destroyed. Thousands have abandoned the hillslopes and moved to the already overcrowded Nepalese lowlands. The landslides add enormous amounts of silt to the rivers downstream, carrying away precious topsoil and aggravating the problem of flooding.

For these reasons, monsoon forecasts have been called the most important predictions in the world. Yet the climatic mechanisms that control the monsoon circulation are still poorly understood. Despite major advances in using satellite pictures and computer models, most meteorologists agree that the onset of the monsoon cannot be accurately predicted more than a few days in advance. For the people of South Asia, the summer monsoons remain one of life's critical uncertainties.

(dharma). The fulfillment of this duty ensured the perpetuation of village life.

The caste system is slowly breaking down as villages become less isolated and the social and economic relations more impersonal. This is particularly true in cities where people from different castes are thrown together

in offices or factories. But change has been slow, painful, and often traumatic. Villages remain strongholds of caste, and despite government legislation outlawing discrimination, outcastes continue to live in segregated areas in most villages. The problems for India are obvious. Conflict between economic development and caste is inevitable since development requires wider interactions and cooperation.

Other religions developed in India in the fourth and fifth centuries B.C. Buddhism emerged as an outgrowth of Hinduism. Missionaries carried the religion to the rest of Asia where it flourished. In South Asia, however, its dominance was short-lived, and today it is significant only in Sri Lanka and in the mountain kingdom of Bhutan.

In the tenth century A.D., Hinduism was challenged by the spread of Islam. Following the traditional route of earlier invaders from the west, it flourished in the Indus Valley and pushed eastward into the Ganges Plain and Delta. Its penetration southward to the peninsula was limited, however. The Islamic invasions continued into the sixteenth century culminating in the Mogul dynasty (1526–1857). This dynasty was centered on the Ganges Plain and at its zenith controlled most of the subcontinent. It produced one of the world's most illustrious empires, fostering major advances in architecture, the arts, and political organization. Yet resistance to Islam was strong. After many centuries of Islamic rule, the heartland of India and the peninsula remained firmly Hindu. Islam had its greatest impact in the Indus Basin and in the Ganges Delta.

By 1700, Mogul control was waning as central authority collapsed. Local rulers assumed control and old rivalries reasserted themselves. It was into this growing power vacuum that the British moved.

Colonial Rule and the British

British presence in India began when Queen Elizabeth I granted a charter to the fledgling British East India Company in 1600, giving it exclusive rights to the spice trade of Asia. The British were not the first Europeans to trade with India. The Portuguese, the Dutch, and later the French had all established trading bases in the region. By the middle of the eighteenth century, however, the British emerged as the exclusive European power operating in South Asia. By 1760, the British East India Company had assumed direct control of several trading stations along the east coast of India and over most of Bihar and Bengal. As Mogul rule deteriorated, the company assumed control over more and more of the region, and by 1856, two thirds of India had come under company rule.

Profit through trade was the objective of the company, but as it attempted to protect this trade, it found itself increasingly embroiled in attempts to maintain po-

litical order. In effect, much of India was governed by a trading company. Resistance to company rule grew, and matters came to a head with the Sepoy Rebellion of 1857. The British government stepped in and assumed control. For the next 90 years, India was administered as a British Crown Colony.

The Colonial Impact

While the British presence never changed the essential "Indianness" of the region, it did bring about major changes in economic and social life. There is a divergence of opinion on the impact of colonial rule. Jawaharlal Nehru, India's first prime minister, felt that colonialism had created a parasitic relationship between the colonial power and the colony. In this view, foreign domination led to a rapid destruction of the self-sufficient economy that India had built up under the Moguls. Others feel that while colonialism had several negative consequences, the British also made positive contributions to Indian life.

One area in which the British made significant contributions to development was in the field of drainage and irrigation projects. The latter were particularly effective in the semiarid regions of the Sind and the Punjab, making agricultural settlement possible over vast tracts of land that had been only sparsely settled. In the Deccan Plateau, east of Bombay, irrigation projects stimulated the commercial production of cotton and wheat. Drainage schemes were particularly effective in creating new agricultural land in the Ganges Delta. All in all, nearly half of the irrigated land in Pakistan, India, and Bangladesh was developed under the British. Another colonial legacy was the extensive network of rail lines and roads radiating from the major port cities. While these lines served British trading and strategic interests, they also provided a well-developed transportation network—an important prerequisite for economic development. The railroads contributed to the spread of a cash economy and made it possible for surplus grain to reach famine areas. The law and order (Pax Britannia) imposed by colonial rule put an end to much of the chronic warfare and civil disorder that characterized the last years of the Mogul dynasty. This fact, together with rudimentary health measures that were introduced, resulted in significant population growth in the nineteenth and twentieth centuries.

The British left behind an efficient system of public administration and bequeathed a legacy of parliamentary democracy, a free press, and impartial law, though these institutions have survived with some continuity only in India and Sri Lanka.

On the other side, many colonial practices had disastrous effects on the region. The Europeans had found an economy that was self-sufficient. Indian cotton textiles, weapons, brassware, and other manufactured goods were renowned, and there was a lively trade between India and Southeastern Asia, the Middle East, and East Africa. The British took over this trade, destroying the pattern of Indian commerce. British rule brought about the near extinction of Indian industry as cheap textiles and other goods from the factories of Manchester, Liverpool, and Birmingham flooded the market. Native weavers, spinners, and craftsmen were unable to compete and were often forced out of business. By the early twentieth century, India formed the biggest single foreign market for traditional British exports. At the same time, cash crops were introduced to be grown for export to Europe. They included cotton, jute, opium, tea, and coffee. Food crops were relegated to inferior lands, and in many places yields fell below precolonial levels. Farmers, in particular, suffered many setbacks as a result of colonial policies. Food production remained overwhelmingly subsistent, and there was a general failure to bring about reforms in agricultural practices and in land tenure. The situation was made worse by rising taxes to pay for the building of railroads, roads, and the British military presence and by the rapid population growth. Landlessness increased along with rural indebtedness. By the 1920s, one-quarter of all taxpayers were also moneylenders.

Some industry did develop during the colonial period, including jute milling, cotton textiles, iron and steel, and coal mining. Production was limited, however, and largely confined to the port cities of Calcutta, Bombay, and Madras. On the eve of independence, factory industry employed only 2 percent of the total work force, and the region was heavily dependent on the import of manufactured goods.

Finally, the British made no attempt to unify or minimize the internal cultural and political divisions of the region, favoring instead a policy of "divide and conquer." At independence, there were some 550 princely states with varying degrees of autonomy, a multitude of dialects and languages, and two dominant and rival religious groups.

The impact of British rule on Sri Lanka was even deeper. At the end of the eighteenth century, almost 90 percent of the island was forested, but was producing spices and enough rice to feed itself. Under colonial rule, the economy was transformed into an export economy producing tea, rubber, and coconuts. As food production was superseded by plantation crops, the country was forced to import large quantities of rice. Tamil workers were brought in from southern India to work on the plantations, changing the cultural makeup of the island. This action sowed the seeds for recurring and often violent conflict between the minority Hindu Tamils (18 percent) and the majority Buddhist Sinhalese.

Independence, Partition, and Religion

The struggle for independence from Britain had been inspired by Mahatma Gandhi's campaign of nonviolent resistance. Tragically, when independence came in 1947, it was marred by violence and bloodshed between Hindus and Muslims. The Muslims formed a substantial minority in British India, and the idea of a separate Muslim state had the overwhelming support of Muslims long before independence. A compromise solution was imposed whereby the Indian subcontinent was partitioned into two separate countries: India and Pakistan (Figure 15.6). Muslim Pakistan comprised two quite different core areas separated by 1000 miles of Indian territory. West Pakistan occupied the more arid basin of the Indus River, while East Pakistan was centered on the humid delta of the Ganges. No one anticipated the violence and dislocation that would follow such a decision. Millions of Muslims and Hindus found themselves on the wrong side of the border, and a massive migration began as Hindus moved to India and Muslims to Pakistan.

Hindu holy men devote their lives to their religion.

FIGURE 15.6

Partition of British India in 1947

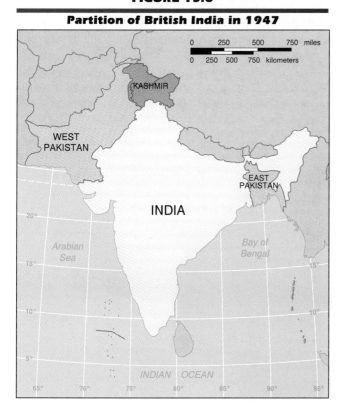

All in all, some 17 million people migrated, and hundreds of thousands were killed or hurt in rioting.

Partition also brought about great economic dislocation. Pakistan, a major food producer, found itself cut off from the iron and steel centers of India. In the Ganges Delta, the jute-producing areas lay in East Pakistan while the jute factories were in neighboring India. Similarly, the cotton fields of Sind were cut off from the textile factories of Bombay. Relations between Pakistan and India since 1947 have been marked by hostility. On two occasions, in 1948 and in 1965, armed conflicts erupted over the disputed territory in Kashmir, a situation that has experienced renewed tension in 1990. In 1971, relations between the two deteriorated further when India supported East Pakistan's decision to secede from Pakistan and form the independent state of Bangladesh.

Significant numbers of Hindus remain in Pakistan (3 percent) and Bangladesh (16 percent). In India, 11 percent of the population is Muslim, but other religious minorities also exist, including *Sikhs* (Sikhism origi-

nated as a reform movement within Hinduism), Jainists, and Christians. Hinduism, however, continues to be the overwhelming binding force that transcends local distinctions in language or custom. Sri Lanka, which gained its independence in 1948, is a predominantly Buddhist country, though it has a significant Hindu minority (15 percent) and smaller numbers of Christians and Muslims.

THE POPULATION DILEMMA

No part of the world faces demographic challenges of the magnitude found in South Asia. In 1986, the combined population of this region passed one billion accounting for one out of every five people on earth (Figure 15.7). The population of India, currently 853 million, adds more than 17 million people each year. It is expected to continue growing until it reaches 1.7 billion sometime late in the next century. In effect, India will add the equivalent of China's population to its current numbers. Bangladesh (115 million), already one of the most crowded lands on earth, is expected to continue to experience population growth well into the next century when its population will be five times greater than its 1982 population. Each year the governments of South Asia must provide additional food, grain, dwellings, schools, jobs, and a host of other services. Gains in industry, agriculture, education, and health care are immediately absorbed by the ever-larger populations. So the region remains poor.

Rapid population growth is a fairly recent phenomenon. In the precolonial era, the population remained fairly steady. Internal strife, disease, and famines kept death rates high. British rule brought with it law and order, rudimentary measures in health care, and improved the transportation system, allowing surplus grain to be moved to famine areas. The death rate fell and the population started to grow. In 1871, after a hundred years of British rule, the population of India (under its present boundaries) stood at 214 million. By 1936 it had risen to 280 million. Today's population is five times greater than at the beginning of British rule. Most of the growth has come from falling death rates, though in recent years the birth rate has also begun to drop (Figure 15.8). In 1940, the crude death rate was approximately 22 per thousand. By 1989 it had fallen to 11 per thousand. Natural increase, the surplus of births over deaths, has been falling in the last decade, and the average for South Asia is 2.3 percent, down from 3 percent in 1970. Nevertheless, the annual growth rates for Pakistan (3.0

FIGURE 15.7

Population Distribution in South Asia

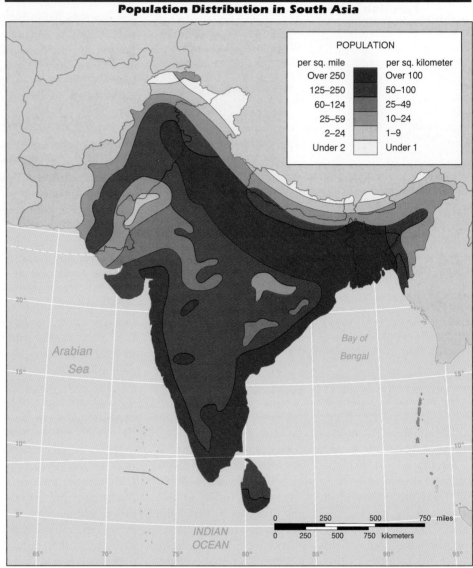

POPULATION

per sq. mile		per sq. kilometer
Over 250		Over 100
125–250		50–100
60–124		25–49
25–59		10–24
2–24		1–9
Under 2		Under 1

Arabian
Sea

Bay of
Bengal

INDIAN
OCEAN

percent) and Bangladesh (2.5 percent) remain high, and only Sri Lanka can be said to have moved into stage three of the demographic transition.

As early as the 1950s, the governments of the region were aware of the problems stemming from rapid population growth. India instituted the world's first national family planning program. Birth control clinics were established in urban and rural areas, and by 1961 more than 4165 clinics were in operation. Unfortunately, with half a million villages to serve, government birth control efforts barely scratched the surface. It was decided to launch a mass education program. Throughout India, posters with the inverted triangle, symbol of the family planning program, carried pictures and slogans like "A small family is a happy family" and "Two children are respectable, three a menace." Teams of family planning workers toured villages showing animated cartoons depicting the disadvantages of large families. The minimum age of marriage was raised to 18 for women and 21 for men. In the 1970s, a vigorous sterilization program was implemented. The program was badly run; there were abuses, and the program met with armed resistance and quite likely brought about the downfall of the Indira Gandhi government. The result was a setback to the family planning program. Today, economic incentives are used to persuade families. Government workers are provided special incentives for small families and penalized if they have more. Small one-time payments of between $11 and $13 are made to men or women who are sterilized, a sum that represents roughly two weeks' wages for rural laborers. In addition, plans are being made to provide a community health worker for each village. By 1985, one-third of India's married couples were practicing some form of contraception, compared with only 10 percent in 1970. Demographers feel that the proportion of users must be above 60 percent by the year 2000 if India is to begin stabilizing its population growth.

All countries in South Asia now have a government-funded family planning program, and there have been some innovative ideas. Sri Lanka and Nepal, for example, have entirely eliminated tax deductions for dependent children. In Bangladesh, family planning and greater economic dependence of women are jointly promoted through credit cooperatives. Most recently,

the country has been considering offering government bonds to sterilization clients with two to three children and to couples who postpone a first pregnancy. Family planning has been most successful in Sri Lanka. With a per capita income of scarcely $400 per year, the country now has one of the lowest mortality and infant mortality rates in the developing world. Life expectancy at birth is 70 years, 15 years longer than in much of the rest of South Asia. Natural increase is now around 1.5 percent and continues to fall as family sizes decline. Sri Lanka was successful because its family planning programs were accompanied by an aggressive social program that included land reform, free primary education, and basic health care provisions.

Despite the success of family planning programs in Sri Lanka, the overall picture is bleak. A major stumbling block is the role children play in overwhelmingly agrarian societies. In urban societies, children are essentially an economic liability, but in rural settings they are viewed as an economic asset. They actively contribute to the family's income and food supply, and their presence makes it unnecessary to hire outside labor. Pension schemes are rare so children provide social security by helping to support and care for aging parents. Another problem is communicating family planning ideas.

FIGURE 15.8

Changes in India's Birth and Death Rates between 1910 and 1980

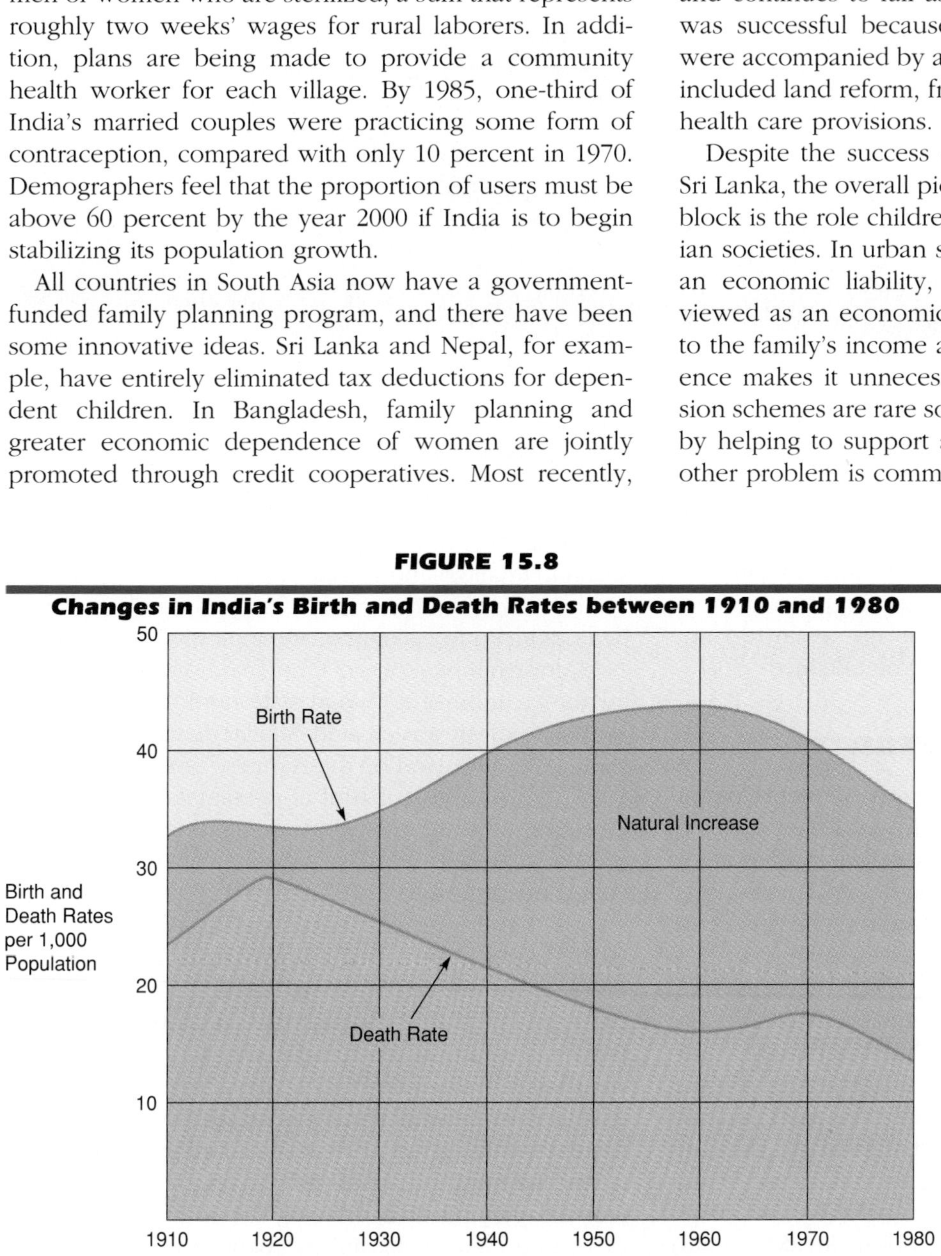

A family planning billboard in South Asia proclaims the advantages and happiness of a small family.

This is especially true in a country like India where there are almost a thousand dialects and languages and only 30 percent of the population is literate. South Asia is a region of many villages, and access to family planning programs is often difficult. It is most restricted in Nepal and Pakistan. About half of currently married women in Nepal are unaware of a modern contraceptive method, and an additional 15 percent who are aware do not know where necessary means can be obtained.

LINGUISTIC DIVERSITY

Religion in South Asia has served both as a **centripetal,** or unifying, **force** and as a **centrifugal force,** dividing people into antagonist groups. Language, on the other hand, is overwhelmingly a divisive force. This is especially true of India, which has several hundred dialects and 15 major languages. Lacking a single national language, Indian leaders have committed themselves to drawing state boundaries along linguistic lines in order to diffuse the language issue (Figure 15.1).

Indian languages fall into two major groups (Figure 15.9). In the south, several Dravidian languages are spoken, including Tamil, Telugu, and Malayalam. They are very different from the languages of northern India, which are Indo-Aryan in origin and related to the Indo-European languages of Europe. They include Hindi, Punjabi, Rajasthani, and Bengali. In 1950, Hindi, a language spoken by about one-third of the population, was adopted as the national language despite resistance from Dravidian South India. Today, 15 languages have official status, and English continues to be used as a second language and lingua franca.

Since independence, many state boundaries have been redrawn to accommodate linguistic aspirations. In 1953, for example, the government yielded to demands for the creation of a Telugu state north of Madras and Andhra Pradesh was created. Similar demands led to the creation of Nagaland on the Burmese border in 1963. In 1981, the Assamese, fearful of cultural submergence by incoming Bengali immigrants, demanded the expulsion of all outsiders while Sikh Punjabis continue to agitate for greater autonomy.

India has one of the most complex cultural patterns in the world, and the centrifugal force of language remains a critical problem. Against this complexity, the preservation of India's territorial integrity since independence is a major accomplishment. It has been possible because the country's leaders have been able to exercise flexibility in accommodating regional demands, thereby avoiding the alternative of balkanization.

Pakistan and Sri Lanka have also experienced problems in unifying their differing ethnic groups. The use of Urdu in West Pakistan and Bengali in East Pakistan was one motive for political separation and the emergence of Bangladesh. In Sri Lanka the conflict between the Tamils and Sinhalese continues, periodically erupting into violent clashes.

POLITICAL CHANGE

Since independence, the history of political development in South Asia has varied considerably from country to country. Despite emergencies and violence, which have included political assassinations, India and Sri Lanka have maintained parliamentary democracies. India, except for a brief period from 1976 to 1977 when a state of emergency was declared, has held general elections regularly. For a country comprising 23 states and several union territories, many with populations exceeding 20 million, and 13 linguistic groups claiming over 10 million speakers, the pressures for regionalism, and even separatism, are considerable. Yet, so far, the Indian Federal Constitution has demonstrated the flexi-

FIGURE 15.9

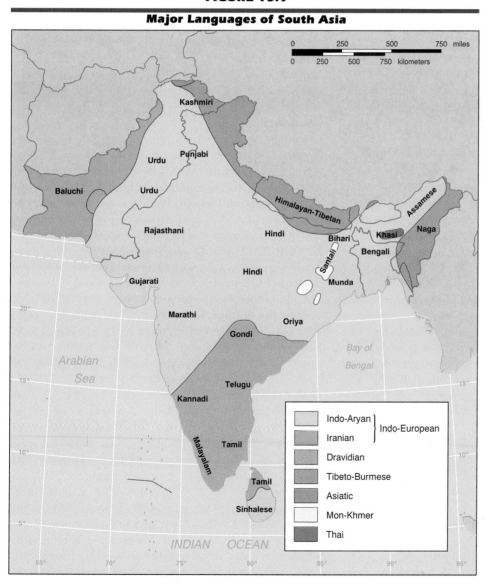

Major Languages of South Asia

bility necessary to contain such separatist tendencies. New states have been created to accommodate various linguistic groups.

Several factors have prevented India from succumbing to the cycle of coups and countercoups found elsewhere in the region. These include a strong dominant party, the Congress party, with broad local support, effective and charismatic leadership by Jawaharlal Nehru (in the years following independence), the constitutional means for dealing with emergencies, and the relatively small size of the armed forces.

For Sri Lanka, insularity has allowed the island to avoid many of the conflicts of the mainland and has given the majority of its people a sense of identity and national purpose. This has allowed the country to develop a democratic system in which political parties of very different shades of opinion have been able to succeed each other relatively peacefully. However, Sri Lanka too has had its share of political violence, and the recent escalation of the Tamil separatist issue is threatening the country's political stability.

The situation has been very different for Pakistan and Bangladesh. Here the story has been one of short-lived parliamentary government alternating with long periods of military rule. Pakistan, unlike India, does not have the advantage of a single, dominant party. Instead, there are many parties, each with a strong local base. The country was deprived of strong leadership in the years following independence with the death of Mohammed Ali Jinnah in 1948 and the assassination of the country's first prime minister, Liaquat Ali Khan, three years later. Since independence, leaders have proclaimed Pakistan to be an Islamic republic. This has produced considerable conflict and compromise resulting from the difficulty of reconciling Islamic concepts of state and divine law with those of parliamentary democracy and with the industrial needs of the twentieth century. Finally, the armed forces have played a much larger role in Pakistan, in terms of both size and influence, than they have in neighboring India.

Bangladesh began its independent existence in 1971 when Sheik Mujbur Rahman and his Awami League took over a country devastated by war. Nevertheless, the country started with a considerable reservoir of goodwill, both internationally and internally. Unfortunately, poor administration, corruption, and rival factions undermined the establishment of a strong central government, and from 1975, there followed a series of coups, countercoups, and political assassinations. Military takeovers, as in Pakistan, have been the rule rather than the exception.

Nepal and Bhutan, which had been monarchies throughout the British period, remain so, although the power exercised by the monarchs and their relationship to other political forces have changed since independence. Both countries have moved toward constitutional monarchies. Successive kings, however, have had an uneasy relationship with politicians who have usually had the support of India.

INDUSTRY

On the eve of independence, South Asia possessed only the skeleton of an industrial framework. What industry did exist was almost exclusively located in India's three port cities, Calcutta, Bombay, and Madras, and employed only 2 percent of all Indian workers. Elsewhere, industrial development was very limited, and the region was heavily dependent on imports of manufactured goods from Europe.

In India, the growth of industry was guided by three important goals: to achieve freedom from reliance on foreign manufactured goods by producing such goods domestically; to decentralize industrial production so that the economic benefits could be spread around; and to provide jobs for the rural poor, many of whom were leaving the villages and moving into the cities in search of work. The methods and approaches by which these goals were to be achieved were, and continue to be, the subject of keen debate among planners and politicians.

Long before independence, Gandhi had urged a return to the **cottage industry** that existed before colonization. He warned of the dangers of a Western technological approach and felt that the factory system could, in the Indian context, lead to mass unemployment. The ruling Congress party, however, felt that freedom from foreign dependence could best be achieved by adopting a centrally planned system in which the state owned heavy industry while private enterprise would concentrate on consumer goods. Industrial goals were to be laid out in a series of five-year plans. Five major industrial regions emerged (Figure 15.10), accounting for two-thirds of the country's manufacturing output. The largest of these is the Calcutta-Jamshedpur region. Calcutta is India's premier manufacturing center. In the 1900s, it had become the center of the jute processing industry. Engineering followed to provide the mills with machinery. Partition saw the city separated

FIGURE 15.10

Major Industrial Centers of South Asia

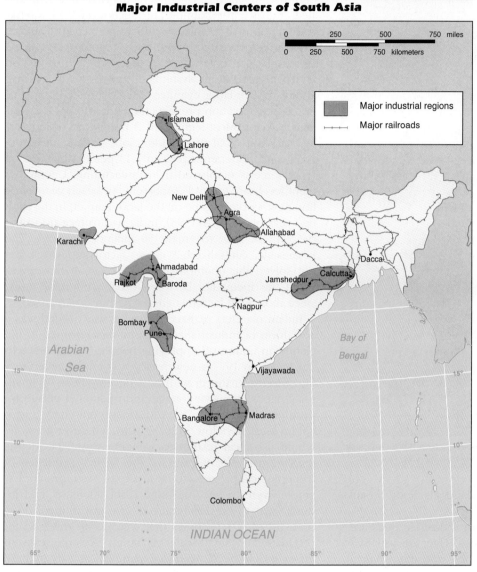

from its source of raw material, which was grown in what is now Bangladesh, and Calcutta was forced to diversify. The city now produces a wide range of goods, including electrical goods, bicycles, and textiles. A hundred and fifty miles to the west lies India's "Ruhr," its major center for heavy industry where everything from nails to locomotives are made. Jamshedpur is the leading iron and steel center, dominated by the giant Tata steel company. The region utilizes coal and high-quality iron ore from the adjacent Chota Nagpur area.

Diversification has also been the trend in India's other major centers. Bombay, Bangalore, and Ahmadabad all originated as textile centers. Today, Bombay is the country's second industrial center. It has the most diversified base, and growth has been greatly stimulated by hydroelectricity from the Western Ghats and the discovery of oil offshore. Petrochemicals, aircraft manufacturing, light industry, and movie making are just some of Bombay's industries. To the south lie the cities of Bangalore and Madras, producing electrical goods, high

A young boy in Ahmedabad makes toys from silk scraps for very low wages.

precision tools, and light industrial goals. Food processing is also significant.

Scattered between these major industrial areas are a number of smaller industrial cities. They are usually located at major transportation hubs, and their main industrial activities are the processing of raw materials. Nagpur, Hyderbad, and Cochin are examples of such cities.

India has achieved its first goal of freedom from reliance on foreign manufacturers. The policy of **import substitution** (producing goods domestically rather than relying on imports) produced a diversified base so that most of the country's basic needs are now met by the manufacturing sector. The country has also fulfilled its second goal of decentralization. Industry is no longer confined to the three major ports. Nevertheless, there have been problems. From 1947 to 1964, industrial production grew by 8 percent annually. After that, production leveled off and, by 1979, actually declined. In the steel industry, for example, production in 1981 was far less than for the 1960s, and India, once self-sufficient in steel, was forced to import large quantities. Problems include lack of investment, deteriorating transportation and communications systems, frequent power failures, and disruptive labor disputes. The continued poverty of the rural population prevents the expansion of the consumer sector. Most significant is the failure of industry to achieve its goal of providing jobs. Today, almost 45 years since independence, only 13 percent of the total work force is employed in manufacturing. Unemployment levels are around 20 percent, and each year 6 million new workers come on the labor market. In the sixth five-year plan (1978–1983), India had to face up to the problem of how to create 50 million new jobs. The large-scale, centralized, Western factory approach has not solved the unemployment problem. It has aggravated the inequality between urban and rural India, since most new factories are located in urban areas.

Indian planners realized that what was needed was a new form of industrialization, one that returned to the Gandhian notion of employing people in preference to machines. In 1977 a new policy was implemented giving priority to the growth and development of small-scale, labor-intensive cottage and traditional village industries. These small-scale industries would have a capital investment below $120,000 and, wherever possible, would use local resources. In some ways, the idea is revolutionary, for it calls for the normal sequence of technological progress to be reversed. Machines, for example, were to become more labor intensive, and industry was to be decentralized. The government drew up a list of 504 products that were to be reserved for production by small-scale industries. No large-scale industry would be allowed to expand its capacity in these goods, which included cycle tires, footwear, clothing, toys, and matches. The government would purchase an additional 241 items exclusively from the small sector. Traditional village crafts such as the making of ploughs, handcarts, or earthenware pots would also receive government aid.

The results have been very encouraging. Planners estimate that for each unit of capital, these types of industries create six times as many jobs as large-scale factory industry. Employment (part-time and full-time) in traditional village industries increased by about 30 percent between 1979 and 1985. For many, these industries provided income during slack periods in the field. The results for small-scale industries are even brighter. In the same time period, employment increased 35 percent, and three times as many workers are now employed in small-scale industries as in the factory sector. The process is leading to decentralization and is rapidly changing the pattern of industry in India.

Sri Lanka has also developed a quite successful program of rural industrialization. A network of divisional development councils was set up throughout the country in 1972. They sponsor, with the help of government funds, a host of small development projects in industry and agriculture. By 1978, these projects had created 80,000 jobs, 60 percent in small-scale industries, which included boat building, brick making, textiles, and the making of two-wheeled tractors. As in India, the average investment required for each job in these industries is considerably lower than for large-scale, state-owned corporations.

Sri Lanka is still an overwhelmingly plantation economy, however, and the bulk of the gross national product comes from tea, rubber, and coconuts. The government is trying to diversify the economy. In addition to rural industrialization, it has created an investment promotion zone outside Colombo and a free trade area within the city along the same lines as Singapore. The venture has been fairly successful in attracting foreign

investment. Electronics, textiles, garments, chemicals, paper milling, and plywood making are some of the industries established here. Yet despite these efforts, the proportion of the population employed in industry was the same in 1981 as it was in 1965 (14 percent).

Pakistan, on the eve of independence, had a far less developed industrial sector than neighboring India. Its major products were cotton, grown in the Sind, and jute, grown in East Pakistan. Partition had cut it off from the textile mills of Bombay and jute factories of Calcutta. The country was further handicapped by a poor mineral resource base with no major deposits of either coal or iron ore. Yet the country later made considerable progress toward industrialization. Like India, it followed the path of import substitution, producing domestically what it previously had to import. The two major centers are Karachi, the former capital, and Lahore in the north. Textiles are Pakistan's leading exports, but there are plants producing iron and steel, petrochemicals, and electronics mainly for domestic consumption. Pakistan's rural industrialization scheme involved the development of small towns (aggrovilles) that would be provided with urban amenities to encourage small-scale industries like cotton ginning, carpet making, and textiles. At a higher level, larger service centers (metrovilles) were built at nodal points to act as growth poles. The results of this scheme have been mixed. Overall, however, Pak-

Tea is Sri Lanka's leading export. Here women pick tea in the south-central highlands.

TABLE 15.1

**Percentage of Labor Force in Agriculture,
Industry, and Services**

	AGRICULTURE		INDUSTRY		SERVICES	
	1965	1981	1965	1981	1965	1981
India	74%	71%	11%	13%	15%	16%
Pakistan	61	57	18	20	21	23
Bangladesh	87	74	3	11	10	15
Sri Lanka	56	54	14	14	30	32
Nepal	95	93	2	2	3	5
Bhutan	95	93	2	2	3	5

SOURCE: World Bank, *World Development Report* (New York: World Bank, 1985).

istan has a higher proportion of its labor force (20 percent) employed in industry than any of the other countries of South Asia (Table 15.1).

The proportion of the labor force involved in industry is much lower in Bangladesh (11 percent). With a population density that is three times greater than that of India, and an economy that remains overwhelmingly agrarian, Bangladesh represents a chronic case of underdevelopment. Nationhood followed years of neglect by West Pakistan, a destructive civil war, and the onslaught of a tropical cyclone that killed nearly half a million people. Aggravated by political instability and a large foreign debt, the economy has never recovered. Much of the country's resources—coal, natural gas, timber, and minerals—go unexploited. Foreign loans are used to buy foodstuffs, and little is left over for development projects. In 1980, a second five-year plan emphasized the development of industry. It has achieved modest success, and the proportion of the population involved in industry rose from 3 percent to 11 percent between 1965 and 1981.

AGRICULTURE

While South Asia has made some progress toward achieving self-sufficiency in the industrial sector, it remains an overwhelmingly rural society. In every country, more than half the population is directly or indirectly engaged in agriculture (Table 15.1). Nepal and Bhutan have the highest proportion (over 90 percent), while in India and Bangladesh, the proportion is above 70 percent.

The village is the cornerstone of agriculture. For centuries, these villages were virtually self-sufficient and had a great deal of autonomy. Village life continues to be traditional and conservative, but improved communications and transportation, the introduction of new, high-yield varieties of rice and wheat (the **Green Revolution**), and the impact of rural industrialization are bringing change. The processes of development are underway in most villages, but the degree of change varies considerably from one region to another. In the Punjab, a full-scale agrarian revolution is in progress while life in many villages in the Ganges Valley is more or less the same as it was a hundred years ago (Figure 15.11).

Because they affect more people, changes in agriculture are probably more significant than those in industry. One of the most important changes has been India's success in achieving food self-sufficiency and eliminating the constant threat of famine. Grain production has increased at an average of 3 percent each year in the last decade, keeping production ahead of population growth. India now has grain reserves and even exports modest amounts to neighboring countries. The change from a food-deficit status to one of food surplus is primarily the result of the Green Revolution.

The Green Revolution

The key element of the Green Revolution was the introduction of high-yielding strains of wheat and rice seed that were highly responsive to chemical fertilizers in areas with reliable water supplies. The government recognized that for the Green Revolution to work, it would have to provide the necessary institutional support. Capital was made available to improve and extend acreage under irrigation. In 1947, 50 million acres (27 million hectares) were irrigated. By 1980, this area had expanded to 120 million acres (48 million hectares). The government invested in roads, markets, rural electrifi-

A farmer uses oxen and a primitive plow to prepare the rice field for another crop.

This patchwork of diked rice fields in various stages of growth illustrates the continuous cropping of rice possible in the Ganges Valley.

After the ploughing and planting of the flooded paddies comes the harvesting and threshing of the rice. Rice cultivation in India is very intensive, but depends almost entirely on human and animal labor.

FIGURE 15.11

Major Agricultural Regions of South Asia

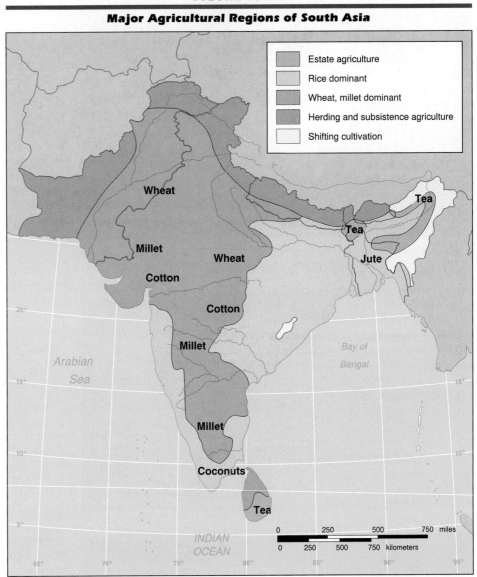

cation, and other supporting infrastructures. Prices were maintained at levels that gave farmers incentives to adopt new practices.

Nowhere was the impact of the Green Revolution greater than in the Punjab. The first high-yielding seeds were introduced in 1966 (Figure 15.12). Wheat yields doubled, and by 1969 two-thirds of the fields were planted with the high-yielding variety. By 1972, farm incomes had doubled and the number of private tube

wells and tractors had increased fivefold. With newly available well water, land that previously lay fallow was brought under cultivation. Wheat acreage increased by 50 percent, and rice was introduced as a secondary crop. With the expansion of farm income, small-scale industry flourished, and many landless farm workers moved to nonfarm jobs. Income per capita has been growing at an average rate of 3 to 3.5 percent each year. Today, the Punjab provides 73 percent of India's wheat

FIGURE 15.12

Agricultural Transformation in Punjab, India

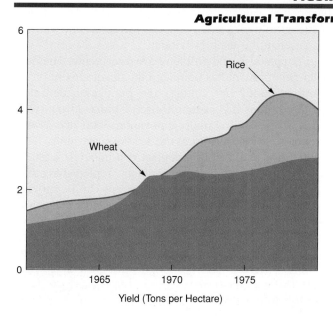

Yield (Tons per Hectare)

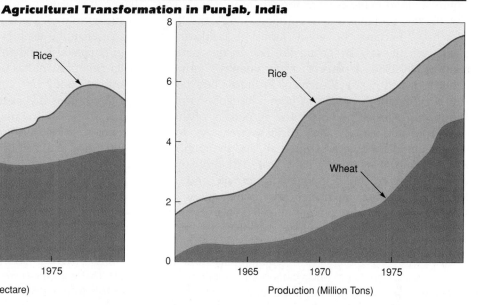

Production (Million Tons)

and 48 percent of its rice. The yield per unit is double the national average for wheat and triple the national average for rice.

Unfortunately, progress has been less dramatic in the major rice-growing areas (Figure 15.11). In the Ganges heartland and delta, farmers are generally poorer, the land more fragmented, irrigation systems less developed, and high-yielding rice seeds more susceptible to pest and disease. Yield increases have been highest along the tropical coasts of Malabar and Coromandel where farmers are more affluent and able to afford the fertilizers and pesticides required. A major drawback of the Green Revolution is that it favors more prosperous farmers and landlords. It does little to help small farmers, dryland farmers, and the cultivators of other staple crops such as pulses and grains like millet, sorghum, and maize. Even within favored areas, small farmers cannot get the credit needed for the extra inputs the seeds require.

A second generation of "miracle" seeds is now emerging that may greatly relieve these farmers' plight. The International Rice Research Institute (IRRI) at Los Banos, Philippines, which produced the original IR-8 "miracle" rice, is now developing strains that will tolerate drought and can flourish in the salty soils of swamps. These strains will need less fertilizers and pes-

ticides and increase the protein content of the rice from 7 to 10 percent. Other cereals are being developed along the same lines.

Persistent Problems

The Green Revolution has been an important element of change in India. It has led to food surpluses and averted the threat of famine. Yet fundamental problems persist in Indian agriculture aggravated by population growth. Each year, food production must increase by more than 2.5 million tons (2.3 million metric tons) just to maintain a minimal consumption level. Despite the surpluses, per capita grain consumption has gone up by only 5 percent in the last generation. Consumption of legumes, a key source of protein in the largely vegetarian diet of most Indians, has fallen by a third during this period. Millions of small farmers and landless laborers exist on the fringes of starvation, many earning less than $75 a year. Unless these agricultural problems can be resolved, the gains of the Green Revolution will have been squandered, and the country will again face the threat of famine.

The most critical problem is India's failure to bring about effective **land reform.** While land reform has always been the goal of the country's leaders, the passage of laws to distribute land was left to the state assemblies where landlord interests were strongly represented. Today, the law prohibits a family from owning more than 18 acres (7 hectares) of "good" farmland. Landowners circumvent the law by giving the title to the land to relatives. In 1980, 4 percent of the country's farming families owned nearly 30 percent of India's best farmland. In contrast, 75 percent of farm households owned less than 5 acres (2 hectares), and 50 percent of these owned as little as 1.25 acres (0.50 hectare). An additional 130 million own no land at all, and the numbers in this group grow rapidly each year. The problem is compounded by rapid population growth and the Hindu and Muslim inheritance laws, which allow the subdivision of property among all sons. This has gradually whittled down the average size of plots, and in many areas the plots are too small to feed a family.

The decline in size makes the small farmers extremely vulnerable. It only takes one year's poor rainfall, an accident, or a serious illness to put the farmer at the mercy of the moneylender. He will have to mortgage his land for a loan and risk forfeiture or sell it off bit by bit to buy a temporary reprieve, ensuring only short-term survival at the risk of long-term ruin. For many it has meant an unavoidable descent down the social ladder from farmer to sharecropper, to landless peasant, to agricultural laborer, and, finally, to exile into the urban areas.

Agriculture is also becoming increasingly polarized. At one end of the scale, small holdings are becoming more fragmented and landlessness is on the rise. At the other, landholding is becoming increasingly concentrated among wealthier farmers and landlords. This concentration has given them considerable economic and political power. Bank loans, for example, go exclusively to the wealthier farmers.

Inefficiency is another problem plaguing Indian agriculture. With a higher proportion of arable land than most of its Asian neighbors, a year-round growing season, and adequate water supplies in most areas, India has the potential to produce large quantities of food. Yet yields remain low. China produces twice as much grain per acre and four times as much rice. India ranks 35th and 42nd in the world in wheat and cotton yields per acre. Much of the country remains unirrigated, and double cropping is not widely practiced. Fertilizer is in short supply and expensive. China, with 30 percent less cropland, uses two and a half times as much fertilizer. Some of India's best farmland is reserved for industrial crops like jute and cotton, rather than for food crops.

Government policies with regard to food production have often aggravated the situation, while overall investment in agriculture continues to represent less than a quarter of the nation's budget.

In Bangladesh, the problems are more acute. The rich delta environment has permitted the development of an intensive paddy rice culture, and population densities are three times higher than those of India. About two-thirds of the land is under cultivation, with rice as the dominant staple, and jute, the leading cash crop. The Ganges and Brahmaputra deposit rich minerals washed down from the Himalayas annually, replenishing the soil's fertility. In good years, three crops can be grown. Yet today, Bangladesh is one of the world's major importers of rice, and the country receives large quantities of food aid. Food production has not kept pace with population growth. Between 1970 and 1980, yields rose by only 5 percent while the population grew by 25 percent. The country suffered an unprecedented series of disasters in the 1970s including a major tropical cyclone (1970), civil war (1971), drought (1972, 1973), and massive floods (1974). Water, the country's great benefactor, is also its tormentor. For four months each year, much of the land is submerged too deeply to grow any kind of crop.

The situation has been compounded by political instability and a failure to bring about much needed land reform. Landlessness, fragmentation of land, and exploitation of small farmers by the increasingly powerful landlords are on the increase. At least two-thirds of all farmers own less than 1.8 acres (0.73 hectares), the amount required to sustain an average family of six at subsistence level. Half of these own less than one acre, and their average holdings amount to 0.23 acres (0.09 hectares). Some farmers are becoming so desperate for land that they have settled in areas of the delta and islands that may be flooded or hit by a cyclone several times a year. Others have moved into neighboring Assam in India, creating resentment and hostility among indigenous Assamese. Meanwhile, the wealthier farmers, who represent only 6 percent of the population, increased their ownership of the land to 47 percent by 1980. Food aid has not been effective in relieving rural hunger. Most of the imported food is sold in ration shops in the cities, providing inexpensive food for urbanites and undercutting the price of food grown in rural areas.

For Pakistan, the problem is too little water rather than too much. Irrigated agriculture is practiced on the fertile alluvial plain of the Indus. Only along the foothills of the Himalayas is there sufficient rainfall to practice agriculture away from the river. At independence, Pakistan inherited most of the great irrigated areas of the Punjab and Sind. These irrigated areas have been expanded, and the two dominant crops are wheat and cotton. Rice has become increasingly important as a secondary crop. In 1965, Pakistan, like India, adopted the use of high-yielding wheat seeds. Production levels have increased, but have not been as high as those of neighboring Indian Punjab. Pakistan continues to import food to supplement local supplies. Land reform has been limited, and agricultural policies continue to favor the landlord. As in India, agriculture is hampered by low yields, inadequate and outdated irrigation systems, landlessness, and poorly developed supporting infrastructure. The situation was aggravated by the Afghan-Soviet crisis when thousands of refugees poured in each year.

Land reform has been most effective in Sri Lanka. In 1972, the ruling Freedom party enacted a series of reforms limiting private holdings to 25 acres (10.1 hectares) of paddy land per family or 50 acres (20.2 hectares) of other land. In 1975, the process was complete when all foreign and nationally controlled estates were nationalized. The goals of land reform were to promote a more equitable distribution of wealth, to increase rural employment, and to boost production of food crops.

The results have been mixed. At independence, Sri Lanka was essentially a colonial export economy growing tea, rubber, and coconuts mainly on large estates. Food crops had a low priority, and the country was forced to import rice. Today, the country is still heavily dependent on exports of tea, rubber, and coconuts. Many nationalized estates were turned into cooperatives, and production of the three major export crops has declined steadily in the last decade. Recent research suggests that the lowly estate laborer benefited very little from land reform.

Production of food crops, on the other hand, has shown an impressive gain and the country is moving toward attaining the goal of self-sufficiency in food production. Production was boosted by the use of high yield seeds and the extension of land under irrigation by more than 100 percent. Since 1965 food production has been gaining on population increase. Rice production has also increased as a result of colonization schemes. The most ambitious of these is the Mahewali-Ganga Project. This largest of Sri Lanka's rivers flows through the "Dry Zone" of the northeast where unpre-

dictable monsoon rains make rice growing difficult. The project will provide year-round irrigation water and hydroelectric power to newly cleared land and to existing farms. Although land reform did not abolish inequities and wealthier farmers continue to receive greater benefits from mechanization and other changes associated with the Green Revolution, the glaring inequities that characterize rural Bangladesh or the central Ganges Valley are absent here.

URBANIZATION

South Asia has always been a land of villages, yet it has many ancient cities. The cities of the precolonial era, like Delhi, Varanasi, or Lahore, were political or religious centers or both. They were generally located inland, and the most important lay in the Indus and Ganges plains. The largest cities today are colonial creations and, with few exceptions, are located on the coast. Founded as military outposts, they later evolved as centers for the import and export traffic between colony and colonizer. These cities include Calcutta, Bombay, and Madras in India; Karachi in Pakistan; Decca in Bangladesh; and Colombo in Sri Lanka. New Delhi, the only major city not a port, was established as the colonial capital in 1911, adjacent to the historic Mogul capital of Old Delhi. Today, these two have coalesced to form a vast metropolitan concentration in the heart of the Ganges Plain (Table 15.2).

Despite the presence of very large cities, the region remains overwhelmingly rural. Pakistan is the least rural county with 28 percent of its population classified as urban. The least urbanized countries are Bangladesh (13 percent), Nepal (7 percent), and Bhutan (5 percent).

TABLE 15.2

Ten Largest Cities in India, 1988 (Estimated)

	POPULATION (MILLIONS)
Calcutta	10.6
Bombay	10.2
Delhi	7.2
Madras	4.9
Bangalore	3.9
Hyderbad	2.8
Ahmadabad	2.4
Kanpur	1.7
Nagpur	1.5

About one-quarter of India's population is urban, and although the percentage is low, it represents over 217 million people. Urbanization throughout the region was fairly slow in the 1950s and 1960s, especially when compared to other developing regions. Ties to villages and families were strong and tended to delay migration. In the last decade, however, rural-urban migration has accelerated, suggesting a loosening or traditional village ties.

India's major urban areas are currently growing more rapidly than the smaller cities. In the 1970s Calcutta grew by 2.2 million and Delhi by 2.1 million, but the most dramatic growth occurred in Bombay, which grew by 3.1 million. Projections for the year 2000 show these three cities will rank among the 10 largest cities in the world with populations of 16.6 million (Calcutta), 16.0 million (Bombay), and 13.3 million (Delhi).

Urban growth has been uncontrolled, creating a planner's nightmare. Population densities are extremely high. In Calcutta, for example, the average population density is 36,000 people per square mile (13,900 per square kilometer) but in poorer sections it approaches 500,000 per square mile (193,050 per square kilometer). New York City, in comparison, has an average density of 24,000 people per square mile (9266 per square kilometer). Rapid growth, which stems from natural increase and rural-urban migration, is creating massive problems in housing, transportation, sanitation, and the provision of jobs. Immigrants from rural areas flock to the outskirts of cities, creating a septic fringe, and invade open areas within the city.

Nowhere are these urban problems more pressing than in Calcutta. As early as the eighteenth century, Robert Clive described Calcutta as "the most wicked place in the Universe." A more recent author referred to the city as the "densest concentration of suffering humanity in the world." Founded by the British East India Company in 1690 on the swamplands of the Hooghly River, the city grew rapidly and became the foremost trading outpost of the British Empire. For 140 years, until 1911, it was the capital of British India, accumulating all the services needed to administer the largest of Britain's colonies. It was the railhead for India's transport network; the financial center for much of Asia; and the country's major port, handling the agricultural products of the Ganges Valley, the Ganges Delta, and Assam and the industrial production of the Damodar Valley. Migrants poured in, and between 1950 and 1985, the city's population almost tripled.

Calcutta probably has one of the lowest urban living standards in the world. The statistics are grim. More than 70 percent of the people live at or below the poverty level, calculated as an average income of less than $8 a month. About a third of the inhabitants live in slum areas called ***bustees*** where there are, on average, one water faucet and one latrine for each 25 dwellings. Others live in shantytowns on the outskirts of the city in dwellings made of rough driftwood planks with roofs of sacking held down by bricks. Squatter settlements are also found along canals where people both wash and excrete. Then there are the street people, some 200,000 of them, who live and die in the streets. In the monsoon, the whole place is awash and some homes are two feet deep in water. Cholera and other water diseases are endemic in these slum areas.

Calcutta's economy has not grown rapidly enough to provide jobs for the rural migrants who continue to pour in. The physical resources of the city continue to break down and are not replaced. The last main sewer, for example, was built in 1896, power blackouts occur regularly, and 20 percent of the city's buildings are classified as unsafe. Yet, despite the bleak picture, visitors, after their initial shock, often comment on the city's vitality and resilience. A new underground railroad is being completed, while in Haldia, 45 miles (73 kilometers) north, a new port, a refinery, and several factory sites are being constructed. A ring of 17 small satellite towns are being planned to relieve congestion in the inner city.

Bombay, in contrast, has always seemed India's most advanced and progressive city. The city developed around an excellent natural harbor surrounded by a more prosperous and less densely settled hinterland than Calcutta. Today, it is India's largest port, handling nearly half the nation's foreign trade. It is also the country's second industrial center with a very diversified industrial base stimulated by the discovery of oil offshore. Some of the industries that have developed around oil refining include petrochemicals, plastics, chemical fertilizers, and synthetic fibers. Bombay is a major international trading center and in recent years has taken over many of the financial functions previously performed by Beirut for the Middle East.

Yet this richest of South Asian cities is strangling itself on its own prosperity. As the economy grows, the city acts as a magnet, and each day 300 to 500 new people pour into the city. Today, 40 to 50 percent of Bombay's population reside in bustees, and the city is experienc-

An estimated 200,000 homeless people in Calcutta use whatever is available for temporary shelter—here, concrete sewer pipes.

ing many of the problems afflicting Calcutta. Current attempts to cope with this massive growth include a new center, New Bombay, which is being built to handle the wholesale transshipment activities of the rest of the nation, leaving the old city to concentrate an international trade. By 1985, long before the completion of the project, some 200,000 inhabitants had already settled in the area that will become New Bombay. In addition, 20 new self-sufficient satellite towns are being built to relieve congestion in the inner city.

SPATIAL CONNECTIVITY

The earth is becoming more and more interdependent as a result of improvements in transportation, communications, and technology. Interdependence brings with it consequences that are both beneficial and detrimental. Capital, foreign aid, arms, resources, raw materials, and manufactured goods move from one hemisphere to another. Giant multinationals exert powerful influences over crucial aspects of development, treating the world as a resource base and a market. The impact of the arms race and superpower rivalry transcends boundary lines drawn by history or established by geography.

In this section we examine the nature and degree of interdependence with South Asia and between South Asia and the rest of the world.

SPATIAL INTERACTION WITHIN SOUTH ASIA

Spatial interaction with South Asia remains weak, and this weakness, in turn, is a contributing factor to the forces of tradition and resistance. Despite advances in industry and commercial agriculture, most people continue to live in villages and practice subsistence agriculture. Regional specialization is limited and interaction minimal. A key element in the development of spatial interaction is the quality of the transportation and communications networks. Studies have shown that the development of integrated transportation and communications systems is vital to modernization. Poorly developed systems aggravate the polarization between city and village.

The transportation network is best developed in India. The country inherited from the British one of the densest road and railway networks in Asia. India has some 63,000 miles of rail, the fourth largest system in the world. But although the network connects most of the country, there are problems. The lines were designed to further British trading and strategic needs rather than to stimulate growth in the country's interior. Consequently the lines focused on the three major port cities, and the system was never fully integrated. Four separate gauges were in use at independence. Despite these drawbacks, railways continue to be cheapest way to move people and goods and have been vital to the growth of industry in India. The system is now undergoing the slow and costly process of standardization and will soon be supplemented by an integrated network of major and minor all-weather roads connecting most settlements.

Elsewhere, however, the situation is very different. The delta nation of Bangladesh is crossed by four major river systems and a host of smaller tributaries. Terrain and climate do not favor transportation development. There are no major road bridges anywhere in the country, and only one railway bridge crosses the Ganges River. Roads and railway lines break off at riverbanks, and crossings are made by ferries, although most of the large towns are connected by roads. Many highways become submerged during the rainy season.

For Nepal and Bhutan, transportation problems are even more formidable. Few nations in the world have such a low density of roads. Nepal has just 424 miles of paved road, and the main means of transportation is a network of footpaths that interlace the mountain terrain. Lack of adequate bridges across the rivers has reduced the usefulness of these trails. Travel for those not fortunate enough to use the meager air service is measured in weeks. Both countries have embarked on road building projects to open up inaccessible areas, but the rugged terrain makes such projects very expensive.

Communications systems in South Asia also remain relatively underdeveloped, particularly for Bangladesh and the mountain kingdoms. India, however, has taken a bold step to bolster communications, one that has the potential to narrow the information gap between rural and urban areas. In 1983, the telecommunications satellite, *INSAT,* was launched. The satellite provides daily weather observations, makes television transmission possible for hundreds of isolated villages, and improves the country's archaic telecommunications system. It allows India to leapfrog a whole era of communications technology, doing away with billions of dollars of ground equipment. Today, about 75 percent of the country is within range of television technology. Unfortunately, many villages still do not have electricity. The primary function of the television system is to provide education programs for rural areas, including school programs, adult education classes, teacher training, health care worker training, family planning counseling, and agricultural information for farmers. In addition, the channels carry national and regional entertainment programs, while a national network of 8000 telephone circuits provides a nationally integrated telecommunications system. This experiment has the potential to be a powerful integrating force in India's diverse and complex cultural landscape.

Another measure of the degree of spatial interaction within a region is trade. Trade within South Asia, like transport and communications, is poorly developed. There are two main reasons for this: the region continues to export a great deal of primary products, and the individual countries are, in fact, in competition with one another. Sri Lanka and India, for example, both export large quantities of tea. After it declared independence in 1971, Bangladesh had high hopes that India would replace West Pakistan as a market for tea and jute. These hopes have not been fulfilled as India already produces those commodities. Where complementarity could exist as, for example, between the food-producing wheat belt of Pakistan and the iron and steel products of the Indian "Ruhr," political animosity has prevented trade from developing, and trade between these two countries is virtually nonexistent.

The conflict between India and Pakistan has overshadowed international relations within South Asia. These two countries have taken up arms against each other on no less than four occasions. Disputes over control of Kashmire remain unresolved, and precious resources have been diverted to defense spending. In 1984, Pakistan spent 4.5 percent of its gross national product on defense while India spent 3.4 percent.

A number of disputes between India and Bangladesh have also occurred despite India's support for secession in 1971. The most serious of these involve Bangladesh's harboring of Naga and Mizo separatists. The Indian government contends that extremists from these tribal groups in northeast India have been using Bangladesh as a base for their operations. A second problem arises from the movement of Bengali Muslims from Bangladesh to neighboring Assam and Meghalaya. There is a great deal of resentment against these immigrants among the indigenous people, and violence has erupted on several occasions.

Relations between India and Sri Lanka are affected by India's concern over the status of Hindu Tamils in that country. Sri Lanka is only 20 miles (32 kilometers) away from India. Although the Sinhalese are a majority in their country, they fear that the tens of millions of Tamils in South India, who regard the Sinhalese as a minority, might make common cause with the Tamils in Sri Lanka. The situation has been aggravated by continuing illicit immigration from India.

Trading has increased between some South Asian countries. Relations between Bangladesh and Pakistan have improved significantly in the last decade, and there has been a corresponding increase in trade. Pakistan is now an important market for Bangladesh's tea and jute, and in 1986 Pakistan's former president Mohammad Zia, welcomed the possibility of an "Islamic state" and sought ties to the Middle East. Landlocked Nepal and Bhutan are beholden to India for their transit trade, and the country is their major trading partner both in purchasing exports and supplying imports.

SOUTH ASIA AND THE GLOBAL ECONOMY

To what extent is South Asia's economy tied to the global economy and what is the nature of its ties? Under colonial rule, the region served as a reservoir of raw materials and as a market for manufactured goods. Tables 15.3 and 15.4 reveal the trade pattern for two time periods, the 1960s and 1987. The figures show that important changes are taking place in the trading pattern of South Asian countries. While exports continue to be dominated by primary commodities, they are far less important today than they were in 1960.

In Sri Lanka, a country that lives by trading its plantation products, the proportion of primary exports dropped from 99 percent in 1960 to 52 percent in 1987. For India, Pakistan, and Bangladesh, the proportion had dropped below 35 percent. In India, manufactured goods including engineering and chemical products now register well ahead of the more traditional exports based on agricultural products, reflecting a measure of economic and industrial maturity. Nevertheless, primary commodities continue to be vital in the region's economy.

Primary commodities are a risky business for a nation to rely on for a livelihood, however. In the unpredictable political and agricultural conditions of the tropics, supplies tend to fluctuate even more than demand in the consumer countries. Prices rise and fall, further destabilized by the commodity dealers and speculators,

and producer nations have little control over the price of their exports. Over the past 25 years, the trend underlying these fluctuations has been heading slowly downward for many South Asian commodities. Over the same period, prices for imports of machinery and other manufactured goods needed for industrialization rose.

There have also been significant changes in the pattern of South Asia's imports. Food imports fluctuate according to the state of domestic crops, although the overall trend is one of decline. The declines have been most significant in India, Sri Lanka, and Nepal. India is now self-sufficient in food grains. In Bangladesh, however, food imports continue to be vital and represent 16 percent of all imports. With increased industrialization, imports of consumer goods have declined in some countries as have capital goods. Intermediate goods, like fertilizers and petroleum, however, have risen rapidly. In India, fuel imports rose from 5 percent of total imports in 1965 to 35 percent in 1983, but dropped to 11 percent by 1987. Pakistan and Sri Lanka showed similar trends. They reflect the impact of OPEC policies on oil prices and the expanded demand for fuel as these nations industrialize. The massive OPEC price hikes of the early 1970s had a devastating effect on the economies of the region. In India, trade fell by 27 percent between 1970 and 1975 while Sri Lanka declined by 35 percent. The region continues to be heavily dependent on the oil-producing nations, especially the Middle East. Recent oil finds off the coast of Bombay should significantly help India's position, but even the most optimistic projections show that the country will continue to be dependent on imported oil.

Since independence, the direction of trade has changed. The outstanding feature is the decline in the region's trading relations with its former colonial ruler. The region has escaped from the colonial relationship and established a more balanced worldwide trading pattern. Table 15.5 shows the major partners for India and Pakistan in 1988. In recent years, Pakistan has increased its trade with its fellow Islamic nations in the Middle East.

A major problem in trading, which was aggravated by the oil crisis, is the growing gap between export earnings and the cost of imports. For Sri Lanka, only six years between 1961 and 1980 showed a positive balance of trade. For Bangladesh, Nepal, and Bhutan, an adverse balance of trade is a chronic condition. Exports have never paid for more than 48 percent of imports in Bangladesh. The remittances sent by South Asians working

TABLE 15.3

South Asia's Leading Exports (as a Percentage of Total Exports), 1960 and 1987

	FUELS, MINERALS, AND METALS		OTHER PRIMARY COMMODITIES		TEXTILES AND CLOTHING		MACHINERY ANY TRANSPORTATION EQUIPMENT		OTHER MANUFACTURED GOODS	
	1960	1987	1960	1987	1960	1987	1960	1987	1960	1987
India	10%	9%	45%	22%	35%	16%	1%	10%	9%	59%
Pakistan	0	1	73	32	23	41	1	3	3	64
Bangladesh	—	16	—	33	—	—	—	17	—	33
Sri Lanka	0	8	99	52	0	25	0	2	0	38
Nepal	—	2	—	26	—	37	—	2	—	70

SOURCE: World Bank, *World Development Report* (New York: World Bank, 1985, 1989).

TABLE 15.4

South Asia's Leading Imports (as a Percentage of Total Imports), 1965 and 1987

	FOOD		FUELS		OTHER PRIMARY COMMODITIES		MACHINERY AND TRANSPORTATION EQUIPMENT		OTHER MANUFACTURED GOODS	
	1965	1987	1965	1987	1965	1987	1965	1987	1965	1987
India	22%	8%	5%	11%	14%	8%	37%	24%	22%	48%
Pakistan	20	16	3	19	5	7	38	31	34	27
Bangladesh	—	16	—	9	—	6	—	28	—	42
Sri Lanka	41	17	8	17	4	3	12	27	34	37
Nepal	22	6	5	8	14	7	37	22	22	51

SOURCE: World Bank, *World Development Report* (New York: World Bank, 1989).

TABLE 15.5

Major Trading Partners of India and Pakistan, 1988

	IMPORTS AS A PERCENTAGE OF TOTAL IMPORTS		EXPORTS AS A PERCENTAGE OF TOTAL EXPORTS	
	Imports from	Percent	Exports to	Percent
India	Japan	12%	United States	18%
	United States	12	Soviet Union	15
	Germany	10	United Kingdom	6
	United Kingdom	8	Japan	9
	Total value of imports = U.S. $19.1 billion		Total value of exports = U.S. $13.3 billion	
Pakistan	Japan	16%	Japan	10%
	United States	11	United States	10
	Kuwait	7	United Kingdom	3
	Saudi Arabia	5		
	Total value of imports = U.S. $6.5 billion		Total value of exports = U.S. $4.5 billion	

SOURCE: *World Almanac* (1991).

abroad are becoming an increasingly important source of income to supplement export earnings. Pakistan now has some 1.5 million workers living abroad, mainly in the OPEC countries of the Middle East. In Nepal, the money sent by Gurkha soldiers working in the Indian and British armies accounts for 11 percent of the country's budget.

South Asian countries have become increasingly dependent on foreign aid to resolve their trade deficits. Foreign aid has also been sought to finance development projects. Increasingly, this aid tends to be provided by specialized international bodies like the World Bank or International Monetary Fund, rather than by individual countries offering bilateral aid. A survey of major creditors for the region shows that aid comes from a wide range of countries representing every shade of political complexion. In addition to the major international agencies, India receives aid from the United States, the Soviet Union, Japan, and Germany, while the list for Pakistan includes the United States, the People's Republic of China, Saudi Arabia, and the Soviet Union. Since the rise in oil prices, Pakistan, as a large Muslim country, has been able to obtain funds from OPEC sources.

The key to the financial crisis plaguing developing countries is the relationship of debt-service payments to export earnings. Traditionally, international banks avoided lending when debt-service payments exceeded

20 percent of export earnings. Table 15.6 shows that South Asia's foreign debt is relatively modest by world standards. India has been following an increasingly careful course and has been able to reduce its debt-servicing costs. Sri Lanka borrowed heavily to finance its Mahewali Project, but debt servicing remains below 20 percent. Donor countries and international agencies, however, have been increasingly concerned about Pakistan's capacity to meet its obligations to creditors. Debt servicing accounts for over 23 percent of its exports, and the consortium of agencies providing assistance has twice had to reschedule debts on condition that the country reduce imports and government spending.

South Asia and Global Politics

International politics in South Asia has been dominated by recurring internal conflict between India and Pakistan. The conflict has also influenced relations between these two countries and the superpowers. India has developed closer relations with the Soviet Union, while Pakistan has moved closer to the People's Republic of China and the United States. Relations between Pakistan and the Soviet Union deteriorated during the Afghan conflict as refugees from that country poured into Pakistan. There is also friction between India and China. In the 1950s, the Chinese invaded and incorporated Tibet and in 1962 invaded India over a border dispute in

TABLE 15.6

South Asia's International Debt, 1988

	TOTAL EXTERNAL DEBT (MILLIONS OF U.S.$)	DEBT SERVICE AS A PERCENTAGE OF EXPORTS
India	$49,695	21.8%
Pakistan	13,944	23.5
Bangladesh	9,330	20.5
Sri Lanka	4,139	17.2
Nepal	1,008	8.5
Bhutan	68	—

SOURCE: World Bank, *World Development Report* (New York: World Bank, 1990).

northeast India. This dispute is still unresolved. The situation was further complicated by the Sino-Soviet rivalry. As China and the Soviet Union drew apart, India and the Soviet Union grew closer. Meanwhile, Chinese links with Pakistan were strengthened with the completion of a strategic road over the Karakoram Mountains in 1978, linking Islamabad with the Tibet-Sinkiang communication system. In recent years, Pakistan has built up links with the Islamic countries of the Middle East.

At times, Sri Lanka and India have been active in the nonaligned movement. Their goal was to turn the Indian Ocean into a "Lake of Peace" from which superpower rivalries would be excluded. Unfortunately, events in the Persian Gulf, the Soviet invasion of Afghanistan, and the creation of a U.S. naval base on the island of Diego Garcia have made this goal increasingly unattainable. The countries of South Asia find it increasingly difficult to remain free of the ever-changing calculations and interventions of the superpowers. The United States views Pakistan as a military ally, strategically placed in relation to the Soviet Union. The Afghanistan crisis meant more American arms for Pakistan and renewed coolness by India toward the United States. The conflict over Kashmir will remain a regional problem for the foreseeable future. Nevertheless, the foreign policies of India and Pakistan have exhibited some convergence in recent years.

PROBLEMS AND PROSPECTS

An analysis of future prospects for South Asia reveals grounds for both pessimism and optimism. The region remains overwhelmingly poor. It is difficult for Westerners to appreciate the extreme physical deprivation and anguish of life in a Calcutta bustee or in rural Bangladesh. Infant mortality rates are high, literacy rates are low, and life expectancy at birth hovers around 55 years. Only in Sri Lanka is the picture less grim. Here education, health, and a variety of social services add up to a quality of life distinctly better than those of its neighbors.

Population growth continues unabated, undermining efforts to improve food supplies, health care, jobs, and education. Population pressure on the land, especially in Bangladesh and Nepal, threatens to destroy the natural resource base, and the entire region is vulnerable to the vagaries of the annual monsoon. Repeated division of small holdings condemns millions of farmers to the inevitable slide into debt, landlessness, and, ultimately, exile to the already overcrowded cities in search of work. The failure of governments to implement meaningful land reform has aggravated the problem and polarized the farming community into "haves" and "have nots." Innovations like those associated with the Green Revolution favor the rich farmers, usually at the expense of the small and landless farmers.

Industrialization, following the Western centralized model, has been successful in providing for the region's domestic needs, but it has failed to solve the problem of employment and has further widened the gap between urban and rural areas. Primary goods continue to represent a substantial proportion of exports, holding the region captive to fluctuating and generally falling world prices. At the same time, the costs of imports of vital fuel and manufactured goods necessary for development remain high, and trade deficits are the norm. The region is forced to seek foreign aid to close the gap between exports and imports. Friction between India and Pakistan continues to dominate political relations in the region and inhibits the development of trade be-

OIL DEPENDENCY AND THE SOUTH ASIAN ECONOMY

South Asia's most serious resource deficiency is oil. The region is heavily dependent on the oil-producing nations, especially the Middle East. South Asia, with a billion people to feed and another billion projected to be added before population growth stops, has less than 1 percent of the world's oil reserves. The Middle East, on the other hand, with only 4 percent of the world's population has 56 percent of global oil reserves. Imports of this commodity, which have risen steadily since independence, impose a major drain on foreign reserves and leave the region's economy vulnerable to price hikes by the oil-producing nations.

Much of the imported oil is used for transportation. In 1985, the proportion of oil used for transportation in India, Pakistan, and Sri Lanka was 45, 49, and 57 percent, respectively. Industry is the second largest user, but the most critical demand for oil has been in agriculture.

Until 1950, South Asian farmers were still largely self-sufficient in energy. They depended on livestock waste for fertilizer and draft animals for tilling. Since then, however, the situation has changed. The adoption of high-yielding, fertilizer-responsive varieties of wheat and rice and the expansion of irrigation using pumps have greatly increased the demand for oil and oil-derived fertilizers and pesticides. Farmers were pushed to use more fertilizers by the extensive use of subsidies. In the early 1980s, these subsidies usually amounted to 50–70 percent of fertilizer cost. Agriculture's increasing reliance on oil has meant that trends in world oil production and prices directly affect efforts to expand food output in South Asia. Furthermore, unlike other sectors of the economy, agriculture does not appear to be reducing its reliance on oil, which makes it even more vulnerable to future price increased by OPEC.

The outlook for the future is not encouraging. Demand for food will continue to rise into the next century, and efforts to increase crop outputs by using more energy-intensive inputs will make agriculture even more oil dependent at a time when world oil supplies are diminishing. Despite the drop in oil prices in the 1980s, they are expected to rise in the 1990s.

The one bright spot in the picture is the discovery of off-shore oil in India. Petroleum production in Assam and Gujarat, which supplied one-third of India's needs, has been supplemented by discoveries on the western continental shelf off Bombay. Six fields are now in production, and it is expected that India will soon be able to meet 70 percent of its domestic needs. The other countries of South Asia, however, have not been so fortunate.

---

tween the two countries. Political instability undermines efforts for development in Pakistan and Bangladesh while the region as a whole finds it increasingly difficult to remain free of the machinations of superpower politics.

On the other hand, there have been some remarkable achievements in the last decade. Population growth rates have fallen, as have infant mortality rates. Life expectancy has improved by nearly 10 years while family planning programs have been increasingly successful in Sri Lanka and India. Nevertheless, the situation remains critical especially for Bangladesh, and population control remains the key to the region's future. Food imports are down in most of the region, and India, once a major importer of food, has been particularly successful in achieving a surplus of food grains. Food self-sufficiency is now an attainable goal for both Sri Lanka and Pakistan. Unfortunately, the situation is much bleaker for Bangladesh where food imports continue to be essential and the country's energies are increasingly devoted to fighting the daily battle against malnutrition. India has been successful in breaking the dependent relationship with its former colonial ruler. The country now is able to produce most of its manufactured goods and has broadened its trade base. Other South Asian nations have not been as successful but are moving in the same direction.

Perhaps the best reason for optimism is the region's awareness of the need for family planning and for an alternative model of development, one that returns to the Gandhian idea of employing people instead of machines. The governments are recognizing the need to shift emphasis away from the cities and large-scale industries and invest in the rural areas. Small-scale industries, rural electrification, commercialization of agriculture, and a host of other projects are part of this trend to close the gap between rural and urban areas and raise rural incomes. This, in turn, would provide a surplus for industrial investment, and domestic markets would be created for the products of manufacturing. In time, the economy as a whole would become productive enough to raise average income levels. Whether this sequence of events will occur remains to be seen. The success of the Punjab farmers and the rapid growth of small-scale industries in the last decade are promising. It is most likely to occur in India and Sri Lanka where democracy has survived and the sinister cycle of coups d'états and civil war has been avoided.

Finally, while the goal of keeping South Asia free from the impact of superpower rivalry may be unattainable, there is a growing recognition throughout the region of the dangers of this rivalry, as shown by the Afghan crisis, and the need to steer a middle course wherever possible.

Suggested Readings

Brush, J. E. "Some Dimensions of Urban Population Pressure in India," in W. Zelinsky et al., eds, *Geography and a Crowding World*. New York: Oxford University Press, 1970.

Dutt, A. "Cities of South Asia," in S. Brunn and J. Williams, eds., *Cities of the World: World Regional Urban Development*. New York: Harper and Row, 1983.

Farmer, B. *An Introduction to South Asia*. London and New York: Methuen, 1984.

———. "The Green Revolution in South Asia." *Geography* 66 (1981).

Hall, A. *The Emergence of Modern India*. New York: Columbia University Press, 1981.

Johnson, B. L. C. *Development in South Asia*. New York: Penguin Books, 1983.

Johnson, B. L. C., and M. Scrivenor. *Sri Lanka: Land, People, Economy*. London: Heinemann, 1981.

Karunatilake, H. N. S. *Economic Development in Ceylon*. New York: Praeger, 1971.

Murton, B. "South Asia," in G. A. Klee, ed., *World Systems of Traditional Resource Management*. London: Edward Arnold, 1980.

National Atlas of India, 8 vols. Calcutta: National Atlas and Thematic Mapping Organization, 1982.

Noble, A. G., and A. K. Dutt, eds. *India: Cultural Patterns and Processes*. Boulder Colo.: Westview Press, 1982.

Saha, K. R., et al. "The Indian Monsoon and Its Economic Impact." *Geo Journal* 3 (1979): 171–78.

Singh, G. *A Geography of India*. Delhi: Atan Ran, 1976.

Sopher, D. E., ed. *An Exploration of India: Geographical Perspectives on Society and Culture*. Ithaca, N.Y.: Cornell University Press, 1980.

Thorner, D. *The Shaping of Modern India*. New Delhi: Allied Publishers, 1980.

CHAPTER 16

Southeast Asia and the Pacific

IMPORTANT TERMS

Endemic

Wallace's line

Circulation

Transmigration

Export processing zones (EPZs)

Irredentism

CONCEPTS AND ISSUES

■ Political and cultural fragmentation of the region resulting in a "shatter zone."

■ Relationship between the formal and informal economies of the region.

■ Evolution of an Asian "common market" via ASEAN.

■ Static economic conditions for many farmers due to involution.

■ Prevalence of urban primacy in the nations of the region.

CHAPTER OUTLINE

PHYSICAL ENVIRONMENT
Climate, Soils, and Natural
    Vegetation
Landform and Drainage Patterns
Mineral Resources

HUMAN ENVIRONMENT
Population
Cultural Environment
Economic and Urban Landscape

SPATIAL CONNECTIVITY
Regional Groupings
Military Alliances and Assistance
Development Assistance
Trade Linkages and Energy
    Dependence
Tourism

PROBLEMS AND PROSPECTS
Common Problems
Conflict

Southeast Asia and the Pacific are really two separate world regions, but they will be treated together here because they share several common characteristics. Southeast Asia has been described as a "shatter belt" because of its very diverse physical and human geography (Figure 16.1). Similarly, the multitude of islands in the vast Pacific region with their varied history of human migration and settlement also qualify as an area marked by great fragmentation and diversity (Figure 16.2).

In understanding Southeast Asia, one must bear in mind two pivotal facts. First, the island and peninsular part of the region is, and has been, very accessible whereas the mainland areas have been accessible only through a series of generally north-south trending river valleys. Second, the region is sandwiched between two great cultural realms: China to the north and the Indian subcontinent to the west. The early migrations of peoples from the subcontinent, and the more recent waves of migrations from China, account for the racial and cultural characteristics of both Southeast Asia's upland and lowland populations. More recent arrivals in Southeast Asia, including Arabs, Europeans, and Americans, have brought further mixing and diversity to the region's population. For centuries Southeast Asia has been a region of considerable conflict. When external participants have been involved, their motive has been in part to maintain the region's accessibility or to control it because of its strategic location. On the other hand, internal conflicts have often resulted because of the great human diversity.

At times treated as a separate world region, classified with Australia and New Zealand, or not considered at all because of its small population, the Pacific region has several characteristics in common with Southeast Asia. First, it is also an extremely fragmented region made up of some 25,000 small islands and islets spread over a vast area; the entire Pacific covers some one-third of the earth's surface. Second, the racial origins of the population can be traced to Southeast Asia and thus originally to China and the Indian subcontinent. Third, like Southeast Asia, the population has also been heavily influenced by external forces of more recent origin, beginning in the sixteenth century with the arrival of the Europeans. Fourth, while individual nations in the two regions have varying types of political and economic systems, they are nearly all developing nations. In short, they are poor and largely dependent upon the wealthy nations as destinations for their primary exports and as origins for technology, military aid,

and development assistance. Finally, like the Southeast Asian nations, independence has brought at least a degree of interregional cooperation and supranationalism.

PHYSICAL ENVIRONMENT

Since both Southeast Asia and the Pacific Island region lie almost entirely between the Tropics of Cancer and Capricorn, they are characterized by tropical, moist climates and the soils and natural vegetation (where it still exists) associated with such climates. Mainland Southeast Asia and the Pacific Islands are characterized by rugged topography, occasionally over 15,000 feet (4572 meters) in elevation. The landforms are a result of recent tectonic activity since a portion of the area is near the

The Golden Buddha located in Bangkok. Buddhism diffused from India into Southeast Asia.

FIGURE 16.1

Southeast Asia

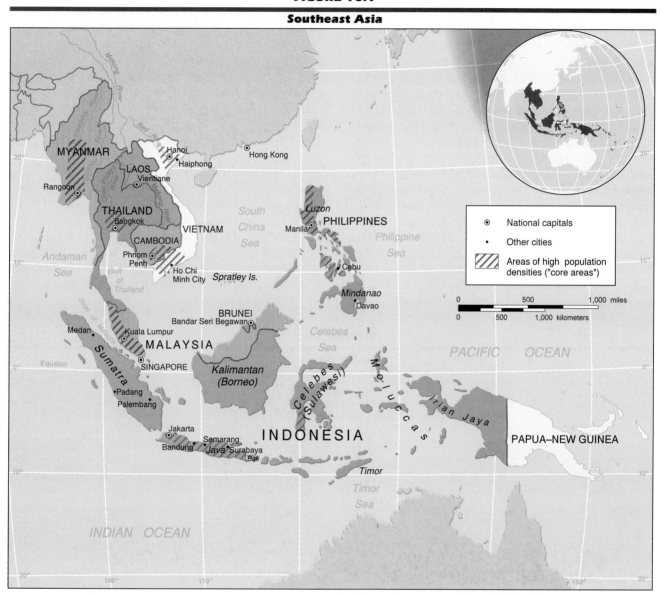

juncture of several tectonic plates. Earthquake and volcanic activity is common since portions of the regions are a part of the circum-Pacific "rim of fire."

CLIMATE, SOILS, AND NATURAL VEGETATION

With the exception of a few highland areas, most notably that of central New Guinea and the northern portion of mainland Southeast Asia, the land areas of both

regions can be classified as tropical wet climate regions. Total annual precipitation amounts are generally high, well over 100 inches (254 centimeters) for many parts of the area, and average temperatures throughout the year are also high (Figure 16.3).

The winds that carry the moisture-laden air are controlled by the seasonal migration of the equatorial low (or Intertropical Convergence Zone) and subtropical highs, which are located to the north and south of the equatorial low. These belts of high and low pressure migrate north and south depending on the season of the year. Much of the Pacific Island region is in the belt of steady trade winds, which blow generally from the east. Southeast Asia, on the other hand, is most affected by the ***monsoon***, winds that carry moisture-laden air in opposite directions during the two seasons of the year. Areas that are on or near the equator or in the belt of trade winds receive a relatively even annual distribution of rainfall (see Figure 16.3a and b). Areas affected by the monsoon, on the other hand, receive the majority of their rainfall in a few short months (Figure 16.4). Rangoon, Myanmar (formerly Burma), for example, receives nearly 90 percent of its annual rainfall in the five-month period from May to September when the summer monsoon winds blow out of the southwest (Figure 16.3c). Such an uneven distribution of rainfall, which affects most of Southeast Asia, is, of course, important from the standpoint of agriculture. Areas such as those near Rangoon are dry for more than half the year and must be irrigated to produce crops during the dry period.

Regions that are warm and moist the year round typically have soils that have undergone the process called laterization. The soils become depleted of nutrients with a relative increase in the iron compounds or laterites; this causes the soil to become hard and impervious when used in intensive agriculture. The tropical forest species have adapted their root systems to the shallow soil and, along with microorganisms, are nurtured by the decaying plant matter almost as rapidly as it drops to the ground. Once the forest cover is burned or cut down, as has happened in so many tropical areas, the soil erodes and is leached of its nutrients much more rapidly than under natural conditions, since it is now exposed directly to the sun's rays and to rain.

There are two major exceptions to the generally infertile soils of the regions. In both Southeast Asia and the Pacific are soils derived from volcanic materials. When they are composed of basic materials (as opposed to acidic), these soils (alfisols) can be extremely fertile.

FIGURE 16.2

Pacific Region

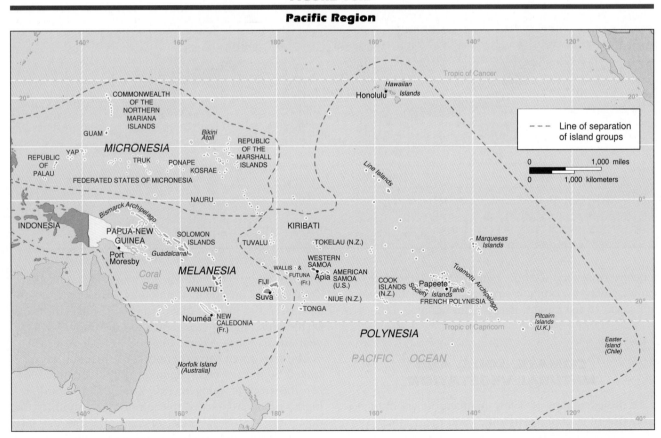

STATISTICAL PROFILE: SOUTHEAST ASIA AND THE PACIFIC ISLANDS

REGION OR COUNTRY	Population Estimate mid-1990 (Millions)	Natural Increase (Annual)	Population Projected to 2000 (Millions)	Infant Mortality Rate	Life Expectancy at Birth (Years)	Urban Population (%)	Per Capita GNP, 1988 (US$)	Area, Thousands of Square Miles (Km²)	Population Density, No./mi² (No./km²)
Southeast Asia									
Brunei	0.3	2.5	0.3	11	71	59	14,120	2.2(6)	115(44)
Cambodia	7.0	2.2	8.5	128	49	11	—	69.9(181)	100(38)
Indonesia	189.4	1.8	223.8	89	59	26	430	735.4(1905)	258(100)
Laos	4.0	2.5	5.0	110	47	16	180	91.4(237)	44(17)
Malaysia	17.9	2.5	21.5	30	68	35	1,870	127.3(330)	140(54)
Myanmar (Burma)	41.3	2.0	49.8	97	55	24	—	261.2(677)	158(61)
Philippines	66.1	2.6	82.7	48	64	42	630	115.8(300)	571(220)
Singapore	2.7	1.5	3.0	6.9	73	100	9,100	0.220(.57)	12,177(4702)
Thailand	55.7	1.5	63.7	39	66	18	1,000	198.5(514)	281(108)
Vietnam	70.2	2.5	88.3	50	66	20	—	127.2(329)	552(213)
Pacific Islands									
Fiji	0.8	2.2	0.9	21	63	39	1,540	7.1(18)	108(42)
French Polynesia	0.2	2.5	0.2	23	68	62	—	1.5(4)	109(42)
New Caledonia	0.2	1.9	0.2	39	67	58	—	7.4(19)	21(8)
Pacific Islands*	0.2	3.1	0.2	29	71	29	—	0.690(2)	269(104)
Papua–New Guinea	4.0	2.7	5.1	59	54	13	770	178.2(462)	23(9)
Solomon Islands	0.3	3.5	0.5	40	61	9	430	11.0(28)	30(12)
Vanuatu	0.2	3.2	0.2	36	69	18	820	5.7(15)	29(11)
Western Samoa	0.2	2.8	0.2	48	66	21	580	1.1(2.8)	155(60)

* Compromising the Federated States of Micronesia, Palau, and the Marshall and Northern Mariana Islands.

One of the classic areas for such rich soils is the central and eastern part of the island of Java, an island that comprises less than 7 percent of Indonesia's land area, yet contains 63 percent of the nation's population. Fertile soils are certainly one reason for such high population densities. The alluvial soils (inceptisols) found in the floodplains of the region's major rivers are the second major exception. The sediments carried by the rivers constantly renew the fertility of the soil in the densely populated rice-producing areas associated with these core regions. Indeed, all of Southeast Asia's most densely populated agricultural areas are located on soils of either alluvial or recent volcanic origin. In addition to Java, these include the Philippines' Central Plain and the valleys of the Red River in northern Vietnam, the lower Mekong in southern Vietnam, the Irrawaddy River of Myanmar (Burma), and Thailand's Chao Phraya (Figure 16.1).

The natural forests associated with tropical wet climate regions are fast being destroyed. As is true of other such regions, people are cutting the forests for timber or to use the land for subsistence agriculture or plantation crops. Only the "Outer" Islands of Indonesia (Java and some nearby islands are known as the "Inner" Islands), some interior regions of mainland Southeast Asia, Mindanao in the Philippines, and New Guinea still have extensive stands of primary forest. It is estimated that perhaps 10 percent of the world's remaining tropical forests are found in Indonesia.

Deforestation is bringing about the destruction of the rich and varied life-forms of the tropical rainforest re-

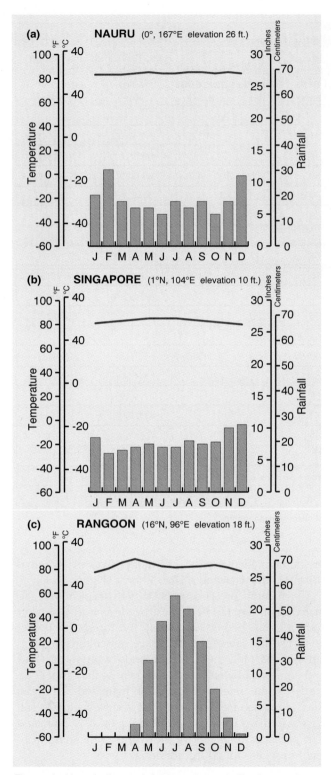

The vertical bars indicate rainfall in each month. The line graphs indicate the temperature variation throughout the year.

FIGURE 16.3 (left)

Climographs of Nauru, Singapore, and Yangon (Rangoon)

gions. Indigenous mammals to Southeast Asia, such as the orangutan, Javan and Sumatran rhinoceros, and tiger, are near extinction. Many of the millions of species of plants and animals (most not yet discovered) in tropical areas may be valuable to humans for medicinal, industrial, and other purposes. Numerous species are **endemic** (local), and therefore deforestation of a small area can mean the disappearance forever of a species. One study estimated that one forested volcano in the Philippines contained more woody plant species, many of them endemic, than the entire United States.

In the eastern part of Indonesia is the dividing line between two great floral and faunal regions. To the west of what is known as **Wallace's line** (modified later by Weber and others) is the Oriental realm and to the east is the Australian realm. Here are found the rather peculiar animals known as marsupials, which include kangaroos, wombats, and koala bears (Figure 16.5). During the Pleistocene epoch (the "Ice Age"), which began about two million years ago and ended only 10,000 years ago, most of island Southeast Asia was attached to the Asian landmass because sea levels were much lower than they are today. Land areas then exposed included the area known generally as the continental shelf (Sunda Shelf in Southeast Asia); today, ocean depths in these areas are less than 650 feet (200 meters) (Figure 16.5). In the area between the lines drawn by Wallace and Weber, however, greater ocean depths kept the areas physically separated by at least 40 miles (65 kilometers). Thus, the water barrier precluded many life-forms from diffusing across the barrier. It is estimated that the first humans, people of the Australoid racial group, did not reach those areas to the east of the barrier (the easternmost islands of Indonesia, New Guinea, and Australia) until perhaps 40,000–50,000 years ago.

LANDFORM AND DRAINAGE PATTERNS

Four different tectonic plates collide in the Southeast Asian and Pacific regions. Near these collision, or subduction, zones, there has been considerable volcanic and earthquake activity, and many volcanoes have erupted in recent memory. The most dramatic eruption occurred in Indonesia in 1887 when Krakatoa, a small island between Java and Sumatra, exploded, carrying tiny dust and ash particles as high as the stratosphere, where they remained in earth orbit for several years. Because the dust partially blocked the sun's rays, the amount of solar radiation received by the earth decreased, temporarily lowering earth temperature. Java alone has over 50 volcanoes that are periodically active. One-sixth (223) of all the major earthquakes that have occurred during this century have been in Southeast Asia and the Pacific, accounting for over 35,000 deaths. The mainland area of Southeast Asia is geologically stable today, but it, too, is mountainous as a result of earlier tectonic activity. The subsequent erosion and deposition that have taken place on the mainland have created a landscape of great complexity.

The islands of the Pacific are of recent volcanic origin or are land areas composed of coral, a material produced by organisms (coral and algae) growing together and secreting rocklike mineral carbonates. As the coral colonies die, limestone is accumulated forming a coral reef or an **atoll.** Usually, atolls are formed on a foundation of extinct volcanic rock beneath the ocean surface. Because coral limestone is the only rock found in such island areas, agriculture is generally not feasible without fertilizers. An exception to this is the coconut palm, a plant native to the area that has considerable commercial importance for its **copra** and coconut oil.

Human populations that reside on low-lying coral islands, such as those in Micronesia, are often in danger of typhoons that are spawned in the western Pacific and move westward into eastern Asia. The Philippines, on the western rim of the Pacific, had a greater number (39) of devastating storms (called cyclones or typhoons) during the 1960s and 1970s than any other single nation.

Understanding the topographic and drainage characteristics of the Southeast Asian region is important to an understanding of human migration, settlement patterns, and population distribution. The densest clusters of population are found in the river valleys and deltas and on lowland areas of volcanic soils. Mainland Southeast Asia contains four major rivers where dense rural populations thrive. The major river of the region, and the sixth longest in the world at 2600 miles (4200 kilometers), is the Mekong (Figure 16.1). It begins in the high mountains of Tibet and reaches its delta in southern Vietnam where each year it discharges a huge amount of water, ranking it fifth among the world's rivers in volume. It is in the delta region in southern Vietnam and in Cambodia where population densities, and rice

FIGURE 16.4

Seasonal Changes in Prevailing Winds and Atmospheric Pressure: January and July

The arrows indicate wind direction. The numbered lines indicate atmospheric pressure in millimeters.

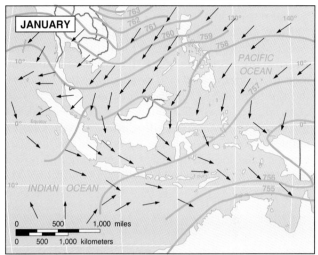

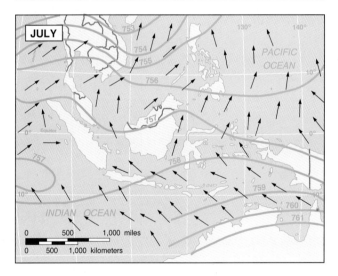

production, have historically been the highest along the Mekong. The other major rivers from the standpoint of human settlement and transportation are the Irrawaddy system of Myanmar (Burma), which, together with its tributaries, provides 5000 miles (8000 kilometers) of navigable waterways in the wet season and 3600 miles (5888 kilometers) in the dry season; the short 200-mile (320-kilometer) Chao Phraya in Thailand, an area noted for intensive rice production that enables Thailand to be ranked among the world's major rice-exporting nations;

and the 500-mile (800-kilometer) long Red River of northern Vietnam, where population densities are among the world's highest. Another major regional river, the Salween (1700 miles, or 2800 kilometers) of Myanmar (Burma), passes through areas of narrow gorges and generally poor soils, causing population densities to be low.

There are no major rivers in Southeast Asia's Malay Peninsula or its island areas, nor in Papua–New Guinea. Because these areas are moist, there are, of course, numerous short rivers, which often drop very rapidly to the sea from the mountainous interior. Where this occurs, falls and rapids are found, and the potential exists for the development of hydroelectric power. One example of such a river is the Agus River and its Maria Cristina Falls, located in Iligan City on the northern coast of the Philippines' second largest island, Mindanao. This

short 22-mile (35-kilometer) river descends very rapidly from about 2300 feet (700 meters) to the sea, creating the falls and associated rapids. It is estimated that about 40 percent of the total hydroelectric potential for the entire nation is found here, only a portion of which has been developed.

MINERAL RESOURCES

While the economies of nearly all the nations of the two regions are dependent primarily on agriculture, there are some mineral resources of importance. Most significant among these is tin; indeed, Malaysia is the world's leading producer and accounts for nearly one-fifth of total production. Indonesia and Thailand are also important producers of tin, and together the three nations

FIGURE 16.5

Biogeographic Regions and the Continental Shelf

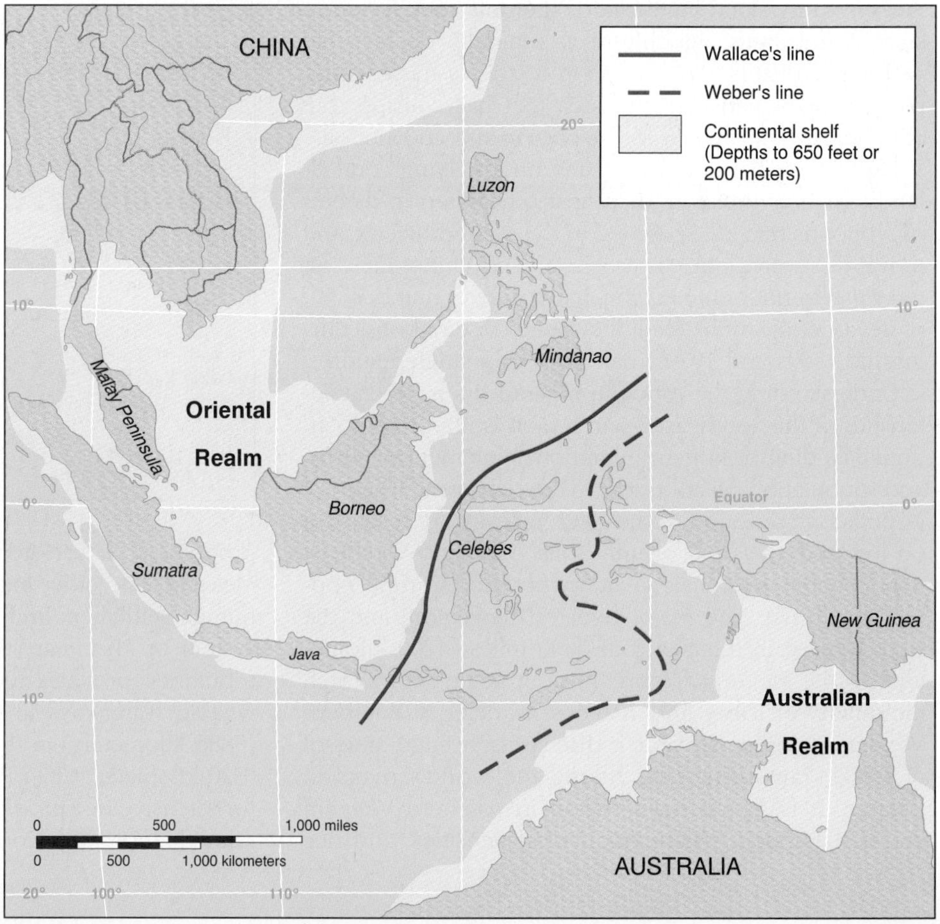

F irst applied to the Balkan area of southeastern Europe, the term *shatter belt* can also be used to describe the Southeast Asian region. Shatter belts are regions characterized by a history of political instability and conflict. Such instability is due to powerful and colliding external cultural, economic, military, and political forces. In the case of Southeast Asia, these have been India to the west, China to the north, and, more recently, the various European colonizers, the Americans, the Japanese, and the Soviets; in the case of the Balkans, such powers have included the Ottoman Empire, Germany, Austria, and Russia.

SHATTER ZONE OR BELT

Such conflict and instability are also created by internal factors. Specifically, the physical environment of Southeast

Asia (and the Balkans) can be described as "fragmented"; that is, the region is divided into literally thousands of islands and islets, and the mainland portion of Southeast Asia is cut up by generally north-south trending mountain and hill ranges that have hindered east-west human interaction. This physical fragmentation has facilitated the historic development of numerous separate upland and lowland ethnic tribal groups. The resultant cultural fragmentation has led to political/military rivalries and political instability within the region.

produce over two-fifths of the world total. Other mineral resources of commercial importance include chromite in the Philippines (which ranked eighth in world production in 1983) and New Caledonia; copper in the Philippines (ranked tenth) and Papua–New Guinea (twelfth); nickel in New Caledonia (the world's third largest producer), Indonesia (fifth), and the Philippines (ninth); and phosphate rock from tiny Nauru (ranked eleventh with over 1 percent of world production), with a total land area of only 8.2 square miles (21.2 square kilometers). Deposits of other mineral ores are limited. In short, except for Southeast Asia's tin and some copper, nickel, and chromite, both regions are poor in mineral resources.

Similarly, Southeast Asia and the Pacific are not particularly well-endowed with fossil fuels. The three major producers are Indonesia, Malaysia, and tiny Brunei. Secondary production occurs in Myanmar (Burma) where Southeast Asia's first commercial oil was produced. Indonesia, an OPEC member, produced over 450 million barrels of oil in 1983 and ranked ninth among the world's producers (with 2.5 percent of world production); it derives over three-quarters of its foreign earnings from mineral fuels. Brunei derives nearly all of its earnings from oil production. Such a dependence upon

one or a few primary commodities, as in the case of Indonesia and Brunei, can severely strain the economy of a nation when the world price for that commodity declines. Oil, of course, is a good example of a commodity whose price has fluctuated enormously, especially since the early 1970s. Except for northern Vietnam, virtually no coal is produced in either region.

Other commercially valuable nonmineral primary resources of Southeast Asia include forests, where a majority of the world's teakwood is produced. Lumber is a primary commodity of great importance to the economies of Laos, Myanmar (Burma), and Malaysia. In Laos, about four-fifths of export earnings come from forest products.

The vast fisheries are another potentially valuable resource, but have not yet been well developed by the nations of the region, in part because of unresolved legal questions. Oceanic territorial claims are an important point for consideration in the development of oceanic resources, especially in archipelagic areas. The United Nations has sponsored "Law of the Sea" conventions to resolve such claims, but the participating nations have met with little success. The United Nations recognizes a 200-mile (320-kilometer) claim for nations even though initially many nations claimed only 3 miles (5 kilometers) of the bordering seas (Figure 16.6).

FIGURE 16.6

Oceanic Claims in the Pacific Ocean

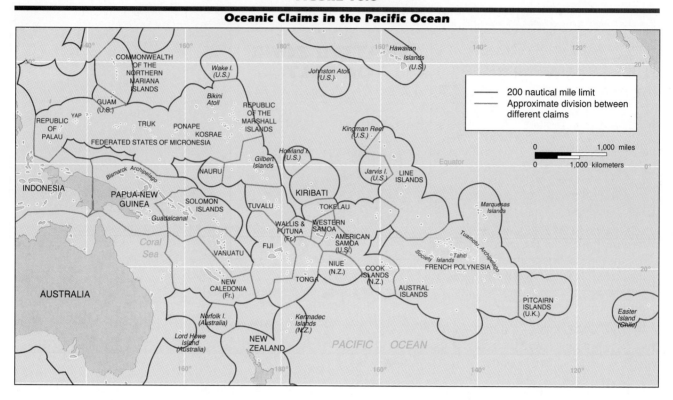

HUMAN ENVIRONMENT

Each of the two regions can be conveniently divided into subregions based on physical and cultural characteristics. Thus, with ten independent nations, Southeast Asia, is composed of the mainland subregion and the peninsular-insular subregion. The Pacific region, sometimes referred to as Oceania or the South Pacific, consists of the three subregions of Melanesia, Polynesia, and Micronesia (Figure 16.2). Politically, the Pacific region consists of nine independent nations and ten dependencies of varying status (Table 16.1). For historical-cultural reasons the Hawaiian Islands and New Zealand are often included with Polynesia. Because of contemporary cultural, racial, political, and economic characteristics that with a few exceptions are quite dissimilar from the rest of the Pacific region, Hawaii and New Zealand have been classified with other regions in this text. The indigenous Hawaiians and New Zealand's Maoris are, of course, true Polynesians.

POPULATION

Southeast Asia has a population in excess of 450 million, or over 8 percent of the world total. By far the most populous nation in the region, and the fifth most populous in the world, is Indonesia with over two-fifths of the regional total. Spread over more than 12,000 islands, Indonesia's total land area also accounts for about two-fifths of the regional total. Vietnam, the Philippines, Thailand, Myanmar (Burma), and Malaysia follow Indonesia in population and in land area; the other nations have notably smaller populations and form a third grouping of nations.

Most of the nations of the region have annual rates of natural population increase of at least 2 percent, which implies a population doubling time of about 35 years. Most of the nations have lowered their birth rates significantly through active family planning programs that have been supported enthusiastically by most governments. Still, the relatively high current population growth rates reflect those of a developing region.

Growth rates, however, are lower in Southeast Asia than in many other developing regions including South Asia, tropical Latin America, the Middle East, and Africa.

Singapore, an independent city-state, is unique in the region and is often classified as a developed nation. Most of Singapore's demographic and socioeconomic indices reflect a higher level of development. In addition to its low population growth rate, life expectancies are 73 years compared to 61 for the region as a whole, and the per capita GNP in 1988 was $9100.

Absolute population densities for the region, which average 255 persons per square mile (99 per square kilometer), are not as high as in other nearby nations, e.g., India with 672 persons per square mile (259 per square kilometer) and China with 302 (117). As the previous discussion implied, however, land for agriculture is limited, and therefore physiological densities are indeed very high. At the extreme is Java, where rural densities attain levels of 2000 persons per square mile (775 per square kilometer). In short, the land available for pioneer agricultural settlement is limited, and therefore many must seek opportunities in cities or in other lands. Since World War II, population redistribution within individual nations has been massive as large numbers of rural people have moved to cities, especially the very largest, or primate, cities of each nation. This movement has been predominantly rural-to-urban migration although there is much evidence that in Southeast Asia and the Pacific Islands, the form of migration known as **circulation** has also become increasingly important. Unlike permanent migration, circulation generally involves movement to cities for short durations, perhaps a few weeks or months. The intent of the circulator is to return to his or her homeplace after work is completed in the city. Due to their mobility, circulators have fewer needs (housing, for example) than do permanent migrants to the overcrowded cities and, thus, urban planners and government officials should consider facilitating circulation as an alternative to permanent migration. There is also considerable circulation to agricultural areas, especially where workers are needed to harvest seasonal crops.

There are also pioneer areas in parts of Southeast Asia to which rural migrants have moved, either spontaneously or through government incentives. An example of the former has been the heavy migration of Filipinos to rural Mindanao Island beginning in the early years of the present century. Filipinos of a variety of ethnic backgrounds, especially Visayans, Ilocanos, and Tagalog-

speaking peoples of central Luzon, have moved in large numbers from their densely populated home areas to Mindanao. Although this movement continues, so many people have settled in rural Mindanao that few areas can still be called pioneer settlement areas. An example of government-encouraged migration is Indonesia's **Transmigration** program. Begun during the Dutch colonial period in 1905 and continued after independence, the purpose of the program was (and remains) to help alleviate population pressures in overcrowded Java through the movement of peoples to the sparsely settled

TABLE 16.1

Political Units of the Pacific

POLITICAL UNIT	POLITICAL STATUS*
MELANESIA	
Fiji	1 (1970)
New Caledonia	3 (1956)
Papua–New Guinea	1 (1975)
Solomon Islands	1 (1978)
Vanuatu	1 (1980)
MICRONESIA	
Guam	3 (1898)
Kiribati	1 (1979)
Nauru	1 (1968)
Trust Territory of the Pacific Islands**	2 (1947)
POLYNESIA	
American Samoa	3 (1899)
Cook Islands	2 (1965)
Easter Island	3 (1888)
French Polynesia	3 (1946)
Niue	2 (1965)
Tokelau	3 (1948)
Tonga	1 (1970)
Tuvalu	1 (1978)
Wallis and Futuna	3 (1961)
Western Samoa	1 (1962)

* Political status and the year that status was attained are given. Units designated "1" are politically independent; "2" indicates units that are internally self-governing or in "free association" with former colonial powers; and "3" indicates a colony with some lesser degree of autonomy.

** Includes the four separate self-governing political entities of the Commonwealth of the Northern Mariana Islands; Federated States of Micronesia; Republic of the Marshall Islands; and Republic of Palau. All have been administered by the United States under a 1947 United Nations trusteeship arrangement.

SOURCES: Frederica Bunge and Melinda W. Cooke, eds., *Oceania: A Regional Study* (Washington, D.C.: U.S. Government Printing Office, 1984); and Population Reference Bureau, *World Population Data Sheet* Washington, D.C.: Population Reference Bureau, (1989).

The skyline of Singapore reflects the affluence of this modern city.

Outer Islands, especially Sumatra and Kalimantan. The program has been deemed unsuccessful by many critics since providing suitable sites has proved costly and only a small fraction of Java's population has been directly affected.

An option for some has been international migration or circulation. In Southeast Asia, the former has been either forced (for example, the considerable refugee movements both intra-and extraregionally) or voluntary (Table 16.2). Forced migration during the past decades has resulted largely from the mass exodus of Vietnamese, Cambodians, and Laotians from Indochina since the mid-1970s. Since then, about 1.6 million Indochinese refugees have sought asylum. The vast majority of these refugees have been resettled. The United States has admitted the largest number of Indochinese—over

750,000. Voluntary movement is generally limited to persons with special skills or those with relatives who are citizens of other nations. For example, during this century many Filipinos have migrated to the United States, where today they number over 2 million and are the second largest Asian ethnic group.

International circulation includes the large number of students who attend colleges and universities in the developed world. Many such students who become doctors or engineers opt to remain in the host country, ultimately gaining citizenship. Such persons are a part of the vast "brain drain" that has unquestionably had a negative impact upon development in the developing world. Another recent example of international circulation is the large number of skilled laborers who take temporary jobs in the oil fields of the Middle East. Many Filipinos, Thais, and other groups have taken such work, and the earnings they send home, called ***remittances,*** can be an important component of a nation's foreign earnings and local incomes.

The total population of the entire Pacific region is just over 6 million, and nearly 63 percent of the total live in Papua–New Guinea, which accounts for over 83 percent of the regional land area. Absolute densities in the entire region are low, only 29 persons per square mile (11 per square kilometer); if Papua–New Guinea is excluded, however, densities average 68 persons per square mile (26 per square kilometer). On some of the smaller islands, densities are very high. Extremely rapid population growth rates, about 3 percent annually for the region, and rapid urbanization are two major problems for the region, especially since the land has a limited capacity to accommodate increased population. Some of the smallest land areas are rapidly becoming overcrowded. The population of Micronesia, for example, has tripled since World War II.

CULTURAL ENVIRONMENT

An understanding of the cultural, political, and economic diversity of contemporary Southeast Asia and the Pacific is not possible without at least a brief examination of the history of the two regions. There is good evidence that Southeast Asia was settled very early in human history, perhaps as early as 500,000 years ago, during the Pleistocene epoch. During that geological time period, most of the Indonesian islands were a part of the mainland since sea levels were much lower than today, exposing much of the continental shelf. For this

TABLE 16.2

Refugees in Need of Protection and Assistance in Southeast Asia

REFUGEE GROUP	TOTAL NUMBER (THOUSANDS)	ASYLUM COUNTRY (THOUSANDS)								
		Thailand	Philippines	Malaysia	Indonesia	Vietnam	Papua–New Guinea	Hong Kong	India	Japan
Cambodian*	333.6	316.7	.2			16.7				
Vietnamese	123.8	12.7	25.8	19.9	8.0			55.4		2
Burman	3.8.8	38.0							.8	
Laotian	69	68.7	.3							
Indonesian	8.1						8.1			
Total	573.3	436.1	26.3	19.9	8.0	16.7	8.1	55.4	.8	2

* Excludes approximately 250,000 Cambodians in Thai border camps.

SOURCE: U.S. Committee for Refugees, *World Refugee Survey 1984* (New York: American Council for Nationalities Service, 1984), pp. 38–41.

reason, archaeologists have found evidence of early humans on Java, who migrated there from the north. Since that time, many different racial and ethnic groups have migrated to, and through, this accessible region. The earliest known racial groups were ancestors of the Australoids, Negritos, and Melanesoids (possibly a mixture of the former two stocks), all today considered minor racial groups because their total numbers are small. Most of these earliest migrants are thought to have originated in the Indian subcontinent. Today, the few relatively pure groups of Negritos that remain in Southeast Asia are found in remote, mountainous areas.

A more recent, and much more numerous, migration to the region has involved peoples of the Mongoloid racial group, who began arriving in the region from China some 3000 to 4000 years ago. Over time this movement has involved large numbers, due partly to increasing population pressures in southern China and the search for new lands on which to settle. These more recent and more sophisticated settlers pushed the earlier inhabitants to less desirable locations, namely, to more remote and hilly interior areas. Historic animosity still persists between the lowland majority and the hill peoples, a wide variety of tribes who comprise the upland minorities. Today, of course, the racial identity of the vast majority of Southeast Asians is Mongoloid. In recent years, migration from China has continued, but Caucasoid peoples from India and, most recently, from Europe and the United States have contributed to an even greater diversity and mixing of the population.

The Indianization of Southeast Asia began more than 2000 years ago when traders, often accompanied by Hindu and Buddhist priests, ventured by sea to nearby Southeast Asian coasts. While their numbers were never very large, these traders/priests brought with them influences that would permanently leave their mark on Southeast Asia. In this manner, the first of the great world religions came to Southeast Asia, as did the first written language (Sanskrit). A new concept of royalty, codes of law, art and architecture, literary forms, and new agricultural techniques including the Indian way of cultivating rice were also borrowed from Indian culture. The first great empires of the region were called "Indianized" states because they adhered closely to these influences. Although not one of the first, perhaps one of the best known of these was the Angkorian Empire in Cambodia, founded in the ninth century A.D. Its capital, Angkor Thom ("City of the Gods"), was one of the largest cities in the world at its zenith in the twelfth century. It is estimated that the city, and the surrounding rural villages that serviced the city, had a population as high as 300,000. This "sacred" city was a replica in miniature of the Indian Hindu-Buddhist cosmology.

Racially and ethnically, the indigenous peoples of the Pacific region are closely related to Southeast Asians since the region's very first inhabitants came from Southeast Asia. New Guinea and Melanesia, the "black islands" (Melanesia derives from the word *melanin,* the skin chemical that accounts for the dark pigmentation), are still predominantly Melanesoid in race. More recently, perhaps 3,000 to 4,000 years ago, Mongoloid peoples from Southeast Asia arrived in their sailing canoes and settled the islands of Micronesia and, especially, Polynesia. The present Polynesian physical type may have developed in the islands of Tonga and Samoa, the former possibly the island group with the longest record of human settlement in Polynesia. The Polynesians, who are known for their great sailing abilities, had discovered and settled even the most remote islands of the Pacific (e.g., Easter Island) by about the fourth century A.D. As in Southeast Asia, the arrival of more recent peoples has changed the racial and ethnic makeup of the indigenous population. Europeans have also tremendously, and forever, altered the political and socioeconomic characteristics of the indigenous population.

The arrival of the Europeans in the two regions can be more precisely dated. In the case of Southeast Asia, the Portuguese arrived in 1509 and two years later had established the first European garrison in the region at Malacca. The earliest Europeans to the Pacific came in 1521 when Magellan landed on Guam during this first European circumnavigation of the globe. The region did not become commercially important to the Europeans, however, until the second half of the eighteenth century after the voyages and discoveries of Dutch, British, and French sailors, the best known of whom was England's Captain James Cook. Following these early navigators came others who represented all of the major European seafaring powers of the time; these powers soon colonized nearly all of the land areas of Southeast Asia and the Pacific (Table 16.3). Only Thailand was never colonized although it was influenced by the West.

It is of utmost importance that we understand some of the broad impacts the West has had upon the regions, since without this background it is not possible to understand the contemporary economic conditions and interrelationships that exist. Each of the colonizers had a somewhat different impact upon its colonies, but all eventually had one overriding reason for control: raw materials and wealth. Thus, the legacies of the colonial period included the development of mineral and forest resources and the development of the plantation system and crops, which in Southeast Asia and the Pacific meant the production of three principal crops: rubber, coconuts, and sugar cane. Concomitant with the development of these resources was the construction of a transportation network to convey these products to port so they could be shipped to the mother country. Thus roads, and in a few instances railroads, were built that connected interior locations where such resources were produced, to a port. Usually, this led to the development of one major, or **primate,** city that served as principal port, government center, cultural and educational cen-

ter, and, eventually, the manufacturing center for the colony. Today, these former colonial ports are the very largest cities in each nation. Wealth is concentrated in these cities, resulting in severe spatial inequalities that can be traced to the colonial period. Indeed, such nations are sometimes referred to as *neocolonies* since the former colonial powers, and some newly developed nations (e.g., Japan), still control much of the economic system.

Other important legacies of the European and American colonial period included the introduction of alien political systems, institutions, and boundaries; the introduction of Western languages through educational systems; and the introduction of Christianity. The latter has been especially significant in the Philippines and in the Pacific region.

Languages

Southeast Asia and the Pacific are incredibly diverse ethno-linguistically. Four major indigenous language families are spoken in the region: Malay-Polynesian (or Austronesian) languages predominate in island and peninsular Southeast Asia and in Polynesia and Micronesia; Sino-Tibetan languages are spoken throughout mainland Southeast Asia; Austro-Asiatic languages are found in Indochina; and the Papuan languages are spoken on New Guinea, in parts of the Moluccas in eastern Indonesia, and in Melanesia. Each family includes hundreds of separate languages and dialects. For example, some 650 Papuan languages are spoken on New Guinea alone, many of which are spoken by only several thousand individuals. Although the languages of individual nations are generally of the same family, many of these languages are mutually unintelligible. In the Philippines, for example, about 70 different major languages and dialects are spoken. And quite frequently, dialects, which are variants of a language, are so unlike that they are essentially separate languages.

Of the Southeast Asian nations, certainly Myanmar (Burma) has had at least as many problems in national development and unity caused by ethno-linguistic differences as any other nation in the region. Although the ethnic Burmans comprise about two-thirds of the total population, various other upland groups are dominant in their own regions. Indeed, upon independence from Britain in 1948, the federal constitution of the Union of Burma (now Myanmar) established six more or less autonomous units based on ethnic considerations. This

TABLE 16.3

Western Colonization of Southeast Asia and the Pacific

COLONY	COLONIZERS	YEAR
American Samoa	United States	1899–present
Brunei	Britain	1888–1984*
Cambodia	France	1863–1954*
Cook Islands	Britain; New Zealand	1892–1901; 1901–1965**
Easter Island	Chile	1888–present
Fiji	Britain	1874–1970*
French Polynesia	France	1842–present
Guam	Spain; United States	1668–1898; 1898–present
Indonesia	Netherlands	1605–1949*
Kiribati***	Britain	1892–1979*
Laos	France	1893–1954*
Malaysia	Britain	1786–1957*
Micronesia	Spain; Germany; Japan; United States	1694–1898; 1885–1918 (Marshalls); 1918–1945; 1945–1980s**
Myanmar (Burma)	Britain	1826–1948*
Nauru	Germany; Australia	1888–1918; 1918–1968*
New Caledonia	France	1853–present
Niue	Britain; New Zealand	1900–1901; 1901–1974**
Papua–New Guinea	Germany; Britain; Australia	1884–1918 (Northeast New Guinea); 1885–1901 (Papua); 1901–1975*
Philippines	Spain; United States	1521–1898; 1898–1946*
Singapore	Britain	1819–1963*
Solomon Islands	Britain	1883–1978*
Thailand	Never colonized	
Tokelau	Britain; New Zealand	1892–1920; 1920–present
Tonga	Britain	1900–1970*
Tuvalu***	Britain	1892–1978*
Vanuatu***	Britain and France	1906–1980*
Vietnam	France	1859–1954*
Wallis and Futuna	France	1842–present
Western Samoa	Germany; New Zealand	1900–1918; 1918–1962*

* Indicates year of independence.
** Indicates year self-governing, "free association" status was attained. Excluding Guam, four separate political entities today comprise Micronesia.
*** Kiribati was formerly the Gilbert Islands; Tuvalu was the Ellice Isands; Vanuatu was the New Hebrides.

was done in an attempt to satisfy the aspirations of the larger minority groups. States emerged for the Karens (9 percent of the population), Shans (7.5 percent), Kachins (2 percent), and Chins (2 percent), in addition to that of Burma proper (Figure 16.7). Occasionally, such groups have even held military control of their states. Local warlords, for example, have been supreme in the Shan State of eastern Myanmar, a part of the notorious "Golden Triangle," in which one-quarter of the world's opium poppy production originates. In addition to these major ethnic groups, scores of smaller groups are found throughout Myanmar, including the Mons, Palaung-Wa,

and Nagas. Ethnic Indians, and Chinese are also important minorities in Myanmar.

Besides indigenous languages, there are others that have been introduced more recently. For example, Chinese is spoken (and taught) by the millions of ethnic Chinese who are found throughout Southeast Asia, especially in cities. Singapore is culturally a Chinese nation in that about three-quarters of its population are of Chinese ancestry; Malaysia's population includes about one-third ethnic Chinese. The other nations of Southeast Asia include lesser, but significant, proportions of Chinese. In all Southeast Asian nations, and in parts of

FIGURE 16.7

Myanmar (Burma): Political Subdivisions and Ethnic Groups

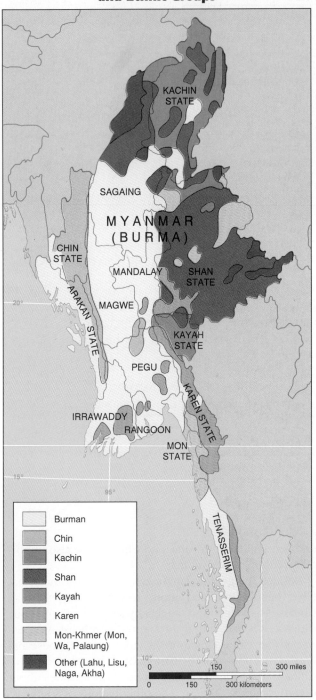

Legend:
- Burman
- Chin
- Kachin
- Shan
- Kayah
- Karen
- Mon-Khmer (Mon, Wa, Palaung)
- Other (Lahu, Lisu, Naga, Akha)

the Pacific, the Chinese control of retailing and whole-saling is disproportionate to their numbers. Today, the Chinese, among the most astute businesspeople in the world, dominate the region's trade and are vital to the success of regional banking, shipping, and manufacturing.

Indians have shown a propensity for commerce as well. Descendants of Indians who came to Malaysia as early as the nineteenth century to work on rubber plantations and in tin mines now own shops and small businesses in the cities of Malaysia and elsewhere in the region. In Fiji, Indians comprise nearly half of the population.

Finally, the Europeans and Americans brought with them their educational systems and schools, and thus also a new overlay of languages. Two of these languages have remained important. Most widely spoken is English, which Britain brought to Malaysia, Myanmar (Burma), and Singapore (where it today is one of four official languages) and the United States introduced into the Philippines during its nearly 50-year rule. Today, English is spoken as a second language by two-fifths of all Filipinos and is often used as the common language because of the great diversity of native languages. All of Manila's major newspapers, and much of the programming on television, are in English. Surprisingly perhaps, even though the Spanish colonial period lasted for some 350 years in the Philippines, Spanish is spoken today by very few Filipinos. French is also important, especially in the Pacific islands that were former colonies and in France's Overseas Territories of French Polynesia and New Caledonia.

Religion

Three major religions are found throughout Southeast Asia and the Pacific today: Buddhism, Islam, and Christianity. The first religions introduced into Southeast Asia were Hinduism and Buddhism, brought by Brahmins (priests) from India beginning about 2000 years ago. Although Hinduism is not a major religion in the region today (it is predominant among the Balinese of Bali and among Indians who are descendants of recent migrants to the region), Buddhism is important on the mainland. The people of Myanmar (Burma), Thailand, Laos, Cambodia, and Vietnam are predominantly Buddhist. In Vietnam, followers of the religion are Mahayana Buddhists in the north and Theravada Buddhists in the south. The former is less orthodox, entering northern Vietnam by way of China whereas the latter diffused into southern Vietnam and the rest of mainland Southeast Asia from the Indian subcontinent. Although Muslim merchants had frequented Southeast Asia since the seventh century A.D., Islam did not become an official religion until the thirteenth century when it was first established in northern Sumatra (Indonesia). Today, it is the predominant religion in Indonesia (80 percent Islam), Malaysia, and Brunei. In the Philippines, whose population is 85 percent Roman Catholic, a 5 percent minority of Muslims is found in the south. A Muslim insurgency has been underway since the early 1970s; some Muslims want independence or, at the very least, a greater degree of autonomy. Muslim minorities are also found in the southern extensions of both Thailand and Myanmar (Burma). Animism is still the dominant belief system of many of the upland tribal groups, especially in the most remote areas of Myanmar (Burma), Laos, and New Guinea.

In the Pacific region, Christianity is the predominant religion, a result of the many missionaries who ventured there. In some areas, notably those influenced by France, Catholicism became most significant whereas Protestant missionaries dominated in others. By the 1860s missionaries were at work in all the island groups except in Melanesia where an absence of ruling chiefs, fragmented and small political units, and a great diversity of difficult languages hindered the missionaries. A few of the island nations, where Indians or other recent immigrant groups make up an important minority, also have other religions (like Hinduism).

Political Characteristics

Southeast Asia can be subdivided into two groups of nations in terms of their political-economic philosophies. The Association of Southeast Asia Nations (ASEAN), an economic "common market" that began in 1967, is comprised of the six free market states of Indonesia, the Philippines, Thailand, Malaysia, Singapore, and Brunei. While most of these nations describe themselves as nonaligned, their economies, trade, and alliances are clearly oriented toward Western Europe, North America, and Japan. On the other hand, the four remaining nations of the region, Myanmar (Burma), Vietnam, Cambodia, and Laos, are socialist states. The latter three have been oriented toward the Soviet or Chinese style of communism since 1975, when the Vietnam War ended, thus concluding 20 years of American involvement in the region. Myanmar (Burma) has been organized as a neutral socialist, one-party state since 1974.

A street scene in Ho Chi Minh City with the rundown City Hall in the background.

Although extremely fragmented, the island nations of the Pacific region are similar in their political systems and outlook. Shared traditions, areal and demographic smallness, relative isolation, and, with a few exceptions, a poverty of resources all have prompted a degree of regional cooperation. Political and economic ties are focused toward the former (and present, in some cases) colonial powers, notably, the United States, Great Britain, France, Australia, and New Zealand.

ECONOMIC AND URBAN LANDSCAPE

Like most other characteristics of the region, the level of economic and technical development differs among the nations (and within the nations). Most would agree that all the nations of both regions, with the exception of Singapore, are less developed. Those indicators most commonly used generally demonstrate this fact; however, it is clear that such indicators also show that differences exist within this group of nations. Because of their high per capita gross national product (GNP), Singapore ($9100 per capita GNP), oil-rich Brunei ($14,120), and several of the Pacific nations are developed. Per capita GNP measures the total domestic and foreign output claimed by residents divided by population; it does not indicate anything about the distribution of such wealth. In fact, in nations like Brunei the actual wealth is distributed very unevenly. The nation's rulers often control the lion's share of the wealth, which, in the case of Brunei, is derived from oil revenues. Within the larger nations of the two regions, Malaysia is

the best off, economically the most developed of the developing. Other large nations that, taken together, have relatively high indicators of development include Thailand, Indonesia, and the Philippines. Clearly, Laos, Myanmar (Burma), Vietnam, Cambodia, and Papau–New Guinea are the least developed in terms of the per capita GNP.

Recent economic growth is an indicator of how nations are progressing. Unfortunately, such information is not always available for all nations so comparisons are difficult. We can, however, conclude that in recent years all of the larger nations in these regions have been growing more rapidly than the average for the developed nations, that is, the industrial market economies. Annual rates of growth in gross domestic product (GDP), a measure of the total final output of goods and services produced by an economy, have been most rapid in Malaysia, Thailand, Indonesia, and Singapore, over 4 percent between 1980 and 1988 (Table 16.4).

Agriculture

The economies of Southeast Asia and the Pacific have historically been based upon agriculture. Today, in most nations, agriculture remains the dominant sector, usually employing over 50 percent of the labor force, but other economic sectors, most notably the services sector, are becoming more important.

Long before the arrival of the Europeans, agriculture flourished throughout Southeast Asia. Indeed, in 1952 the geographer Carl Sauer proposed in his classic work

TABLE 16.4

Economic Growth Indicators in Southeast Asian Nations

NATION	AVERAGE ANNUAL GROWTH IN GDP (PERCENT)	
	1965–1980	1980–1988
Cambodia	N.A.	N.A.
Indonesia	8.0	5.1
Malaysia	7.3	4.6
Myanmar (Burma)	N.A.	N.A.
Papua–New Guinea	4.1	3.2
Philippines	5.9	0.1
Singapore	10.1	5.7
Thailand	7.2	6.0
Vietnam	N.A.	N.A.
High-Income economies	3.7	2.8

SOURCE: World Bank, *World Development Report, 1990* (New York: World Bank, 1990), pp. 220–21.

Agricultural Origins and Dispersals that sedentary agriculture was first practiced in Southeast Asia (in northern Indochina), rather than in the Middle East. When the Europeans arrived, two major agricultural systems were evident. One may be described as intensive subsistence cultivation, which usually meant the production of rice along with various minor crops such as fruits or vegetables.

Shifting cultivation, known locally as *caingin* in the Philippines and *ladang* in parts of Indonesia and in Malaysia, is very widespread; it is found where intensive subsistence cultivation is not possible due to soil depletion. Upland areas are often in shifting cultivation. Although Westerners once believed shifting cultivation was harmful to the environment, it has since become clear that as long as population densities do not reach a point where land is used for too long a period, this practice is not detrimental to the environment. In Southeast Asia, the practice involves cutting away and burning a small patch of forest and then planting it with a wide variety of crops (e.g., upland or "dry" rice, maize, tubers, mango, banana) for several years. The land is eventually abandoned, and the forest is allowed to replenish itself. If the same land is used for too long a period, or too frequently, then the soil and forest are destroyed and may be replaced by a grass vegetative cover. The grass is often too coarse and sharp edged to be used even for grazing, however.

Sedentary agriculture, though not as widespread as shifting, involves far greater numbers of people. Rice was, and still is, the principal subsistence crop of the region. Production is most intensive in those areas where physical conditions are best suited to rice production: broad river valleys and deltas. At least one-half of Southeast Asia's cultivated land is in rice, a greater proportion than in any other world region. Literally hundreds of varieties of lowland, or "wet," rice have been developed over thousands of years. Rice as cultivated in the region is very labor intensive. First, the wet-field (*sawah* in Indonesian), referred to as a rice paddy, must be prepared. This often means leveling the land and constructing a dike around the paddy so the field can be flooded, either by water from irrigation or from rain. Although practices may vary somewhat, the next step usually involves planting seeds close together in nursery beds; several weeks later, the seedlings are transplanted to the main fields. By then the main fields have been flooded and the soil worked into a soft mud through repeated plowing and harrowing, most often

A woman replants rice seedlings in a flooded paddy in Malaysia.

with the aid of small tractors and water buffalo, the work animal of monsoon Asia. Following transplanting, water in the paddy has to be carefully regulated so that the upper portion of the plant is not submerged. As the rice nears maturity, the paddy is drained and the rice is harvested, usually by hand. In some upland and even mountainous areas, such wet rice is grown in terraces that have been painstakingly built into the hillsides. The most dramatic example of this is found in the mountain areas of northern Luzon, in the Philippines, where such practices began thousands of years ago. The construction of the paddy in both lowland and terraced areas provides protection against excessive soil erosion. New flood waters in the paddy each growing season mean that mineral matter is replaced; as a result, some areas have produced rice for centuries without serious loss of soil fertility.

TABLE 16.5

Change in Agricultural Production, Cereal and Tuber Yields, and Population, 1965–1983

NATION	AGRICULTURAL PRODUCTION (%)	YIELD OF CEREALS (%)	YIELD OF ROOTS AND TUBERS (%)	POPULATION CHANGE (%)
Cambodia	−33%	−16%	−24%	−2%
Fiji	44	19	3	40
Indonesia	101	119	34	48
Laos	73	106	54	38
Malaysia	131	27	7	60
Myanmar (Burma)	89	92	165	53
Papua–New Guinea	62	−39	3	48
Philippines	106	64	16	65
Solomon Islands	93	67	17	50
Thailand	103	9	26	65
Vietnam	70	32	−9	62

SOURCES: World Resources Institute, *World Resources 1986* (New York: Basic Books, 1986), pp. 262–63. Percentages are derived from an index of agricultural production and crop yields in kilograms per hectare. Population growth rates are derived from United Nations, *1970 Demographic Yearbook* (New York: United Nations, 1970) and the Population Reference Bureau, *World Population Data Sheet* (Washington, D.C.: Population Reference Bureau, 1983).

Beginning in the 1960s, the Green Revolution has brought about the development of new, higher-yielding rice varieties so that today most of the Southeast Asian nations produce sufficient rice to feed their own populations. Both total production and yields and cereals and tuberous crops have increased dramatically, in most cases more rapidly than population (Table 16.5).

Thailand is the world's major exporter of rice, producing far more than needed for its own people in the fertile alluvial basin of the Chao Phraya. The lower Mekong region of Cambodia and southern Vietnam has also historically been a rice-exporting region, but since the 1960s continued military conflict has greatly reduced production. In Cambodia, beset by years of internal turmoil, agricultural production and yields have actually declined, as has the population (Table 16.5). Agricultural production has increased most dramatically in the ASEAN as a result of the new agricultural technology and the expansion of cropped areas. It is perhaps not surprising that the "miracle" rice has been adopted in most of the ASEAN since the new varieties were developed at the International Rice Research Institute, located in Los Baños, not far from Manila in the Philippines.

Subsistence and shifting cultivation are also the traditional agricultural systems in the Pacific region. Although cereals, especially rice, are also important here, root and tuber crops are the mainstays in the diets of many peoples.

The Europeans brought plantation agriculture to the rural landscape of Southeast Asia and the Pacific. In so doing, the colonizers, particularly the British and Dutch, introduced a wide range of new commercial plants from other tropical lands. Indeed, except for rice, sugar, and coconuts, all of the major commercial crops produced in Southeast Asia today were introduced during the colonial period. The Europeans introduced maize, coffee, tea, cinchona (quinine is obtained from the bark of the tree), tobacco, oil palm, and, perhaps most importantly, rubber.

Rubber was brought to Southeast Asia by the British. Indigenous to the Amazon jungle, the rubber tree was introduced into Malaya (the name for Malaysia until after independence) around the turn of the twentieth century. The fortuitous development of the automobile about the same time soon led to a great demand for rubber. By the 1920s, rubber was by far the most important export of Malaya, and soon it became the leading export of the Dutch East Indies as well. Myanmar (Burma), Thailand, and southern Indochina also produced significant amounts of the crop. In many ways rubber was an ideal crop for the region because it thrived even on laterized soils; it also provided more or less permanent employment for workers on the rubber estates, since the latex (which drains from incisions made in the tree) had to be collected throughout the year. Today, much of the rubber and other Malaysian "estate" crops like oil palm (whose production increased dramatically after the 1960s) are produced on smallholder operations, rather than on large plantations.

Coconut, an indigenous crop, also became more important after the European period although copra and coconut oil had been used for a considerable period in traditional recipes of both regions. The copra trade became a major economic activity in the mid-nineteenth century when a large demand for tropical vegetable oils developed in Europe. The Germans launched the early copra trade as a result of their advances in the technology of extracting oil from copra. They had established an oil-processing plant in Western Samoa by 1856. By the early twentieth century, large-scale production had begun in Southeast Asia. Today, the major producing areas are along the coasts of the Philippines and Indonesia. Because it requires minimal care, the coconut is also a popular crop among small landholders and is often grown as a cash supplement to subsistence crops.

Plantation crops are of major commercial importance in the island and peninsular areas of Southeast Asia and in some of the Pacific nations; in mainland Southeast Asia on the other hand, commercial agriculture is dominated by rice, a crop ideally suited to the monsoon conditions and the wide deltas and floodplains of the large rivers of the mainland. The colonial powers contributed little to rice cultivation techniques, which are practiced today much as they were in the past. However, the colonizers did introduce new drainage, irrigation, and flood control technologies.

These early advances, as well as the higher-yielding varieties and technologies developed during the Green Revolution, have increased food production, allowing higher population densities to be supported. Unfortunately, higher yields have not always meant improvements in rural living standards. Rather, rural people often continue to survive at bare subsistence levels. In other words, scientific advances in agriculture may not necessarily lead to improvements in life-style, but rather to maintenance of the status quo. Such a "no change" situation has been called "agricultural involution" by the anthropologist Clifford Geertz, who based his ideas on findings from Java, one of the world's most densely populated rural areas. Geertz proposed that as population increased, the economy remained static since wet rice agriculture was able to respond to the rising population by absorbing increasing numbers of workers. As a result, the agricultural system became internally more elaborate, ornate, and complicated, "all in an effort to provide everyone with a niche, however small, in the over-all system." * What is tragic, according to Geertz, is that this involution need not have occurred except that

* Clifford Geertz, *Agricultural Involution: The Processes of Ecological Change in Indonesia* (Berkeley: University of California Press, 1963), p. 82.

A worker taps a rubber tree on a plantation in Malaysia.

the Dutch imposed their agricultural system on Java. The Javanese could have made the transition to modernism around the 1830s much more easily than is possible today. A basic question, of course, is how long such an involutionary system can continue to absorb ever greater numbers and simply maintain the status quo. The implication is that at a crisis point some form of revolution may occur.

Urbanization and Industrialization

As is true of agriculture, the colonial period greatly influenced the urban and industrial character of the nations of the two regions. Since the end of World War II, urbanization has been very rapid in most of the developing world as millions have moved from rural to urban areas where jobs in industry or the service sector were perceived to exist. Changes in the percentage of the population residing in urban areas between 1950 (15 percent) and 1990 (40 percent) clearly demonstrate this. The United Nations estimates that 36 percent of Southeast Asia's population will live in cities by the year 2000.

The metropolitan areas that have grown the most rapidly are the very largest cities in each nation, especially the primate cities. Those nations where the largest city completely overshadows all other cities in size are said to have a high primacy index, an index derived simply by dividing the population of the very largest city by that of the next largest metropolitan area. Furthermore, it appears that some primate cities, like Manila and Bangkok, have continued their very rapid growth relative to secondary urban places. In the two largest nations of the region, Indonesia and Vietnam, this two-city primacy index is not as high as might be expected. In the case of Indonesia, this is due in part to the large population and areal size of the nation. This large size has enabled Indonesia to support several cities with populations over one million; after Jakarta (Southeast Asia's largest city), these are Surabaya, Bandung, and Medan.

Ho Chi Minh City (formerly Saigon) is Vietnam's largest city, but Hanoi in the north may surpass Ho Chi Minh City by the year 2000 (Table 16.6). Vietnam has been divided into at least two parts at various periods throughout the last millennium. Between 1954, the year in which the French forces were defeated at Dien Bien Phu by Ho Chi Minh's guerilla forces, and 1975, there were in fact two nations: Hanoi was the capital of communist North Vietnam, and Saigon was the capital of American-supported South Vietnam. It appears that the region's large cities will continue their rapid growth, and at least three of them, Jakarta, Manila, and Bangkok, will rank among the 30 largest in the world by the year 2000 (Table 16.6).

Vietnamese farmer relies on animate energy to plow a soggy field in preparation for planting.

TABLE 16.6

Major Metropolitan Areas of Southeast Asia: 1950, 1980, and 2000

1980 RANK	METROPOLITAN AREA	POPULATION (THOUSANDS)			PERCENTAGE GROWTH
		1950	1980	2000 (Projected)	1950–1980
1	Jakarta	1,565	6,503	16,933	316%
2	Manila	1,532	5,926	12,683	287
3	Bangkok	964	4,960	11,030	415
4	Ho Chi Minh City	970	2,701	5,066	178
5	Singapore	484	2,410	3,029	398
6	Yangon (Rangoon)	686	2,200	7,372	221
7	Surabaya	679	2,080	5,038	206
8	Bandung	511	1,463	4,126	186
9	Medan	245	1,379	2,270	463
10	Semarang	371	1,027	1,858	177
11	Kuala Lumpur	208	938	1,468	351
12	Hanoi	225	820	5,109	264
13	Palembang	277	787	1,876	274
14	Ujung Pandang	228	709		211
15	Cebu	178	667		275

SOURCE: United Nations, *Global Review of Human Settlements, Statistical Annex* (New York: United Nations, 1976), Table 6 and various other sources.

As during the colonial period, the major share of the wealth, financial and cultural activity, national political institutions, manufacturing establishments, and commercial activities has remained in the primate, port cities. Given these circumstances it is not surprising that so many migrate to these cities in search of economic opportunities. Cities in the region cannot support the vast numbers of poor who come to them, however, and new migrants are often forced to reside with relatives and friends before moving out on their own. When they do start their own households, they are often in the ubiquitous slum and **squatter** areas that account for up to one-third, and occasionally over one-half, of the city population. These residential communities are built where vacant land exists, and today this, of course, means on land in the peripheral areas. This type of land use in urban areas has proliferated in Southeast Asia since World War II when the cities began experiencing the heavy rural-urban migration that continues to the present day. Ideally, migrants hope to live near potential jobs, like those that exist in the port zones and the central business district, but usually land is not available for housing these areas. As Southeast Asian cities have expanded spatially, new industrial and commercial land uses have emerged in peripheral areas, as is suggested in the model of the Southeast Asian port city developed by the geographer T. G. McGee (Figure 16.8).

Often the migrants do not find jobs and are forced to return to their origins, join the long list of unemployed in the city, or secure work in what is called the urban "informal" sector. The *informal sector* includes many different types of service-related jobs that do not provide a regular wage or salary. People who work as vendors in public markets or on the streets, beggars, drivers of certain types of vehicles like the pedicab in Indonesia and the jeepney in the Philippines, and itinerant stevedores are all examples of informal sector workers. In-

Squatter settlements in the Philippines are often built over the sea because land is scarce in urban areas.

FIGURE 16.8

Model of Land Use in Southeast Asian Cities

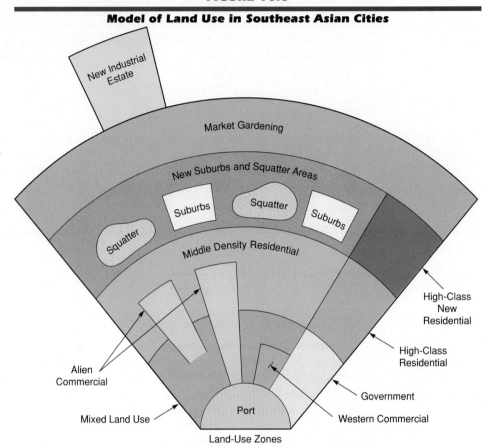

SOURCE: T. G. McGee, *The Southeast Asian City* (New York: Frederick A. Praeger, 1967). Used by permission.

formal sector workers are often described as underemployed; that is, they do not work full-time but rather only when work is available. On the other hand, many informal sector workers do work full-time and quite frequently earn more than those in the "formal" sector. The latter include employees who receive a regular wage or salary, as, for example, government workers, clerks in large department stores and supermarkets, professionals like lawyers, doctors, and teachers, and factory workers. Today in many cities in the developing world, informal jobs account for a greater percentage of urban employment than do those in the formal sector. Most new employment in Southeast Asia has emerged in the service sector, which includes both formal and informal workers.

Thus, the urban informal sector plays a very important role in cities in that it provides employment for the many low-income, often poorly educated migrants who come from the nation's densely populated rural areas. Some have argued that these cities, like the rural areas, are also places of involution because the informal sector eventually absorbs many of those who are not successful in finding formal sector jobs but choose to remain in the city. Again, the question of how long such places can continue to absorb migrants without some kind of more radical transformation must be raised. Clearly, a rapid decline in the rate of natural population increase coupled with rapid improvements in economic conditions will do much to forestall or eliminate such a possibility in both rural and urban areas.

The Western model for modernization, which has been the model followed in most former colonial areas, has included industrialization. In the early 1980s, manufacturing typically accounted for about three-quarters of the total industrial sector (which comprises mining, manufacturing, construction, and electricity, water, and gas) except in Indonesia and Malaysia where the mining (tin and oil) subsector is most important. In the structural transformation of economies, the industrial sector is expected to increase its share of the GDP substantially, and growth of manufacturing is also the main vehicle of technological progress. In some nations, notably Myanmar (Burma), the Philippines, and Thailand, food and agricultural manufacturing industries account for a large share of total manufacturing output.

Indonesia has a larger-than-average share of chemical industries related to its oil production. The Indonesian government is committed to industries that make the best use of its natural resources and abundant labor. Like many developing nations, it is opting to reduce capital goods imports, which use up needed foreign exchange, by encouraging industries that do not require such imports.

In Singapore, over half the output of manufacturing industries is derived from machinery and transportation equipment, a pattern more typical of a developed economy. The government of Singapore is continuing to upgrade manufacturing away from labor-intensive, low-productivity industries toward high-technology, capital-intensive, and export-oriented types of manufacturing. In the early 1980s, investments were mainly for projects in such industries as computers, fabricated metals, and machinery.

In Malaysia, the government is currently restructuring its manufacturing sector in line with its "New Economic Policy." For Malaysia this means the development of industries aimed at producing goods for export rather than goods for internal consumption that would otherwise have to be imported. Industrial output has also increased in Vietnam in recent years, especially in such industries as food processing, textile yarn, and various kinds of light industry. Shortages of raw materials, fuel, and spare parts are among the major problems for industrialization in Vietnam.

Thus, these nations are utilizing different strategies in the development of their manufacturing industries. Optimally, the strategy selected should be based upon the natural resources of the nation, the composition of its labor force, its trading relationships with other nations, and the size of its domestic market. To protect their own industries Southeast Asian nations have instituted measures, such as import tariffs or import surcharges, to protect certain domestic activities against foreign competition. Generally, such policies discriminate against primary agricultural and mineral commodities since manufactured goods receive the highest protection. In short, this shifts the internal terms of trade in favor of the finished goods. Such policies, of course, have a

Many people work in the informal economic sector in urban areas, such as the drivers of these small family-owned buses in the Philippines.

number of implications, one of which is that consumers must pay higher prices for competing imported products. It appears that a number of Southeast Asia's nations, such as Thailand and the Philippines, are becoming increasingly aware of some of the adverse implications of high protection levels and have lowered their tariff structures recently. The need for increased foreign exchange earnings has caused several nations to opt for other industrial incentives such as preferential duties on imports of machinery required for the production of exports in Fiji and Tonga and the creation of **export processing zones (EPZs)** in Singapore, Malaysia, and the Philippines. Approximately 175,000 workers are employed in EPZs, mostly in electrical and electronic goods and textile and apparel manufacturing establishments owned by transnational corporations.

The majority of the nations of the two regions have adopted a rather liberal policy toward foreign economic involvement and, indeed, have actively promoted such foreign investment through a variety of fiscal and other incentives. For example, the inflow of capital has been rapid in ASEAN because corporate taxes are much lower in Southeast Asia (15–20 percent) than in other regions, such as South Asia (55–65 percent). Thus, multinational corporations are being welcomed as a way to increase output, capital, technological levels, exports, and, of course, employment. These corporations have often been criticized for excessive repatriation of profits (returning profits made abroad to their home country) and "neocolonialism." Nevertheless, foreign direct investment in these nations has grown rapidly—by more than four times during the 1970s. Japan has been the most significant foreign investor in Indonesia and Malaysia, which received the largest volume of foreign direct investment among all developing Asian countries.

Multinational corporations accounted for about 20 percent of total manufacturing exports from nations with foreign investment. The proportion of such exports varies widely with Singapore having by far the largest, over 70 percent of its total (Table 16.7). With the exception of Singapore, however, multinationals play a proportionately much smaller role in the developing Asian nations than they do elsewhere in the world. Nevertheless, the Asian economies have developed a heavy dependence on imported technology and on securing access to foreign markets through trade that is wholly controlled from decision-making headquarters abroad. The overall employment impact of the multinationals, except in Singapore where 105,000 workers

are employed in manufacturing in EPZs, does not reach more than a small fraction of the total labor force. One relatively new development is the growing trend for multinational firms originating in developing Asian nations like Korea, Hong Kong, Singapore, India, and the Philippines to invest in other Asian nations, especially Indonesia, Malaysia, and Thailand.

SPATIAL CONNECTIVITY

In varying degrees, the political entities of Southeast Asia and the Pacific are dependent economically and militarily upon developed nations outside the region. The degree of such dependence can be measured by the amount of foreign investment, military alliances and aid assistance, economic development assistance, the direction and share of foreign trade, and tourist flows. Examination of the linkages between the nations of the two regions and those outside the regions (that is, extraregional linkages) and among nations within each region (intraregional) clearly supports the existence of two distinct political-economic groupings of nations. Extraregionally, the centrally planned economies of Vietnam, Cambodia, and Laos are oriented toward the Soviet Union or The People's Republic of China whereas the ASEAN and Pacific nation-states are oriented toward the West and Japan.

REGIONAL GROUPINGS

The first regional grouping of nations in Southeast Asia emerged in 1961, when Malaya (and Singapore, politically a part of Malaya until 1965), the Philippines, and Thailand joined together in the Association of South-East Asia (ASA). This organization, and another that followed briefly in 1963 called "Maphilindo," rather quickly disappeared in large part because of long-standing boundary disputes. The dispute between Malaya and the Philippines over the latter's claim to Sabah led to the end of ASA. Maphilindo, an organization that was to unite Malaya, the Philippines, and Indonesia, disappeared almost as soon as it was formed because of differences between Malaya and Indonesia over their boundary in Kalimantan (then Borneo).

The most successful example of regional cooperation in Southeast Asia has been the Association for Southeast Asian Nations (ASEAN). While the six member nations—

TABLE 16.7

Foreign Direct Investment in Selected Nations of Southeast Asia and the Pacific, Late 1970s

| NATION | FOREIGN DIRECT INVESTMENT FROM OECD* | | DISTRIBUTION OF FOREIGN AFFILIATES BY HOME COUNTRY (%) | | | | | | |
	Annual Average (Millions of Dollars)	As a Percentage of Total Flow of Resources	Australia–New Zealand	United States	Japan	United Kingdom	Netherlands	France	West Germany
Cambodia	0.3	<1	—	—	—	—	—	—	—
Fiji	12.4	21	63	3	2	19	N.A.	N.A.	N.A.
Indonesia	111.7	9	5	38	21	13	7	1	4
Malaysia	104.5	20	7	15	7	40	3	1	1
Myanmar (Burma)	0.1	<1	—	—	—	—	—	—	—
Papua–New Guinea	30.9	9	70	3	1	19	1	0	1
Philippines	200.4	19	2	62	13	5	1	1	3
Singapore	383.9	74	6	34	8	29	2	1	3
Thailand	98.1	11	2	36	22	15	3	1	4
Tonga	<0.1	<1	—	—	—	—	—	—	—
Vanuatu	4.2	11	35	8	2	12	0	9	0
Vietnam	0	0	—	—	—	—	—	—	—

* OECD = Organization for Economic Cooperation and Development.

SOURCE: United Nations Centre on Transnational Corporations (UNCTC), *Transnational Corporations in World Development, Third Survey* (New York: United Nations, 1983), Tables II.14 and II.19.

DOMINO THEORY

The domino theory, which was first applied to Southeast Asia during the 1950s under the Eisenhower administration, was an expression of the fear that one by one, the Southeast Asian nations would fall under communist domination and influence like dominoes in a line. To some extent, this has been the case as Vietnam, Cambodia, and Laos have become communist countries. The theory was a major justification for the American entry into Vietnam and the subsequent war there. Although the term is not often used today, it is still a guiding principle for military, political, and economic interrelations and external relations in the region. The establishment of ASEAN, the foreign policies of Indonesia and Thailand, and the American bases in the Philippines are but a few examples of the continuing concern that communism will spread from nearby communist countries. Although the theory is not adhered to as rigidly as it was in the 1960s and 1950s, it continues to influence U.S. policy in Central America and other regions of the world.

Underlying the domino theory is the notion that the spatial diffusion of an ideology is more rapid in situations where nations are in close proximity or adjacent to each other. As a practical matter, an adjacent country can more easily support guerrilla wars or insurgency movements to foster the adoption of an ideology in its neighbor. Nicaraguan shipment of arms to rebels in nearby El Salvador during the 1980s is an example of how proximity provides a logistical advantage supporting an ideological struggle.

In most cases, the domino theory has been applied to the spread of communism in Southeast Asia and Latin America. Most recently, however the domino effect has been very apparent in the toppling of communist regimes one after the other in Eastern Europe. The enthusiasm for increased freedoms in each nation was fueled by ideas and actions in adjacent nations. The "dominoes" of Eastern Europe fell much more easily and more rapidly than anyone could have imagined.

Indonesia, the Philippines, Thailand, Malaysia, Singapore, and Brunei—differ somewhat in their economies, they are in basic agreement with regard to their political-strategic outlook. Created in Bangkok in 1967, ASEAN was open for membership to all Southeast Asian nations. Beginning in 1972, the member nations began to bargain collectively with trading blocs such as the European Economic Community (EEC; now European Community, or EC). The heads of state agreed in 1976 to establish the ASEAN Industrial Projects (AIP), and in 1977 they agreed on a Preferential Trading Arrangement (PTA) that reduced some tariffs. Ownership of industrial projects was to be shared by the member nations. As of the early-1980s, only projects in Indonesia and Malaysia (both urea fertilizer plants) were close to completion. Other examples of regional cooperation include an arrangement whereby 50,000 tons of rice are stockpiled to cover emergency requirements anywhere in the region and the construction of a telecommunications network that will eventually tie together the member nations with more than 4000 nautical miles of submarine cable.

In summary, while progress has been slower than planned, an important beginning has been made toward regional cooperation and ultimately, perhaps, regional integration. Certainly, the Vietnam War played a major role in the early cooperation of the ASEAN members, and, since 1978, a central preoccupation of ASEAN has been a search for political settlement in Cambodia. The great success of the European Common Market has also served as a catalyst for cooperation among ASEAN nations. The differences in the economies and varying levels of development of the member nations present a major obstacle to future cooperation, however. Because individual nations still maintain a basically nationalistic view of costs and benefits, each fears that it will not get its fair share.

Had it not been for the Vietnam conflict, the Mekong Valley Scheme would almost certainly have greatly furthered regional cooperation among nearly all the nations of mainland Southeast Asia. An idea that grew out of studies completed in the 1950s, most notably those by the United Nations' Economic Committee on Asia and the Far East (ECAFE), the scheme involved the damming of the Mekong for hydroelectric, irrigation, improved fishing, flood control, and navigation purposes. While several of the smaller dams planned have been built on some of the Mekong's tributaries in Thailand, most of the work on the project came to a halt as the Vietnam conflict escalated (Figure 16.9). Although currently differences are too great among the nations for significant progress to be made on the scheme, there is potential for the future.

Two major regional organizations are found in the Pacific region. The first, established in 1947, is the South Pacific Commission (SPC). Originally, its members consisted of Australia and New Zealand and the four other colonial powers of the region at the time: Great Britain, France, the Netherlands, and the United States. Its functions then, as now, were restricted to matters such as economic development, education, health, and human welfare. Given the fragmented nature of the region, one of the commission's major concerns today is regional integration and planning. Gradually, the island states of the Pacific became more involved and ultimately gained full membership in the commission. This process began in 1950 when delegates from the various political entities met for the first time at the first South Pacific Conference. Perhaps it was here that a regional identity, that of the "Pacific Islander," emerged. The Pacific states gained a greater voice in the commission, especially after 1962 when Western Samoa became the region's first independent nation (Table 16.1). As the other Pacific states gained independence or a greater degree of autonomy, they, too, joined the commission, and by 1984 there were 27 members. The Netherlands ceased being a member in 1962 when it lost its only Pacific colony, West New Guinea, to Indonesia. The former colony is now Indonesia's easternmost province, Irian Jaya.

The apolitical nature of the commission was irritating to the independent states of the region, which by 1970 numbered four: Nauru, Tonga, and Fiji in addition to Western Samoa. These nations were concerned about a number of issues including the decolonization of other Pacific islands and the testing, dumping, and deploying of nuclear weapons in the region. This led to the founding of a separate organization in 1971, the South Pacific Forum. In addition to the region's newly independent nations, membership was extended to Australia and New Zealand as these were recognized as being inextricably linked with the region. The forum has established a joint venture shipping line and the Forum Fisheries Agency to coordinate regional fishing policies. In the 1980s, the 13 full members of the forum focused on gaining the independence of several colonies (especially French Oceania), halting French nuclear testing and American storage of nuclear materials in the Pacific, and settling oceanic claims with the United States.

Nuclear questions have been at the forefront of regional problems since the first test of a nuclear bomb in 1946. In American-controlled Micronesia, Bikini and Eniwetok atolls have been the most directly affected. Even today, after the cessation of nuclear testing here, over one thousand Bikinians are forced to live some 500 miles south of their home island because Bikini is still dangerously radioactive.

One other minor but interesting attempt at regional cooperation has been the establishment of the University of the South Pacific. This regional institution is supported by Australia and New Zealand, and 11 English-speaking Pacific states send students to it for higher education.

There are other less well-known examples of extraregional, regional, and bilateral agreements, established mostly for economic or research purposes. For example, Laos is a signatory to the 1976 Bangkok Agreement on Trade Expansion along with Korea, India, Bangladesh, and Sri Lanka; the Southeast Asian Tin Research and Development Centre (SEATRADC) coordinates efforts in tin mining in Malaysia, Indonesia, and Thailand; the Asian and Pacific Coconut Community, with 11 members, assists in the promotion and marketing of exports; and Malaysia has assisted the Philippines in port development.

MILITARY ALLIANCES AND ASSISTANCE

Today, formal military alliances among groups of nations are rare in the region, but this was not always the case. Thailand and the Philippines, along with Pakistan, the United States, Great Britain, New Zealand, Australia, and originally France, were members of the Southeast Asia Treaty Organization (SEATO). SEATO was a defensive and anticommunist alliance that was created as a response to what was viewed as communist expansion in Southeast Asia. This fear grew out of the creation of a communist state in North Vietnam in 1954 and the

FIGURE 16.9

The Mekong Valley Scheme

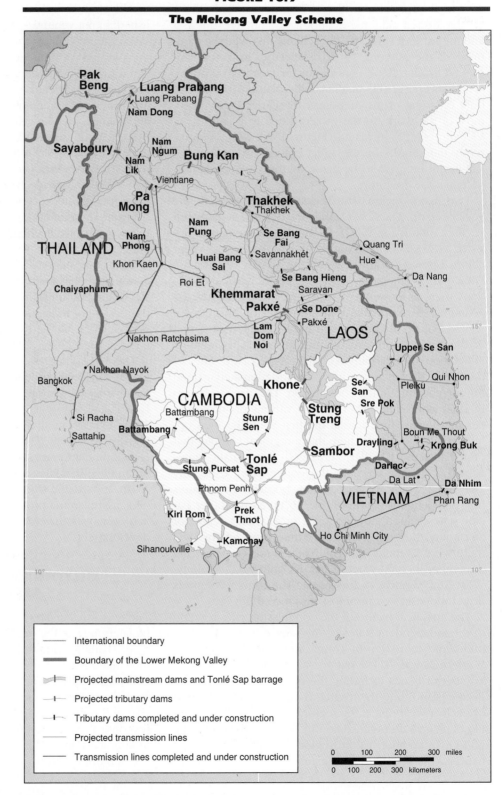

likelihood of a united North and South Vietnam. It was feared that a united communist Vietnam would ultimately lead to communist domination throughout the entire region. SEATO, never particularly cohesive, ceased to exist in 1975, but the geopolitical philosophy among the ASEAN members in the mid-1980s was still based largely on this perception of communist expansionism.

Since 1975, there have been some changes in bilateral military alliances. One basic change has been related to the Sino-Soviet split. Vietnam and the Soviet Union have emerged as allies against the People's Republic of China. Since 1979, Vietnam has even provided the Soviet Union with two military bases at the sites of the former American bases at Cam Ranh Bay and Danang. China and Vietnam have clashed along their common border several times, due largely to Hanoi's tilt toward Moscow and its invasion and occupation of Cambodia in 1978. Neighboring Thailand is very much concerned over the Vietnamese occupation and has attempted, with little success, to negotiate a Vietnamese withdrawal. On the other hand, strongly anti-Chinese Indonesia tends to favor a Soviet-backed Vietnam as a barrier to Chinese influence in the region. In short, current policies in noncommunist nations of the region are still a response to communist expansionism, but now such policies must take into account two opposing players, China and the Soviet Union.

One of the longest-standing bilateral military agreements has been that between the United States and the Philippines whereby the United States maintains two huge military installations in the Philippines; Clarke Air Force Base in central Luzon north of Manila and Subic Bay Naval Base, also on Luzon Island.

One way in which the strength of military ties can be measured is by examining who supplies the region's nations with arms and other forms of military assistance (Table 16.8). Vietnam received almost as many arms in total value as did all of the other Southeast Asian nations combined between 1978 and 1982. Nearly all the arms were supplied by the Soviet Union, as was also true for Laos. Myanmar (Burma), reflecting its postindependence posture of neutrality, is more diversified insofar as arms suppliers are concerned. The United States supplies arms to the Philippines, Thailand, Singapore, Indonesia, and Malaysia. In short, arms transfers in the region reflect recent events and power relationships.

Bilateral defense agreements in the Pacific are wholly with former colonial powers, including Australia and New Zealand. However, Soviet diplomatic, economic,

and technical inroads are beginning to be made on a minor scale in the Pacific. Thus, Kiribati and Vanuatu have recently entered into fishing agreements with Moscow.

DEVELOPMENT ASSISTANCE

Another way in which to measure both the level of development and dependency of nations is through examination of economic aid for development. All nations of the region receive bilateral and multilateral development assistance; the amount received through such arrangements from the developed nations is tabulated and reported by the United Nations. As we have noted, one way to measure level of development is by using per capita GNP, although due to differences in political-economic systems this measure is by no means ideal. Nevertheless, if we assume for the moment that such a measure is indicative of development level, then we might further propose that those nations with the lowest per capita income would receive the greatest per capita development assistance. By using this latter measure, rather than, for example, the total amount of assistance, we are holding population constant; that is, we are accounting for the differences in the populations of the nations. Figure 16.10 suggests there may be some relationship between level of development as measured by per capita GNP and development assistance as measured by per capita development assistance, but it is by no means clear and, at best, the relationship is very weak. As might be expected, very poor nations like Papua–New Guinea and Cambodia received a greater share of assistance, and Singapore and Malaysia, nations with higher incomes, received less assistance.

TRADE LINKAGES AND ENERGY DEPENDENCE

Another way to assess spatial connectivity and dependency among nations is through examination of foreign trade linkages. Such linkages usually are closely related to other aspects of foreign relations. Nations with a large trade relative to total GNP are more dependent on, and thus more sensitive to, external relationships. Trade, of course, is essential because no nation is self-sufficient. While a nation may be able to produce a certain commodity, it may opt to import that commodity because it can be bought less expensively elsewhere. In short, exports are necessary so that a nation can obtain imports. The difference between the two, the net trade balance,

TABLE 16.8

Arms Transfers and Major Suppliers, 1978–1982

RECIPIENT NATION	TOTAL VALUE (MILLION U.S.$)	SUPPLIER (PERCENTAGE OF TOTAL)					
		Soviet Union	United States	People's Republic of China	Western Europe	Eastern Europe	Other
Cambodia	80	25%	0%	63%	0%	0%	12%
Fiji	10	0	0	0	0	0	100
Indonesia	1300	0	19	0	26	0	55
Laos	130	100	0	0	0	0	0
Malaysia	525	0	36	0	32	0	30
Myanmar (Burma)	110	0	9	0	36	0	55
Papua–New Guinea	60	0	0	0	0	0	100
Philippines	290	0	75	0	7	0	18
Singapore	320	0	70	0	9	0	21
Thailand	1100	0	75	0	17	0	8
Vietnam	3700	99	0	0	0	1	0

NOTE: Western Europe includes France, Britain, West Germany, and Italy; Eastern Europe includes Czechoslovakia, Poland, and Romania; "other" includes, among others, Australia and New Zealand.

SOURCE: U.S. Arms Control and Disarmament Agency, *World Military Expenditures and Arms Transfers, 1972–1982* (Washington, D.C.: Arms Control and Disarmament Agency, 1984), pp. 96–98.

FIGURE 16.10

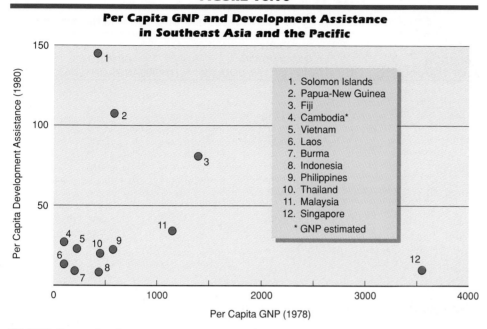

Per Capita GNP and Development Assistance in Southeast Asia and the Pacific

1. Solomon Islands
2. Papua-New Guinea
3. Fiji
4. Cambodia*
5. Vietnam
6. Laos
7. Burma
8. Indonesia
9. Philippines
10. Thailand
11. Malaysia
12. Singapore

* GNP estimated

SOURCES: Data are from Population Reference Bureau, *1980 World Population Data Sheet* (Washington, D.C.: Population Reference Bureau, 1980) (for GNP); and United Nations *Statistical Yearbook 1982* (New York: United Nations, 1982), pp. 482–83 (for Development Assistance).

is negative when the value of imports exceeds the value of exports. In Southeast Asia, Singapore is the most notable example of a nation heavily dependent upon foreign trade.

Often a major share of payments for imports must be used to meet domestic energy requirements since the majority of the world's nations do not produce enough energy to meet their needs. In Southeast Asia and the Pacific, only three nations—Brunei, Indonesia, and Malaysia—produce sufficient quantities of energy to meet their domestic requirements (Table 16.9). Energy in the form of petroleum is a major export for these nations.

As we have already noted, most Southeast Asian and Pacific nations are heavily dependent upon the production and export of primary commodities, most notably, oil, tin, forest products, fish, and a variety of agriculturally derived goods, especially from coconuts, sugar, and rubber. It is necessary to sell these commodities so that needed machinery, spare parts, electronic goods, transportation equipment, and other manufactured items not produced locally can be obtained. In short, the mix of imports of such nations is clearly oriented toward industrial products. Unstable primary commodity prices

related to events such as changes in demand, changes in preferential agreements or subsidies, and the introduction of substitutes (e.g., aluminum for tin in canning, synthetic rubber for natural rubber) have often hurt the developing nations, especially those dependent on one or a few such commodities. Brunei is almost totally dependent on oil for export revenues, and in Indonesia over three-quarters of exports relate to fossil fuels. Food-related items make up the vast majority of exports in many of the regions' nations, for example, about 90 percent in Fiji (75 percent for sugar alone), Western Samoa, and Vanuatu. During the 1970s, most of the regions' nations did not succeed in decreasing their dependency on primary commodities as had been hoped. Clearly, diversification of exports is a strategy that must be used to strengthen national economies. Furthermore, since about 1970, the growth of exports has undergone a marked slowdown brought about by increased protectionism and global recession. The average annual growth rate among ASEAN nations, for example, stood at a healthy 40 percent between 1970 and 1975, declined to 20 percent in the 1975–82 period, and was a dismal 2 percent in the 1982–84 period.

TABLE 16.9

Production and Consumption of Commercial Energy in Southeast Asia and the Pacific, 1988

NATION	ENERGY PRODUCTION (THOUSANDS OF TERAJOULES)	ENERGY CONSUMPTION (THOUSANDS OF TERAJOULES)	DIFFERENCE (THOUSANDS OF TERAJOULES)
SOUTHEAST ASIA			
Brunei	689	109	580
Cambodia	0	6	−6
Indonesia	4192	1502	2690
Laos	4	4	0
Malaysia	1610	644	946
Myanmar (Burma)	81	76	5
Philippines	79	487	−408
Singapore	0	572	−572
Thailand	391	886	−495
Vietnam	169	215	−46
PACIFIC			
Papua–New Guinea	2	32	−30
All other	3	114	−111

SOURCE: United Nations, *1988 Energy Statistics Yearbook* (New York: United Nations, 1990), Table 3.

Diversification in the geographical patterns of trade is also desirable as this means that nations are less dependent upon one export destination or import origin. Nations of the two regions have to some extent maintained their historic patterns of trading with former colonial rulers, but there is some evidence of changing directions of trade. Japan has become a major trading partner for many of the nations, and there is also significant intraregional trade. As a number of the nations in Southeast Asia, most notably Singapore, Malaysia, and the Philippines, have increased their manufacturing output, the amount of intraregional trade can be expected to increase.

TOURISM

One final indicator of the interrelationships among nations is tourism, which has grown rapidly in some of the nations of the regions, especially those in the Pacific. As in assistance and trade, the geographical origins of tourists to the region reflect in part the colonial past (Table 16.10). Thus, the United States, Great Britain, France, Australia, and New Zealand are among the major origins for international travelers. In addition, the Japanese have become the largest single group of extraregional visitors in recent years. This is not surprising

given their high income levels, economic interests, and close proximity to Southeast Asia and the Pacific. Certainly, Japanese travel to the two regions, as well as financial investment, can be expected to continue to increase in the near future. In many ways, no other single nation outside the region has had a greater influence than Japan during the 1980s and 1990s.

PROBLEMS AND PROSPECTS

COMMON PROBLEMS

Nearly all the individual nations of both regions are classed as developing and thus have common problems. As is the case throughout the developing world, many of today's difficulties can be traced to the legacies of the Western colonial period. Certainly, the economies of most nations in the two regions have been and still are dependent upon a relatively small number of primary commodities that were introduced during the colonial period. The colonial powers generally discouraged industrialization as this would have meant competition for industries in the mother country. Whereas some nations have attempted to diversify their economies, through

TABLE 16.10

Tourist Arrivals in Southeast Asia and the Pacific by Country of Origin, 1981

NATION	NUMBER OF VISITORS (THOUSANDS)	JAPAN	UNITED STATES	UNITED KINGDOM	AUSTRALIA AND NEW ZEALAND	FRANCE	WEST GERMANY	ASEAN
Brunei	286.5	82%	1%	4%	2%	*	*	1%
French Polynesia	96.9	1	43	1	13	14%	4	*
Indonesia	600.2	11	9	8	17	5	5	21
Malaysia	1679.4	8	3	6	6	*	*	60
Papua–New Guinea	35.1	6	8	5	58	1	3	3
Philippines	939.0	21	19	2	7	1	3	10
Singapore	2828.6	10	5	4	11	2	3	38
Thailand	2015.6	11	6	8	4	4	4	31
Tonga	12.6	2	18	2	44	*	7	*

* Indicates <0.5%.

SOURCE: United Nations, *1982 Statistical Yearbook* (New York: United Nations, 1985), Table 172, pp. 928–46.

manufacturing, for example, this has been extremely difficult because of the keen competition from the already well-entrenched manufacturing sector of the developed world. In short, virtually all the national economies are still dependent upon the developed nations, often the former colonial power, and hence neocolonialism is said to prevail. Southeast Asia and the Pacific can be described as a part of the global "periphery."

Another legacy of the colonial period has been the continuing economic dominance of the primate city. The economic development that has taken place, such as the establishment of new industries, has usually occurred in the very largest metropolitan areas. This, of course, has exacerbated the already unequal economic opportunities between urban and rural areas. In fact, the disparities are increasing between the primate urban centers and the hinterlands. As a result, the much poorer rural and village populations are located in areas that have become the testing and training grounds for revolutionary movements. In short, poverty (some would argue this is also a relict of the colonial period) has spurred insurgency.

While fertility levels have declined in a few nations, rapid population growth remains a major problem for nearly all the nations of the two regions with the exception of Singapore, Thailand, and Indonesia where growth rates have declined significantly through strong government support of family planning programs. On the other hand, the Malaysian government has advocated increased population growth! In some of the smaller Pacific nations, notably those in Micronesia and Polynesia, growth rates remain at very high levels, and population densities are fast increasing.

Another problem common to both regions relates at least partially to a geographical characteristic common to both, fragmentation. The numerous islands and separated valleys, mountain basins, and peninsulas on the Southeast Asian mainland have made travel and communication between places difficult. If economic development and regional integration are to be facilitated, then the degree of relative isolation between places (especially in the Pacific) must be lessened. Clearly, this is possible only through continued investments in inter- and extraregional transportation and communication networks.

CONFLICT

For several reasons, Southeast Asia has had a long history of conflict. First, issues of national identity have been a major source for recent and earlier conflicts. These issues have arisen from long-standing ethnic differences, from resentment toward alien minorities (most notably the Chinese), and, most recently, from revolutionary social challenge. Conflict has historically been especially problematic for Southeast Asia where ethnic animosities have long been the rule. Thus, the nations of the region often include ethnic minorities unwilling to reconcile themselves to political dominance by other groups. Although many conflicts have emerged, some of which are still in progress, attempts by minorities to secure separate political status have not generally been successful, nor have major regional conflicts emerged as a result. In short, separatism, though endemic to Southeast Asia as a source of conflict, has only had a limited impact. Nevertheless, it exists and continues to have a negative impact upon national unity.

A source of conflict that has had a much greater impact on the region since decolonization has been that brought about by revolutionary social challenge. Stark poverty and gross disparities of wealth in most of the region have facilitated revolutionary movements. Typically, this type of conflict has pitted organized armed guerilla movements against the elites who have succeeded the colonial rulers. In most cases, an insurgent Communist party has proffered the major challenge to postcolonial governments in the region. In only one political unit, North Vietnam, did the Communist party succeed to power immediately after colonial withdrawal. Since 1975 the communists have controlled all of Indochina. In every other nation of the region, the Communist party has played an active political role at various times. In the Philippines, the Hukbalahap movement was formed in the 1940s as a response to the Japanese. It is the origin of the Communist party of the Philippines, and today its armed wing, the New People's Army, is believed to include over 10,000 active guerillas operating throughout the nation.

When the colonial powers drew political boundaries, they failed to take into account ethnic realities. As a result, such boundaries have been the source for numerous disputes between nations leading to **irredentist** claims. Virtually every international boundary created has been in dispute at one time or another since independence. Some boundary disputes are still not resolved, and occasionally military actions are the result. Differences still exist between Brunei and Malaysia, Malaysia and Thailand, Indonesia and Papua–New Guinea, and Vietnam and China. Boundary disputes have also been the source, if not the root, of conflict between Vietnam and Cambodia. Future boundary disputes will involve control of offshore ocean areas in the South China Sea and the Pacific Ocean.

Today, the most important trouble spot in the region is Cambodia. The issues involved in this conflict go beyond the simply political; bitter differences between Khmers and Vietnamese existed long before the French colonized the area. In the 1830s, for example, the Vietnamese attempted to eradicate the traditional culture of the area and replace it with their own. The Cambo-

dians, and for that matter the Thais as well, have historically feared the expansionist Vietnamese. Thailand regards the new, Vietnam-dominated status of both Cambodia and Laos, traditionally considered buffer states against Vietnam, as a major threat to its own security. The possible return to power of the Khmer Rouge as the Vietnamese leave Cambodia is also a concern due to their murderous past. It is thus not surprising that Thailand, as well as other ASEAN states that harbor similar fears, have encouraged (and even supplied with arms) anti-Vietnamese groups in Cambodia to oppose the new colonizers. Until this issue is resolved, there is little hope for any kind of agreement between the communist Indochinese states and those that belong to ASEAN or, for that matter, between the major external powers of regional influence.

Extraregional major powers continue to be extensively involved in Southeast Asia. This is partly because of past ties, partly because of present economic interrelationships, and partly because of the strategic significance of Southeast Asia. For evidence of the latter point, we can again turn to Cambodia since the differences between Vietnam and the ASEAN states over that country have been the catalyst that have brought the great powers back into regional politics. The strong Soviet backing of Vietnam's position in Cambodia has forged an unexpected alliance between the United States and the People's Republic of China, at least with regard to Vietnamese domination of Cambodia. This, of course, has pleased Thailand but, on the other hand, has given cause for concern to Indonesia and Malaysia, both of which fear China's expansionist desires in the region.

Another major player in the region, Japan, cannot be ignored. Indeed, economically Japan is a major force in the region, and its importance will surely increase. Already it is ASEAN's major trading partner. It may be that trade and aid relationships between Japan and Vietnam will become more important, thereby ultimately decreasing Vietnam's dependence on the Soviet Union. Increased trade with, and aid from, Japan (and the United States and ASEAN, for that matter) may eventually induce Vietnam toward an agreement for a nonaligned Cambodia. The region's current dilemma, one of many rooted in the past, is an extremely competitive and complex situation, whose outcome will probably be stalemated for some time. In summary, Southeast Asia remains one of the world's potential centers of regional conflict.

Suggested Readings

Brookfield, H. C., ed. *The Pacific in Transition: Geographical Perspectives on Adaptation and Change.* New York: St. Martin's Press, 1973.

Bunge, F., and M. W. Cooke, eds. *Oceania: A Regional Study* Washington, D.C.: U.S. Government Printing Office, 1984. Area Handbook Series.

Dwyer, D. J. (ed.). *South East Asian Development: Geographical Perspectives.* New York: Wiley/Longman, 1990.

Fisher, C. A. *South-East Asia: A Social, Economic and Political Geography.* London: Methuen, 1964.

Fryer, D. W. *Emerging Southeast Asia: A Study in Growth and Stagnation,* 2d ed. New York: John Wiley & Sons, 1979.

Geertz, C. *Agricultural Involution: The Processes of Ecological Change in Indonesia.* Berkeley: University of California Press, 1963.

Hauser, P. M., D. B. Suits, and N. Ogawa, eds. *Urbanization and Migration in ASEAN Development.* Tokyo: National Institute for Research Advancement, 1985.

Leifer, M. *Conflict and Regional Order in South-East Asia.* London: International Institute for Strategic Studies, Adelphi Papers No. 162, 1981.

Leinbach, T. R., and R. Ulack, "Cities of Southeast Asia," in S. D. Brunn and J. F. Williams, *Cities of the World: World Regional Urban Development.* New York: Harper & Row, 1983.

Leng, L. Y. *The Razor's Edge: Boundaries and Boundary Disputes in Southeast Asia.* Singapore: Institute of Southeast Asian Studies, Discussion Paper No. 15, 1980.

_____ . *Southeast Asia and the Law of the Sea: Some Preliminary Observations on the Political Geography of Southeast Asian Seas,* 2d ed. Singapore: Singapore University Press, 1980.

McCloud, D. G. *System and Process in Southeast Asia: The Evolution of a Region.* Boulder, Colo.: Westview Press, 1986.

McGee, T. G. *The Southeast Asian City.* New York: Praeger. 1967.

Morgan, J. R. and M. J. Valencia (eds.). *Atlas for Marine Policy in Southeast Asian Seas.* Berkeley: University of California Press, 83.

Ulack, R., and G. Pauer. *Atlas of Southeast Asia.* New York: MacMillan, 1989.

United Nations, Centre on Transnational Corporations. *Transnational Corporations in World Development, Third Survey.* New York: United Nations, 1983.

United Nations, Economic and Social Commission for Asia and the Pacific (ESCAP). *Economic and Social Survey of Asia and the Pacific 1985.* Bangkok: United Nations, 1986.

United States Committee for Refugees. *World Refugee Survey 1984.* New York: American Council for Nationalities Service, 1984.

Selected National Statistics

REGION OR COUNTRY	Population Estimate mid-1990 (Millions)	Birth Rate (per 1,000 pop.)	Death Rate (per 1,000 pop.)	Natural Increase (Annual, %)	Population "Doubling Time" in Years (at current rate)	Population Projected to 2000 (Millions)	Infant Mortality Rate[a]	Total Fertility Rate[b]	% Population Under Age 15/65+	Life Expectancy at Birth (Years)	Urban Population (%)	Per Capita GNP, 1988 (US$)
World	5,321	27	10	1.8	39	6,292	73	3.5	33/6	64	41	$ 3,470
More Developed	1,214	15	9	0.5	128	1,274	16	2.0	22/12	74	73	15,830
Less Developed	4,107	31	10	2.1	33	5,018	81	4.0	36/4	61	32	710
Less Devel. (Excl. China)	2,987	35	11	2.4	29	3,738	91	4.6	40/4	59	36	870
Africa	661	44	15	2.9	24	884	109	6.2	45/3	52	31	600
Northern Africa	144	38	10	2.8	25	183	87	5.2	43/4	59	41	1,110
Algeria	25.6	40	9	3.1	22	32.7	74	6.1	46/4	60	43	2,450
Egypt	54.7	38	9	2.9	24	69.0	90	4.7	41/4	60	45	650
Libya	4.2	38	7	3.1	22	5.6	69	5.5	44/3	66	76	5,410
Morocco	25.6	35	10	2.6	27	31.4	82	4.8	42/4	61	43	750
Sudan	25.2	45	16	2.9	24	33.6	108	6.4	45/3	50	20	340
Tunisia	8.1	28	7	2.0	34	10.1	59	4.1	39/4	65	53	1,230
Western Sahara	0.2	49	23	2.5	28	0.2	—	—	—	—	—	—
Western Africa	206	47	17	3.0	23	279	119	6.6	46/2	48	30	340
Benin	4.7	51	19	3.2	22	6.6	110	7.0	47/3	47	39	340
Burkina Faso	9.1	50	18	3.2	21	12.5	126	7.2	48/4	51	8	230
Cape Verde	0.4	38	10	2.8	25	0.5	66	5.2	42/5	61	27	—
Côte d'Ivoire	12.6	51	14	3.7	19	18.5	96	7.4	49/2	53	43	740
Gambia	0.9	47	21	2.6	27	1.1	143	6.4	44/3	43	21	220
Ghana	15.0	44	13	3.1	22	20.4	86	6.3	45/3	55	32	400
Guinea	7.3	47	22	2.5	28	9.2	147	6.2	43/3	42	22	350
Guinea-Bissau	1.0	41	20	2.1	33	1.2	132	5.4	41/4	45	27	160
Liberia	2.6	45	13	3.2	22	3.7	83	6.4	46/3	56	43	450
Mali	8.1	52	22	3.0	23	10.7	117	7.2	47/3	45	18	230
Mauritania	2.0	46	19	2.7	25	2.7	127	6.5	44/3	46	35	480
Niger	7.9	51	21	3.0	23	11.1	135	7.1	47/3	45	16	310
Nigeria	118.8	46	17	2.9	24	160.8	121	6.5	45/2	48	31	290

REGION OR COUNTRY	Population Estimate mid-1990 (Millions)	Birth Rate (per 1,000 pop.)	Death Rate (per 1,000 pop.)	Natural Increase (Annual, %)	Population "Doubling Time" in Years (at current rate)	Population Projected to 2000 (Millions)	Infant Mortality Rate[a]	Total Fertility Rate[b]	% Population Under Age 15/65+	Life Expectancy at Birth (Years)	Urban Population (%)	Per Capita GNP, 1988 (US$)
Western Africa (continued)												
Senegal	7.4	46	19	2.7	26	9.7	128	6.4	44/3	46	36	630
Sierra Leone	4.2	48	23	2.5	28	5.4	154	6.5	44/3	41	28	240
Togo	3.7	50	14	3.6	19	5.2	114	7.2	49/2	55	22	370
Middle Africa	**68**	**45**	**16**	**3.0**	**23**	**91**	**118**	**6.1**	**45/3**	**50**	**37**	**420**
Angola	8.5	47	20	2.7	26	11.1	137	6.4	45/3	45	25	—
Cameroon	11.1	42	16	2.6	26	14.5	125	5.8	44/3	50	42	1,010
Central African Republic	2.9	44	19	2.5	27	3.7	143	5.6	42/3	46	35	390
Chad	5.0	44	20	2.5	28	6.2	132	5.9	43/4	46	27	160
Congo	2.2	44	14	3.0	23	3.0	113	6.0	45/3	53	40	930
Equatorial Guinea	0.4	43	17	2.6	27	0.5	120	5.5	43/4	50	60	350
Gabon	1.2	39	16	2.2	31	1.6	103	5.0	33/6	52	41	2,970
Sao Tome and Principe	0.1	36	9	2.7	25	0.2	61.7	5.4	42/5	65	38	280
Zaire	36.6	47	14	3.3	21	50.3	108	6.2	46/3	53	40	170
Eastern Africa	**199**	**47**	**17**	**3.0**	**23**	**273**	**116**	**6.7**	**47/3**	**50**	**18**	**230**
Burundi	5.6	48	15	3.2	22	7.7	114	7.0	45/3	51	5	230
Comoros	0.5	47	13	3.4	20	0.7	94	7.1	48/3	55	23	440
Djibouti	0.4	47	18	3.0	23	0.6	122	6.6	46/3	47	78	—
Ethiopia	51.7	44	24	2.0	34	70.8	154	6.2	46/4	41	11	120
Kenya	24.6	46	7	3.8	18	35.1	62	6.7	50/2	63	20	360
Madagascar	12.0	46	14	3.2	22	16.6	120	6.6	45/3	54	22	180
Malawi	9.2	52	18	3.4	20	11.8	130	7.7	48/3	49	14	160
Mauritius	1.1	19	7	1.3	54	1.2	25.2	2.0	30/5	68	41	1,810
Mozambique	15.7	45	19	2.7	26	20.4	141	6.4	44/3	47	19	100
Reunion	0.6	24	6	1.8	39	0.7	14	2.4	32/5	71	98	—
Rwanda	7.3	51	17	3.4	20	10.4	122	8.3	49/2	49	6	310

(-) data unavailable or inapplicable
[a] Infant deaths per 1,000 live births
[b] Average number of children born to a woman during her lifetime

REGION OR COUNTRY	Population Estimate mid-1990 (Millions)	Birth Rate (per 1,000 pop.)	Death Rate (per 1,000 pop.)	Natural Increase (Annual, %)	Population "Doubling Time" in Years (at current rate)	Population Projected to 2000 (Millions)	Infant Mortality Rate[a]	Total Fertility Rate[b]	% Population Under Age 15/65+	Life Expectancy at Birth (Years)	Urban Population (%)	Per Capita GNP, 1988 (US$)
Eastern Africa (continued)												
Seychelles	0.1	25	8	1.7	41	0.1	17.0	2.7	36/6	70	52	3,800
Somalia	8.4	51	20	3.1	23	10.4	132	6.6	47/3	45	33	170
Tanzania	26.0	51	14	3.7	19	36.5	106	7.1	49/2	53	19	160
Uganda	18.0	52	17	3.6	20	25.1	107	7.4	49/2	49	9	280
Zambia	8.1	51	14	3.8	18	11.6	80	7.2	49/2	53	45	290
Zimbabwe	9.7	42	10	3.2	22	13.1	72	5.8	45/3	58	25	660
Southern Africa	45	36	9	2.7	26	59	61	4.7	40/4	62	53	2,150
Botswana	1.2	40	11	2.9	24	1.6	64	5.3	46/3	59	22	1,050
Lesotho	1.8	41	12	2.8	24	2.4	100	5.8	43/4	56	17	410
Namibia	1.5	44	12	3.2	22	2.1	106	6.1	45/3	56	51	—
South Africa	39.6	35	8	2.7	26	51.5	55	4.5	40/4	63	56	2,290
Swaziland	0.8	46	15	3.1	22	1.1	130	6.2	47/2	50	26	790
Asia	**3,116**	**27**	**9**	**1.9**	**37**	**3,718**	**74**	**3.5**	**34/5**	**63**	**29**	**1,430**
Asia (Excl. China)	**1,997**	**31**	**10**	**2.1**	**33**	**2,438**	**88**	**4.1**	**38/4**	**60**	**33**	**2,140**
Western Asia	**132**	**36**	**8**	**2.8**	**24**	**175**	**71**	**5.2**	**41/4**	**64**	**58**	**2,860**
Bahrain	0.5	27	3	2.3	30	0.7	24	4.2	33/2	67	81	6,610
Cyprus	0.7	19	9	1.0	67	0.8	11	2.4	26/10	76	62	6,260
Gaza	0.6	50	7	4.3	16	0.8	55	7.0	50/3	65	—	—
Iraq	18.8	46	7	3.9	18	27.2	67	7.3	45/3	67	68	—
Israel	4.6	23	7	1.6	43	5.4	10.0	3.1	32/9	75	89	8,650
Jordan	4.1	41	6	3.5	20	5.7	54	5.9	46/2	69	64	1,500
Kuwait	2.1	27	2	2.5	28	2.9	15.6	3.7	37/1	73	94	13,680
Lebanon	3.3	28	7	2.1	33	4.1	49	3.7	40/5	68	80	—
Oman	1.5	46	13	3.3	21	2.1	100	7.2	46/3	55	9	5,070
Qatar	0.5	25	2	2.3	30	0.7	25	4.5	29/1	69	88	11,610
Saudi Arabia	15.0	42	8	3.4	20	22.0	71	7.2	45/3	63	73	6,170

REGION OR COUNTRY	Population Estimate mid-1990 (Millions)	Birth Rate (per 1,000 pop.)	Death Rate (per 1,000 pop.)	Natural Increase (Annual, %)	Population "Doubling Time" in Years (at current rate)	Population Projected to 2000 (Millions)	Infant Mortality Rate[a]	Total Fertility Rate[b]	% Population Under Age 15/65+	Life Expectancy at Birth (Years)	Urban Population (%)	Per Capita GNP, 1988 (US$)
Western Asia (continued)												
Syria	12.6	45	7	3.8	18	18.0	48	6.8	49/4	65	50	1,670
Turkey	56.7	29	8	2.1	32	69.0	74	3.6	36/4	64	53	1,280
United Arab Emirates	1.6	23	4	1.9	36	2.0	26	4.8	31/2	71	81	15,720
Yemen, North	7.2	52	17	3.5	20	10.0	129	7.6	50/3	49	20	650
Yemen, South	2.6	48	14	3.4	20	3.6	110	7.0	48/3	52	40	430
East Asia	**1,336**	**20**	**6**	**1.3**	**52**	**1,510**	**35**	**2.2**	**26/6**	**69**	**29**	**2,460**
China	1,119.9	21	7	1.4	49	1,280.0	37	2.3	27/6	68	21	330
Hong Kong	5.8	13	5	0.8	82	6.3	74.4	1.4	22/8	77	93	9,230
Japan	123.6	10	6	0.4	175	127.5	4.8	1.6	20/11	79	77	21,040
Korea, North	21.3	26	5	2.1	32	24.9	33	2.5	34/4	70	64	—
Korea, South	42.8	16	6	1.0	72	46.0	30	1.6	27/5	68	70	3,530
Macao	0.5	18	3	1.5	47	0.5	9	—	23/8	76	97	—
Mongolia	2.2	36	8	2.8	25	2.8	50	4.8	41/4	65	52	—
Taiwan	20.2	17	5	1.2	57	22.1	17	1.8	28/6	74	71	—
Southern Asia	**1,192**	**35**	**12**	**2.3**	**30**	**1,485**	**101**	**4.6**	**40/3**	**57**	**26**	**310**
Afghanistan	15.9	48	22	2.6	27	25.4	182	7.1	46/4	41	18	—
Bangladesh	114.8	39	14	2.5	28	146.6	120	4.9	43/3	54	13	170
Bhutan	1.6	38	17	2.1	32	1.9	128	5.5	38/4	48	5	150
India	853.4	32	11	2.1	33	1,042.5	95	4.2	39/3	57	26	330
Iran	55.6	45	10	3.6	20	75.7	91	6.3	45/3	63	54	—
Maldives	0.2	46	9	3.7	19	0.3	76	6.6	45/2	61	26	410
Nepal	19.1	42	17	2.5	28	24.3	112	6.1	42/3	52	7	170
Pakistan	114.6	44	13	3.0	23	149.1	110	6.7	44/4	56	28	350
Sri Lanka	17.2	21	6	1.5	47	19.4	22.5	2.3	35/4	70	22	420
Southeast Asia	**455**	**29**	**8**	**2.1**	**34**	**547**	**70**	**3.6**	**38/4**	**61**	**27**	**—**
Brunei	0.3	29	3	2.5	27	0.3	11	3.6	37/3	71	59	14,120

(-) data unavailable or inapplicable
[a] Infant deaths per 1,000 live births
[b] Average number of children born to a woman during her lifetime

REGION OR COUNTRY	Population Estimate mid-1990 (Millions)	Birth Rate (per 1,000 pop.)	Death Rate (per 1,000 pop.)	Natural Increase (Annual, %)	Population "Doubling Time" in Years (at current rate)	Population Projected to 2000 (Millions)	Infant Mortality Rate[a]	Total Fertility Rate[b]	% Population Under Age 15/65+	Life Expectancy at Birth (Years)	Urban Population (%)	Per Capita GNP, 1988 (US$)
Southeast Asia (continued)												
Cambodia	7.0	39	16	2.2	31	8.5	128	4.5	36/3	49	11	—
East Timor	0.7	44	22	2.2	31	0.9	166	5.4	35/3	43	12	—
Indonesia	189.4	27	9	1.8	38	223.8	89	3.3	38/3	59	26	430
Laos	4.0	41	16	2.5	28	5.0	110	5.5	43/3	47	16	180
Malaysia	17.9	30	5	2.5	28	21.5	30	3.6	38/4	68	35	1,870
Myanmar (Burma)	41.3	33	13	2.0	34	49.8	97	4.2	37/4	55	24	—
Philippines	66.1	33	7	2.6	27	82.7	48	4.3	39/3	64	42	630
Singapore	2.7	20	5	1.5	47	3.0	6.9	2.0	23/6	73	100	9,100
Thailand	55.7	22	7	1.5	45	63.7	39	2.6	35/4	66	18	1,000
Viet Nam	70.2	33	8	2.5	28	88.3	50	4.2	42/4	66	20	—
North America	**278**	**16**	**9**	**0.7**	**93**	**298**	**9**	**2.0**	**22/12**	**75**	**74**	**19,490**
Canada	26.6	14	7	0.7	96	29.3	7.3	1.7	21/11	77	77	16,760
United States	251.4	16	9	0.8	92	268.3	9.7	2.0	22/12	75	74	19,780
Latin America	**447**	**28**	**7**	**2.1**	**33**	**535**	**54**	**3.5**	**38/5**	**67**	**69**	**1,930**
Central America	**118**	**32**	**6**	**2.5**	**27**	**145**	**51**	**4.1**	**42/4**	**67**	**61**	**1,640**
Belize	0.2	37	6	3.1	22	0.3	36	5.0	44/4	69	50	1,460
Costa Rica	3.0	29	4	2.5	28	3.8	17.4	3.3	36/5	76	45	1,760
El Salvador	5.3	35	8	2.7	26	6.5	54	4.4	45/4	62	43	950
Guatemala	9.2	40	9	3.1	23	11.8	59	5.6	46/3	63	40	880
Honduras	5.1	39	8	3.1	23	6.8	63	5.3	47/3	63	42	850
Mexico	88.6	30	6	2.4	29	107.2	50	3.8	42/4	68	66	1,820
Nicaragua	3.9	42	9	3.3	21	5.1	69	5.5	47/4	62	57	830
Panama	2.4	27	5	2.2	32	2.9	23	3.1	36/5	72	52	2,240
Caribbean	**34**	**25**	**8**	**1.7**	**40**	**38**	**57**	**3.1**	**33/6**	**68**	**55**	**—**
Antigua and Barbuda	0.1	15	5	1.0	71	0.1	24	1.7	27/6	71	58	2,800
Bahamas	0.2	20	5	1.5	46	0.3	21.7	2.3	34/5	71	75	10,570

REGION OR COUNTRY	Population Estimate mid-1990 (Millions)	Birth Rate (per 1,000 pop.)	Death Rate (per 1,000 pop.)	Natural Increase (Annual, %)	Population "Doubling Time" in Years (at current rate)	Population Projected to 2000 (Millions)	Infant Mortality Rate[a]	Total Fertility Rate[b]	% Population Under Age 15/65+	Life Expectancy at Birth (Years)	Urban Population (%)	Per Capita GNP, 1988 (US$)
Caribbean (continued)												
Barbados	0.3	16	9	0.7	100	0.3	16.2	1.8	28/11	75	32	5,990
Cuba	10.6	18	7	1.2	60	11.6	11.9	1.9	25/8	75	72	—
Dominica	0.1	26	5	2.1	33	0.1	14	2.7	34/7	75	—	1,650
Dominican Republic	7.2	31	7	2.5	28	8.6	65	3.8	39/3	66	52	680
Grenada	0.1	37	7	3.0	23	0.1	30	4.9	39/7	71	—	1,370
Guadeloupe	0.3	20	7	1.4	51	0.4	18.0	2.2	31/7	73	90	—
Haiti	6.5	35	14	2.2	32	7.8	122	5.1	40/5	53	26	360
Jamaica	2.4	22	5	1.7	41	2.7	16	2.4	37/6	76	49	1,080
Martinique	0.3	19	6	1.3	54	0.4	11	2.1	30/7	74	82	—
Netherlands Antilles	0.2	18	5	1.3	55	0.2	9	2.0	30/7	73	53	—
Puerto Rico	3.3	20	7	1.2	57	3.4	14.2	2.2	30/9	72	67	5,540
St. Kitts-Nevis	0.04	23	11	1.3	55	0.04	39.7	2.6	34/9	68	45	2.770
Saint Lucia	.02	28	6	2.2	31	0.2	21.5	3.8	44/6	71	46	1,540
St. Vincent and the Grenadines	0.1	25	6	1.9	37	0.1	24.7	2.8	44/6	72	21	1,100
Trinidad and Tobago	1.3	27	7	2.0	34	1.7	13.7	3.1	34/6	70	64	3,350
Tropical South America	**247**	**29**	**8**	**2.1**	**33**	**298**	**59**	**3.5**	**37/4**	**65**	**71**	**2,020**
Bolivia	7.3	38	12	2.6	27	9.3	110	5.1	43/4	53	49	570
Brazil	150.4	27	8	1.9	36	179.5	63	3.3	36/4	65	74	2,280
Colombia	31.8	28	7	2.0	34	38.0	46	3.4	36/4	66	68	1,240
Ecuador	10.7	33	8	2.5	27	13.6	63	4.3	42/4	65	54	1,080
Guyana	0.8	25	5	1.9	36	0.8	30	2.8	37/4	67	33	410
Paraguay	4.3	35	7	2.8	25	5.5	42	4.6	41/4	67	43	1,180
Peru	21.9	32	8	2.4	29	26.4	76	4.1	38/4	65	69	1,440
Suriname	0.4	27	6	2.0	34	0.5	40	3.0	34/4	68	66	2,450
Venezuela	19.6	28	5	2.3	30	24.1	33	3.5	39/4	70	83	3,170

REGION OR COUNTRY	Population Estimate mid-1990 (Millions)	Birth Rate (per 1,000 pop.)	Death Rate (per 1,000 pop.)	Natural Increase (Annual, %)	Population "Doubling Time" in Years (at current rate)	Population Projected to 2000 (Millions)	Infant Mortality Rate[a]	Total Fertility Rate[b]	% Population Under Age 15/65+	Life Expectancy at Birth (Years)	Urban Population (%)	Per Capita GNP, 1988 (US$)
Temperate South America	49	21	8	1.4	51	55	28	2.8	33/8	71	85	2,320
Argentina	32.3	21	9	1.3	54	36.0	32	3.0	34/9	71	85	2,640
Chile	13.2	22	6	1.7	41	15.3	18.5	2.5	31/6	71	84	1,510
Uruguay	3.0	18	10	0.8	87	3.2	22.3	2.4	26/12	71	87	2,470
Europe	501	13	10	0.3	266	515	12	1.7	20/13	74	75	12,170
Northern Europe	84	14	11	0.2	286	86	9	1.8	19/15	75	85	14,300
Denmark	5.1	12	12	0.0	(-)	5.2	7.8	1.6	18/15	75	84	18,470
Finland	5.0	13	10	0.3	239	5.0	5.9	1.7	19/13	75	62	18,610
Iceland	0.3	19	7	1.1	61	0.3	6.2	2.3	25/10	78	89	20,160
Ireland	3.5	15	9	0.6	108	3.5	9.7	2.2	28/11	74	56	7,480
Norway	4.2	14	11	0.3	231	4.3	8.4	1.8	19/16	76	71	20,020
Sweden	8.5	14	11	0.2	311	8.8	5.8	2.0	18/18	77	83	19,150
United Kingdom	57.4	14	12	0.2	301	59.1	9.5	1.8	19/15	75	90	12,800
Western Europe	159	12	10	0.2	326	164	8	1.6	18/14	76	83	17,270
Austria	7.6	12	11	0.1	1,155	7.7	8.1	1.4	18/15	75	55	15,560
Belgium	9.9	12	11	0.2	462	9.9	9.2	1.6	18/14	74	95	14,550
France	56.4	14	9	0.4	157	57.9	7.5	1.8	20/14	77	73	16,080
Germany, West	63.2	11	11	-0.0	(-)	65.7	7.5	1.4	15/15	76	94	18,530
Luxembourg	0.4	12	10	0.2	346	0.4	8.7	1.4	17/13	75	78	22,600
Netherlands	14.9	13	8	0.4	165	15.3	7.6	1.5	18/13	77	89	14,530
Switzerland	6.7	12	9	0.3	231	6.8	6.8	1.6	17/15	77	61	27,260
Eastern Europe	113	14	11	0.3	215	115	16	2.0	24/11	71	64	—
Bulgaria	8.9	13	12	0.1	630	9.0	13.5	2.0	21/12	72	67	—
Czechoslovakia	15.7	14	11	0.2	289	16.3	11.9	2.1	24/11	71	75	—
Germany, East	16.3	13	13	0.0	6,930	15.5	8.1	1.7	19/13	73	77	—
Hungary	10.6	12	13	-0.2	(-)	10.6	15.8	1.8	21/13	70	60	2,460
Poland	37.8	16	10	0.6	122	38.9	16.2	2.1	26/10	71	61	1,850
Romania	23.3	16	11	0.5	141	24.5	25.6	2.3	25/9	70	54	—

REGION OR COUNTRY	Population Estimate mid-1990 (Millions)	Birth Rate (per 1,000 pop.)	Death Rate (per 1,000 pop.)	Natural Increase (Annual, %)	Population "Doubling Time" in Years (at current rate)	Population Projected to 2000 (Millions)	Infant Mortality Rate[a]	Total Fertility Rate[b]	% Population Under Age 15/65+	Life Expectancy at Birth (Years)	Urban Population (%)	Per Capita GNP, 1988 (US$)
Southern Europe	145	12	9	0.3	250	150	14	1.5	21/12	75	68	8,650
Albania	3.3	26	6	2.0	34	3.8	28	3.2	35/5	71	35	—
Greece	10.1	11	9	0.2	408	10.2	11.0	1.5	20/13	77	58	4,790
Italy	57.7	10	9	0.1	1,155	58.6	9.5	1.3	18/14	75	72	13,320
Malta	0.4	16	8	0.8	87	0.4	8.0	2.0	24/10	75	85	5,050
Portugal	10.4	12	10	0.2	301	10.7	14.9	1.6	22/13	74	30	3,670
Spain	39.4	11	8	0.3	247	40.7	9.0	1.5	22/13	77	91	7,740
Yugoslavia	23.8	15	9	0.6	114	25.1	24.5	2.0	23/9	71	46	2,680
USSR	291	19	10	0.9	80	312	29	2.5	25/9	69	66	—
Oceania	27	20	8	1.2	57	31	26	2.6	27/9	72	70	9,550
Australia	17.1	15	7	0.8	90	19.1	8.7	1.8	22/11	76	86	12,390
Fiji	0.8	27	6	2.2	32	0.9	21	3.3	38/3	63	39	1,540
French Polynesia	0.2	31	6	2.5	27	0.2	23	3.9	37/4	68	62	—
New Caledonia	0.2	25	6	1.9	36	0.2	39	3.0	33/4	67	58	—
New Zealand	3.3	17	8	0.8	82	3.5	10.0	2.0	24/11	74	84	9,620
Pacific Islands[c]	0.2	36	5	3.1	22	0.2	29	5.0	47/4	71	29	—
Papua-New Guinea	4.0	39	12	2.7	26	5.1	59	5.7	41/2	54	13	770
Solomon Islands	0.3	41	5	3.5	20	0.5	40	6.3	47/3	61	9	430
Vanuatu	0.2	37	5	3.2	22	0.2	36	5.5	45/3	69	18	820
Western Samoa	0.2	34	7	2.8	25	0.2	48	4.6	40/4	66	21	580

(-) data unavailable or inapplicable
[a] Infant deaths per 1,000 live births
[b] Average number of children born to a woman during her lifetime
[c] Comprising the Federated States of Micronesia, Palau, and the Marshall and N. Mariana Islands

Definitions

Mid-1990 Population: Estimates are based upon a recent census or official national data or upon UN, U.S. Bureau of the Census, or World Bank projections. The effects of refugee movements and population shifts due to contemporary political events have been taken into account to the extent possible.

Birth and Death Rate: These rates are often referred to as "crude rates" since they do not take a population's age structure into account. Thus, crude death rates in more developed countries (MDCs) are often higher than those in less developed countries (LDCs) because the MDCs have a larger proportion of older persons.

Rate of Natural Increase (RNI): Birth rate minus the death rate, implying the annual rate of population growth without accounting for net immigration or emigration.

Population "Doubling Time": The number of years in which a population would double assuming a **constant** rate of natural increase. This column is included to provide an indication of the potential growth associated with a particular RNI. It is not intended to forecast the doubling of any population; for a more realistic expectation of future growth or decline, see the columns on projected population in 2000 and 2020.

Population in 2000: Projected populations based upon reasonable assumptions about the future course of fertility, mortality, and migration. Data are based upon official country projections or upon series issued by the UN, the U.S. Census Bureau, or the World Bank.

Infant Mortality Rate: The annual number of deaths to infants under age 1 year per 1,000 live births. Rates shown with one decimal place are those generally considered to be completely registered. Some differences in definition of an infant death affect comparability of rates from country to country and some "complete" rates are for year of registration rather than occurrence.

Total Fertility Rate (TFR): The average number of children a woman will bear in her lifetime assuming that current age-specific birth rates will remain constant throughout her child-bearing years (usually considered to be ages 15–49).

Population Under Age 15/Age 65+: The percent of the total population in those two age groups, which are often considered the "dependent ages."

Life Expectancy at Birth: The average number of years a newborn infant can be expected to live under *current* mortality levels.

Urban Population: Percent of total population in areas termed urban by that country.

Per Capita GNP: Gross National Product data are from the *World Bank Atlas, 1989*. GNP includes the value of all domestic and foreign output. Figures in italics refer to 1987.

Sources

Data compiled by Population Reference Bureau, Inc., 777 14th Street, NW, Suite 800, Washington, D.C. 20005 (202) 639-8040. Rates and figures are primarily compiled from the following sources: statistical bulletins and demographic yearbooks of the individual countries; the UN *Demographic Yearbook, 1988* (forthcoming) and *Population and Vital Statistics Report, Data Available as of 1 April 1990* (forthcoming) of the UN Statistical Office; *World Population Prospects as Assessed in 1988* of the UN Population Division; the data files of the Center for International Research, U.S. Bureau of the Census; data from the publications of the Council of Europe and the European Communities; and long-term projections of the World Bank. Other sources include recent data from demographic surveys, direct communication with demographers and statistical bureaus in the U.S. and abroad, and special studies. Specific data sources may be obtained by contacting the authors of the *Data Sheet.*

For countries with complete registration of births and deaths (indicated by an infant mortality rate shown to a decimal place), rates are those most recently available. For developed countries, nearly all birth, death, infant mortality, and total fertility rates refer to 1988 or 1989. For less developed countries, those rates generally refer to the latter half of the 1980s. Other measures, unless otherwise indicated, are for the year most recently available in the sources cited above.

GLOSSARY

Absolute location. Location of a place by use of a mathematical grid system, such as latitude and longitude.

Acid precipitation. Rain, snow, sleet, and fog with a lowered pH (more acid) due to the presence of atmospheric pollution by sulfur and nitrogen compounds from industrial sources.

Agglomeration. The spatial grouping of people and activities to improve interaction and minimize the cost and effort of overcoming distance.

Alluvial soils. Soils deposited by water; usually, they are rich and fertile.

Altiplano. A plain or high plateau, usually referring to the high-elevation inter-Andean basin of Peru and Bolivia.

Americas. All the nations in the western hemisphere; a combination of North America, Central America, South America, and the Caribbean.

Andean Pact. A 1969 agreement that established an Andean subgroup consisting of Columbia, Ecuador, Peru, Bolivia, and Chile within LAFTA. *see also* LAFTA.

Animism. The belief that a spiritual life exists in natural phenomena and the universe itself and that this spirit is capable of having an influence on human beings.

Antecedent boundaries. Boundaries that are drawn prior to the development of the cultural landscape.

Anthropogeographic boundaries. Boundaries that are drawn in an effort to conform to the spatial extent of a specific culture.

Apartheid. Meaning "apartness"; a system of racial separation practiced in South Africa.

Aquiculture (aquaculture). The artificial cultivation of marine products; "sea farming."

Aquifer. A stratum of rock or sand containing water.

Arabic. The language of close to 200 million people in Southwest Asia and North Africa. It is also the ritual language of Islam.

Arable land. Land that can be cultivated.

Archipelago. A geologically related group of islands.

Aridisols. A soil order applied to soils of dry areas having little organic matter and shallow depth.

Arithmetic density. Population density in a geographic area; usually expressed as the number of persons per unit area (e.g., 70 would mean that there are 70 persons per square kilometer).

Atoll. A coral island consisting of a reef surrounding a lagoon.

Autonomous republic. An administrative unit for an ethnic group within one of the union republics of the Soviet Union.

Balance of payments. A statistical record of all economic transactions between the reporting country and all other countries.

Balkanization. The process of creating many new states from a larger political unit. The term was used initially to describe the changes that occurred in Eastern Europe as a result of the breakup of the Ottoman Empire and the Austro-Hungarian Empire.

Bandeirantes. "Flag bearers." Members of semimilitary expeditions out of São Paulo, the bandeirantes set off in search of gold, but instead found Indians who were subsequently sold as slaves for the coastal plantations. The bandeirantes traveled over the interior of the South American continent and expanded the borders of Brazil to the west and south. Intermarriage with Indian women spawned the onset of the racial category known as mamelucos.

Biomes. A classification of the earth's surface based on the prevalent vegetation and climatic conditions.

Buffer zone. A country or group of countries separating ideological or political adversaries.

Bustee. A term used in India to describe shantytowns constructed from cast-off materials on vacant land in and near cities. They are usually occupied by people new to urban life who are unable to find jobs or housing.

Caatinga. A vegetative region in northeast Brazil that is characterized by recurring droughts and floods. The wetter area of the caatinga is dense, with trees as high as 30 feet. In drier areas, the drought-resistant trees are low and widely spaced. The most important commercial crop in the caatinga is cotton.

Camino réal. Literally, the "royal highway"; this term was generally applied to official routes of transportation throughout the Spanish colonial empire, e.g., the camino réal from Lima to Caracas.

Campos. Vegetation cover of south central Brazil characterized by tropical grassland interspersed with varying amounts of tree cover, ranging from dense (campo cerrado) to sparse (campo limpo).

Capitalism. A belief in the private and corporate ownership of capital goods with the production, prices, and distribution of goods being determined by the free market.

Caribbean Basin. All Middle American countries that border on the Caribbean Sea; all the islands in the Caribbean and all Central American countries except El Salvador; (although the east coast of Mexico's Yucatán Peninsula touches the Caribbean Sea, it is *not* considered part of the basin).

Caribbean Middle America. All the islands in the Caribbean Sea.

CARICOM. A common market grouping of nations now including Antigua, Barbados, Belize, Dominica, Grenada, Guyana, Jamaica, Montserrat, Saint Kitts–Nevis, Saint Lucia, Saint Vincent, Suriname, and Trinidad and Tobago, which was formed in 1973 to promote intraregional economic development.

Carreterra Marginal de la Selva. Based on a 1957 idea of the architect and former president of Peru Fernando Belaúnde, a highway that will eventually connect Venezuela, Colombia, Ecuador, Peru, and Bolivia along the eastern flanks of the Andes. Penetration roads from the major national settlement cores are to connect with the Marginal Forest Highway, which began construction in 1968 to open millions of acres in the upper Amazon Basin for colonization by poor highland farmers. The latest scheduled date of completion is 1995.

Carrying capacity. In terms of human population, the potential number of inhabitants that can be supported within a given area of land under the prevailing level of technology without degradation of the environment.

Caste system. The hereditary and hierarchial social classes of Hindu society and religion. Members of each caste are usually restricted in their occupation and limit their association with members of other castes.

Centralization. The concentration of almost all political and economic power in one dominant city, usually the capital.

Centrally planned economics. Economies in which the decision making for industry, agriculture, and other sectors is done by a central government authority; the decision-making authority sets prices and determines output levels rather than allowing them to be determined by supply and demand as in a free market economy.

Centrifugal force. A force that tends to destroy or weaken the unity and cohesion of the people of a country or region.

Centripetal force. A force that brings about unity within a country or region. Religion and language are examples of forces that can enhance cohesion among people.

Chuché. The North Korean policy of self-reliance.

Circulation. A temporary form of human mobility falling somewhere between commuting and migration, the other two forms of mobility. Circulators leave their homeplace for a period of time (a few days or even a few years) but intend to return. There are many reasons for circulation, but the most common is for employment purposes.

Climate. The long-term temporal and geographic pattern of atmospheric and weather characteristics.

CMEA. *See* Council for Mutual Economic Assistance.

Collectivization. The socialization of agriculture in which collective farms (kolkhozes in the Soviet Union) and state farms (sokhozes in the Soviet Union) replace individual holdings. All land is owned by the state, and workers either are paid wages, as on a state farm, or receive a share of the farm income, as on a collective farm.

COMECON. *See* Council for Mutual Economic Assistance.

Commercial economies. Free market, capitalist economies such as those found in the United States, Canada, Western Europe, Japan, Australia, and New Zealand.

Comparative advantage. The relative superiority of a location over other possible locations for certain activities.

Confucianism. The teachings of Confucius (551–479 B.C.); it served as a basis for social conduct, order, and obligation in Chinese culture.

Continentality. Used in a climatic sense, a term to describe remoteness from marine influences. Generally drier, continental regions also experience great daily temperature fluctuations, and—at higher latitudes—greater degrees of seasonality.

Conurbation. The growing together of several large cities into one multicentered urban complex. The Ruhr Valley of Germany is probably the best example. The same as a megalopolis.

Copra. The dried meat of the coconut, which yields coconut oil.

Coqueo. The Andean cultural tradition, surviving primarily in Peru and Bolivia, of chewing coca leaves. The mildly narcotic stimulant was used as a means to alleviate the harsh climate and working conditions, often in the mines.

Cordillera. A system or chain of mountains, usually referring to the Andes.

Core-periphery. A term used to describe the difference between the centers of decision making and financial power on the one hand and the poor and powerless fringe areas on the other.

Cottage Industry. Any industry in which part or all of the production process is carried on in the home, rather than in a factory.

Council for Mutual Economic Assistance (COMECON or CMEA). An organization formed to regulate economic relations between the Soviet Union and Eastern European countries.

Cradle to the grave. A social welfare system that aims to guarantee the well-being of every individual from birth to death. It is most clearly associated with the Scandinavian countries.

Daimyo. The most powerful landholding military lords in premodern Japan.

Demilitarized zone (DMZ). A zone two and one-half miles (4 kilometers) wide separating North and South Korea.

Demographic transition. The change over time from high birth rates and death rates to low birth rates and death rates. The change is associated with modernization.

Demokratizatsiya (democratization). A term used to refer to changes and reforms in the Soviet government and economy.

Dependency ratio. The number of unproductive individuals as a percentage of productive individuals. Conventionally, the cutoff ages are those under 15 and those 65 or older.

Desertification. The expansion of desert areas into steppe areas due to the combined effects of drought, overgrazing, vegetation clearance, and improper cultivation.

Developed world. The broad region of the earth characterized by heavy reliance on technology, a high degree of mechanization, the consumption of large amounts of inanimate energy, intensely interconnected transportation and communication networks, and a large amount of wealth as measured by gross national product and per capita income.

Developing world. The broad region of the earth in which the majority of the people follow a traditional way of life separated from the conveniences of the developed world and its level of living.

Dharma. The Hindu concept of an individual's duty and obligation. It is related to the caste system since fullfilment of this duty follows from observing the customs and laws of one's particular caste.

Diet. The legislative branch of the Japanese government; according to the 1947 constitution, it is the "highest organ of state power."

Distance-decay. A term used to express the decline in development and prosperity that accompanies the increase in distance from the economic center of a country or region.

Double cropping. The practice of planting and harvesting two crops a year.

Doubling time. The number of years required for a population to double; it can be estimated by using the "law of 70" (i.e., dividing the annual natural increase into 70).

Dravidians. One of the earliest cultural groups to occupy the Indian subcontinent. From about 1500 B.C., they were pushed southward onto the Deccan Plateau by successive invasions of Aryans (IndoEuropeans) from Central Asia. Today, Dravidian languages dominate southern India.

Dynasty. One of a number of families that ruled China in succession for almost four thousand years until the revolution in 1911.

Economies of scale. Cost savings per unit of production derived from increases in the volume of output.

Ecumene. Permanently inhabited areas of the earth.

Ejidos. Rural farming communities in Mexico in which parcels of land are distributed to the peasants, but ownership of the land is retained by the ejido.

El Dorado. "The Gilded One". Traditionally, the ruler of the Chibchas would be sprinkled with gold dust and bathe in the cold waters of Lake Guatavita in highland Columbia annually. Stories of this event became stretched so much that El Dorado came to signify mythical cities of gold that lured many adventurers into the interior of northern South America.

El Niño. Every few years along the north coast of Peru, winds from across the equator bring southward a coast-hugging countercurrent of warm water, which spreads over the surface of the cold water. Since this freak warm countercurrent arrives near Christmas, it is called El Niño, or the Christ Child.

Emulation. The ambition to equal or surpass another; to adopt and adapt from another, producing a product superior to the original.

Endemic. Restricted to a locality or a region, as in a plant species or endemic disease.

Entrepôt. A place, usually a port city, where goods (perhaps including human immigrants) arrive via one type of carrier (e.g., ship) and are transshipped via another (e.g., rail or highway).

Environmental system. The combination and interaction of all natural living and nonliving components, including humans.

Export processing zones (EPZs). Areas set aside by governments in an attempt to encourage the location of foreign industries. Industries that locate in such zones are given incentives, such as minimum taxes and tariffs, to do so.

Exurban. Residential and commercial development occurring in outlying, nonurban areas, but often tied functionally to an urban area.

Federalism/federal state. A form of political structure in which a national government coexists with the governments of the constituent political units. Australia and the United States are examples of federal states.

Fertile Crescent. An area in Southwest Asia at the outer fringes of the desert where cultivation is possible. It encompasses the Tigris-Euphrates Valley and the eastern Mediterranean borderlands. Some authors include the Nile Valley in Egypt.

Fjord (fyord). A long, narrow, steep-sided inlet of the ocean. Fjords are glacial valleys whose outlets are below sea level.

Formal region. A part of the earth delimited by common geographic characteristics.

Fossil fuels. Coal, petroleum, and natural gas.

Four modernizations. A strategy for balanced growth and development of agriculture, industry, science and technology, and defense; initiated by Deng Xiaoping in the People's Republic of China.

Frontier migration. Human migration into a frontier area, usually stimulated by a combination of push (poverty at home) and pull (the lure of plenty of cheap land) factors. In South America, frontier migration usually consists of migration from temperate coastlands or highlands into humid tropical lowlands.

Functional region. An operational areal unit that functions around some political, socioeconomic, or other geographic activity.

"Gang of four". A group of four individuals led by Mao's wife, Chiang Qing, who were accused of being responsible for the radical policies of Mao's later years. After Mao's death they were purged by the Chinese Communist party, clearing the way for a more moderate and pragmatic approach to development.

Gaucho. A cowboy of mixed Spanish and Indian ancestry, living on the South American pampas. As with the North American cowboy, the gaucho culture has become somewhat romanticized.

Geometric (delimitation of boundaries). The practice of delimiting boundaries by using latitude and longitude or other mathematical measurements.

Glasnost. A Russian term meaning "openness." It refers to a more open policy in the Soviet Union regarding political, social, and economic matters.

Global interdependence.. Interdependence and connectivity between people, places, things, and ideas.

Gondwana. Term applied to the southern landmass of the supercontinent Pangaea. It included what is now Australia, Antarctica, Africa, South America, and India.

Gradational processes. The erosional, weathering, and depositional modifications that work to wear down the features of the earth's surface.

Greater Antilles. All the Caribbean islands west of Puerto Rico, including Puerto Rico and the Bahamas.

Greenhouse effect. The "blanket" created by the atmosphere by its absorption of outgoing terrestrial long-wave radiation and consequent accumulation of heat. The effect has been enhanced by continued atmospheric carbon dioxide buildup due to air pollution, chlorofluorocarbons, and rampant destruction of the world's forests, the prime consumers of carbon dioxide and producers of oxygen.

Green Revolution. A term used to describe recent agricultural advances that have resulted in increased agricultural productivity. These advances include new high-yielding varieties of seeds and improved fertilization and cultivation techniques.

Gross domestic product (GDP). The total monetary value of goods and services produced domestically by the citizens and businesses of a country, typically in one year.

Gross national product (GNP). The total value of goods and services produced by a country, typically in one year.

This value reflects only the monetary exchanges in an economy.

Growth pole. A city or town selected to receive substantial developmental aid in the expectation that economic prosperity will spread out from the growing center to the area around it.

Guomindang. The Chinese Nationalist party, now in exile on Taiwan.

Haciendas. Large self-sufficient land holdings, frequently used for stock raising.

Hard technology. The equipment, facilities, and other tangible items meant to increase wealth, production, or the quality of life.

Hercynian. A period of mountain building in Europe; the ranges that emerged during this period such as the Harz Mountains.

Hinterland. The area surrounding an activity center, such as a city, upon which the center is dependent for food, resources, or labor.

Homogeneous region. *See* formal region.

Human environment. The features of an area that are related to human beings, e.g., culture, settlement patterns, and economic activities.

Hydrosphere. The earth's water environment, including atmospheric water vapor, oceans, rivers, and groundwater.

Hylea. The regularly inundated floodplain of the Amazon River and its tributaries. The renourishment of the alluvial soils of the hylea via flooding accounts for a high agricultural productivity, and wet-rice agriculture has been expanding in this area.

Import substitution. A strategy used in developing countries to foster development of domestic industry and reduce reliance on imported goods.

Industrial core. The area within a region having the greatest amount of industrial activity.

Industrial infrastructure. The basic components of industry, including utilities, transportation networks, communications systems, and associated commercial and industrial enterprises.

Industrial revolution. The name given to the changeover of manufacturing from human and animal power to mechanical power based on the burning of fuels. It began in Great Britain around 1770 and led to the construction of large factories dependent on coal and iron.

Internal migration. Movement of people from one part of the country to another.

Irredentism. A policy directed toward the incorporation of a territory historically or ethnically related to another political unit.

Islam. A monotheistic religion with about 900 million adherents, found mostly in Southwest Asia and North Africa. Islam stresses the oneness of God (Allah) and that Muhammad is his messenger.

Japanese Core. The region along the Pacific coast of Honshu, extending from Tokyo to Osaka and Kobe; the political, economic, and cultural center of Japan with 40 percent of the country's population and 70 percent of its industrial output.

Kami. A term used to refer to the divine in the Shinto religion; it is translated as divinity, deity, god, spirit, or supernatural force.

Karst. Terrain consisting of white limestone rocks with fissures, eroded by rainwater.

Kolkhoz. A collective farm in the Soviet Union. Although all the land is owned by the state, part is divided into private plots for the individual families. The remainder is farmed by the workers as a group.

Koran. See Quran.

Kray. A large administrative unit in Siberia.

Ladino. A Central American Indian who has been assimilated into the mainstream, Spanish-speaking culture.

LAFTA. The Latin American Free Trade Association that was formed in 1960 in response to calls for a common market approach to regional development. It included Mexico, Venezuela, Colombia, Ecuador, Peru, Bolivia, Chile, Paraguay, Argentina, Uruguay, and Brazil. *See also* Andean Pact.

LAIA. The Latin American Integration Association that was formed in 1980 as a restructured LAFTA; it retains the original 11-nation membership. This treaty recognized differing levels of economic development among the 11 nations and established three different levels of tariff rates.

Land reform. The redistribution of farmland from large landholders to landless peasants in order to achieve a more equitable distribution of agricultural land.

Laterization. A soil-forming process that involves the decay and removal of the organic layer and the leaching of minerals by excessive rainfall. As nutrients are leached out, oxides of iron and aluminum are often left behind, and the soil takes on characteristic red or yellowish colors. The process is associated with hot and humid climates and produces soils such as oxisols and ultisols.

Leaching. The process in which percolating water dissolves and removes many of the organic nutrients, especially minerals, from the upper layer of a soil.

Leeward. "Downwind"; often used to describe the "rainshadow" side of a mountain or mountain range.

Lesser Antilles. All the Caribbean islands that lie east and south of Puerto Rico; the northern portion is sometimes called the Leeward Islands; the southern portion is frequently called the Windward Islands.

Lingua franca. A common, well-established language that is used to conduct business, international trade, or governmental affairs so as to enable persons to communicate with each other in a common language.

Lithosphere. The uppermost solid layer of the earth's crust.

Llanos. The savanna grasslands of the Orinoco River basin in Venezuela and Columbia.

Loess. A fertile soil composed of thick layers of fine silt de-

posited by the wind. Loess is common around the southern edges of glaciated areas.

Maghrib (Maghreb). Arabic for the west. It includes Morocco, Algeria, and Tunisia. Western Libya, essentially from Tripoli westward, has sometimes been included.

Mainland Middle America. All the countries of Central America plus Mexico.

Marxism. A belief that the abuses of the capitalist system will result in the uprising of the working classes to establish a classless society.

Mediterranean climate. A climate type characterized by hot dry summers and mild wet winters. It occurs in both hemispheres, so that the reference is to local summers and winters, and is found on the west coasts of continents at about 30°–40° of latitude.

Megalopolis. A large urban region comprised of several urban centers that have spread and merged together.

Meiji Restoration. A series of political, social, and economic changes in Japan in the latter part of the nineteenth century that resulted in the end of feudalism and Japan's development into a modern state.

Mesoamerica. A term referring to the "high-culture" area of Mexico and Central America where the native peoples had well-developed agriculture, which included domesticated plants and animals.

Mesopotamia. The land between the two rivers, specifically the Tigris and the Euphrates. Today, the term is associated with Iraq.

Mestizo. From the Latin for "mixed"; a person of mixed ancestry, specifically a combination of European and American Indian.

Metropolitan area. A central city and the adjacent built-up area.

Migrant worker (guest worker). Someone who moves temporarily or seasonally to another country to find work.

Migration. the act of moving from one area or country to another.

Mineral reserves. The estimated amounts of a mineral that can be obtained at a specified price with existing technology.

Monroe Doctrine. A message delivered to the U.S. Congress by President James Monroe in 1823, warning Europeans not to meddle in the affairs of the western hemisphere; it is still used by the United States as a justification to eliminate foreign influences in the region.

Monsoon. A regional wind pattern that varies with the season, often accompanied by heavy summer rains.

Montaña. The forested region of Peru along the eastern border valleys of the Andes and the eastern plains. Historically a very isolated region, this long and narrow region has come to be perceived as a utopia in terms of climate and productivity of soils by in-migrants from the Andean highlands and desert coasts.

Mulatto. A person of mixed white (Caucasoid) and black

(Negroid) ancestry.

Multinational corporation (Transnational). A corporation based in one country, but having subsidiaries and operations in many countries.

Nation. A culturally distinctive group of people occupying a particular region that are bound together by a commonality of shared traits such as language, religion, ethnicity, and other attributes.

Nation-state. A state whose population has the characteristics of a nation.

National okrug. An administrative unit for small ethnic groups in the northern part of the Soviet Union.

Nationalism. The strong love felt by entire nations for their country. It has often led to feelings of superiority and to excuses for aggression.

Negritude. A philosophy that promotes black cultural pride; its roots were in former French-controlled countries in West Africa and the West Indies.

Neocolonialism. The dependence of former colonies upon Western economic and financial institutions after achieving independence from the colonial power.

Nile Valley. The lifeline of Egypt. The Nile River originates in Ethiopia and in the Lake Victoria basin in East Africa. It is also an important river in Sudan.

Nodal region. *See* Functional region.

Nomadism. The migratory cyclic movement of animals and people.

Nonecumene. The uninhabited areas of the earth.

Nutritional density. *See* Physiological density.

Oblast. A Soviet administrative unit somewhat similar to a U.S. county.

Orographic. Pertaining to mountains; often used to describe the rains that are triggered by forced ascent of moisture-laden clouds over mountains.

Oxisols. A soil order consisting of highly weathered, extensively lateralized soils in tropical latitudes that usually contain various minerals such as iron, manganese, and bauxite in the upper horizons.

Pan-Africanism. A movement that presses for the spiritual unity and equality of all black people in the world.

Pan-American Highway. The highway linking the Americas, which stemmed from a plan formulated by the U.S. Congress in the 1920s. One segment of the highway (sometimes referred to as the Inter-American Highway) extends from Laredo, Texas, to Panama. In South America the highway follows the western axis of the continent, although branches extend out to Rio de Janeiro and Buenos Aires. Only one gap—the Darién Gap in Panama—remains.

Pan-Americanism. A late nineteenth-century movement toward unity in the Americas in which the United States perceived its role to be a "big brother" to the newly independent republics south of its borders.

Pangaea. A term applied to the former supercontinent when most of the present continents consisted of a singular landmass.

Partition. The division of British India at independence into Hindu India and Muslim Pakistan.

Paulista. An inhabitant of São Paulo. Sometimes the term was synonymous with bandeirante, as many of the "flag bearers" hailed from São Paulo.

Peninsula. A narrow stretch of land extending from the mainland into the sea.

People's Liberation Army (PLA). The military forces of the People's Republic of China

Perestroika. The restructuring of the Soviet economy; it involves changes in management practices and more sensitivity to market forces in decision making.

Permafrost. Permanently frozen subsoil.

Physical environment. The physical features of an area, e.g., climate, vegetation, soils, landforms, water resources, energy resources and minerals, and wildlife.

Physiographic features. Physical features of the earth such as rivers, lakes, and mountain ranges; sometimes used to delimit boundaries.

Physiological density. The number of persons or density of persons per unit (e.g., acres, hectares) of land in agricultural use.

Placer mining. The "mining" of gold or other valuable minerals by washing, sifting, dredging, or hydraulic sluicing of streamborne sediments.

Plantation. A large family- or corporate-owned estate organized to produce a cash crop, often a singular cash crop such as sugar or bananas. Most plantations were established within the tropics or subtropics, primarily along coastal or riverine sites. Since 1950 many have been divided into small parcels or reorganized as cooperatives.

Podzolization. A process of soil formation in which the upper layers are leached and material accumulates in the lower layers.

Population density. The number of individuals per unit area (usually given as so many people per square mile or square kilometer). It does not take into account any human and environmental factors and must therefore be used with care.

Postindustrial era. The period following the evolution of a developed economy from a concentration on extractive (primary) and industrial (secondary) economic activities to a reliance on the production of selected goods and services (tertiary economic activity).

Precipitation effectiveness. The water that is available for human use after evaporation (and transpiration). Thus, an area that receives 25 inches (63.5 centimeters) of rainfall annually may have a lower precipitation effectiveness than a place that receives 20 inches (50.8 centimeters).

Primary commodities. Goods or materials derived from primary economic activities such as farming, fishing, or mining.

Primate city. The largest city in a country; usually, it is twice as large as the next largest city; frequently, it is the capital city.

Private plots. Small plots of an acre or less located on collective and state farms. Workers are allowed to grow crops on the plots for private production and income.

Push-pull migration factors. The factors in their homeland that encourage people to leave their homes (push) and emigrate to an area they believe will provide better opportunities (pull) for them.

Quechua. A language spoken by a number of Andean Indian tribes within the realm of the former Incan Empire. The term is sometimes used to denote the highland Andean culture.

Quran (Qur'an, Koran). The holy book of Islam. It contains revelations believed to have been made to Muhammad by the angel Gabriel. It must be read in the original Arabic by believers.

Rain shadow effect. The leeward side of a physiographic feature (mountain, plateau) that receives less precipitation due to the "wringing out" of the moisture as it is uplifted over the feature on the windward side (orographic precipitation).

Rayon. A Soviet administrative unit somewhat similar to a U.S. township.

Reduction-in-bulk. Quite literally, the reduction in mass of a good, often a primary commodity such as bauxite or iron ore. This process takes place at a favorable transshipment point, such as for shipment overseas, or at an industrial site.

Reexport. An imported good that is then exported to another country; also, to export something that has been imported from elsewhere.

Relative location. Location of a place in relation to another place.

Remittances. Money sent to the homeplace by a temporary or permanent migrant who is working elsewhere. Many remittances come from foreign workers, for example, Filipinos who work in Middle Eastern oilfields or Mexicans working in the United States.

Remoteness. The condition of being far away; usually used to mean far away from urban centers or other human settlements.

Renaissance. Rebirth; the name given to the revival of interest in the art and literature of ancient Greece and Rome, which began in Italy around 1400. It resulted in one of history's greatest periods for painting, sculpture, and architecture.

Replicative society. A society that borrows the culture of another society and attempts to reproduce it in its own society.

Salinization. A soil-forming process occurring in low-lying desert regions. The resulting soils are unsuitable for agriculture and are characterized by high concentrations of soluble salts as a result of the evaporation of surface water.

Samurai. The warrior elite of premodern Japan that formed the real ruling class from the late twelfth century until the Meiji Restoration. They were dissolved in the 1870s, but by that time, former samurai were leaders in government and other sectors of modernizing Japan.

Savanna. A vegetation zone characterized by shrubs and bushes scattered among grasses occurring in areas with a tropical climate.

Sclerophytic (xerophytic). The kind of vegetation whose leaves and bark retard evaporation. Such vegetation is common in deserts and in areas with long dry seasons.

Selva. True closed-canopy rainforest, most common in the Amazon Basin. Unlike jungle, which is characterized by dense underbrush, the lack of sunlight penetration to the rainforest floor precludes lower-story vegetation growth.

Shatter belt. An area that is broken apart or shattered by an intervening power or state.

Shia. A Muslin who believes that succession to Muhammad must be by direct descent through Ali, Muhammad's cousin and son-in-law. About 10 percent of the world's Muslims are Shia, most of them in Iran.

Shield. A vast area of ancient hard rocks. Most shields are located around the Arctic Ocean and have been heavily glaciated.

Shinto. Japan's indigenous religion, appearing as a loosely structured set of practices, creeds, and attitudes; it had become a state religion by the end of the nineteenth century, but this status was abolished in 1945.

Shogunate. Any of three military governments that ruled Japan during most of the period between 1192 and 1867.

Sikhism. A reform movement within Hinduism founded in India around A.D. 1500 as a protest against the perceived excesses of Hinduism, especially the caste system.

Site. The local physical and human characteristics of a place.

Situation. Relationship of a place to other places or geographic areas.

Slash and burn agriculture. A system of shifting agriculture that involves the cutting down and burning of a parcel of rainforest for purposes of temporary (normally one to two seasons) cultivation of crops.

Soft technology. The science, education, and management skills necessary to increase wealth and production and utilize hard technology.

Solidarity. A workers' labor movement for political reform in Poland.

Sovereign state. A country that controls its internal affairs and is not governed by a foreign power.

Sovkhoz. A state farm in the Soviet Union. Workers are paid wages rather than receiving a share of the farm income as on a collective farm.

Spatial connectivity. Spatial linkages between geographic phenomena.

Spatial distribution. Distribution and arrangement of geographic phenomena such as physical features and socioeconomic activities on the earth's surface.

Spatial interaction. Connections and movements of people, ideas, and goods within and between places.

Squatter settlement. An area within or on the fringes of a city in which people illegally establish ramshackle residences on land they do not own. In time, squatter settlements may gain some legitimacy, and some basic services may be provided and structures may also be upgraded.

Squatters. Persons who settle in an area in which they have no legal right to settle. Usually applied to urban squatters in cities of the developing world.

State. A geographical area that is organized into a political unit.

Steppe. A semiarid climatic area characterized by short grass vegetation.

Subduction. In plate tectonics, the descent of the edge of a crustal plate under the edge of an adjoining crustal plate. This contact zone is normally characterized by extensive tectonic activity and volcanism.

Subsequent boundaries. Boundaries that are drawn after the cultural landscape has developed.

Subsistence agriculture. Refers to the production of crops for family or local consumption.

Sunni. An orthodox Muslim who believes in the teachings of the Quran and Mohammad. Succession to Muhammad is by consensus of the religious leaders. About 90 percent of the world's Muslims are Sunni.

Superimposed boundaries. Boundaries that are imposed on societies by other powers without taking into account the spatial extent of existing ethnic groups; such boundaries have often been imposed as a result of colonialism or wars and frequently have resulted in the division of ethnic groups between two or more states.

Taiga. Coniferous forest.

Talkie-talkie. The lingua franca, or de facto national language, of Suriname. Its origins lie in the polyglot, polyethnic character of the country's inhabitants, which include descendants of Amerindians, British, Chinese, Dutch, East Indians, French, Indonesians, and Portuguese.

Technology diffusion. The geographic spread of technology from one area to another; it is governed by the presence or absence of cultural, physical, and economic barriers.

Tectonic processes. The modifications of the earth's crust by movement, folding, faulting, and volcanism, which build up the earth's surface.

Terminal moraine. A ridge of glacial materials deposited at the furthest edge of advance of a continental ice sheet.

Transhumance. A form of agriculture based on the seasonal movement of herds or flocks of animals. Most commonly, it refers to the shifting of animals up to mountain pastures in the summer and back down into the valleys in the winter.

Transmigration. A program in Indonesia, begun in the early twentieth century by the Dutch, whereby families were encouraged and offered incentives to migrate from densely populated Java to Indonesia's sparsely populated "Outer" Islands, especially Sumatra, Kalimantan, and Sulawesi.

Transnational corporation. See Multinational.

Trans-Amazon Highway. An east-west, mostly dirt highway running south of and roughly parallel to the Amazon River. The highway, which began construction in the early 1970s, was intended to allow settlement in Brazil's "last frontier" and to stimulate development.

Treaty ports. Ports along the sea and rivers of East Asia, where foreigners were allowed to trade as a result of the "unequal treaties."

Tropical cyclone. An intense cyclonic storm beginning in the tropics in late summer and early autumn. Called a typhoon or a hurricane in common usage.

Tsunami. A large wave produced by an earthquake at sea.

Tundra. A zone of sparse herbaceous vegetation occurring in polar climatic areas.

Unequal treaties. A series of treaties signed between the countries of East Asia and Western powers, by which trade was expanded and special privileges were granted to Westerners.

Union republic. One of the 15 republics making up the Union of Soviet Socialist Republics.

Untouchables. The lowest group in the caste system, so called because it was believed (and still is by many Indians) that this group has the ability to defile members of a higher caste with whom they come in contact.

Urbanization. The movement of people from rural areas to towns and cities; usually measured as a ratio of urban to rural.

Veranillo. "Little summer"; in the humid tropics of the Americas, this refers to a short (two to three week) dry season, which often occurs in the midst of the rainy season.

Vertical (altitudinal) zonation. A term referring to the climatic gradation encountered as one changes elevation in tropical mountain areas. In Latin America, the vertical zones include the "tierra caliente" (0–3000 feet), the "tierra templada" (3000–7000 feet), the "tierra fría" (7000–12,000 feet), and the "tierra helada" (above 12,000 feet). Each of the zones is characterized by specific types of climate, natural vegetation, and cultural adaptations. In some aspects, it is the altitudinal equivalent of latitudinal climatic zones.

Wallace's line. The boundary, located in eastern Indonesia, between the Oriental and Australian floral and faunal regions.

Watershed. The water collection area for a river or lake system.

Windward. The direction from which the wind blows; often referring to the slopes of highlands that face the prevailing winds and are thus subject to higher rainfall.

Workers' management. The system used in Yugoslavia by which workers run their own enterprises.

World city. A large city that serves the whole world in terms of finance, management, communications, and culture. London, Paris, and New York are the best examples.

Xerophytic. See Sclerophytic.

Zaibatsu. The industrial and financial conglomerates that dominated the Japanese economy from the Meiji Restoration to World War II.

Zambo. A person of mixed African and Indian origin. The zambo is less common in Latin America than the mulatto or mestizo, due to the fact that African slaves were not imported in great numbers to places where a large labor pool of Indians was available.

Zero population growth (ZPG). A stabilized population where the number of deaths and births are equal.

Zionism. A movement to promote Jewish nationalism by colonizing Jews in present-day Israel.

INDEX

Boldface page numbers indicate figures and maps. Italicized page numbers indicate tables.

A

Aborigines, 304–305, **304**, 316
Absolute location, 16
Acadians, in United States, 113–114
Acid precipitation, 131, 185
  impact areas, **133**
Adelaide, Australia, 312
Aden, Gulf of, 233
Adidjan, Côte d'Ivoire (Ivory Coast), 512
Adriatic Sea, 147
Aegean Sea, 147, 243
Aerial photographs and images, 15
Afghanistan, 469
  national cohesion in, 468
  population of, *437*, 439
  Soviet aid to, 232, 233
  Soviet invasion of, 611, 612
Africa. *See countries in;* Southwest Asia–
    North Africa; Sub-Saharan Africa
African coast, Soviet naval presence on,
    234
African plate, 26
Afroasiatic languages, 501
Age-gender structure, 60–62, **60, 61, 63**
Agency for International Development
    (AID), 135
Agglomeration, 121, 122
Agriculture
  in Australia, 308–310, **310**
  and biomes, 38–39
  in Canada, 117, **118, 119**
  commercial, 403, 506, 508
  in Eastern Europe, 251–254
  in floodplain, **403**
  in Ireland, **169**
  in Japan, 284, **285**
  in Middle America, 351–355, **352, 353**
  in People's Republic of China,
    559–560, **559**
  plantation, 403
  slash and burn, 401, 403
  in South America, 403, **403**
  in South Asia, 600–605, *600*, **602**
  in Southeast Asia and Pacific Islands,

635, *636*, 637–638, **637, 638**
  in Southwest Asia–North Africa, **456**,
    458–460
  in Soviet Union, 196, 201, **201**,
    205–208, 223
  in Sub-Saharan Africa, 506, **507**, 508,
    508
  subsistence, **401**, 403
  in United States, 117, **118**, 119–121,
    120, *122*
  in Western Europe, 167–169, **168**
Agus River, 624
Ahmadabad, India, 597
Air transportation
  in People's Republic of China, 562
  in South America, 421
  in Soviet Union, 227
  in Sub-Saharan Africa, 527
Akuja, Nigeria, 512
Alaska Range, 97
Albania
  agriculture in, 253–254
  economy of, 264
  foreign trade of, 266
  industry in, 264
  population of, 249, 251
  post–World War II, 246, 247
  transportation in, 264
  urbanization of, 249
Alfisol(s), *30*
  in Anglo-America, 104
  in Eastern Europe, 242
  in Middle America, 336
  in New Zealand, 300
  in Southeast Asia and Pacific Islands,
    620
  in Southwest Asia–North Africa, 436
  in Western Europe, 156
Algeria, 463, 472
  national cohesion in, 465
  population of, *437*, 438, 440
Aliens, problem of illegal, in U.S., 116
Alluvial soil(s)
  in Eastern Europe, 242
  in Japan, 274

in South Asia, 582
in Southeast Asia and Pacific Islands,
    621
Alpine region, 171–172
Alps, 153, 155, 176
Altay Mountains, 542
Altiplano, 381
Altitude, correlation between
    precipitation and, 433
Alumina, in Middle America, 336
Alumina shale, in China and East Asia,
    544, 546
Aluminum
  in Poland, 257
  in Romania, 259
  in Soviet Union, 216, 219
Alunite, in Soviet Union, 219
Amazon Basin, 383
  frontier in, 396, 398
  soil in, 398
Amazonian development program, 397,
    398
Amazon River, 381, 386, 392
Americas, 328. *See also countries in*
Amerindian peoples, 339
Amharic language, 501
Amin, Idi, 514
Amish people, 113–114
Amsterdam, 149
Amu Dar'ya, 195, 197
Amundsen, Roald, 323
Amur River, 197, 209, 543
Andalusian Basin, 149
Andean cordillera, 381
  climate in, 383
Andean Pact (1969), 417
Angara River, 197, 214
Angel Falls, Venezuela, 381
Anglican Church, in Australia, 306
Anglo-America, 96. *See also* Canada;
    United States
  adjacency and trade, 129, 131
  agricultural production, 117, **118**,
    119–121
  boundaries of, 97, **98**

PHOTO CREDITS

CHAPTER 6

188 ©Peter Turnley/Black Star
200 ©Peter Turnley/Black Star
202 (left) TASS/SOVFOTO
202 (right) TASS/SOVFOTO
206 TASS/SOVFOTO

208 TASS/SOVFOTO
215 TASS/SOVFOTO
220 TASS/SOVFOTO
222 TASS/SOVFOTO
235 AP/Wide World Photos

CHAPTER 7

236 ©J. Langevin-Sygma
243 Courtesy of Romanian National Tourist Office
247 ©F. Zecchin/Magnum Photos, Inc.
248 AP/Wide World Photos
252 ©Mario Rossi/Photo Researchers, Inc.

253 K. Scholz/Superstock
256 ©Owen Franken/Stock, Boston
261 Courtesy of Balkan Holidays
263 David Maenza/The Image Bank

CHAPTER 8

270 Tony Stone Worldwide
275 Courtesy of Robert Cramer, National Council for Geographic
        Education (NCGE)
283 (left) Courtesy of Robert Cramer, NCGE

283 (right) ©1987 Japan National Tourist Organization
285 Courtesy of Robert Cramer, NCGE
286 ©Patrick Ward/Stock, Boston
289 Courtesy of Robert Cramer, NCGE

CHAPTER 9

294 ©Brian Brake/Photo Researchers, Inc.
301 ©Northern Territory Tourist Commission
302 Frank/Waterhouse/H. Armstrong Roberts
304 ©Northern Territory Tourist Commission

305 Embassy of Australia
311 K. Scholz/Superstock
312 New Zealand Tourism
313 Smith/ZEFA/H. Armstrong Roberts

SUPPLEMENT   Polar Regions

319 H. Armstrong Roberts

322 ©Francois Gohier/Photo Researchers, Inc.

PART THREE

324–25 ©David Robert Austen/Stock, Boston

CHAPTER 10

326 ©Rob Crandall/Stock, Boston
331 A. Upitis/Superstock
333 ©David Woo/Stock, Boston
335 Courtesy of William V. Davidson
338 ©John Cancalosi/Stock, Boston
341 Courtesy of William V. Davidson

342 M. Thonig/H. Armstrong Roberts
345 ©Owen Franken/Stock, Boston
351 J. Messerschmidt/H. Armstrong Roberts
354 ©Elizabeth Harris/TSW
355 ©Robert Frerck/TSW
358 Courtesy of William V. Davidson

CHAPTER 11

376 Tony Stone Worldwide
378 Courtesy of John O'Hear
381 Courtesy of John Saunders
401 Courtesy of John Saunders
403 Courtesy of John Saunders

409 ©Robert Frerck/The Stock Market
411 Courtesy of John Saunders
414 Courtesy of John O'Hear
415 Courtesy of Klaus Meyer-Arendt
416 Courtesy of Martha Works

CHAPTER 12

426 H. Armstrong Roberts
432 ©Jim Holland/Stock Boston
444 (left) Courtesy of Basheer K. Nijim
444 (right) R. Opfer/H. Armstrong Roberts
446 Courtesy of Basheer K. Nijim
456 (left) ©P. Zachmann/Magnum Photos, Inc.
456 (right) Courtesy of Basheer K. Nijim

457 (left) ©Michael R. Schneps/The Image Bank
457 (right) ©Owen Franken/Stock, Boston
459 ©Hugh Sitton/TSW
460 ©Fulvio Roiter/The Image Bank
464 ©Paul Rickenback/Photo Researchers, Inc.
477 AP/Wide World Photos

CHAPTER 13

480 D. Degnan/H. Armstrong Roberts
488 Patrick Clarke/KTONY
490 ©Luis Villota/The Stock Market
494 Kenya Tourist Office, New York
508 (top) UNESCO/A. Vorontzoff

508 (bottom) UNESCO/Dominique Roger
511 ©Alex Webb/Magnum Photos, Inc.
515 ©Robert Caputo/Stock, Boston
516 Superstock
519 I. Ingram/Superstock

CHAPTER 14

536 ©Margaret Gowan/TSW
538 ©Paolo Koch/Photo Researchers, Inc.
543 ©Fong Siu Nang/The Image Bank
550 Eastphoto
559 Superstock

562 ©Stephanie Stokes/The Stock Market
565 Korea National Tourism Corporation
569 ©Nathan Benn/Stock, Boston
574 ©Chris Bensley/Stock, Boston

CHAPTER 15

576 ©Suzanne and Nick Geary/TSW
578 ©Adrian Murrell/TSW
582 ©Ted Kerasote/Photo Researchers, Inc.
590 ©Sheldan Collins/The Stock Market
594 ©David R. Austen/Stock, Boston
598 ©Courtesy of Melinda Meade

599 ©Aslaby Kirchane/TSW
601 (top) Camerique/H. Armstrong Roberts
601 (center) Courtesy of Melinda Meade
601 (bottom) Courtesy of Melinda Meade
607 ©Bruno Barbey/Magnum Photos, Inc.

CHAPTER 16

616 ©Jean-Mare Truchet/TSW
618 Courtesy of Richard Ulack
628 ©Tim Bieber, Inc./The Image Bank
633 ©Tom McHugh/Photo Researchers, Inc.
635 ©Hugh Sitton/TSW

637 ©Bill Wassman/The Stock Market
638 ©Philip Jones Griffiths/Magnum Photos, Inc.
639 Courtesy of Richard Ulack
641 ©Owen Franken/Stock, Boston